THEORY AND
APPLICATIONS OF
OPTICAL REMOTE
SENSING

WILEY SERIES IN REMOTE SENSING

Jin Au Kong, Editor

THEORY AND APPLICATIONS OF OPTICAL REMOTE SENSING

Edited by

Ghassem Asrar
Headquarters
National Aeronautics and Space Administration
Washington, D.C.

WILEY

A WILEY-INTERSCIENCE PUBLICATION
JOHN WILEY & SONS
New York • Chichester • Brisbane • Toronto • Singapore

Library of Congress Cataloging in Publication Data:

Theory and applications of optical remote sensing/edited by Ghassem
  Asrar.
      p.   cm. -- (Wiley series in remote sensing)
    ''A Wiley-Interscience publication.''
    Includes bibliographies and index.
    ISBN 0-471-62895-6
    1. Remote sensing.   I. Asrar, Ghassem.   II. Series.
  G70.4.T47   1989
  621.36'78--dc19   88-35182
                                                          CIP

Printed in the United States of America

10 9 8 7 6 5 4 3 2 1

CONTRIBUTORS

GHASSEM ASRAR Headquarters, National Aeronautics and Space Administration, Washington, DC 20546

BHASKAR J. CHOUDHURY Hydrological Sciences Branch, Goddard Space Flight Center, National Aeronautics and Space Administration, Greenbelt, MD 20771

DONALD W. DEERING Earth Resources Branch, Goddard Space Flight Center, National Aeronautics and Space Administration, Greenbelt, MD 20771

JEFF DOZIER Center for Remote Sensing and Environmental Optics, University of California, Santa Barbara, CA 93106, and Jet Propulsion Laboratory, California Institute of Technology, Pasadena, CA 91109

NARENDRA S. GOEL Department of Systems Science, State University of New York, Binghamton, NY 13901

ALEXANDER F. H. GOETZ Center for the Study of Earth from Space/CIRES, University of Colorado at Boulder, Boulder, CO 80309-0449

MICHAEL F. GROSS College of Marine Studies, University of Delaware, Newark, DE 19716

MICHAEL A. HARDISKY Department of Biology, University of Scranton, Scranton, PA 18510

ALFREDO R. HUETE Department of Soil and Water Science, The University of Arizona, Tucson, AZ 85721

JAMES R. IRONS Earth Resources Branch, Goddard Space Flight Center, National Aeronautics and Space Administration, Greenbelt, MD 20771

RAY D. JACKSON United States Water Conservation Laboratory, Agricultural Research Service, United States Department of Agriculture, Phoenix, AZ 85040

EDWARD T. KANEMASU Evapotranspiration Laboratory, Waters Annex, Kansas State University, Manhattan, KS 66506

YORAM J. KAUFMAN University of Maryland, College Park, MD 20742, and Goddard Space Flight Center, National Aeronautics and Space Administration, Greenbelt, MD 20771

VYTAUTAS KLEMAS College of Marine Studies, University of Delaware, Newark, DE 19716

WILLIAM P. KUSTAS Hydrology Laboratory, Beltsville Agricultural Research Center, Agricultural Research Service, United States Department of Agriculture, Beltsville, MD 20705

RANGA B. MYNENI Institüte für Bioklamatologie, D-3400 Göttingen, Federal Republic of Germany

GARY W. PETERSEN Department of Agronomy, College of Agriculture, and Office for Remote Sensing of Earth Resources, Environmental Resources Research Institute, The Pennsylvania State University, University Park, PA 16802

DAVID L. PETERSON Ames Research Center, National Aeronautics and Space Administration, Moffett Field, CA 94035

JOHN C. PRICE Beltsville Agricultural Center, Agricultural Research Service, United States Department of Agriculture, Beltsville, MD 20705

STEVEN W. RUNNING School of Forestry, University of Montana, Missoula, MT 59812

PIERS J. SELLERS Center for Ocean-Land-Atmosphere Interactions, Department of Meteorology, University of Maryland, College Park, MD 20742

RICHARD A. WEISMILLER Department of Agronomy, College of Agriculture, University of Maryland, College Park, MD 20742

STEPHEN W. WHARTON Earth Resources Branch, Goddard Space Flight Center, National Aeronautics and Space Administration, Greenbelt, MD 20771

DIANE E. WICKLAND Terrestrial Ecosystems Program, Headquarters, National Aeronautics and Space Administration, Washington, DC 20546

PREFACE

The recent developments in the field of remote sensing have resulted in its increased application in a wide range of scientific disciplines. These advances have been made in several different areas including sensor design, measurement techniques, data analysis methods, and modeling activities. The results of progress in these areas have been reported in a wide variety of scientific journals that are discipline specific. The remote sensing books that are currently available either discuss the basic physical principles in depth with limited emphasis on applications or treat the subject of interest to a particular discipline in depth without reference to its broader applications in other disciplines. In our view, future remote sensing books should help to overcome these shortcomings so that the interested reader can obtain enough information to answer the basic questions of how remotely sensed data are obtained, what the factors affecting such measurements are, and how these data are used in different disciplines.

In this book, we intend to cover both principles and applications of remote sensing over the spectral region of 0.40 to 16 μm in a wide range of disciplines that are concerned with land-surface processes. The objective is to demonstrate how the same basic measurements are used in different disciplines to obtain quantitative information about the land-surface conditions and processes. The motivation for taking this interdisciplinary approach is the general awareness among scientific communities of the observation and monitoring of land-surface processes and renewable resources on a global scale. This can be accomplished in a timely manner by remote sensing. The main concern, however, is our limited knowledge of the interrelationship between the land surfaces and the atmosphere. This limitation is the result of past research activities that were primarily intradisciplinary.

Understanding and proper interpretation of remotely sensed data require a priori knowledge of the underlying physical, biological, and chemical processes that contribute to a particular spectral reflectance and/or emission spectra. This knowledge can be obtained only through cooperation and research among scientists from pertinent disciplines. Therefore, it is imperative to encourage interdisciplinary research in this field of science toward the attainment of a common goal: using remote sensing to understand the earth as a system. To this end I have invited scientists who are actively involved in research in this area to write chapters in their fields of interest. This book therefore presents our current knowledge of the principles and applications of visible and infrared remote sensing in the study of land-surface processes. The chapters are self contained but their interrelationship is discussed in Chapter 1.

This book is not intended to be the final word in our knowledge in this area of science, but only a start in establishing open communication and future cooperation among scientific disciplines using optical remote sensing as a tool to study land-surface processes. The challenge remains to further our understanding of remotely sensed data exhibiting high spectral and spatial resolution and to devise broader applications of the data which will be acquired by the Polar Orbiting Earth Observing Platforms during the 1990s.

GHASSEM ASRAR

Washington, D.C.

CONTENTS

CHAPTER 4

SOIL INFLUENCES IN REMOTELY SENSED VEGETATION-CANOPY SPECTRA   **107**

Alfredo R. Huete

CHAPTER 5

THE THEORY OF PHOTON TRANSPORT IN LEAF CANOPIES   **142**

Ranga B. Myneni, Ghassem Asrar, and Edward. T. Kanemasu

CHAPTER 6

INVERSION OF CANOPY REFLECTANCE MODELS FOR ESTIMATION OF BIOPHYSICAL PARAMETERS FROM REFLECTANCE DATA   **205**

Narendra S. Goel

CHAPTER 7

ESTIMATION OF PLANT-CANOPY ATTRIBUTES FROM SPECTRAL REFLECTANCE MEASUREMENTS **252**

Ghassem Asrar, Ranga B. Myneni, and Edward. T. Kanemasu

CHAPTER 8

VEGETATION-CANOPY SPECTRAL REFLECTANCE AND BIOPHYSICAL PROCESSES **297**

Piers J. Sellers

CHAPTER 9

THE ATMOSPHERIC EFFECT ON REMOTE SENSING AND ITS CORRECTIONS **336**

Yoram J. Kaufman

1

INTRODUCTION

GHASSEM ASRAR

Headquarters
National Aeronautics and Space Administration
Washington, D.C.

I INTRODUCTION

Remote sensing is defined as acquisition of information about the condition and/or the state of a target by a sensor that is not in direct physical contact with it. The sensors that are currently used for this purpose are divided into two groups: active and passive systems. The active sensors generate and transmit a signal toward the target and receive and record the returned signal after its interaction with the target. The relationship between the transmitted and received signal is used to characterize the condition or state of the target. Active microwave sensors and lidars are examples for this category. The passive sensors do not generate or transmit a signal, however; they detect and record the natural electromagnetic energy reflected and/or emitted from a target. The magnitude and shape of the signal are indicators of the condition/state of the target. Cameras, radiometers, and scanners are some examples for this category.

In this book, we consider the measurements, analyses, and applications of the remotely sensed data acquired by the passive-sensor systems in the visible to thermal infrared ($0.4-16$ μm) regions of the electromagnetic spectrum. The following chapters were prepared by a group of scientists who are actively conducting research on these topics. We have organized these chapters to provide continuity throughout the book. In the following sections, we present an overview of the topics covered throughout the book to establish the relationship among different chapters. We also devote some space to the topics of types of sensors, radiometric calibration, and data-reduction techniques in this chapter, since they are mentioned frequently throughout the book but not treated as separate topics.

II SENSORS

The sensor systems currently used in optical remote sensing of the Earth's surface can be classified in different ways. This classification can be based on the mode of operation of the sensors, their spectral characteristics, and/or the platform on which the sensor is

1

deployed. If classified according to the last approach, the optical sensors can be divided into ground-based, airborne, and spaceborne systems.

There is a wide variety of ground-based radiometers with different bandpass filters that mimic the operational airborne and spaceborne radiometers and scanners. Most of these sensors operate basically on the same principles, except that they measure the reflected or emitted energy from the surface over different wavelength regions of the electromagnetic spectrum. The history of development of ground-based sensors along with some examples are discussed in Chapter 2. The airborne sensors are generally designed to serve as a prototype for the future spaceborne sensor systems. Some of the currently operational airborne sensors are summarized in Table 1. The spaceborne sensors have provided the global coverage of the Earth's surface conditions at different spatial and temporal resolutions. Some of the past and present satellites with optical sensors and their characteristics are summarized in Tables 2 and 3. The following chapters describe in detail how data obtained by these satellites are used to assess the land-surface condition/state by scientists from different disciplines. A large number of future spaceborne sensor systems are being proposed in connection with the Polar Orbiting Platforms. A summary of these activities, including the characteristics of the proposed sensors, is presented in Chapter 18.

III ABSOLUTE RADIOMETRIC CALIBRATION

The first step in analysis of remotely sensed data obtained by a radiometer is to convert the instrument output units, typically counts or volts, to energy units, typically radiance. The relationship between the instrument output and incident radiance is called absolute radiometric calibration of the sensor. The following discussion only pertains to the visible and short-wavelength (0.4–3.0 μm) region of the spectrum.

The output of a sensor, Q (volt), is a function of the spectral radiance (L_λ) at the

TABLE 1 Some Examples of Airborne Optical Sensor Systems

Sensor	Number of Channels	Spectral Bandwidth (μm)	Spectral Sampling Interval
Airborne Ocean Color Imager (AOCI)	10	0.44–12.28	20-nm visible & near-IR 60-nm shortwave–IR 3.86-μm thermal–IR
Advanced Solid-State Array Spectrometer (ASAS)	30	0.45–0.88	14 nm
Airborne Visible-Infrared Imaging Spectrometer (AVIRIS)	220	0.41–2.45	9.4–9.7 nm
Thermal Infrared Multispectral Scanner (TIMS)	8	8.20–12.2	0.40–1.4 μm
Thematic Mapper Simulator (TMS-NS001)	8	0.45–12.3	0.06–1.4 μm
Thematic Mapper Simulator (TMS-Deadalus)	11	0.42–14.0	0.02–5.5 μm

TABLE 2 Past Satellites with Optical Sensor Systems

Satellite	Sensor	Spectral Regions (μm)	Orbital Characteristics	Repeat Cycle	Time of Data Acquisition (hour)	Data Availability
ITOS-1 and NOAA-1	SR (Scanning Radiometer)	0.45–0.65, 0.50–1.00 / 10.5–12.5	Near-polar sun-synchronous	12 hours		April 1970 to June 1971 / April to June 1971
NOAA-2 NOAA-5	VHRR (Very High Resolution Radiometer)	0.60–0.70 / 10.5–12.5	Near-polar, sun-synchronous	12 hours	08.30 and 20.30	November 1972 to March 1978
TIROS-N	AVHRR (Advanced Very High Resolution Radiometer)	0.55–0.90, 0.725–1.1 / 3.55–3.93, 10.5–11.5	Near-polar, sun-synchronous	12 hours, every 9.2 days	15.00 (ascending, node) and 03.00 (descending node)	February 1979 to November 1980
NOAA-8	AVHRR-2	0.58–0.68, 0.725–1.1 / 3.55–3.93, 10.5–11.5, 11.5–12.5	Near-polar, sun-synchronous	12 hours, every 9.2 days	19.30 (ascending node) and 07.30 (descending node)	March 1983 to June 1984
SEASAT-A	VIR (Visible and Infrared Radiometer)	0.47–0.94 / 10.5–12.5	Near-polar, nonsun-synchronous	Every 36 hours	Alternating, day and night	June to October 1978; only selected areas covered
LANDSAT-3	MSS (Multispectral Scanner System)	0.50–0.60, 0.60–0.70, 0.70–0.80, 0.80–1.10 / 10.4–12.6	Near-polar, sun-synchronous	18 days	09.31 (descending node)	March 1978 to March 1979
HCMM	HCMR (Heat Capacity Mapping Radiometer)	0.55–1.10 / 10.5–12.5	Sun-synchronous, circular between 85°N and 85°S	12 hours, every 16 days	14.00 (ascending node) and 02.00 (descending node) at equator; 13.30 and 02.30 in Northern Hemisph.; 14.30 and 01.30 in Southern Hemisph.	April 1978 to September 1980

TABLE 3 Present Satellites with Optical Sensor Systems

Satellite	Sensor	Spectral Regions (μm)		Orbital Characteristics	Repeat Cycle	Time of Data Acquisition (hour)	Data Availability
SMS-1 and 2 GOES-1 to GOES-5	VISSR (Visible and Infrared Spin Scan Radiometer)	0.55–0.70	10.5–12.6	Geostationary	Stationary	Every 30 min	June 1974 to August 1981 January 1976 to present
METEOSAT	VISSR	0.40–1.10 5.70–7.10	10.5–12.5	Geostationary	Stationary	Every 30 min	March 1977 to present
GMS	VISSR	—	10.5–12.5	Geostationary	Stationary	Every 30 min	April 1978 to present
NIMBUS-7	CZCS (Coastal Zone Color Scanner)	0.433–0.453 0.510–0.530 0.540–0.560 0.660–0.680 0.700–0.800	10.5–12.5	Near-polar, sun-synchronous	—	12.00	September 1978 to present
DMSP Block 5 F1 to F4	OLS (Operational Linescan System)	0.40–1.10	8.0–13.0 F4: 10.5–12.0	Near-polar, sun-synchronous	12 hours	—	March 1979 to present
NOAA-6	AVHRR	0.58–0.68 0.725–1.1	3.55–3.93 10.5–11.5	Near-polar, sun-synchronous	12 hours, every 9.2 days	19.30 (ascending node) and 07.30 (descending node)	June 1979 to present Taken out of service March 1983 Reinstated June 1984

Satellite	Sensor	Spectral bands (μm)		Orbit	Repeat cycle	Equator crossing time	Period of operation
NOAA-7	AVHRR-2	0.58–0.68 0.725–1.1	3.55–3.93 10.5–11.5 11.5–12.5	Near-polar, sun-synchronous	12 hours, every 9.2 days	14.30 (ascending node) and 02.30 (descending node)	June 1981 to present
NOAA-9	AVHRR-2	0.58–0.68 0.725–1.1	3.55–3.93 10.5–11.5 11.5–12.5	Near-polar, sun-synchronous	12 hours, every 9.2 days	19.30 (ascending node) and 07.30 (descending node)	November 1984 to present
LANDSAT-4 LANDSAT-5	TM (Thematic Mapper)	0.45–0.52 0.52–0.60 0.63–0.69 0.76–0.90 1.55–1.75 2.08–2.35	10.4–12.5	Near-polar, sun-synchronous	16 days	09.45 ± 15 min	July 1982 to present March 1984 to present
SPOT-1*	HRV (High Resolution Visible)	0.50–0.59 0.62–0.66 0.77–0.87		Near-polar, sun-synchronous	26 days, and pointing capability provide shorter cycles	10.30	March 1986

*SPOT-2 is scheduled for launch in 1989.

sensor aperture and the absolute spectral responsivity (R_λ) of the sensor:

$$Q = F(L_\lambda, R_\lambda) \tag{1}$$

The output from most of the currently operational systems are linear or nearly linear with input radiance across their usable dynamic range, and Eq. 1 is a valid representation. The actual relationship is more complicated since it should represent the spatial, temporal, spectral, and environmental dependence and characteristics of the instrument and the target. In Eq. 1, it is assumed that the radiance is spatially uniform and depolarized, and the target and sensor responses are time-independent. Based on these assumptions, Eq. 1 reduces to

$$Q = k \int_{d\lambda} L_\lambda r_\lambda \, d\lambda + O \tag{2}$$

where O is the sensor offset, k is the instrument peak responsivity, r_λ is the relative spectral responsivity of the instrument, and $d\lambda$ is the wavelength band. The desired quantity is typically the average spectral radiance ($\bar{L}_\lambda$) over the spectral wavelength bandpass ($d\lambda$) or an integrated radiance (L') over the bandpass:

$$\bar{L}_\lambda = \frac{\int_{\lambda_1}^{\lambda_2} L_\lambda \, d\lambda}{\lambda_2 - \lambda_1} = \frac{L'}{\lambda_2 - \lambda_1} \tag{3}$$

where λ_1 and λ_2 are the lower and upper bandpass limits, respectively, in micrometers. An exact solution to Eq. 3 exists only if the bandpass is well defined and the spectral radiance from the target as a function of wavelength is known (Kostkowski and Nicodemus, 1978). In most cases, a weighted average spectral radiance (L'_λ) can be used if a slowly varying radiance can be expected at the sensor aperture:

$$\bar{L}'_\lambda \cong \bar{L}_\lambda = \frac{\int_{d\lambda} L_\lambda r_\lambda \, d\lambda}{\int_{d\lambda} r\lambda \, d\lambda} \tag{4}$$

In this case, the moments method (Palmer and Tomasko, 1980) is an accurate representation of the bandpass:

$$\lambda_c = \frac{\int_{d\lambda} \lambda r_\lambda \, d\lambda}{\int_{d\lambda} r\lambda \, d\lambda} \tag{5}$$

and

$$\sigma^2 = \frac{\int_{d\lambda} \lambda r_\lambda \, d\lambda}{\int_{d\lambda} r\lambda \, d\lambda} - \lambda_c^2 \tag{6}$$

where

$$\lambda_1 = \lambda_c - 3^{1/2}\sigma \tag{7}$$

$$\lambda_2 = \lambda_c + 3^{1/2}\sigma \tag{8}$$

In sensors that have one detector per spectral wavelength band (e.g., the Advanced Very High Resolution Radiometer), this is a reasonable approach; however, even in this case, some radiometric errors can still result from spectral dependence of the target. In sensors with multiple detectors within a given wavelength bandpass, this approach may result in spectral striping of the image, especially for targets that do not have a flat spectral response (e.g., vegetation).

Equation 2 can now be written as

$$Q = GL'_\lambda + O \tag{9a}$$

or

$$L'_\lambda = \frac{Q - O}{G} \tag{9b}$$

where G is the gain and defined as

$$G = \frac{\int_{d\lambda} r_\lambda \, d\lambda}{k} \tag{10}$$

Therefore, to compute a mean in-bandpass radiance, one has to know the gain, the offset, and the wavelength spectral bandpass.

Sensors are usually calibrated by exposing them to at least two levels of known spectral radiance and calculating the gains based on the Eqs. 2 or 10. There are several sources of information on the gains and offsets for the operational Earth-observing satellites. These include (1) prelaunch measurements, (2) on-board or extraterrestrial measurements, and (3) measurements made on the known targets at the Earth's surface at the time of satellite overpass. The group 1 and 2 data are normally included in the instrument documents and/or as header information on the data tapes provided by the operational organization(s) or the vendor of the data in case of commercial satellites. A summary of these data is published by Malila and Anderson (1986). Data from group 3 activities are available in a variety of sources in the open literature. There are some coordinated efforts to compile most of this information in one place for ease of access by the interested reader. The results from some of these activities are published in a special issue (Vol. 22, No. 1, 1987) of *Remote Sensing of Environment*.

IV METHODS OF DATA ANALYSIS

There is a variety of methods for deriving the information content of the spectral-reflectance and thermal-emittance data acquired by remote-sensing sensors. The choice of method depends on the type of information sought and the complexity of the underlying processes to be depicted with the data. The simplest method combines the spectral data from different wavelength bandpasses to produce a photographic product called an image. An example is a false-color infrared image in which the reflected energy from the red wavelength band is usually replaced with the reflected energy from the near-infrared region. In this case, a target such as green vegetation looks mostly red due to its strong reflectance in the near-infrared region. The utility of such image products is limited to their qualitative or semiquantitative applications, at best.

The spectral-reflectance and emittance data can be also used in conjunction with radiative transfer models to assess quantitatively the surface biological and geophysical parameters. This approach is discussed in detail in some of the following chapters, especially Chapter 6. The most popular method of using remotely sensed spectral-reflectance data is based on the spectral indices. An index is formulated to depict a feature of the target based on its spectral characteristics in certain regions of the electromagnetic spectrum. For example, the ratio index (RI),

$$RI = \frac{\rho_r}{\rho_{ir}} \tag{11}$$

and normalized difference vegetation index (NDVI),

$$NDVI = \frac{\rho_{ir} - \rho_r}{\rho_{ir} + \rho_r} \tag{12}$$

were proposed to study the vegetative surfaces. Other similar vegetation indices have been suggested by many scientists. The two-dimensional vegetation indices correlate well with plant canopy attributes, such as green leaf area and phytomass. A detailed discussion of this topic is presented in Chapters 7 and 8. The main disadvantage of the two-dimensional spectral indices is their dependence on the spectral characteristics of the background features, such as soil and senescent vegetation for plant canopies. This topic is treated in detail in Chapter 4.

A second group of spectral indices are computed from linear combinations of the multispectral-reflectance data. This approach is particularly useful for the sensors that have a large number of wavelength bandpasses that are highly correlated. Kauth and Thomas (1976) first introduced such an approach for the analysis of the Multispectral Scanner System (MSS) data for plant canopies. They discovered four orthogonal indices relating to soil and vegetation conditions. The method is based on the spectral contrast between the green vegetation and the background feature(s). Jackson (1983) extended this approach to allow analysis of data from sensors with any number of wavelength bandpasses. This extended procedure requires only three multispectral data points, representing dry and wet soil, and green vegetation. Miller et al. (1983) proposed a modified principle component analysis (PCA) procedure that allows using complete multispectral and multitemporal data sets for deriving the orthogonal indices. Badhwar et al. (1982) proposed an automatic method of identifying the stages of plant-canopy development and their subsequent classification based on the orthogonal greenness vegetation indices.

Future remote-sensing sensor systems will have higher spectral resolution and larger numbers of spectral bandpasses, as compared with the current sensor systems. This will result in data sets of higher dimensions that contain additional information about the land-surface condition/processes. A summary of data-analysis techniques commonly used in geology is presented in Chapter 12. The current methods of multispectral data analysis do not provide the needed capability for deriving the full information content of these high-resolution multispectral sensor systems. There is a need for the development of more innovative approaches for deriving quantitative information offered by the current and future generation of high-resolution multispectral sensors.

V OUTLINE OF THE FOLLOWING CHAPTERS

The radiant solar energy that reaches the Earth's surface is composed of direct solar-beam and diffuse-sky radiation. The diffuse component results from propagation through the atmosphere. A portion of the total (direct + diffuse) incident solar energy that reaches the Earth's surface is absorbed by the surface and the rest is reflected back toward the atmosphere. The absorbed energy is used in biological, chemical, and/or physical processes such as photosynthesis, evaporation, and heating of the Earth's surface. The quantity and quality of the reflected component depends on the condition and state of the surface. This component is the main focus of passive remote-sensing studies in the visible to short-wavelength (0.4–2.5 μm) region of the spectrum. The reflected radiation from the surface is measured as a function of wavelength by radiometers, scanners, and spectrometers that are mounted on ground-based, airborne, or spaceborne platforms. This information is then used to characterize the condition/state of the surface at the time of measurements. A large number of factors may affect these measurements.

The solar energy reflected from the surface must propagate through the atmosphere before reaching the sensor. In case of ground-based measurements, the small volume of the atmosphere between the surface and the sensor does not play a significant role in attenuation and scattering of the reflected upwelling solar energy; however, for aircraft- and satellite-based measurements, the atmospheric effects are significant and they should be resolved for proper assessment of the surface conditions/state from remotely sensed data. The ground-based measurement techniques are discussed in depth in Chapter 2. The effects of atmosphere on aircraft- and satellite-based measurements and some methods for removing such effects are described in detail in Chapter 9.

The magnitude of reflected solar energy from the Earth's surface depends also on the surface-cover type and condition. For example, the spectral-reflectance characteristics of a bare-soil surface is distinctly different from the same soil if it is partially covered with vegetation; see Figure 1. The unique spectral features observed at different regions of the spectrum for different soil types may be used to distinguish among them; see Figure 2. In Chapter 3, a detailed discussion of the interaction of solar energy with bare soils along with some examples of applications of this information in soil classification

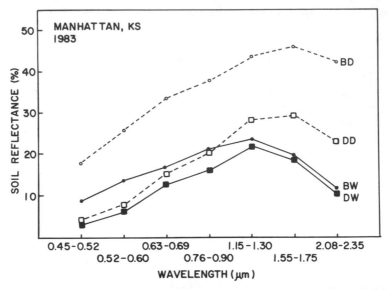

Figure 1 Visible to short-wavelength infrared bidirectional reflectance of dark (*D*) and bright (*B*) soils under dry (*D*) and wet (*W*) conditions. [After Asrar et al. (1987).]

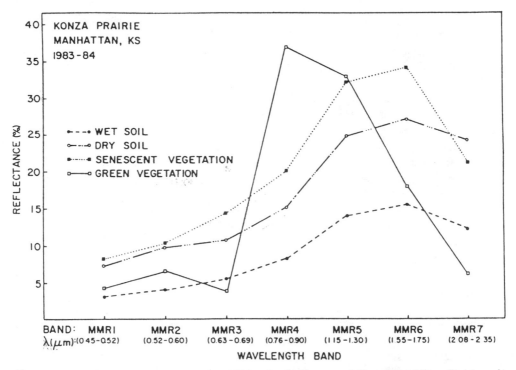

Figure 2 Visible to short-wavelength infrared bidirectional reflectance of dry and wet soils, and green and senescent tallgrasses. [Reprinted by permission of the publisher from Asrar et al. (1986). Copyright, 1986 by the Elsevier Scientific Publishing Co., Inc.]

is presented. In remote sensing of vegetative surfaces, the underlying soil also plays a major role, especially under partial vegetation cover conditions. In this case, the reflected solar energy from the surface is composed of the contributions from both soil background and vegetation. To characterize the vegetative surfaces based on the remotely sensed data, the influence of soil background on composite canopy reflectance has to be minimized. Chapter 4 is devoted to a detailed discussion of this subject.

The interaction of solar energy with vegetation canopies can be modeled based on the radiative transfer equations, subject to some simplifying assumptions. A wide variety of methods is used to solve these equations, based on the complexity of the assumptions made in solving them. A general discussion of such models along with some examples are presented in Chapter 5. The radiative transfer models can be used in the forward direction to simulate the interaction of solar energy with the Earth's surface. This is an effective method of studying the role of factors that contribute to a given composite surface-reflectance characteristic. Radiative transfer models can be also used along with measured surface-reflectance data in an inverse mode, to infer some surface or near-surface biological or geophysical characteristics. The conceptual approach and some examples for inversion of plant-canopy radiative transfer models are presented in Chapter 6.

Remotely sensed spectral-reflectance data can be used to estimate plant-canopy attributes such as green-leaf area, green phytomass, and photosynthetic capacity. Chapters 7 and 8 are devoted to estimation of plant-canopy attributes and physiological processes from remotely sensed visible and near-infrared reflectance data. These two chapters present a few case examples for retrieving some quantitative information about the condition/state of plant canopies from remotely sensed spectral-reflectance data, based on some basic physical and physiological principles. Chapter 9 is devoted to a detailed discussion of atmospheric effects on remotely sensed visible to short-wavelength infrared reflectance and some procedure for correction of such effects. Chapters 10–13 are devoted to applications of remotely sensed data in forestry, wetlands, geology, and snow studies. The case studies presented in these four chapters demonstrate how the same basic information is used in different disciplines to assess the Earth's land-surface condition/state. Chapter 14 contains description of a knowledge-based spectral-classification model. This is becoming a very popular approach in conjunction with Geographic Information Systems (GIS) for identification and classification of land-surface cover types and conditions.

Absorption of the solar energy by the Earth's surface results in its increased internal kinetic energy and, hence, an increase in its temperature. This increased temperature leads to a thermal-gradient difference between the surface and the surrounding environment. During daylight hours, this gradient is normally positive toward the surface, but at night, it reverses, and the surface usually emits the extra internal energy. In addition, the surface may gain or lose energy through exchange of latent and sensible heat until it comes to equilibrium with the surrounding environment. The land-surface cover type and the type of materials immediately below the surface play a significant role in the rate of gain/release of the energy from the surface. Therefore, the thermal energy released from the surface can be used as an indicator of the state/condition of the surface at the time of measurement. Similar to the visible and short-wavelength infrared regions of the spectrum, this upwelling thermal energy must travel through the atmosphere before reaching the airborne and spaceborne remote-sensing sensor systems. The atmosphere has a pronounced influence on the emitted thermal energy from the surface, es-

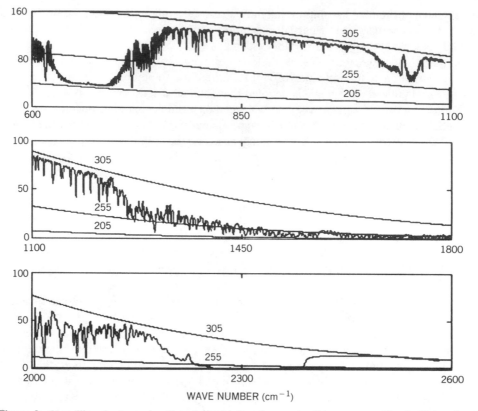

Figure 3   Upwelling short-wavelength to thermal infrared spectral radiance measured by the University of Wisconsin High Resolution Infrared Sounder (HIS) aboard a NASA U2 aircraft at an altitute of 20 km over Huntsville, Alabama, on June 15, 1986. Curves of Plank radiance for equivalent blackbody temperatures of 205, 255, and 305 K are superimposed. The upper panel displays the absorption lines due to attenuation of upwelling radiance by carbon dioxide (600–700 cm^{-1}), water vapor (750–1000 cm^{-1}), and ozone (1000–1100 cm^{-1}). The middle panel shows absorption due mainly to water vapor and the lower panel illustrates absorption by water vapor (2000–2200 cm^{-1}) and nitrous oxide and carbon dioxide (2200–2400 cm^{-1}).

pecially in a certain water-absorption region of the spectrum; see Figure 3. Some knowledge of the atmospheric gaseous composition at the time of observation is required for proper interpretation and applications of the thermal-emission data. Chapter 15 is devoted to a detailed discussion of thermal infrared measurements and modeling. A brief description of the atmospheric effects on thermal infrared data is also presented.

The incident solar energy plays a significant role in biological, chemical, and physical processes that take place at the Earth's surface. Remotely sensed radiometric measurements can be used to assess and estimate the components of surface-energy budget. This is an attractive approach, since operational Earth-observing satellites provide a synoptic coverage of extended areas over a very short period, repeatedly. An overview of the principles involved in using remotely sensed data to estimate the components of land-surface energy budget and a few examples are presented in Chapter 16. Chapter 17 is devoted to the modeling of evapotranspiration and carbon assimilation by plant canopies, and the application of remotely sensed data in such models. Both one- and two-dimensional models along with some case examples are discussed. The final chapter contains a summary of the past accomplishments and current remote-sensing research activities in the field of terrestrial ecology. Some future directions for research activities in this field of science are also presented in this chapter.

ACKNOWLEDGMENT

The section on radiometric calibration was prepared with the help of Mr. Brian Markham, NASA/Goddard Space Flight Center, Greenbelt, Maryland.

REFERENCES

Asrar, G., R. L. Weiser, D. E. Johnson, E. T. Kanemasu, and J. M. Killeen (1986). Distinguishing among tallgrass prairie cover types from multispectral reflectance. *Remote Sens. Environ.* **19:**159–169.

Asrar, G., R. D. Martin, G. P. Miller, and E. T. Kanemasu (1987). Remote sensing of near-surface soil conditions. *Proc. Int. Conf. Infiltr. Dev. Appl.*, pp. 466–474.

Badhwar, G. D., J. G. Carnes, and W. W. Austin (1982). Use of Landsat-derived temporal profile for corn-soybean features extraction and classification. *Remote Sens. Environ.* **12:**57–79.

Jackson, R. D. (1983). Spectral indices in *n*-space. *Remote Sens. Environ.* **13:**401–429.

Kauth, R. J., and G. S. Thomas (1976). The tassled cap—A graphic description of the spectral-temporal development of agricultural crops as seen by Landsat. In *Machine Processing of Remotely Sensed Data.* Purdue University, West Lafayette, Indiana, pp. 4B/41–51.

Kostkowski, H. J., and Nicodemus, F. E. (1978). An introduction to measurement equation. *NBS Tech. Note (U.S.)* **910-2,** p. 105.

Malila, W. A., and D. M. Anderson (1986). *Satellite Data Availability and Calibration Documentation for Land Surface Climatology Studies*, Rep. No. 180300-1-f. Environmental Research Institute, University of Michigan, Ann Arbor.

Miller, G. P., M. Fuchs, M. J. Hall, G. Asrar, E. T. Kanemasu, and D. E. Johnson (1983). Analysis of seasonal multispectral reflectance of small grains. *Remote Sens. Environ.*, **14:**153–167.

Palmer, J. M., and M. G. Tomasko (1980). Broadband radiometry with spectrally selective filter. *Opt. Lett.* **5**(5):208.

2

FIELD MEASUREMENTS OF BIDIRECTIONAL REFLECTANCE

DONALD W. DEERING

Earth Resources Branch
Laboratory for Terrestrial Physics
Goddard Space Flight Center
National Aeronautics and Space Administration
Greenbelt, MD

I INTRODUCTION

Since the beginnings of Earth resources remote sensing from satellites, the general expectation has been that field measurements or observations taken directly at the Earth's surface should ultimately become unnecessary. Progress toward that expectation has been elusive, however, as field research programs have evolved through several stages and philosophies. The philosophy expressed by many (but not all) involved in remote sensing from space in the early 1970s was overly optimistic. The rationale was that measurements from sophisticated spaceborne sensors encompassing a variety of wavelength regimes should eventually provide all of the information required to correct for perturbations in the intervening atmosphere and permit the extraction of the desired surface parameters. One example of the kind of thought that was prevalent in those years was a preoccupation with "spectral signatures" (e.g., Holter, 1970), which was thought to be the key to unique discrimination of major plant species or assemblages, soil and rock types, etc. from satellite remote sensors.

Major remote-sensing research programs, such as the multi-U.S. agency sponsored Large Area Crop Inventory Experiment, or LACIE (Bauer et al., 1978; MacDonald and Hall, 1980), and the Agriculture and Resources Inventory Surveys Through Aerospace Remote Sensing (AgRISTARS) program (Bauer et al., 1986), which contained major field measurement efforts, were launched in order to develop operational satellite-use capabilities for government and eventually private industry. Most of the Earth resources remote-sensing investigations throughout the world for more than a decade and a half beginning about 1970 were focused on such applications. They ranged in scope from geological mapping and exploration to vegetation mapping and crop yield assessment to wildlife census and used primarily the Landsat series of satellites, including, initially, the Multispectral Scanner System (MSS) and, later, the Thematic Mapper (TM). Uses of other satellite sensors, such as the Advanced Very High Resolution Radiometer (AVHRR), and somewhat later the French Systeme Probatoire pour l'Observation de la Terre (SPOT) satellite sensors had become important by the mid-1980s.

The intensive field measurement programs of the 1970s and early 1980s were fruitful and contributed much to satellite data-processing and thematic information capability development (e.g., "signature-extendable technology" for global space-based crop recognition; Hall and Badhwar, 1987). But they were found to have yielded less than expected in the way of fundamental understanding of the physical relationships between electromagnetic (E-M) radiation and surface variables and processes, which was needed to substantially advance the state of the art. Much of the earlier field work was limited to scene-classification verification. As a consequence, when funding institutions began to suffer budget cuts and rescope their research programs, most of the field program activities were winnowed out such that the mid-1980s saw little field measurement research being conducted. Field measurements in support of fundamental remote-sensing research activities and satellite ground truth had become rather diffused.

Recently, a greatly heightened global awareness of the need to repetitively measure and model the Earth, its components and processes, as a single system has ushered in a new era of global "Earth Systems Science." This auspicious global monitoring and study approach is opportunely coupled with the prospects for more advanced, multiple-sensor Earth Observing Satellite (EOS) systems for the 1990s, incorporating instruments from many nations [National Aeronautics and Space Administration (NASA), EOS Steering Committee, 1987]. And unlike the earlier philosophy, this new research thrust gives a clear recognition that ground-based measurement programs are critical for both the short and long term.

> Observations *in situ* will probably contribute a large part of the information needed for process studies designed both to identify the key global variables that define the states of the various subsystems of the Earth system and to establish the understanding required for models of the connections among these subsystems. In addition, almost all measurement systems for global variables will continue to include an *in situ* element. . . . The real power of the approach comes from a judiciously chosen blend of space-based remote-sensing and *in situ* techniques, combining the global coverage from space with the measurements-validation and calibration-reference capabilities of *in situ* approaches [National Aeronautics and Space Administration (NASA) Advisory Council, 1988].

The message from the current scientific community is that space observations will have to be rather routinely complemented by field measurement studies of Earth properties. And it seems likely that the increase in understanding of the physical processes involved in the interactions of radiation with the atmosphere and Earth's surfaces will be followed by an increased demand for ground-based information of greater precision and detail to calibrate and validate remote-sensing data and products. Thus, the requirement for field measurements should continue. In fact, even more comprehensive field campaigns are being undertaken in the late 1980s than have previously been conducted, such as the First ISLSCP Field Experiment (FIFE) under the auspices of the International Satellite Land Surface Climatology Project (ISLSCP).

Prior to FIFE, there had been no significant attempts to take simultaneous land-surface observations of meteorological and biophysical parameters at sufficient temporal and spatial resolution to allow for the effects of surface heterogeneity and over a large enough area to be observable from moderate-resolution satellite sensors. Such data would permit legitimate comparison of satellite-derived quantities with actual surface conditions. The *two problems* central to FIFE, which are *generically relevant* to most "ground truth" studies, are *(1) whether our understanding of processes* (e.g., biological) *on the small scale, of microns to meters, can be rigorously integrated over space to describe interactions appropriate to atmospheric length scales and (2) whether such processes* (as photosynthesis and evapotranspiration) *or associated states*, (such as chlorophyll density, soil moisture, and reflectance), *can be quantified sufficiently by remote sensing techniques*. FIFE was also seen as an opportunity to refine ground-measurement techniques and improve the experimental methodology to allow the execution of further, simplified experiments for the wide variety of land-surface types that occur throughout the world (Sellers et al., 1988).

I.A   Purposes and Predicaments

Field measurements are typically taken to serve as (1) "ground truth" for spacecraft or aircraft remote-sensing data calibrations, validations, and interpretation and/or (2) for studying the relationships between surface physical properties or process phenomena and the impinging E-M radiation field. In most instances, the purpose of the so-called ground truth data "is to standardize the [satellite or aircraft] data and to remove atmospheric effects, to define the surface context and to determine the relations between the spectral signature and the object of inquiry" (Steven, 1987). Ground truth involves the collection of measurements and observations about the type, size, distribution, condition, and any other physical or chemical properties that may be important concerning the materials on the Earth's surface or the intervening atmosphere being remotely sensed.

Remote-sensing studies of the second type may or may not have coincident aircraft or satellite data acquisitions. Often the more fundamental studies of the physical relationships between E-M radiation and surface properties and processes are conducted with strictly ground-based radiometers. The major concern here is understanding the physics of the interaction between solar radiation and the Earth's surface elements (e.g., plant canopies, their component parts and structure, and soils).

A third potential use for field measurements is to evaluate the potential for the application of remote sensing using different spectral bands for a specific task (Milton, 1980), such as the discrimination of mineralized zones (Goetz et al., 1975; see also Chapter 12), although this could be the logical product of the second objective of field measurements (fundamental studies).

In the first stage of developing an understanding of the surface-energy relationships, field measurements are typically gathered in conjunction with the remote-sensing data in order to try to establish statistical correlations between the observed E-M measurements and the properties of the irradiated surface.

> Once statistical relationships have been established, the next stage is to try to understand the physics underlying the relationship, both because of its intrinsic interest and because of the need to optimize system performance for particular applications. This places a second role on ground truth, since it must be used to validate the physical theories stemming from it (Lamont et al. 1987).

In other words, the physical understanding is incorporated into a model(s) that relates the surface variables to the spectral reflectance (or emittance). Then the model is validated with spectral measurements for which the surface characteristics are known, and, ultimately, the model is used to determine the unknown characteristics of a surface from measured spectral reflectances. The research approach in the new era of Earth system science will appropriately consist of four *cyclic* steps: observations, analysis and interpretation, models (i.e., physical theories), and verification and prediction (NASA Advisory Council, 1988).

The knowledge base or understanding that has been acquired over the past couple of decades had clearly proven useful in Earth resources remote sensing, but the methods and results have both been subject to criticism. Most of the remote-sensing knowledge generated in recent years has been developed through inductive methodology, with the emphasis on image interpretation (or image-data analysis) through ground control (e.g., correlations with ground variables) and verification. This methodology usually emphasizes manipulation and generalization of data.

Curran (1987) relates that inductive methodology uses presumably true and unbiased observations to build up an objective description of facts or phenomena that are then shaped or ordered to derive theory and thereby knowledge. He provides some insight into the justifications for some of the criticisms through his appraisal that

> in the context of remote sensing the inductive methodology can be criticized on four counts. *Firstly*, no apparently confirming observations can ever show that a theory is true [that is, logical error]; *secondly*, the observations upon which the theory is based are theory-laden [that is, contain human bias and theory] and therefore unsuitable; *thirdly*, the ability to construct theory from observations alone can result in the theories being used, without understanding of process, as instruments of prediction or description; and *fourthly*, data fixation can lead to an emphasis on remotely-sensed data [as if the accumulation of data was an end in itself] rather than on ideas.

Deductive methodology, where the emphasis is not on observation but on the formulation of theory first and then testing of hypotheses, is clearly the much preferred *modus operandum* in remote sensing today, as it is in most other scientific circles. Unfortunately, in the context of remote sensing, the scientist still has a predicament as Curran (1987) concludes that

> the deductive methodology can also be criticized on four counts. *Firstly*, it is very difficult to falsify a theory as the observations rather than the theory are usually the more dubious; *secondly*, [remote-sensing scientists] tend to use the traditional but less powerful route of verification [by testing inferential rather than factual hypotheses]; *thirdly*, the deductive process is inherently subjective as a result of the ever-changing nature of theory acceptance [and a "protective belt" of hypotheses that includes auxiliary, restrictive hypotheses and assumptions]; and *fourthly*, there is evidence of pseudo-deduction where inductive work is made to wear deductive clothes.

With a new and worldwide interest in Earth system science, with increasing global experience in already proven remote-sensing capabilities, and with recent and soon-to-be-realized improvements in technology for field and space use, we can restate afresh Holz's (1973) optimism that ". . . we are on the threshold of a significant breakthrough in methods of understanding the environment in which we live."

Major advancements in remote sensing from space will require a thorough understanding of the way electromagnetic energy is transferred and how it interacts when it encounters matter in the atmosphere, at the Earth's surface and at the remote-sensing instrument. Well-designed field measurement programs will continue to play an integral role in this development process, especially for achieving plant-canopy level and whole-scene understanding for pushing ahead the state of the art, and then for repetitively checking the accuracy and stability of spaceborne sensors.

The remainder of this chapter will discuss necessary considerations for making field measurements through use of examples, discussions, and references to published literature on these topics. The goal is to provide the reader with an overview of field research that has been performed, the terminology that is used, the calibration procedures that are performed, and some precautions that must be considered, as well as a descriptive examination of the variety of instrumentation that has been employed in ground-based remote-sensing studies—on the backdrop of current technology and knowledge. A discussion of some of the observations that are acquired in remote-sensing field research is included, along with some useful factors to be considered in conducting a field measurements program. Some attention is given to a variety of salient topics, but pedantic treatment is avoided, especially for those subjects where a more detailed treatment is provided in other chapters in this book.

The primary emphasis in this chapter is given to field measurements in the optical-reflective or "total solar" regime of the E-M spectrum, which is nominally from 0.3 to about 3.0 μm. Additionally, most of the discussion will focus on plant-canopy or vegetation measurements, as soils and geology are treated extensively in Chapters 3 and 12, respectively. Field measurement instrumentation and its more recent historical evolution is given considerable synoptic emphasis. Thermal measurements are given only brief mention here as they are examined in detail in Chapter 15. Presentation of graphs and charts of radiance and reflectance data is minimized, as literally hundreds of internationally published journal articles (a number of which are cited herein) are replete with such data; however, some are given as illustrative aids to the reader.

II FUNDAMENTAL CONSIDERATIONS

At the most basic level, remote sensing is accomplished through the transfer of energy that is either emitted or reflected from a radiating object to a sensor. The sensor, in turn, reconstitutes this energy either into an image that can be detected by human sense organs or an electronic signal that can be recorded and subsequently displayed and analyzed. In Earth resources remote sensing, spectral radiances that are reflected from the surface and received by the sensor are a function of many factors. These include the continuously varying incident solar irradiance (e.g., geometry and spectral distribution), atmospheric conditions (e.g., water vapor and aerosols), meteorological conditions (e.g., temperature, wind, dew), reflectance properties of the surface (e.g., spatial, spectral, and biophysical conditions), and sensor viewing conditions (geometry, time of observation).

In the specific case of vegetation remote sensing, there are basically four factors that determine canopy reflectance: incoming solar flux, spectral properties of the vegetation elements (especially, leaves), architecture of the canopy (i.e., spatial distribution of vegetated and nonvegetated areas and vegetation characteristics, such as leaf-angle distribution and leaf-area index), and scattering from the soil (Goel, 1987).

II.A BRDFs and Reflectance Factors

The fundamental and intrinsic physical property governing the reflectance behavior of a scene element is its bidirectional reflectance distribution function, or BRDF (Nicodemus, 1982). The BRDF is useful as an underlying concept, but as it is geometrically and mathematically defined, ". . . it can never be measured directly because truly infinitesimal elements of solid angle do not include measurable amounts of radiant flux" (Nicodemus et al., 1977). The integration of this quantity over finite solid angles of incidence and exitance yields the "reflectance factor" that is actually estimated in most field measurements (Robinson and Biehl, 1979). However, "omnidirectional" (or multidirectional) field radiance measurements of sufficient angular density can also provide useful estimates of the BRDF.

The quantities that are directly estimated by radiometers are radiances, which are a complex combining of the BRDF with the irradiance field and instrument transducing characteristics. Because of the time dependence of radiance data, reflectance factors are frequently calculated to facilitate comparison of information concerning surface features from multitemporal data (Jackson, et al., 1987). A reflectance factor is defined as the ratio of the radiant flux actually reflected by a sample surface to that which would be reflected into the same reflected-beam geometry by an ideal (lossless, that is, 100% reflectivity) perfectly diffuse (Lambertian) standard surface irradiated in exactly the same way as the sample (Nicodemus et al., 1977).

The biconical reflectance factor is considered the standard reflectance term as defined fully by Nicodemus et al. (1977), although in field measurements, such terms as the hemispherical–conical reflectance factor (hemispherical incidence–conical reflection or collection) may be more appropriate. Usually, however, the term bidirectional reflectance factor, sometimes simply called the reflectance factor, is most often used to describe the field measured reflectance, whereby the target surface radiance is divided by the radiance of a level reference surface standard irradiated by the sun. For small fields of view ("angle of acceptance" or aperture $<20°$ full angle), this term is appropriate—with one direction being associated with the viewing angle (usually $0°$ from normal) and the other direction being associated with the solar zenith and azimuth angles (Robinson and Biehl, 1979). The term "bidirectional reflectance" is frequently used rather loosely to refer to the bidirectional reflectance factor(s) measured over targets from one or more nadir and off-nadir viewing angles (e.g., to sample the BRDF).

The bidirectional reflectance factor field calibration procedure of Robinson and Biehl (1979) is designed to obtain a property of the scene that is nearly independent of the incident irradiation and atmospheric conditions at the time of the measurement. It consists of the measurement of the response V_s of the instrument viewing the subject and measurement of the response V_r of the instrument viewing a level reference surface to produce an approximation to the bidirectional reflectance factor of the subject:

$$R_s(\theta_i, \phi_i; \theta_r, \phi_r) = \frac{V_s}{V_r} R_r(\theta_i, \phi_i; \theta_r, \phi_r) \qquad (1)$$

where $R_r(\theta_i, \phi_i; \theta_r, \phi_r)$ is the bidirectional reflectance factor of the reference surface; θ_i and θ_r are the zenith angles of incidence and reflection, respectively; and ϕ_i and ϕ_r are the azimuth angles of the incident and reflected rays, respectively. R_r is required to correct for its nonideal reflectance properties described later.

The assumptions in the approach are (1) that the incident radiation is dominated by its directional component (i.e., clear sky), (2) that the instrument responds linearly to entrant flux, (3) that the reference surface is viewed in the same manner as the subject and that the conditions of illumination are the same, (4) that the entrance aperture is sufficiently distant from the subject and that the angular field of view is small with respect to the hemisphere of reflected beams, and (5) that the reflectance properties of the reference surface are known.

Duggin and Cunia (1983) have further pointed out that large errors can arise in the reflectance factor due to changes in irradiance on the target between the time of measuring radiance reflected from the target and measuring the irradiance at the target (i.e., reference standard) if the "sequential" method is used. Duggin (1981) reported errors of up to 100% or more that were caused by sequential measurements in conditions of variable illumination, such as can occur with high, thin cirrus cloud cover. Their recommendation is to simultaneously measure target radiance and irradiance on the target. Such simultaneous measurements of radiance reflected from a target and from a reference standard panel can be made with two different instruments of the same type that are intercalibrated. A variant to this approach would be to use a cosine receptor on an inverted (sky-looking) instrument for the irradiance instead of the reference panel. Milton (1982) suggests that this additional expense and effort should not be necessary with replication of each measurement and careful data screening, which he believes are the best safeguards against the effects of short-time irradiance fluctuations.

II.B Reference Standards for Reflectance Computation

Perfect lossless Lambertian surfaces for use in computing reflectance factors cannot be achieved in practice; consequently, reference surface panels that are used as standards must be properly calibrated to account for their "loss" and non-Lambertian properties. Kimes and Kirchner (1982) discovered in a laboratory goniometer study of a typical spray-painted barium sulfate reference panel that errors in total spectral irradiance (at 0.68 μm) for a clear sky were 0.6, 6.0, 13.0, and 27.0% for solar zenith angles of 0°, 45°, 60° and 75°, respectively. Jackson et al. (1987) developed a relatively simple field goniometer procedure whereby reference panels can be calibrated using the sun as the radiation source. The radiometer that is used for field measurements is also used as the calibration instrument; and the component due to diffuse flux from the atmosphere is measured with the aid of an occluding disk and subtracted from the total irradiance. With cloud-free sky conditions, the accuracy of the method is estimated to be 1%.

Reference surfaces for field use have generally been fabricated by coating an aluminum panel with a highly reflective material such as barium sulfate ($BaSO_4$) or halon (polytetrafluoroethylene) (Schutt et al., 1981). However, inconsistencies in preparation and normal field use have been shown to cause substantial differences in painted-panel reflectance properties (Jackson et al., 1987). Weidner and Hsia (1981) provide instruction for the preparation of pressed halon surfaces according to National Bureau of Standards (NBS) specifications. Weidner et al. (1985) found that such reference standards were reproducible to within ±0.005 in reflectance for the 0.3–2.0 μm spectral interval and ±0.01 for the 2.0–2.5 μm interval. Most likely because of the difficulties in producing durable, consistent reflectance standards, a private research laboratory has recently marketed a product called Spectralon[1] (Labsphere, Inc.) that is a relatively hard, "baked" material that is declared to be durable, chemically inert, and washable, and therefore "ideal" for remote-sensing field use.

Another source of spectral radiance or reflectance error, the presence of nearby objects (e.g., the researcher's body), is of special concern in hand-held radiometry but

---

[1]The use of commercial product or company names in this chapter is not intended as an endorsement of any products or companies, but are included for descriptive purposes as an aid to the reader.

should also be considered with truck or other field operation configurations. Typically, a researcher positions himself on the side of the target point opposite the sun. The presence of his body (or truck) blocks a portion of the incoming diffuse solar radiance and ground exitance onto the target (or reference panel). Additionally, his body reflects incoming diffuse and direct solar irradiance and ground exitance onto the target point. As an example of the effect, for a solar zenith angle of 75°, the red and near-infrared (NIR) spectral bands radiances were found to be in error by 15 and 18%, respectively, with a person wearing white clothing at 0.5 m from the target (Kimes et al., 1983). Wearing black clothing resulted in only a 2% measurement error in both spectral bands. The effects are diminished with increased distance between the person and the target and with higher solar elevation angles. Measuring a reference panel in a different position than that of the target (e.g., positioned on top of a truck) can also introduce similar kinds of errors in the reflectance factor.

II.C Radiometric Calibration and Accuracy

Accuracy refers to confidence in the correlation between measurements in one location and another or between a measurement and a recognized standard. Precision, on the other hand, implies careful measurement under controlled conditions that can be repeated again and again with similar results. It also means that small differences can be detected and measured with confidence. Precision is related, then, to confidence in successive measurements with the same equipment and operating conditions. While it is not possible for a measurement to be highly accurate without being precise as well, accuracy and precision are not necessarily closely related (Collier, 1978). Precision refers to the repetitiveness of measurement, whereas accuracy refers to its correct relationship to a standard.

Remote-sensing field researchers have generally been more concerned with precision (although they may have called it calibration accuracy) and whether two or more similar instruments respond alike. The requirements for making high-accuracy field measurements for sensing the Earth's surface and near surface have not advanced as rapidly as they have in some other Earth science disciplines (Guenther, 1987), but the needs for high accuracy are increasing with more advanced space sensors being developed and the scientific community becoming more involved in intercomparison studies.

For acceptable calibration, instruments for field use must be characterized in comparison with standards maintained by the National Bureau of Standards (NBS) and related to an international, scientifically based set of standard units. The absolute radiometric scale for field measurements is obtained by using the field instrument as a transfer device to relate the energy from a scene to the energy emitted at a calibration standard. Interpretation of field measurements in terms of the output of the calibration source typically requires knowledge of several instrument operating characteristics, including relative linearity of the system over its dynamic range, system sensitivity to out-of-field-of-view sources, and system "crosstalk" between the several spectral and spatial channels (Guenther, 1987). Other details of instrument performance are important, such as definitive spectral resolution, sharpness of the field of view, temperature stability, and polarization sensitivity, and are receiving more critical attention as attempts to improve absolute accuracies and to push the interpretation of results increase. These details are mentioned to alert the reader but are not examined here further.

A laboratory calibration standard that has been shown to be useful for field as well

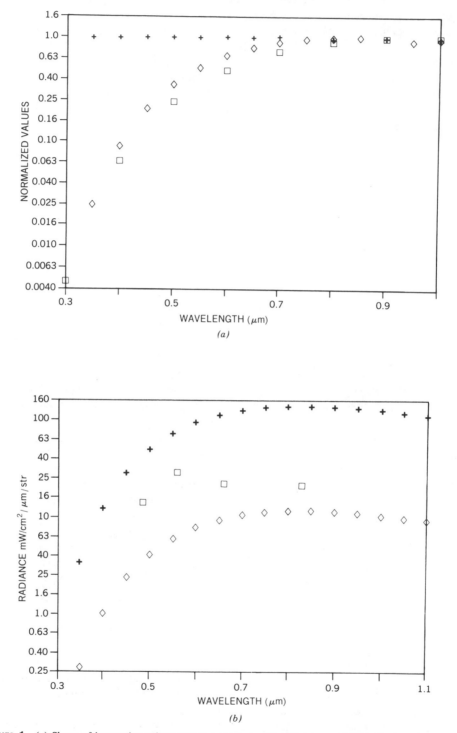

Figure 1 (a) Shape of integrating sphere output compared with reflectance and lamp shades. Normalized radiance of integrating sphere (diamonds) compared to reflectance of BaSO$_4$ paint (+) and blackbody source at 2850 K (squares). (b) Comparison of radiances for Thematic Mapper Bands 1–4 (squares) and radiances of integrating sphere with 1 (diamonds) and 12 (+) lamps. [From Guenther (1987).]

as aircraft instruments, and many other types of radiometers, is a large integrating sphere. Lovell (1984) describes an approach to relate the output of an integrating sphere to NBS maintained standards. At the NASA/Goddard Space Flight Center (GSFC) a 1.8 m diameter integrating sphere, which is equipped with twelve 200 W tungsten filament lamps mounted on the inside of the sphere, is used for instrument calibrations that are traceable to the NBS standard. The sphere output as compared with $BaSO_4$ paint and a 2850 K blackbody source is given in Figure 1(a), and a comparison of two different lamp level outputs that ''bracket'' typical Thematic Mapper spectral radiances in the visible and near infrared regions is illustrated in Figure 1(b). A modified integrating sphere, an integrating hemisphere, has also been developed and is shown in Figure 2; it is being used to calibrate a field radiometer.

Currently, the uncertainties in absolute calibration with the integrating sphere are no better than 5–6% (sometimes rising to over 10%) with the uncertainty of the lamps contributing nearly 3% of this in typical use. However, with improved technology, such as that based on the physics of silicon photodiode detectors, which is transduced to the candidate field instrument through the use of feedback-controlled dye laser systems, absolute calibrations with uncertainties smaller than 0.3% are feasible (Guenther, 1987).

II.D Geometric Parameters

In the early studies of Earth surface reflectivities with aircraft scanners, it was shown that reflectance anisotropy was a factor that must be understood in remote sensing. For

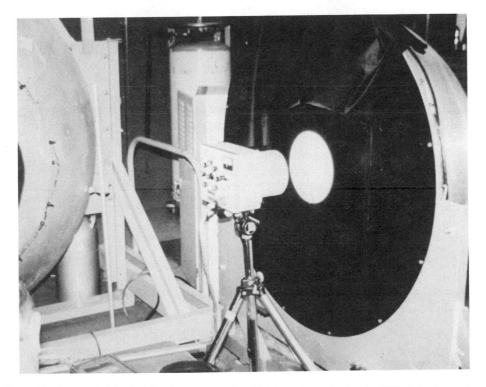

Figure 2 Barnes Modular Multiband Radiometer (MMR) being calibrated using the GSFC 1.2 m integrating hemisphere. Note the 1.8 m integrating sphere partially visible at left. (Photo courtesy F. G. Hall.)

example, Salomonson and Marlatt (1971) showed from aircraft scanner radiometry that reflectance increases sharply with view zenith angle. It may be that "geometric signatures" can be generalized for some broad classes of Earth surfaces (e.g., related to surface-structure and structural-element reflectance properties) from studies of their anisotropy characteristics (i.e., analogous to the spectral-signatures concept). Field measurements afford the opportunity to study such basic reflectance properties at angles out to the horizon, whereas aircraft sensors are often limited by instrument or aircraft hardware constraints, as well as atmospheric path-length problems (Horvath et al., 1970). And although large angular measurements are not practical from satellites, they *are* *crucial* to plant-canopy model formulation (Otterman and Weiss, 1984; Goel, 1987; see also Chapter 6). Such models are then useful for forming strategies for satellite operation and for developing satellite data-analysis techniques.

Understanding the geometric relationships of the target, sensor, and sun to each other, and the associated terminology, is essential for measuring and reporting the results of field measurements, particularly for multiple-viewing-angle bidirectional reflectance-factor measurements, which are becoming more common today. Unfortunately, the literature evidences little consistency in the choice of coordinate systems used by remote-sensing researchers. Even the same authors have used different coordinate systems in different papers.

In the early 1970s, when field measurement programs began to proliferate, the principal concern was acquiring data for the same view and illumination conditions under which the aircraft and satellite sensors were acquiring their measurements. It was generally recognized that changes in solar elevation could confound the measurements, so some attempt to standardize the data-acquisition procedures was needed. In addition, a considerable amount of time was required to acquire a set of measurements on any given day for a variety of crops or targets and treatments. Thus, a sort of universal operating procedure was "adopted" throughout the Earth resources remote-sensing community wherein the measurements were taken primarily from a position normal to the horizontal surface (nadir-view angle) and during a nominal time interval of ± 2 hours of solar noon. During this period of time, the sun changes in altitude the least during its daily trajectory across the sky (variations in atmospheric path length and illumination geometry effects, including row orientation effects, are minimized), the illumination of the target by the sun is at a maximum, and the measurements were considered to be comparable to the mid- to late-morning (e.g., Landsat MSS, TM) and early afternoon (e.g., NOAA/ AVHRR) satellite overpass data.

With nadir-view-only measurements, the only significant geometric concerns for the field researcher are (1) the instantaneous field of view (IFOV) and its relationships to the size and distribution of the target elements and (2) the orientation of the sun azimuth relative to any preferred orientations of the target, such as crop row direction. Therefore, as many researchers began to implement field measurement programs that included off-nadir viewing, the coordinate systems that they used were often defined relative to either the crop row direction or a map coordinate system (i.e., cardinal directions; e.g., Kimes et al., 1984a) or more commonly both (crops being planted with North–South or East–West row orientations; e.g., Ranson et al., 1985). Likewise, the rotation of the view azimuths was clockwise from the North, as would be expected with a map coordinate system frame of reference (i.e., rotation from North to East to South to West).

Those field scientists who were involved in making measurements of polarization, as well as others interested in phase-angle considerations, reported their data in coordinate reference systems that are constant relative to the azimuth of the sun (e.g., Vanderbilt and Grant, 1985). In fact, most of the literature that presents off-nadir viewing geometry can be found to use the convention of defining the coordinate system with the primary axis "rotating" relative to the dynamic solar azimuth position. What is not consistent, however, is the definition of the coordinate system azimuths relative to the sun position. In cases where authors have only taken (or reported) data for the solar principal plane, the sensor-viewing azimuth rotation is usually not given.

Some authors fix the solar azimuth (direction to the sun) at $0°$ ($\theta_s = 0°$) and rotate the sensor-viewing azimuths counterclockwise, as is common in mathematics (Kriebel, 1978, aircraft data; Nicodemus et al., 1977, theoretical; Walthall et al., 1985, empirical model), whereas others place the sun at $0°$ azimuth, but rotate the sensor-view azimuths clockwise (e.g., Shibayama and Wiegand, 1985). Still others have the solar azimuth at $180°$ and rotate the sensor-view azimuths counterclockwise (e.g., Coulson et al., 1965), and, finally, some position the solar azimuth at $180°$ and rotate the sensor-view azimuths in a clockwise direction, as shown in Figure 3 (e.g., Kimes et al., 1984b).

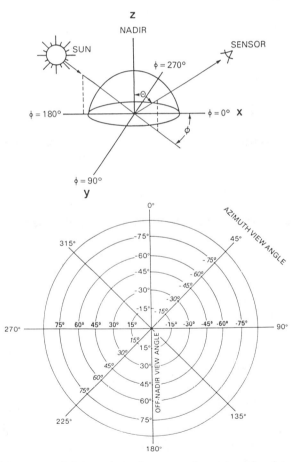

Figure 3 A coordinate system defining solar and sensor angles, and a polar plot showing a scheme for plotting bidirectional reflectance factors. The solar azimuth is always $180°$. The sensor azimuths and off-nadir angles are shown as ϕ and θ, respectively. The spectral bidirectional reflectance factors of azimuth planes are defined in the polar plot, where the distance from the origin represents the off-nadir-view angle of the sensor, and the angle from $\phi = 0°$ represents the sensor's azimuth. [From Kimes et al. (1984b).]

This last configuration is somewhat comparable to the map or geographic reference and to the crop geometry coordinate system reference discussed before. Traditional field scientists relate to these coordinates since the sun is at the "South" (180°) position (i.e., sun position at solar noon in the northern hemisphere) and the rotation is clockwise like the logical geographic North–East–South–West progression. This coordinate system ($\phi_s = 180°$, clockwise rotation) may have merit in the larger context of relating field data sets to satellite-acquired data whose angular characteristics are referenced to geographic coordinates. This is becoming increasingly important as truly global data-base systems for ingesting, assimilating, archiving, and distributing the massive quantities of diverse satellite, aircraft, and ground data will be required for future Earth system science studies.

However, since the BRDF is a property of the surface, for some study objectives, the coordinate systems that continuously rotate to "track" the sun for orienting the solar azimuth at 0° or 180° may be less desirable than those that fix on the surface. For example, when studying plant geometry and wind-direction effects, the instrument geometry variation may make it more difficult to sort out the biophysical variable effects. It should be clear from the preceding that the researcher must accurately report the orientation of the measurements in coordinate space in published works.

III FIELD INSTRUMENTS

III.A Early Measurements

The most extensive *in situ* measurements of the spectral reflectance of natural surfaces is undoubtedly that of Krinov (1947). Krinov used eight different types of spectroradiometers and acquired over 10,000 spectra in the course of five years. As was typical of such measurements taken from then and on through the 1960s, the instruments were essentially laboratory instruments with a small solid angle of reception and that were modified for field use. Since it is now common knowledge that the sensor viewing angles that lie along the solar principal plane offer the most dynamic and probably most useful features of the bidirectional reflectance distribution, it is interesting to note that Krinov's measurements that were not taken in the nadir direction were mostly taken at an azimuth of 90° from the sun (and at a view zenith angle of 45°).

Almost two decades later, simple adaptations to what were essentially laboratory instruments similar to those of Krinov were still being utilized by Chen and Rao (1968), Coulson et al. (1965), Gates et al. (1965), Kondratyev et al. (1964), and many others. These early measurements, however, provided good evidences of wavelength, sun-angle, and viewing-angle dependencies of reflectance for a wide variety of Earth surface materials, including grass canopies, soils, and geologic materials.

As recently as 1972, rather primitive "field instruments" were being used. One interesting approach included two simple camera light (exposure) meters that were adapted for measuring the angular reflectance properties of grass and asphalt under different solar illumination angles (Egbert and Ulaby, 1972). One light meter had a 1° "angle of acceptance" or aperture for the higher incidence angles (view zenith angles), and a second meter had a 10° aperture for the lower incidence angles. Spectral discrimination was accomplished with gelatin filters taped over the apertures, and the measurements were referenced to an Eastman Kodak 18% grey card to obtain reflectances. An "ideal" instrument for such experiments was envisioned by the authors to be "a light meter with a zoom lens" (Egbert and Ulaby, 1972).

At approximately this same time, Coulson and Reynolds (1971) had developed an instrument for field use based on an S-20 photomultiplier tube radiometer that was fitted with an integrating sphere as the primary receiver. Bihemispherical reflectance measurements were made in six spectral bands that ranged from the ultraviolet to the near-infrared and that were defined by narrow-band (0.01 μm) interference filters. Their measurements were of considerable interest to agriculturists as they included several soils and vegetation types, including alfalfa, bluegrass, rice, sugar beets, and sorghum. They concluded that surfaces of a complex structure generally show a decrease of reflectance with increasing sun angle and that at the shorter wavelengths, the reflectances are nearly Lambertian for sun elevations greater than 20°.

III.B The "Modern" Era

The late 1960s and early 1970s saw a rapid advancement in sensor technology with the development of new satellite and aircraft multispectral scanners, which spawned the development of significantly advanced field instruments based on the same general concepts. The launch of the first Earth Resources Technology Satellites (or ERTS-1; the name was later changed to Landsat-1) was an unqualified success on an international scale. At that time, there existed very little ground control data with which to interpret the satellite imagery. The highly publicized aircraft-sensor-based "Corn Blight Watch Experiment" in 1970 and 1971 (MacDonald et al., 1972) had also captured the imaginations of agriculturists and natural resource scientists, as well as the general public. These two events unleashed the age of contemporary remote sensing.

Although there were no ground sensors involved in the corn-blight study (only field observers to classify the extent of the fungus damage at sample fields), a 12-channel multispectral scanner complemented the color infrared aerial photography and proved to more accurately detect the blight damage levels than the photointerpreters. More importantly, however, the experiment provided a major impetus to the development of Earth resources remote sensing and identified numerous areas in need of further research, including field studies that would yield more accurate multispectral image interpretation.

Simultaneous to these events, sophisticated mobile field spectrometer laboratories were being developed, such as an EG&G model 580-585 spectroradiometer that was fitted with a reflective telescope, interfaced to a Hewlett-Packard minicomputer and housed in a small mobile home trailer. The International Biological Program (IBP) studies on the Pawnee grasslands of Colorado provided the framework for the reflectance studies of the blue grama grasslands. From these studies came the basic understanding of the close statistical relationships of red and near-infrared (NIR) reflectance with plant biomass and other biophysical variables (Pearson and Miller, 1972; Tucker et al., 1975; Tucker, 1977a,b).

The IBP biomass studies were reinforced by those of Colwell (1974) at the University of Michigan. He used a field spectroradiometer that was sensitive in the spectral region from 300 to 800 μm (in 8 μm wavelength intervals), and was pieced together with "building blocks" from Gamma Scientific, Inc., and some additional miscellaneous hardware (like a PVC pipe baffle tube) to adapt it to field use. Colwell's original plan was to measure the reflectance of monospecific flats of oats and timothy, over light and dark soils, in the laboratory using artificial lighting. But the photomultiplier detector was not sufficiently sensitive to make accurate reflectance measurements. Thus, it became a "field" experiment, using the sun as the source of illumination. Colwell's work added the additional dimensional component of view-angle variation to the quantitative radiometric study of biophysical plant canopy variables, and he included an analytical modeling approach to the data interpretation.

III.C   Hand-Held Radiometers

From the relatively simple red and near-infrared reflectance versus biomass relationships (Figure 4) came the idea that a useful, very simple, and very portable field instrument could be constructed. The hand-held two-channel digital radiometer or "biometer" was the result (Pearson et al., 1976). The prototype unit was a Tektronics Model J-16 portable radiometer that was modified to use two probes. Each silicon-detector unit was fitted with an interference filter to measure radiances of approximately 0.680 and 0.800 μm. A pocket calculator, which was interfaced to the unit, automatically computed the ratio of the two spectral radiances. Then, using stored regression coefficients, the estimated biomass was calculated and displayed in g/m^2. An inexpensive "production" unit that was developed from the prototype design, but which displayed the radiances (voltages) rather than biomass, is shown in Figure 5. Units of this type were successfully used on a variety of ecosystems around the world (Tucker et al., 1981).

The concept of "spectral mapping" of biomass (Pearson et al., 1976) or ratio-regression biomass analysis caught on quickly in the remote-sensing community. This concept formed the singular principle upon which quantitative applications of the four-band Landsat MSS satellite sensor were based for more than a decade. And, in fact, the two-band concept has persisted to form the basis for continental and global biomass mapping

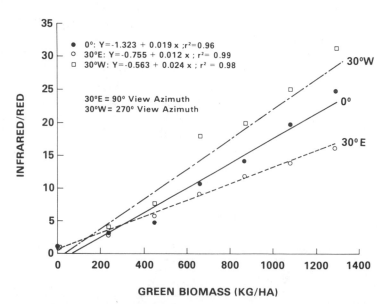

Figure 4   Linear relationships of the simple NIR/red radiance ratio to wheat dry green biomass at the nadir-view zenith angle (0°) and for two 30° off-nadir angles in the solar principal plane, one viewing in the forward-scatter direction (30° E) and the other in the backscatter direction (30° W); see also Figure 5(a). The sun zenith illumination angle is 50°.

Figure 5 A two-channel "Biometer" being used to measure (a) the directional radiances of wheat growing in greenhouse flats and (b) nadir radiances for pasture grass at red and near-infrared wavelengths.

being performed routinely today using the AVHRR visible (VIS) and near-infrared (NIR) sensors on board the NOAA satellites (Holben, 1986; Justice, 1986).

The simple ratio of the two bands is seldom used today, however, in preference to several other ratio techniques (Tucker, 1979) such as the Normalized Difference (ND), which is the ratio of the NIR and red bands (or VIS in the case of the AVHRR) radiances (Deering et al., 1975). Thus, for the AVHRR, $ND = (NIR - VIS)/(NIR + VIS)$. This ratio gives a measure of photosynthetic capacity such that the higher the ratio, the more photosynthetically active the cover type (Sellers, 1985) and, in general, the higher the quantity of green plant biomass or leaf-area index.

The early ground-based spectrometers, which include the two previously mentioned and the NIWARS (Netherlands Interdepartmental Working Community for the Application of Remote Sensing Technologies)-field spectrometer in The Netherlands (Bunnik and Verhoef, 1974), the Exotech Model 20-C used at Purdue University, the S-191H or Field Signature Acquisition System operated by the NASA/Johnson Space Center (Figure 6), and others, were quite simply very cumbersome for *in situ* studies. In addition, most of these field instruments or field-adapted instruments were very expensive to build, operate, and maintain. These characteristics, along with their lack of portability (Tucker et al., 1981), were significant and often prohibitive limitations for a great many Earth resources scientists who wished to conduct remote-sensing research.

Further, remote sensing is being increasingly applied in less-developed areas of the world, which presents several extra demands upon radiometric instrumentation (Milton, 1980). The instruments must be rugged, reliable, and easy to use by semi-skilled personnel; furthermore, repairs should be possible with a minimum of specialized assistance. And, finally, the instruments must be sufficiently inexpensive so as to be used in relatively large numbers—thus to extend over a wide area the data derived from a small

Figure 6 The Field Signature Acquisition System spectroradiometer, equivalent to the Skylab S-191 sensor, on a "cherry-picker" boom with a van housing the data-processing computer equipment.

number of conventional spectroradiometers operating at fixed locations. Consequently, other simple instruments, but more elaborate than the two-band biometer, were designed with the special demands and constraints of field work in mind.

The development of satellite remote-sensing systems that recorded data in a limited number of spectral bands gave a tremendous boost to the development and use of field-portable multiband instruments that were matched (in spectral response at least) to the satellite sensors (e.g., Landsat MSS and TM; NOAA AVHRR). A commercial "Landsat radiometer," the Exotech-100 (Figure 7) was marketed in 1972 near the time of the launch of Landsat-1. The radiometer has a 15° IFOV and acquires data in the 0.5–0.6, 0.6–0.7, 0.7–0.8, and 0.8–1.1 μm spectral wavelength regions. These radiometers were used extensively during the decade encompassed by the LACIE and AgRISTARS programs. But their application for simple field spectral reflectance measurements has gone beyond those programs into virtually all of the Earth science disciplines, particularly cropland agriculture, but including geologic "ground truth" studies (Marsh and Lyon, 1980). Other similar instruments have been developed, such as the Milton multiband radiometer (four bands—three silicon photodiodes and one cadmium-sulphide cell). A couple of interesting features of the Milton instrument are the selectable fields of view (using built-in adjustable-length hoods in front of the sensors) and selectable internal filters (Milton, 1980).

The Landsat Thematic Mapper (TM) sensor offered additional, narrower spectral wavelength bands and covered spectral regions beyond the NIR that were not available on the MSS sensor (or the Exotech-100), including shortwave infrared (SWIR; 1.55–1.75 μm) and thermal infrared channels (TIR; 10.4–12.5 μm). Consequently, another self-contained hand-held field radiometer (Figure 8) was designed to collect radiance data in the TM spectral bands, which were numbered 3, 4, and 5 at the 0.63–0.69, 0.76–0.90, and 1.55–1.75 μm spectral wavelengths, respectively (Tucker et al., 1981).

Figure 7 Exotech 100 A four-band MSS radiometer being used to acquire ground-truth measurements on prairie grasslands in support of a satellite overpass.

Figure 8   (*a*) A three-band hand-held TM band radiometer (with an inclinometer attached) being used in a variety of ways—to acquire reflectances of rangeland vegetation, including (*b*) native grass species and (*c*) a juniper bush, and (*d*) to measure soybean-canopy reflectances, using a specially designed pole mount, for studying crop-water stress relationships to TM-band reflectances.

The three- and four-band hand-held radiometers met a certain need in the remote-sensing community, but they did not even provide all of the spectral channels that the TM satellite sensor afforded. Additionally, data-handling hardware or software systems were not commercially available for these instruments, and relatively few investigators were using the instruments in a manner that permitted comparable and repeatable measurements. Consequently, a significant effort was undertaken by NASA and Purdue University, primarily under the auspices of the LACIE and AgRISTARS programs, to develop an eight-band (spanning 0.4–12.5 μm) multiband radiometer that would be (1) relatively inexpensive to acquire and maintain, (2) simple to operate and calibrate, (3) complete with data-handling hardware and software, and (4) well-documented for use by researchers (Robinson et al., 1979).

The eight-band radiometer became known commercially as the Barnes Modular Multiband Radiometer, or MMR. The MMR, which was interfaced to commercially available 12-bit data loggers, was intended to be

> well suited to the needs of remote sensing field researchers who required portability, spectral accuracy, well defined fields of view, appropriate dynamic range, and excellent radiometric performance in the ranges 0.4 to 2.5 μm and 10 μm to 15 μm (Robinson et al., 1981).

Figure 8 (*Continued*)

The designers anticipated that the MMR would be suitable for hand-held operation, but at a weight of about 6.4 kg and physical dimensions of 26.4 × 20.5 × 22.2 cm, it is more practical to use with a tripod, truck, or other somewhat more substantial mounting platform. The prototype MMR was extensively tested and some modifications were made before commercial production. Fifteen MMRs were purchased for use at various widely scattered sites in the AgRISTARS program alone (Bauer et al., 1986). The MMR has undoubtedly become the most fully characterized and widely disseminated of field instruments to date (Figure 9; see also Figure 2).

The possibilities that polarized light may provide additional information related to the chemical and physical properties of plant leaves has led to the development of a very specialized field instrument called a polarization photometer. Vanderbilt and Grant (1986) developed an instrument for rapidly determining the bidirectional, polarized, and diffuse light-scattering properties of individual leaves illuminated and measured *in situ* at an angle of 55° (approximately Brewster's angle) in six wavelength bands in the visible and near-infrared E-M regions. They demonstrated the potential for discriminating between three species of oak using polarization data.

Figure 8   *(Continued)*

Figure 9 Barnes Modular Multiband Radiometer in a variety of different applications: (*a*) over a BaSO4 reference panel in Vermont forest, (*b*) mounted on a truck boom over simulated forest stand, (*c*) on a helicopter over prairie grassland, and (*d*) on a long pole for taking off-nadir measurements. (Photos *a* and *b* courtesy D. L. Williams.)

III.D Finer Spectral Resolution

The future hopes of spaceborne systems were that even greater technological remote-sensing capabilities than those of the TM or SPOT sensors could be realized, including higher spectral resolution, more wavelength bands, and off-nadir pointing capabilities. Thus, the early 1980s saw the development of still additional field instruments that could deepen our understanding of Earth surface and atmospheric radiative transfer characteristics and aid in defining the requirements for the next generation of space instruments.

A considerable number of new field spectrometers emerged in the 1980s—each with its own unique and worthwhile features. One of the earlier of these instruments was a Barnes Model 12-550 spectroradiometer (Zweibaum and Chapelle, 1979). Its distinguishing feature was the capability to measure real-time reflectance (as opposed to re-

Figure 9   (*Continued*)

flected radiance) of the target using a unique optical chopper and reflectometer/comparator optical system, which made the normally required frequent measurements of a reference standard panel unnecessary. The instrument covered the full spectral range from 0.4–2.5 μm, as illustrated in Figure 10, using a four-stage thermoelectrically cooled (to −50°C) lead-sulfide detector and a three-segment continuously variable filter. All operations were automatic through a preprogrammed microprocessor unit. The instrument was field-portable, Figure 11(*a*), but the physical dimensions and weight of the radiometer head and the microprocessor unit made it most amenable to use with at least a small vehicle, Figure 11(*b*).

The very compact field-portable data-logging spectroradiometer called the Spectron Engineering SE590 instrument consists of a lightweight (1 kg) and small (5 × 13 × 15 cm) photodiode array detector head that simultaneously acquires a continuous spectrum in 256 wavelength bands in less than one second. The spectral output goes to a 3 kg, 9 × 23 × 23 cm microprocessor-controlled and miniature-tape recording data logger. The spectral range for the primary system is 0.4–1.1 μm. Additional radiometer

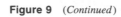

Figure 9 (*Continued*)

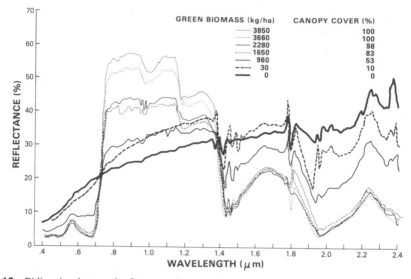

Figure 10   Bidirectional spectral reflectances for seven dry green biomass levels for a selectively ''thinned'' mature alfalfa canopy. Vertical foliage projection cover percentages for each biomass class are indicated. Solar zenith angle is 55°.

heads are available, however, which enable sampling the spectral ranges from 0.2–0.4 μm and 1.0–2.5 μm (lead-sulfide detector).

The SE590 instrument's keen advantage over some similar systems (such as the one just discussed) offering high spectral resolution is its physical size and weight (Figure 12). The SE590 has been used at sites ranging from southwestern U.S. desert scrub vegetation to the Canadian subarctic. In the subarctic, Petzold and Goward (1988) found some unusual spectral reflectance patterns in the lichen communities of boreal forests and low-arctic tundra that they conclude will enable accurate monitoring of the boreal forest-tundra ecotone and detection of its vigor and movement (spread or decline).

Numerous other field spectroradiometers are being used today by remote-sensing scientists, including the Barringer Hand Held Reflectance Radiometer (TM bands in VIS, NIR, and SWIR) and the NASA/Jet Propulsion Lab Portable Field Reflectance Spectrometer, or PFRS, which can be reconfigured into a thermal-emittance spectrometer (Figure 13). Another field-worthy instrument is the Programmable Multispectral Radiometer called PROBAR (0.4–2.5 μm) manufactured by Moniteq Ltd. of Canada. A field-portable spectrometer that costs approximately the same as the SE590 and is approximately of the same physical size is the LI-COR LI-1800 instrument. However, the optical detector unit is mounted within the primary microprocessor unit, making the ''optical head'' somewhat more bulky. Also the scan time is on the order of minutes, depending upon the number of spectral bands selected for sampling. The salient special feature is the several adaptive optional units that, among other things, allow the use of a hand-held integrating sphere for leaf transmittance and reflectance measurements.

Geological remote-sensing scientists have long clamored for very-high-resolution spectral instruments, particularly in the spectral regions from about 0.86–2.5 μm, where important mineral absorptive features occur. One need has been to make *in situ* spectral measurements with a resolution comparable to or higher than the new class of imaging

Figure 11 (*a*) Barnes Model 12-550 Spectroradiometer, and (*b*) the instrument mounted on a pickup for small plot measurements. (Photo courtesy E. W. Chappelle.)

Figure 12 Spectron Engineering SE590 instrument mounted on a special tripod boom for off-nadir measurements; reference panel measurements being taken.

spectrometers such as the Airborne Imaging Spectrometer (AIS) and the Airborne Visible-Infrared Imaging Spectrometer (AVIRIS) (Goetz, 1987). The Portable Instant Display and Analysis Spectrometer (PIDAS) was developed to solve these problems in a field instrument. The PIDAS field instrument acquires 872 data points within the range from 0.4–2.5 μm in 2 s. It can acquire spectra every 8 s and stores up to 288 spectra in bubble memory, with the capability to display the current spectrum instantly on a hand-held liquid-crystal display unit, and in superimposition with another previously stored spectrum, if desired.

The PIDAS is portable, but is rather heavy, with the backpack weighing more than 30 kg (see Figure 14). The visible sampling interval from 0.425–0.922 μm has a declared resolution of 0.88 μm (actually, about 2 μm), and the infrared sampling interval from 0.856–2.490 μm has a resolution of 4.7 μm (actually, about 10 μm). PIDAS measurements in the field have been compared with laboratory spectra of the same samples, and the PIDAS has been shown to be superior to the Beckman 5240 laboratory instrument (25 μm resolution) in discriminating absorptive features important in geological remote sensing (Goetz, 1987).

III.E Improved BRDF Measurement Capability

The Portable Apparatus for Rapid Acquisition of Bidirectional Observations of the Land and Atmosphere, or PARABOLA, instrument (Deering and Leone, 1986) was designed to overcome important limitations of existing field instruments for obtaining the adequate sampling needed to analyze the bidirectional reflectance angular distributions of Earth surface targets and the sky. The previous major instrumentation limitation was deter-

Figure 13 Field spectral radiometers used by NASA/JPL scientists for geologic and vegetation remote-sensing studies: (*a*) Barringer Associates Hand Held Reflectance Radiometer, or HHRR, and (*b*) the NASA/JPL-built Portable Field Reflectance Spectrometer, or PFRS.

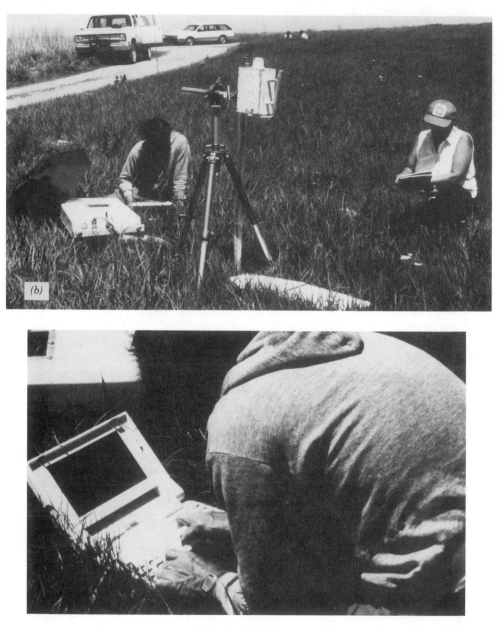

Figure 13   (*Continued*)

Figure 14 Portable Instant Display and Analysis Spectrometer, or PIDAS, high-resolution spectral field spectrometer for geologic mineral analysis (foreground). (Photo courtesy Brian Curtiss.)

mined to be the difficulty in obtaining multiple viewing-angle measurements of ground surfaces in a very short period of time to eliminate, or at least sufficiently minimize, the effects of changing sun position, sky conditions, and the vegetation's dynamic biophysical conditions during the sampling period. A second important previous limitation was the capability to measure the downwelling sky radiance distribution concurrent with measurements of the ground target.

The PARABOLA is a battery-powered two-axis scanning head three-channel (visible, near-infrared, and midinfrared; 0.650–0.670, 0.810–0.840, and 1.620–1.690 μm, respectively) motor-driven radiometer, Figures 15(a) and (b), that enables the acquisition of radiance data for almost the complete (4π) sky- and ground-looking hemispheres in 15° instantaneous field-of-view sectors in only 11 s. The two silicon and one germanium detectors are temperature-regulated (by cooling or heating) through thermoelectric proportional-control circuits. Also, due to the tremendous range in target brightness that can be expected in scanning a 4π sr field of view, an autoranging amplifier is used to switch the gain levels back and forth by factors of 1, 10, and 100 to maintain maximum radiometric sensitivity.

The PARABOLA system has been deployed using a variety of mounting platforms, including (1) a large van boom over agricultural fields and pasturelands; (2) four-wheel-drive pickup trucks over rugged rangeland areas using a lightweight collapsible-boom apparatus, Figure 15(c); (3) various large tripods over lake ice and snow, Figure 15(d); (4) a 30 m high tower tram over a forest and beneath the forest canopy, (Figures 15(e) and (f); and (5) a hot-air balloon over pine forest and wetlands.

Three-dimensional graphs of the bidirectional reflectance angular distributions (BRDF estimates) from PARABOLA measurements show considerably different angular-reflection properties for several natural surface types. A bare-soil surface (Figure 16), for example, features continuously increasing reflectance factors when shifting from the

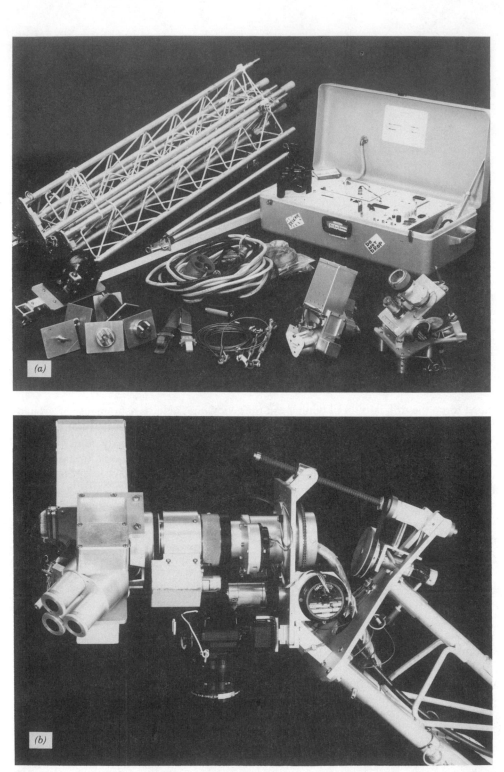

Figure 15 PARABOLA multidirectional field radiometer: (*a*) primary sensor system components and Transportable Pickup Mount System (TPMS) dismantled, (*b*) radiometer head and camera mounted on the leveling head of the end of the TPMS, (*c*) in operation on rangeland, (*d*) mounted on a tripod for lake ice and snow measurements, (*e*) mounted on a tram at the forest floor beneath the tree canopy, and (*f*) above the forest canopy.

Figure 15 (*Continued*)

Figure 15 *(Continued)*

X–Y vs POLAR COORDINATES (AZIMUTHS POSTED)

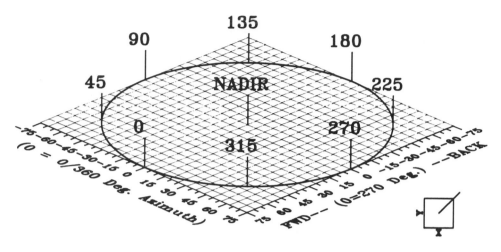

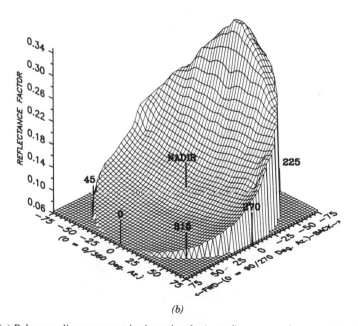

Figure 16 (*a*) Polar-coordinate system viewing azimuths (posted) represented on a cartesian (*x–y*) plane as used in three-dimensional data plotting. View zenith angles are shown out to ±75°, as shown on the *x*- and *y*-axis label. (Note that the coordinate system is the same as in Figure 3.) (*b*) Recently plowed bare-soil field bidirectional reflectance surface at 65° solar zenith angle for 0.662 μm.

forward-scatter direction view zenith angles ($\phi_v = 0°$) to the backscatter direction view zenith angles ($\phi_v = 180°$). A rangeland brush species, shinnery oak (Figure 17), has a prominent "ridge" along the solar principal plane, but the red and near-infrared spectral bands show differing responses for the larger-view zenith angle reflectances. A tallgrass prairie grassland (Figure 18) exhibits a very strong "deep bowl" anisotropy, with the

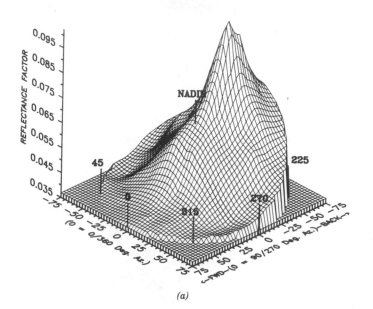

(a)

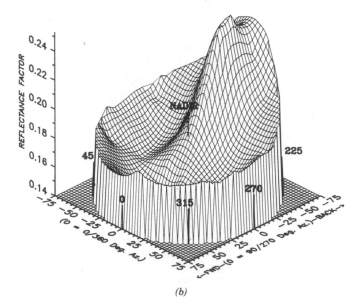

(b)

Figure 17   Sand shinnery oak-rangeland community bidirectional reflectance surfaces for (*a*) 0.662 μm and (*b*) 0.826 μm at 31° solar zenith angle (see also Figure 15c).

near-infrared spectral-band forward-scatter reflectances at approximately the same magnitude as the backscatter reflectances. Deering and Eck (1987) and Wardley (1984) point out that these differing effects of viewing geometry on the red and near-infrared spectral reflectances also affect the various "vegetation indices" (e.g., the Normalized Difference).

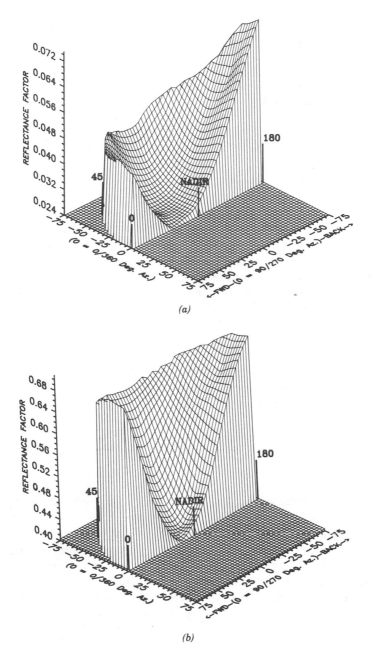

(a)

(b)

Figure 18 Tallgrass prairie grassland community bidirectional reflectance surfaces for (*a*) 0.662 μm and (*b*) 0.826 μm at 48° solar zenith angle (see also Figure 23).

Another recently developed optical spectroradiometer whose field of view can be mechanically scanned over the sky hemisphere is the Skyscanner by Moniteq, Ltd. Ten filters and two detectors allow the wavelength range of 0.25–2.50 μm to be covered. Each complete spectral scan and recording sequence requires only about 2 s. The spatial sampling is achieved by mechanically stepping the field of view (selectable at 5°, 20°, or 30°) through an azimuth and elevation sequence (i.e., about 10 min for a reasonable spatial sample).

III.F Ancillary Data

Frequently, to satisfy satellite overpass ground-truth requirements and in field studies of the bidirectional reflectance properties of Earth surfaces, ground measurements other than the surface bidirectional spectral reflectances are needed. Solar radiometers, or sun photometers (Figure 19), for example, can provide important information on the state of the atmosphere (which the irradiant flux must pass through to strike the field target or reference panel), such as the atmospheric aerosol optical thickness (Figure 20).

Photographic recording of the sky conditions (Figure 21) and the state of the ground target at the time that the radiometric measurements are taken can often be very helpful in determining data quality and interpreting the data. McArthur and Hay (1981) have described a technique for mapping the distribution of diffuse solar radiation over the sky hemisphere using all-sky photographs, supplemented with actinometric measurements. Hamalainen et al. (1985) devised a rather novel "multipyranometer," which is a conglomeration of 25 radiation sensors mounted at different angles on a metal hemisphere, for measuring the direct component and angular distribution of solar radiation.

Knowledge of the proportion of diffuse to total radiation impinging on the subject surface can be important in interpreting bidirectional spectral reflectances. Figure 22 shows the effects that an increase in the amount of haze in the atmosphere (increases the proportion of total irradiance that is diffuse) can have on the surface-reflectance properties of a "clumped" orchardgrass pasture under cloudless sky conditions (Deering and Eck, 1987). In this case, the reduction in the amount of dark shadow being cast on the light-colored bare-soil surfaces between the clumps of green plants caused the bidirectional reflectance factors to increase by approximately 30 and 10% for the red and NIR spectral bands, respectively. Without the photographic record, it may have been difficult to explain why the reflectance factors changed so dramatically when the biophysical variable measurements (biomass, plant cover percentage, surface-soil moisture) showed no change.

Figure 23 shows the shadowing characteristics on a uniform tallgrass prairie site for three different solar zenith angles (see also Figure 18). An assessment of the proportions of shadowed plant and soil area at different viewing angles (different distances from the center of the photograph) could prove useful in analyzing reasons for variations in off-nadir bidirectional reflectances. A plant-canopy reflectance model has been developed that can, in essence, utilize the "shadow information" contained within multiple viewing-angle bidirectional spectral reflectance measurements. The model has been successfully used to derive a useful plant-canopy parameter for desert scrub vegetation, and an approach has been suggested for utilizing ground fisheye photographs for assessing the canopy structure of vegetation over bright soil (Otterman et al., 1987).

Acquiring radiometric measurements under other than very clear sky conditions can potentially cause major problems in the calculation of bidirectional reflectance factors. As a large cumulus cloud approaches the solar disc, a sensor of global irradiance will

Figure 19 Two types of sun photometers being used to acquire atmospheric optical properties data: (*a*) West German Mainz II in hand-held operation and (*b*) "Reagan sun photometer" (built at the University of Arizona) in use with a tripod.

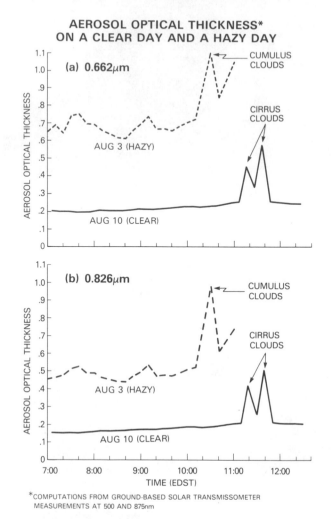

AEROSOL OPTICAL THICKNESS*
ON A CLEAR DAY AND A HAZY DAY

*COMPUTATIONS FROM GROUND-BASED SOLAR TRANSMISSOMETER
MEASUREMENTS AT 500 AND 875nm

Figure 20 Aerosol optical thickness at 0.662 μm (top) and 0.826 μm (bottom), as computed from ground-based solar transmissometer measurements (taken at 0.500 μm and 0.875 μm) for clear and hazy days (see Figure 22).

indicate increased intensity (as much as 30% is typical). Then, as the cloud begins to cover the solar disk, the intensity decreases markedly. Thus, any measurement made during these events will be subject to error due to intensity changes with time. More importantly, a sizeable error can be introduced by the marked difference in direction of the light from the cloud. Bands of cloud covering half the horizon to an elevation of 20° (zenith of 70°) will, however, produce little effect on the measurement of reflectance factor. But the effects of thin cirrus clouds are quite frustrating and can change not only the proportion of "skylight," but also the intensity of the direct component (Robinson and Biehl, 1979).

Among the numerous biophysical and physical variables that have been measured in conjunction with vegetation field radiometry, probably the more frequently measured are the above-ground plant biomass (Figure 24), leaf-area index (or LAI), proportion of green and brown plant material, plant and soil moisture contents, plant-canopy coverage,

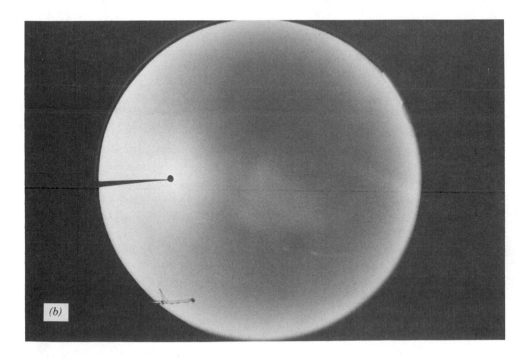

Figure 21 (*a*) Sky camera with an 8 mm full-fisheye lens (capped). Photographs of three sky conditions: (*b*) cloudless, Raleigh sky, (*c*) bright sky with variable thin cloud development, and (*d*) sky containing cumulus cloud cover in an otherwise clear atmosphere.

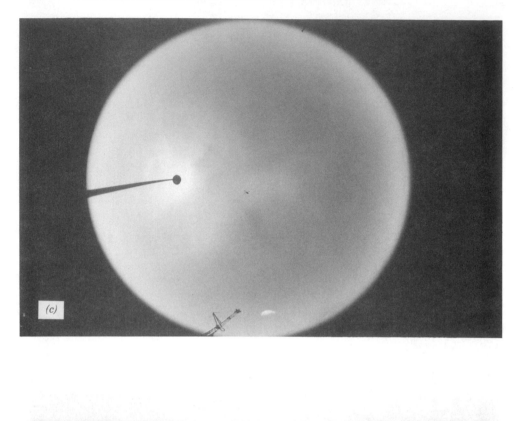

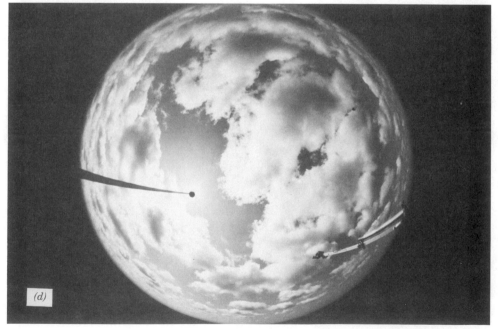

Figure 21   (*Continued*)

CLEAR SKY
AUGUST 10
9:05 a.m. EDST
($\Theta_z = 59°$, $\Phi_z = 95°$)

$(mW/cm^2/\mu m/ster.)$

(a)

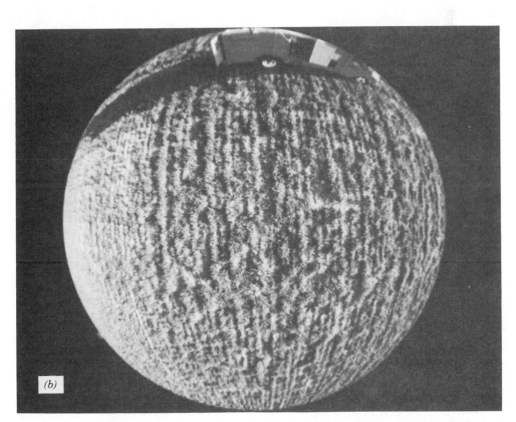

(b)

Figure 22 For a cloudless, clear day (Aug. 10): (*a*) the PARABOLA-derived sky irradiance distribution, and (*b*) a bird's-eye view photograph of an orchardgrass pasture. For a cloudless hazy day (Aug. 3): (*c*) the sky irradiance distribution and (*d*) a bird's-eye view photograph of the same pasture site. [From Deering and Eck (1987).]

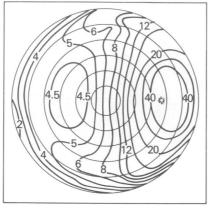

HAZY SKY
AUGUST 3

9:16 a.m. EDST
($\Theta_z = 56°$, $\Phi_z = 96°$)

(mW/cm^2/μm/ster.)

(c)

(d)

Figure 22   (*Continued*)

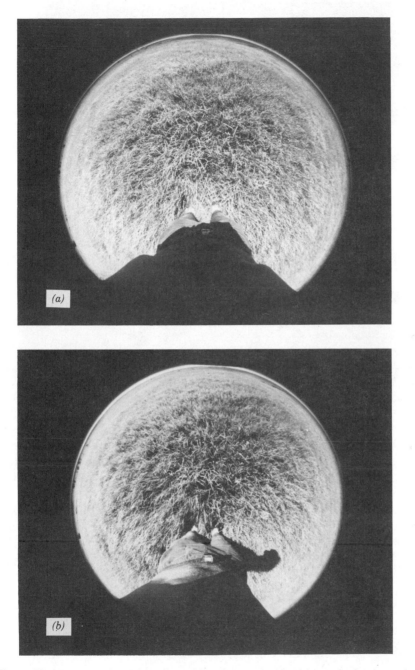

Figure 23 Ground vertical fisheye (8 mm lens) photographs of prairie grassland at three solar zenith angles of illumination: (*a*) 27°, (*b*) 48°, and (*c*) 68°.

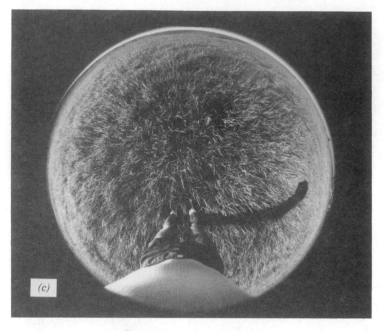

Figure 23 (*Continued*)

and plant height. Of these, LAI is the single most useful structural characteristic of vegetation for the quantitative analyses of interest in global ecology (Botkin and Running, 1984). LAI provides a means of estimating carbon fixation, evapotranspiration, and elemental accumulations of the vegetation directly.

Some ecologists, therefore, have considered that the first priority of a global vegetation study must be to accurately measure LAI from a satellite data base. The work of Sellers (1985) suggests, however, that the spectral reflectance data are good indicators of the *rates* associated with the vegetation (i.e., photosynthesis, transpiration), but unreliable indicators of the *state* of the vegetation (LAI, biomass). Weiser et al. (1986) found that direct relationships between the spectral reflectance indices and grass biophysical parameters (LAI and phytomass) were site-specific and year-specific. Wiegand and Richardson (1987) found that absorbed photosynthetically active radiation (APAR) could be reliably estimated from spectral vegetation indices; thus, a direct tie between spectral reflectance and photosynthesis seems plausible. Detailed discussions of these topics are given in Chapters 7 and 8.

Field studies of vegetation reflectance should also consider the possible need to acquire other structural information about plant canopies, such as leaf-angle distribution. Kirchner et al. (1982) found that distinct changes occurred in the pattern of radiance observed by a sensor as structural characteristics and biomass changed. Kimes and Kirchner (1983) concluded that the three-dimensional distribution of leaf orientations have to be routinely collected in diurnal sun angle studies and then incorporated into mathematical models of sensor response. There is also some evidence that wind may affect the measured reflectances. Schutt and Newcomb (1986) found a drop in red and NIR reflectances with a 7 m/s wind and that the greatest drop in reflectance occurred in the morning when leaf turgor was at a maximum, enabling normally horizontal leaves on the windward side of the canopy to assume stable vertical positions.

Figure 24   Clipping circular plot for above-ground phytomass estimates of rangeland grass production to correlate with spectral reflectance measurements; see Figure 8(*b*).

IV   FIELD MEASUREMENT PRINCIPLES AND CONSIDERATIONS

The kinds of observations that should be taken in a field measurement program can vary significantly depending on the objectives of the remote-sensing activity. Conceptually, there is a hierarchy of information types or classes for characterizing the Earth's surface features, and ground observations should be sought at the appropriate level or "scale." Remote-sensing applications such as land classification that are concerned only with the separation of discrete classes simply require information at the thematic end of the scale, but more fundamental studies may have to span the hierarchy from the physical end (e.g., radiation interactions with the surface and biophysical variable measurements) of the scale toward the thematic (Steven, 1987).

For example, some satellite or aircraft remote-sensing studies, such as the Corn Blight Watch Experiment, may not necessarily require ground radiometry, but may involve simple visual observations of surface type or condition. Ground-truth measurements designed primarily to relate radiometric measurements acquired by a space or aircraft sensor to surface reflectances will often differ substantially in the spatial sampling needs from those necessary in fundamental studies of radiative transfer properties and their relationships to biophysical variables or process phenomena.

Carefully designed and conducted field measurements of surface radiances may be needed simply to accurately analyze and utilize space-acquired Earth surface radiances. For example, the restrictions imposed by the geometry of a satellite–sun–surface config-uration, the construction of satellite scanning radiometers, and physical constraints placed upon their viewing angles necessitate making a number of basic assumptions in order to estimate surface albedo from satellite observations—including assumptions that surface reflectivity is Lambertian (Hughes and Henderson-Sellers, 1982).

Otterman and Tucker (1985) and Kimes and Sellers (1985) report, however, that errors in inferring albedo or hemispherical reflectance from nadir reflectance can be as high as 40–45% for a wide variety of cover types and a wide range of solar zenith angles. Middleton et al. (1987) found that errors up to 55% could occur in estimating the hem-ispheric spectral reflectance with the assumption of isotropy (i.e., that only a nadir view measurement could accurately estimate the hemispheric spectral reflectance).

Eaton and Dirmhirn (1979) report that it is only with accurate angular ground mea-surements that a means is provided to (a) determine the actual "indicatrix" (geometrical or angular reflection characteristics) of different natural surfaces unaffected by atmo-spheric conditions, (b) derive accurate space data-scattering functions by considering the accurate ground values, and (c) determine ground albedo patterns from space data (with the assumption of knowing the atmospheric scattering functions).

Field sampling sites used for ground truth must often be selected for uniformity on a scale of several space sensor pixels to ensure an uncontaminated value, and they must often be chosen to span the range of radiometric values that occur across the landscape. Greater attention must also be placed on "matching" the field radiometer's spectral bands and possibly viewing angle to those of the space or aircraft sensor in this type of field study.

The required or optimum number of sample sites and number of measurement repli-cates at a given site must be determined. These factors will once again depend consid-erably on the scale of the study and its objectives, that is, whether intended as ground truth for a remote sensing overflight or an independent fundamental investigation. At the local site scale (as contrasted with the several space-sensor-pixels scale referred to be-fore), the optimum number of sample sites and replicates will, to a large extent, be related to the configuration of the radiometer being used and the spatial heterogeneity of the surfaces to be represented by the data. Variations in the field of view and height of the instrument above the surface will affect the degree to which variations in the spectral reflectance of a surface can be generalized.

What the field researcher desires to know is, "How many observations or measure-ments must be acquired to be reasonably confident of (1) detecting differences between different surface types or conditions being measured or (2) sufficiently characterizing the surface bidirectional reflectance properties?" Daughtry et al. (1982) conducted an ex-periment to determine how canopy reflectance varies as a function of sensor altitude above the crop and what minimum altitude (height of the sensor above the canopy or soil surface) is needed to acquire repeatable reflectance measurements with a desired precision.

For corn and soybean row crop canopies, the number of measurements required for a given level of precision decreased with increasing sensor altitude and as the sensor's field of view contained a more representative sample of the canopy. The coefficients of variation of reflectance decreased exponentially as the sensor altitude increased. At the lower altitudes (<3 m), more measurements were required because reflectance measurements tended to be erratic as the sensor was moved across the rows. For example, to detect 20% differences in red-band-canopy reflectance factors of two soybean canopies with approximately 70% soil cover using a simple random sampling scheme, at least 39 measurements were required when the sensor was 2.0 m above the soil. Nineteen measurements were required at 3.0 m, and only five were needed at 7.0 m above the soil. Sampling schemes that employed *a priori* knowledge of the canopy row spacing were more efficient (required fewer measurements for a given level of precision) than simple random-sampling schemes (Daughtry et al., 1982).

The timing of ground-truth measurements may be critical, and it is almost always important for logistics purposes to determine in advance which variables must be measured at the time of a spacecraft or aircraft overpass and which variables can be assumed to be relatively constant (and thus could be sampled at some earlier or later time). One of the most important aspects of remote-sensing research involves preplanning to determine the type of surface observations that must be obtained, when they are to be obtained, and the methods to be used in collecting the data, as is more fully discussed by Hoffer (1971).

The basic purpose of field measurements in remote sensing is to establish useful relationships between two kinds of data, namely, digital radiances and user (whether scientist or resource manager) requirements. But, in addition to the many practical problems that may arise in ground measurements and surveys, there is also the conceptual problem of correlating the two kinds of data, that is, converting the data into information.

The problem is that the relationships that are sought are usually indirect, for example, the photosynthetic capacity of vegetation and its relationship to the Normalized Difference spectral-band ratio parameter (discussed before). The variables to be measured at the surface (including radiometer and surface variables) must be chosen to reasonably ensure establishing a connection between the measured radiometric characteristics of the surface and the surface parameter or process under study. For example, in studying the relationships between plant-canopy reflectance and phytomass for a specific plant-cover type, it may not be necessary to measure the temperature, humidity, and wind speed, and the soil-moisture observation needs may be limited to knowing whether the surface is wet or dry. But if the objective is to study the canopy reflectance relationships to leaf photosynthesis, then measurements of the complete soil-moisture profile as well as micrometeorological variables will likely be essential. Steven (1987) suggests that there is an *art* to taking remote-sensing field measurements or ground truth—and that is to

> tune the models of radiation interactions with well-chosen surface variables, so as to optimize the information content of the remotely sensed data.

REFERENCES

Bauer, M. E., M. C. McEwen, W. A. Malila, and J. C. Harlan (1978). Design, implementation, and results of LACIE field research. In *Large Area Crop Inventory Equipment*, Vol. III. NASA/ Johnson Space Center, Houston, Texas, pp. 1037–1066.

Bauer, M. E., C. S. T. Daughtry, L. L. Biehl, E. T. Kanemasu, and F. G. Hall (1986). Field spectroscopy of agricultural crops. *IEEE Trans. Geosci. Remote Sens.* **GE-24**(1):65–75.

Botkin, D. B., and S. W. Running (1984). Role of vegetation in the biosphere. In *Machine Processing Remotely Sensed Data*, Purdue University, West Lafayette, Indiana, pp. 326–332.

Bunnik, N. J. J., and W. Verhoef (1974). *The Spectral Directional Reflectance of Agricultural Crops: Measurements on Wheat and a Grass Canopy for Some Stages of Growth*, NIWARS Publ. No. 33. Interdepartmental Working Community for the Applications of Remote Sensing Techniques, Delft, The Netherlands.

Chen, H., and C. R. N. Rao (1968). Polarization of light on reflection by some natural surfaces. *Br. J. Appl. Phys.* **2**(1):1191–1203.

Collier, R. D. (1978). Accuracy and precision. *Ind. Res. Dev.* **4**:81–83.

Colwell, J. E. (1974). Grass canopy bidirectional spectral reflectance. *Proc. 9th Int. Symp. Remote Sens. Environ.*, University of Michigan, Ann Arbor, pp. 1061–1085.

Coulson, K. L., and D. W. Reynolds (1971). The spectral reflectance of natural surfaces. *J. Appl. Meteorol.* **10**:1285–1295.

Coulson, K. L., G. M. Bouricius, and E. L. Gray (1965). Optical reflection properties of natural surfaces. *J. Geophys. Res.* **70**(18):4601–4611.

Curran, P. J. (1987). Remote sensing methodologies and geography. *Int. J. Remote Sens.* **8**(9):1255–1275.

Daughtry, C. S. T., V. C. Vanderbilt, and V. J. Pollara (1982). Variability of reflectance measurements with sensor altitude and canopy type. *Agron. J.* **74**:744–751.

Deering, D. W., and T. F. Eck (1987). Atmospheric optical depth effects on angular anisotropy of plant canopy reflectance. *Int. J. Remote Sens.* **8**(6):893–916.

Deering, D. W., and P. Leone (1986). A sphere-scanning radiometer for rapid directional measurements of sky and ground radiance. *Remote Sens. Environ.* **19**:1–24.

Deering, D. W., Rouse, J. W., R. H. Haas, and J. A. Schell (1975). Measuring forage production of grazing units from Landsat MSS data. *Proc. 10th Int. Symp. Remote Sens. Environ.*, University of Michigan, Ann Arbor, pp. 1169–1178.

Duggin, M. J. (1981). Simultaneous measurement of irradiance and reflected radiance in field determination of spectral reflectance. *Appl. Opt.* **20**:3816–3818.

Duggin, M. J., and T. Cunia (1983). Ground reflectance measurement techniques: A comparison. *Appl. Opt.* **23**:3771–3777.

Eaton, F. D., and I. Dirmhirn (1979). Reflected irradiance indicatrices of natural surfaces and their effect on albedo. *Appl. Opt.* **18**(7):994–1008.

Egbert, D. D., and F. T. Ulaby (1972). Effect of angles on reflectivity. *Photogramm. Eng. Remote Sens.* **38**(6):556–564.

Gates, D. M., H. J. Keegan, J. C. Schleter, and V. R. Weidner (1965). Spectral properties of plants. *Appl. Opt.* **4**:11–20.

Goel, N. S. (1987). Models of vegetative canopy reflectance and their use in estimation of biophysical parameters from reflectance data. *Remote Sens. Rev.*, **3**:1–212.

Goetz, A. F. H., F. C. Billingsley, D. Elston, I. Lucchitta, E. M. Shoemaker, M. J. Abrams, A. R. Gillespie, and R. L. Squires (1975). Portable field reflectance spectrometer. In *Application of ERTS Images and Image Processing to Regional Geologic Problems and Geologic Mapping in Northern Arizona*, JPL Tech. Rep. 32. California Institute of Technology, Jet Propulsion Laboratory, Pasadena, p. 1597.

Goetz, A. F. H. (1987). The Portable Instant Display and Analysis Spectrometer (PIDAS). *JPL Publ.* **87-30**:8–17.

Guenther, B. (1987). Practical aspects of achieving accurate radiometric field measurements. *Remote Sens. Environ.* **22**:131–143.

Hall, F. G., and G. D. Badhwar (1987). Signature-extendable technology: Global space-based crop recognition. *IEEE Trans. Geosci. Remote Sens.* **GE-25**(1):93–103.

Hamalainen, M., P. Nurkkanen, and T. Slaen (1985). A multisensor pyranometer for determination of the direct component and angular distribution of solar radiation. *Sol. Energy* **35**(6):511–525.

Hoffer, R. M. (1971). The importance of "ground truth" data in remote sensing. *Aerofotogeografia*, No. 7. Instituto de Geografia, Universidade de Sao Paulo, Sao Paulo, Brazil. 12 p.

Holben, B. N. (1986). Characteristics of maximum-value composite images from temporal AVHRR data. *Int. J. Remote Sens.* **7**(11):1417–1434.

Holter, M. R. (1970). Research needs: The influence of discrimination, data processing and system design. In *Remote Sensing with Special Reference to Agriculture and Forestry*. National Academy of Sciences, Washington, D. C., pp. 354–421.

Holz, R. K. (Ed.) (1973). *The Surveillant Science: Remote Sensing of the Environment*. Houghton Mifflin, Boston, Massachusetts, p. 38.

Horvath, R., J. G. Braithwaite, and F. C. Polcyn (1970). Effects of atmospheric path on airborne multispectral sensors. *Remote Sens. Environ.* **1**:203–215.

Hughes, N. A., and A. Henderson-Sellers (1982). System albedo as sensed by satellites: Its definition and variability. *Int. J. Remote Sens.* **3**(1):1–11.

Jackson, R. D., M. S. Moran, P. N. Slater, and S. F. Bigger (1987). Field calibration of reference reflectance panels *Remote Sens. Environ.* **22**:145–158.

Justice, C. O. (Ed.) (1986). Monitoring the grasslands of semi-arid Africa using NOAA-AVHRR data. *Int. J. Remote Sens.* **7**(11), Spec. Issue:1383–1622.

Kimes, D. S., and J. A. Kirchner (1982). Irradiance measurement errors due to the assumption of a Lambertian reference panel. *Remote Sens. Environ.* **12**:141–149.

Kimes, D. S., and J. A. Kirchner (1983). Diurnal variations of vegetation canopy structure. *Int. J. Remote Sens.* **4**(2):257–271.

Kimes, D. S., and P. J. Sellers (1985). Inferring hemispherical reflectance of the Earth's surface for global energy budgets from remotely sensed nadir or directional radiance values. *Remote Sens. Environ.* **18**:205–223.

Kimes, D. S., J. A. Kirchner, and W. W. Newcomb (1983). Spectral radiance errors in remote sensing ground studies due to nearby objects. *Appl. Opt.* **22**:8–10.

Kimes, D. S., W. W. Newcomb, J. B. Schutt, P. J. Pinter, Jr., and R. D. Jackson (1984a). Directional reflectance factor distributions of a cotton row crop. *Int. J. Remote Sens.* **5**(2):263–277.

Kimes, D. S., B. N. Holben, and C. J. Tucker (1984b). Optimal directional view angles for remote sensing missions. *Int. J. Remote Sens.* **5**(6):887–908.

Kirchner, J. A., D. S. Kimes, and J. E. McMurtrey, III (1982). Variation of directional reflectance factors with structural changes of a developing alfalfa canopy. *Appl. Opt.* **21**(20):3766–3774.

Kondratyev, K. Ya., F. Mironova, and A. N. Otto (1964). Spectral albedo of natural surfaces. *Pure Appl. Geophys.* **59**:207–216.

Kriebel, K. T. (1978). Measured spectral bidirectional reflection properties of four vegetated surfaces. *Int. J. Remote Sens.* **17**(2):253–259.

Krinov, E. L. (1947). Spectral reflectance properties of natural surfaces. *NRC Tech. Transl.* **439**:1–268.

Lamont, J., S. Quegan, and I. A. Ward (1987). Ground truth requirements for radar observations over land and sea. *Int. J. Remote Sens.* **8**(7):1057–1067.

Lovell, D. J. (1984). Theory and applications of integrating sphere technology. *Laser Focus/Electro-Opt. Mag.* **20**(5):86–96.

MacDonald, R. B., and F. G. Hall (1980). Global crop forecasting. *Science* **208**:670–679.

MacDonald, R. B., M. E. Bauer, R. D. Allen, J. W. Clifton, J. D. Erickson, and D. A. Landgrebe (1972). Results of the 1971 corn blight watch experiment. *Proc. 8th Int. Symp. Remote Sens. Environ.*, University of Michigan, Ann Arbor.

Marsh, S. E., and R. J. P. Lyon (1980). Quantitative relationships of near-surface spectra to Landsat radiometric data. *Remote Sens. Environ.* **10**:241–261.

McArthur, L. J. B., and J. E. Hay (1981). A technique for mapping the distribution of diffuse solar radiation over the sky hemisphere. *J. Appl. Meteorol.* **20**:421–429.

Middleton, E. M., D. W. Deering, and S. P. Ahmad (1987). Surface anisotropy and hemispheric reflectance for a semiarid ecosystem. *Remote Sens. Environ.* **23**:193–212.

Milton, E. J. (1980). A portable multiband radiometer for ground data collection in remote sensing. *Int. J. Remote Sens.* **1**(2):153–165.

Milton, E. J. (1982). Field measurement of reflectance factors: A further note. *Photogramm. Eng. Remote Sens.* **48**(9):1474–1476.

National Aeronautics and Space Administration (NASA), Advisory Council, Earth System Sciences Committee (1988). *Earth System Science, A Closer View.* Office of Interdisciplinary Earth Studies, University Corporation for Atmospheric Research, Boulder, Colorado.

National Aeronautics and Space Administration (NASA), EOS Steering Committee (1987). *From Pattern to Process: The Strategy of the Earth Observing System*, NASA EOS Steering Comm. Rep., Vol. II. NASA, Washington, D. C.

Nicodemus, F. E. (1982). Reflectance nomenclature and directional reflectance emissivity. *Appl. Opt.* **9**(6):1474–1475.

Nicodemus, F. E., J. C. Richmond, J. J. Hsia, I. W. Ginsberg, and T. Limperis (1977). Geometrical considerations and nomenclature for reflectance. *NBS Monogr. (U.S.)* **160**:1–52.

Otterman, J., and C. J. Tucker (1985). Satellite measurements of surface albedo and temperatures in semi desert. *J. Clim. Appl. Meteorol.* **24**:228–233.

Otterman, J., and G. H. Weiss (1984). Reflections from a field of randomly located vertical protrusions. *Appl. Opt.* **23**(12):1931–1936.

Otterman, J., D. Deering, T. Eck, and S. Ringrose (1987). Techniques of ground-truth measurements of desert-scrub structure. *Adv. Space Res.* **7**(11):153–158.

Pearson, R. L., and L. D. Miller (1972). Remote mapping of standing crop biomass for estimation of the productivity of the shortgrass prairie. *Proc. 8th Int. Symp. Remote Sens. Environ.*, University of Michigan, Ann Arbor, pp. 1357–1381.

Pearson, R. L., L. D. Miller, and C. J. Tucker (1976). A hand-held spectral radiometer to estimate gramineous biomass. *Appl. Opt.* **15**(2):416–418.

Petzold, D. E., and S. N. Goward (1988). Reflectance spectra of subarctic lichens. *Remote Sens. Environ.* **24**:481–492.

Ranson, K. J., C. S. T. Daughtry, L. L. Biehl, and M. E. Bauer (1985). Sun-view angle effects on reflectance factors of corn canopies. *Remote Sens. Environ.* **18**:147–161.

Robinson, B. F., and L. L. Biehl (1979). Calibration procedures for measurement of reflectance factor in remote sensing field research. *Proc. Soc. Photo-Opt. Instrum. Eng.* **196**:16–26.

Robinson, B. F., M. E. Bauer, D. P. DeWitt, L. F. Silva, and V. C. Vanderbilt (1979). Multiband radiometer for field research. *Proc. Soc. Photo-Opt. Instrum. Eng.* **196**:8–15.

Robinson, B. F., R. E. Buckley, and J. A. Burgess (1981). Performance evaluation and calibration of a modular multiband radiometer for remote sensing field research. *Proc. Soc. Photo-Opt. Instrum. Eng.* **308**:147–157.

Salomonson, V. V., and W. E. Marlatt (1971). Airborne measurements of reflected solar radiation. *Remote Sens. Environ.* **2**:1–8.

Schutt, J. B., and W. W. Newcomb (1986). On the behavior of a stressed cotton canopy in a direct air stream. *Int. J. Remote Sens.* **7**(10):1251–1262.

Schutt, J. B., B. N. Holben, C. M. Shai, and J. H. Henninger (1981). Reflectivity of TFE—a washable surface—compared with that of BaSO₄. *Appl. Opt.* **20**:2033–2035.

Sellers, P. J. (1985). Canopy reflectance, photosynthesis and transpiration. *Int. J. Remote Sens.* **6**(8):1335–1372.

Sellers, P. J., F. G. Hall, G. Asrar, D. E. Strebel, and R. E. Murphy (1988). The First ISLSCP Field Experiment (FIFE). *Bull. Am. Meteorol. Soc.* **69**(1):22–27.

Shibayama, M., and C. L. Wiegand (1985). View azimuth and zenith, and solar angle effects on wheat canopy reflectance. *Remote Sens. Environ.* **18**:91–103.

Steven, M. D. (1987). Ground truth, an underview. *Int. J. Remote Sens.* **8**(7):1033–1038.

Tucker, C. J. (1977a). Asymptotic nature of grass canopy spectral reflectance. *Appl. Opt.* **16**(5):1151–1156.

Tucker, C. J. (1977b). Resolution of grass canopy biomass classes. *Photogramm. Engr. Remote Sens.* **43**(8):1059—1067.

Tucker, C. J. (1979). Red and photographic infrared linear combinations for monitoring vegetation. *Remote Sens. Environ.* **8**:127–150.

Tucker, C. J., L. D. Miller, and R. L. Pearson (1975). Shortgrass prairie spectral measurements. *Photogramm. Eng. Remote Sens.* **41**(9):1157–1162.

Tucker, C. J., W. H. Jones, W. A. Kley, and G. J. Sundstrom (1981). A three-band hand-held radiometer for field use. *Science* **211**(16):281–283.

Vanderbilt, V. C., and L. Grant (1985). Plant canopy specular reflectance model. *IEEE Trans. Geosci. Remote Sens.* **GE-23**(5):722–730.

Vanderbilt, V. C., and L. Grant (1986). Polarization photometer to measure bidirectional reflectance factor R (55°, 0°; 55°, 180°) of leaves. *Opt. Eng.* **25**(4):566–571.

Walthall, C. L., J. M. Norman, J. M. Welles, G. Campbell, and B. L. Blad (1985). Simple equation to approximate the bidirectional reflectance from vegetative canopies and bare soil surfaces. *Appl. Opt.* **24**:383–387.

Wardley, N. W. (1984). Vegetation index variability as a function of viewing geometry. *Int. J. Remote Sens.* **5**(5):861–870.

Weidner, V. R., and Hsia, J. J. (1981). Reflection properties of pressed polytetrafluoroethylene powder. *J. Opt. Soc. Am.* **71**:856–861.

Weidner, V. R., J. J. Hsia, and B. Adams (1985). Laboratory intercomparison study of pressed polytetrafluoroethylene powder reflectance standards. *Appl. Opt.* **24**:2225–2230.

Weiser, R. L., G. Asrar, G. P. Miller, and E. T. Kanemasu (1986). Assessing grassland biophysical characteristics from spectral measurements. *Remote Sens. Environ.* **20**:141–152.

Wiegand, C. L., and A. J. Richardson (1987). Spectral components analysis: Rationale, and results for three crops. *Int. J. Remote Sens.* **8**(7):1011–1032.

Zweibaum, F. M., and E. M. Chapelle (1979). Making real-time sun reflectance measurements with a microprocessor-based spectroradiometer. *Proc. Soc. Photo-Opt. Instrum. Eng.* **180**:242–255.

3

SOIL REFLECTANCE

James R. Irons

Earth Resources Branch
Goddard Space Flight Center
National Aeronautics and Space Administration
Greenbelt, Maryland

Richard A. Weismiller

Department of Agronomy
College of Agriculture
University of Maryland
College Park, Maryland

Gary W. Petersen

Department of Agronomy
College of Agriculture
and
Office of Remote Sensing of Earth Resources
Environmental Resources Research Institute
The Pennsylvania State University
University Park, Pennsylvania

I INTRODUCTION

Remote observations of soil reflectance in the visible and shortwave infrared regions of the electromagnetic spectrum are important for at least two reasons. First is the ubiquitous presence of soil over the land surface of the Earth. With the exception of areas covered by ice, snow, or man-made materials, soil intercepts much of the solar energy incident on the land surface. Even a portion of the solar radiation incident on dense plant canopies penetrates to the underlying soil. Soils thus influence the reflectance of the composite land surface, with this influence increasing in regions of sparse or senescent vegetation and dominating in areas devoid of vegetation. Clearly, many climatologic and biologic land processes involve the absorption and reflection of solar radiation by soils.

A second reason for studying soil reflectance is pedologic. Pedology is the study of soils as three-dimensional natural bodies resulting from weathering processes, as conditioned by climate, biota, and topography, acting on parent material over time. Pedology includes the description, quantitative characterization, classification, and mapping of soils. Since the characteristics of radiation reflected from a material are a function of material properties, observations of soil reflectance can provide information on the properties and state of the soil. The most familiar application of this concept is the observation of soil color to describe and help classify soils. Spectrometers, radiometers, and polarimeters provide more quantitative measurements of reflected energy, which have found applications in soil characterization, classification, and mapping.

The purpose of this chapter is to discuss the interaction of shortwave (solar) electromagnetic energy (0.3–3.0 μm) with bare-soil surfaces and to present some examples of applications of remote-sensing observations in pedology. Although productive research has been conducted on the reflection and emission of thermal and microwave energy from soils, particularly for the remote sensing of soil moisture (Salomonson et al., 1983; Schmugge et al., 1980), much of the research on remote sensing for pedology has focused on observation of reflected solar energy. The focus is likely due to the past predominance of instruments that passively measure reflected solar energy as compared to thermal and microwave sensors. In addition, the influence of soil on the reflectance of vegetated land is discussed in Chapter 4. The emphasis of this chapter is, therefore, on the interaction of shortwave energy with soil materials and bare-soil surfaces.

II   THE COMPOSITION AND PHYSICAL PROPERTIES OF SOILS

This discussion of soil reflectance will be couched within the pedologic concept of soils. Almost the entire land surface is mantled by unconsolidated materials overlying bedrock. As discussed by Brady (1984), this material is known as the regolith, and the depth of the regolith may vary from negligibly shallow to tens of meters in depth. The upper portion of the regolith is closest to the atmosphere and is thus exposed to physical, chemical, and biological weathering processes. The usual consequence of these processes is the development of horizontal layers called horizons that are distinguished by variations in composition and physical properties. Only this upper portion of weathered and layered materials is considered soil by pedologists. Soils are distinguished from the underlying material by the presence of organic matter, plant roots, soil organisms, and horizons. Thus, soils are not just unconsolidated lithologic materials, but contain mineral and organic products of weathering in association with water and air and with varying degrees of consolidation, structure, and layering. The properties, relative concentrations, arrangements, and interrelationships of the soil materials all affect the reflection of solar energy from soils.

The major soil components are inorganic solids, organic matter, air, and water. The inorganic component of most soils consists primarily of crystalline minerals, but an appreciable amount of noncrystalline substances also develop in soils (Jackson et al., 1986). In common usage, the term *soil mineral* encompasses both the crystalline and noncrystalline inorganic materials. Soil organic matter consists of roots, decaying plant residue, and living soil organisms. The pore space left between particles of these solid soil materials is occupied by either soil water or soil air in varying relative concentrations. The soil water can be more completely described as a solution containing a variety of dissolved compounds. Water molecules can also be found as a structural component in some crystal lattices. Note that all three phases of matter, solid, liquid, and gas, are considered integral components of soil.

II.A   Soil Minerals

Soil minerals are derived from the weathering of rocks. The minerals are classified as either primary minerals or secondary minerals, depending on origin. Primary minerals are components of igneous and metamorphic rocks and are formed at temperatures and pressures greater than those found at the surface of the Earth. Primary minerals are incorporated into soils when rocks disintegrate as a result of mechanical weathering. Secondary minerals are formed by the chemical weathering of primary minerals. Pedologists further divide soil minerals into particle size fractions or separates (Table 1). The general distribution of mineral classes as a function of particle size is shown in Figure 1.

Primary minerals occur predominately in the sand and silt fractions of a soil. Although quartz comprises only about 12% of the minerals in igneous rock (Johnson, 1979), it is the most abundant primary mineral in soil due to its high resistance to chemical weathering. Table 2 lists other common primary minerals found in soils.

TABLE 1   Soil Separates as Defined by the U.S. Department of Agriculture

Soil Separate	Particle-Size Range (mm)
Gravel	>200
Very coarse sand	1.00–2.00
Coarse sand	0.50–1.00
Medium sand	0.25–0.50
Fine sand	0.10–0.25
Very fine sand	0.05–0.10
Silt	0.002–0.05
Clay	<0.002

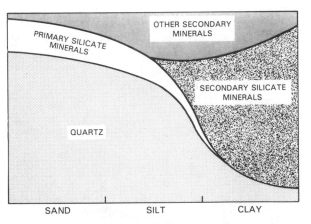

Figure 1   General distribution of soil minerals as a function of particle size. Quartz is the most abundant mineral in the sand and coarse-silt separates. Other primary silicates are found in the sand fraction but become less common in the silt fraction. Secondary silicate minerals are the most abundant type of mineral in the clay fraction. (Adapted with permission of Macmillian Publishing Company from *The Nature and Properties of Soils* by N. C. Brady, Fig. 2.4, Copyright © 1984, Macmillan Publishing Company.)

TABLE 2 Primary Minerals Commonly Found in Soils

Name	Elemental Composition	Crystal System
1. Quartz	SiO_2	Trigonal
2. Feldspars		
a. orthoclase	$KAlSi_3O_8$	Monoclinic
b. plagioclase		
albite	$NaAlSi_3O_8$	Triclinic
anorthite	$CaAl_2Si_2O_8$	Triclinic
3. Layer Silicates		
a. mica		
muscovite	$KAl_3Si_3O_{10}(OH)_2$	Monoclinic
biotite	$KAl(Mg, Fe)_3Si_3O_{10}(OH)_2$	Monoclinic
b. chlorite	$(Mg, Fe, Al)_6(Si, Al)_4O_{10}(OH)_8$	Monoclinic
4. Ferromagnesian Silicates		
a. amphibole	$(Na, Ca, Mg, Fe)_{2\text{-}3}(Mg, Fe)_{5\text{-}7}$ $(Al_{2-x}Si_{6+x})O_{22}(OH)_2$	Monoclinic
b. pyroxene	$(Ca, Mg, Fe)_2Si_2O_6$	Monoclinic or orthorhombic
c. olivene	$(Mg, Fe)_2SiO_4$	Orthorhombic

Discussion of the reflectance properties of these minerals is complicated by isomorphous substitution and impurities (Hunt and Salisbury, 1970). Isomorphous substitution occurs when different atoms occupy the same site within a crystal lattice. Either calcium, magnesium, or ferrous iron, for example, can occupy the same site in the mineral pyroxene, $(Ca,Mg,Fe)_2Si_2O_6$. The ratio of the total number of calcium, magnesium, and iron atoms to either silicon atoms or oxygen atoms remains constant, but the relative proportions of the three interchangeable elements can vary in nature, resulting in an isomorphous series of pyroxenes. Isomorphous substitution occurs in many of the other primary soil minerals, as indicated by the chemical formulas (Table 2), and different members of the same isomorphous series can have distinct reflectance properties (Hunt and Salisbury, 1970). In addition, minerals in nature are rarely pure, since trace elements are often trapped in crystal lattices during crystallization. These impurities affect the color and other reflectance properties of minerals. The purity, elemental composition, and crystalline structure of the primary minerals are all factors in determining soil reflectance.

The primary minerals are decomposed into secondary soil minerals by chemical weathering, either by the chemical alteration of the original primary minerals or by the recrystallization of the decomposition products. Water usually plays a role in chemical weathering, either as a reactant or by providing the reacting ionic species in solution. The most abundant products of chemical weathering are clay minerals and oxides and hydroxides of iron, aluminum, silicon, and titanium (Table 3). The processes also produce carbonates, sulfates, and phosphates (Table 3). These secondary soil minerals tend to be stable in soils because the minerals are formed in the temperature and pressure environment of the soil. The clay fraction of soils is predominately composed of secondary minerals.

Clay minerals are layer aluminosilicates, also called crystalline phyllosilicates (Jackson et al., 1986), in which atomic sheets of silicon tetrahedra, Figure 2(a), are bound to sheets of aluminum octahedra, Figure 2(b). The most abundant clay minerals in soils are the kaolins having a 1:1 layered structure (Figure 3) and the smectites and illites

TABLE 3 Secondary Minerals Commonly Found in Soils

Name	Elemental Composition	Crystal System
1. Clay minerals (layer aluminosilicates)		
a. Kaolin		
kaolinite	$Al_2Si_2O_5(OH)_4$	Triclinic or monoclinic
b. Smectite		
montmorillonite	$Na_{0.33}(Al_{1.67}Mg_{0.33})Si_4O_{10}(OH)_2$	Monoclinic
c. Illite	$Al_2(Si_{3.85}Al_{0.15})O_{10}(OH)_2$	Monoclinic
d. Vermiculite	$Mg_3(Al, Si)_4O_{10}(OH)_2 \cdot 4H_2O$	Monoclinic
e. Chlorite	$(Mg, Fe, Al)_6(Si, Al)_4O_{10}(OH)_8$	Monoclinic
2. Oxides and hydroxides of Al and Fe		
a. gibbsite	$Al(OH)_3$	Monoclinic
b. goethite	$FeO(OH)$	Orthorhombic
c. hematite	Fe_2O_3	Trigonal
3. Carbonates		
a. calcite	$CaCO_3$	Trigonal
b. dolomite	$CaMg(CO_3)_2$	Trigonal
4. Sulfates		
a. gypsum	$CaSO_4 \cdot 2H_2O$	Monoclinic
5. Phosphates		
a. apatite	$Ca_5(PO_4)_3(Cl, F, OH)$	Hexagonal

(A)

(B)

Figure 2 The fundamental units of clay mineral structure: (*a*) A silicon tetrahedron is formed by four oxygen atoms (O^{4-}) surrounding a silicon atom (Si^{4+}). Each oxygen has an unbalanced valence bond that may be satisfied by bonding to another silicon atom, to two aluminum atoms, or to a hydrogen atom. The sharing of oxygen atoms between two silicon atoms is the basis of silicon sheet formation and the sharing of oxygen atoms between a silicon atom and two aluminum atoms bonds silicon sheets to aluminum octahedra sheets. (*b*) An aluminum octahedron consists of six oxygen atoms surrounding an aluminum atom (Al^{3+}). Each oxygen has one and one-half unbalanced valence bonds. The half bond is usually satisfied by another aluminum atom, which is the basis of aluminum sheet formation. The remaining full bond is satisfied by a silicon atom or a hydrogen atom. (Adapted with permission from *Physical Edaphology* by S. A. Taylor and G. L. Ashcroft, Fig. 6.4 and Fig. 6.5, Copyright © 1974, W. H. Freeman and Company.)

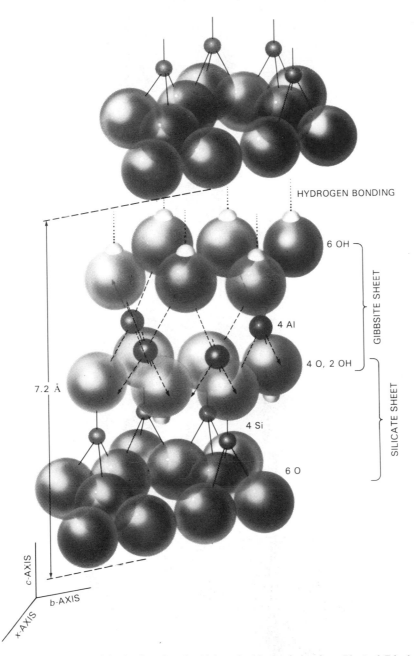

HYDROGEN BONDING

6 OH

4 Al

4 O, 2 OH

4 Si

6 O

GIBBSITE SHEET

SILICATE SHEET

7.2 Å

c-AXIS

b-AXIS

a-AXIS

Figure 3 The 1:1 structure of the kaolin minerals. (Adapted with permission from *Physical Edaphology* by S. A. Taylor and G. L. Ashcroft, Fig. 6.6, Copyright © 1972, W. H. Freeman and Company.)

having a 2:1 layered structure (Figure 4). Each layer in the 1:1 structure consists of a silicon tetrahedra sheet bound to an aluminum octahedra sheet. Each layer in the 2:1 structure is formed by an aluminum octahedra sheet sandwiched between two silicon tetrahedra sheets. Other less abundant clay minerals in soils include chlorite and vermiculite.

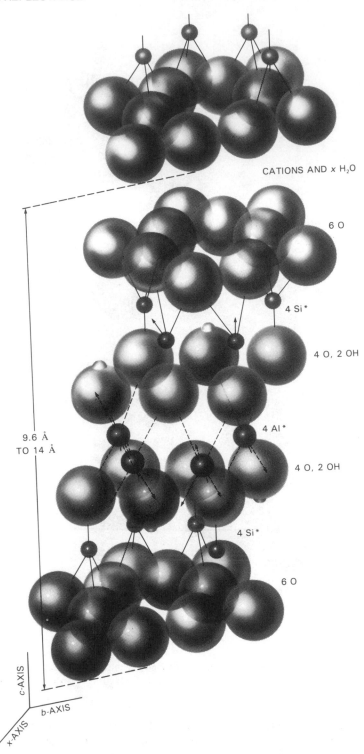

CATIONS AND x H₂O

6 O

4 Si*

4 O, 2 OH

9.6 Å
TO 14 Å

4 Al*

4 O, 2 OH

4 Si*

6 O

c-AXIS

b-AXIS

x-AXIS

Figure 4   The 2:1 structure of the smectite minerals. (Adapted with permission from *Physical Edaphology* by S. A. Taylor and G. L. Ashcroft, Fig. 6.7, Copyright © 1972, W. H. Freeman and Company.)

Cations of comparable ionic radii can also fill some of the atomic sites normally occupied by silicon ions (Si^{4+}) and aluminum ions (Al^{3+}) in clay mineral sheets. The cations commonly involved in this isomorphous substitution include ferric iron (Fe^{3+}), ferrous iron (Fe^{2+}), magnesium (Mg^{2+}), and zinc (Zn^{2+}). The specific minerals within the kaolin, smectite, and illite groups are defined on the basis of the degree of isomorphous substitution and on the substituting ionic species.

Unsatisfied valences occur on the surfaces and at the broken edges of clay mineral particles. The exposed oxygens and hydroxyls of the silica and alumina sheets act as negatively charged sites, which attract hydrogen ions (protons) and other cations. The hydroxyl groups formed at the edges of clay particles cause absorption features in soil reflectance spectra.

Crystalline particles of both primary and secondary minerals are often coated with noncrystalline materials in soils, which consist of both organic and inorganic substances. The inorganic substances are primarily hydroxides and hydrous oxides of metals and noncrystalline forms of the layer aluminosilicates. The noncrystalline aluminosilicates exhibit only local and nonrepetitive short-range structural order in comparison to crystalline particles (Jackson et al., 1986). The metal hydroxides and hydrous oxides also occur in soils as both particle coatings and as crystalline mineral particles such as goethite and hematite. The various particle coatings often serve as cementing agents that bind particles into aggregates. The noncrystalline soil materials affect reflectance by both coating soil particles and by aggregating particles.

II.B   Organic Matter

Soil organic matter consists of decomposing plant and animal residue, substances derived from the products of decomposition, and microorganisms and small animals that dwell in the soil. These materials are categorized as either nonhumic substances or as humus. Nonhumic substances are still recognizable as physical or chemical components of plant or animal tissue and include proteins, peptides, amino acids, fats, waxes, and organic acids (Schnitzer, 1982). These are readily decomposed by soil microorganisms and are usually broken down quickly in the soil. The decomposition leaves both resistant compounds of higher plant origin (e.g., oils, fats, waxes, and lignins) and new compounds synthesized by microorganisms and retained as a part of their tissue (e.g., polysaccharides and polyuronides). These products of decomposition form a complex and resistant mixture of brown or dark brown amorphous and colloidal substances called humus (Brady, 1984). Humus constitutes approximately 65–75% of the organic matter in mineral soils (Schnitzer, 1982) and can occur as a discrete substance in the soil or as coatings on mineral particles. It can also act as a binder between particles in aggregates.

Mineral soils by definition contain less than 20% organic matter and cover over 95% of the world land surface (Steila, 1976). Mineral soils in the United States typically contain 1–6% organic matter (Johnson, 1979). Despite this generally low content, organic matter exerts a profound influence on mineral soil properties such as structure, tilth, fertility, water-holding capacity, and in particular, reflectance. Soils developed in semiarid grassland environments (i.e, mollisols), for example, contain abundant humus, which imparts a very dark pigmentation on the soils. That pigmentation is less intense in soils of humid-temperate regions and is least apparent in the soils of the tropics and semitropics (Brady, 1984). Soil organic matter can thus strongly influence soil reflectance, but the influence is highly dependent on climate and environment.

II.C Soil Water and Soil Air

The pore spaces left between the particles and aggregates of solid soil materials are filled with soil water and soil air in varying concentrations. Soil water is actually a solution containing dissolved salts. Many of the salts are needed for plant growth, and a constant exchange of ions occurs between the soil solids and the soil water and between the water and plant roots. The soil air differs from the atmosphere in several respects. The relative humidity of soil air is usually near 100%, and due to microbial activity, a higher concentration of carbon dioxide and a lower concentration of oxygen is found relative to the atmosphere (Brady, 1984). Soil air is forced out of pores when water from precipitation, irrigation, or surface runoff infiltrates into and percolates through the soil under the forces of gravity and capillary attraction. Soil air diffuses back into the pores when soil water is depleted by evapotranspiration and gravitational drainage. This natural fluctuation of soil water and soil air concentrations causes frequent temporal variations of soil reflectance.

Water is retained in soil pores by the forces of capillary attraction. Due to the polar nature of the water molecule, soil solids attract water molecules, resulting in the adsorption of water onto particle surfaces. A mutual attraction (cohesion) also exists between water molecules, and the water molecules adsorbed onto particle surfaces attract other molecules further removed from the surfaces. The tenacity by which the water molecules are held in the pores decreases as the distance from a particle surface increases. These are the same type of attractive forces that cause water to rise in a capillary tube.

The affinity of the soil for soil water is characterized by the water potential, or water tension. For an incremental volume of soil water, the water potential represents the energy status of the soil water relative to a hypothetical volume of pure free water at the same location. The potential energy of soil water is less than the energy of the pure free water because of the attraction of soil solids for water and because of the salts dissolved in the soil water. Since water tends to flow from a region of high potential to a region of lower potential, the water potential is often treated as a positive suction or tension responsible for the attraction and retention of water. The water tension thus increases as the actual potential energy of the soil water decreases. Water tension is usually quantified in units of either energy per unit weight, which reduces to a unit of height (e.g., cm), or in units of energy per unit volume, which reduces to a unit of pressure (e.g., bar).

Although water exists at a continuum of potentials within a particular soil volume, soil water can be categorized for descriptive purposes on the basis of water tension. Water that readily drains from the soil following saturation is referred to as gravitational water and generally occupies large pores at tensions of 0.1–0.2 bar. Capillary water is held in the smaller pores by the forces of capillary attraction at tensions between 0.1 and 31 bars. It is the primary source of water for growing vegetation. The water adsorbed tightly by the soil solids at tensions greater than 31 bars is hygroscopic water (Brady, 1984). In dryer soils, the soil water can be envisioned as forming a film of hygroscopic and capillary water that surrounds soil particles or occupies the wedges between adjoining particles.

II.D Texture and Structure

Soil reflectance, as well as most other important soil properties, is not only influenced by the chemical composition of the soil constituents, but also by the size and arrangement of the soil particles in relation to the soil air and water. Soil texture refers to the size distribution of the soil mineral particles. The physical arrangement and aggregation of these particles provides a soil with structure. Texture and structure determine the amount of pore space available in a soil for occupation by water and air. Porosity refers to the relative volume and the size distribution of the pores. Soils of similar mineral composition and texture can have distinct structures and porosities.

Farmers and pedologists can often be observed reaching down to grab a handful of soil and rubbing the soil between a thumb and forefinger. The rubbing breaks down aggregates into primary particles and the feel, or texture, of the broken-down soil is determined to a large degree by the particle size distribution of the soil minerals. The continuum of particle sizes in a soil sample can range over three to four orders of magnitude, and this large range has been divided into the particle size fractions shown in Table 1. Although continuous mathematical distribution functions might characterize soil texture more completely, texture is more easily and more commonly described by determining the relative proportions of the particle size fractions on a mass basis. Texture descriptions are further facilitated by the definition of textural classes on the basis of relative proportions of sand-, silt-, and clay-sized particles (Figure 5). Laboratory methods are used for objective measurements of the relative proportions of the particle size classes, but an experienced pedologist can generally identify the textural class of a soil sample by feel.

A measurement or feel for soil texture is fundamental to understanding the characteristics of a soil. The mineralogy of the inorganic solids, for example, is related to particle size as previously shown in Figure 1. Sandy soils contain a high proportion of quartz and other primary minerals, whereas clayey soils contain a higher proportion of secondary minerals. The soil texture and related mineralogy have a strong influence on

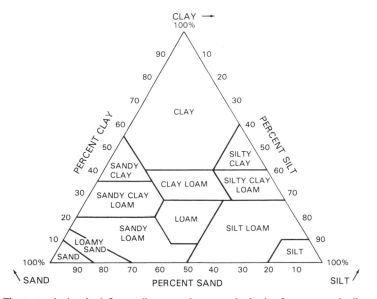

Figure 5 The textural triangle defines soil-texture classes on the basis of percent sand, silt, and clay. For any point within the triangle, the percent clay can be determined by drawing a line through the point and parallel to the base (sand axis) of the triangle. The intersection of the line with the clay axis indicates the percent clay. Similarly, the percent silt is determined by drawing a line parallel to the clay axis and the percent sand is determined by drawing a line parallel to the silt axis.

properties such as tilth, fertility, bulk density, thermal and hydraulic properties, and reflectance. In particular, sandy soils tend to be brighter than clayey soils. Additionally, the presence of rock fragments (i.e., particles greater than 2 mm in size) modifies texture and hence affects the physical properties and reflectance of soils. Since the size distribution of the mineral particles is not generally subject to rapid alteration in soils, texture is considered a relatively constant soil property and its determination is a basic requirement for any soil description.

Soil structure refers to the physical arrangement and aggregation of the soil particles. An undisturbed volume of soil is not a random arrangement of unconsolidated particles. Almost all adjoining soil particles adhere to some degree (Kemper and Rosenau, 1986), and particles often cluster into coherent aggregates that are distinctly separated from adjacent aggregates by surfaces of weak adherence. The formation of aggregates in a soil generally opens up larger pores and thereby promotes drainage and aeration. Although the size, shape, and stability of aggregates are difficult to quantify, some common types of aggregates and soil structures are recognized by pedologists and used in soil description. These soil structures are described as platelike, prismlike, blocklike, and spheroidal (Brady, 1984). Soil structure in conjunction with texture and composition are the primary soil properties controlling soil water movement, heat transfer, porosity, aeration, bulk density, and reflectance.

II.E Soil Variation

The natural processes that form soils produce both temporal and spatial variations of soil characteristics, and the spatial variations occur in both the vertical and horizontal dimensions. A vertical section of soil from the ground surface to the underlying unweathered material is called a soil profile. They are typically divided into distinct horizontal layers, called horizons, which are distinguished by more-or-less abrupt changes in soil characteristics between horizons. Figure 6 shows the horizons of a representative soil profile. This characteristic layering of soil materials forms the basis of soil description and classification. The characteristics of each horizon are usually included in a soil description, and profiles having the same sequence of horizons are classified as belonging to the same soil series. The function of a soil map is to describe the horizontal variation of soil profiles over the landscape.

Soil maps typically divide the landscape into discrete soil mapping units with distinct boundaries. Although the delineation of mapping units is useful for descriptive purposes, abrupt soil boundaries are rare due to the nature of the soil-forming processes (Petersen and Calvin, 1986). More often, a gradual transition in soil properties and profiles occurs over the landscape. This variation is amenable to observation from remote platforms, which can afford a synoptic perspective of a geographic area. Remotely acquired images show variations in surface reflectance, which can indicate corresponding variations in the underlying soil profiles. The routine use of aerial photography for soil mapping is the most familiar application of remote sensing to the description of soil spatial variation.

The soil-forming processes are continuous and ongoing. Thus, soil profiles continue to develop and change over the years, but the changes are generally gradual. The sequence and character of horizons in a profile often persist for decades or even centuries. The rapid and frequent variations of surface-soil conditions are of perhaps more relevance to remote-sensing applications. Soil-surface roughness, structure, moisture content, and other properties are readily altered by weather, cultivation, erosion, plant

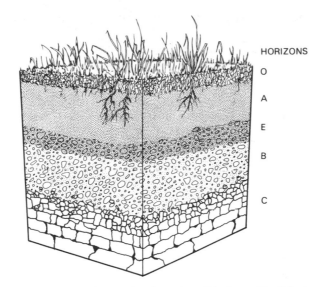

Figure 6 A representative soil profile showing the major soil horizons. The O horizon is a layer of plant litter and residue that accumulates above the mineral soil. The A horizon is a soil layer occurring at or near the surface and is characterized by the leaching, or eluviation, of clays and iron and aluminum oxides and by the presence of humidified organic matter that imparts a darker color relative to lower horizons. The E horizon is a zone of maximum eluviation with a concentration of sand and silt particles of quartz and other resistant minerals. The B horizon is a zone of illuviation that contains an accumulation of materials such as clays and iron and aluminum oxides. The C horizon is a layer of unconsolidated lithologic materials that have been relatively unaffected by the soil-forming processes.

growth, and other surface phenomena. These temporal variations affect the surface reflectance properties. Changes in these reflectance properties can complicate the use of remote sensing for soil mapping and characterization, but on the other hand, observations of reflectance variation over time can also indicate characteristics of the underlying soil properties. Thus, an appreciation of spatial and temporal soil variability is needed to understand soil reflectance and the applications of remote sensing to pedology.

II.F Soil Color

The eye perceives the wavelength spectrum of reflected light as color. Soil color is a readily observable soil property that is routinely used to describe soils and to distinguish between soil types. As discussed by Baumgardner et al. (1985), the United States Department of Agriculture (USDA) has standardized color designations for soil description by the use of the "Munsell Soil Color Charts" (Munsell Color, 1975). These charts contain standard color chips with colors specified by designations for hue, value, and chroma as well as by descriptive names.

In the field, a pedologist determines soil color by a direct comparison between a soil sample and the color chips. The solar radiation reflected from soils can also be observed by remote-sensing instruments that offer measurement capabilities beyond those of the naked eye. In either case, an understanding of the interaction of light with soil is needed to infer soil properties and processes from the observed reflectance. The following sections discuss the relationships between the soil properties previously discussed and the characteristics of the reflected solar radiation.

III   THE INTERACTION OF SOLAR ENERGY WITH SOIL

Soil is a semiinfinite medium relative to solar radiation. In other words, solar radiation incident on a bare soil is either absorbed within the soil profile or is reflected from the soil surface. The absorbed radiation heats the soil, evaporates soil water, drives chemical reactions and biologic processes, or otherwise raises the energy of the soil system. The radiation that is not absorbed is reflected and can be observed by remote-sensing instruments.

As defined by Nicodemus et al. (1977), "reflection is the process by which electromagnetic flux (power), incident on a stationary surface or medium, leaves that surface or medium from the incident side without change in frequency; reflectance is the fraction of the incident flux that is reflected." To precisely discuss the observation of reflected radiation and reflectance, the geometry of the beam of incident flux (i.e., the direction and solid angle of the beam) and the geometry of the beam in which reflected flux is collected and detected must be specified. This chapter follows a nomenclature designated by Nicodemus et al. (1977) for specifying the beam geometries. In particular, bidirectional reflectance refers to the idealized case, where both the incident beam and the collected beam are infinitely small and the direction of each beam is specified by angular parameters (e.g., azimuth and zenith angles). Bihemispherical reflectance refers to the case where both beams subtend a 2π sr solid angle. Nicodemus et al. (1977) provide a more complete discussion of the nomenclature and the concept of reflectance.

Several characteristics of reflected solar radiation can be remotely observed and quantified. Spectral composition (i.e., flux or reflectance as a function of wavelength) is the most commonly observed characteristic of radiation reflected from soils. Another observable characteristic is the directional distribution of the reflected radiation. Due to the rough nature of soil surfaces, solar radiation is reflected in all directions within the 2π sr solid angle above an element of soil surface, but the reflectance is generally not equal in all directions. In addition, the fraction of incident solar flux reflected over all directions (i.e., the albedo) can be observed. A third observable characteristic is the polarization state of reflected radiation, which can be observed as a function of wavelength and direction. The objective of remote sensing for pedology is to relate observed characteristics of reflected radiation to soil properties.

III.A   Microscopic Interactions

To relate the remote observations to soil properties, a quantum mechanical viewpoint of the interaction of light with matter is useful. According to the principles of quantum mechanics, a system composed of atoms and molecules can possess energy in only specific discrete amounts called energy levels. Electromagnetic energy is either emitted or absorbed when an atom or molecule changes from one of the characteristic energy levels to another. Such a change is referred to as an energy transition, and the frequency of the emitted or absorbed radiation is given by

$$|\Delta E| = h\nu \tag{1}$$

where $|\Delta E|$ is the difference in energy between the two levels, h is Planck's constant (6.626×10^{-34} Js), and ν is the frequency of the radiation.

The basis of the quantum mechanical model of matter is Schrödinger's equation:

$$E\Psi = \mathbf{H}\Psi = (-\hbar^2/2m)\nabla^2\Psi + \mathbf{V}\Psi \tag{2}$$

where E is the total energy of a particle, $\mathbf{H}$ is the Hamiltonian operator, Ψ is the wave function of the particle, $\hbar$ is Planck's constant divided by 2π ($= h/2\pi$), m is the mass of the particle, ∇^2 is the Laplacian operator, and $\mathbf{V}$ is a potential-energy operator (note that Eq. 2 is the time-independent form of the Schrödinger equation). Mathematical solutions to this partial differential equation are only possible for certain discrete values of E. In other words, the allowable energy levels for an atomic particle are the eigenvalues of the Schrödinger equation (Hunt, 1980). Physicists have found the equation to accurately and quantitatively describe a wide range of atomic phenomena. In practice, Schrödinger's equation is too complicated to solve exactly for complex atomic systems. The important concept, however, is that a validated mathematical description exists for the quantum nature of matter and energy. The microscopic interactions of light with matter are quantized.

Transitions in the energy levels of molecular systems involve changes in the motions of atomic nuclei or in the energy state of electrons. The motions of the nuclei can be translational, rotational, or vibrational. In soils, interactions with solar radiation primarily involve either vibrational motions or electronic energy states, since molecular translations and rotations are restricted in most soil materials.

The vibrational motions consist of oscillations in the relative positions of bonded atomic nuclei. The oscillations either stretch molecular bond lengths or bend interbond angles. Transitions between vibrational states result in the emission or absorbtion of radiation only if the vibration is accompanied by an oscillation in the magnitude of the dipole moment of a molecule (Castellan, 1983). Transitions in vibrational energy states are generally associated with the emission or absorption of radiation within the infrared portion of the spectrum.

Transmission spectra of free water, for example, show three sharp and strong absorption bands between 2.5 and 7.0 μm (Table 4). Each absorption band is associated with a distinct vibrational mode of the polar water molecules. Two of the modes involve stretching of the oxygen-hydrogen bonds, and the remaining mode involves bending of the angle between the two oxygen-hydrogen bonds, as illustrated in Figure 7. A set of discrete energy levels is associated with each vibrational mode. The frequency of each absorption band corresponds to the difference in energy between the ground state of the associated vibrational mode and the next allowable energy level of that mode. These three transitions in vibrational energy levels are referred to as the fundamental vibration transitions of water.

In addition to the fundamental absorption bands, several weaker absorption bands occur in the transmission spectra of water. These bands are referred to as overtone bands and combination bands. Overtone bands are due to the transition of one vibrational mode from a ground state to an energy level two or more levels above the ground state. Combination bands are attributed to the splitting of an absorbed quantum of radiation to raise the energy level of more than one vibrational mode (Table 4). The overtone and combination tone bands are weaker than the fundamental bands because the fundamental vibrational transitions occur more frequently.

As with all matter, the possible vibrational energy levels of the molecular soil components depend on the atomic composition and structure of the molecules. The funda-

TABLE 4   Absorption Bands for Water

Wavelength (μm)	Frequency (s^{-1})	Wavenumber (cm^{-1})	Energy State of Each Vibrational Mode			Remarks
			$v1$	$v2$	$v3$	
6.270	4.782E + 13	1595	0	1	0	Fundamental, $v2$
3.173	9.450E + 13	3152	0	2	0	First overtone, $2v2$
2.738	1.095E + 14	3652	1	0	0	Fundamental, $v1$
2.662	1.126E + 14	3756	0	0	1	Fundamental, $v3$
1.876	1.598E + 14	5331	0	1	1	Combination, $v2 + v3$
1.135	2.640E + 14	8807	1	1	1	Combination, $v1 + v2 + v3$
0.942	3.182E + 14	10613	2	0	1	Combination, $2v1 + v3$
0.906	3.307E + 14	11032	0	0	3	Second overtone, $3v3$
0.823	3.643E + 14	12151	2	1	1	Combination, $2v1 + v2 + v3$
0.796	3.767E + 14	12565	0	1	3	Combination, $v2 + 3v3$
0.652	4.601E + 14	15348	3	1	1	Combination, $3v1 + v2 + v3$

[a] After Castellan (1983; Table 25.2).

mental transitions of vibrational energy levels for most soil mineral molecules involve the emission or absorption of radiation in the infrared portion of the spectrum between 2.5 and 50 μm (White, and Roth, 1986). For this reason, infrared spectroscopy is a useful tool for laboratory soil analyses. For remote sensing within the solar spectrum, however, observations of vibrational phenomena are generally limited to overtone and combination bands (Hunt, 1980).

The interactions of light with soil materials also include transitions in the energy levels of the electrons in soil atoms and molecules. Electronic transitions may be due to charge-transfer modes, crystal field transitions, and transitions into the conduction band (Smith, 1983). These electronic transitions generally involve higher energy than vibrational transitions. Thus, the absorption and emission of radiation due to electronic tran-

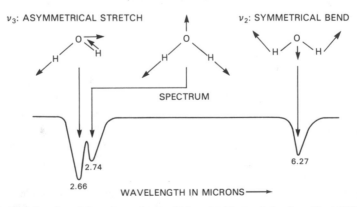

Figure 7   Vibrational modes of the water molecule. [Adapted with permission from Hunt (1980). Copyright © 1980, John Wiley & Sons, Inc.]

sitions usually occur at the higher frequencies within the ultraviolet and visible portion of the spectrum. The absorption bands associated with electronic transitions are also typically much broader than the bands corresponding to vibrational transitions (Hunt and Salisbury, 1970).

III.B Macroscopic Interactions

The processes of reflection, absorption, transmission, and refraction by a macroscopic medium are perhaps more familiar than microscopic quantum mechanical interactions. To quantify these processes, the optical properties of a medium (e.g., the index of refraction) are defined and measured for calculations of quantities such as reflectance, absorptance, and the angle of refraction. The optical properties, however, depend on the atomic and molecular composition and structure of the medium. The processes of reflection, absorption, etc. are actually the composite effects of the multitude of quantum mechanical interactions involving the enormous number of atoms and molecules in real media. We resort to the use of macroscopic optical properties for calculations because it is impossible to follow or account for the numerous microscopic interactions in most real media such as soils.

Many of the fundamental optical properties of a medium are embodied in its complex index of refraction. This index is a complex number and is a function of wavelength:

$$m = n - ik \tag{3}$$

The real part, n, is called the real refractive index, and it is equal to the ratio of the speed of light in vacuum, c, to the speed of light in the material, v:

$$n = c/v \tag{4}$$

The imaginary part, k, of the complex index of refraction is called the extinction coefficient. The attenuation of radiation travelling through a finite thickness of the material is determined by the extinction coefficient. According to the Lambert–Beer law, as a beam of radiation travels through a slab of uniform material, the relative decrease in the intensity of the beam is directly proportional to the thickness of the slab:

$$-\delta E/E = a\,\delta x \tag{5}$$

Consequently,

$$E_x = E_0 e^{-ax} \tag{6}$$

where E_0 is the flux density of the beam incident of the slab, x is the thickness of the slab, E_x is the flux density of the beam leaving the slab, and a is a constant of proportionality called the absorption coefficient of the material. The extinction coefficient k is related to the absorption coefficient a by the following equation (Egan and Hilgeman, 1979):

$$k = a\lambda/4\pi \tag{7}$$

where λ is the wavelength of the radiation.

Although fundamental to the comprehension of light interactions with soil, data on the complex indices of refraction of soil constituents are sparse in the literature (Egan and Hilgeman, 1979). Real refractive indices are provided for many soil minerals in texts on optical crystallography, but these values are not usually given as a function of wavelength. The indices are generally average values for the white light typically used for illumination in petrographic microscopes. Other sources of data on soil mineral indices include literature addressing atmospheric attenuation due to suspended soil particles (e.g., Grams et al., 1974) or literature addressing lunar and planetary observations (e.g., Egan et al., 1975; Egan and Hilgeman, 1979). Otherwise, the optical properties of soil constituents have not been well studied.

Part of the reason for the sparsity of data on soil complex indices of refraction is the physical nature of some soil constituents. Many secondary soil minerals, for example, occur only as powders consisting of small particles. Measurements of transmittance or angles of refraction cannot be made with large crystals of these minerals. Measurements are further complicated by the optical anisotropy of many minerals. Due to crystal structure, the indices of refraction of an anisotropic mineral depend on the direction of radiation propagation relative to the crystal axes. Some indices have been derived for secondary minerals from measurements of compressed powders and from particles compressed in a transparent matrix such as potassium bromide (Egan and Hilgeman, 1979), but interpretation of these measurements is complicated by the anisotropy and nonspherical shape of the minerals (Bohren and Huffman, 1983). Specifying the indices of amorphous soil minerals and organic matter is also difficult due to the variable and indefinite molecular composition and structure of these materials. Regardless of the difficulties, further data on the indices of soil materials are needed to advance understanding of soil reflectance.

Knowledge of the indices of refraction alone is not sufficient for the comprehension of light interactions with soil. Solar radiation incident on a soil surface does not strike a smooth, flat surface consisting of a homogeneous mixture of materials having some well-known composite index of refraction. Instead, the soil presents a rough, irregular surface of variable composition. A photon of incident radiation may undergo multiple reflections by a variety of materials before ultimately being absorbed or reflected from the surface. These multiple interactions depend on physical soil characteristics as well as the complex indices of refraction. Relevant characteristics include the structure, aggregation, and physical associations of the soil constituents; the size distribution and shapes of soil particles and aggregates; and surface roughness due to clods, furrows, rock fragments, or other large-scale surface structures. Any attempt to realistically model soil reflectance must account for the physical characteristics of soil as well as the intrinsic optical properties of the soil materials.

IV MODELS OF SOIL REFLECTANCE

Models of soil reflectance are generally developed from the concept of multiple scattering by soil particles. As explained by Bohren and Huffman (1983), the electric charges within any particle of matter (i.e., the atomic electrons and protons) are set into oscillatory motion by the electric field of an incident electromagnetic wave. Note that the incitation of the oscillations on an atomic level is a quantum mechanical process (i.e., is quantized), as previously discussed. The accelerated electric charges either transform the incident radiation into other forms of energy (i.e., absorb the radiation) or reradiate the energy at the same frequency but in different directions. This secondary radiation from a particle is referred to as scattered radiation. Multiple scattering occurs when particles are in close proximity and are affected not only by the electromagnetic field from an external source, but also by the electromagnetic fields scattered from nearby particles. Given the close proximity of soil particles, reflection from soil surfaces is often modeled as a manifestation of multiple scattering by soil particles.

Computational methods have been developed to quantitatively describe the multiple scattering of light by collections of particles. The general methods have been reviewed by a number of authors (Egan and Hilgeman, 1979; Hanson and Travis, 1974; van de Hulst, 1980) and have been applied to modeling reflectance from and transmission through a variety of media such as Earth and planetary atmospheres, oceans, snow, plant canopies, and, in particular, soils. Valid models offer the ability to predict reflectance from the media and offer the possibility of model inversion for the estimation of media characteristics from remote observations. Further, models aid the comprehension of remote observations by quantitatively describing the physics of the light-scattering process. Several multiple-scattering models applicable to the study of soil reflectance are discussed here.

IV.A Single Scattering

The problem of light scattering by a single particle is well studied. The concepts of geometric optics can be applied to scattering by large particles and the technique of ray tracing can be used to obtain numerical results describing the scattered energy (Hansen and Travis, 1974). Ray tracing involves the use of the Fresnel equations to calculate coefficients of reflection and transmission and the use of Snell's law to determine the direction of the refracted rays. The scattering of light by a spherical particle of arbitrary radius is also rigorously described by the Mie theory, which provides a formal solution to Maxwell's equations with appropriate boundary conditions for spheres. Computational methods and computer code for single-scattering calculations using the Mie theory are provided by Dave (1968), Wiscombe (1980), Bohren and Huffman (1983), and Egan and Hilgeman (1979). The Mie theory is often applied to scattering by smaller particles having radii approaching the wavelengths of the scattered radiation. In-depth treatments of the Mie theory and single-particle scattering are provided by Bohren and Huffman (1983) and van de Hulst (1957).

Soil particles are not typically spherical, but the Mie theory is often applied to irregular particles by treating the particles as spheres of equivalent radii, volumes, or surface areas. Bohren and Koh (1985) have shown that this approach can lead to unrealistic representations of scattering by irregular particles. Mugnai and Wiscombe (1986) have calculated numerical results for nonspherical particles described by Chebyshev polynomials, but exact solutions for single scattering by small nonspherical particles are not generally available.

The objective of theoretical work on single scattering is to compute parameters describing the radiation scattered by particles (Hansen and Travis, 1974). The parameters of interest include the scattering cross section (σ_{sca}, dimension of area), the single-scattering albedo (ω, dimensionless), and the phase function (P). The scattering cross section describes the magnitude of the energy scattered in all directions relative to the energy of the incident radiation. The incident energy falling on to a hypothetical area of

σ_{sca} is equal to the total energy of the scattered radiation. The single-scattering albedo is the ratio of the scattered energy to the total energy either scattered or absorbed by the particle. The phase function describes the angular distribution of the energy scattered by the particle. These parameters are applicable to calculations of multiple scattering if the particles can be treated as independent scatters (see Chapter 5 for more details).

IV.B Multiple Scattering

Computational methods applicable to multiple scattering by soils generally have a basis in radiative transfer theory. The theory involves the development of conservation of energy equations that describe the radiation field within a medium that absorbs, emits, and scatters radiation (Chandrasekhar, 1960). The time-dependent radiative transfer equation for a plane parallel media may be written as follows (Smith, 1983):

$$\mu \frac{\delta I}{\delta \tau} (\tau; \mu, \phi) = -I(\tau; \mu, \phi) + J(\tau; \mu, \phi) \tag{8}$$

where μ is the cosine of the angle between the direction of propagation and a normal to the plane parallel media; ϕ is the azimuth angle of the propagation direction; τ is the optical depth of the media; I is radiation intensity, or radiance; and J is a source term comprised of the scattering of surrounding flux and attentuation of an external source. The source term is given by

$$J(\tau; \mu, \phi) = \frac{\omega}{4\pi} \int_0^{2\pi} \int_{-1}^{1} P(\mu', \phi'; \mu, \phi) I(\tau; \mu', \phi') \, \delta\mu' \, \delta\phi'$$

$$+ \frac{\omega}{4\pi} P(\mu_0, \phi_0; \mu, \phi) \pi F_0 e^{-(\tau - \tau_0)/\mu} \tag{9}$$

where ω is the single-scattering albedo, τ_0 is the total optical depth of the media, P is the phase function, and F_0 is the incident radiance from an external source. To paraphrase Smith (1983), the problem in modeling soil reflectance is not in writing down this general formulation of the multiple-scattering process. The difficulty is in developing abstractions (i.e., simplified representations) of the soil that realistically relate soil properties to the scattering parameters and processes represented within the radiative transfer equation.

Many of the applicable methods require the abstraction of soils as mineral powders. Vincent and Hunt (1968), for example, investigated the relative contributions of specular reflectance and multiple scattering to the bihemispherical reflectance of mineral powders. Specular reflectance was modeled by the Fresnel equations and multiple scattering was modeled by the Kubelka–Munk theory, which is based on the simultaneous solution of two first-order differential equations in one dimension (Egan and Hilgeman, 1979):

$$\delta I = -(\alpha + \sigma) a_0 I \, \delta x + \sigma b_0 J \, \delta x \tag{10}$$

$$\delta J = (\alpha + \sigma) b_0 J \, \delta x + \sigma a_0 I \, \delta x \tag{11}$$

where I is the radiant flux in the positive x direction through a uniform layer of scatterers, J is the radiant flux in the negative x direction, α is the fraction of radiation absorbed per unit path length, σ is the fraction of radiation scattered per unit path length, and a_0 and b_0 are constants relating δx to the average path lengths of the radiation. For an infinitely thick layer, the solution of a differential equations yields a value of zero for transmission and the following expression for reflectance, ρ (Egan and Hilgeman, 1979):

$$\rho = \frac{1 - \left[\alpha/(\alpha + 2\sigma)\right]^{1/2}}{1 + \left[\alpha/(\alpha + 2\sigma)\right]^{1/2}} \tag{12}$$

In comparing the Fresnel equations to the Kubelka-Munk theory, Vincent and Hunt (1968) concluded that specular reflectance should increase and multiple scattering should decrease as the absorption coefficient increases.

Emslie and Aronson (1973) also considered light scattering by mineral powders. They treated scattering by large mineral particles differently than that by small particles. Absorption and scattering coefficients were primarily derived from geometrical optics for particles significantly larger than the wavelength of the scattered radiation. Corrections were made for additional absorption due to the edges and asperities (i.e., small projections or roughness on a surface) of large particles. Powders consisting of particles smaller than the scattered wavelength were modeled as assemblies of randomly oriented ellipsoids having a range of shapes and the same volumes as the actual particles. This abstraction allowed the use of established optical methods to calculate absorption and scattering coefficients. The applied wave optics accounted for both coherent and noncoherent scattering by closely spaced fine particles. A bridging formula between large and small particles was also given for the calculation of coefficients for intermediate-size particles.

For any size of particle, Emslie and Aronson (1973) represented scattering by six discrete beams. Six simultaneous differential equations were formulated to describe the balance of radiation flux in the transverse directions as well as the forward and backward directions within a particulate medium. The solution of these equations with appropriate boundary conditions provided an equation for bihemispherical reflectance.

An important feature of this method is the ability to model mixtures of different mineral particles by appropriate summations over particle volume fractions, particle sizes, and complex indices of refraction (Smith, 1983). Aronson and Emslie (1973) found good agreement between their model and measurements of infrared reflectance from a mixture of fine quartz and corundum particles.

Egan and Hilgeman (1978) compared the model of Emslie and Aronson to spectral reflectance observations of various particulate samples. Good agreement was not found between the model and observations, which ranged from the ultraviolet to the infrared. The discrepancy was attributed to scattering by asperities (i.e., small projections or roughness on a surface) on coarse particles (Egan and Hilgeman, 1978); the model included only absorption, not scattering, due to asperities. Scattering by asperities was known to be inversely proportional to wavelength and was thus believed to be more significant at wavelengths shorter than the infrared region observed by Aronson and Emslie (1973).

To account for asperity scattering as well as absorption, Egan and Hilgeman (1978) developed a Monte-Carlo approach to formulate scattering by coarse particles. The computational method assigned probabilities to reflection, refraction, and internal and external scattering as a function of the asperities on single spherical particles. Radiation fluxes in six directions were then determined stochastically. The six-direction radiative transfer method of Emslie and Aronson (1973) was used to describe multiple scattering and to calculate the reflectance of particulate surfaces. Egan and Hilgeman (1978) found good agreement between this refined model and visible to near-infrared reflectance observations of multiconstituent particulate samples.

Another approach to computing the reflectance of mineral powders was proposed by Conel (1969). He abstracted mineral powders as semiinfinite cloudy atmospheres consisting of a plane-parallel layer of independent scatters. On the basis of this abstraction, a numeric solution to the radiative transfer equation was derived by the method of discrete ordinates (Hansen and Travis, 1974). The resulting equation was a function of two single-scattering parameters:

$$\rho = \frac{u - 1}{u + 1} \tag{13}$$

where

$$u^2 = \frac{1 - \omega(g/3)}{1 - \omega} \tag{14}$$

and ρ is reflectance, ω is the single-scattering albedo, and g is a factor that quantifies the anisotropy of light scattered by a single particle. The anisotropy factor g was determined by expressing the phase function as a truncated Legendre polynomial series:

$$P(\theta) = 1 - g \cos \theta \tag{15}$$

where θ is the phase angle between incident and scattered radiation. These single-scattering parameters were derived from the Mie theory as functions of wavelength and mineral particle size and complex index of refraction. Conel (1969) applied his model to study the spectral emissivity of quartz particles in the infrared region.

The models discussed up to this point all calculate hemispherical reflectances for mineral powders. Remote soil observations, however, are often directional. Hapke (1981) has developed a model for the bidirectional reflectance of mineral powders as well as for hemispherical reflectances. Hapke derived an approximate analytic solution to the radiative transfer equation that took mutual shadowing by closely spaced particles into account. The resulting equation for bidirectional reflectance follows:

$$\rho(\mu_0, \mu, \theta) = \frac{\omega}{4\pi} \frac{\mu_0}{\mu_0 + \mu} \left\{ [1 + B(\theta)] P(\theta) + H(\mu_0) H(\mu) - 1 \right\} \tag{16}$$

where

μ cosine of the zenith angle of illumination;

μ_0 cosine of the zenith angle of reflection;

ω single-scattering albedo;

θ phase angle between directions of reflection and illumination;

$P(\theta)$ single-scattering phase function;

$H(\mu)$ approximation to the H-function of Chandrasekhar (1960);

$B(\theta)$ function that accounts for retroreflectance.

Hapke and Wells (1981) compared reflectances calculated by Eq. 16 to measured bidirectional reflectances of cobalt glass powder with close agreement. Hapke (1981) also suggested that Eq. 16 could be applied to powders containing multiple constituents by computing a weighted average of the single-scattering albedos of the constituents.

Retroreflectance refers to a large sharp peak in bidirectional reflectance near zero phase (i.e.; the reflectance radiation scattered directly back in the antiillumination direction). It is a common phenomenon in the observation of natural materials and is also known as the opposition effect in planetary observations and the hot spot in remote sensing of plant canopies. To account for retroreflectance in Eq. 16, Hapke (1986) derived the following retroreflectance function:

$$B(\theta) = \frac{B_0}{1 + \tan(\theta/2)/h} \tag{17}$$

where B_0 is the ratio of the radiation scattered by surface particles to the total radiation scattered at zero phase ($\theta = 0$), and h is a parameter that controls the angular width of the retroreflectance. Retroreflectance from particulate surfaces has also been considered by Lumme and Bowell (1981) in their derivation of particle surface phase functions from the radiative transfer equation. Lumme and Bowell (1981) concluded that retroreflectance from low-albedo surfaces was primarily attributable to the porosity of the surface.

Besides abstracting soils as mineral powders, several models deal with the effects of large-scale surface roughness on soil reflectance. These models abstract the soil surface as either distributions of geometric objects or periodic geometric surfaces. The models then quantitatively describe light scattering and shadowing by the objects or structures.

Hapke (1984), for example, represented rough planetary surfaces as an assembly of tilted facets that had no preferred orientation in azimuth but a Gaussian distribution of zenith angles. By assuming large facets relative to wavelength, the proportions of visible versus hidden and illuminated versus shadowed surface areas were analytically related to the mean slope angles of the facets and the directions of illumination and viewing. These relationships were used to derive a shadowing function that was multiplied by the Hapke (1981) bidirectional reflectance function (Eq. 16) to model the bidirectional reflectance of rough particulate surfaces.

A similar shadowing function was developed by Cierniewski (1987) for soil surfaces represented by equal-sized spheres arranged so that the sphere centers formed a square grid on a freely sloping plane. Geometry was used to express the proportion of shadowed surface area as a function of sphere radius, sphere spacing, slope angle, and the directions of illumination of viewing. The reflectance of a rough soil surface was then related to the reflectance of a smoothed soil surface by an exponential function of the shadowed proportion. Cierniewski (1987) used empirical observations of smoothed soil surfaces and his shadowing function to predict rough-soil reflectance. The predictions agreed closely with field observations of rough soils.

Cooper and Smith (1985) used a Monte-Carlo approach to allow the representation of a soil surface by any function $f(x, y)$ that was periodic in x and y with a period of 1. This approach involved the tracing of photons incident on arbitrary points across the periodic surface. As each photon was reflected from a point on the surface, the new direction of the reflected photon was determined stochastically under the assumption of Lambertian reflectance; in other words, reflection was equally probable in any direction within the 2π sr solid angle (hemisphere) surrounding a point on the surface. Since a point of reflection could occur on a sloping increment of the surface, reflection into the hemisphere bounded by the surface increment could result in a subsequent intersection of the photon with another point on the surface. Each photon was followed in this manner through, potentially, multiple reflections until the photon escaped past the top of the surface. The direction of the escaping photon was then stored. After following a number of photons, a directional distribution of escaping photons was established. Bidirectional reflectance was expressed as the ratio of photons escaping in a particular direction to the number of incident photons.

In summary, current multiple-scattering models applicable to soil reflectance generally require the abstraction of soils as either mineral powders or surfaces formed by large geometric structures. These models provide valuable insight into the interactions of solar radiation with soil, particularly with respect to the effects of mineral composition, particle size, and rough surfaces. Actual soils, however, are far more complex than these simple abstractions. Current models do not incorporate such fundamental soil properties as moisture content, organic matter content, and structure. Further theoretical developments are needed to advance our comprehension and modeling of light interactions with soils.

V OBSERVATIONS OF SOIL REFLECTANCE

The interactions of light with soils have also been studied empirically. The most thoroughly observed characteristic of light scattered by soils has been spectral reflectance. A more limited amount of data on the directional distribution and polarization state of scattered light can be found in the literature. These empirical observations have demonstrated general relationships between soil properties and the characteristics of scattered energy.

V.A Spectral Reflectance

Observations of spectral soil reflection have typically been quantified by taking the ratio of spectral radiant flux reflected from a soil surface to the spectral radiant flux reflected from a reference material illuminated and viewed in the same manner as the soil surface. If the reference material can be considered a lossless perfectly diffuse (Lambertian) surface, then this ratio is precisely defined as a spectral reflectance factor, instead of spectral reflectance, by Nicodemus et al. (1977). Barium sulfate and polytetrafluoroethylene (halon) are commonly used as reference materials because these materials approximate lossless Lambertian surfaces in the visible and near-infrared portions of the spectrum.

A substantial number of researchers have measured spectral reflectance factors from soil samples in the laboratory and from soil surfaces in the field (Beck et al., 1976; Bowers and Hanks, 1965; Condit, 1970; Mathews et al., 1973; Montgomery and Baumgardner, 1974; Obukhov and Orlov, 1964; Orlov, 1966). Perhaps the most comprehensive study of reflectance from United States soils was conducted by Stoner and Baumgardner (1980, 1981) and Stoner et al. (1979). In the laboratory, they rigorously measured spectral bidirectional reflectance factors from samples of over 240 soil series. Measurements were made over the 0.52–2.32 μm wavelength region of the spectrum. The samples were selected by a stratified sampling of soil series within 17 temperature and moisture regimes across the contiguous United States. Several samples of tropical soils from Brazil were included. The samples were prepared in a manner that ensured a uniform soil-moisture tension (0.1 bar) in all samples. Empirical studies of soil spectral reflec-

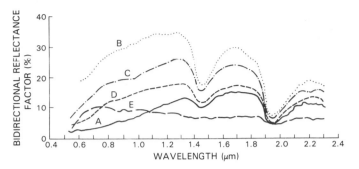

Figure 8 The characteristic soil bidirectional reflectance spectra of Stoner and Baumgardner (1981). Curve A: soils having high (>2%) organic-matter content and fine texture. Curve B: soils having low (<2%) organic-matter content and low (<1%) iron-oxide content. Curve C: soils having low (<2%) organic-matter content and medium (1 to 4%) iron-oxide content. Curve D: soils having high organic-matter content (>2%), low iron-oxide content (<1%), and moderately coarse texture. Curve E: soils having high iron-oxide content (>4%) and fine texture. (Reproduced from "Characteristic Variations in Reflectance of Surface Soils" by E. R. Stoner and M. F. Baumgardner, Fig. 1, *Soil Science Society of America Journal*, Volume 45, No. 6, pages 1161–1165 by permission of the Soil Science Society of America, Inc.)

tance are well reviewed by Baumgardner et al. (1985) with an emphasis on the work of Stoner and Baumgardner (1981).

Stoner and Baumgardner (1981) defined five spectral curves (Figure 8) that were characteristic of the observed soil reflectance spectra. In other words, they believed each of the observed spectra resembled one of the five curves shown in Figure 8 and the resemblance was determined by soil properties, especially organic matter and iron-oxide content. The discriminating features of the five curves are shape and the presence or absence of absorption bands. The strategy here will be to discuss observed relationships between soil properties and spectral reflectance with reference to the five characteristic spectra of Stoner and Baumgardner (1981).

V.A.1 Moisture Content

Reflectance spectra of moist soils include prominent absorption bands centered at 1.4 and 1.9 μm (Fig. 8). These bands, along with weaker absorption bands at 0.97, 1.20, and 1.77 μm, are attributable to overtones and combinations of the fundamental vibrational frequencies of water molecules in the soil. The bands at 1.4 and 1.9 μm are typically broad, indicating an unordered arrangement of water molecules at various sites in the soil (Baumgardner et al., 1985).

In addition to the absorption bands, increasing moisture content generally decreases soil reflectance across the entire shortwave spectrum. Wet soils usually appear darker to the eye than dry soils for this reason. The data of Bowers and Hanks (1965) are frequently cited to demonstrate decreasing spectral reflectance as a function of increasing moisture content for a silt loam soil (Figure 9). As reviewed by Planet (1970), Angstrom (1925) attributed the decrease in reflectance to internal reflections within the film of water covering soil surfaces and particles.

V.A.2 Iron and Iron-Oxide Content

Some iron is usually found in a soil. Iron ions are readily soluble and are therefore widespread. Iron ions can also easily substitute into octahedral Al^{3+} sites and less easily into tetrahedral Si^{4+} sites and are thus retained in soils (Hunt, 1980). Iron also com-

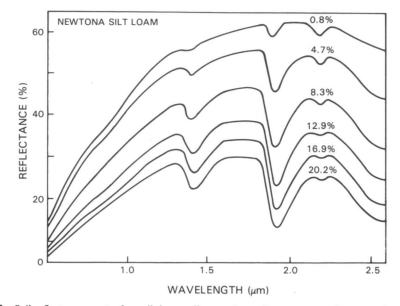

Figure 9 Soil reflectance spectra for a silt loam soil at varying moisture contents. Percent moisture content by weight is shown above each curve. (Reproduced with permission from ''Reflection of Radiant Energy from Soils'' by S. A. Bowers and R. J. Hanks, Fig. 1, *Soil Science*, Volume 100, pages 130–138, Copyright © 1965, Williams and Wilkins.)

monly occurs as a principal constituent of some soil minerals. Many of the absorption features in soil reflectance spectra are due to the presence of iron in some form.

These absorption features are caused by either crystal field effects or charge transfers involving iron ions. For example, the steep decrease in reflectance toward the blue and ultraviolet wavelengths is a characteristic of almost all soil reflectance spectra (Figure 8). This decrease is due to a strong iron–oxygen charge transfer band that extends into the ultraviolet (Hunt, 1980). Other absorption bands often occur near 0.7 and 0.87 μm (Stoner and Baumgardner, 1980) due to electronic transitions of ferric (Fe^{3+}) iron (Figure 8, curve E). Electronic transitions involving ferrous (Fe^{2+}) iron can cause strong absorption bands near 1.0 μm (Figure 8, curves C and E) (Hunt and Salisbury, 1970; Stoner and Baumgardner, 1980). Several weaker absorption bands between 0.4 and 0.55 μm can be found in some spectra due to one or the other iron ion (Hunt and Salisbury, 1970; Mulders, 1987). Even trace amounts of iron can result in the appearance of iron absorption bands in soil reflectance spectra (Stoner and Baumgardner, 1980).

Additional absorption in the middle infrared wavelengths can be attributed to ferrous iron in disordered octahedral sites (Hunt and Salisbury, 1970; Mulders, 1987). Curve E of Figure 8 represents the spectra of soils with high iron-oxide content (greater than 4%) such as the tropical soils (oxisols) observed by Stoner and Baumgardner (1980). Iron absorption in the middle infrared by these soils can be strong enough to obliterate the water-absorption band at 1.4 μm (Stoner and Baumgardner, 1981).

V.A.3 Organic-Matter Content

Organic matter has a strong influence on soil reflectance. Spectral reflectance generally decreases over the entire shortwave region as organic matter content increases (Stoner and Baumgardner, 1980). At organic-matter contents greater than 2%, the decrease due to organic matter may mask other absorption features in soil spectra (Baumgardner et al., 1970). The spectra of soils with organic-matter contents greater than 5% often have a concave shape between 0.5 and 1.3 μm (Figure 8, curve A) as compared to the convex shape of spectra for soils with lower organic-matter content (Stoner and Baumgardner, 1981).

The reflectance spectra of organic soils (i.e., organic-matter content greater than 20%) depend on the decomposition of the organic material. Spectra of fully decomposed (sapric) materials resemble curve A of Figure 8, whereas spectra of partially decomposed (hemic) materials resemble curve D of Figure 8 (Stoner and Baumgardner, 1981). The spectral reflectance of minimally decomposed (fibric) organic matter is high in the near-infrared and is similar to the spectral reflectance of senescent leaves (Stoner and Baumgardner, 1981).

V.A.4 Mineralogy

Most soils are predominantly composed of minerals. The principal atomic constituents of soil minerals, however, are silicon, aluminum, and oxygen, which do not possess energy levels having permissible transitions within the visible and near-infrared portions of the spectrum. Therefore, soil minerals primarily effect shortwave spectra in an indirect manner. They impose their crystal structures on the energy levels of ions (e.g., ferrous iron and hydroxyls) bound to the structures (Hunt, 1980).

The most comprehensive collection of mineral spectra from particulate samples was acquired and published by Hunt and Salisbury (1970, 1971) and Hunt et al. (1971a,b, 1972, 1973). The relevance of this body of work to soil reflectance is concisely reviewed by Mulders (1987). Briefly, quartz has high reflectance throughout the shortwave region, and shortwave quartz spectra do not contain absorption features unless impurities are present. Other primary minerals are less reflective and have spectra containing absorption features due to electronic iron transitions (e.g., amphiboles display an absorption feature near 1.0 μm) or due to the vibrations of hydroxyl ions (e.g., muscovite displays absorption bands at 1.4 μm and between 2.2 and 2.6 μm).

Spectra of the secondary layer silicates also display absorption features due to electronic iron transitions and hydroxyl ion vibrations. Hydroxyl bands near 1.4 and 2.2. μm are characteristic of layer silicates. Stoner and Baumgardner (1980) found the 2.2 μm hydroxyl band difficult to identify in most of their soil spectra, but the band was apparent in the spectra of a few soils having clay contents greater than 20%. The hydroxyl band at 1.4 μm could not be distinguished due to the strong water band also at 1.4 μm.

The spectra of other secondary minerals also contain characteristic features. Calcite spectra display absorption bands between 1.8 and 2.5 μm due to carbonate. Gypsum spectra exhibit absorption bands at 1.8 and 2.3 μm due to overtones and combinations of water-molecule vibrational frequencies (Mulders, 1987).

In the work of Stoner and Baumgardner (1980), soils with gypsic mineralogy were found to have the highest spectral reflectances on average for all observed wavelengths. Montmorillonitic soils had the lowest average spectral reflectances between 0.52 and 1.0 μm. Kaolinitic soils generally displayed a wide absorption band near 0.9 μm due to the common presence of free iron oxides.

V.A.5 Particle Size

Bidirectional reflectances of particulate soil minerals generally increase and the contrasts of absorption features decrease as particle size decreases (Bowers and Hanks, 1965; Hunt, 1980; Stoner and Baumgardner, 1980). This behavior is characteristic of transparent materials and most silicate minerals behave transparently in the shortwave region

(Salisbury and Hunt, 1968). In contrast, the bidirectional reflectances of opaque materials decrease as particle size decreases (Hunt, 1980; Salisbury and Hunt, 1968). Decreasing spectral reflectance with a decrease in particle size was observed for some oxides and for metal sulfides by Hunt et al. (1971a,b).

In common experience, clayey soils often appear darker to the eye than sandy soils even though primary clay particles are much smaller than sand grains. The difference may be explained in part by the different mineralogies of clay and sand particles, but may also be due to the tendency of clay particles to aggregate. That aggregation into agglomerates and clods larger than sand grains can contribute to the darker appearance of clayey soils.

V.A.6 *Surface Conditions*

Changes in bare-soil surface conditions complicate the remote sensing of soils. Conditions such as roughness, moisture content, and the presence of plant residue are easily and frequently altered by weather and tillage. Changes in these conditions have been found to affect the spectral reflectance of soils.

Latz et al. (1984) studied the effects of erosion on Alfisols. Reflectances from the A horizons were found to be low and reflectance spectra had the concave curve shape between 0.5 and 0.8 μm, typical of soils having high organic-matter contents (Figure 8, curve A). Erosion of the Alfisols exposed B horizons that had higher iron contents and lower organic-matter contents than the upper A horizons. Reflectances from the B horizons were higher and the spectra had the convex shape between 0.5 and 0.8 μm, typical of soils with low organic-matter and high iron contents (Figure 8, curve C). Latz et al. (1984) concluded that the shape of reflectance spectra could be used to discriminate degrees of erosion in Alfisols. Weismiller et al. (1985) have reviewed other research on the use of spectral reflectance data for soil-erosion investigations.

Surface roughness and the formation of dry surface crusts have also been observed to affect spectral reflectance. Recently tilled soils are generally rougher, with larger clods and higher surface-moisture contents, than soil left to the effects of weather. Coulson and Reynolds (1971) found that hemispherical reflectance from dry smooth soil to be about 50% higher than reflectance from soil after disking. Cipra et al. (1971) observed higher spectral reflectance values between 0.43 and 0.73 μm from a crusted soil relative to the same soil with the crust broken.

The effects of moisture and plant residue on the reflectance spectra of two soils, an Alfisol and a Mollisol, were investigated by Stoner et al. (1980). Reflectance spectra were obtained both in the field and from moist samples in the laboratory. For each soil, the spectrum from the laboratory generally retained the same shape and form as the spectra from the bare-field plots. Increasing the moisture content lowered reflectance, but did not change the shape of the spectra. The presence of a corn residue did flatten the concave shape of Mollisol spectra, which, when taken from a totally bare plot, were typical for soils of high organic-matter content (Figure 8, curve A). The corn residue also increased reflectance relative to totally bare soil at comparable moisture contents. Flattened sugar-cane residue was also found to increase soil reflectance by Gausman et al. (1975), but standing sugar-cane litter reduced reflectance. Shadowing by the standing litter may have caused this reduction.

V.B Directional Distribution of Soil Bidirectional Reflectance

In most studies of soil reflectance, soil surfaces were both illuminated from a single direction and viewed from a single direction. Soil surfaces, however, are not Lambertian surfaces. Soil reflectance is not only a function of wavelength, but also depends on the directions of illumination and viewing. A limited amount of soil bidirectional reflectance data for multiple illumination or view directions can be found in the literature.

The most extensive data set is provided by Coulson (1966) and Coulson et al. (1965). They observed spectral bidirectional reflectance in the laboratory from samples of crushed minerals, beach and desert sands, and loamy and clayey soils. Observations were acquired using multiple-illumination zenith angles, view zenith angles, and view azimuth angles. Coulson (1966) found that particulate materials having low absorption, such as desert sand (gypsum) and beach sand (quartz), strongly scattered light in the forward direction (i.e., away from the direction of illumination) with a maximum reflectance at a view zenith angle greater than the zenith angle of specular reflectance. Forward scattering was less pronounced and retroreflectance (i.e., backscattering in the antiillumination direction) was more pronounced in observations of reflectance from highly absorbing materials such as clayey and loamy soils. Coulson (1966) also observed a broad reflectance minimum at view zenith angles near 0° (i.e., nadir) for all samples. These results are consistent with the bidirectional reflectance measurements of Irons et al. (1987b) for quartz and clay minerals (Figure 10).

Coulson (1966) attributed the observed retroreflectance maxima to the effects of mutual shadowing among particles. A sensor does not view shadows if and only if the sensor views a particulate surface from the antiillumination direction. The proportion of shadowed area in view increases as the sensor moves away from the antiillumination direction. The increase in viewed shadow is rapid at first and then slows. In materials having low absorption, such as quartz, shadows are softened by light that is multiply scattered or transmitted through particles. This softening of shadows mutes retroreflectance by low-absorption materials.

Soil reflectance has also been measured as a function of illumination direction and view direction in the field. Coulson and Reynolds (1971) observed soil hemispherical reflectance under varying solar zenith angles. Hemispherical reflectance maxima were found at solar zenith angles between 70° and 80° (i.e., low solar elevation angles). Kimes (1983), Kimes and Sellers (1985), and Kimes et al. (1985) have observed the bidirectional reflectance of bare fields from multiple-view zenith and azimuth angles (Figure 11) in their studies of plant-canopy reflectance. Irons et al. (1987a) remotely observed bare-soil bidirectional reflectance using a pointable airborne sensor. A rough recently plowed bare-soil surface was found to more strongly scatter light back toward the antisolar direction than a smooth-soil surface. This study demonstrated the importance of surface roughness in determining the directional distribution of soil reflectance in the field.

V.C Albedo

The ratio of the total solar flux reflected in all directions to the incident solar flux is albedo. Idso et al. (1975) observed the albedo of a loamy Entisol as a function of solar zenith angle, time of year, and water content. The albedos varied between approximately 12 and 38%, depending primarily on water content; albedo increased as the soil dried. If the water content remained fairly constant over a day, the variation in albedo was approximately symmetric about solar noon. The albedo minimum occurred near the smallest solar zenith angle at solar noon, whereas albedo maxima occurred at the greatest solar zenith angles encountered in the morning and afternoon. Idso et al. (1975) normalized the observed albedos for solar zenith-angle effects and found a strong linear relationship between normalized albedo and the volumetric water content of a thin soil

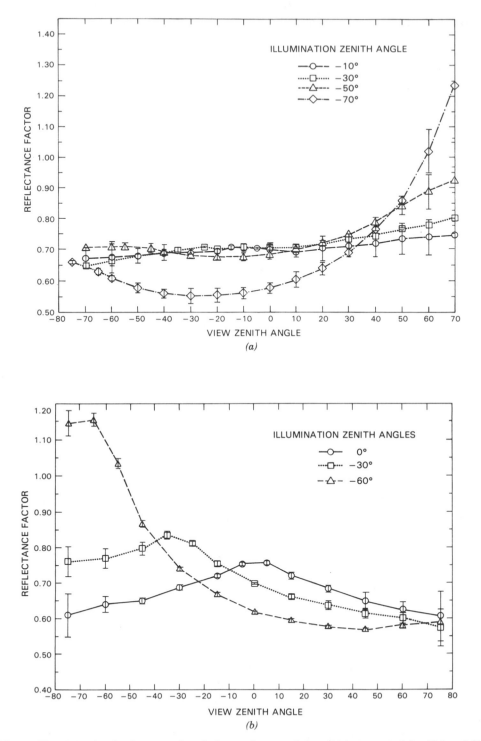

Figure 10 Bidirectional reflectance of particulate surfaces consisting of (*a*) quartz particles (0.1 to 0.25 mm) and (*b*) kaolinite particles (approximately 2 μm) at varying view and illumination directions. The view direction was always in the principal plane of illumination. The wavelength of illumination was 0.6328 μm (Irons et al., 1987b).

BARE SOIL - VISIBLE BAND

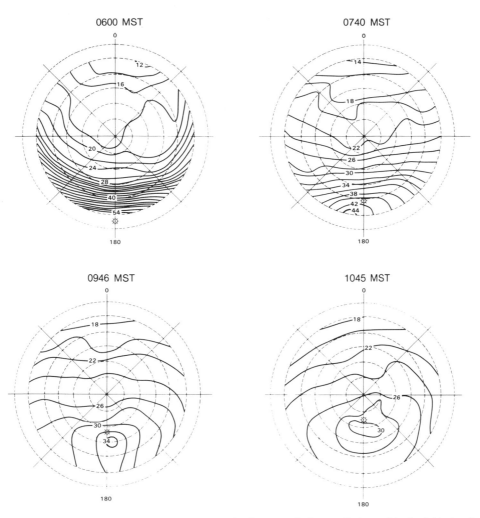

Figure 11 Directional distribution of bidirectional reflectance of a loam soil observed in the field. A polar coordinate system is used to plot the view direction. The polar coordinate θ is used to plot the view azimuth angle relative to the solar azimuth and the polar coordinate r is used to plot the view zenith angle. The solar direction is shown by the X at a relative azimuth of $\theta = 180°$. The observed spectral band was 0.58–0.68 μm. [Reprinted from Kimes (1983).]

surface layer less than 2 mm thick. The relationship held for a water-content range of 0 to 18% in the loamy soil. In addition, they found that raking to roughen the soil surface reduced albedo by approximately 2% (absolute) for a wet soil and by approximately 4% (absolute) for a dry soil.

V.D Polarization

The degree of linear polarization of light scattered by soil and particulate mineral samples has been observed by several investigators. The degree of linear polarization is usually determined by measuring the intensity of light passing through a polarizing filter as the filter is rotated around the axis defined by the direction of light propagation (Coulson, 1966). The desired quantity is then calculated by

$$P = \frac{I_0 - I_{90}}{I_0 + I_{90}} \tag{18}$$

where P is the degree of linear polarization, and I_0 and I_{90} are intensities of the light passing through the polarizing filter at two orthogonal orientations. The polarizing filter is conventionally oriented with the axis of polarization parallel to the observed surface (horizontally oriented) to measure I_0 and perpendicular to the surface (vertically oriented) to measure I_{90}.

Polarization data for soils have been acquired using multiple-illumination and -view directions. The degree of polarization and bidirectional reflectance were measured simultaneously by Coulson (1966) and Coulson et al. (1965) when they observed the soil samples from multiple directions as previously discussed. They provided polar plots of the degree of polarization observed from multiple-view directions (Figure 12). They found that darker particulate surfaces, such as loamy soil, polarized light to a greater degree than did highly reflective surfaces, such as desert sand. This observation is consistent with the polarization measurements made by Egan (1985) of several soil and mineral samples. Coulson (1966) also concluded the phase angle (i.e., the angle formed by the illumination direction and the view direction) is the primary geometric variable controlling the degree of polarization by soils, which is the same conclusion reached by Woessner and Hapke (1987) in a study of polarization by soil surfaces and plant canopies. Finally, Coulson (1966) found the degree of polarization by soils to decrease as the observed wavelength increased from 0.4 to 1.0 μm. This trend was also observed by Genda and Okayama (1978) in measurements of polarization by a sample of loamy

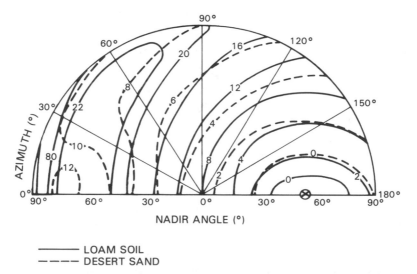

Figure 12 Directional distribution of the degree of linear polarization of radiation scattered from a loam soil and a desert sand in the laboratory. A polar coordinate system is used to plot the view direction. The observed wavelength was 0.492 μm. The illumination direction is shown by the X at a zenith angle of $r = 53°$ and an azimuth angle of $\theta = 180°$. [Reprinted from Coulson (1966).]

soil and a sample of river sand. Genda and Okayama (1978) also found an increase in the degree of polarization when they increased the moisture content of their soil and sand samples.

VI   REMOTE-SENSING APPLICATIONS TO PEDOLOGY

Observations of soil reflectance have been used in pedologic studies for many years. Visual assessment of soil color, with reference to the Munsell color chips, is a standard element of soil descriptions, and color is also used as a criteria in the taxonomic identification and classification of soils. Black-and-white aerial photographs provide a pictorial record of surface reflectance and have been used for decades as an aid to soil mapping and a cartographic base for published soil maps. Remote-sensing capabilities were more recently extended with the advent of satellite-borne sensors in the early 1970s, and the application of these capabilities to pedology have been extensively explored.

Remotely sensed data in the form of black-and-white aerial photographs were first used to prepare base maps for a soil survey in Jennings County, Indiana, in 1929 (Bushnell, 1929). The results represented a vast improvement over the use of plane tables to draw base and soil maps, and by 1938, most soil surveys in the United States used black-and-white aerial photographs as the base map (U.S. Department of Agriculture, 1951). In the early 1960s, soil researchers reported that boundaries between soil mapping units could be differentiated more accurately from color photography than from black-and-white photography (Dominiquez, 1960). Color and color infrared photography also proved useful in identifying soil-drainage characteristics, slopes, and organic-matter content (Peterson et al., 1975). Because of color photography's high cost, however, black-and-white panchromatic photography has remained the major soil-mapping aid.

Success in discerning objects with aerial photography prompted analysts to investigate more sophisticated remote-sensing techniques. Scientists reduced the early images collected with optical-mechanical scanners to a photographic film format for analysis. Tonal qualities in a given scene were matched with individual Earth surface features using techniques such as the grey-scale step-wedge method. Investigations of crop tonal responses revealed spectral response to be a combination of vegetative cover and soil background reflectance. In general, soils were characterized by a high spectral response in the thermal infrared, a rather low response in the reflective infrared, and a varied response in the visible portions of the spectrum (Hoffer, 1967).

Development of computer technologies in the 1960s made possible many advances in remote-sensing analysis. Among them was the overlaying of optical-mechanical multiple-aperture images. This technique provided investigators with more information about the scenes being studied and allowed a more accurate recognition of Earth surface features. By 1965, the University of Michigan had improved upon the multiple-aperture scanner by devising a single-aperture 12-channel scanner (0.32 to 13.5 μm) with an internal calibration. Although capabilities devised for overlaying multiple-aperture images were no longer necessary, the same techniques could be used to analyze multidate and multiimage data.

Researchers later combined numerical-analysis techniques and data collected by the single-aperture 12-channel scanner to study further the application of remote-sensing techniques to soil investigations. Early studies showed that the Alfisol and Mollisol soil orders were separable spectrally despite the influence of cultural practices that contributed to shadowing and reduced reflectance of surface soils (Cipra et al., 1971; Stoner and Horvath, 1971). Soil parameters, such as type, texture, color, moisture relationships, and organic-matter content, proved to be distinguishable using numerical analysis of aircraft scanner data (Baumgardner et al., 1970; Kristof et al., 1978). Some researchers noted that identification of soil series was not feasible, however, because both surface and subsurface properties differentiate series (Kristof and Zachary, 1971).

Prior to the launch of Landsat, scientists hypothesized that multispectral scanner data could provide the capability to map soils at the family level of taxonomic classification. Simulated spacecraft imagery suggested that accurate delineation of soil associations could be accomplished in grassland areas and areas of minimal tree cover (Simonett, 1965). May and Petersen (1975) compared airborne MSS data, corrected for solar illumination and atmospheric differences, to laboratory spectral reflectance data of soils. They found that soil-map unit boundaries generated from classification of MSS and laboratory data compared exceedingly well with field survey map unit boundaries

Launch of the first Landsat satellite in July 1972 began a new era in the acquisition and availability of remotely sensed data. Data from the Landsat Multispectral Scanner System (MSS) provided information in the visible and near-infrared portion of the spectrum with a synoptic view over a much greater area than possible with aerial photography. Early research with Landsat MSS data proved that gross variations in soil features could be identified. The synoptic view enabled the observation and delineation of soil patterns, land use, slope effects, and drainage patterns (Westin, 1973). Image interpretation of simulated infrared and individual single-band black-and-white imagery produced soil-association maps of single counties (Langlois et al., 1976; Steinhardt et al., 1975). Image interpretation of Landsat data was used to produce a soil-association map for the entire state of South Dakota (Westin and Frazee, 1976).

Early research using computer-aided analysis of Landsat MSS data to aid the national soil survey prompted specific investigations related to soil survey and soil identification. Spectral analysis of a narrow strip of Mollisols and adjacent areas located in an area of predominately Alfisols in Indiana indicated that predominantly mollic areas and mollic inclusions could be separated successfully from the Alfisols (Peterson et al., 1975). Other studies nearby correlated soil-moisture characteristics with spectral soil classes, showing that soil boundaries could be readily identified and soil-map units easily quantified (Kirschner et al., 1978). Seubert et al. (1979) collected laboratory spectral data of eroded Alfisols that suggested that eroded soils could be delineated from uneroded soils. Latz et al. (1984) later showed that various degrees of soil erosion could be delineated in Alfisol regions both under laboratory conditions and in the field using Landsat MSS data.

A spectral soil map produced for Chariton County, Missouri, provided results that stimulated new data-analysis techniques. Because Landsat multispectral scanner data portray only surface reflectance properties, soils that vary widely with respect to subsurface properties may exhibit the same surface spectral properties. Ancillary data in the form of physiographic boundaries provided additional information and allowed for sorting and correlation of spectral soil classes and soil series (Weismiller et al., 1977). Davidson (1981) in a similar study, used polaroid photographs of Landsat MSS visible and infrared bands to aid in delineation of soil boundaries.

Analytical procedures developed in conjunction with the soil survey efforts in Indiana and Missouri were extended to Jasper County in northwestern Indiana. Previous success in preparing soil-association maps and methods of using ancillary data boundaries prompted the initial preparation of a parent-material map that would be digitized and overlayed with Landsat multispectral scanner data to produce a spectral soil map with designated soil series. Classification techniques were accomplished in areas limited to individual parent materials, thus increasing the correlation of soil spectral classes to soil properties, particularly drainage class, and thus pertinent soil series. Spectral maps correlated to soil series for individual field sheets were prepared at a scale of 1 : 15,840 and were used to aid in map-unit delineation and map-unit homogeneity assessment. These spectral maps along with standard field mapping techniques increased the efficiency and accuracy of the soil survey of Jasper County, Indiana (Weismiller et al., 1979). Techniques similar to those developed and tested in Jasper County, Indiana, were later extended to soil-survey efforts in Ford County, Illinois, the Big Desert Area of Idaho, Federal Bureau of Land Management lands, and Tipton County, Indiana (Kiefer et al., 1980; Lund et al., 1980; Harrison, 1980; DiPaolo, 1979; Dipaolo and Hill, 1983). This approach of stratifying data proved a useful method in generating general soil maps and in providing preliminary mapping and map-unit design. The spectral data also aided in increasing the accuracy and confidence of the soil survey.

Imhoff et al. (1980, 1982; Imhoff and Petersen, 1980) working in the Green River area of Utah, overlayed spectral data with thematic maps and enhanced images and other forms of georegistered data to enhance the classification of soil data for soil-survey efforts. These techniques increased the speed and accuracy of the manual process of soil unit boundary delineation and provided a means to evaluate soil-map unit variability as related to soil spectral variability. Research in arid and semiarid regions have found Landsat MSS data useful in providing information on soil-vegetation site characteristics. These combined soil-vegetation spectral responses proved most valuable in soil-map unit delineation (Horvath et al., 1980, 1984).

In 1984, a significant advance in resolution of satellite data came about with the launch of an advanced sensor, the Thematic Mapper (TM), aboard Landsat 4. The TM offered a higher spatial resolution (30 m) and additional spectral bands (six bands in the visible and shortwave infrared plus a thermal band) relative to the four spectral bands and 80 m resolution of the Landsat MSS sensors. Several researchers have investigated the usefulness of Thematic Mapper (TM) data to soil investigations. Thompson et al. (1984) found that the increased spectral and spatial characteristics of TM data could indeed provide information pertinent to soil-boundary delineation even under intensively cropped acres in central Iowa. Seeley et al. (1984) have studied the effect of tillage, residue, and weathering on spectral reflectance in the TM spectral bands. Residue type and amount were found to have the greatest influence on spectral reflectance in the TM bands, particularly when large quantities of residue were present on the surface.

Computer-aided analyses of satellite multispectral scanner data can aid field soil scientists. Specifically, soil maps generated by computer analysis of Landsat MSS and TM data can help in the preparation of soil-association maps; the delineation of parent-material boundaries; the refinement of soil-map unit boundaries; the quantification of soil-map units; the determination of percentages of inclusions within soil-map units; and the preparation of single-feature maps, such as drainage and organic matter content (Weismiller and Kaminsky, 1978).

As summarized by Baumgardner et al. (1985), research reported in the literature over the past 15–20 years demonstrates the utility of the spectral and spatial data collected by Landsat MSS and TM sensors as an aid to soil survey, soil inventory, and soil management. The sensors acquire digital image data that afford a synoptic view of the area of interest, quantitative assessment of map-unit homogeneity, near-orthographic quality of imagery, and multispectral and multitemporal data. Furthermore, Landsat data are acquired in a digital format that allows for convenient incorporation into geographic information systems and ease of computer manipulation. While computer-aided analysis of multispectral scanner data alone cannot produce a soil survey, analysis of Landsat MSS and TM data can effectively aid the preparation of all orders of soil surveys.

Sources of remote-sensing data for soil survey and other pedologic studies are not limited to the Landsat satellites. Although beyond the scope of this chapter, thermal and microwave sensors also generate useful data for soil studies, particularly with respect to soil moisture (Salomonson et al., 1983; Schmugge et al., 1980). Also, as reviewed by Mulders (1987), several countries have launched or planned satellite remote-sensing systems that are potentially applicable to soil studies. In particular, the French Centre National d'Etudes Spatiales launched in 1986 a satellite called the Systeme Probatoire pour l'Observation de la Terre (SPOT). This satellite carries "high-resolution visible (HRV)" sensors that acquire digital image data having either 10 m spatial resolution in a single panchromatic band or 20 m resolution in three bands in the green, red, and near-infrared portions of the spectrum. The HRV sensors can also be pointed off-nadir for the acquisition of stereoscopic imagery. Simulated SPOT data along with thermal data and Landsat TM data have been evaluated by Petersen et al. (1987) for the identification of desert soils and geomorphic features. As they observed, the high spatial resolution and stereoscopic capabilities of the HRV sensors may render SPOT data especially useful for the discrimination of geomorphic features related to surface hydrology, soil-erosion potential, and landscape stability.

VII CONCLUDING REMARKS

Despite the demonstrated utility of advanced remote-sensing techniques for pedologic studies, neither Landsat data nor data from other satellite sensors are routinely used for soil mapping in the United States. Black-and-white aerial photography remains the standard type of remote-sensing data used in the U.S. national soil-survey program. The use of aerial photographs has been established over several decades and pedologists have become familiar and comfortable with the techniques of visual photointerpretation. More advanced analyses of digital remote-sensing data generally require an investment in automated data-processing capabilities (i.e. computer hardware and software) and an understanding of image processing and spectral analysis. These factors, however, should not be allowed to hinder the adoption of advanced remote-sensing techniques for soil surveys. The advantages of satellite remote sensing, as discussed, are too substantial to ignore.

Obstacles to the adoption of remote-sensing methods for pedologic studies are being reduced. Remote-sensing data from the Landsat, SPOT, and other satellites are commercialized and are readily available to the general public. Computer hardware and image-processing software are increasingly accessible and economical. The use of microcomputers for digital image analyses is now common. Educational opportunities are also increasing as courses in remote sensing are offered by most major universities and even by small colleges. With increasing access to data, computer technology, and education, the advantages of satellite remote sensing for soil mapping may prove especially attractive in countries that are not mapped and developed to the extent of the United States. These countries may lead the way in the application of remote sensing to pedology.

The use of remote sensing for soil studies can certainly be further advanced by an enhanced understanding of the process of light scattering by soil surfaces. Current models provide some insight into this process, but further developments are needed to incorporate significant soil properties such as moisture content, organic-matter content, metal-oxide coatings, aggregation, and structure into the theory and models. Theoretical developments are also needed to predict the directional distribution and polarization state of light scattering by soils and to more fully evaluate the observation of these radiation characteristics for soil studies. Theoretical advancements would additionally aid in the recognition and representation of the soil contribution to the radiation balance and to the reflectance of vegetated surfaces. An understanding of light scattering by soils is critical to the comprehension and application of remote observations of terrestrial land surfaces.

REFERENCES

Angstrom, A. (1925). The albedo of various surfaces of ground. *Geogr. Ann.* **7**:323.

Aronson, J. R., and A. R. Emslie (1973). Spectral reflectance and emittance of particulate materials. 2. Application and results. *Appl. Opt.*, **12**:2573–2584.

Baumgardner, M. F., S. J. Kristof, C. J. Johannsen, and A. L. Zachary (1970). Effects of organic matter on the multispectral properties of soils. *Proc. Indiana Acad. Sci.* **79**:413–422.

Baumgardner, M. F., L. F. Silva, L. L. Biehl, and E. R. Stoner (1985). Reflectance properties of soils. *Adv. Agron.* **38**:1–44.

Beck, R. H., B. F. Robinson, W. W. McFee, and J. B. Peterson (1976). *Spectral Characteristics of Soils Related to the Interaction of Soil Moisture, Organic Carbon, and Clay Content*, LARS Inf. Note 081176. Purdue University, West Lafayette, Indiana.

Bohren, C. F., and D. R. Huffman (1983). *Absorption and Scattering of Light by Small Particles*. Wiley, New York.

Bohren, C. F., and G. Koh (1985). Forward-scattering corrected extinction by nonspherical particles. *Appl. Opt.* **24**:1023–1029.

Bowers, S. A., and R. J. Hanks (1965). Reflection of radiant energy from soils. *Soil Sci.* **100**:130–138.

Brady, N. C. (1984). *The Nature and Properties of Soils*, 9th Ed. Macmillan, New York.

Bushnell, T. M. (1929). Aerial photographs of Jennings County. *Proc. Indiana Acad. Sci.* **39**:229–230.

Castellan, G. W. (1983). *Physical Chemistry*, 3rd ed. Addison-Wesley, Reading, Massachusetts.

Chandrasekhar, S. (1960). *Radiative Transfer*. Dover, New York.

Cierniewski, J. (1987). A model for soil surface roughness influence on the spectral response of bare soils in the visible and near-infrared range. *Remote Sens. Environ.* **23**:97–115.

Cipra, J. E., M. F. Baumgardner, E. R. Stoner, and R. B. MacDonald (1971). Measuring radiance characteristics of soil with a field spectroradiometer. *Soil Sci. Soc. Am. Proc.* **35**:1014–1017.

Condit, H. R. (1970). The spectral reflectance of American soils. *Photogramm. Eng.* **36**:955–966.

Conel, J. E. (1969). Infrared emissivities of silicates: Experimental results and a cloudy atmosphere model of spectral emission from condensed particulate mediums. *J. Geophys. Res.* **74**:1614–1634.

Cooper, K., and J. A. Smith (1985). A Monte Carlo reflection model for soil surfaces with 3D structure. *IEEE Trans. Geosci. Remote Sens.* **GE-23**:668–673.

Coulson, K. L. (1966). Effects of reflection properties of natural surfaces in aerial reconnaissance. *Appl. Opt.* **5**:905–917.

Coulson, K. L., and D. W. Reynolds (1971). The spectral reflectance of natural surfaces. *J. Appl. Meteorol.* **10**:1285–1295.

Coulson, K. L., G. M. Bouricius, and E. L. Gray (1965). Optical reflection properties of natural surfaces. *J. Geophys. Res.* **70**:4601–4611.

Dave, J. V. (1968). *Subroutines for Computing the Parameters of the Electromagnetic Radiation Scattered by a Sphere*, Rep. No. 320-3237, IBM Scientific Center, Palo Alto, California.

Davidson, S. E. (1981). *Use of Landsat Data to Define Soil Boundaries in Carroll County, Missouri. The 1981 NASA ASEE Summer Faculty Fellowship Program*, Vol. I, Tech. Final Rep., Johnson Space Center, Houston, Texas.

DiPaolo, W. D. (1979). *An Analysis of Landsat Data for Soils Investigations on Federal Lands in Southwest Idaho*, LARS Contract Rep. 082879. Purdue University, West Lafayette, Indiana.

DiPaolo, W. D. and L. R. Hill (1983). *The Use of Remote Sensing for Soils Investigations on Bureau of Land Management Lands*, TN-361. U.S. Department of Interior, Bureau of Land Management, Denver, Colorado, pp. 1–62.

Dominiquez, O. A. (1960). A comparative analysis of black and white aerial photographs as aids in the mapping of soils in wild land area. In *Manual of Photographic Interpretation*. American Society of Photogrammetry, Washington, D.C., pp. 398–402.

Egan, W. G. (1985). *Photometry and Polarization in Remote Sensing*. Elsevier, New York.

Egan, W. G., and T. W. Hilgeman (1978). Spectral reflectance of particulate materials: A Monte-Carlo model including asperity scattering. *Appl. Opt.* **17**:245–252.

Egan, W. G., and T. W. Hilgeman (1979). *Optical Properties of Inhomogeneous Materials; Applications to Geology, Astronomy, Chemistry, and Engineering*. Academic Press, New York.

Egan, W. G., T. W. Hilgeman, and K. Pang (1975). Ultraviolet complex refractive index of Martian dust: Laboratory measurements of terrestrial analogs. *Icarus* **25**:344–355.

Emslie, A. G., and J. R. Aronson (1973). Spectral reflectance and emittance of particulate materials. 1. Theory. *Appl. Opt.* **12**:2563–2566.

Gausman, G. W., A. H. Gerbermann, C. L. Weigand, R. W. Leamer, R. R. Rodriguez, and J. R. Noriega (1975). Reflectance differences between crop residues and bare soils. *Soil Sci. Soc. Am. Proc.* **39**:752–755.

Genda, H., and H. Okayama (1978). Estimation of soil moisture and components by measuring the degree of spectral polarization with a remote sensing simulator. *Appl. Opt.* **17**:3439–3443.

Grams, G. W., I. H. Blifford, D. A. Gillette, and P. B. Russell (1974). Complex index of refraction of airborne soil particles. *J. App. Meteorol.* **13**:459–471.

Hansen, J. E., and L. D. Travis, (1974). Light scattering in planetary atmospheres. *Space Sci. Rev.* **16**:527–610.

Hapke, B. (1981). Bidirectional reflectance spectroscopy. 1. Theory. *JGR, J. Geophys. Res.* **86**:3039–3054.

Hapke, B. (1984). Bidirectional reflectance spectroscopy. 3. Correction for macroscopic roughness. *Icarus* **59**:41–59.

Hapke, B. (1986). Bidirectional reflectance spectroscopy. 4. The extinction coefficient and the opposition effect. *Icarus* **67**:264–280.

Hapke, B., and E. Wells (1981). Bidirectional reflectance spectroscopy. 2. Experiments and observations. *JGR, J. Geophys. Res.* **86**:3055–3060.

Harrison W. D. (1980). Application of Landsat data on a low order soil survey in south-central Idaho. In *Machine Processing of Remotely Sensed Data*. Purdue University, West Lafayette, Indiana, p. 68.

Hoffer, R. M. (1967). *Interpretation of Remote Multispectral Imagery of Agricultural Crops*. Agricultural Experiment Station, Purdue University, West Lafayette, Indiana.

Horwath, E. H., D. F. Post, W. M. Lucas, and R. A. Weismiller (1980). Use of Landsat digital data to assist in mapping soils on Arizona rangelands. In *Machine Processing of Remotely Sensed Data*. Purdue University, West Lafayette, Indiana, pp. 235–240.

Horvath, E. H., D. F. Post, and J. B. Kelsey (1984). The relationships of Landsat digital data to the properties of Arizona rangelands. *Soil Sci. Soc. Am. J.* **48**:1331–1334.

Hunt, G. R. (1980). Electromagnetic radiation: The communications link in remote sensing. In *Remote Sensing in Geology* (B. S. Siegal and A. R. Gillespie, Eds). Wiley, New York, pp. 5–45.

Hunt, G. R., and J. W. Salisbury (1970). Visible and near-infrared spectra of minerals and rocks. I. Silicate minerals. *Mod. Geol.* **1**:283–300.

Hunt, G. R., and J. W. Salisbury (1971). Visible and near-infrared spectra of minerals and rocks. II. Carbonates. *Mod. Geol.* **2**:23–30.

Hunt, G. R., J. W. Salisbury, and C. J. Lenhoff (1971a). Visible and near-infrared spectra of minerals and rocks. III. Oxides and hydroxides. *Mod. Geol.* **2**:195–205.

Hunt, G. R., J. W. Salisbury, and C. J. Lenhoff (1971b). Visible and near-infrared spectra of minerals and rocks. IV. Sulphides and sulphates. *Mod. Geol.* **3**:1–4.

Hunt, G. R., J. W. Salisbury, and C. J. Lenhoff (1972). Visible and near-infrared spectra of minerals and rocks. V. Halides, phosphates, arsenates, vanadates, and borates. *Mod. Geol.* **3**:121–132.

Hunt, G. R., J. W. Salisbury, and C. J. Lenhoff (1973). Visible and near-infrared spectra of minerals and rocks. VI. Additional silicates. *Mod. Geol.* **4**:85–106.

Idso, S. B., R. D. Jackson, R. J. Reginato, B. A. Kimball, and F. S. Nakayama (1975). The dependence of bare soil albedo on soil water content. *J. Appl. Meteorol.* **14**:109–113.

Imhoff, M. L., and G. W. Petersen (1980). *Role of Landsat Products in Soil Surveys*, Final Rep. for Contract No. NAS 5-25667. NASA/Goddard Space Flight Center, Greenbelt, Maryland.

Imhoff, M. L., G. W. Petersen, and J. R. Irons (1980). Delineation of soil boundaries using image enhancement and spectral signature classification of Landsat data. In *Machine Processing of Remotely Sensed Data*. Purdue University, West Lafayette, Indiana, p. 64.

Imhoff, M. L., G. W. Petersen, S. G. Sykes, and J. R. Irons (1982). Digital overlay of cartographic information on Landsat MSS data for soil surveys. *Photogramm. Eng. Remote Sens.* **48**:1337–1342.

Irons, J. R., B. L. Johnson, Jr., and G. H. Linebaugh (1987a). Multiple-angle observations of reflectance anisotropy from an airborne linear array sensor. *IEEE Trans. Geosi. Remote Sens.* **GE-25**:372–383.

Irons, J. R., J. M. Kestner, H. W. Leidecker, and N. A. Horning (1987b). Goniometric measurement of soil mineral reflectance. In *Agronomy Abstracts. 1987 Annual Meetings*. American Society of Agronomy, Madison, Wisconsin, p. 159.

Jackson, M. L., C. H. Lim, and L. W. Zelazny (1986). Oxides, hydroxides, and aluminosilicates. In *Methods of Soil Analysis* (A. Klute, Ed.), Part 1. 2nd ed., *Agronomy 9*. American Society of Agronomy, Madison, Wisconsin, pp. 101–150.

Johnson, L. J. (1979). *Introductory Soil Science*. Macmillan, New York.

Kemper, W. D., and R. C. Rosenau (1986). Aggregate stability and size distribution. In *Methods of Soil Analysis*, (A. Klute, Ed.), Part 1. 2nd ed., Agronomy 9. American Society of Agronomy, Madison, Wisconsin, pp. 425–442.

Kiefer, L. M., E. E. Voss, F. R. Kirschner, R. A. Weismiller, S. J. Kristof, and L. J. Lund (1980). Application of multispectral data in developing a detailed soil survey of Ford County, Illinois. In *Machine Processing of Remotely Sensed Data*. Purdue University, West Lafayette, Indiana, p. 66.

Kimes, D. S. (1983). Dynamics of directional reflectance factor distributions for vegetation canopies. *Appl. Opt.* **22**:1364–1372.

Kimes, D. S., and P. J. Sellers (1985). Inferring hemispherical reflectance of the earth's surface for global energy budgets from remotely sensed nadir or directional radiance values. *Remote Sens. Environ.* **18**:205–223.

Kimes, D. S., W. W. Newcomb, C. J. Tucker, I. S. Zonneveld, W. Van Wijngaarden, J. De Leeuw, and G. F. Epema (1985). Directional reflectance factor distributions for cover types of northern Africa. *Remote Sens. Environ.* **18**:1–19.

Kirschner, F. R., S. A. Kaminsky, R. A. Weismiller, H. R. Sinclair, and E. J. Hinzel (1978). Soil map unit composition and assessment using drainage classes by Landsat data. *Soil Sci. Soc. Am. J.* **42**:768–771.

Kristof, S. J. and A. L. Zachary (1971). Soil types from multispectral scanner data. *Proc. 7th Int. Symp. Remote Sens. Environ.*, University of Michigan, Ann Arbor, pp. 1427–1434.

Kristof, S. J., M. F. Baumgardner, R. A. Weismiller and S. Davis (1978). Application of multispectral reflectance studies of soils. Pre-Landsat. In *Machine Processing or Remotely Sensed Data*. Purdue University, West Lafayette, Indiana, pp. 52–61.

Langlois, K. H., L. C. Osterholz, and F. R. Kirschner (1976). Use of Landsat imagery as a base for making a general soil map. *Proc. Indiana Acad. Sci.* **85**:126.

Latz, K., R. A. Weismiller, G. E. Van Scoyoc, and M. F. Baumgardner (1984). Characteristic variations in spectral reflectance of selected eroded Alfisols. *Soil Sci. Soc. Am. J.* **48**:1130–1134.

Lumme, K., and E. Bowell (1981). Radiative transfer in the surfaces of atmosphereless bodies. I. Theory. *Astron. J.* **86**:1694–1704.

Lund, L. J., R. A. Weismiller, S. J. Kristof, F. R. Kirschner, and W. D. Harrison (1980). Development of spectral maps for soil-vegetation mapping in the Big Desert Area, Idaho. In *Machine Processing of Remotely Sensed Data*. Purdue University, West Lafayette, Indiana, p. 67.

Mathews, H. L., R. L. Cunningham, and G. W. Petersen (1973). Spectral reflectance of selected Pennsylvania soils. *Soil Sci. Soc. Am. Proc.* **37**:421–424.

May, G. A. and G. W. Petersen (1975). Spectral signature selection for mapping unirrigated soils. *Remote Sens. Environ.* **4**:211–220.

Montgomery, O. L., and M. F. Baumgardner (1974). *The Effects of the Physical and Chemical Properties of Soil on the Spectral Reflectance of Soils*, LARS Inf. Note 112674. Purdue University, West Lafayette, Indiana.

Mugnai, A., and W. J. Wiscombe (1986). Scattering from nonspherical Chebyshev particles. I: Cross sections, single-scattering albedo, asymmetry factor, and backscattered fraction. *Appl. Opt.* **25**:1235–1244.

Mulders, M. A. (1987). *Remote Sensing in Soil Science*. Elsevier, Amsterdam.

Munsell Color (1975). *Munsell Soil Color Charts*. MacBeth Division of Kollmurgen Corporation, Baltimore, Maryland.

Nicodemus, F. E., J. C. Richmond, J. J. Hsia, I. W. Ginsberg, and T. Limperis (1977). Geometrical considerations and nomenclature for reflectance. *NBS Monog. (U.S.)* **160**:1–52.

Obukhov, A. I., and D. S. Orlov (1964). Spectral reflectivity of the major soil groups and possibility of using diffuse reflection in soil investigations. *Sov. Soil Sci. (Engl. Transl.)* **2**:174–184.

Orloy, D. S. (1966). Quantitative patterns of light reflection by soils. I. Influence of particle (aggregate) size on reflectivity. *Sov. Soil Sci. (Engl. Transl.)* **13**:1495–1498.

Petersen, G. W., K. F. Connors, D. A. Miller, R. L. Day, and T. W. Gardner (1987). Aircraft and satellite remote sensing of desert soils and landscapes. *Remote Sens. Environ.* **23**:253–271.

Petersen, R. G., and L. D. Calvin (1986). Sampling In *Methods of Soil Analysis*, (A. Klute, Ed.) Part 1, 2nd ed., Astronomy 9. American Society of Agronomy, Madison, Wisconsin, pp. 33–51.

Peterson, J. B., F. E. Goodrick, and W. N. Melhorn (1975). Delineation of the boundaries of a buried pre-glacial valley with Landsat 1 data. *Proc. 1st NASA Earth Resour. Surv. Symp.*, Vol. IA, Houston, Texas, pp. 97–103.

Planet, W. G. (1970). Some comments on reflectance measurements of wet soils. *Remote Sens. Environ.* **1:**127–129.

Salisbury, J. W., and G. R. Hunt (1968). Martian surface materials; effect of particle size on spectral behavior. *Science* **161:**365–366.

Salomonson, V. V., T. J. Jackson, J. R. Lucas, G. K. Moore, A. Rango, T. Schmugge, and D. Scholz (1983). Water resources assessment. In *Manual of Remote Sensing* (R. N. Colwell, J. E. Estes, and G. A. Thorley, Eds.), 2nd ed., Vol. II. American Society of Photogrammetry, Falls Church, Virginia, pp. 1497–1570.

Schmugge, T. J., T. J. Jackson, and H. L. McKim (1980). Survey of methods for soil moisture determination. *Water Resourc. Res.* **16:**961–979.

Schnitzer, M. (1982). Organic matter characterization. In *Methods of Soil Analysis* (A. L. Page, Ed.), Part 2, 2nd ed. Agronomy 9. American Society of Agronomy, Madison, Wisconsin, pp. 581–594.

Seeley, M., E. Larson, E. Schroder, and D. Lindren (1984). Changes in TM response of soil scenes due to tillage, residue and weathering. *Proc. Int. Geosci. Remote Sens. Symp. (IGARSS' 84)*, Strasbourg, Germany, pp. 27–30.

Seubert, C. E., M. F. Baumgardner, R. A. Weismiller, and F. R. Kirschner (1979). Mapping and estimating real extent of severely eroded soils of selected sites in Northern Indiana. In *Machine Processing of Remotely Sensed Data.*, Purdue University, West Lafayette, Indiana, pp. 234–238.

Simonett, D. S. (1965). Pedologic and ecologic applications of orbital radar imagery for studies of tropical environments in use of imaging radar and two frequencies altimeter scatterometer on manned spacecraft to investigate lunar and terrestrial surfaces. Ph.D. Thesis, University of Kansas, Lawrence.

Smith, J. A. (1983). Matter-energy interaction in the optical region. In *Manual of Remote Sensing* (R. N. Colwell, D. S. Simonett, and F. T. Ulaby, Eds.), 2nd ed., Vol. I. American Society of Photogrammetry, Falls Church, Virginia, pp. 115–161.

Steila, D. (1976). *The Geography of Soils*. Prentice-Hall, Englewood Cliffs, New Jersey.

Steinhardt, G. C., D. P. Franzmeier, and J. E. Cipra (1975). Indiana soil associations compared to Earth Resources Technology Satellite imagery. *Proc. Indiana Acad. Sci.* **84:**463–468.

Stoner, E. R., and M. F. Baumgardner (1980). *Physicochemical, Site, and Bidirectional Reflectance Factor Characteristics of Uniformly Moist Soils*, LARS Tech. Rep. 111679. Purdue University, West Lafayette, Indiana.

Stoner, E. R., and M. F. Baumgardner (1981). Characteristic variations in reflectance of surface soils. *Soil Sci. Soc. Am. J.* **45:**1161–1165.

Stoner, E. R. and E. H. Horvath (1971). The effect of cultural practices on multispectral response from surface soil. *Proc. 7th Int. Symp. Remote Sens. Environ.*, University of Michigan, Ann Arbor, pp. 2109–2113.

Stoner, E. R., M. F. Baumgardner, L. L. Biehl, and B. F. Robinson (1979). *Atlas of Soil Reflectance Properties*, LARS Tech. Rep. 111579. Purdue University, West Lafayette, Indiana.

Stoner, E. R., M. F. Baumgardner, R. A. Weismiller, L. L. Biehl, and B. F. Robinson (1980). Extension of laboratory-measured soil spectral to field conditions. *Soil Sci. Soc. Am. J.* **44:**572–574.

Taylor, S. A., and G. L. Ashcroft (1972). *Physical Edaphology. The Physics of Irrigated and Nonirrigated Soils.* Freeman, San Francisco, California.

Thompson, D. R., K. E. Henderson, A. G. Houston, and D. E. Pitts, (1984). Variations in alluvial derived soils as measured by Landsat thematic mapper. *Soil Sci. Soc. Am. J.* **40:**137–142.

U.S. Department of Agriculture (1951). *Soil Survey Manual,* Agric. Handb. No. 18. USDA, Soil Survey Staff, Washington, D.C.

van de Hulst, H. C. (1957). *Light Scattering by Small Particles.* Dover, New York.

van de Hulst, H. C. (1980). *Multiple Light Scattering, Tables, Formulas, and Applications,* Vol. 1. Academic Press, New York.

Vincent, R., and G. Hunt (1968). Infrared reflectance from mat surfaces. *App. Opt.* **7:**53–59.

Weismiller, R. A. and S. A. Kaminsky (1978). Application of remote-sensing technology to soil survey research. *J. Soil Water Conserv.* **33:**287–289.

Weismiller, R. A., I. D. Persinger, and O. L. Montgomery (1977). Soil inventory prepared from digital analysis of satellite multispectral scanner data and digitized topographic data. *Soil Sci. Soc. Am. J.* **41**(6):1,116–1,170.

Weismiller, R. A., S. K. Kast, M. F. Baumgardner, and F. R. Kirschner (1979). Landsat MSS data as an aid to soil survey—an operational concept. In *Machine Processing of Remotely Sensed Data.* Purdue University, West Lafayette, Indiana, p. 240.

Weismiller, R. A., G. E. Van Scoyoc, S. E. Pazar, K. Latz, and M. F. Baumgardner (1985). Chapter 10. Use of soil spectral properties for monitoring soil erosion. In *Soil Erosion and Conservation.* Soil Conservation Society of America, Ankeny, Iowa, pp. 119–127.

Westin, F. C. (1973). ERTS 1 imagery: A tool for identifying soil associations. In *Earth Survey Problems Through the Use of Space Techniques.* General Assembly Committee of Space Research, Ronstanz, West Germany.

Westin, F. C. and C. J. Frazee (1976). Landsat data, its use in a soil survey program. *Soil Sci. Soc. Am. J.* **40:**81–89.

White, J. L. and C. B. Roth (1986). Infrared spectrometry. In *Methods of Soil Analysis* (A. Klute, Ed.), Part 1, 2nd ed., Agronomy 9. American Society of Agronomy, Madison, Wisconsin, pp. 291–330.

Wiscombe, W. J. (1980). Improved Mie scattering algorithms. *Appl. Opt.* **19:**1505–1509.

Woessner, P., and B. Hapke (1987). Polarization of light scattered by clover. *Remote Sens. Environ.* **21:**243–261.

4

SOIL INFLUENCES IN REMOTELY SENSED VEGETATION-CANOPY SPECTRA

Alfredo R. Huete

Department of Soil and Water Science
The University of Arizona
Tucson, Arizona

I INTRODUCTION

The spectral composition of the radiant flux emanating from the Earth's surface provides information about the biological, chemical, and physical properties of soil, water, and vegetation features in terrestrial ecosystems. Remote-sensing techniques, models, and indices are designed to convert this spectral information into a form that is readily interpretable. However, the fundamental interactions of radiant energy with the Earth's surface must be understood for remote sensing to be efficiently applied. The scope, applicability, and limitations of remote-sensing measurements must be realized in solving ecological problems.

Remote-sensing studies of vegetation abundance, composition, productivity, and health from satellite optical measurements have been hampered by the atmosphere medium through which ground-reflected radiance must traverse and the soil background from which the vegetation signal must be discriminated. Atmospheric scattering and absorption of radiant flux have been carefully analyzed in several studies (Kondratyev et al., 1974; Pitts et al., 1974; Otterman and Fraser, 1976; Slater, 1980; Fraser and Kaufman, 1985) and is discussed in detail by Kaufman in Chapter 9. Atmosphere turbidity generally inhibits reliable measures of vegetation and may delay the detection of an onset of stress in canopies (Slater and Jackson, 1982; Jackson et al., 1983). These studies have shown the importance of characterizing the atmosphere in order to extend ground-based studies to satellite sensor data and vice versa. By contrast, soil influences on measured canopy spectra and vegetation dynamics over the Earth's surface have received less attention.

I.A Spatial and Temporal "Global" Soil Variation

The soil background contribution to plant-canopy spectral response is significant in partial canopy situations, especially where there may occur significant spatial or temporal

107

soil-surface variations. Partial or incomplete canopies dominate the majority of the Earth's terrestrial surface with the exception of dense tropical rain forests and barren deserts. In arid and semiarid environments, complex associations of soil types and sparse canopy covers limit the extraction of reliable vegetation information and often inhibit vegetation detection below 30% green cover (Colwell, 1974; Pearson et al., 1976; Elvidge and Lyon, 1985; Huete et al., 1984). Rangelands and woodlands also possess large temporal and spatial sources of environmental variability attributed to species diversity, dry-matter accumulation and decomposition, and topographic influences. Living and dead plant materials vary spatially, whereas growth and decomposition overlap in time. Incomplete canopies are also prevalent in agricultural areas for most of the growing season. Irrigated agricultural soils undergo numerous wetting and drying cycles and exhibit irregular drying patterns.

Spectral data collected over partial canopies represent a complex mixture and arrangement of individual plant, soil-background, and shadow contributions. Soil-surface spectral contributions to canopy responses vary with the amount of soil exposed, surface-moisture content, organic-matter content (decomposed and undecomposed), particle-size distribution, soil mineralogy, soil structure, surface roughness, coarse fragments, crusting, presence of shadow, and cultural practices (Angstrom, 1925; Bowers and Hanks, 1965; Stoner and Baumgardner, 1981; Rao et al., 1979; Jackson et al., 1979; Kollenkark et al., 1982a). Thus, even within a given soil type, surface conditions may vary greatly. Consequently, spectral responses associated with differences in vegetation-related parameters (leaf-area index, biomass, cover, stress, disease) are difficult to detect amidst a variable soil-background substrate since canopy response is sensitive to both forms of variation.

I.B   Factor-Variation Approach

The spectral properties of a vegetated canopy are particularly difficult to model and are a function of an infinitely large number of variables ranging from (1) microscopic soil mineralogical composition and leaf-pigment composition to (2) macrovariables such as soil texture, stone content, leaf angles, and plant height. In addition, canopy spectral response will vary with extrinsic factors such as sun angle, sensor view angle, and atmospheric condition. Since many of these properties are interrelated, it is not necessary to define and quantify all of them to predict canopy reflectance; thus, canopy spectral behavior may be analyzed in terms of a limited number of state factors. These are variables that can be independently varied while holding the other state variables constant (factor variation).

Numerous multispectral band indices have been developed in order to isolate the green-vegetation parameters from the soil background and other nongreen components. They have been widely shown to be valuable techniques in the measurement and interpretation of vegetation conditions and are important toward the establishment of simple ''global'' models that can adequately describe dynamic of soil-vegetation systems. Radiative transfer models have also been developed to characterize the spectral behavior of vegetation canopies (Suits, 1972; Kimes et al., 1985; Chance, 1977; Norman and Welles, 1983), but they are generally complex, difficult to apply in global studies, and they commonly require input of soil optical properties.

The spectral responses of partial vegetation covers may be treated as a function of four factors: (1) the structural and biophysical makeup of the vegetation, (2) the soil-surface background, (3) solar illumination/ and sensor viewing conditions, and (4) the atmosphere medium. Factor-variation studies are most easily applied to the last two extrinsic variables by measuring canopy spectral response under different sun angle, view angle, or atmospheric conditions. Atmosphere radiative transfer programs may be used to vary atmosphere conditions in ground-based studies (Herman and Browning, 1975; Iqball, 1983; Fraser and Kaufman, 1985). Separation of soil and vegetation properties, however, is much more difficult since in nature they covary. Soil properties affect the type of vegetation growing over it, and vegetation influences soil development.

In this chapter, the role of the soil background on canopy response is critically examined for a range of soil types, surface-moisture conditions, vegetation densities, solar zenith angles, and atmospheric conditions. The sensitivity and limitations of spectral vegetation indices are analyzed as to their ability to resolve the vegetation signal amidst variable soil-background conditions and under different sun-angle and atmospheric conditions. It is the goal of this analysis to determine how soils and vegetation mix to produce the remotely sensed canopy signal. With this understanding, it is hoped that simple global models may be developed to better approximate the spectral dynamics of soil-vegetation behavior.

To study soil and vegetation interactions, a series of "controlled" experiments conducted by Huete et al. (1985) and Huete and Jackson (1987) are utilized here. In these studies, ground-measured spectra were collected over a series of cotton (*Gossypium hirsutum* L. var DPL-70) and range-grass (*Eragrostis lehmanniana* Nees) canopies with different soil backgrounds alternately inserted underneath the canopy, via the use of soil trays. In this way, (1) reflected radiances from a "constant" vegetation cover with variable soils were collected for specific days, and (2) reflected radiances of a "constant" soil with varying vegetation densities were collected between days. Canopy characteristics for the cotton experiment are summarized in Table 1.

The soils, which were used in a variety of soil-moisture conditions, included a (i) bright yellowish-brown Superstition sand (sandy, mixed, hyperthermic Typic Calciorthid), (ii) a high-iron red Whitehouse sandy clay loam (fine, mixed thermic Ustollic Haplargid), (iii) a brown Avondale loam (fine-loamy, mixed, hyperthermic Typic Torrifluvent), and (iv) a dark, organic-rich Cloversprings loam (fine-silty, mixed Cumulic

TABLE 1   Canopy Characteristics for Cotton from Bare Soil to Full Cover

Day of Year	Green Cover (%)	Leaf-Area Index (m^2/m^2)	Canopy Height (m)	Dry Biomass (kg/m^2)	Wet Biomass (kg/m^2)
170	0	0	0	0	0
196	20	0.5	0.15	0.03	0.15
198	25	0.7	0.22	0.06	0.30
208	40	1.0	0.28	0.10	0.50
215	55	1.5	0.36	0.14	0.60
217	60	1.7	0.39	—	—
223	75	2.8	0.47	0.25	1.40
228	90	3.0	—	—	—
235	95	3.3	0.69	0.30	1.70
242	100	3.6	0.83	0.33	1.80

Source: Reprinted by permission of the publisher from Huete et al. (1985). Copyright 1985 by Elsevier Science Publishing Co., Inc.

TABLE 2   Physical Characteristics of the Four Soil-Background Types

Soil Series	Textural Class	Organic Matter (%)	Extractable Iron (%)	CaCO₃ (%)	Munsell Color Dry	Moist
Cloversprings	Loam	5.7	1.8	0	10YR3/2	10YR2/1
Whitehouse	Sandy clay loam	1.5	2.5	0	2.5YR4/6	2.5YR3/6
Avondale	Loam	0.9	0.8	5	7.5YR5/4	7.5YR4/4
Superstition	Sand	0.2	0.2	3	10YR7/4	10YR5/4

Source: Reprinted by permission of the publisher from Huete et al. (1985). Copyright 1985 by Elsevier Science Publishing Co., Inc.

TABLE 3   Red and Near-Infrared Extinction and Path Radiance ($Wm^{-2}sr^{-1}$) Spectral Properties for the Two "Simulated" Atmospheres at Two Solar Zenith Angles

Atmosphere Condition	$\theta_z = 25°$ Red	NIR	$\theta_z = 60°$ Red	NIR
Clear atmosphere				
Path radiance	10.13	3.57	7.12	2.50
Extinction coeff.	0.160	0.125	0.160	0.125
Turbid atmosphere				
Path radiance	18.91	8.86	14.09	6.76
Extinction coeff.	0.492	0.425	0.492	0.425

Source: Reprinted by permission of the publisher from Huete and Jackson (1988). Copyright 1988 by Elsevier Science Publishing Co., Inc.

Cryoboroll) (Table 2). Reflected radiances were measured simultaneously in seven spectral bands (0.45–0.52, 0.52–0.60, 0.63–0.69, 0.76–0.90, 1.15–1.30, 1.55–1.75, and 2.08–2.30 μm) using a Barnes Modular Multispectral Radiometer (MMR) (Barnes Engineering Co., Stanford, CT), with a 15° field of view. Reflectance factors were calculated by ratioing spectral responses of the canopy with those obtained over a barium-sulfate reference panel. Parameters describing the two "simulated" atmospheres (clear and turbid) used in this chapter are given in Table 3.

II SOIL SPECTRAL CHARACTERISTICS

The spectral properties and behavior of bare-soil surfaces are important to understand, since the soil serves as a useful reference from which we attempt to discriminate and measure green-vegetation growth. Knowledge of soil spectral properties are also a requisite to understanding their spectral influences on vegetation as well as to develop effective soil-normalization techniques. Soil spectral curves or signatures, as used in remote sensing, are a function of the number of wavelengths and their spectral resolutions (bandwidths). For a given set of wavelengths, soil signatures may vary in two distinct manners; there are (1) brightness differences associated with the magnitude (amplitude) of reflected radiance, and (2) spectral curve-shape variations attributed to mineralogic, organic, and water-absorption features. Each of the two types of soil spectral variations has important and distinct influences on the spectral interpretation of vegetation parameters as will be discussed here.

II.A Soil-Brightness Variations

The major difference among soil spectral signatures is the amplitude or magnitude of reflected radiance. This represents the primary source of soil reflectance variation with light-colored soils possessing higher amplitudes than darker soils. In the Munsell color decimal notation (U.S. Department of Agriculture, 1975), the "value" of a soil describes its relative brightness and it is proportional to the amplitude of the soil reflectance curve in the visible part of the spectrum. Moisture and surface roughness will alter the amplitude of the soil signal, resulting in a "lighter" or "darker" soil color value. In satellite remote sensing, brightness differences are also present due to solar irradiance variations caused by sun angle, atmosphere, and surface inclination.

Kauth and Thomas (1976) analyzed soil spectral variance in four-dimensional Landsat MSS signal (Euclidean) space and found most of the variability of bare-soil signals to be attributed to brightness, since nearly all spectral data fell along a line extending from the origin. The four-space diagonal line has been named the "soil line," or the soil-brightness vector (Kauth and Thomas, 1976). Richardson and Wiegand (1977) also developed a "soil line" concept utilizing red and near-infrared wavelength combinations. A one-dimensional soil line in the NIR–red, four-space, or any other number of wavelength regions, implies that soil spectral signatures vary only in amplitude across the 2, 4, or n wavelengths. In this case, the amplitude of the signature determines the position of the soil on this line. The higher the amplitude, the brighter the soil, and the further away from the origin the soil falls on the soil line. If the soils had equal responses across the n wavelengths, then the Euclidean slope of the soil line would equal 1.

The soil-line concept has become widely accepted in the interpretation and analysis of multispectral data (Fukuhara et al., 1979; Jackson et al., 1980; Thompson and Wehmanen, 1980; Wiegand and Richardson, 1982; Miller et al., 1984). Although the soil line may shift with atmosphere conditions and satellite calibration factors, it has generally been assumed that there exists one, "global" soil line encompassing a wide range of soil types and soil-surface conditions. The "global" soil-line concept, however, may not be a proper representation of bare-soil spectral variance. Jackson et al. (1980) plotted a wide range of soil types and reported a curvilinear relationship. Huete et al. (1984) found that the "global" soil line actually consists of numerous nonparallel soil lines that represent soil-moisture variations for specific soil types. Consequently, the soil line becomes soil-specific and scene-dependent.

II.B Spectral Curve-Shape Variations

Aside from amplitude differences, there are curve-shape variations associated with soil-absorption properties. These variations result in soil-color (Munsell hue and chroma) differences in the visible wavelengths as well as mineralogic, organic, and water-absorption effects in the longer optical wavelengths. Hue is related to the wavelength composition of reflected visible radiation, and chroma refers to the purity of color. These create "unique" spectral signatures unrelated to the amplitude of soil brightness. Condit (1970), utilizing a small range of optical wavelengths (0.3–1.0 μm), found only three unique soil curve forms among 160 soil samples. Stoner and Baumgardner (1981) analyzed 485 moist soils and identified five soil curve forms in the spectral region from 0.52–2.32 μm. These were associated mainly with the organic matter and iron contents of the soils they studied. Soil curve-shape differences are more difficult to assess since currently operational optical sensors have a limited number of "broad" wavelength bands that limit the amount of spectral detail available.

Soil spectral curve-shape differences are responsible for deviations away from the principal axis of soil variance (soil line). These secondary soil (color) variations provide the "width" of a soil line and depending on the wavelength axes, the soil line may resemble an ellipsoid or a plane. A soil plane concept was used by Kauth and Thomas (1976) to describe secondary soil spectral behavior in four-band MSS space. They attributed this second soil component to red and yellow colored soils. The magnitude of these deviations becomes critical to vegetation analysis if these sources of spectral variance overlap with those normally attributable to vegetation. This secondary axis of soil variation in wavelength space may add a noise component to vegetation studies and renders soil spectra more difficult to normalize. Additional factors responsible for soil spectral behavior are covered in Chapter 3.

III   SOIL INFLUENCES ON VEGETATION SPECTRA

The interaction of radiant energy with vegetation is generally more complex than with soils. In converting water (from the soil) and carbon (atmosphere CO_2 gas) to carbohydrates, plants selectively absorb visible radiation, and under the influence of chlorophyll and carotene pigments, they will absorb slightly greater amounts of red and blue light leaving a "green" appearance to actively photosynthesizing leaves. The spongelike mesophyll tissue in healthy leaves is responsible for strong water-related absorptions in the middle-infrared region (1.3–3.0 μm) and a strong near infrared reflectance peak. Vegetation optical behavior is also related to the macroscopic arrangement of plants above a soil. This includes canopy architecture, structure, leaf morphology, leaf area, and plant density. As a result of the vegetation spectral features cited, a decrease in reflected red radiation (Figure 1) and an increase in reflected near-infrared radiation (Figure 2) generally occurs, with the emergence of green vegetation over a soil. One might also anticipate blue, green, and middle-infrared reflectances to decrease with the vegetation growth due to leaf pigment and water absorption in these wavelengths.

III.A   Soil Type and Moisture Variations

The change in composite canopy response with the development of vegetation is dependent on the optical properties of the soil background as well as those of vegetation. Significant decreases in red reflectance with increasing amounts of green cover occur with lighter-colored soil backgrounds, such as the Superstition sand and Avondale loam shown in Figure 1. Red-canopy reflectances, however, may vary little over darker soils, as seen with the Cloversprings loam in wet- and dry-surface conditions (Figures 1 and 3). Near-infrared reflectances, on the other hand, generally increase with increasing vegetation amounts (Figure 2), although in some very bright desert soils, canopy NIR response may decrease with the presence and growth of vegetation (Figure 4). Depending on the optical properties of both the vegetation and soil background, composite canopy spectral response may increase, decrease, or remain invariant to changes in green-vegetation amounts.

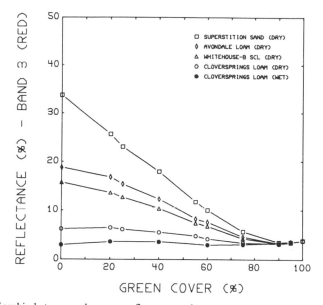

Figure 1 Relationship between red-canopy reflectance and green-canopy cover for one wet- and four dry-soil backgrounds: (□) Superstition sand (dry), (◇) Avondale loam (dry), (△) Whitehouse s.c.l. (dry), (●) Cloversprings loam (dry), (●) Cloversprings loam (wet). [Reprinted by permission of the publisher from Huete et al. (1985). Copyright 1985 by Elsevier Science Publishing Co., Inc.]

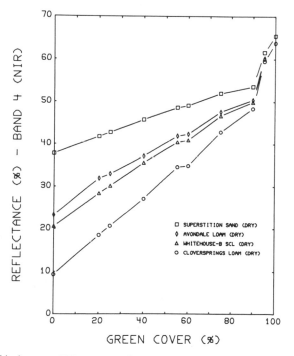

Figure 2 Relationship between NIR-canopy reflectance and green-canopy cover for four dry-soil backgrounds. See Figure 1 caption for symbols. [Reprinted by permission of the publisher from Huete et al. (1985). Copyright 1985 by Elsevier Science Publishing Co., Inc.]

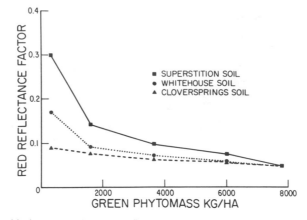

Figure 3 Relationship between red-canopy reflectance and Lehmann lovegrass phytomass for three soil backgrounds. [Reprinted by permission of the publisher from Huete and Jackson (1987). Copyright 1987 by Elsevier Science Publishing Co., Inc.]

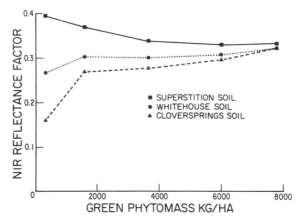

Figure 4 Relationship between NIR-canopy reflectance and Lehmann lovegrass phytomass for three soil backgrounds. [Reprinted by permission of the publisher from Huete and Jackson (1987). Copyright 1987 by Elsevier Science Publishing Co., Inc.]

The reflectance data of the grass and cotton canopies (Figures 1–4) may also be depicted in NIR–red wavelength plots to demonstrate combined soil-vegetation behavior (Figure 5). The resulting triangular shape with apices defined by dense green vegetation (100% green cover; LAI = 3.6) and dark and bright bare soils (soil line) forms the principal plane of the tasseled cap concept (Kauth and Thomas, 1976). The emergence of vegetation over the soil causes composite response to shift away from the soil line. The characterization of this vegetation-induced shift in Euclidean space becomes crucial to the interpretation and discrimination of vegetation. For any specific amount of vegetation, a new soil-varying spectral line is formed that represents an isoline of constant greenness. These isolines have (i) shifted away from the soil line in the direction of vegetation variance, and (ii) they are shorter in length compared with the bare-soil line and eventually converge to a point where soil influences disappear entirely as a result of dense canopy closure. The individual cotton and grass spectral isolines followed a pat-

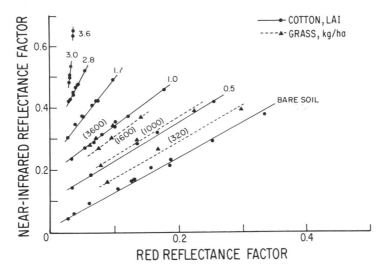

Figure 5 Observed vegetation isolines in NIR–red wavelength space for various canopy densities of cotton and grass with differing soil-background conditions. [Reprinted by permission of the publisher from Huete (1988). Copyright 1988 by Elsevier Science Publishing Co., Inc.]

tern of increasing slopes and NIR intercepts with higher vegetation densities (Figure 5 and Table 4). This same soil-vegetation spectral behavior has been verified with various canopy radiant transfer models including the SAIL model (S. Goward, personal communication), the two-stream hemispherical model of Sellers (1985), and the directional radiative transfer model utilized by Choudhury (1987).

III.B Solar-Angle Interactions

Understanding the spectral behavior of partial vegetation canopies at different solar zenith angles is necessary in many remote-sensing applications such as seasonal and phenological vegetation analyses and productivity studies (Kimes et al., 1980). The change

TABLE 4 Geometric Properties of Cotton-Canopy Isolines in NIR–Red Wavelength Space

Green Cover (%)	Slope	NIR Intercept	Length	n	Correlation Coefficient (r)
0	1.06	0.026	0.396	8	0.997
20	1.24	0.106	0.301	5	0.998
25	1.30	0.131	0.278	4	0.999
40	1.54	0.186	0.224	8	0.997
55	2.01	0.253	0.156	4	0.999
60	2.50	0.240	0.153	8	0.997
75	3.93	0.299	0.094	8	0.994
90	15.26	0.006	0.052	4	0.977
95	2.19	0.005	0.021	4	0.134
100	55.33	−0.015	0.017	2	—

Source: Reprinted by permission of the publisher from Huete et al. (1985). Copyright 1985 by Elsevier Science Publishing Co., Inc.

in vegetation-canopy spectra with sun angle is dependent upon numerous factors, including canopy composition, architectural differences, canopy closure, leaf optical properties, and the optical properties of the soil background (Coulson and Reynolds, 1971; Jackson et al., 1979; Kollenkark et al., 1982b; Pinter et al., 1985; Huete, 1987a). These studies not only show diurnal reflectance trends to be highly complex and variable in vegetation canopies, but also stress the importance of physically describing the plant geometry and sunlit/shaded soil components in order to develop appropriate sun-angle correction techniques.

In Figure 6, the red and NIR canopy reflectances of a 40% green cotton cover (LAI = 1) are shown as a function of underlying bare-soil reflectance (measured without the presence of a canopy) for two solar zenith-angle conditions. In this experiment, the soil between the North–South cotton rows was mainly sunlit at the smaller solar zenith angle (25°) and shaded at the larger solar zenith angle (60°). As expected, red and NIR canopy reflectances increased with higher reflecting soil backgrounds at both sun angles. The sensitivity in canopy response (slopes in Figure 6) to soil background, however, was higher in the NIR and at the smaller solar zenith angle when most of the soil was sunlit.

The red canopy reflectances were higher, and NIR reflectances were lower at the smaller solar zenith angle for all soil backgrounds considered here (Fig. 6). The difference in red-canopy reflectance between the two solar zenith angles was greater for higher reflecting soils, whereas the sun-angle difference in NIR-canopy response was greater for the darker soils. Thus, red reflectance is relatively insensitive to sun angle in dark-soil canopies, whereas NIR reflectance is insensitive to sun angle in light-soil canopies. This means that diurnal changes in red-canopy response is primarily a soil effect since there are no response changes for a hypothetical "zero" reflecting soil. NIR diurnal response is due to vegetation since the addition of higher reflecting soil spectra tends to nullify the diurnal response. Figure 6 illustrates that canopy reflectances may increase, decrease, or remain invariant with solar angle, depending on the optical properties of the underlying soil as well as the wavelength used to measure the canopy. Consequently,

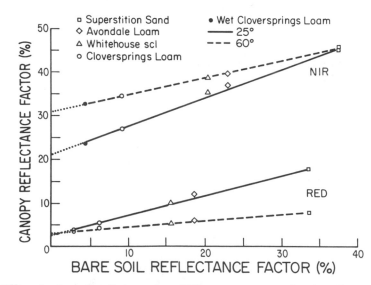

Figure 6 NIR- and red-canopy reflectances for a 40% cotton canopy as a function of bare-soil reflectance for two solar zenith angles. [From Huete (1987a).]

one cannot generalize on expected diurnal trends in canopy reflectances without consideration of soil optical properties in addition to those of the vegetation.

III.C Atmosphere Influences

Both soil and atmosphere conditions influence the application of remote-sensing techniques to the study of soil-vegetation dynamics over the Earth's surface. There are two atmosphere influences of concern in interpreting satellite remote-sensing data (Slater, 1980). Atmosphere scattered "path radiances" represent an added flux component, scattered onto the sensor detectors, which have not been reflected by the Earth's surface. A second atmosphere influence concerns the optical thickness or extinction properties of the atmosphere medium, which results in an attenuation (scattering and absorption) of ground-reflected radiance. Atmospheric path radiances are generally most influential over "darker" surface targets, whereas extinction properties primarily affect "brighter" surface targets (Slater and Jackson, 1982; Fraser and Kaufman, 1985); but both are most influential at the shorter optical wavelengths.

Atmosphere extinction and path radiance effects on surface radiances are counteracting, i.e., depending on the brightness of the ground surface, the added path radiance may be equivalent to all extinction losses. Fraser and Kaufman (1985) reported a "critical surface reflectance," whereby the radiance remained constant as the optical thickness varied. Since the soil background strongly influences the overall brightness of partially vegetated canopies, atmospheric modification of canopy reflected radiances will be strongly dependent on the optical properties of the soil. In this section, the role of the soil background on atmospheric modification of canopy spectra is carefully examined.

The combined influences of atmosphere extinction, path radiance, and soil background on the red (0.63–0.69 μm) and NIR (0.76–0.90 μm) radiances sensed from a 40% cotton canopy are illustrated in Figure 7 for two "simulated" atmospheres at a 25°

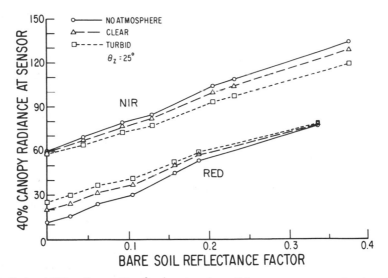

Figure 7 Red and NIR radiance ($Wm^{-2}sr^{-1}$) values for a 40% cotton canopy as a function of soil-background reflectance and atmosphere condition. [Reprinted by permission of the publisher from Huete and Jackson (1988). Copyright 1988 by Elsevier Science Publishing Co., Inc.]

solar zenith angle (Table 3). In order of decreasing reflectance, the soils may be identified as Superstition sand, Avondale loam, Whitehouse s.c.l., wet Avondale loam, Cloversprings loam, wet Cloversprings loam, and an extrapolated "zero soil" (see Huete, 1987b). Both red- and NIR-canopy radiances increased with higher reflecting soil substrates, but in a manner dependent on atmospheric condition. In comparison to the "no" atmosphere case (ground radiance values), canopy radiances calculated at the top of clear (100 km meteorological range) and turbid (10 km meteorological range) atmospheres increased in the red and decreased in the near-infrared for a given soil. This was a result of atmospheric red path radiance being dominant over atmospheric red extinction, whereas NIR extinction was stronger in magnitude than the NIR-path radiance contribution.

Soil spectral contributions toward canopy radiance played a significant role in determining the net effect of atmosphere extinction and path radiance. The increase in canopy red radiance with atmosphere turbidity was greatest with the darkest soils and became nearly invariant with the lighter-colored soils. Thus, the brighter the soil background, the greater was the influence of atmosphere extinction in offsetting the additive red-path radiance. Atmosphere extinction effects controlled NIR radiance values with maximum decreases in response occurring over "bright" soils with invariant results over dark soil backgrounds (Figure 7). The combined influence of the two atmosphere factors was dependent on the underlying substrate in partial-canopy covers since the spectral characteristics of soil background influenced canopy radiance values. NIR reflected radiances from the 40% cotton canopy cover doubled between the darkest and brightest soil backgrounds ($68-130$ $Wm^{-2}sr^{-1}$) and red reflected radiances varied by a factor of 4 ($15-70$ $Wm^{-2}sr^{-1}$) due to soil influences.

Also shown in Figure 7 are the red and NIR radiance values for the cotton canopy with a "zero" soil-background condition. This hypothetical situation shows that the reflected NIR radiance from only the vegetation (40% cotton cover; "zero" soil) was not significantly influenced by the atmosphere and that it was the soil contribution, related to the amount of soil as well as the soil-surface condition, that created atmosphere-induced variability. On the other hand, the red-vegetation radiance component was very sensitive to the atmosphere condition and the soil contribution acted to minimize or offset such effects.

Figure 8 is a NIR-red wavelength plot of the 40% canopy relative to the soil line (radiances of four soils in wet and/or dry conditions) and full cotton (100% green cover). Each line represents a 40% canopy, measured with different underlying soil substrates, at a particular atmosphere condition. Compared with the "no" atmosphere situation, increasingly turbid atmospheres tended to shift the 40% canopy isoline toward the bare-soil line. The individual 40% canopy isolines were nearly parallel and shifted so that for a bright soil situation, there were mainly changes in NIR radiances with smaller variations in red radiance, whereas for darker soils, red radiances were most sensitive. In contrast to the 40% canopy, the bare soils themselves tended to cluster along a soil line. With increasing atmosphere turbidity, light-colored soils moved down the soil line, darker soils shifted upward along the soil line, whereas soils in between were minimally affected. The solid symbols in Fig. 8 represent bare-soil, full-cotton, and partial-canopy radiances measured with a 60° solar zenith angle, where the soils between the cotton rows were mostly shaded. In relation to the smaller solar zenith angle (25°), total canopy red and NIR radiances were much lower due to less solar irradiance as well as reduced soil contributions.

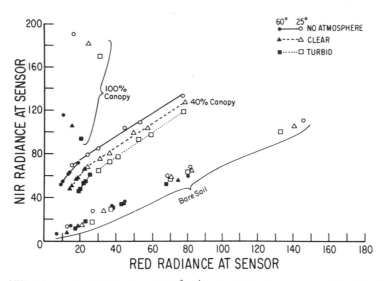

Figure 8 NIR–red wavelength behavior ($Wm^{-2}sr^{-1}$) of a 40% cotton canopy with different soil backgrounds relative to bare soil and full canopy for various atmosphere, sun-angle, and soil-background conditions. [Reprinted by permission of the publisher from Huete and Jackson (1988). Copyright 1988 by Elsevier Science Publishing Co., Inc.]

In summary, the soil background plays an important role in determining the reflectance characteristics of partially vegetated canopies. In comparison to canopies underlain by light-colored soils, dark-soil canopies are more sensitive to atmosphere condition in the red and less sensitive to sun-angle changes and vegetation-density changes. In the NIR wavelengths, the dark-soil canopies will become more dependent on sun angle and vegetation density and less sensitive to atmosphere condition.

IV SOIL INFLUENCES ON VEGETATION SPECTRAL INDICES

IV.A Vegetation-Index Theory

Numerous spectral vegetation indices have been developed to characterize vegetation canopies. These indices generally attempt to enhance the spectral contribution of green vegetation while minimizing those from soil background, solar irradiance, sun angle, senescent vegetation, and atmosphere (Tucker, 1979; Colwell, 1974; Kauth and Thomas, 1976; Richardson and Wiegand, 1977; Ashley and Rea, 1975). The most common of these indices utilize red- and NIR-canopy reflectances or radiances in the form of ratios (normalized difference and ratio vegetation indices: see Tucker, 1979) or in linear combination (perpendicular-vegetation index, Richardson and Wiegand, 1977; and green-vegetation index, Kauth and Thomas, 1976). These indices have been found to be well correlated with various vegetation parameters including green-leaf area, biomass, percent green cover, productivity, and photosynthetic activity (Colwell, 1974; Hatfield et al., 1984; Asrar et al., 1984; Sellers, 1985). A final requirement of spectral indices is that they should not be site-dependent, but instead be (globally) applicable across space and time.

IV.A.1   Ratio-Based Indices

Vegetation indices or greenness measures developed thus far can be classified into two broad categories: the ratio indices and orthogonal indices. The ratio vegetation index (RVI = NIR/red) and normalized difference vegetation index [NDVI = (NIR − red)/(NIR + red) = (RVI − 1)/(RVI + 1)] are the most common of the ratio transformations used as vegetation measures. They involve ratioing a linear combination of the NIR and red bands by another linear set of the same bands. In the two-dimensional NIR–red space, these indices are graphically displayed by vegetation isolines of increasing slopes diverging out from the origin (Figure 9).

The ratio indices essentially rely on the existence of the soil-line concept in normalizing soil behavior and discriminating vegetation spectra. Since most soil spectra fall on or close to the soil line, and since the intercept of such a line is close to the origin, RVI and NDVI values of bare soils (ratios) will be nearly identical for a wide range of soil conditions. These indices have been found effective in normalizing soil-background spectral variations (Colwell, 1974), and variations in irradiance conditions (Tucker, 1979). Tucker (1977) and Ripple (1985) found that the NDVI was the best estimator of low amounts of blue grama (*Bouteloua gracilis*) and tall fescue (*Festuca arundinacea*) grass phytomass. Colwell (1974) and Pearson et al. (1976) found that the RVI was unreliable in grass canopies with low green covers (<30%). Weiser et al. (1984) reported a direct correlation between RVI and tall grass prairie phytomass, however, they found their results to be site-dependent, year-specific, and sensitive to the presence of senescent vegetation.

IV.A.2   Orthogonal-Based Indices

The second broad category of spectral vegetation measures are orthogonal transformations that include the two-band perpendicular-vegetation index (PVI) of Richardson and Wiegand (1977) and the four-band green-vegetation index (GVI) of Kauth and Thomas

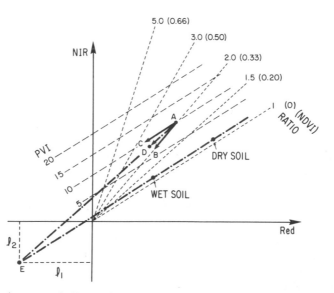

Figure 9   Vegetation spectra isolines and convergence points in NIR–red wavelength space as predicted by the ratio, normalized difference, and perpendicular-vegetation indices. [Reprinted by permission of the publisher from Huete (1988). Copyright 1988 by Elsevier Science Publishing Co., Inc.]

(1976). The six-band (Crist and Cicone, 1984) and n-wavelength band (Jackson, 1983) GVI also represent orthogonal indices. The orthogonal indices are distinct from the ratio indices in that isolines of equal ''greenness'' do not converge at the origin, but instead remain parallel to the principal axis of soil spectral variation, i.e., the soil line (Figure 9). A greenness vector, orthogonal to the soil line, is computed to maximally include green-vegetation signals while holding soil background constant. The projection of composite spectra onto this vector is subsequently used as the measure of vegetation.

The explanation offered by the two types of indices are contradictory with each other in describing soil-vegetation spectral behavior. As an example, a partial canopy over a dry-soil background (A) is shown in Figure 9. If the soil background were to become wet, a vegetation isoline bounded by dry- and wet-soil conditions would be formed. In order for the RVI and NDVI to effectively normalize such a background change, the vegetated pixel would have to shift directly toward the origin (B), following an isoline of constant RVI and NDVI values. The PVI, however, would require the pixel to shift along an isoline parallel to the soil line (C), so that both the wet- and dry-soil vegetated pixels maintain a constant PVI value (equidistant to the soil line). Another vegetation index, called the soil-adjusted vegetation index $[SAVI = [(NIR - red)/(NIR + red + 0.5)]1.5]$ was developed by Huete (1988) to describe observed grass- and cotton-vegetation isolines (Figure 5). The SAVI assumes a shift (D) that lies between those predicted by the ratio and orthogonal approaches (Figure 9). This index takes on properties of both the NDVI and PVI and is more fully described later in this chapter.

There is a lack of detailed analyses concerning the limitations of these vegetation indices in assessing greenness and in monitoring of the plant canopy (Tucker, 1979). For the most part, quantitative information regarding the performance of various spectral measures have been collected over uniform soil backgrounds with ground-based radiometers. Multispectral data collected from space- or airborne sensors, on the other hand, will quite often include different soil types under several soil-moisture conditions and with varying quantities of dead organic material. The usefulness of vegetation spectral indices depend, in part, on how well they minimize these soil-background spectral variations. In wavelength space, it is the migration of a vegetated pixel away from the soil line in relation to vegetation density that must be adequately modeled.

IV.B Soil-Brightness Effects

Soil brightness influences have been noted in numerous studies, where, for a given amount of vegetation, darker- or lighter-soil substrates resulted in higher vegetation index values when the NDVI, RVI, PVI, and GVI were used as vegetation measures. Colwell (1974) found dark-soil canopies to have the highest RVI values at low- to intermediate-percent oat covers. Elvidge and Lyon (1985) additively mixed vegetation spectra with a series of arid-region soil substrates and reported higher RVI and NDVI greenness values for darker substrates, but found no effect on the PVI. Jackson et al. (1983) found the PVI and GVI to be sensitive to soil-moisture condition on a uniform soil type. Huete et al. (1985) found RVI and NDVI values to decrease while PVI and GVI values increased with increasing soil brightness under a constant amount of vegetation. All of these studies show spectral indices to be partially correlated with soil brightness over certain ranges of vegetation density. Thus, in areas where there are considerable soil-brightness variations resulting from moisture differences, roughness variations, shadow, or organic-matter differences, there are soil-induced influences on the vegetation index values.

Soil-brightness influences are prevalent in partially vegetated canopies because the ratio-based and orthogonal-based vegetation indices fail to predict the behavior of vegetated pixels as they migrate away from the soil line. If vegetation isoline behavior does not agree with that predicted by spectral vegetation indices, then different soil backgrounds under constant vegetation amounts will produce different spectral index values. The isolines presented in Figure 5 neither converged at the origin (as required by the RVI and NDVI) nor maintained slopes identical with the soil line (a PVI requirement). At low-vegetation densities, the isolines are nearly parallel with the soil line, while at very high densities, they nearly approach "zero" intercepts. Over most of the range of vegetation densities, the isolines do not obey the pattern expected from the ratio and orthogonal indices.

Figure 10 shows the soil-brightness problem inherent in the NDVI. Decreases in red-canopy reflectance, due to darker-soil substrates, cause significant increases in the NDVI. The NDVI appears to be as sensitive to soil darkening as to vegetation development. Thus, a very low amount 320 kg/ha, of grass phytomass on a dark-soil substrate has the same NDVI (~0.3) as 1000 kg/ha on a bright sandy substrate. The NDVI values for a 20% green cotton cover (LAI = 0.5) ranged from 0.24–0.60 and approached the values for a 60% green cover (LAI = 1.7). The NDVI also does not account for the orientation of the bare-soil line (Figure 10). Since the soil line has a positive NIR intercept in NIR–red wavelength space, darker soils have higher NDVI values than the lighter-colored soils because their vector slope to the origin is steeper. As a result, the NDVI would not be able to differentiate these darker (bare) soils from a 320 kg/ha grass canopy over lighter soils.

Figure 11 demonstrates the nonparallel nature of cotton GVI isolines. From these data one can see: (1) soil-brightness influences, where brighter soils produce the highest GVI values for equivalent vegetation densities, and (2) secondary soil variations within the isolines. Secondary soil noise variations increase with the additional use of more wavelengths in the computation of orthogonal indices. In Figure 11, the GVI values of

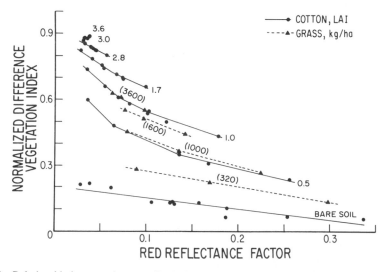

Figure 10   Relationship between the normalized difference and bare-soil red reflectance for various canopy densities of cotton and grass. Numbers in parentheses denote grass biomass. [Reprinted by permission of the publisher from Huete (1988). Copyright 1988 by Elsevier Science Publishing Co., Inc.]

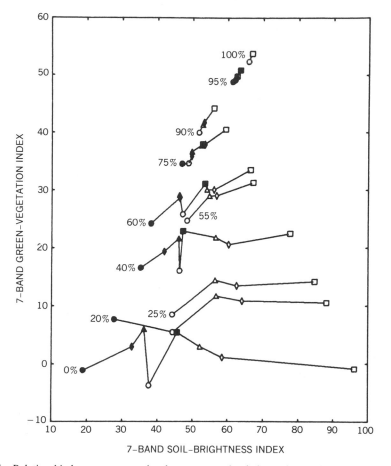

Figure 11 Relationship between a seven-band green vegetation index and seven-band soil-brightness index for various green cotton covers. See Figure 1 caption for symbols; solid symbols denote wet-soil backgrounds. [Reprinted by permission of the publisher from Huete et al. (1985). Copyright 1985 by Elsevier Science Publishing Co., Inc.]

bare soil overlap with those from the 20% green canopy (LAI = 0.5). As expected, secondary soil influences decrease in magnitude with increases in vegetation density due to the decreased soil signal.

Irrespective of how bare-soil spectra are normalized (ratios or rotation), soil-brightness influences become greater with increasing vegetation densities up to 40–60% green-canopy cover (LAI = 1–2). The spectral index values from the 75% canopy cover (LAI = 2.8) are as sensitive to soil background as those from the 20% canopy covers (LAI = 0.5) and soil-induced variations in the NDVI for a constant canopy of LAI = 1.0 encompassed half the total NDVI dynamic range (zero to full cover) of the green canopy.

The SAVI, which was designed to minimize soil-brightness influences, produces vegetation isolines more nearly independent of the soil background (Figure 12). Soil variations are reduced by the SAVI in both the narrow-leaf grass (erectophile) and the broad-leaf cotton (planophile) canopies. In Figure 13, soil-induced spectral index variations are shown as a function of green cotton LAI for the NDVI, PVI, and SAVI. The relationship between NDVI and PVI with LAI is very soil-dependent, as seen by the con-

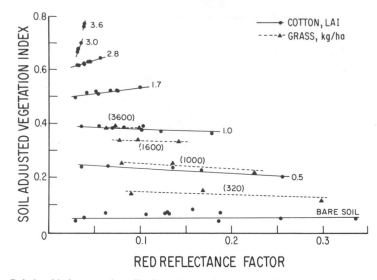

Figure 12 Relationship between the soil-adjusted vegetation index and bare-soil red reflectance for various canopy densities of cotton and grass. Numbers in parentheses denote grass biomass. [Reprinted by permission of the publisher from Huete (1988). Copyright 1988 by Elsevier Science Publishing Co., Inc.]

siderable range in NDVI or PVI values for a constant vegetation density. Note the opposite soil-brightness effects between these two indices. The PVI values for a constant amount of vegetation are greatest with the light-soil background, whereas it is the darker soils that result in highest NDVI values. In both cases, the soil-brightness influence becomes more serious at intermediate levels of vegetation density than at either very

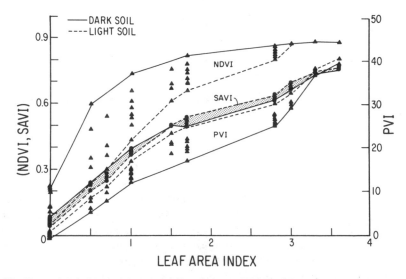

Figure 13 Vegetation index response and soil variations of the NDVI, SAVI, and PVI as a function of cotton leaf-area index. [Reprinted by permission of the publisher from Huete (1988). Copyright 1988 by Elsevier Science Publishing Co., Inc.]

low or very high densities (Figure 13). The SAVI, by comparison, substantially reduces soil-induced variations, and it improves the linearity between the index and LAI.

IV.C   Solar-Angle Influences

Jackson et al. (1980) showed how sunlit- and shaded-soil backgrounds might affect spectral indices, primarily through differences in red and NIR flux penetration through the canopy. The changes in red- and NIR-canopy reflectances with sun angle (Figure 6) were mostly a result of differences in sunlit- and shaded-soil contributions. Figure 14 shows the NDVI of a 40% cotton canopy plotted against bare-soil NIR reflectance for the two solar zenith angles. As discussed in the previous section, NDVI values increase as the soil becomes darker for constant vegetation conditions (Figure 10). The NDVI values are also higher at the larger solar zenith angle, partly due to a greater proportion of plant-material irradiance and partly as a result of greater soil darkening (shadow) at this angle. Not only are NDVI values greater, but the soil-induced effects are also smaller at the larger sun angle. The NDVI of the 40% green canopy is greatest with darker-soil types and larger solar zenith angles (shaded soil). This corresponded to minimal soil-component contributions (darkest soil surface) and maximum vegetation irradiance conditions.

Figure 14 shows that NDVI differences with solar zenith angle are primarily a soil-induced effect since (a) they become greater with lighter-colored soils and (b) they are minimal with very dark soils. With the extrapolated "zero" soil background, the 40% canopy only varies from 0.79–0.77 units. The soil-dependent nature of the NDVI is of concern to temporal data sets due to NDVI sensitivity to differences in sunlit- and shaded-soil components resulting from variations in solar zenith angle. Huete (1987a) analyzed sun-angle influences on the PVI and found a strong soil-component influence on diurnal canopy spectral responses.

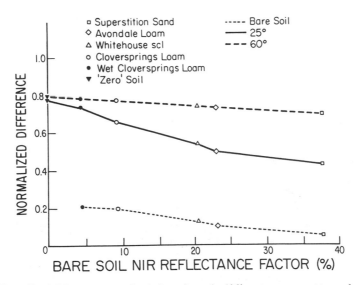

Figure 14   Normalized difference vegetation index values of a 40% cotton canopy at two solar zenith angles and different soil backgrounds. [From Huete (1987a).]

IV.D Atmosphere Influences

Slater and Jackson (1982) found "simulated" turbid atmospheres to modify the NIR–red ratios of a wheat canopy by 5.6% for a dry-soil background and 29.5% for a wet-soil background. Figure 15 illustrates the influence of clear and turbid simulated atmospheres on the NDVI for varying densities of cotton over light- and dark-soil backgrounds (Huete and Jackson, 1988). In general, increasing atmosphere turbidity degrades spectral index responses for all vegetation densities. The relative influences of soil background and atmosphere, however, varies for the two spectral indices. The distance between the two soil curves for a particular atmosphere represents the soil influence on the spectral index, and variations among the three atmosphere curves for a particular soil background represent the atmosphere effect on the spectral index. Figure 15 shows the NDVI to be most sensitive to atmosphere influences when underlain by dark soils. The NDVI of the light-colored soil with 40% or less green cover is invariant to atmosphere condition, whereas the darker-soil NDVI response decreases considerably with increasing turbidity. The atmosphere modified the NDVI signal of the light-soil–20% cotton canopy by only 0.04 units compared with a 0.28 unit dark-soil shift. With the exception of the highest densities, darker-soil backgrounds resulted in greater sensitivity to atmospheric condition. It may also be noted that, under turbid atmospheres, soil influences on the NDVI are much smaller. The atmosphere greatly reduces the soil effect on canopy NDVI values, but at the expense of significantly lowering the NDVI of the green covers. NDVI sensitivity to soil background is maximal at the lower green-canopy covers, particularly with 40% green-canopy covers and below. Soil influences on the NDVI also exceed those created by the atmosphere over this green-canopy range. It is only beyond 60% canopy covers that the influence of the atmosphere dominates over that of the soil in determining NDVI values.

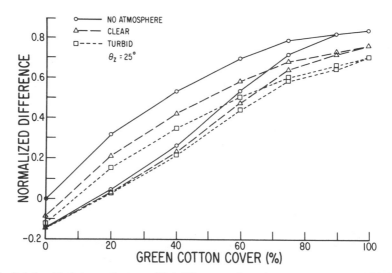

Figure 15 Relationship between the normalized difference and percent green cotton cover with dark- and a light-soil backgrounds at three atmosphere conditions. The upper curve at each atmosphere is for the dark soil. [Reprinted by permission of the publisher from Huete and Jackson (1988). Copyright 1988 by Elsevier Science Publishing Co., Inc.]

IV.E   Bare-Soil Noise Problems

Soil spectral variations have been known to hinder the discrimination of low densities of vegetation over croplands and arid/semiarid environments. The soil response is particularly strong in sparsely vegetated areas and the weak vegetation signal is hard to decipher amidst the strong and variable soil background. In this section, the lower boundary condition (bare-soil spectra and the soil line) of canopy spectral behavior is investigated. A sensitivity analysis is performed to determine to what extent low-vegetation responses can be discriminated from bare-soil responses. A vegetation index should not confuse secondary bare-soil deviations from the soil line with those associated with green vegetation. Figure 11 shows the projection of the eight bare-soil spectra onto a seven-band (MMR) GVI. The GVI is sensitive to bare-soil type and soil-moisture condition, and the width of these secondary variations is equivalent to a 20% green cotton GVI response. Bare-soil spectral variations establish a threshold level of vegetation discrimination below which one cannot reliably detect vegetation from bare soil.

A "vegetation index" signal to "soil noise" ratio (S/N) may be utilized to test the sensitivity of vegetation spectral indices. This is obtained by ratioing the dynamic range of vegetation index values (0–100% green) to that obtained with a series of bare soils. This provides a measure of how well vegetation may be discriminated from bare soil; however, it would not establish a lower limit of "reliable" vegetation detection due to variable spectral index response sensitivities. An alternative approach is to project bare-soil values onto a vegetation index measure and directly obtain the level of green cover discriminable from the scatter "cloud" of bare-soil values (as in Figure 11). This produces a "vegetation equivalence" value that represents a range of spectral index values attributed to soil variations.

The vegetation equivalence values of secondary soil variations are presented in Table 5 for six vegetation indices. The best spectral indices for discriminating low amounts of vegetation from the eight bare soils were the PVI and SAVI, where they could be used to discriminate approximately 5% green cover from bare soil. The RVI and NDVI values

TABLE 5   Bare-Soil Spectral Variations on Vegetation Index Measures Using 8 and 40 Soil Backgrounds

Vegetation and Soil Variations	RVI	NDVI	SAVI	PVI	GVI-4	GVI-7
Dynamic range (cotton)	15.79	0.783	0.702	0.394	0.424	0.487
Soil range (8)	0.42	0.157	0.042	0.017	0.050	0.090
% Vegetation Equivalence (8)	12.8	15.3	5.4	4.6	13.3	20.8
% Vegetation Equivalence (40)	14.7	17.1	9.1	10.9	18.7	39.5
Dynamic range (range grass)	5.46	0.610	0.392	0.161	0.157	—
Green grass Equivalence (kg/ha) (40)	203	487	179	163	288	—

of the eight bare soils had vegetation equivalences of 13 and 15%, respectively. The four- and seven-band GVI (13 and 21%) proved less reliable than the two-band PVI (5%), as a result of the extra wavelength bands that introduced more soil noise than vegetation information. The vegetation equivalence values shown in Table 5 is scene-dependent and can vary in several respects: (i) the spectral properties of the soil surfaces present in an image (or data set), (ii) the spectral characteristics of the vegetation in an image; and (iii) external factors such as atmosphere condition, and sun-view angle geometry. These are now discussed below.

IV.E.1 Soil Composition

The level of soil noise is dependent on the spectral variance of the soils present in an image. Thus, a wider range of soil types may increase the magnitude of soil noise and raise the threshold of vegetation detection. In Table 5, the spectral index values of 40 bare soils (see Huete et al., 1984) are used to establish a ''global'' range of bare-soil variations. The ''global'' bare-soil spectral ranges increased the magnitude of soil noise, almost doubling the threshold of vegetation detection for certain indices. The SAVI and PVI still produced the best capability for discriminating low-vegetation densities (9.1 and 10.9%, respectively). The seven-band GVI produced spectral index values for the 40 bare soils that were equivalent to a green cotton cover of nearly 40%.

On the other hand, a more restrictive or uniform group of soils will reduce the magnitude of bare-soil spectral index values; however, if data sets are to be transferable to other locations and compared with other data sets, then the global soil range must be considered. Otherwise, individual soil spectra or soil-specific soil lines would have to be utilized and reported in vegetation studies. Huete et al. (1984) found a substantial reduction in bare-soil noise problems when vegetation spectra were measured in relation to ''individual'' soil lines. In either case, the characterization of soil-background spectra is user dependent. In certain situations, a general scene-independent set of soil reference spectra (global soil line) is the only requirement for vegetation classification. In other situations, scene- or temporal-dependent soil optical properties may be needed for improved vegetation discrimination. The level to which vegetation can be measured, however, may become limited by other factors such as sensor noise, atmospheric conditions, and spatial and radiometric resolution.

IV.E.2 Vegetation Characteristics

The second factor that affects the level of vegetation discrimination is the spectral properties of the vegetation community in an image. The green cotton canopy utilized in the previous analysis resulted in fairly large spectral index dynamic ranges since cotton is a relatively broad-level lush green crop. In establishing the limitations of spectral indices in rangeland areas, one would have to adjust the green dynamic range to that expected in rangeland areas. Usually a ''pure'' (100%) vegetation signal in natural ecosystems such as rangelands is much lower than that found in managed agricultural ecosystems due to the lower amount of green-leaf foliage and complex (nonuniform) canopy structures in the natural ecosystems. A lower vegetation dynamic range results in a higher threshold of vegetation detection since soil variations may remain the same. In Table 5, bare-soil spectral influences relative to the reflectance spectra of a dense stand of Lehmann lovegrass are shown. The vegetation dynamic range decreased by factors of nearly 3 (RVI) to 1.3 (NDVI). This data suggest that in rangelands, it would be more difficult to detect low amounts of green-vegetation cover with the RVI, NDVI, and four-band GVI. The PVI and SAVI enabled the lowest levels of vegetation discrimination (160 and 180 kg/ha, respectively), which represents approximately 10% of the maximum biomass (2000 kg/ha) encountered in arid and semiarid rangelands.

IV.E.3 External Factors

Atmosphere conditions will also alter the soil noise in spectral indices. In Figure 8, the 40 and 100% cotton spectra were drawn closer to a relatively stable soil line. Thus, there are greater atmosphere influences in vegetated canopies than in bare soils. This reduces the dynamic range of vegetation signals and amplifies the soil-noise problem. This same trend was observed for the 25° and 60° solar zenith angles (Figure 8). Solar- and view-angle conditions alter the relative proportion of plant and soil signal reaching the sensor, and as a result, soil-noise problems will similarly be changed. Soil spectral variations are reduced at larger solar zenith angles due to the increased amount of soil shading and greater amount of irradiance on the vegetation component. Also, for larger view angles (i.e., off-nadir measurements of canopy response), a smaller proportion of soil background is "seen" by the sensor and the soil-noise problems are reduced.

V MODELING SOIL–VEGETATION SPECTRAL INTERACTIONS

In the preceeding sections, we described how soil-background influences complicate the interpretation of remotely sensed vegetation information. In this section, the interactions and contributions of soil and vegetation in forming composite-canopy response are examined. This requires an understanding of the spectral composition of the irradiance at the soil surface as well as that of the soil-reflected radiance. Differences in red and NIR flux transfers (Kimes et al., 1985; Sellers, 1985; Choudhury, 1987) through a canopy are dependent on the density and optical properties of the overlying vegetation and results in a complex soil–vegetation interaction that makes it difficult to subtract or correct for soil-background influences. A vegetated canopy will scatter and transmit a significant amount of NIR flux toward the soil surface, irradiating the soil underneath as well as in between individual plants. Allen and Richardson (1968) and Lillesaeter (1982) reported that NIR radiation can be transmitted through eight layers of leaves. The soil subsequently reflects part of this transmitted and scattered flux back toward the sensor in a manner dependent upon the optical properties of the soil surface. By contrast, red light is strongly absorbed by the uppermost leaf layers of the canopy, and irradiance at the soil surface is limited to that received directly from the sun and sky through the canopy gaps. Heilman and Kress (1987) and Huete (1987b) investigated the differential rates of radiant flux penetration in incomplete cotton canopies and found the spectral response reflected from the soil surface to mimic that of green vegetation.

Several studies have dealt with radiant flux dynamics in partially vegetated canopies (Jackson et al., 1979; Norman and Welles, 1983; Verhoef, 1984; Kimes et al., 1985). These models are oriented toward predicting canopy reflected flux by characterizing the complex geometrical, structural, and optical leaf properties of a vegetation canopy. The purpose of this section is not to model canopy optical behavior, but to examine radiant transfer at the soil surface. This is done through an anlaysis of the canopy-attenuated spectral irradiance at the soil surface.

V.A Soil- and Vegetation-Component Analysis

Radiant flux transfer through the canopy medium may be evaluated by Lambert-Bouguer's law:

$$E_s(\lambda) = E_o(\lambda) \exp\left[-k(\lambda)m\right] \tag{1}$$

or

$$T(\lambda) = \exp\left[-k(\lambda)m\right] \tag{2}$$

where E_o is the global (diffuse + direct) irradiance entering the canopy medium; E_s is the irradiance at the soil surface after passing through a canopy optical path length m; k is the extinction coefficient; km is the extinction optical thickness (dimensionless); T is the transmittance through the canopy medium; and λ is the wavelength (Iqball, 1983). By using LAI as the optical depth parameter (m), canopy extinction becomes the slope obtained in plotting $-\ln(T)$ against LAI.

A simple first-order interactive model of radiant transfer through a canopy may be used to study soil influences and separate measured canopy response into a vegetation spectral component and a soil-dependent spectral component (Figure 16):

$$D_m(\lambda) = E_o(\lambda)R_c(\lambda) + E_o(\lambda)R_s(\lambda)T^2(\lambda) \tag{3}$$

where the measured canopy signal D_m is equal to the sum of (i) the vegetation component E_oR_c, which is the radiance directly reflected by the vegetation, and (ii) the soil-dependent component $E_oR_sT^2$, which is the product of global irradiance E_o times soil reflectance R_s times the downward and upward global transmittance through the canopy T^2. All radiant energy interacting with the soil background forms the soil-dependent signal reflected from the canopy. The vegetation component represents the reflected canopy radiance that would be measured if a "zero" reflecting soil background were underneath the canopy. This spectral response is free of soil influences, and it may be related to LAI through some constant B that encompasses the scattering coefficients and inclination angles in canopy radiant transfer models.

Equation 3 is based on single-order spectral interactions between soil and vegetation components. Second-order spectral interactions between soil and vegetation would re-

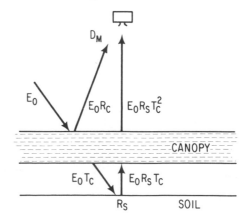

Figure 16 Canopy radiative transfer model involving first-order soil–vegetation spectral interactions. [Huete (1987b). Reproduced from *Agronomy Journal*, Volume 79, pages 61–68 by permission of the American Society of Agronomy, Inc.]

quire an additional product term. Huete (1978b) utilized factor analytic inversion techniques (Malinowski and Howery, 1980) to decompose composite-canopy spectra into soil and vegetation spectral components and to determine the number of significant "order" terms needed to solve Eq. 3. Decomposition of NIR-canopy spectra via eigenfunction analysis resulted in a two-component model that accounted for 99.996% of the spectral variability with a residual standard deviation of 0.37% reflectance. This was comparable with the root-mean-square experimental uncertainty (0.30%) encountered in the ground-based cotton experiment. Despite high NIR flux scattering within vegetation canopies, a two-component (single-order) soil–vegetation model (Eq. 3) was sufficient in explaining soil–vegetation spectral interactions. The suitability of such a model can be seen in Figure 17, where measured NIR-canopy reflectance (%) is plotted as a function of soil-background reflectance (%). The slopes represent the two-way global-canopy transmittance terms in Eq. 3, and the intercepts are equal to the vegetation responses free of soil influences (zero soil reflectance). The slopes, which range from 1 (bare soil; 100% transmittance) to 0 (dense vegetation; no transmittance), represent the limiting conditions.

The spectral composition of soil-surface irradiance and soil- and vegetation-reflected radiances (normalized by the irradiance at the top of the canopy) is derived through factor decomposition of each wavelength band in accordance with Eq. 3. The transmitted canopy flux reaching the soil surface through the various canopy densities resembled the spectral signature of green vegetation, Figure 18(a). With a 90% green cover, only 10% of the canopy incident red light reached the soil surface as compared to 38% NIR flux penetration. Multiplication of the square of canopy transmittance by the reflectance of the underlying soil, $R_s T^2$, produces the soil-component spectral contribution toward composite response, Figure 18(b). The upper spectral curve is the bare-soil reflectance for the light-colored Superstition sand. The presence of the green canopy lowers the soil-component response; however, the lower soil responses become wavelength-dependent as the red response decreases more, relative to NIR response. The soil-component spectral curves consequently took on the optical properties of both the soil and overlying vegetation. The vegetation-component response toward canopy reflectance, Figure 18(c),

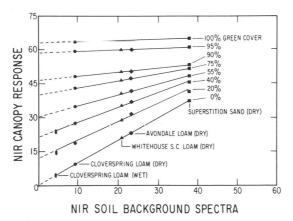

Figure 17 Relationship between NIR-canopy reflectance and underlying bare-soil NIR reflectance for various levels of percent green cotton cover. [Huete (1987b). Reproduced from *Agronomy Journal*, Volume 79, pages 61–68 by permission of the American Society of Agronomy, Inc.]

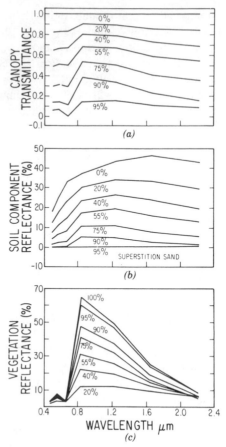

Figure 18   Spectral curves for canopy transmittance and decomposed soil- and vegetation-component spectra for various levels of green cover. [Huete (1987b). Reproduced from *Agronomy Journal*, Volume 79, pages 61–68 by permission of the American Society of Agronomy, Inc.]

represents vegetation spectral signatures free from soil-background influences (zero soil intercepts). The addition of any vegetation spectral curve, Figure 18(*c*), with the corresponding soil-component curve, Figure 18(*b*), reproduces measured spectral response to within experimental error.

The decomposed soil- and vegetation-component responses may then be inserted into the various spectral indices to determine soil-component influences. An optimal vegetation index would be invariant to the soil-dependent component, yet remain sensitive to the vegetation component. The RVI and PVI values for the separated component spectra as well as for the experimental spectra are plotted using light- and dark-soil backgrounds (Figure 19). The soil-filtered vegetation component is linearly related to the RVI and PVI values. The RVI values for the soil component, however, are only invariant at low green covers, Figure 19(*a*). With increased amounts of vegetation, soil-component RVI values increased curvilinearly and introduced a soil-dependent RVI signal that mixed with that from the vegetation component. The composite canopy curve, situated between the two separated component curves, increased curvilinearly as a result of the soil influence. The soil-component RVI values for the darker soil is greater than for the lighter soil, resulting in the soil-brightness influence in which darker soils produced higher canopy RVI values.

The decomposed PVI curves differed from the RVI curves in that composite-canopy

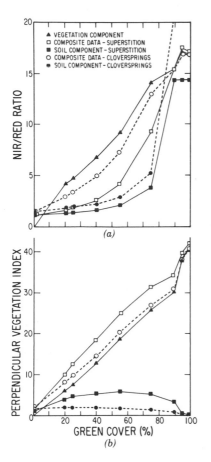

Figure 19 NIR/red ratio and perpendicular vegetation index values for composite and decomposed soil- and vegetation-components for two soil backgrounds at various levels of green cover. [Huete (1987b). Reproduced from *Agronomy Journal*, Volume 79, pages 61–68 by permission of the American Society of Agronomy, Inc.]

values were the sum of the PVI values of the two separated components, Figure 19(*b*). The orthogonal-based spectral indices assume additive mixing of fractional amounts of soil and vegetation spectra. Consequently, the PVI of canopy spectra was always higher than that of either component. Furthermore, soil-component PVI values were higher with lighter-soil backgrounds, causing a soil-brightness effect whereby brighter soil produced higher PVI values for the same density of vegetation.

Thus, for both the ratio- and orthogonal-based spectral indices, the index (greenness) signal from the soil component is indistinguishable from that of green vegetation. This spectral component renders these vegetation indices "soil-dependent" because the magnitude of the index signal it produces varies with the reflectance properties of the underlying soil. Furthermore, soil-component spectral index values increase with green cover until an intermediate vegetation density is attained. At low-vegetation densities, there is not enough vegetation to impart a soil-dependent (canopy-scattered) index signal, whereas at very high densities, the soil response is too attenuated to be of significance. At intermediate levels of vegetation, however, there is significant scattering and transmission of NIR flux through the canopy in relation to absorbed visible and middle-infrared flux to produce soil-component spectra strongly resembling vegetation spectra.

The separated soil components for the 40% green canopy at two solar zenith angles are shown in Figure 20 for the bright Superstition sand and dark Cloversprings loam

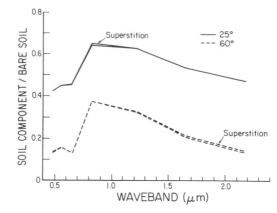

Figure 20 Normalized soil-component response at two solar zenith angles from a 40% cotton canopy with Superstition sand and Cloversprings loam backgrounds. [From Huete (1987a).]

backgrounds. The soil-reflected responses are normalized by their respective bare-soil reflectances to highlight the vegetation-imparted signal in the soil component. The top two curves represent the "normalized" soil response at a 25° solar zenith angle (sunlit soil), and the two lower curves are normalized responses for 60° solar zenith angle (shaded soil). At both sun angles, the soil-reflected contribution toward composite-canopy response resembles green vegetation.

V.B Correcting for Soil Background

The removal of soil-background effects on measured canopy spectra may be accomplished with (1) factor-analytic inversion models as described before, (2) canopy radiant transfer models that require specific input parameters, and (3) spectral vegetation indices that require minimal information. The factor-analysis techniques are most suited to multitemporal data sets, where the primary source of temporal variance is due to vegetation dynamics with soils primarily varying spatially. Since soil–vegetation interactions are wavelength-dependent, factor decomposition must be accomplished one band at a time. Factor decomposition, or principal-components analysis of individual scenes, does not remove the soil-dependent "greenness" signal, as this technique assumes simple additive mixing of ground-reflecting features (Huete, 1986).

The application of canopy radiant transfer models to incomplete canopy situations generally requires prior information on soil reflectance properties as well as leaf optical and plant structural properties. These models, thus far, have limited applications in "global" terrestrial ecosystem studies. Vegetation spectral indices are more versatile, require minimal information for their use, and are widely used in most vegetation-related studies. Unfortunately, their sensitivity to soil background, sun angle, and atmosphere condition limit their accuracy and reliability. However, soil influences were shown to be greatly reduced with the SAVI (Figures 12 and 13). The SAVI recognizes first-order soil–vegetation interactions by allowing for vegetation isoline slope changes (in NIR–red wavelength space) that match those obtained with experimental soil–grass and soil–cotton spectral mixtures. If these isoline slope changes can be physically related to canopy radiant transfer processes, then the complexity in modeling soil–vegetation dynamics may be considerably reduced.

The slope and intercept of vegetation isolines in wavelength space are related to the optical and structural properties of the canopy medium. Utilizing Eqs. 2 and 3, one can derive the slope behavior of isolines in NIR–red wavelength space:

$$D_m(\lambda)/E_o(\lambda) = R_m(\lambda) = B(\lambda)\text{LAI} + R_s(\lambda) \exp\left\{-2k(\lambda)\text{LAI}\right\} \qquad (4)$$

where measured canopy reflectance ($R_m = D_m/E_o$) is the sum of the vegetation- and soil-dependent components. Combining the red and NIR versions of Eq. 4 with a soil-line equation (Eq. 5) results in an equation for the slope of vegetation isolines (M_{vi}) in NIR–red wavelength space:

$$r_{s(\text{NIR})} = r_{s(\text{red})}M_s + I_s \qquad (5)$$

and

$$M_{vi} = M_s \exp\left[2(k_{\text{red}} - k_{\text{NIR}})\text{LAI}\right] \qquad (6)$$

M_s and I_s are the slope and NIR intercept, respectively, of the soil-line equation. Equation 6 shows that the slope of a vegetation isoline is dependent on the slope of the soil line, LAI, and the difference in red- and NIR-canopy extinction. The following generalizations can be made:

1. If LAI = 0, Eq. 6 reduces to the soil-line slope.
2. If $k_{\text{red}} = k_{\text{NIR}}$, the slope of the isolines remain constant (parallel) and equal to the soil-line slope.
3. If $k_{\text{red}} > k_{\text{NIR}}$ (a photosynthetically active canopy), the flux transfer difference is positive and the slope of the isoline becomes greater than the soil-line slope.

Vegetation isolines increase in slope with increasing density because the soil reflects a high amount of vegetation-scattered NIR flux whereas red light is mostly absorbed by the canopy medium. The extinction (k) values for the cotton experimental data set utilized here were 0.46 for red light and 0.23 for NIR flux (Figure 21). Measured canopy

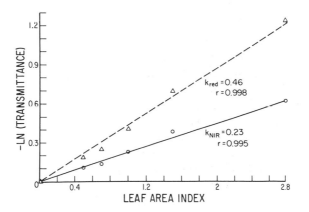

Figure 21 Relationship among cotton-canopy transmittance, extinction, and leaf-area index in the red and NIR wavelengths. [Reprinted by permission of the publisher from Huete (1988). Copyright 1988 by Elsevier Science Publishing Co., Inc.]

response could be corrected for soil background with the use of Eq. 4, however, the reflectance of the soil and LAI would have to be known.

An alternative approach, which would not require prior knowledge of soil background or LAI, is to graphically model vegetation isoline behavior (Figure 9). Vegetation isolines do not actually converge to one common point (Figure 5), as convergence is partly a function of vegetation density (LAI). However, a first-order approximation may be obtained by shifting the origin to a point where intermediate densities of vegetation converge with the bare-soil line. Inserting the k values from Figure 21 into Eqs. 4–6 results in a vegetation isoline slope (LAI = 1) that converges with the soil line at $(-0.26, -0.24)$. This results in approximately equal x- and y-axis shifts (0.25 reflectance units). Shifting the origin negative values is equivalent to adding the constants l_1 and l_2 to the entire red and NIR reflectance data set. By utilizing the NDVI format, a soil-adjusted version (SAVI) is derived as follows:

$$[(NIR + l_2) - (red + l_1)]/[(NIR + l_2) + (red + l_1)] \tag{7}$$

which simplifies to

$$(NIR - red)/(NIR + red + 0.5) \tag{8}$$

In order to maintain the bounded conditions of the NDVI equation (NDVI can vary from -1 to $+1$), Eq. 8 is multiplied by 1.5:

$$SAVI = [(NIR - red)/(NIR + red + 0.5)]1.5 \tag{9}$$

A sensitivity analysis by Huete (1988) revealed that a substantial reduction in soil influences may be attained for a wide range in shifted origins. Soil influences were effectively minimized in both cotton (planophile) and range grass (erectophile) isolines (Figure 12). Consequently, only the relative difference in spectral extinction through a canopy is needed to describe and model soil-brightness effects. Data on spectral radiant flux transfers through vegetation biomes, however, are not readily available.

VI   SUMMARY AND CONCLUSIONS

Remote sensing of agricultural, forest, and rangelands frequently involves the measurement of reflected energy from two or more components (vegetation, soil, atmosphere, etc.) in the presence of each other. The development and usefulness of vegetation spectral models, transformations, and indices are dependent upon the degree to which the spectral contribution of nonvegetation components can be isolated from measured canopy response. Although vegetation indices have been widely shown to be valuable "global" tools in the measurement and interpretation of vegetation conditions, several limitations have been identified of which two are soil-related problems. These two factors are the focus of this chapter and they are (1) soil-brightness effects and (2) secondary soil spectral deviations that protrude into spectral vegetation measures.

Detection of low amounts of vegetation is restricted by soil spectral characteristics that partially overlap the spectral behavior of vegetation. The "width" of the global soil line inhibits the discrimination of low amounts of vegetation, since their respective pixels occupy similar positions in spectral index space. Soils with high chromas, such as red- or yellow-colored soils, often deviate the most from the soil line and project themselves into "green vegetation" space. These secondary variations are partly responsible for the soil "artifact" areas appearing in NDVI imagery of arid and semiarid portions of Africa, especially when the role of the atmosphere in enhancing soil problems is considered. Although minimal over NIR–red wavelengths, they become greater in magnitude with the inclusion of more wavelengths. Soil spectral deviations affect all spectral indices, including the NDVI, PVI, and SAVI. The spectra of the "global" soils data set resulted in a range of SAVI values equivalent to a green cotton cover of 9%. The respective index values for the NDVI, PVI, and four-band GVI were 17, 11, and 19% green cover. This soil problem is not easily removed without prior knowledge of specific soil optical properties or prior stratification of soil spectral units. Canopy radiant transfer models are also affected since they also make use of reference soil spectra and generalized soil spectral behavior.

In arid and semiarid ecosystems, numerous soils may be present in complex spatial patterns and the vegetation densities may be also low. Soil-noise limitations are far more crucial to these areas than to agricultural ecosystems, where the vegetation signals are high and the soil surfaces are generally more uniform. However, in agricultural areas, varying soil-moisture conditions and tillage practices generate soil spectral variations that may limit vegetation studies. The use of site-specific soil lines reduces soil-moisture influences and may also minimize variations attributed to roughness and tillage operations. In either case, the development and use of appropriate soil lines or soil reference spectra will depend on the desired intensity of vegetation analysis or level of vegetation classification.

Secondary soil variations are minor relative to soil-brightness effects. However, they are more difficult to model since they depend not on macroscopic radiant energy interactions (which are responsible for most soil-brightness variations), but depend instead on microscopic (absorption-related) energy interactions related to specific soil mineralogic, organic, and water-absorption features. Since this falls in the realm of spectroscopy, it is hoped that future sensors that contain more wavelengths to characterize soils will enable such spectroscopic effects to be modeled and understood.

Vegetation spectral indices and models generally assume that soil influences decrease with vegetation development, since the relative fraction of soil exposed is reduced. Consequently, soil-background influences on canopy reflectance would approach a maximum level at low vegetation densities. However, as illustrated in this chapter, the most important soil influence on spectral vegetation measures is the soil-brightness effect, which is most pronounced at intermediate levels of vegetation density due to first-order soil–vegetation spectral interactions. Differential transmission of spectral fluxes through a canopy imparts a soil-component signature that resembles that spectral signature of green vegetation. Since the soil-component signal also varies with the optical properties of the soil background, a soil-brightness influence occurs as lighter or darker soils reflect different amounts of canopy-transmitted signal back to the sensor. As the soil background becomes brighter, much more of the vegetation transmitted/scattered NIR signal is reflected in relation to the nontransmitted red light.

Soil–vegetation interactions may be studied via factor–analytic inversion models, which allow composite canopy spectra of soil–plant mixtures to be separated into soil and vegetation component spectra. The soil-component reflected flux as well as the canopy transmitted flux strongly resembles the spectral response of green vegetation. Jordan

(1969) found the spectral composition of radiant flux at the forest floor to be more useful in assessing leaf-area index than the spectral response measurements made above the canopy. The spectral component separation shown here documents how soil and vegetation mix to produce canopy spectra and also shows how the soil signal mixes into spectral indices to create the soil-brightness problem. The factor models verify that first-order soil–vegetation interactions are adequate in describing experimental canopy measurements. Thus, the spectral response from a partially vegetated canopy consists of a primary vegetation-reflected response and a secondary vegetation-backscattered response (vegetation-scattered, soil-reflected). Soil-correction methodologies would have to be based on removing this secondary vegetation signal. Since plants are efficient NIR scatterers, there is very little "pure" soil response reflected from pixel-size areas as this secondary vegetation signal would be fairly ubiquitous in remote-sensing imagery.

Soil-brightness effects complicate the study of atmosphere and sun-angle influences on vegetation spectra. As shown here, similar vegetation densities over certain soils are minimally degraded by the atmosphere, whereas the same vegetation amounts over other soils become very sensitive to atmosphere condition. Similarly, diurnal canopy responses are also dependent on the underlying soil background. The removal of soil-brightness effects should facilitate atmosphere and sun-angle correction techniques.

The SAVI is only a preliminary step in modeling soil–vegetation interactions at the "global" level, i.e., without detailed input canopy parameters. The SAVI is also a method by which spectral indices may be refined or "calibrated" so that soil-substrate variations are effectively normalized and they are not spectrally influencing vegetation measures. Graphically, the transformation involves a shifting of the origin of spectra plotted in NIR–red wavelength space to account for first-order soil–vegetation interactions and differential red and NIR extinction through vegetated canopies. Although the SAVI successfully minimized soil-brightness problems in the two architecturally different canopies tested here (broad-leaf cotton and narrow-leaf grass) and in various canopy models, further field studies would be needed to ensure SAVI applicability on a global basis. Radiant flux transfers through the natural ecosystems involve a mixture of several plant species. The arrangement of different plant species, along with their morphologies, architectures, and densities will determine the composition and magnitude of soil-surface irradiance and hence the contribution and influence of the soil-dependent canopy signal. The limitations of the SAVI thus need to be analyzed over several vegetation biomes and agricultural situations. However, since the SAVI minimizes soil-induced variations, canopy response sensitivity to sun angle and atmosphere is not complicated by soil-background variations.

It is hoped that factor variation studies and spectral separation studies, as presented here, will aid in the development and refinement of spectral models, so that they can adequately describe the spectral dynamics of Earth surface components. The soil background has been shown to be a major surface component controlling the spectral behavior of all but the most dense vegetation canopies. Since soil surfaces are temporally dynamic (because of moisture, cultivation, erosion, organic inputs), an understanding of soil spectral properties, its behavior and interactions with plant life and water, is crucial to the development of global Earth models.

REFERENCES

Allen, W. A., and A. J. Richardson (1968). Interaction of light with a plant canopy. *J. Opt. Soc. Am.* **58**:1023–1028.

Angstrom, A. (1925). The albedo of various surfaces of ground. *Geogr. Ann.* **7**:323–342.

Ashley, M. D., and J. Rea (1975). Seasonal vegetation differences from ERTS imagery. *Photogramm. Eng. Remote Sens.* **41**: 713–719.

Asrar, G., M. Fuchs, E. T. Kanemasu, and J. L. Hatfield (1984). Estimating absorbed photosynthetic radiation and leaf area index from spectral reflectance in wheat. *Agron. J.* **76**:300–306.

Bowers, S. A., and R. J. Hanks (1965). Reflection of radiant energy from soils. *Soil Sci.* **100**:130–138.

Chance, J. E. (1977). Applications of Suits spectral model to wheat. *Remote Sens. Environ.* **6**:147–150.

Choudhury, B. J. (1987). Relationship between vegetation indices, radiation absorption, and net photosynthesis evaluated by a sensitivity analysis. *Remote Sens. Environ.* **22**:209–233.

Colwell, J. E. (1974). Vegetation canopy reflectance. *Remote Sens. Environ.* **3**:175–183.

Condit, H. R. (1970). The spectral reflectance of American soils. *Photogramm. Eng.* **36**:955–966.

Coulson, K. L., and D. W. Reynolds (1971). The spectral reflectance of natural surfaces, *J. Appl. Meteorol.*, **10**:1285.

Crist, E. P., and R. C. Cicone (1984). Application of the tasseled cap concept to simulated thematic mapper data. *Photogramm. Eng. Remote Sens.* **50**:343–352.

Elvidge, C. D., and R. J. P. Lyon (1985). Influence of rock-soil spectral variation on assessment of green biomass. *Remote Sens. Environ.* **17**:265–279.

Fraser, R. S., and Y. J. Kaufman (1985). Relative importance of aerosol scattering and absorption in remote sensing. *IEEE Trans. Geosci. Remote Sens.* **GE-23**:625–633.

Fukuhara, M., S. Hayashi, Y. Yasuda, et al. (1979). Extraction of soil information from vegetated areas. In *Machine Processing of Remotely Sensed Data*, Purdue University, West Lafayette, Indiana, pp. 242–252.

Hatfield, J. L., G. Asrar, and E. T. Kanemasu (1984). Intercepted photosynthetically active radiation estimated by spectral reflectance. *Remote Sens. Environ.* **14**:65–75.

Heilman, J. L., and M. R. Kress (1987). Effects of vegetation on spectral irradiance at the soil surface. *Agron J.* **79**:765–768.

Herman, B. M., and S. R. Browning (1975). The effect of aerosols on the earth-atmosphere albedo. *J. Atmos. Sci.* **30**:1430–1445.

Huete, A. R. (1986). Separation of soil-plant spectral mixtures by factor analysis. *Remote Sens. Environ.* **19**:237–251.

Huete, A. R. (1987a). Soil and sun angle interactions on partial canopy spectra. *Int. J. Remote Sens.* **8**:1307–1317.

Huete, A. R. (1987b). Soil-dependent spectral response in a developing plant canopy. *Agron. J.* **79**:61–68.

Huete, A. R. (1988). A soil adjusted vegetation index (SAVI). *Remote Sens. Environ.* **25**:295–309.

Huete, A. R., and R. D. Jackson (1987). The suitability of spectral indices for evaluating vegetation characteristics on arid rangelands. *Remote Sens. Environ.* **23**:213–232.

Huete, A. R., and R. D. Jackson (1988). Soil and atmosphere influences on the spectra of partial canopies. *Remote Sens. Environ.* **25**:89–105.

Huete, A. R., D. F. Post, and R. D. Jackson (1984). Soil spectral effects on 4-space vegetation discrimination. *Remote Sens. Environ.* **15**:155–165.

Huete, A. R., R. D. Jackson, and D. F. Post (1985). Spectral response of a plant canopy with different soil backgrounds. *Remote Sens. Environ.* **17**:37–53.

Iqball, M. (1983). *An Introduction to Solar Radiation*. Academic Press, New York.

Jackson, R. D. (1983). Spectral indices in n-space. *Remote Sens. Environ.* **13**:409–421.

Jackson, R. D., R. J. Reginato, P. J. Pinter, Jr., and S. B. Idso (1979). Plant canopy information extraction from composite scene reflectance of row crops. *Appl. Opt.* **18**:3775–3782.

Jackson, R. D., P. J. Pinter, Jr., R. J. Reginato, and S. B. Idso (1980). *Hand-held Radiometry*, Agric. Rev. Manuals W-19. U.S. Department of Agriculture, Science and Education Administration, Oakland, California.

Jackson, R. D., P. N. Slater, and P. J. Pinter, Jr. (1983). Discrimination of growth and water stress in wheat by various vegetation indices through clear and turbid atmospheres. *Remote Sens. Environ.* **13**:187–208.

Jordan, C. F. (1969). Derivation of leaf-area index from quality of light on the forest floor, *Ecology* **50**:663–666.

Kauth, R. J., and G. S. Thomas (1976). The tasseled cap-a graphic description of the spectral-temporal development of agricultural crops as seen by Landsat. In *Machine Processing of Remotely Sensed Data*. Purdue University, West Lafayette, Indiana, pp. 41–51.

Kimes, D. S., J. A. Smith, and K. J. Ranson (1980). Vegetation reflectance measurements as a function of solar zenith angle. *Photogramm. Eng. Remote Sens.* **46**:1563.

Kimes, D. S., J. M. Norman, and C. L. Walthall. (1985). Modeling the radiant transfers of sparse vegetation. *IEEE Trans. Geosci. Remote Sens.* **GE-23**:695–704.

Kollenkark, J. C., C. S. T. Daughtry, M. E. Bauer, and T. L. Housley (1982a). Effects of cultural practices on agronomic and reflectance characteristics of soybean canopies. *Agron J.* **74**:751–758.

Kollenkark, J. C., V. C. Vanderbilt, C. S. T. Daughtry, and M. E. Bauer (1982b). Influence of solar illumination angle on soybean canopy reflectance. *Appl. Opt.* **21**:1179.

Kondratyev, K. Ya., L. N. Dyachenko, and N. P. Piatovskaya (1974). On the relationship between the earth-atmosphere system albedo and the earth's surface albedo. In *Earth Survey Problems*. Akademie-Verlag, Berlin, pp. 473–482.

Lillesaeter, O. (1982). Spectral reflectance of partly transmitting leaves: Laboratory measurements and mathematical modeling. *Remote Sens. Environ.* **12**:247–254.

Malinowski, E. R., and D. G. Howery (1980). *Factor Analysis in Chemistry*. Wiley, New York.

Miller, G. P., M. Fuchs, M. J. Hall, G. Asrar, E. T. Kanemasu, and D. E. Johnson (1984). Analysis of seasonal multispectral reflectances of small grains. *Remote Sens. Environ.* **14**:153–167.

Norman, J. M., and J. M. Welles (1983). Radiative transfer in an array of canopies. *Agron. J.* **75**:481–488.

Otterman, J., and R. S. Fraser (1976). Earth-atmosphere system and surface reflectivities in arid regions from Landsat MSS data. *Remote Sens. Environ.* **5**:247–266.

Pearson, R. L., C. J. Tucker, and L. D. Miller (1976). Spectral mapping of short grass prairie biomass. *Photogramm. Eng. Remote Sens.* **42**:317–324.

Pinter, P. J., Jr., R. D. Jackson, C. E. Ezra, and H. W. Gausman (1985). Sun-angle and canopy-architecture effects on the spectral reflectance of six wheat cultivars. *Int. J. Remote Sens.* **6**:1813.

Pitts, D. E., W. E. McAllum, and A. E. Dillinger (1974). The effect of atmospheric water vapor on automatic classification of ERTS data. *Proc. 9th Int. Symp. Remote Sens. Environ.*, University of Michigan, Ann Arbor, pp. 483–497.

Rao, V. R., E. J. Brach, and A. R. Mack (1979). Bidirectional reflectance of crops and the soil contribution. *Remote Sens. Environ.* **8**:115–125.

Richardson, A. J., and C. L. Wiegand (1977). Distinguishing vegetation from soil background information. *Photogramm. Eng. Remote Sens.* **43**:1541–1552.

Ripple, W. J. (1985). Asymptotic reflectance characteristics of grass vegetation. *Photogramm. Eng. Remote Sens.* **51**:1915–1921.

Sellers, P. J. (1985). Canopy reflectance, photosynthesis and transpiration. *Int. J. Remote Sen.* **6**:1335–1372.

Slater, P. N. (1980). *Remote Sensing: Optics and Optical Systems.* Addison-Wesley, Reading, Massachusetts, p. 575.

Slater, P. N., and R. D. Jackson (1982). Atmospheric effect on radiation reflected from soil and vegetation as measured by orbiting sensors using various scanning directions. *Appl. Opt.* **21**:3923–3931.

Stoner, E. R., and M. F. Baumgardner (1981). Characteristic variations in reflectance of surface soils. *Soil Sci. Soc. Am. J.* **45**:1161–1165.

Suits, G. H. (1972). The calculation of directional reflectance of a vegetative canopy. *Remote Sens. Environ.* **2**:117–125.

Thompson, D. R., and O. A. Wehmanen (1980). Using Landsat digital data to detect moisture stress in corn-soybean growing regions. *Photogramm. Eng. Remote Sens.* **46**:1087–1093.

Tucker, C. J. (1977). Asymptotic nature of grass canopy spectral reflectance. *Appl. Opt.* **6**:1151–1156.

Tucker, C. J. (1979). Red and photographic infrared linear combinations for monitoring vegetation. *Remote Sens. Environ.* **8**:127–150.

U.S. Department of Agriculture (1975). *Soil Taxonomy*, Agric. Handb. No. 436. USDA, Soil Survey Staff, Washington, D.C.

Verhoef, W. (1984). Light scattering by leaf layers with application to canopy reflectance modeling: the SAIL model. *Remote Sens. Environ.* **16**:125–141.

Weiser, R. L., G. Asrar, G. P. Miller, and E. T. Kanemasu (1984). Assessing grassland biophysical characteristics from spectral measurements. *Proc. 10th Int. Symp. Mach. Process. Remote Sens. Data*, Purdue University, West Lafayette, Indiana, pp. 357–361.

Wiegand, C. L., and A. J. Richardson (1982). Comparisons among a new soil index and other two- and four-dimensional vegetation indices. *Proc. 48th Annu. Meet. Am. Soc. Photogramm.*, pp. 211–227.

5

THE THEORY OF PHOTON TRANSPORT IN LEAF CANOPIES

RANGA B. MYNENI

Institut für Bioklimatologie
George-August-Universität Göttingen
Göttingen, Federal Republic of Germany

GHASSEM ASRAR

Headquarters
National Aeronautics and Space Administration
Washington, D.C.

and

EDWARD T. KANEMASU

Evapotranspiration Laboratory
Kansas State University
Manhattan, Kansas

I INTRODUCTION

The computation of the radiant-energy distribution in leaf canopies has two principal motivations. First, the leaf-projected scalar fluxes are needed to compute canopy photosynthetic rates (deWit, 1965; Monteith, 1965; Duncan et al., 1967; Gutschick and Weigel, 1984; amongst many others) and the distribution of the angular flux escaping the canopies is needed for remote-sensing applications (Gerstl and Simmer, 1986). Second, the inversion of the mathematical formulations that predict emergent radiation intensities allows estimation of canopy attributes such as green leaf-area index and absorbed radiant energy, from remote measurements of canopy reflectance (Goel and Strebel, 1983; Asrar et al., 1984).

Past mathematical formulations developed for the canopy problem differ considerably in details, depending primarily on the purpose for which they were formulated. If one is interested solely in computing canopy photosynthetic rates, then leaves may be considered black, since leaves scatter less than 10% of the incident photosynthetically active radiation. With such a simplification, it is straightforward to compute the uncollided propagation of direct sunlight and diffuse skylight into the canopy and, in conjunction with a leaf light-photosynthesis response function, canopy photosynthetic rates may be calculated. The topic will not be further pursued here; instead, we refer the interested reader to the excellent paper by Norman (1980), where a step-by-step discussion of the computational procedure is given.

With recent technological advances in spectral and spatial resolution of space-borne radiometric sensors, and with continuing refinement of observational techniques for greater measuremental accuracy, the problem of quantitatively assessing the quality of remotely sensed radiometric data gathered over agricultural and natural vegetated surfaces has drawn considerable attention in recent years. Specifically, in the remote sensing of vegetation canopies with sensors that are highly sensitive in narrow spectral bands and with smaller footprints, it becomes increasingly important to understand the physics of photon propagation through the vegetated canopies. Hence, an accurate solution of the transport equation is sought in studies concerning applications in remote sensing.

The general plan of this chapter is as follows. In Section II, we derive the governing equation for photon transport in leaf canopies and show that it reduces to the standard form only under certain idealistic conditions. In Section III, the G function and the scattering transfer functions [in both one and two angles] are derived and explicit analytical results for the scattering kernel are presented. In Sections IV, V, and VI, exact analytical, numerical, and approximate methods of solving the transport equation will be discussed in detail. The emphasis in these sections is on a concise presentation of the method, and no effort is made to reference all the important papers dealing with that particular method. Instead, a conscious effort is made to refer to papers that detail the method and that should help the reader if further reading is necessary.

II THE TRANSPORT EQUATION FOR LEAF CANOPIES

Consider a flat horizontal leaf canopy of depth T, which is perpendicular to the z axis and which is illuminated spatially uniformly on top. Then, the steady-state radiance distribution function [or the *specific energy intensity*] $I(z, \Omega)$ at any given wavelength in the absence of polarization, frequency-shifting interactions, and radiation sources in the canopy is given by the transport equation in plane geometry (Davison, 1958),

$$-\mu \frac{\partial I}{\partial z}(z, \Omega) + \sigma_e(z, \Omega) I(z, \Omega)$$
$$= \int_{4\pi} d\Omega' \, \sigma_s(z, \Omega' \rightarrow \Omega) I(z, \Omega') \qquad 0 < z < T \qquad (1)$$

where σ_e is the *extinction coefficient*, and σ_s is the *differential scattering coefficient* for photon scattering from direction Ω' into a unit solid angle about direction Ω. The *unit vector* $\Omega(\mu, \phi)$ has an *azimuth angle* ϕ and a *polar angle* $\theta = \cos^{-1} \mu$ with respect to the outward normal.

Equation 1 has been studied extensively for the case of constant σ_e and σ_s. Many numerical approximations have been used to obtain numerical solutions (Chandrasekhar, 1960; Duderstadt and Martin, 1979). However, in the canopy problem, σ_e is not necessarily independent of the direction of photon travel and σ_s is generally *not rotationally invariant* (i.e., it generally depends upon the absolute directions of photon travel Ω' and

Ω and not just on the scattering angle, $\cos^{-1}(\Omega.\Omega')$, as is usually the case). Hence, the standard methods of treating Eq. 1 for most neutron and photon-transport problems must be somewat modified, or certain approximations must be made about both the canopy and Eq. 1 to justify the standard treatments. In this section, we identify those features of the photon-transport equation for a leaf canopy that requires special attention.

To obtain an explicit expression for the extinction coefficient σ_e, assume that the canopy consists of plane leaves with a *leaf-area density* $u_L(z)$, defined to be the total one-sided leaf area per unit volume of the canopy at depth z. Although the leaf-area density may vary with depth z, we assume that there is no lateral variation (as implied by the plane symmetry assumption inherent in Eq. 1). The probability that a leaf has a normal $\Omega_L(\mu_L, \phi_L)$ (directed away from the top surface) in a unit solid angle about Ω_L is given by the *leaf normal distribution function* $g_L(z, \Omega_L)$, which is normalized as

$$\frac{1}{2\pi} \int_0^{2\pi} d\phi_L \int_0^1 d\mu_L \, g_L(z, \Omega_L) = 1 \tag{2}$$

Here, all leaves are assumed to face upwards, so that all normals are confined to the upper hemisphere.

Now consider photons at a depth z traveling in direction Ω. The extinction coefficient is then the probability, per unit pathlength of travel, that a photon hits a leaf, i.e., the probability that a photon while traveling a distance ds along Ω is intercepted by a leaf divided by the distance ds. Mathematically,

$$\sigma_e(z, \Omega) = G(z, \Omega) u_L(z) \tag{3}$$

where the *geometry factor* $G(z, \Omega)$ is the fraction of total leaf area, per unit volume of the canopy, that is perpendicular to Ω (Ross, 1981), namely,

$$G(z, \Omega) = \frac{1}{2\pi} \int_0^{2\pi} d\phi_L \int_0^1 d\mu_L \, g_L(z, \Omega_L) |\Omega_L.\Omega| \tag{4}$$

The differential scattering coefficient $\sigma_s(z, \Omega' \to \Omega)$ may be expressed in terms of the *leaf scattering distribution function* $f(\Omega' \to \Omega; \Omega_L)$. For a leaf with outward normal Ω_L, this transfer function is the fraction of intercepted energy (from photons initially traveling in direction Ω') that is reradiated (scattered) into a unit solid angle about direction Ω. Since the rate at which energy in direction Ω' is intercepted by leaves of orientation Ω_L, per unit volume at depth z, is $u_L(z) |\Omega_L.\Omega'| I(z, \Omega')$, the rate at which photons traveling in all initial directions are scattered into a unit solid angle about Ω is

$$\int_{4\pi} d\Omega' \, u_L(z) |\Omega_L.\Omega'| f(\Omega' \to \Omega; \Omega_L) I(z, \Omega') \tag{5}$$

Finally, integration of this expression over all leaf orientations, weighted by $g_L(z, \Omega_L)$, gives the total volumetric rate at which photons are scattered by leaves of all orientation into a unit solid angle about Ω as

$$\frac{u_L(z)}{2\pi} \int_{2\pi} d\Omega' \left[\int_{2\pi} d\Omega_L \, g_L(z, \Omega_L) |\Omega_L.\Omega'| f(\Omega' \to \Omega; \Omega_L) \right] I(z, \Omega') \tag{6}$$

This quantity must equal the right-hand-side of Eq. 1, and, thus, the differential scattering coefficient $\sigma_s(z, \Omega' \rightarrow \Omega)$ is given by

$$\sigma_s(z, \Omega' \rightarrow \Omega) = \frac{u_L(z)}{2\pi} \int_{2\pi} d\Omega_L \, g_L(z, \Omega_L) \, |\Omega_L . \Omega'| \, f(\Omega' \rightarrow \Omega; \Omega_L) \quad (7)$$

In general, it is seen that the scattering coefficient depends on both the initial and scattered photon directions. However, under various approximations for the leaf normal distribution functions $g_L(z, \Omega_L)$ and the leaf scattering distribution functions $f(\Omega' \rightarrow \Omega; \Omega_L)$, σ_s may depend on fewer variables (see the next section).

If it now is assumed that the angular distribution of leaves, $g_L(z, \Omega_L)$, is independent of depth z, then the photon-transport equation, Eq. 1, can be written as

$$-\mu \frac{\partial I}{\partial \tau}(\tau, \Omega) + G(\tau, \Omega) I(\tau, \Omega) = \frac{1}{\pi} \int_{4\pi} d\Omega' \, \Gamma(\Omega' \rightarrow \Omega) I(\tau, \Omega') \quad (8)$$

where the *optical depth* $\tau(z)$ is defined as

$$\tau(z) \equiv \int_0^z dz \, u_L(z) \quad (9)$$

and the *area scattering transfer function* $\Gamma(\Omega' \rightarrow \Omega)$ as

$$\frac{1}{\pi} \Gamma(\Omega' \rightarrow \Omega) \equiv \frac{1}{2\pi} \int_{2\pi} d\Omega_L \, g_L(\Omega_L) \, |\Omega_L . \Omega'| \, f(\Omega' \rightarrow \Omega; \Omega_L) \quad (10)$$

The leaf scattering transfer function $f(\Omega' \rightarrow \Omega; \Omega_L)$, when integrated over all exit photon directions, yields the *single-scatter albedo* (per unit leaf area), ω, i.e.,

$$\int_{4\pi} d\Omega \, f(\Omega' \rightarrow \Omega; \Omega_L) = \omega \quad (11)$$

In general, ω may depend on both the initial photon direction Ω' and the leaf orientation Ω_L; however, for many scattering models, ω is independent of Ω' and Ω_L, so that from Eqs. 4 and 11, it is seen that $\Gamma(\Omega' \rightarrow \Omega)$ is normalized as

$$\frac{1}{\pi} \int_{4\pi} d\Omega \, \Gamma(\Omega' \rightarrow \Omega) = \omega G(\Omega') \quad (12)$$

Under this restrictive condition for $f(\Omega' \rightarrow \Omega; \Omega_L)$, the photon-transport equation, Eq. 8, may be simplified by defining a *normalized transfer function*,

$$P(\Omega' \rightarrow \Omega) = 4 \frac{\Gamma(\Omega' \rightarrow \Omega)}{\omega G(\Omega')} \quad (13)$$

which is normalized to unity, i.e.,

$$\frac{1}{4\pi} \int_{4\pi} d\Omega \, P(\Omega' \to \Omega) = 1 \tag{14}$$

With this transfer function, the photon-transport equation, Eq. 8, becomes

$$-\mu \frac{\partial I}{\partial \tau} (\tau, \Omega) + G(\tau, \Omega) I(\tau, \Omega) = \frac{\omega}{4\pi} \int_{4\pi} d\Omega' \, P(\Omega' \to \Omega) \, G(\Omega') I(\tau, \Omega')$$

$$\tag{15}$$

II.A   Approximations and Simplifications

Equation 15 may be considerably simplified if scattering by a leaf is assumed to be isotropic, i.e.,

$$f(\Omega' \to \Omega; \Omega_L) = \omega/4\pi \tag{16}$$

where ω is the fraction of intercepted energy that is scattered. For this special case,

$$\frac{1}{\pi} \Gamma(\Omega' \to \Omega) = G(\Omega') \frac{\omega}{4\pi} \tag{17}$$

If, in addition, it is assumed that the orientation in the canopy is completely random, i.e., $g_L(\Omega_L) = 1$, then from Eq. 4 it is found that $G(\Omega) = 0.5$, so that Eq. 15 reduces to

$$-\mu \frac{\partial I}{\partial \tau'} (\tau', \Omega) + I(\tau', \Omega) = \frac{\omega}{4\pi} \int_{4\pi} d\Omega' \, I(\tau', \Omega') \tag{18}$$

with $\tau' = \tau/2$. Equation 18 can be reduced to a one-angle problem by simply averaging over the azimuth angle ϕ to give

$$-\mu \frac{\partial I}{\partial \tau'} (\tau', \mu) + I(\tau', \mu) = \frac{\omega}{2} \int_{-1}^{1} d\mu' \, I(\tau', \mu') \tag{19}$$

where the *azimuthally averaged radiation intensity* is defined as

$$I(\tau', \mu) = \frac{1}{2\pi} \int_{0}^{2\pi} d\phi \, I(\tau', \mu, \phi) \tag{20}$$

Equations 18 and 19 are forms of the one-speed transport equation, which have been well studied in many photon- and neutron-transport problems and used in earlier canopy investigations (see Gerstl and Zardecki, 1985b; Gutschick and Weigel, 1984). However, for the canopy problem, it is seen to apply only for a very restrictive case.

The elimination of the azimuthal angles in the canopy transport equation with anisotropic scattering, Eq. 15 can also be effected by azimuthal averaging under two conditions. Specifically, $G(\Omega)$ must be independent of ϕ and

$$P(\mu' \to \mu) = \frac{1}{2\pi} \int_0^{2\pi} d\phi \, P(\Omega' \to \Omega) = 4 \frac{\Gamma(\mu' \to \mu)}{\omega G(\mu)} \tag{21}$$

must be independent of ϕ'. If these conditions are satisfied, the azimuthal average of Eq. 15 yields

$$-\mu \frac{\partial I}{\partial \tau}(\tau, \mu) + G(\mu) I(\tau, \mu) = \frac{\omega}{2} \int_{-1}^{1} d\mu' \, P(\mu' \to \mu) G(\mu') I(\tau, \mu') \tag{22}$$

The condition that $G(\mu)$ be independent of ϕ holds whenever the leaves are distributed randomly in azimuth, i.e., $g_L(\Omega_L)$ is independent of ϕ_L. The condition that $P(\mu' \to \mu)$ be independent of ϕ' holds for many different scattering transfer functions $P(\Omega' \to \Omega)$; for example, if the scattering is rotationally invariant (i.e., $P(\Omega' \to \Omega)$ depends only on $\Omega' . \Omega$) or, for importance to the canopy problem, if $P(\Omega' \to \Omega)$ depends only on $(\Omega_L . \Omega)$ and $(\Omega_L . \Omega')$.

II.B   Boundary Conditions for the Canopy Problem

To specify a unique solution of Eq. 15 or Eq. 22, it is necessary to specify the incident-radiation intensity at the canopy boundaries, i.e., at $\tau = 0$ and $\tau = \tau(T) \equiv \tau_T$. The canopy is illuminated from above by both a direct monodirectional solar component (in the direction $\Omega_0(\mu_0, \phi_0)$, $\mu_0 < 0$) as well as by diffuse sky radiation. Thus, at $\tau = 0$, the radiation field is

$$I(0, \Omega) = I_d(\Omega) + I_0 \delta(\Omega - \Omega_0) \qquad \mu < 0 \tag{23}$$

where $I_d(\Omega)$ is the diffuse component (often assumed to be isotropic), and I_0 is the intensity of the direct solar radiation that is incident in direction Ω_0. At the ground surface, it is assumed that a fraction r_s of the energy reaching the ground through the canopy is reradiated isotropically back into the canopy, i.e.,

$$I(\tau_T, \Omega) = \frac{r_s}{\pi} \int_0^{2\pi} d\phi' \int_{-1}^0 d\mu' \, |\mu'| \, I(\tau_T, \Omega') \qquad \mu > 0 \tag{24}$$

For the azimuthally averaged problem, these boundary conditions become

$$I(0, \mu) = \left[\frac{1}{2\pi} \int_0^{2\pi} d\phi \, I_d(\Omega) \right] + \left[\frac{I_0}{2\pi} \delta(\mu - \mu_0) \right] \qquad \mu < 0 \tag{25}$$

and

$$I(\tau_T, \mu) = 2r_s \int_{-1}^0 d\mu' \, |\mu'| \, I(\tau_T, \mu') \qquad \mu > 0 \tag{26}$$

II.C   Separation of Uncollided and Collided Intensities

Because of the delta function in the incident intensity (see Eq. 23 or Eq. 25], it is computationally advantageous to separate analytically the uncollided intensity I^0 (which also contains a delta function) from the intensity of photons I^s that have scattered one or more times in the canopy. Consider the azimuthally dependent problem (Eqs. 15, 23, and 24]. Let

$$I(\tau, \boldsymbol{\Omega}) \equiv I^0(\tau, \boldsymbol{\Omega}) + I^s(\tau, \boldsymbol{\Omega}) \tag{27}$$

where the intensity of uncollided photons I^0 for $\mu < 0$ is

$$I^0(\tau, \boldsymbol{\Omega}) = I(0, \boldsymbol{\Omega}) \exp\left[\frac{G(\boldsymbol{\Omega})\,\tau}{\mu}\right] \tag{28a}$$

and for $\mu > 0$ is

$$I^0(\tau, \boldsymbol{\Omega}) = \frac{r_s}{\pi} \exp\left[\frac{-G(\boldsymbol{\Omega})\,(\tau_T - \tau)}{\mu}\right]$$

$$\cdot \left\{\int_0^{2\pi} d\phi' \int_{-1}^0 d\mu'\,|\mu'|\,I(0, \boldsymbol{\Omega}') \exp\left[\frac{G(\boldsymbol{\Omega}')\,\tau_T}{\mu'}\right]\right\} \tag{28b}$$

Substitution of Eqs. 27 and 28 into the transport equation, Eq. 15, then gives

$$-\mu \frac{\partial I^s}{\partial \tau}(\tau, \boldsymbol{\Omega}) + G(\boldsymbol{\Omega})\,I^s(\tau, \boldsymbol{\Omega})$$

$$= \frac{\omega}{4\pi} \int_{4\pi} d\boldsymbol{\Omega}'\,P(\boldsymbol{\Omega}' \to \boldsymbol{\Omega})\,G(\boldsymbol{\Omega}')\,I^s(\tau, \boldsymbol{\Omega}') + Q(\tau, \boldsymbol{\Omega}) \tag{29}$$

where the *first collision source* $Q(\tau, \boldsymbol{\Omega})$ is

$$Q(\tau, \boldsymbol{\Omega}) = \frac{\omega}{4\pi} \int_{4\pi} d\boldsymbol{\Omega}'\,P(\boldsymbol{\Omega}' \to \boldsymbol{\Omega})\,G(\boldsymbol{\Omega}')\,I^0(\tau, \boldsymbol{\Omega}') \tag{30}$$

The boundary conditions for the modified transport equation, Eq. 29, are

$$I^s(0, \boldsymbol{\Omega}) = 0 \qquad \mu < 0 \tag{31}$$

$$I^s(\tau_T, \boldsymbol{\Omega}) = \frac{r_s}{\pi} \int_0^{2\pi} d\phi' \int_{-1}^0 d\mu'\,|\mu'|\,I^s(\tau_T, \mu') \qquad \mu > 0 \tag{32}$$

The numerical solution of Eq. 29 subject to the boundary conditions of Eqs. 31 and 32 is usually far easier than attempting to solve Eq. 15 directly.

II.D The Standard Problem

In this section, we describe a *standard problem* that we will later solve using the various analytical and numerical methods. In Section V.C we consider a more realistic problem (diffuse sky incidence, soil boundary, nonuniform leaf normal orientation, etc.) and its solution is described in some detail.

Consider a horizontally homogeneous source-free leaf canopy of thickness τ'_T, illuminated by a monodirectional beam and surrounded by vacuum. Assume that the leaves are uniformly inclined in this canopy, so that the transport equation and the scattering transfer function are rotationally invariant. For such a leaf canopy, we can now define the optical depth τ as just one-half the leaf-area index, i.e., $\tau = \tau'/2$ and $\tau_T = \tau'_T/2$ [see Section II.A]. After separation of uncollided and collided fluxes, the governing transport equation is of the form

$$-\mu \frac{\partial I}{\partial \tau}(\tau, \mathbf{\Omega}) + I(\tau, \mathbf{\Omega}) = \frac{\omega}{4\pi} \int_{4\pi} d\mathbf{\Omega}' \, P(\mathbf{\Omega}' \to \mathbf{\Omega}) I(\tau, \mathbf{\Omega}')$$

$$+ \frac{\omega}{4} I_0 P(\mathbf{\Omega}_0 \to \mathbf{\Omega}) \exp\left(\frac{\tau}{\mu_0}\right) \qquad (33)$$

where I_0 is the intensity of the monochromatic external collimated source perpendicular to its direction of propagation $\mathbf{\Omega}_0$. The associated boundary conditions are

$$I(0, \mathbf{\Omega}) = 0 \qquad \mu < 0 \qquad (34)$$

and

$$I(\tau_T, \mathbf{\Omega}) = 0 \qquad \mu > 0 \qquad (35)$$

For the azimuthally averaged (one-angle) problem, the governing transport is

$$-\mu \frac{\partial I}{\partial \tau}(\tau, \mu) + I(\tau, \mu)$$

$$= \frac{\omega}{2} \int_{-1}^{1} d\mu' \, P(\mu' \to \mu) I(\tau, \mu') + \frac{\omega}{4} I_0 P(\mu_0 \to \mu) \exp\left(\frac{\tau}{\mu_0}\right) \qquad (36)$$

The boundary conditions are now

$$I(0, \mu) = 0 \qquad \mu < 0 \qquad (37)$$

and

$$I(\tau_T, \mu) = 0 \qquad \mu > 0 \qquad (38)$$

We emphasize here that our convention is that $\mu \in (-1, 0)$ corresponds to downward directions and conversely, $\mu \in (0, 1)$ corresponds to upward directions. When we depart from this convention, we will state this explicitly.

III OPTICAL MODELS FOR A LEAF CANOPY

The orientation of leaves in a canopy can greatly affect the radiation field, and selection of a model for $g_L(\Omega_L)$ must retain essential features of plant structure and yet be sufficiently simple to allow subsequent analysis. In this section, we discuss the G function and the scattering transfer function and give numerical examples for some model leaf canopies.

III.A The G Function

It is often reasonable to assume that the polar and azimuthal angles of the leaf normals are independent, i.e., $g_L(\Omega_L) = g_L(\mu_L) h_L(\phi_L)$. The geometry factor $G(\Omega)$ thus can be written as

$$G(\Omega) = \int_0^1 d\mu_L \, g_L(\mu_L) \, \psi(\mu, \mu_L) \tag{39}$$

where

$$\psi(\mu, \mu_L) = \frac{1}{2\pi} \int_0^{2\pi} d\phi_L \, h_L(\phi_L) \, |\Omega_L . \Omega| \tag{40}$$

is the azimuthal average of the projected leaf area perpendicular to Ω. There is ample experimental evidence (see Ross, 1981) to suggest a random leaf normal distribution in azimuth, i.e., $h_L(\phi_L) = 1$. With this assumption, ψ and G become independent of ϕ. Explicitly, one obtains

$$\psi(\mu, \mu_L) = |\mu\mu_L| \quad \text{if } |\cot\theta \cot\theta_L| > 1 \tag{41a}$$

otherwise,

$$\psi(\mu, \mu_L) = \mu\mu_L[2\phi_t(\mu)/\pi - 1] + (2/\pi)(1 - \mu^2)^{1/2}(1 - \mu_L^2)^{1/2} \sin\phi_t \tag{41b}$$

In the above,

$$\phi_t(\mu) = \cos^{-1}[-\cot\theta \cot\theta_L] \tag{42}$$

This result agrees with that reported by Gutschick and Weigel (1984) and is shown as a contour plot in Figure 1. For a canopy whose leaves all have the same polar inclination (μ_0) but are distributed uniformly in azimuth, the geometry factor $G(\mu) = \psi(\mu, \mu_0)$. For leaves whose normals are distributed uniformly in all directions, $G(\Omega) = 0.5$.

III.B Scattering Transfer Functions for a Leaf Canopy

There is presently only a small amount of experimental information about scattering from different types of leaves (see Breece and Holmes, 1971; Woolley, 1971). Consequently, scattering has often been assumed to be isotropic or, for anisotropic problems, to be represented by some idealized scattering transfer function, such as the Henyey–Green-

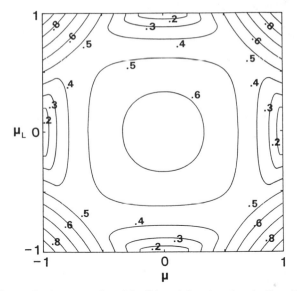

Figure 1   A contour plot of the $\Psi(\mu, \mu_L)$ function given by Eqs. 41.

stein kernel (Gerstl and Zardecki, 1985a,b) or various Legendre polynomial expansions (see Van de Hulst, 1980). Such treatments are based on rotationally invariant scattering models and, hence, are not entirely satisfactory for the leaf-canopy problems. A simple but realistic leaf scattering model has been proposed by Ross and Nilson (1968) and recently suggested for the general canopy problem by Gutschick and Weigel (1984). In this model, a fraction r_L of the intercepted energy is reradiated as a cosine distribution about the leaf normal (i.e., Lambertian reflectance). Similarly, a fraction t_L is transmitted in a Lambertian distribution on the opposite side of the leaf, so that the single-scatter albedo of the leaf becomes $\omega = r_l + t_L$. This *bi-Lambertian* scattering model can be described mathematically as

$$f(\Omega' \to \Omega; \Omega_L) = \begin{cases} r_L|\Omega_L.\Omega|/\pi & (\Omega_L.\Omega)(\Omega_L.\Omega') < 0 \\ t_L|\Omega_L.\Omega|/\pi & (\Omega_L.\Omega)(\Omega_L.\Omega') > 0 \end{cases} \tag{43}$$

The associated area scattering transfer function becomes

$$\Gamma(\Omega' \to \Omega) = r_L\Gamma^-(\Omega' \to \Omega) + t_L\Gamma^+(\Omega' \to \Omega) \tag{44}$$

where

$$\Gamma^\pm(\Omega' \to \Omega) = \pm\frac{1}{2\pi} \int_0^1 d\mu_L \int_0^{2\pi} d\phi_L \, g_L(\Omega_L)(\Omega_L.\Omega)(\Omega_L.\Omega') \tag{45}$$

The $\pm$ in the last definition indicates that the ϕ_L integration is over that portion of the 0–2π range for which the integrand is either positive ($+$) or negative ($-$).

For an isotropic canopy, i.e., one with a uniform leaf normal distribution, $g_L(\Omega_L)$ = 1, Eqs. 44 and 45 may be evaluated analytically to give the anisotropic rotationally invariant scattering transfer function

$$\Gamma(\Omega' \to \Omega) = \frac{\omega}{3\pi} (\sin \beta - \beta \cos \beta) + \frac{t_L}{\pi} \cos \beta \tag{46}$$

where $\beta \equiv \cos^{-1}(\Omega . \Omega')$, the angle between Ω and Ω'. In general, however, $\Gamma(\Omega' \to \Omega)$ is not rotationally invariant; moreover, numerical methods must be used to evaluate Eqs. 44 and 45. The area scattering transfer function of Eq. 46 is shown in Figure 2, and it is seen that scattering becomes increasingly anisotropic as r_L/t_L increases.

From the definition of $\Gamma(\Omega' \to \Omega)$, it is seen that the bi-Lambertian scattering transfer function imbues the area scattering transfer function with a very useful symmetry property, namely,

$$\Gamma(\Omega' \to \Omega) = \Gamma(\Omega \to \Omega') = \Gamma(-\Omega' \to -\Omega) \tag{47}$$

For the special case of equal leaf reflectance and transmittance, $r_L = t_L$, the following additional symmetry holds, namely,

$$\Gamma(\Omega' \to \Omega) = \Gamma(-\Omega' \to \Omega) \tag{48}$$

These symmetry properties considerably reduce the computational effort when the elements of the transfer matrix are calculated from Eqs. 44 and 45.

As noted by Ross (1981), the general lack of rotational invariance in $\Gamma(\Omega' \to \Omega)$ precludes the use of Legendre polynomial expansions and the addition theorem that are traditionally used to treat anisotropically scattering media. Consequently, most of the available computer codes for solving anisotropic transport problems cannot be directly applied to the canopy problem. Modifications must be made to use the transfer function

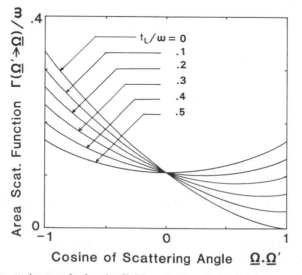

Figure 2 The area scattering transfer function $\Gamma(\Omega' \to \Omega)$ for an isotropic leaf canopy (Eq. 46). Each leaf is assumed to scatter according to the bi-Lambertian model (Eq. 43).

(or its discretized transfer matrix) directly and to incorporate the cross section $G(\Omega)$ that is direction-dependent.

Even though a Legendre polynomial expansion of $\Gamma(\Omega' \rightarrow \Omega)$ for the bi-Lambertian scattering function cannot be used, the azimuthally dependent transport equation can still be reduced to the azimuthally indpendent form of Eq. 22. For an azimuthally random leaf normal distribution and the bi-Lambertian scattering distribution, Eq. 21 gives the azimuthally averaged transfer function as

$$\Gamma(\mu' \rightarrow \mu) = \int_0^1 d\mu_L \, g_L(\mu_L) \left[t_L \Psi^+(\mu, \mu', \mu_L) + r_L \Psi^-(\mu, \mu', \mu_L) \right] \quad (49)$$

where

$$\Psi^\pm(\mu, \mu', \mu_L) = \pm \frac{1}{4\pi^2} \int_0^{2\pi} d\phi \int_0^{2\pi} d\phi_L \, (\Omega_L . \Omega)(\Omega_L . \Omega') \quad (50)$$

It can be shown that the double integration over ϕ and ϕ_L for this particular scattering distribution also eliminates ϕ', so that the azimuthally averaged transport equation is given by Eq. 22, an equation involving only the cosines of the polar angles. In fact, this elimination of ϕ' and the subsequent reduction of the transport equation to a one-angle equation occurs for any scattering distribution $f(\Omega' \rightarrow \Omega; \Omega_L)$ that depends only on $(\Omega_L . \Omega)$ and $(\Omega_L . \Omega')$.

Evaluation of the double integral in Eq. 50 gives

$$\Psi^\pm(\mu, \mu', \mu_L) = H(\mu, \mu_L) H(\pm\mu', \mu_L) + H(-\mu, \mu_L) H(\mp\mu', \mu_L) \quad (51)$$

where the H function is (with $\theta = \cos^{-1} \mu$)

$$H(\mu, \mu_L) = \mu\mu_L \quad \text{if } (\cot\theta \cot\theta_L) > 1 \quad (52a)$$

$$H(\mu, \mu_L) = 0 \quad \text{if } (\cot\theta \cot\theta_L) < -1 \quad (52b)$$

$$H(\mu, \mu_L) = \frac{1}{\pi} \left[\mu\mu_L \phi_t(\mu) + (1 - \mu^2)^{1/2} (1 - \mu_L^2)^{1/2} \sin\phi_t(\mu) \right] \quad (52c)$$

otherwise

and

$$\phi_t(\mu) \equiv \cos^{-1}(\cot\theta \cot\theta_L) = \pi - \phi_t(-\mu) \quad (53)$$

This $H(\mu, \mu_L)$ function appears frequently in the transport description based on the bi-Lambertian scattering function and is illustrated in Figure 3. From Eqs. 49 and 51, it can be seen that the azimuthally averaged scattering transfer function also possesses symmetry properties, namely,

$$\Gamma(\mu' \rightarrow \mu) = \Gamma(\mu \rightarrow \mu') = \Gamma(-\mu' \rightarrow -\mu) \quad (54)$$

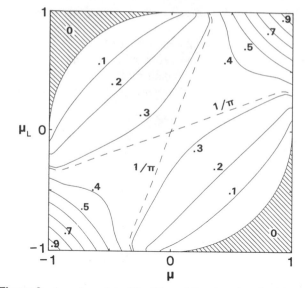

Figure 3   A contour plot of the $H(\mu, \mu_L)$ function given by Eqs. 52.

Finally, for the special case of equal leaf reflectance and transmittance (i.e., $r_L = t_L$), Eqs. 49 and 51 reduce to

$$\Gamma(\mu' \to \mu) = \frac{\omega}{2} \int_0^1 d\mu_L \, g_L(\mu_L) \left[H(\mu, \mu_L) + H(-\mu, \mu_L) \right] \\ \cdot \left[H(\mu', \mu_L) + H(-\mu', \mu_L) \right] \tag{55a}$$

which reduces to

$$\Gamma(\mu' \to \mu) = \frac{\omega}{2} \int_0^1 d\mu_L \, g_L(\mu_L) \, \psi(\mu, \mu_L) \, \psi(\mu', \mu_L) \tag{55b}$$

and an additional symmetry occurs, namely, $\Gamma(\mu' \to \mu) = \Gamma(\mu' \to -\mu)$.

For the two limiting cases of horizontal ($\theta_L = 0°$) and vertical leaves ($\theta_L = 90°$), Eqs. 44 and 45 may be evaluated analytically. For horizontal leaves, one obtains

$$\Gamma(\Omega' \to \Omega) = \begin{cases} t_L \mu\mu' & \mu\mu' > 0 \\ r_L |\mu\mu'| & \mu\mu' < 0 \end{cases} \quad \text{for } \theta_L = 0° \tag{56}$$

whereas for vertical leaves,

$$\Gamma(\Omega' \longrightarrow \Omega) = \Gamma_1(\beta)(1 - \mu^2)^{1/2}(1 - \mu'^2)^{1/2} \quad \text{for } \theta_L = 90° \tag{57}$$

where $\beta \equiv \phi - \phi'; 0 \le \beta \le \pi$; and

$$\Gamma_1(\beta) = \frac{\omega}{2\pi}(\sin \beta - \beta \cos \beta) + \frac{t_L}{2} \cos \beta \tag{58}$$

The same results were reported by Ross (1981). In the case of horizontal leaves, the $\Gamma(\mathbf{\Omega}' \to \mathbf{\Omega})$ is seen to be independent of the exit azimuth, whereas in the case of erect leaves, this is not so. To obtain the corresponding azimuthally independent function $\Gamma(\mu' \to \mu)$, average Eqs. 56 and 57 over all ϕ from $0-2\pi$ (or, equivalently, over all β from $0-\pi$). The result is

$$\Gamma(\mu' \to \mu) = \begin{cases} r_L |\mu\mu'| & \mu\mu' < 0 & \theta_L = 0° \\ t_L \mu\mu'/2 & \mu\mu' > 0 & \theta_L = 0° \\ \dfrac{2\omega}{\pi^2} (1 - \mu^2)^{1/2} (1 - \mu'^2)^{1/2} & & \theta_L = 90° \end{cases} \tag{59}$$

The same results can be obtained directly from Eqs. 49 and 51. For horizontal leaves, $\theta_L = 0°$, $g_L(\mu_L) = \delta(1 - \mu_L)$, and from Eq. 51, $H(\mu, 1) = \mu$ if $\mu > 0$ and zero otherwise. For vertical leaves, $\theta_L = \pi/2$, $g_L(\mu_L) = \delta(\mu_L)$ and $H(\mu, 0) = (1 - \mu^2)^{0.5}/\pi$. With these results, Eqs. 49 and 51 reduce to Eq. 59.

To illustrate realistic canopies whose leaves are oriented in different directions, models for the leaf-inclination function $g_L(\mu_L)$ must be used. Based on the data of deWit (1965), Bunnik (1978) gave empirical expressions for a variety of leaf canopies. For example, the leaves in a *planophile* canopy (mainly horizontal leaves) were found to be distributed as

$$g_L(\theta_L) = \frac{2}{\pi} (1.0 + \cos 2\theta_L) \tag{60}$$

The normalized transfer function $P(\mathbf{\Omega}' \to \mathbf{\Omega})$ for leaf canopies with the planophile distribution function is shown in Figure 4 for a near-normal incident direction. The

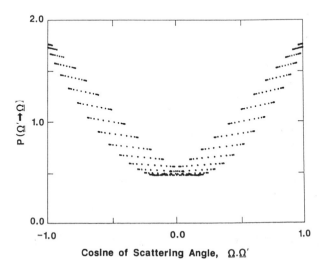

Figure 4   The normalized azimuthally dependent scattering transfer function $P(\mathbf{\Omega}' \to \mathbf{\Omega})$ for a planophile canopy (predominantly horizontal leaves) given by Eq. 60. The leaf transmittance and reflectance are both equal to 0.5, and $\mathbf{\Omega}'$ is fixed at $\theta' = 170°$ and $\phi' = 0°$. Each point is the value of the transfer function for a discrete value of $\mathbf{\Omega}_{ij} = (\mu_i, \phi_j)$ (the nearly horizontal row of points is for a fixed μ_i as ϕ_j varies). These results illustrate that the transfer function is not just a function of the scattering angle, $\cos^{-1}(\mathbf{\Omega}'\mathbf{\Omega})$, i.e., it is not rotationally invariant.

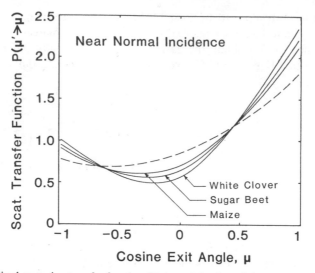

Figure 5   Normalized scattering transfer function $P(\mu' \rightarrow \mu)$ for three field crops illuminated almost normally ($\theta' = 170°$). The original data on the leaf normal inclination distribution were reported by deWit (1965) and are combined with the bi-Lambertian transfer function. Also, shown [dashed-line curve] is the result for a uniform leaf normal distribution, $g_L(\Omega_L) = 1$. For all curves, $t_L/\omega = 0.3$.

transfer function for these as well as other theoretical and measured leaf distributions are not rotationally invariant, since they depend on both the incident and scattered directions. Nevertheless, it is apparent that there is generally more scattered radiation in the direction close to their incident direction (i.e., around $(\Omega \cdot \Omega') \approx \pm 1$). Measurements of directional canopy reflectances made at right angles to the incident (solar) direction show lower reflectances than at larger (or smaller) angles from the incident direction. This effect can also be observed visually. Most leaf canopies (where specular reflection is unimportant, as in grasslands) appear duller in the antisolar-view directions, $(\Omega \cdot \Omega') \approx 0$, than in near retrosolar directions, $(\Omega \cdot \Omega') \approx -1$. This effect should not be confused with the *Hot-Spot* effect, which is a local bright spot centered around the retrosolar direction and is purely a shading effect.

Examples of the azimuthally integrated bi-Lambertian transfer function based on measured leaf normal-orientation distribution for three agricultural crops (deWit, 1965) are shown in Figures 5 and 6. The dashed lines in these figures are for a canopy with uniform leaf normal distribution $g_L(\Omega_L)$. From these examples, it is seen that field crops may be poorly described by an isotropic scattering model (i.e., $P(\Omega' \rightarrow \Omega) = 1$), and even when a bi-Lambertian scattering model is used, a uniform leaf normal distribution may not be a justifiable assumption. Extensive discussion on the transfer functions in one and two angles can be found in Myneni et al. (1988a,c).

IV   ANALYTICAL METHODS FOR SOLVING THE TRANSPORT EQUATION

Analytical methods are those that are capable of producing results in a closed analytical form. In most cases, these methods result in equations that can be solved only numerically, and in many cases, straightforward numerical methods may be just as accurate.

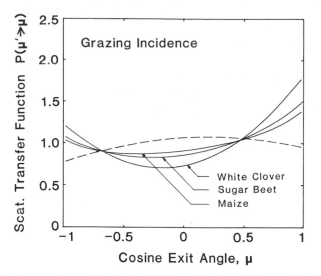

Figure 6 Normalized scattering transfer function $P(\mu' \rightarrow \mu)$ for three field crops illuminated almost tangentially ($\theta' = 100°$). The original data on the leaf normal inclination distribution were reported by deWit (1965) and are combined with the bi-Lambertian transfer function. Also, shown [dashed-line curve] is the result for a uniform leaf normal distribution, $g_L(\Omega_L) = 1$. For all curves, $t_L/\omega = 0.3$.

For instance, in the singular eigenfunction and the Wiener–Hopf methods, the end result of the initial analysis is a reduction to singular integral equations that can be reduced to Fredholm integral equations in the case of a finite slab, and these equations can be solved only numerically. The X and Y functions, from which the emergent flux distributions can be obtained, are solved by numerical solution of various pairs of coupled equations: nonlinear integral equations, nonlinear integro-differential equations, linear singular integral equations, and regular Fredholm equations. The primary usefulness of these methods is then the understanding of the mathematical structure and behavior of solutions of the transport equation. Also, analytical solutions serve as computational standards, benchmarks that one can use to test the accuracy of more practical approximate and numerical methods.

Under this category, we discuss only the singular eigenfunction analysis and the method of reduction to X and Y functions. Discussion on the Wiener–Hopf method and the transfer matrix method can be found in Lenoble (1985). For the more mathematically oriented reader, the spectral methods and the monograph by Kaper et al. (1982) are recommended.

IV.A   Singular Eigenfunction Method

We consider the standard one-angle problem described in Section II.D. The governing equation for the azimuthally averaged intensity is

$$\mu \frac{\partial I}{\partial \tau}(\tau, \mu) + I(\tau, \mu) = \frac{\omega}{2} \int_{-1}^{1} d\mu' \, P(\mu' \rightarrow \mu) \, I(\tau, \mu') \qquad 0 < \tau < \tau_T \quad (61)$$

The associated boundary condition are

$$I(0, \mu) = \delta(\mu - \mu_0) \qquad 0 < \mu\mu_0 < 1 \tag{62a}$$

$$I(\tau_T, \mu) = 0 \qquad -1 < \mu < 0 \tag{62b}$$

Note that $+\mu$ now corresponds to the downward direction and $-\mu$ corresponds to the upward direction. If the transfer function is expanded in a finite sum of Legendre polynomials of the scattering angle, then Eq. 61 can be written as

$$\mu \frac{\partial I}{\partial \tau}(\tau, \mu) + I(\tau, \mu) = \frac{\omega}{2} \sum_{n=0}^{N} b_n P_n(\mu) \int_{-1}^{1} d\mu' \, P_n(\mu') \, I(\tau, \mu') \tag{63}$$

where the b_n are the coefficients of the $N + 1$ term Legendre expansion.

The complete solution can be uniquely expanded as (Kaper, 1966)

$$I(\tau, \mu) = \sum_{i=1}^{M} a_{+i} \xi(\nu_i, \mu) e^{-\tau/\nu_i} + \sum_{i=1}^{M} a_{-i} \xi(-\nu_i, \mu) e^{\tau/\nu_i}$$

$$+ \int_{-1}^{1} d\nu \, A(\nu) \, \xi(\nu, \mu) e^{-\tau/\nu} \tag{64}$$

where the $a_{\pm i}$ and $A(\nu)$ are the expansion coefficients, and ξ and ν are the *eigenfunctions* and *eigenvalues* of the homogeneous transport equation. For $|\nu| > 1$, the eigenfunctions are given by

$$\xi(\pm\nu_i, \mu) = \frac{\omega\nu_i}{2} \frac{D(\nu_i, \pm\mu)}{\nu_{i\mp}\mu} \qquad i = 1, 2, \cdots, M \tag{65}$$

where the M pairs of discrete eigenvalues, $\pm\nu_i$, are given by the dispersion relation

$$1 - \frac{\omega\nu}{2} \int_{-1}^{1} d\mu \frac{D(\mu, \mu)}{\nu - \mu} = 0 \tag{66}$$

The function D is given by

$$D(\nu, \mu) = \sum_{n=0}^{N} b_n h_n(\nu) P_n(\mu) \tag{67}$$

where the function h is defined by the recurrence relation

$$(n + 1)h_{n+1} - \nu(2n + 1 - b_n\omega)h_n(\nu) + nh_{n-1}(\nu) = 0 \qquad n = 0, 1, 2, \ldots \tag{68}$$

and $h_0 \equiv 1$.

On the other hand, for $\nu \in (-1, 1)$, the eigenfunctions are given by

$$\xi(\nu, \mu) = \frac{\omega\nu}{2} \frac{D(\nu, \mu)}{\nu - \mu} + \lambda(\nu) \delta(\nu - \mu) \tag{69}$$

where the generalized function $(\nu - \mu)^{-1}$ and $\delta(\nu - \mu)$ are defined by the functionals

$$\langle (\nu - \mu)^{-1}, \zeta(\mu) \rangle = \int_{-1}^{1} d\mu \, \frac{\zeta(\nu)}{\nu - \mu} \tag{70}$$

and

$$\langle \delta(\nu - \mu), \zeta(\mu) \rangle = \zeta(\nu) \quad \text{if } \nu \in (-1, 1) \tag{71}$$
$$= 0 \quad \text{otherwise}$$

respectively, for all test functions ζ. The integration in Eq. 70 refers to integration in the Cauchy principal-value sense. Substituting Eq. 69 in the following normalization (since the eigenvalue problem is homogeneous),

$$\int_{-1}^{1} d\mu \, \xi(\nu, \mu) = 1 \tag{72}$$

gives the λ function as

$$\lambda(\nu) = 0.5 [\Lambda^{+}(\nu) + \Lambda^{-}(\nu)] \tag{73}$$

where Λ^{+} and Λ^{-} are the boundary values of a function that is analytic in the complex plane cut along the real axis between -1 and $+1$,

$$\Lambda(z) = 1 - \frac{\omega z}{2} \int_{-1}^{1} d\mu \, \frac{D(z - \mu)}{z - \mu} \quad z \neq \notin (-1, 1) \tag{74}$$

and given by the expressions

$$\Lambda^{\pm}(\nu) \equiv \Lambda(\nu \pm i0) = 1 - \frac{\omega \nu}{2} \int_{-1}^{1} d\mu \, \frac{D(\nu, \mu)}{\nu - \mu} \pm 0.5\pi i \omega \nu D(\nu, \nu) \quad \nu \in (-1, 1) \tag{75}$$

The integration in Eq. 75 also refers to integration in the Cauchy principal-value sense. For $\omega > 1$, the discrete roots are not necessarily real and, hence, many of the functions in the ensuing analysis become complex. However, for the canopy photon-transport problem, from physical arguments, $\omega < 1$ and we shall restrict our further analysis to this constraint.

The eigenfunctions ξ satisfy an orthogonality relations that may be symbolically stated as

$$\int_{-1}^{1} d\mu \, \xi(\nu, \mu) \, \xi(\nu', \mu) = 0 \quad \text{if } \nu \neq \nu' \text{ for } \nu, \nu' \in (-1, 1) \text{ or } \nu^{\pm} \tag{76}$$

$$\int_{-1}^{1} d\mu \, \mu \zeta^{2}(\pm \nu_i, \mu) = \pm N_i^{-1} \tag{77}$$

$$\int_{-1}^{1} d\mu \, \mu \zeta(\nu, \mu) \, \zeta(\nu', \mu) = N^{-1}(\nu) \, \delta(\nu - \nu') \tag{78}$$

where

$$N_i^{-1} \equiv \frac{\omega \nu_i^2}{2} \int_{-1}^{1} d\mu \, \frac{D^2(\nu_i, \mu)}{(\nu_i - \mu)^2} \tag{79}$$

and

$$N(\nu) \equiv \frac{1}{\nu \left\{ \lambda^2(\nu) + \left[(\omega \pi \nu / 2) D(\nu, \nu) \right]^2 \right\}} \tag{80}$$

From the previous orthogonality relations, expressions for the expansion coefficients of Eq. 64 can be obtained in terms of the emergent distributions. For instance, let $\tau = 0$ in Eq. 64, multiply through by $[\,\mu \xi(\nu_i, \mu)\,]$, integrate over μ, and use the orthogonality properties together with the first boundary condition in Eqs. 62 to obtain

$$a_{+i} = N_i \left[\mu_0 \xi(\nu_i, \mu_0) - \int_0^1 d\mu \, \mu \xi(-\nu_i, \mu) R(\mu) \right] \tag{81}$$

Similarly, by multiplying Eq. 64 by $[\xi(\nu, \mu), \nu > 0]$ instead of $\xi(\nu_i, \mu)$, one obtains

$$A(\nu) = N(\nu) \left[\mu_0 \xi(\nu, \mu_0) - \int_0^1 d\mu \, \mu \xi(-\nu, \mu) R(\mu) \right] \qquad \nu > 0 \tag{82}$$

Likewise, one obtains

$$a_{-i} = -N_i e^{-\tau_T/\nu_i} \int_0^1 d\mu \, \mu \xi(-\nu_i, \mu) T(\mu) \tag{83}$$

and

$$A(-\nu) = -N(\nu) e^{-\tau_T/\nu} \int_0^1 d\mu \, \mu \xi(-\nu, \mu) T(\mu) \tag{84}$$

In the preceding, $R(\mu) \equiv I(0, -\mu)$ and $T(\mu) \equiv I(\tau_T, \mu)$.

Substitution of these results into the expression of Eq. 64 gives the complete solution $I(\tau, \mu)$ in terms of the reflected and transmitted distributions. In particular, this result can be evaluated for the reflected and transmitted distributions by setting $\tau = 0$ and $\tau = \tau_T$, respectively. This yields a pair of coupled integral equations of the form

$$R(\mu) = \mu_0 \sum_{i=1}^{M} N_i \xi(\nu_i, \mu_0) \, \xi(-\nu_i, \mu) + \mu_0 \int_0^1 d\nu \, N(\nu) \, \xi(\nu, \mu_0) \, \xi(-\nu, \mu)$$

$$- \int_0^1 d\mu' \, \mu' K(\mu, \mu') R(\mu') - \int_0^1 d\mu' \, \mu' G(\mu, \mu') T(\mu') \qquad \mu < 0 \tag{85}$$

and

$$T(\mu) = \mu_0 \sum_{i=1}^{M} N_i \xi(\nu_i, \mu_0) \, \xi(\nu_i, \mu) e^{-\tau_T/\nu_i} + \mu_0 \int_0^1 d\nu \, N(\nu) \, \xi(\nu, \mu_0) \, \xi(\nu, \mu) e^{-\tau_T/\nu}$$

$$- \int_0^1 d\mu' \, \mu' G(\mu, \mu') R(\mu') - \int_0^1 d\mu' \, \mu' K(\mu, \mu') \, T(\mu') \qquad \mu < 0$$

$$(86)$$

where the kernel functions K and G are defined as

$$K(\mu, \mu') = \sum_{i=1}^{N} N_i \xi(-\nu_i, \mu') \, \xi(-\nu_i, \mu) + \int_0^1 d\nu \, N(\nu) \, \xi(-\nu, \mu') \, \xi(-\nu, \mu) \quad (87)$$

and

$$G(\mu, \mu') = \sum_{i=1}^{M} N_i \xi(-\nu_i, \mu') \, \xi(\nu_i, \mu) e^{-\tau_T/\nu_i}$$

$$+ \int_0^1 d\nu \, N(\nu) \, \xi(-\nu, \mu') \, \xi(\nu, \mu) e^{-\tau_T/\nu} \qquad (88)$$

Equations 85 and 86 are difficult to solve, and it is necessary to separate the singular uncollided intensity from the continuous collided distribution. The singularity may be removed by rewriting these equations in terms of the collided transmitted distribution $T'(\mu)$:

$$T'(\mu) = T(\mu) - \delta(\mu - \mu_0) \exp\left[-\tau_T/\mu_0\right] \qquad (89)$$

and

$$T'(\mu) = g(\mu) - \int_0^1 d\mu' \, \mu' G(\mu, \mu') \, R(\mu') - \int_0^1 d\mu' \, \mu' K(\mu, \mu') \, T'(\mu') \qquad \mu > 0$$

$$(90)$$

Substituting Eq. 89 into Eq. 85 gives

$$R(\mu) = k(\mu) - \int_0^1 d\mu' \, \mu' K(\mu, \mu') \, R(\mu') - \int_0^1 d\mu' \, \mu' G(\mu, \mu') \, T'(\mu') \qquad \mu < 0$$

$$(91)$$

where $k(\mu)$ and $g(\mu)$ are defined as

$$k(\mu) \equiv \mu_0 \sum_{i=1}^{M} N_i \Big\{ \xi(\nu_i, \mu_0) \, \xi(-\nu_i, \mu) - \xi(-\nu_i, \mu_0)$$

$$\cdot \xi(\nu_i, \mu) \exp\left[-\left(\frac{\tau_T}{\mu_0}\right) - \left(\frac{\tau_T}{\nu_i}\right)\right] \Big\}$$

$$+ \mu_0 \int_0^1 d\nu \, N(\nu) \Big\{ \xi(\nu, \mu_0) \, \xi(-\nu, \mu) - \xi(-\nu, \mu_0)$$

$$\cdot \xi(\nu, \mu) \exp\left[-\left(\frac{\tau_T}{\mu_0}\right) - \left(\frac{\tau_T}{\mu}\right)\right] \Big\} \qquad (92)$$

and

$$g(\mu) \equiv \mu_0 \sum_{i=1}^{M} N_i \xi(\nu_i, \mu_0) \, \xi(\nu_i, \mu) \left[\exp \left(\frac{-\tau_T}{\nu_i} \right) - \exp \left(\frac{-\tau_T}{\mu_0} \right) \right]$$

$$+ \mu_0 \int_0^1 d\nu \, N(\nu) \, \xi(\nu, \mu_0) \, \xi(\nu, \mu) \left[\exp \left(\frac{-\tau_T}{\nu} \right) - \exp \left(\frac{-\tau_T}{\mu_0} \right) \right] \quad (93)$$

The functions $k(\mu)$, $g(\mu)$, $K(\mu, \mu')$, and $G(\mu, \mu')$ for $[\mu, \mu' \epsilon (0, 1)]$ are continuous, and, thus, Eqs. 90 and 91 are a pair of regular Fredholm equations of the second kind for emergent distributions. The solution of these equations is quite tedious and will not be discussed here. Excellent reviews on this method can be found in Case and Zweifel (1967) and in McCormick and Kuscer (1973).

IV.B The Method of Reduction to X and Y Functions

We consider the standard problem described in Section II.D. After the separation of uncollided and collided fluxes, the governing equation is Eq. 33 and the associated boundary conditions are given by Eqs. 34 and 35. Expand the transfer function in Legendre polynomials with a finite number of terms

$$P(\mathbf{\Omega}' \to \mathbf{\Omega}) = \sum_{n=0}^{N} b_n P_n(\mathbf{\Omega}' \to \mathbf{\Omega}) \quad (94)$$

where the b_n are the $N + 1$ expansion coefficients. The Legendre polynomials for the argument in Eq. 94 can be expanded by the addition theorem for spherical harmonics as (since the transfer function is rotationally invariant)

$$P(\mathbf{\Omega}' \to \mathbf{\Omega}) = \sum_{m=0}^{N} \sum_{n=m}^{N} b_n^m P_n^m(\mu) \, P_n^m(\mu') \cos m(\phi' - \phi) \quad (95)$$

where

$$b_n^m = (2 - \delta_{0,m}) b_n \frac{(n - m)!}{(n + m)!} \quad n = m, \ldots, N; \quad 0 \le m \le N \quad (96)$$

and

$$\delta_{0,m} = \begin{cases} 1 & \text{if } m = 0 \\ 0 & \text{otherwise} \end{cases} \quad (97)$$

and P_n^m are the associated Legendre polynomials.

We now expand the intensity and the source terms in the form

$$I(\tau, \mathbf{\Omega}) = \sum_{m=0}^{N} I^m(\tau, \mu) \cos m(\phi_0 - \phi) \quad (98)$$

$$J(\tau, \boldsymbol{\Omega}) = \sum_{m=0}^{N} I^m(\tau, \mu) \cos m(\phi_0 - \phi) \qquad (99)$$

where $J(\tau, \boldsymbol{\Omega})$ is the right-hand side of Eq. 33.

Substituting Eqs. 98 and 99 into the transport equation (Eq. 33) shows that each azimuthal component I^m satisfies a separate transport equation with the source function J^m. Note that for $m = 0$, the intensity expressed in Eq. 98 corresponds to the azimuth independent case and, hence, the transport equation with I^0 and J^0 is the corresponding one-angle transport equation. We may now integrate the $N + 1$ independent transport equations using an integrating factor, subject to the boundary conditions given in Eqs. 34 and 35, to obtain

$$I^m(\tau, \mu) = \int_{\tau}^{\tau_T} \frac{d\tau'}{\mu} J^m(\tau', \mu) \exp\left[\frac{-(\tau' - \tau)}{\mu}\right] \qquad \mu > 0 \qquad (100)$$

and

$$I^m(\tau, \mu) = -\int_0^{\tau} \frac{d\tau'}{\mu} J^m(\tau', \mu) \exp\left[\frac{(\tau - \tau')}{\mu}\right] \qquad \mu < 0 \qquad (101)$$

The radiation field is thus determined if we can find the source function $J^m(\tau, \mu)$.

We limit our discussion here to the emergent distributions and define the reflection and transmission functions as

$$R(\boldsymbol{\Omega}, \boldsymbol{\Omega}_0) = I(0, \boldsymbol{\Omega})/I_0 \qquad (102)$$

$$T(\boldsymbol{\Omega}, \boldsymbol{\Omega}_0) = I(\tau_T, \boldsymbol{\Omega})/I_0 \qquad (103)$$

Expand the above in the form

$$R(\boldsymbol{\Omega}, \boldsymbol{\Omega}_0) = R^0(\mu, \mu_0) + 2 \sum_{m=1}^{N} R^m(\mu, \mu_0) \cos m(\phi_0 - \phi) \qquad \mu > 0 \quad (104)$$

and

$$T(\boldsymbol{\Omega}, \boldsymbol{\Omega}_0) = T^0(\mu, \mu_0) + 2 \sum_{m=1}^{N} T^m(\mu, \mu_0) \cos m(\phi_0 - \phi) \qquad \mu < 0 \quad (105)$$

Each azimuthal component can be expressed in terms of *auxilliary functions* Φ and Ψ as

$$R^m(\mu, \mu_0) = \frac{\omega}{4} \sum_{n=m}^{N} b_n(-1)^{n+m} \frac{\Phi_n^m(\mu) \Phi_n^m(\mu_0) - \Psi_n^m(\mu) \Psi_n^m(\mu_0)}{\mu + \mu_0} \qquad (106)$$

$$T^m(\mu, \mu_0) = \frac{\omega}{4} \sum_{n=m}^{N} b_n \frac{\Phi_n^m(\mu_0) \Psi_n^m(\mu) - \Phi_n^m(\mu) \Psi_n^m(\mu_0)}{\mu - \mu_0} \qquad (107)$$

The auxilliary functions can be expressed in terms of the X and Y functions as

$$\Phi_n^m(\mu) = \left[X^m(\mu) r_n^m(\mu) + (-1)^{n+m} Y^m(\mu)\right] P_m^m(\mu) \qquad (108)$$

$$\Psi_n^m(\mu) = \left[X^m(\mu) s_n^m(\mu) + (-1)^{n+m} Y^m(\mu) r_n^m(-\mu)\right] P_m^m(\mu) \qquad (109)$$

where r_n^m and s_n^m are polynomials that must be determined by solving a system of linear algebraic equations (see Sobolev, 1975). The X^m and Y^m functions are determined by nonlinear equations of the form (cf. Chandrasekhar, 1960, Chapter VIII)

$$X^m(\mu) = 1 + \mu \int_0^1 d\mu' \, \frac{Y^m(\mu')}{\mu + \mu'} \left[X^m(\mu') \, X^m(\mu) - Y^m(\mu) \, Y^m(\mu') \right]$$

and

$$Y^m(\mu) = \exp\left(\frac{-\tau}{|\mu|}\right) + \mu \int_0^1 d\mu' \, \frac{Y^m(\mu')}{\mu - \mu'} \left[Y^m(\mu) \, X^m(\mu') - Y^m(\mu') \, X^m(\mu) \right]$$

$$(111)$$

where the *characteristic function* $\mathrm{Y}^m(\mu')$ is an even polynomial in μ' and generally satisfies the condition

$$\int_0^1 d\mu' \, \mathrm{Y}^m(\mu') \leq 0.5 \qquad (112)$$

The X^m and Y^m functions can also be determined from linear singular equations (cf. Leonard and Mullikin, 1964; Carlstedt and Mullikin, 1966). In any case, appropriate constraints are necessary to ensure uniqueness of the solution.

Thus, the method of X and Y functions involves three main steps: (1) determination of the characteristic function $\mathrm{Y}^m(\mu')$, (2) solution of the integral equations for $X^m(\mu)$ and $Y^m(\mu)$, and (3) computation of the reflection R^m and transmission T^m functions from X and Y functions. The main attraction of this method is the reduction of the solution to functions of a single angle. The primary drawback is the extensive amount of algebra that has to be employed for most transfer functions. In Section V.E we will use the principles of invariance to derive a basic set of four integral equations for the reflection and transmission functions. Using the expansions developed here, we show that the integral equations can be solved by reducing them to the X and Y functions.

V NUMERICAL METHODS FOR SOLVING THE TRANSPORT EQUATION

The complexity of the equations describing particle-transport processes usually force one to implement numerical (i.e., computer-based) methods of solution. Some methods introduce approximations that convert the integro-differential (or integral) form of the transport equation into a system of linear algebraic equations that are most amenable to solution by a computer (e.g., the discrete ordinates finite difference method). Depending on the method, there may be a good deal of analytical treatment before the numerical procedures are implemented (as in methods based on the principles of invariance). These methods, in general, can account for strongly anisotropic scattering, realistic boundary conditions, and can give highly accurate results when solved rigorously on minicomputer systems. We have recently developed a discrete ordinates/exact kernel method that gave good results for the general canopy problems. This method is discussed in detail in Section V.C. The main emphasis here is to present the method and a few important equations along with some results whenever they are available.

V.A   Successive Orders of Scattering Approximations

The method of Successive Orders of Scattering Approximations (SOSA) is one of the oldest and simplest in concept of all the different solutions to the multiple-scattering problem. Various methodologies of SOSA have been developed over the years, but the common denominator is the computation of photons scattered once, twice, three times, etc., with the total scattered intensity obtained as the sum over all orders (Dave, 1964; Irvine, 1965; Hansen and Travis, 1974). Thus, the total emergent intensities may be obtained as

$$I(0, \boldsymbol{\Omega}) = \sum_{n=1}^{\infty} I_n(0, \boldsymbol{\Omega}) \tag{113a}$$

$$I(\tau_T, \boldsymbol{\Omega}) = \sum_{n=1}^{\infty} I_n(\tau_T, \boldsymbol{\Omega}) \tag{113b}$$

where the subscript n denotes the order of scattering. Similarly, the reflection and transmission matrices are given by

$$R(\boldsymbol{\Omega}, \boldsymbol{\Omega}_0) = \sum_{n=1}^{\infty} R_n(\boldsymbol{\Omega}, \boldsymbol{\Omega}_0) = - \sum_{n=1}^{\infty} \frac{I_n(0, \boldsymbol{\Omega})}{\mu_0 I_0} \tag{114a}$$

$$T(\boldsymbol{\Omega}, \boldsymbol{\Omega}_0) = \sum_{n=1}^{\infty} T_n(\boldsymbol{\Omega}, \boldsymbol{\Omega}_0) = - \sum_{n=1}^{\infty} \frac{I_n(\tau_T, \boldsymbol{\Omega})}{\mu_0 I_0} \tag{114b}$$

We consider the standard two-angle problem described in Section II.D. The governing equation may be written as

$$-\mu \frac{\partial I}{\partial \tau} + I(\tau, \boldsymbol{\Omega}) = S(\tau, \boldsymbol{\Omega}) + J(\tau, \boldsymbol{\Omega}) \tag{115}$$

where

$$S(\tau, \boldsymbol{\Omega}) = \frac{\omega}{4\pi} \int_{4\pi} d\boldsymbol{\Omega}' \, P(\boldsymbol{\Omega}' \to \boldsymbol{\Omega}) \, I(\tau, \boldsymbol{\Omega}') \tag{116}$$

$$Q(\tau, \boldsymbol{\Omega}) = \frac{\omega}{4} I_0 P(\boldsymbol{\Omega}_0 \to \boldsymbol{\Omega}) \exp\left(\frac{\tau}{\mu_0}\right) \tag{117}$$

The associated boundary conditions are given by Eqs. 34 and 35. There are two methods of SOSA that are different enough to warrant separate treatment and these will be discussed in detail in what follows.

V.A.1   The Classical Method of τ Integration

The formal solution of the transport equation (Eq. 115) subject to the boundary conditions specified by Eqs. 34 and 35 is

$$I(\tau, \boldsymbol{\Omega}) = \int_{\tau}^{\tau_T} \frac{d\tau'}{\mu} J(\tau', \boldsymbol{\Omega}) \exp\left[\frac{-(\tau' - \tau)}{\mu}\right] \qquad \mu > 0 \tag{118}$$

and

$$I(\tau, \Omega) = -\int_0^\tau \frac{d\tau'}{\mu} J(\tau', \Omega) \exp\left[\frac{(\tau - \tau')}{\mu}\right] \qquad \mu < 0 \qquad (119)$$

where $J(\tau, \Omega)$ is the source term. For the first order of scattering, the source function $J(\tau, \Omega) = Q(\tau, \Omega)$, since $S(\tau, \Omega) = 0$. Thus, substitution of Eq. 117 into Eqs. 118 and 119, and solving for the emergent flux distribution gives

$$I(0, \Omega) = \frac{\omega |\mu_0| I_0}{4(\mu + |\mu|)} P(\Omega_0 \rightarrow \Omega) \left[1 - \exp\left[-\tau_T\left(\frac{1}{\mu} - \frac{1}{\mu_0}\right)\right]\right] \qquad \mu > 0$$

$$(120)$$

$$I(\tau_T, \Omega) = \frac{\omega |\mu_0| I_0}{4(|\mu| - |\mu_0|)} P(\Omega_0 \rightarrow \Omega) \left[\exp\left(\frac{\tau_T}{\mu}\right) - \exp\left(\frac{\tau_T}{\mu_0}\right)\right]$$

$$\mu < 0 \text{ and } \mu \neq \mu_0 \qquad (121)$$

and if $\mu = \mu_0$, then

$$I(\tau_T, \Omega) = -\frac{\omega \tau_T I_0}{4\mu_0} P(\Omega_0 \rightarrow \Omega) \exp\left(\frac{\tau_T}{\mu_0}\right) \qquad \mu < 0 \qquad (122)$$

Now, the source functions and intensity for each higher order scattering can be obtained successively by means of recursion principles:

$$J_{n+1}(\tau, \Omega) = \frac{\omega}{4\pi} \int_{4\pi} d\Omega' \, P(\Omega' \rightarrow \Omega) I_n(\tau, \Omega') \qquad (123)$$

$$I_n(\tau, \Omega) = \int_\tau^{\tau_T} \frac{d\tau'}{\mu} J_n(\tau', \Omega) \exp\left[\frac{-(\tau' - \tau)}{\mu}\right] \qquad \mu > 0 \qquad (124)$$

$$I_n(\tau, \Omega) = -\int_0^\tau \frac{d\tau'}{\mu} J_n(\tau', \Omega) \exp\left[\frac{\tau - \tau'}{\mu}\right] \qquad \mu < 0 \qquad (125)$$

In the previous equations, $n \geq 1$ and the zeroth-order intensity is

$$I_0(\tau, \Omega) = \pi I_0 \exp(\tau_T/\mu_0) \delta(\mu - \mu_0) \delta(\phi - \phi_0) \qquad (126)$$

One method of numerically solving these equations is to expand the scattering transfer function in Legendre polynomials and the azimuthal dependence of the intensity and the source function is a cosine series [as in Section IV.B] and evaluate the emergent distributions successively for each azimuthal component. The other method is to perform the τ and Ω integrals numerically using Gauss quadrature and constructing a transfer matrix for scattering from a direction to all possible directions in the quadrature set. This method has been used by Myneni et al. (1987a, b) to compute the profiles of scattered flux and the emergent distributions from a soybean canopy. The usefulness of this method as

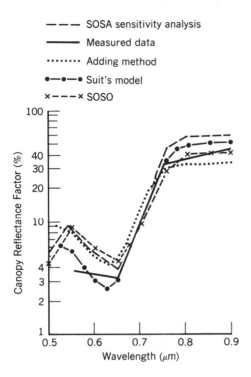

- - - - SOSA sensitivity analysis
——— Measured data
········· Adding method
•—•—• Suit's model
×----× SOSO

Figure 7 Soybean-canopy reflectance factors at different wavelengths for a nadir-view direction and for a solar zenith angle of 145°. The results of the adding method and the Suits model are from Cooper et al. (1982). The measured data are from Ranson and Biehl (1982). For a definition of the reflectance factor, see Eq. 181.

compared with other methods of computing the reflectance factors of a soybean canopy can be seen in Figures 7 and 8 (the measured data are of Ranson and Biehl, 1982).

V.A.2   The Method of SOSA Based on an Invariance Principle

For a homogeneous vegetation canopy, the numerical τ integration (Eqs. 124 and 125) can be avoided in a manner somewhat more elegant than that previously described. This method is based on an invariance principle: *if a layer having the same optical properties is added to the top of a homogeneous layer, the reflection and the transmission matrices are the same as they would be if the second layer was instead added at the bottom.* Hence, the reflection and the transmission matrices for the nth order of scattering ($n >$ 1) may be obtained from the same matrices but for lower orders of scattering. This hybrid formulation was first suggested by Uesugi and Irvine (1970) and was later developed by Hansen and Travis (1974). Recently, this method was applied to the canopy transport problem by Myneni et al. (1987c), where a complete derivation of the equations for the reflection and transmission matrices is presented.

Consider a homogeneous canopy of optical thickness τ (for a problem description, see Section II.D). A thin layer of optical thickness $\Delta\tau(\Delta\tau \ll 1)$ having the same optical properties is added to the top of this canopy (Figure 9). Furthermore, let $\Delta\tau$ be so small that $\Delta\tau^2$ can be safely neglected and at most only single scattering is possible within $\Delta\tau$. There are four possible scattering events that may augment or diminish the reflection matrix (A_1, B_1, C_1, and D_1) and similarly four events that affect the transmission matrix (A_2, B_2, C_2, and D_2). Let ΔR_{ij} and ΔT_{ij} denote changes in the reflection and the transmission matrices, respectively, where i denotes if the added layer is at the top ($i = 1$) or bottom ($i = 2$), and j denotes the scattering events A through D. It is possible

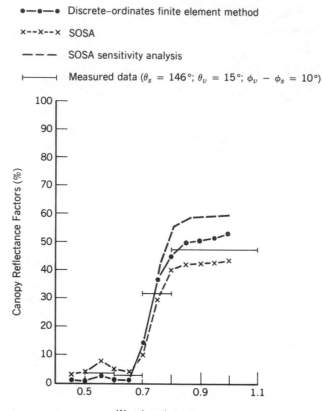

Figure 8 Soybean-canopy reflectance factors at different wavelengths. The sun and view zenith angles are $\theta_0 = 146°$ and $\theta = 15°$, respectively. The azimuthal difference between the sun and the sensor $(\phi - \phi_0)$ is $10°$. The results of the discrete-ordinates finite element method are from Gerstl and Zardecki (1985b). The measured data are from Ranson and Biehl (1982). For a definition of the reflectance factor, see Eq. 181.

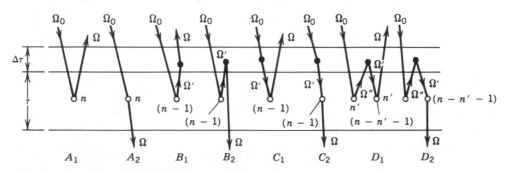

Figure 9 Schematic representation of nth-order diffuse reflection and transmission when an optically thin layer is added at the top of the principal layer.

to write analytical expressions for the eight scattering events shown in Figure 9. For instance, ΔR_{1D} can be written as

$$\Delta R_{1D} = \frac{\omega \Delta \tau}{4\pi} \sum_{n'=1}^{n-2} \int_{2\pi} d\Omega' \, R_{n-n'-1}(\Omega, \Omega')$$
$$\cdot \left[\frac{1}{\pi} \int_{2\pi} d\hat{\Omega} \, P(\hat{\Omega} \to \Omega') \, R_{n'}(\hat{\Omega}, \Omega_0) \right] \quad n > 2 \quad (127)$$

Likewise, adding a layer at the bottom of the principal layer results in eight scattering events, A_1 through D_2 and the corresponding analytical expressions $\Delta R_{2A}, \Delta T_{2A}, \ldots, \Delta T_{2D}$. From the invariance principle stated earlier, we have

$$(\Delta R_{1A} + \Delta R_{1B} + \Delta R_{1C} + \Delta R_{1D}) - (\Delta R_{2A} + \Delta R_{2B} + \Delta R_{2C} + \Delta R_{2D}) = 0$$
$$(128a)$$

$$(\Delta T_{1A} + \Delta T_{1B} + \Delta T_{1C} + \Delta T_{1D}) - (\Delta T_{2A} + \Delta T_{2B} + \Delta T_{2C} + \Delta T_{2D}) = 0$$
$$(128b)$$

After substituting the expressions for ΔR_{ij} and ΔT_{ij}, and neglecting terms of $\Delta \tau^2$, one obtains the final expressions for computing the nth-order reflection and transmission matrices:

$$\left[\frac{1}{\mu} - \frac{1}{\mu_0} \right] R_n(\Omega, \Omega_0) = \frac{\omega}{4\pi\mu} \int_{2\pi} d\Omega' \, P(\Omega' \to \Omega) \, R_{n-1}(\Omega', \Omega_0)$$

$$- \frac{\omega}{4\pi\mu_0} \int_{2\pi} d\Omega' \, R_{n-1}(\Omega, \Omega') \, P(\Omega_0 \to \Omega')$$

$$+ \frac{\omega}{4\pi\mu_0} \exp\left(\frac{\tau_T}{\mu_0} \right) \int_{2\pi} d\Omega' \, T_{n-1}(\Omega, \Omega') \, P(\Omega_0 \to \Omega')$$

$$- \frac{\omega}{4\pi\mu} \exp\left(\frac{-\tau_T}{\mu} \right) \int_{2\pi} d\Omega' \, P(\Omega' \to \Omega) \, T_{n-1}(\Omega', \Omega_0)$$

$$+ \sum_{n'=1}^{n-2} \frac{\omega}{4\pi} \int_{2\pi} d\Omega' \, R_{n-n'-1}(\Omega, \Omega')$$

$$\cdot \left[\frac{1}{\pi} \int_{2\pi} d\hat{\Omega} \, P(\hat{\Omega} \to \Omega') \, R_{n'}(\hat{\Omega}, \Omega_0) \right]$$

$$- \sum_{n'=1}^{n-2} \frac{\omega}{4\pi} \int_{2\pi} d\Omega' \, T_{n-n'-1}(\Omega, \Omega')$$

$$\cdot \left[\frac{1}{\pi} \int_{2\pi} d\hat{\Omega} \, P(\hat{\Omega} \to \Omega') \, T_{n'}(\hat{\Omega}, \Omega_0) \right] \quad (129)$$

and

$$\left[\frac{1}{\mu} + \frac{1}{\mu_0}\right] T_n(\mathbf{\Omega}, \mathbf{\Omega}_0) = \frac{\omega}{4\pi\mu} \int_{2\pi} d\mathbf{\Omega}' \, P(\mathbf{\Omega}' \to \mathbf{\Omega}) \, T_{n-1}(\mathbf{\Omega}', \mathbf{\Omega}_0)$$

$$+ \frac{\omega}{4\pi\mu_0} \int_{2\pi} d\mathbf{\Omega}' \, T_{n-1}(\mathbf{\Omega}, \mathbf{\Omega}') \, P(\mathbf{\Omega}_0 \to \mathbf{\Omega}')$$

$$- \frac{\omega}{4\pi\mu_0} \exp\left(\frac{\tau_T}{\mu_0}\right) \int_{2\pi} d\mathbf{\Omega}' \, R_{n-1}(\mathbf{\Omega}, \mathbf{\Omega}') \, P(\mathbf{\Omega}_0 \to \mathbf{\Omega}')$$

$$- \frac{\omega}{4\pi\mu} \exp\left(\frac{-\tau_T}{\mu}\right) \int_{2\pi} d\mathbf{\Omega}' \, P(\mathbf{\Omega}' \to \mathbf{\Omega}) \, R_{n-1}(\mathbf{\Omega}', \mathbf{\Omega}_0)$$

$$+ \sum_{n'=1}^{n-2} \frac{\omega}{4\pi} \int_{2\pi} d\mathbf{\Omega}' \, R_{n-n'-1}(\mathbf{\Omega}, \mathbf{\Omega}')$$

$$\cdot \left[\frac{1}{\pi} \int_{2\pi} d\hat{\mathbf{\Omega}} \, P(\hat{\mathbf{\Omega}} \to \mathbf{\Omega}') \, T_{n'}(\hat{\mathbf{\Omega}}, \mathbf{\Omega}_0)\right]$$

$$- \sum_{n'=1}^{n-2} \frac{\omega}{4\pi} \int_{2\pi} d\mathbf{\Omega}' \, T_{n-n'-1}(\mathbf{\Omega}, \mathbf{\Omega}')$$

$$\cdot \left[\frac{1}{\pi} \int_{2\pi} d\hat{\mathbf{\Omega}} \, P(\hat{\mathbf{\Omega}} \to \mathbf{\Omega}') \, R_{n'}(\hat{\mathbf{\Omega}}, \mathbf{\Omega}_0)\right] \qquad (130)$$

Note that for $n = 2$, the last two terms in Eqs. 129 and 130 should be neglected. As is apparent, the formulation involves integrals over the angular coordinates only, and these may be evaluated accurately using Gauss quadrature. This method was applied to study the anisotropy in the emergent flux distributions in soybean canopies by Myneni et al. (1987c).

The main advantage of the method of SOSA is that it provides insight useful for interpreting multiple-scattering results. Also, one obtains for as many depths (in the classical formulation) and directions as the order of the quadrature set used in evaluating the integrals. The chief drawback is that the convergence is slow in strongly scattering layers ($\omega \approx 1$; e.g., scattering by leaf canopies at the near-infrared wavelengths).

V.B Method of Gauss–Seidel Iteration

We now consider the standard two-angle problem described in Section II.D. The governing equation is Eq. 33 and the associated boundary conditions are specified by Eqs. 34 and 35. The complete formal solution of the transport equation (Eq. 33) subject to the boundary conditions is

$$I(\tau, \mathbf{\Omega}) = I(\tau_T, \mathbf{\Omega}) \exp\left[\frac{-(\tau_T - \tau)}{\mu}\right] + \int_\tau^{\tau_T} \frac{d\tau'}{\mu} J(\tau', \mathbf{\Omega})$$

$$\cdot \exp\left[\frac{-(\tau' - \tau)}{\mu}\right] \qquad \mu > 0 \qquad (131)$$

$$I(\tau, \mathbf{\Omega}) = I(0, \mathbf{\Omega}) \exp\left(\frac{\tau}{\mu}\right) - \int_0^\tau \frac{d\tau'}{\mu} J(\tau', \mathbf{\Omega}) \exp\left(\frac{\tau - \tau'}{\mu}\right) \qquad \mu < 0 \quad (132)$$

The vegetation canopy is now split into N contiguous layers, each of thickness $\Delta\tau = \tau_T/N$. We also discretize the angular variable by choosing a finite set of M directions in the unit sphere, Ω_j, and denote the associated weights as $w_j(j = 1, \ldots, M)$. For the spatial interval (τ_i, τ_{i+2}), Eqs. 131 and 132 can be approximated as

$$I(\tau_i, \Omega_j) = I(\tau_{i+2}, \Omega_j) \exp\left(\frac{-2\,\Delta\tau}{\mu_j}\right) + \int_{\tau_i}^{\tau_{i+2}} \frac{d\tau'}{\mu_j} J(\tau', \Omega_j)$$

$$\cdot \exp\left[\frac{-(\tau' - \tau_i)}{\mu_j}\right] \qquad u_j > 0 \tag{133}$$

$$I(\tau_{i+2}, \Omega_j) = I(\tau_i, \Omega_j) \exp\left(\frac{2\,\Delta\tau}{\mu_j}\right) - \int_{\tau_i}^{\tau_{i+2}} \frac{d\tau'}{\mu_j} J(\tau', \Omega_j)$$

$$\cdot \exp\left(\frac{\tau_{i+2} - \tau'}{\mu_j}\right) \qquad \mu_j < 0 \tag{134}$$

where J is the source term. We now introduce the following approximations, all of which are valid so long as $\Delta\tau$ is small:

$$\int_{\tau_i}^{\tau_{i+2}} \frac{d\tau'}{\mu_j} J(\tau', \Omega_j) \exp\left[\frac{(\tau' - \tau_i)}{\mu_j}\right] = \int_{\tau_i}^{\tau_{i+2}} \frac{d\tau'}{\mu_j} J(\tau', \Omega_j) \exp\left(\frac{\tau_{i+2} - \tau'}{\mu_j}\right)$$

$$\equiv J(\tau_{i+1}, \Omega_j)\left[1 - \exp\left(-2\,\Delta\tau/|\mu_j|\right)\right]$$

$$\tag{135}$$

and for higher orders of scattering,

$$J(\tau_{i+1}, \Omega_j) = \frac{\omega}{4\pi} \sum_{k=1}^{M} w_k P(\Omega_k \to \Omega_j) I(\tau_{i+1}, \Omega_k) + Q(\tau_{i+1}, \Omega_j) \tag{136}$$

where P is now the transfer matrix for scattering from a direction to all other directions in the quadrature set (of dimensions $M \times M$).

In this fashion, the formal solution (Eqs. 131 and 132) is replaced by a system of linear algebraic equations with the unknowns $I(\tau_i, \Omega_j)$, $i = 1, \ldots, N$ and $j = 1, \ldots, M$. These equations can be easily solved by the one-step Gauss–Seidel iteration procedure. The downward fluxes are computed from layer to layer for every iteration step. For instance, in the nth iteration step, the values of the downward fluxes of the layer $i + 2$ are computed from the downward fluxes of layers i and $i + 1$ of the same iteration step and from the upward flux values of the layer $i + 1$ of the iteration step $n - 1$. For a more detailed discussion of this method, the reader is referred to Herman and Browning (1965). Dave (1970). and of Dave and Gazdag (1970). This method, according to Dave, is about 50% faster than the classical SOSA method (cf. Hansen and Travis, 1974, p. 595).

The advantages of this method are that the internal radiation field is readily obtained without additional labor and that it is possible to treat vertical inhomogeneities. The chief limitations is that it becomes too tedious for deep plant canopies (ca. LAI > 6).

V.C   The Method of Discrete Ordinates

In the discrete-ordinates method (also called the S_n method in neutron-transport theory), photons are restricted to travel in $2n$ discrete directions and the angular integrals are evaluated very accurately by Gaussian quadrature. The limit $n = 1$, corresponds to the two-stream method with one upward and one downward stream. For large n, the method in principle becomes quite accurate. By taking the limit $n \to \infty$, the method of eigenvalues is obtained (see Section IV.A). The classical discrete ordinates method of Wick (1943) and Chandrasekhar (1944, 1960) will not be discussed here (for the general case of anisotropic scattering, see Lenoble, 1985, pp. 36–39); instead, we will discuss a variant that we will call the *discrete-ordinates exact-kernel method*.

V.C.1   Problem Description

Consider a flat horizontal leaf canopy of depth T, which is illuminated spatially uniformly on the top and bounded by an isotropically reflecting soil at the bottom. The governing equation after separation of collided and uncollided intensities is [Section II.C]

$$-\mu \frac{\partial I}{\partial \tau}(\tau, \mathbf{\Omega}) + G(\mathbf{\Omega}) I(\tau, \mathbf{\Omega}) = \frac{\omega}{4\pi} \int_{4\pi} d\mathbf{\Omega}' \, P(\mathbf{\Omega}' \to \mathbf{\Omega}) \, G(\mathbf{\Omega}') \, I(\tau, \mathbf{\Omega}')$$
$$+ \frac{\omega}{4\pi} \int_{4\pi} d\mathbf{\Omega}' \, P(\mathbf{\Omega}' \to \mathbf{\Omega}) \, G(\mathbf{\Omega}') \, I^0(\tau, \mathbf{\Omega}')$$

(137)

where I^0 is the intensity of uncollided photons. For $\mu < 0$,

$$I^0(\tau, \mathbf{\Omega}) = I(0, \mathbf{\Omega}) \exp\left[G(\mathbf{\Omega})\tau/\mu\right]$$

(138)

and for $\mu > 0$,

$$I^0(\tau, \mathbf{\Omega}) = \frac{r_s}{\pi} \exp\left[\frac{-G(\mathbf{\Omega})(\tau_T - \tau)}{\mu}\right] \int_0^{2\pi} d\phi' \int_{-1}^0$$
$$\cdot \, d\mu' \, |\mu'| I(0, \mathbf{\Omega}') \exp\left[\frac{G(\mathbf{\Omega}')\tau_T}{\mu'}\right]$$

(139)

where r_s is the hemispherical soil reflectance. The canopy is illuminated from above by both a direct monodirectional solar component (in direction $\mathbf{\Omega}_0(\mu_0, \phi_0)$, $\mu_0 < 0$) and by diffusion radiation from the sky. Thus,

$$I(0, \mathbf{\Omega}) = I_d(\mathbf{\Omega}) + I_0 \delta(\mathbf{\Omega} - \mathbf{\Omega}_0)$$

(140)

where I_d is the diffuse component (often assumed isotropic), and I_0 is the intensity of the direct solar radiation perpendicular to its direction of propagation. The optical depth τ here is also commonly refered to as the cumulative leaf-area index (one-sided green leaf area per unit ground area) and differs from the "mean free path" used in neutron-transport theory. The geometry factor $G(\mathbf{\Omega})$ is physically the fraction of the total leaf area (per unit volume of the canopy) that is perpendicular to the direction $\mathbf{\Omega}$ [see Section

III.A]. Finally, the boundary conditions are

$$I(0, \Omega) = 0 \quad \mu < 0 \tag{141}$$

and,

$$I(\tau_T, \Omega) = \frac{r_s}{\pi} \int_0^{2\pi} d\phi' \int_{-1}^0 d\mu' \, |\mu'| I(\tau_T, \Omega') \quad \mu > 0 \tag{142}$$

For a more detailed discussion on the canopy problem, see Section II of this chapter. Evaluation of the $G(\Omega)$ function is discussed in Section III.A.

V.C.2 Angular Discretization of the Transport Equation

We approximate the angular dependence of the transport equation by the discrete-ordinates method, where the angular variables μ and ϕ are discretized into a set of $N \times M$ discrete directions at which the angular flux is to be evaluated. The scattering source terms are evaluated by numerical quadrature, where (μ_i, ϕ_j) values are the quadrature ordinates. The set of corresponding weights is denoted by w_i and w_j. With this approximation, Eq. 137 can be written as

$$-\mu_i \frac{\partial I}{\partial \tau} (\tau, \Omega_{ij}) + G(\Omega_{ij}) I(\tau, \Omega_{ij})$$
$$= S(\tau, \Omega_{ij}) + Q(\tau, \Omega_{ij}) \quad i = 1, \ldots, N \text{ and } j = 1, \ldots, M \tag{143}$$

where the *distributed source* term is

$$S(\tau, \Omega_{ij}) = \frac{\omega}{4\pi} \sum_{m=1}^M w_m \sum_{n=1}^N w_n P(\Omega_{nm} \to \Omega_{ij}) G(\Omega_{nm}) I(\tau, \Omega_{nm}) \tag{144}$$

and the first collision source term is

$$Q(\tau, \Omega_{ij}) = \frac{\omega}{4\pi} \sum_{m=1}^M w_m \sum_{n-1}^N w_n P(\Omega_{nm} \to \Omega_{ij}) G(\Omega_{nm}) I^0(\tau, \Omega_{nm}) \tag{145}$$

The exact-kernel method of evaluating the scattering source terms will be discussed later. The accuracy that can be achieved in solving the discrete-ordinates equation when Eqs. 144 and 145 are used to evaluate the scattering source term is largely dependent on the quadrature set used. In general, one would like to use a set that is large enough to adequately describe the angular detail in the fluxes, yet small enough so that excessive computational effort is not involved. The choice of such an optimum set is typically problem-dependent, especially when anisotropic redistribution is involved. Unfortunately, there is no standard procedure for choosing a priori an optimum set. The choice is usually made by trial and error. However, as a general rule, the following criteria should be met (Duderstadt and Martin, 1979);

(a) *Projection Invariance:* For one-dimensional slab geometry with azimuthal symmetry, the discrete directions should be symmetric about $\mu = 0$. Physically, this corresponds to treating photons traveling downwards ($\mu < 0$) the same as photons traveling upwards ($\mu > 0$). However, if one knows that the angular flux is peaked near a certain direction μ_i, it may be advantageous to tailor a nonsymmetric quadrature set with several directions clustered near μ_i.

(b) *Positivity of the Scalar Flux:* The zeroth moment of the angular flux, the scalar flux,

$$
I'(\tau) = 2\pi \int_{-1}^{1} d\mu\, I(\tau, \mu) \approx 2\pi \sum_{i=1}^{N} w_n I(\tau, \mu_i) \tag{146}
$$

should always be positive. Choosing $w_i > 0$ will ensure positivity, as long as the angular flux is positive. It is possible in the case of iteration to obtain negative fluxes. In such a case, a negative flux fix-up procedure must be employed (Section V.C.5).

(c) *Accurate Evaluation of the Angular Integral:* The scalar flux and the scattering sources should be evaluated accurately with a minimum of quadrature ordinates and weights.

In view of the preceding, Gauss quadrature is considered ideal. Because of the discontinuity of the angular flux at the plane interface, $\mu = 0$, even-order quadrature sets are choosen. In many cases, to avoid this discontinuity, one splits the angular integration range into two parts, $-1 \leq \mu < 0$ and $0 \leq \mu < 1$, and performs Gaussian quadrature separately over each subrange. This approach is called the double-P_n method (DP_n).

V.C.3 *Spatial Discretization*

There are two methods of effecting a numerical solution to the angle-discretized transport equation (Eq. 143). The conventional method involves approximating the spatial derivative in the transport equation by a finite difference scheme and using the diamond difference relations to relate cell-edged fluxes to the cell-centered fluxes. The other method involves using a finite element formulation based on a unique implementation of the Galerkin integral law. The two methods are discussed in what follows.

The Finite Difference Method. The transport equation is numerically solved by forming a set of finite difference equations obtained by discretizing the spatial variable τ into a set of spatial nodes $[\tau_k]$. Let the upper and the lower boundaries of the vegetation canopy be τ_1 and τ_{K+1}, respectively. The spatial derivative of the angular flux in Eq. 143 can be approximated as

$$
\frac{\partial}{\partial \tau} I(\tau_{k+0.5}, \Omega_{ij}) = \frac{I(\tau_{k+1}, \Omega_{ij}) - I(\tau_k, \Omega_{ij})}{\Delta_{k+0.5}} \tag{147}
$$

where

$$
\tau_{k+0.5} = \frac{\tau_k + \tau_{k+1}}{2} \tag{148}
$$

and

$$\Delta_{k+0.5} = \tau_{k+1} - \tau_k \tag{149}$$

We thus obtain the finite difference form of the discrete-ordinates equations as

$$-\mu_i \frac{I(\tau_{k+1}, \mathbf{\Omega}_{ij}) - I(\tau_k, \mathbf{\Omega}_{ij})}{\Delta_{k+0.5}} + G(\mathbf{\Omega}_{ij}) I(\tau_{k+0.5}, \mathbf{\Omega}_{ij})$$
$$= S(\tau_{k+0.5}, \mathbf{\Omega}_{ij}) + Q(\tau_{k+0.5}, \mathbf{\Omega}_{ij})$$
$$k = 1, \ldots, K; \, i = 1, \ldots, N, \text{ and } j = 1, \ldots, M \tag{150}$$

Before the set of equations represented by Eq. 150 can be solved, it is necessary to reduce the number of unknowns by assuming a relation between cell-edged and cell-centered fluxes. This can be done by using the diamond difference scheme with a weighting factor α:

$$I(\tau_{k+0.05}, \mathbf{\Omega}_{ij}) = (1 - \alpha) I(\tau_k, \mathbf{\Omega}_{ij}) + \alpha I(\tau_{k+1}, \mathbf{\Omega}_{ij}) \qquad \mu < 0 \tag{151}$$

$$I(\tau_{k+0.5}, \mathbf{\Omega}_{ij}) = (1 - \alpha) I(\tau_{k+1}, \mathbf{\Omega}_{ij}) + \alpha I(\tau_k, \mathbf{\Omega}_{ij}) \qquad \mu > 0 \tag{152}$$

Note that here if $\alpha = 0.5$, we obtain the standard diamond difference relations. The weighted diamond difference relations are used whenever a negative flux is encountered.

Equation 150 can be solved for $I(\tau_{k+1}, \mathbf{\Omega}_{ij})$ in terms of $I(\tau_k, \mathbf{\Omega}_{ij})$ after substituting the relation in Eq. 151 as

$$I(\tau_{k+1}, \mathbf{\Omega}_{ij}) = a_{ij} I(\tau_k, \mathbf{\Omega}_{ij}) - b_{ij} J(\tau_{k+0.5}, \mathbf{\Omega}_{ij}) \qquad \mu < 0 \tag{153}$$

or for $I(\tau_k, \mathbf{\Omega}_{ij})$ in terms of $I(\tau_{k+1}, \mathbf{\Omega}_{ij})$ after substituting the relation in Eq. 152 as

$$I(\tau_k, \mathbf{\Omega}_{ij}) = c_{ij} I(\tau_{k+1}, \mathbf{\Omega}_{ij}) - d_{ij} J(\tau_{k+0.5}, \mathbf{\Omega}_{ij}) \qquad \mu > 0 \tag{154}$$

In the previous equations,

$$a_{ij} = [1 + f_{ij}(1 - \alpha)]/(1 - f_{ij}\alpha) \tag{155}$$

$$b_{ij} = f_{ij}/(1 - f_{ij}\alpha) \tag{156}$$

$$c_{ij} = [1 - f_{ij}(1 - \alpha)]/(1 + f_{ij}\alpha) \tag{157}$$

$$d_{ij} = f_{ij}/(1 + f_{ij}\alpha) \tag{158}$$

$$f_{ij} = [G(\mathbf{\Omega}_{ij})\Delta_{k+0.5}]/\mu_i \tag{159}$$

$$J(\tau_{k+0.05}, \mathbf{\Omega}_{ij}) = S(\tau_{k+0.5}, \mathbf{\Omega}_{ij}) + Q(\tau_{k+0.05}, \mathbf{\Omega}_{ij}) \tag{160}$$

These equations are of the standard form (Duderstadt and Martin, 1979) except for a directionally dependent geometry factor (or cross section) $G(\mathbf{\Omega}_{ij})$ in the coefficients a_{ij} through d_{ij}. Nevertheless, this difference presents no computational difficulties, and Eqs.

153 and 154 may be solved by the standard inward-outward technique. Use Eq. 138 to obtain the incident intensity at τ_1 in the inward direction ($\mu < 0$), and use Eq. 153 to evaluate I at successively lower canopy levels for all Ω_{ij} with $\mu < 0$. Now use the boundary condition (Eq. 139) to evaluate the upward intensities $\mu > 0$ at the ground and sweep upwards using Eq. 154 to find I at successively higher levels for all Ω_{ij} with $\mu > 0$. Once this inward-outward sweep has been completed, reevaluate J from Eqs. 144 and 145, using the average of the cell-edged values to estimate the cell-centered values $J(\tau_{k+0.5}, \Omega_{ij})$. This sweep and source reestimation is repeated until convergence is achieved.

The Finite Element Method. The application of the finite element method in the solution of the discrete-ordinates equations is based on a unique implementation of the Galerkin integral law formulation of the transport equation (Duderstadt and Martin, 1979). The solution algorithm of traditional discrete-ordinates codes can be retained by applying the finite element method locally on individual mesh cells, and typically one solves the system of linear equations in the direction of photon flight.

We begin by defining the *finite element basis functions*:

$$\xi_k(\tau) = \frac{\tau_{k+1} - \tau}{\Delta\tau} \qquad \tau_k \le \tau \le \tau_{k+1} \tag{161}$$

and zero otherwise. Similarly,

$$\xi_{k+1}(\tau) = \frac{\tau - \tau_k}{\Delta\tau} \qquad \tau_k \le \tau \le \tau_{k+1} \tag{162}$$

and zero otherwise. In the previous equations, $\Delta\tau = \tau_{k+1} - \tau_k$. Expanding the intensity $[I(\tau, \Omega_{ij})]$ and the source term $[J(\tau, \Omega_{ij}); J(\tau, \Omega_{ij}) = S(\tau, \Omega_{ij}) + Q(\tau, \Omega_{ij})]$ as

$$I(\tau, \Omega_{ij}) = I(\tau_k, \Omega_{ij})\,\xi_k(\tau) + I(\tau_{k+1}, \Omega_{ij})\,\xi_{k+1}(\tau) \tag{163}$$

$$J(\tau, \Omega_{ij}) = J(\tau_k, \Omega_{ij})\,\xi_k(\tau) + J(\tau_{k+1}, \Omega_{ij})\,\xi_{k+1}(\tau) \tag{164}$$

and substituting the explicit forms of $\xi_k(\tau)$ and $\xi_{k+1}(\tau)$, we obtain

$$I(\tau, \Omega_{ij}) = \frac{1}{\Delta\tau}\left[(\tau_{k+1} - \tau)\,I(\tau_k, \Omega_{ij}) + (\tau - \tau_k)\,I(\tau_{k+1}, \Omega_{ij})\right] \tag{165}$$

$$J(\tau, \Omega_{ij}) = \frac{1}{\Delta\tau}\left[(\tau_{k+1} - \tau)\,J(\tau_k, \Omega_{ij}) + (\tau - \tau_k)\,J(\tau_{k+1}, \Omega_{ij})\right] \tag{166}$$

Introducing these expansions in the angle-discretized transport equation (Eq. 143), we obtain

$$-\mu_i \frac{\partial I}{\partial \tau}(\tau, \Omega_{ij}) + \frac{G(\Omega_{ij})}{\Delta t}\left[(\tau_{k+1} - \tau)\,I(\tau_k, \Omega_{ij}) + (\tau - \tau_k)\,I(\tau_{k+1}, \Omega_{ij})\right]$$

$$= \frac{1}{\Delta\tau}\left[(\tau_{k+1} + \tau)\,J(\tau_k, \Omega_{ij}) + (\tau - \tau_k)\,J(\tau_{k+1}, \Omega_{ij})\right] \tag{167}$$

If the weighting functions for the finite element method are chosen to be the basis functions $[\xi_k(\tau), \xi_{k+1}(\tau)]$, the resulting sweep equations will be quite complicated. Hence, we choose the alternative weighting functions (Hill, 1975),

$$I_1(\tau) = 1 \quad \text{and} \quad I_2(\tau) = \tau - \tau_k \quad \text{for } \mu_i < 0 \qquad (168)$$

and

$$I_1(\tau) = 1 \quad \text{and} \quad I_2(\tau) = \tau_{k+1} - \tau \quad \text{for } \mu_i > 0 \qquad (169)$$

Hill (1975) shows that the use of these weighting functions is equivalent to Galerkin weighting and, more importantly, the resulting sweep equations are considerably simpler.

The Galerkin formulation of the finite element method consists of multiplying Eq. 167 by the finite element basis functions (Eqs. 168 and 169) and integrating over the mesh cell. Consider the case when $\mu_i < 0$:

$$-\mu_i \int_{\tau_k}^{\tau_{k+1}} d\tau \, 1 \, \frac{\partial I}{\partial \tau}(\tau, \mathbf{\Omega}_{ij}) + \frac{G(\mathbf{\Omega}_{ij})}{\Delta \tau} \int_{\tau_k}^{\tau_{k+1}} d\tau \, 1$$

$$\cdot \left[(\tau_{k+1} - \tau) I(\tau_k, \mathbf{\Omega}_{ij}) + (\tau - \tau_k) I(\tau_{k+1}, \mathbf{\Omega}_{ij})\right]$$

$$= \frac{1}{\Delta \tau} \int_{\tau_k}^{\tau_{k+1}} d\tau \, 1 \left[(\tau_{k+1} - \tau) J(\tau_k, \mathbf{\Omega}_{ij}) + (\tau - \tau_k) J(\tau_{k+1}, \mathbf{\Omega}_{ij})\right] \qquad (170)$$

The streaming term can be integrated and evaluated by imposing the following boundary conditions:

$$\lim_{\tau \to \tau_k} I(\tau, \mathbf{\Omega}_{ij}) = I^B(\mathbf{\Omega}_{ij}) \qquad \lim_{\tau \to \tau_{k+1}} I(\tau, \mathbf{\Omega}_{ij}) = I(\tau_{k+1}, \mathbf{\Omega}_{ij}) \qquad \mu_i < 0 \quad (171)$$

and

$$\lim_{\tau \to \tau_k} I(\tau, \mathbf{\Omega}_{ij}) = I(\tau_k, \mathbf{\Omega}_{ij}) \qquad \lim_{\tau \to \tau_{k+1}} I(\tau, \mathbf{\Omega}_{ij}) = I^B(\mathbf{\Omega}_{ij}) \qquad \mu_i > 0 \quad (172)$$

where $I^B(\mathbf{\Omega}_{ij})$ is the angular flux on the boundary due to the adjacent cell. So the outgoing angular flux from a mesh cell is treated as an incoming source for the adjacent mesh cell. The angular flux just inside the boundary is not constrained to equal the incoming angular flux and, hence, the angular flux can be discontinuous at the cell boundaries. Evaluation of Eq. 170 gives

$$\left[G(\mathbf{\Omega}_{ij}) \frac{\Delta \tau}{2}\right] I(\tau_k, \mathbf{\Omega}_{ij}) + \left[-\mu_i + G(\mathbf{\Omega}_{ij}) \frac{\Delta \tau}{2}\right] I(\tau_{k+1}, \mathbf{\Omega}_{ij})$$

$$= \frac{\Delta \tau}{2} \left[J(\tau_k, \mathbf{\Omega}_{ij}) + J(\tau_{k+1}, \mathbf{\Omega}_{ij})\right] - \mu_i I^B(\mathbf{\Omega}_{ij}) \qquad \mu_i < 0 \qquad (173)$$

The second equation for $\mu_i > 0$ is obtained by weighting Eq. 167 with $I_2(\tau) = (\tau - \tau_k)$:

$$\left[3\mu_i + G(\mathbf{\Omega}_{ij})\,\Delta\tau\right]I(\tau_k, \mathbf{\Omega}_{ij}) + \left[-3\mu_i + 2G(\mathbf{\Omega}_{ij})\,\Delta\tau\right]I(\tau_{k+1}, \mathbf{\Omega}_{ij})$$
$$= 2\,\Delta\tau\left[0.5J(\tau_k, \mathbf{\Omega}_{ij}) + J(\tau_{k+1}, \mathbf{\Omega}_{ij})\right] \qquad \mu_i < 0 \qquad (174)$$

Similarly, for $\mu_i > 0$, we obtain two equations:

$$\left[G(\mathbf{\Omega}_{ij})\frac{\Delta\tau}{2}\right]I(\tau_{k+1}, \mathbf{\Omega}_{ij}) + \left[\mu_i + G(\mathbf{\Omega}_{ij})\frac{\Delta\tau}{2}\right]I(\tau_k, \mathbf{\Omega}_{ij})$$
$$= \frac{\Delta\tau}{2}\left[J(\tau_k, \mathbf{\Omega}_{ij}) + J(\tau_{k+1}, \mathbf{\Omega}_{ij})\right] + \mu_i I^B(\mathbf{\Omega}_{ij}) \qquad \mu_i > 0 \qquad (175)$$

$$\left[-3\mu_i + G(\mathbf{\Omega}_{ij})\,\Delta\tau\right]I(\tau_{k+1}, \mathbf{\Omega}_{ij}) + \left[3\mu_i + 2G(\mathbf{\Omega}_{ij})\,\Delta\tau\right]I(\tau_k, \mathbf{\Omega}_{ij})$$
$$= 2\,\Delta\tau\left[J(\tau_k, \mathbf{\Omega}_{ij}) + 0.5J(\tau_{k+1}, \mathbf{\Omega}_{ij})\right] \qquad \mu_i > 0 \qquad (176)$$

The four equations, Eqs. 173–176, permit evaluation of four unknowns [$I(\tau_k, \mathbf{\Omega}_{ij})$ and $I(\tau_{k+1}, \mathbf{\Omega}_{ij})$ for $\mu_i < 0$; and $I(\tau_k, \mathbf{\Omega}_{ij})$ and $I(\tau_{k+1}, \mathbf{\Omega}_{ij})$ for $\mu_i > 0$] for any mesh cell. Unlike the conventional finite difference formulation, Section V.C.3, we now solve for four unknowns at each mesh cell: two cell-edged angular fluxes in the downward sweep and two in the upward sweep. The disadvantage is that there are more operations per mesh cell and, hence, computing times are large. However, for large optical depths, the number of spatial nodes required for adequate flux convergence is considerably less than in the conventional finite difference formulation. Furthermore, since the diamond difference relations are not used, there is no problem of negative fluxes.

V.C.4 The Exact-Kernel Method

The accuracy that can be achieved in the solution of the discrete-ordinates equations is dependent on, among other factors, the accuracy obtained in the evaluation of the distributed and the first collision source terms. Any numerical technique that purports to accurately evaluate the scattering-source terms relies on some finite expansion methodology, the degree of which is dictated by the anisotropy in the scattered flux. Highly anisotropic scattering situations might arise from a highly peaked scattering transfer function in the monoenergetic case. The most common method of treating such highly anisotropic scattering situations is with a Legendre polynomial expansion approximation of the scattering-source terms (Odom, 1975). For very highly anisotropic scattering kernels, a Legendre polynomial expansion requires a very large number of expansion terms (Hunt, 1964). If the moments of the flux are calculated, the number of moments calculated must be equal to the order of the expansion needed to represent the scattering transfer function, since any lower-order truncation of the moments will neglect the higher-order expansion terms of the scattering transfer approximation. Such higher orders of expansion render the corresponding discrete-ordinates solution infeasible. Odom (1975) analyzed the Legendre expansion approximation and several other expansion techniques (such as the small-angle approximation, the truncated-peak technique, transport approximation, forward-backward approximation) that have been developed to evaluate the scattering-source terms in neutron-transport theory. He concluded that such approximations in practical calculations can sometimes yield accurate results (accuracy $> 95\%$)

for integral quantities such as the scalar flux. However, the results are generally inaccurate (greater than 25% inaccuracy) for detailed angular fluxes, especially at grazing angles.

As shown earlier (Section III.B), the scattering transfer functions in vegetation canopies are generally not rotationally invariant (i.e., they do not depend on just the scattering angle). This lack of rotational invariance precludes the use of Legendre polynomial expansions and the addition theorem for the spherical harmonics expansion of the argument of the Legendre polynomials, which are traditionally used to treat anisotropic media. Consequently, most of the available computer codes for solving anisotropic transport problems cannot be directly applied to canopy problems. Modifications must be made to use the transfer function or its discretized matrix directly and to incorporate the cross section $G(\mathbf{\Omega})$ that is direction-dependent.

The exact-kernel method (after Odom, 1975) involves the computation of a transfer matrix composed of scattering cross sections for every $\mathbf{\Omega}_{nm} \rightarrow \mathbf{\Omega}_{ij}$ transfer. Use of the exact-kernel method guarantees positivity of the computed fluxes, provided certain other conditions are satisfied, and it has been shown to provide accurate results for both neutron- and photon-transport problems (cf. Risner, 1985). As the name implies, the exact-kernel forms of the distributed and the first collision source terms are simply as given in Eqs. 144 and 145, respectively, where $P(\mathbf{\Omega}_{nm} \rightarrow \mathbf{\Omega}_{ij})$ is the exact-kernel cross section for transfer from the direction $\mathbf{\Omega}_{nm}$ to $\mathbf{\Omega}_{ij}$. A blessing in disguise is the exact analytical solutions of the scattering transfer functions in general cases that permit a rapid evaluation of any transfer value.

It should be noted, however, that the exact-kernel technique has two serious drawbacks. The most obvious of the two is the problem of cross-section storage. In the Legendre expansion method, the only cross-section terms required are the $M + 1$ expansion coefficients for the Mth order cross-section expansion. In the exact-kernel method, however, it is necessary to store an $N \times N$ matrix of transfer values (N being the total number of discrete ordinates). In the one-angle problem, storing a 16×16 matrix for DP_8 quadrature may not be a problem even for small computers, but in the two-angle problem, a matrix of $16 \times 16 \times 16 \times 16$ might drastically increase the computation time even on advanced minicomputers. That is why it is so important to exploit the symmetries in the transfer matrix.

The second drawback associated with the exact-kernel technique is that of angular redistribution of scattered photons or the zeroth-order ray effects. This problem can be best illustrated by considering a highly anisotropic transfer function, which is nonzero only over a small portion of the scattering angle (assume that the transfer function is rotationally invariant). For such a transfer function, many of the exact-kernel transfer values will be practically zero. Whether or not a particular exact-kernel cross section is zero depends on the spacing between adjacent $\mathbf{\Omega}_{ij}$ values of the quadrature set. If the spacing between the ordinates is too large, particles traveling in the $\mathbf{\Omega}_{ij}$ direction will never scatter into other directions. This inability of particles to redistribute angularly remains even after multiple scatters. But in our case, the transfer function most generally encountered are not that peaked. The remedy for this malady is to choose more ordinates or tailor a quadrature set to accurately approximate the peaks in the transfer function.

V.C.5 Negative-Flux Fix-Up Procedure

The choice of the number of spatial nodes K is critical, since a coarse mesh will easily generate either negative angular fluxes or the converged solution may actually be a poor approximation of the correct solution. Consider the positivity of the diamond difference scheme represented by Eqs. 151 and 152, where the coefficients contain the weighting factor α, and say $\alpha \equiv 0.5$. Examination of Eq. 154 indicates that for $J(\tau_{k+0.5}, \Omega_{ij}) > 0$ and $I(\tau_{k+1}, \Omega_{ij}) > 0$, the condition

$$\frac{G(\Omega_{ij}) \Delta_{k+0.5}}{2\mu_i} \leq 1 \tag{177}$$

will guarantee that $I(\tau_k, \Omega_{ij}) > 0$. In other words, the diamond difference scheme in one-dimensional slab geometry is positive if Eq. 177 is satisfied. In practice, however, it may be difficult to meet this constraint, because the quadrature point with smallest magnitude might be too close to zero. For instance, for a Gaussian quadrature set of order 16, the smallest direction cosine $\mu = 0.095$. The minimal value of $\Delta_{k+0.5}$ is 0.2 when $G(\Omega_{ij}) = 1$, i.e., a mesh spacing of less than one-fifth of the mean free path. Furthermore, in two-dimensional geometries, it can be shown that the diamond difference scheme cannot be guaranteed to be positive (Duderstadt and Martin, 1979). Hence, we need some sort of negative-flux fix-up prescription to treat negative fluxes if they appear during a calculation.

The simplest flux fix-up scheme is to set the offending flux to zero. This may seem to be somewhat artificial, but actually it should be kept in mind that a zero flux is probably the best approximation to the correct solution, since the computed negative flux is based on the accurate diamond difference scheme. An alternative flux fix-up procedure involves switching to a positive difference scheme whenever negative fluxes are encountered (Engle, 1973). Let us generalize by writing the diamond difference scheme with a weighting factor (as in Section V.C.3)

$$I(\tau_{k+0.5}, \Omega_{ij}) = (1 - \alpha) I(\tau_k, \Omega_{ij}) + \alpha I(\tau_{k+1}, \Omega_{ij}) \qquad \mu < 0 \tag{178}$$

$$I(\tau_{k+0.5}, \Omega_{ij}) = (1 - \alpha) I(\tau_{k+1}, \Omega_{ij}) + \alpha I(\tau_k, \Omega_{ij}) \qquad \mu > 0 \tag{179}$$

Now, if we assume $S(\tau_{k+0.5}, \Omega_{ij}) = 0$, for the sake of argument, and substitute the correct modified difference scheme (Eq. 179) in Eq. 150, we obtain

$$I(\tau_k, \Omega_{ij}) = \left[\frac{1 - f_{ij}(1 - \alpha)}{1 + f_{ij}\alpha} \right] I(\tau_{k+1}, \Omega_{ij}) \tag{180}$$

Clearly, $I(\tau_k, \Omega_{ij}) > 0$, if $\alpha = 1$ and $I(\tau_{k+1}, \Omega_{ij}) > 0$. Therefore, the weighted diamond difference relations are strictly positive for $\alpha = 1$. This step-function scheme is substituted for the standard diamond difference scheme ($\alpha = 0.5$) whenever $I(\tau_k, \Omega_{ij}) < 0$. Once a positive $I(\tau_k, \Omega_{ij})$ is obtained, we can switch back to the standard diamond difference relation. There is a disadvantage, however, in that the resulting difference scheme is no longer of second-order accuracy in the mesh spacing. Therefore, we must sacrifice accuracy for positivity, a common trade-off in the application of the discrete-ordinates method. This scheme is attractive in two-angle transport calculations, where a large memory requirement is often stifling and, hence, there is a need for fewer mesh nodes while ensuring positivity of the computed fluxes.

V.C.6 *Results*

Gerstl and Zardecki (1985a,b) used the discrete-ordinates finite element method for studying photon transport in a coupled medium of atmosphere and vegetation (a result from their work is plotted in Fig. 8). However, they assumed isotropic scattering in the canopy, which for realistic canopies can give poor approximations for the reflected angular fluxes. In a later study (Simmer and Gerstl, 1985; Gerstl and Simmer, 1986), they used the Henyey–Greenstein transfer function with a negative (backscattering) aniso-tropic scattering parameter as an approximation to the actual scattering transfer function of the vegetation canopy. However, the anisotropic parameter in this transfer function must be estimated heuristically, and there are presently no experimental data from which it may be even empirically inferred. Moreover, as discussed earlier, such rotationally invariant transfer functions that permit expansion techniques are generally not charac-teristic of vegetation canopies.

Shown in Figure 10 is a comparison for a maize canopy of measured values for red and near-infrared wavelengths, with values computed using the bi-Lambertian transfer function, an isotropic leaf normal distribution, and the discrete-ordinates method as a function of the view zenith angle ($\theta = \cos^{-1} \mu$) (the measured data are of Ranson and Biehl, 1983). The azimuthal dependence of the reflectance factors of a soybean canopy is shown in Figure 11 (the measured data are of Ranson and Biehl, 1982). It should be emphasized here that there were no free adjustable parameters in these calculations, and that all canopy properties were based on measured values.

The directional reflectance factor is the ratio of canopy radiance to that of a calibration panel (conservative Lambertian reflector) radiance under similar viewing and illumina-

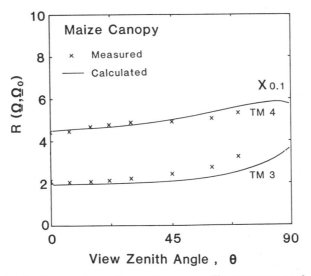

Figure 10 Directional reflectance factors for a maize canopy. The parameters are $\theta_0 = 140°$ and $\phi_0 = 100°$, the zenith and azimuth angles, respectively, of the sun; θ and $\phi = 0°$, the zenith and azimuth angles, respectively, of the detector's view direction; $\tau_T = 4$, optical depth (leaf-area index), $\omega = 0.11$ and 0.92, single-scatter albedo of the leaves in the red (TM band 3, 630–690 nm) and near-infrared (TM band 4, 760–900 nm) wavelengths, respectively; $r_s = 0.10$ and 0.20, soil reflectance in the TM bands 3 and 4, respectively; $\beta = 0.8$, fraction of direct solar radiation in the total incident flux density; $g_L(\Omega_L) = 1$, isotropic leaf normal orientation distribution function; DP_{10} quadrature set used for both μ_i and ϕ_j in the discrete-ordinates calcu-lations. The measured data are from Ranson and Biehl (1983).

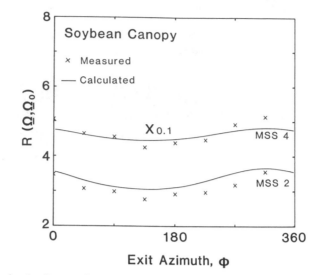

Figure 11   Directional reflectance factors for a soybean canopy. The parameters are the same as for Figure 10, with the following exceptions: incident direction $\theta_0 = 142°$, $\phi_0 = 136°$; view zenith angle $\theta = 22°$; $\tau_T = 2.9$; $\omega = 0.137$ and 0.972 in the red (MSS band 2, 600–700 nm) and near-infrared (MSS band 4, 800–1100 nm), respectively. The measured data are from Ranson and Beihl (1982).

tion conditions:

$$R(\boldsymbol{\Omega}, \boldsymbol{\Omega}_0) = \pi I(0, \boldsymbol{\Omega}) \left[\pi \left| \mu_0 \right| I_0 + \int_0^{2\pi} d\phi' \int_{-1}^{0} d\mu' \left| \mu' \right| I_d(\boldsymbol{\Omega}') \right]^{-1} \quad (181)$$

Besides the good agreement between measured and calculated values, it is apparent that there is lower reflectance in the antisolar directions (i.e., $\phi_0 - \phi \rightarrow \pi$). This effect is expected from the nature of the scattering transfer functions (for instance, see Figure 4), which preferentially scatter photons into directions close to the incident directions. It is for this reason that most vegetated canopies look duller when viewed in the antisolar directions than when viewed from near retrosolar directions (if specular reflectance is not predominant).

Finally, the influence of the leaf normal distribution $g_L(\boldsymbol{\Omega}_L)$ on the reflected intensity is shown in Figure 12, as a function of the cosine of the exit polar angle μ, for both near-perpendicular and tangential illuminations. Curve 1 is for a planophile leaf normal distribution, curve 2 is for a plagiophile distribution, curve 3 is for a isotropic distribution, $g_L(\boldsymbol{\Omega}_L) = 1$, and curve 4 is a pure Lambertian (cosine) profile shown for comparison. It is seen that for near-normal incidence (right half of Figure 12), the shape of the angular reflectance from the leaf canopy is rather insensitive to the leaf normal distribution function. For grazing incidence (left half of Figure 12), however, the reflected angular fluxes are very much affected by the angular orientation of the leaves in the canopy.

The azimuthal dependence of the reflectance factors is shown in Figure 13 for the same three-leaf normal distribution functions and for both near-normal and tangential incidence. Again, it is seen that grazing incidence results in greater anisotropy in the

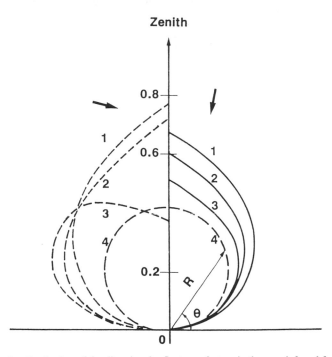

Figure 12 Angular distribution of the directional reflectance factors in the near-infrared for vegetation canopies with three different leaf normal distribution functions: curve 1 is for a planophile distribution, curve 2 is for a plagiophile distribution, and curve 3 is for an isotropic distribution. The dashed-line curve 4 is a Lambertian (cosine) distribution. The parameters are $\tau_T = 3$; $\omega = 1.0$; $r_s = 0.2$; DP_8 quadrature sets for both μ and ϕ; view azimuth $\phi = 6.08°$; no diffuse sky flux in the incident radiation ($\beta = 1$). Incident solar direction is $\mu_0 = -0.9848$ (nearly normal) and $\phi_0 = 0°$ for the right-hand plot, whereas $\mu_0 = -0.1736$ (grazing) and $\phi_0 = 0°$ for the left-hand plot.

azimuthal variation, with the reflectance being greatest in the incident azimuthal plane. More details on this method can be found in Myneni et al. (1988,a,b,c,d), where the derivations and more results are given.

V.D Doubling or the Adding Method

The principle of this method is simple: *if reflection and transmission is known from each of the two layers, the reflection and transmission from the combined layer can be obtained by computing the successive reflections back and forth between the two layers.* If the two layers are identical, then this method is referred to as the doubling method. The current form of the adding method was developed by Van de Hulst (1963). It has been extended to include polarization and extensively used by Hansen (cf. Hansen, 1971a,b) and by Hovenier (1971) for applications in planetary radiative transfer. The procedure used by Grant and Hunt (1969a,b,c), called the matrix-operator method, is similar to the adding method as far as the actual computer calculations are concerned.

Let τ_1 and τ_2 be the optical thickness of the two layers to be added, and, further, let the subscripts 1 and 2 refer to the top and the bottom layers, respectively. Let R_i and T_i denote the reflection and the transmission functions, respectively, of the ith layer. Assume the top layer is illuminated by collimated flux (πI_0 along $\Omega_0(\mu_0, \phi_0)$); see Figure

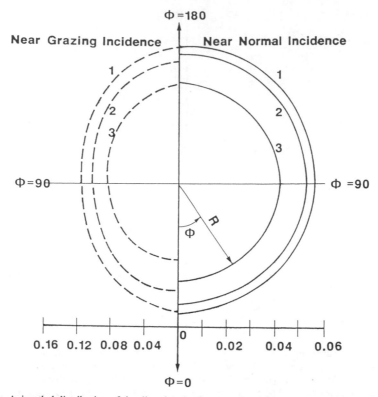

Figure 13 Azimuthal distribution of the directional reflectance factors in the red wavelength for vegetation canopies with three different leaf normal distribution functions: curve 1 is for a planophile distribution, curve 2 is for a plagiophile distribution, and curve 3 is for an isotropic distribution. Parameters are $\tau_T = 3$; $\omega = 0.2$; $r_s = 0.1$; DP_8 quadrature sets for both μ and ϕ; view zenith cosine $\mu_0 = 0.2372$; no diffuse sky flux in the incident radiation ($\beta = 1$). Incident solar direction is $\mu_0 = -0.9848$ (nearly normal) and $\phi_0 = 180°$ for the right-hand plot, whereas $\mu_0 = -0.1736$ (grazing) and $\phi_0 = 180°$ for the left-hand plot. Note that the right-hand and the left-hand plots have different scales.

14. The recipe for the adding method is then

$$S = R_1^* R_2 (1 - R_1^* R_2)^{-1} \tag{182}$$

$$D = T_1 + S \exp\left(-\frac{\tau_1}{|\mu_0|}\right) + ST_1 \tag{183}$$

$$U = R_2 \exp\left(-\frac{\tau_1}{|\mu_0|}\right) + R_2 D \tag{184}$$

$$R(\tau_1 + \tau_2) = R_2 \exp\left(-\frac{\tau_1}{|\mu|}\right) U + T_1^* U \tag{185}$$

$$T(\tau_1 + \tau_2) = \exp\left(-\frac{\tau_2}{|\mu|}\right) D + T_2 \exp\left(-\frac{\tau_1}{|\mu_0|}\right) + T_2 D \tag{186}$$

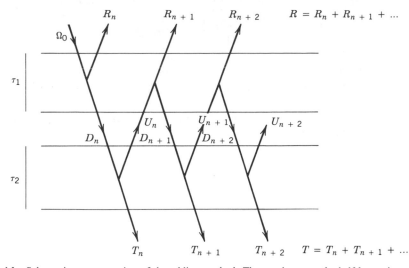

Figure 14 Schematic representation of the adding method. The gap between the half layers is an artifice. Each fork represents any number of successive scatterings that may occur within that half layer.

In the previous expressions, the exponential terms refer to uncollided transmission through layers 1 or 2; T is collided transmission. $T(\tau_1 + \tau_2)$ is the collided transmission at the bottom of the two layers, and the projected uncollided transmission through both the layers is simply $|\mu_0| \pi I_0 \exp[(\tau_1 + \tau_2)/\mu_0]$. $|\mu_0| I_0 D$ and $|\mu_0| I_0 U$ are collided downward and upward intensities, respectively, at the interface. Finally, R_i^* and T_i^* are the reflection and the transmission functions, respectively, when the relevant layer is illuminated from below.

The numerical computations can be handled efficiently by means of a Fourier-series expansion for the azimuthal dependence and numerical quadrature for the integration. A prescription for computing the R and T matrices in case of a horizontally homogeneous azimuthally random leaf canopy with Lambertian scatters is given by Cooper et al. (1982) (for some results from their work, see Figure 7).

V.E Principles of Invariance

This method, first developed by Ambartsumian (1943, cf. Chandrasekhar, 1960), involves determining the radiation field of a principal layer, when a thin layer of similar optical properties is added or removed. Mathematically, it involves partial derivatives of functions characterizing the emergent distributions, with respect to the optical depth τ, that result in integral equations that may be solved by reducing them to the X and Y functions.

We consider the standard two-angle problem detailed in Section II.D. The governing transport equation is Eq. 33 and the boundary conditions are given by Eqs. 34 and 35. To express the diffuse reflection and transmission functions, we shall use the Chandrasekhar's S and T functions:

$$I(0, \mathbf{\Omega}) = \frac{I_0}{4\mu} S(\tau_T; \mathbf{\Omega}, \mathbf{\Omega}_0) \qquad \mu > 0 \tag{187}$$

$$I(\tau_T, \Omega) = \frac{I_0}{4|\mu|} T(\tau_T; \Omega, \Omega_0) \qquad \mu < 0 \qquad (188)$$

which differ from the standard reflection and transmission functions by a factor of $4\mu\mu_0$. Note also that there is an explicit dependence on the optical depth of the principal layer, τ_T.

Consider Figure 15(a). We could formulate the first invariance principle as *the intensity $I(\tau, \Omega)$ in the outward direction ($\mu > 0$) at any depth τ results from the reflection of the uncollided incident flux, $\pi I_0 \exp(\tau/\mu_0)$, and the collided flux, $I(\tau, \Omega')$ ($\mu < 0$), incident on the surface τ by the underlying layer of thickness, $\tau_T - \tau$.* Mathematically, this principle is

$$I(\tau, \Omega) = \frac{I_0}{4\mu} \exp\left(\frac{\tau}{\mu_0}\right) S(\tau_T - \tau; \Omega, \Omega_0) + \frac{1}{4\pi\mu} \int_{2\pi} d\Omega'$$

$$\cdot S(\tau_T - \tau; \Omega, \Omega') I(\tau, \Omega') \qquad \mu > 0, \mu' < 0 \qquad (189)$$

Likewise, from Figures 15(b)–(d), it is possible to formulate three invariance principles and the associated mathematical expressions:

$$I(\tau, \Omega) = \frac{I_0}{4|\mu|} T(\tau; \Omega, \Omega_0)$$

$$+ \frac{1}{4\pi|\mu|} \int_{2\pi} d\Omega' \, S(\tau; \Omega, \Omega') I(\tau, \Omega') \qquad \mu < 0, \mu' > 0$$

$$(190)$$

$$\frac{I_0}{4\mu} S(\tau_T; \Omega, \Omega_0) = \frac{I_0}{4\mu} S(\tau; \Omega, \Omega_0) + \exp\left(\frac{-\tau}{\mu}\right) I(\tau, \Omega)$$

$$+ \frac{1}{4\pi\mu} \int_{2\pi} d\Omega' \, T(\tau; \Omega, \Omega') I(\tau, \Omega') \qquad \mu > 0, \mu' > 0$$

$$(191)$$

$$\frac{I_0}{4|\mu|} T(\tau_T; \Omega, \Omega_0) = \frac{I_0}{4|\mu|} \exp\left(\frac{\tau}{\mu_0}\right) T(\tau_T - \tau; \Omega, \Omega_0) + \exp\left(\frac{\tau_T - \tau}{\mu}\right) I(\tau, \Omega)$$

$$+ \frac{1}{4\pi|\mu|} \int_{2\pi} d\Omega' \, T(\tau_T - \tau; \Omega, \Omega') I(\tau, \Omega')$$

$$\mu < 0, \mu' < 0 \qquad (192)$$

The importance of these four principles arises from the fact that they can be used to derive a basic set of four integral equations for the scattering and the transmission functions. Differentiating Eqs. 189–192 with respect to τ, and passing either to the limit $\tau = 0$ (in Eqs. 189 and 192) or to the limit $\tau = \tau_T$ (in Eqs. 190 and 191) and using the

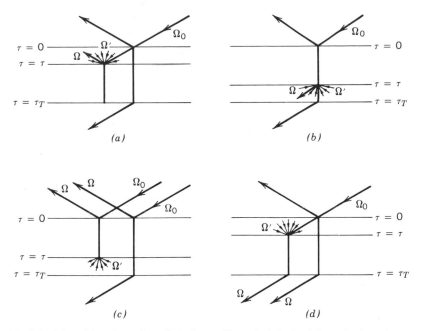

Figure 15 Principles of invariance for a finite layer. The optical depth of the entire layer is τ_T. Incident and exit directions are denoted as Ω_0 and Ω, respectively. Ω' refers to the diffuse internal field radiation.

boundary conditions specified by Eqs. 34 and 35, we obtain

$$\left[\frac{dI}{d\tau}(\tau, \Omega)\right]_{\tau=0} = \frac{I_0}{4\mu}\left[\frac{1}{\mu_0}S(\tau_T; \Omega, \Omega_0) - \frac{\partial S}{\partial \tau_T}(\tau_T; \Omega, \Omega_0)\right]$$

$$+ \frac{1}{4\pi\mu}\int_{2\pi}d\Omega'\, S(\tau_T; \Omega, \Omega')\left[\frac{dI}{d\tau}(\tau, \Omega')\right]_{\tau=0}$$

$$\mu > 0,\ \mu' < 0 \tag{193}$$

$$\left[\frac{dI}{d\tau}(\tau, \Omega)\right]_{\tau=\tau_T} = \frac{I_0}{4|\mu|}\frac{\partial T}{\partial \tau_T}(\tau_T; \Omega, \Omega_0)$$

$$+ \frac{1}{4\pi|\mu|}\int_{2\pi}d\Omega'\, S(\tau_T; \Omega, \Omega')\left[\frac{dI}{d\tau}(\tau, \Omega')\right]_{\tau=\tau_T}$$

$$\mu < 0,\ \mu' > 0 \tag{194}$$

$$0 = \frac{I_0}{4\mu}\frac{\partial S}{\partial \tau_T}(\tau_T; \Omega, \Omega')\exp\left(\frac{-\tau_T}{\mu}\right)\left[\frac{dI}{d\tau}(\tau, \Omega)\right]_{\tau=\tau_T}$$

$$+ \frac{1}{4\pi\mu}\int_{2\pi}d\Omega'\, T(\tau_T; \Omega, \Omega')\left[\frac{dI}{d\tau}(\tau, \Omega')\right]_{\tau=\tau_T}$$

$$\mu > 0,\ \mu' > 0 \tag{195}$$

$$0 = \frac{I_0}{4|\mu|}\left[\frac{1}{\mu_0} T(\tau_T; \mathbf{\Omega}, \mathbf{\Omega}_0) - \frac{\partial T}{\partial \tau_T}(\tau_T; \mathbf{\Omega}, \mathbf{\Omega}_0)\right]$$

$$+ \exp\left(\frac{\tau_T}{\mu}\right)\left[\frac{dI}{d\tau}(\tau, \mathbf{\Omega})\right]_{\tau=0}$$

$$+ \frac{1}{4\pi|\mu|}\int_{2\pi} d\mathbf{\Omega}'\, T(\tau_T; \mathbf{\Omega}, \mathbf{\Omega}')\left[\frac{dI}{d\tau}(\tau, \mathbf{\Omega}')\right]_{\tau=0}$$

$$\mu < 0,\ \mu' < 0 \qquad (196)$$

The derivatives of intensity can be found from the equation of transfer; hence, we have (cf. Eqs. 33–35, 187, and 188)

$$\left[\frac{dI}{d\tau}(\tau, \mathbf{\Omega})\right]_{\tau=0} = \frac{1}{\mu}\left[\frac{I_0}{4\mu} S(\tau_T; \mathbf{\Omega}, \mathbf{\Omega}_0) - J(0, \mathbf{\Omega})\right] \qquad \mu > 0 \qquad (197)$$

$$\left[\frac{dI}{d\tau}(\tau, \mathbf{\Omega})\right]_{\tau=0} = \frac{1}{|\mu|} J(0, \mathbf{\Omega}) \qquad \mu < 0 \qquad (198)$$

$$\left[\frac{dI}{d\tau}(\tau, \mathbf{\Omega}')\right]_{\tau=\tau_T} = -\frac{1}{\mu} J(\tau_T, \mathbf{\Omega}) \qquad \mu > 0 \qquad (199)$$

$$\left[\frac{dI}{d\tau}(\tau, \mathbf{\Omega})\right]_{\tau=\tau_T} = \frac{1}{\mu}\left[\frac{I_0}{4|\mu|} T(\tau_T; \mathbf{\Omega}, \mathbf{\Omega}_0) - J(\tau_T, \mathbf{\Omega})\right] \qquad \mu < 0 \qquad (200)$$

where J is the source term. From Eqs. 33–35, 187, and 188, we have (set $\omega = 1$; case of conservative scattering)

$$J(0, \mathbf{\Omega}) = 0.25 I_0\left[P(\mathbf{\Omega}_0 \to \mathbf{\Omega}) + \frac{1}{4\pi}\int_{2\pi}\frac{d\hat{\mathbf{\Omega}}}{\hat{\mu}} P(\hat{\mathbf{\Omega}} \to \mathbf{\Omega}) S(\tau_T; \hat{\mathbf{\Omega}}, \mathbf{\Omega}_0)\right]$$

$$\hat{\mu} > 0 \qquad (201)$$

$$J(\tau_T, \mathbf{\Omega}) = 0.25 I_0\left[\exp\left(\tau_T/\mu_0\right) P(\mathbf{\Omega}_0 \to \mathbf{\Omega})\right.$$

$$\left. + \frac{1}{4\pi}\int_{2\pi}\frac{d\hat{\mathbf{\Omega}}}{|\hat{\mu}|} P(\hat{\mathbf{\Omega}} \to \mathbf{\Omega}) T(\tau_T; \hat{\mathbf{\Omega}}, \mathbf{\Omega}_0)\right] \qquad \hat{\mu} < 0 \qquad (202)$$

Substituting the expressions in Eqs. 201 and 202 into Eqs. 197–200, and resubstituting the resulting expressions into Eqs. 193–196, one obtains the following four integral equations for diffuse reflection and transmission by a finite plane-parallel canopy:

$$\left(\frac{1}{\mu} - \frac{1}{\mu_0}\right) S(\tau_T; \mathbf{\Omega}, \mathbf{\Omega}_0) + \frac{\partial S}{\partial \tau_T}(\tau_T; \mathbf{\Omega}, \mathbf{\Omega}_0) = P(\mathbf{\Omega}_0 \to \mathbf{\Omega})$$

$$+ \frac{1}{4\pi}\int_{2\pi}\frac{d\hat{\mathbf{\Omega}}}{\hat{\mu}} P(\hat{\mathbf{\Omega}} \to \mathbf{\Omega}) S(\tau_T; \hat{\mathbf{\Omega}}, \mathbf{\Omega}_0)$$

$$+ \frac{1}{4\pi} \int_{2\pi} \frac{d\hat{\boldsymbol{\Omega}}}{|\hat{\mu}|} S(\tau_T; \boldsymbol{\Omega}, \boldsymbol{\Omega}') P(\boldsymbol{\Omega}_0 \to \boldsymbol{\Omega}')$$

$$+ \frac{1}{4\pi} \int_{2\pi} \frac{d\hat{\boldsymbol{\Omega}}}{|\hat{\mu}|} S(\tau_T; \boldsymbol{\Omega}, \boldsymbol{\Omega}') \left[\frac{1}{4\pi} \int_{2\pi} \frac{d\hat{\boldsymbol{\Omega}}}{\hat{\mu}} P(\hat{\boldsymbol{\Omega}} \to \boldsymbol{\Omega}) S(\tau_T; \hat{\boldsymbol{\Omega}}, \boldsymbol{\Omega}_0) \right]$$

$$\mu > 0, \mu' < 0, \hat{\mu} > 0 \tag{203}$$

$$\frac{\partial S}{\partial \tau_T}(\tau_T; \boldsymbol{\Omega}, \boldsymbol{\Omega}_0) = P(\boldsymbol{\Omega}_0 \to \boldsymbol{\Omega}) \exp\left[-\tau_T \left(\frac{1}{\mu} - \frac{1}{\mu_0} \right) \right]$$

$$+ \exp\left(\frac{-\tau_T}{\mu} \right) \frac{1}{4\pi} \int_{2\pi} \frac{d\hat{\boldsymbol{\Omega}}}{|\hat{\mu}|} P(\hat{\boldsymbol{\Omega}} \to \boldsymbol{\Omega}) T(\tau_T; \hat{\boldsymbol{\Omega}}, \boldsymbol{\Omega}_0)$$

$$+ \exp\left(\frac{\tau_T}{\mu_0} \right) \frac{1}{4\pi} \int_{2\pi} \frac{d\boldsymbol{\Omega}'}{\mu'} T(\tau_T; \boldsymbol{\Omega}, \boldsymbol{\Omega}') P(\boldsymbol{\Omega}_0 \to \boldsymbol{\Omega}')$$

$$+ \frac{1}{4\pi} \int_{2\pi} \frac{d\boldsymbol{\Omega}'}{\mu'} T(\tau_T; \boldsymbol{\Omega}, \boldsymbol{\Omega}')$$

$$\cdot \left[\frac{1}{4\pi} \int_{2\pi} \frac{d\hat{\boldsymbol{\Omega}}}{|\hat{\mu}|} P(\hat{\boldsymbol{\Omega}} \to \boldsymbol{\Omega}') T(\tau_T; \hat{\boldsymbol{\Omega}}, \boldsymbol{\Omega}_0) \right]$$

$$\mu > 0, \mu' > 0, \hat{\mu} < 0 \tag{204}$$

$$\frac{1}{|\mu|} T(\tau_T; \boldsymbol{\Omega}, \boldsymbol{\Omega}_0) + \frac{\partial T}{\partial \tau_T}(\tau_T; \boldsymbol{\Omega}, \boldsymbol{\Omega}_0) = \exp\left(\frac{\tau_T}{\mu_0} \right) P(\boldsymbol{\Omega}_0 \to \boldsymbol{\Omega})$$

$$+ \frac{1}{4\pi} \int_{2\pi} \frac{d\hat{\boldsymbol{\Omega}}}{|\hat{\mu}|} P(\hat{\boldsymbol{\Omega}} \to \boldsymbol{\Omega}) T(\tau_T; \hat{\boldsymbol{\Omega}}, \boldsymbol{\Omega}_0)$$

$$+ \exp\left(\frac{\tau_T}{\mu_0} \right) \frac{1}{4\pi} \int_{2\pi} \frac{d\boldsymbol{\Omega}'}{\mu'} S(\tau_T; \boldsymbol{\Omega}, \boldsymbol{\Omega}') P(\boldsymbol{\Omega}_0 \to \boldsymbol{\Omega}')$$

$$+ \frac{1}{4\pi} \int_{2\pi} \frac{d\boldsymbol{\Omega}'}{\mu'} S(\tau_T; \boldsymbol{\Omega}, \boldsymbol{\Omega}') \left[\frac{1}{4\pi} \int_{2\pi} \frac{d\hat{\boldsymbol{\Omega}}}{|\hat{\mu}|} P(\hat{\boldsymbol{\Omega}} \to \boldsymbol{\Omega}') T(\tau_T; \hat{\boldsymbol{\Omega}}, \boldsymbol{\Omega}_0) \right]$$

$$\mu < 0, \mu' > 0, \hat{\mu} < 0 \tag{205}$$

$$\frac{1}{|\mu_0|} T(\tau_T; \boldsymbol{\Omega}, \boldsymbol{\Omega}_0) + \frac{\partial T}{\partial \tau_T}(\tau_T; \boldsymbol{\Omega}, \boldsymbol{\Omega}_0) = \exp\left(\frac{\tau_T}{\mu_0} \right) P(\boldsymbol{\Omega}_0 \to \boldsymbol{\Omega})$$

$$+ \exp\left(\frac{\tau_T}{\mu} \right) \frac{1}{4\pi} \int_{2\pi} \frac{d\hat{\boldsymbol{\Omega}}}{|\hat{\mu}|} P(\hat{\boldsymbol{\Omega}} \to \boldsymbol{\Omega}') S(\tau_T; \hat{\boldsymbol{\Omega}}, \boldsymbol{\Omega}_0)$$

$$+ \frac{1}{4\pi} \int_{2\pi} \frac{d\boldsymbol{\Omega}'}{|\mu'|} T(\tau_T; \boldsymbol{\Omega}, \boldsymbol{\Omega}') P(\boldsymbol{\Omega}_0 \to \boldsymbol{\Omega}')$$

$$+ \frac{1}{4\pi} \int_{2\pi} \frac{d\boldsymbol{\Omega}'}{|\mu'|} T(\tau_T; \boldsymbol{\Omega}, \boldsymbol{\Omega}') \left[\frac{1}{4\pi} \int_{2\pi} \frac{d\hat{\boldsymbol{\Omega}}}{\hat{\mu}} P(\hat{\boldsymbol{\Omega}} \to \boldsymbol{\Omega}') S(\tau_T; \hat{\boldsymbol{\Omega}}, \boldsymbol{\Omega}_0) \right]$$

$$\mu < 0, \mu' < 0, \hat{\mu} > 0 \tag{206}$$

Now, by simple subtraction, we can eliminate $\partial S/\partial \tau_T$ from Eqs. 203 and 204; likewise we can eliminate $\partial T/\partial \tau_T$ from Eqs. 205 and 206. The resulting pair of integral equations, between S and T only, may be regarded as the expression of *the invariance of the laws of diffuse reflection and transmission to the addition (or removal) of layers of arbitrary thickness to (or from) the principal layer at the top and the simultaneous removal (or addition) of layers of equal thickness from (or to) the principal layer at the bottom* (cf. Chandrasekhar, 1960).

Since the transfer function is rotationally invariant, for the particular problem considered here, it may be expanded as a series (see Section IV.B):

$$P(\mathbf{\Omega}' \rightarrow \mathbf{\Omega}) = \sum_{m=0}^{N} \sum_{n=m}^{N} b_n^m P_n^m(\mu) P_n^m(\mu') \cos m(\phi' - \phi) \tag{207}$$

where b_n^m are the expansion coefficients and are given by Eq. 96. The scattering and the transmission functions also have analogous expansions:

$$S(\tau_T; \mathbf{\Omega}, \mathbf{\Omega}_0) = \sum_{m=0}^{N} S^m(\tau_T; \mu, \mu_0) \cos m(\phi_0 - \phi) \tag{208}$$

$$T(\tau_T; \mathbf{\Omega}, \mathbf{\Omega}_0) = \sum_{m=0}^{N} T^m(\tau_T; \mu, \mu_0) \cos m(\phi_0 - \phi) \tag{209}$$

Note that S^m and T^m are now functions of τ_T, μ, and μ_0 only. Substituting these expansions for S and T in Eqs. 203–206, we find that the equations for the various Fourier components separate. After some algebraic manipulation, we obtain

$$\left(\frac{1}{\mu} - \frac{1}{\mu_0}\right) S^m(\tau_T; \mu, \mu_0) = \sum_{n=m}^{N} (-1)^{n+m} b_n^m \times \left[\Phi_n^m(\tau_T; \mu) \Phi_n^m(\tau_T; \mu_0)\right.$$
$$\left. - \Psi_n^m(\tau_T; \mu) \Psi_n^m(\tau_T; \mu_0)\right] \quad \mu > 0, \mu_0 < 0 \tag{210}$$

$$\left(\frac{1}{\mu} - \frac{1}{\mu_0}\right) T^m(\tau_T; \mu, \mu_0) = \sum_{n=m}^{N} b_n^m \times \left[\Psi_n^m(\tau_T; \mu) \Phi_n^m(\tau_T; \mu_0)\right.$$
$$\left. - \Phi_n^m(\tau_T; \mu) \Psi_n^m(\tau_T; \mu_0)\right] \quad \mu < 0, \mu_0 < 0 \tag{211}$$

The auxiliary functions in the last equations are

$$\Phi_n^m(\tau_T; \mu) = P_n^m(\mu) + 0.5\mu \sum_{k=m}^{N} (-1)^{k+n} b_n^m \int_0^1 d\mu' \frac{P_n^m(\mu')}{\mu + \mu'}$$
$$\times \left[\Phi_n^m(\tau_T; \mu) \Phi_n^m(\tau_T; \mu') - \Psi_n^m(\tau_T; \mu) \Psi_n^m(\tau_T; \mu')\right]$$
$$\mu = |\mu|, \mu' = |\mu'| \tag{212}$$

and

$$\Psi_n^m(\tau_T; \mu) = \exp(-\tau_T/\mu) P_n^m(\mu) + 0.5\mu \sum_{k=m}^{N} b_n^m \int_0^1 d\mu' \frac{P_n^m(\mu')}{\mu - \mu'}$$

$$\times \left[\Psi_n^m(\tau_T; \mu) \Phi_n^m(\tau_T; \mu') - \Phi_n^m(\tau_T; \mu) \Psi_n^m(\tau_T; \mu') \right]$$

$$\mu = |\mu|, \mu' = |\mu'| \qquad (213)$$

and $m = 0, 1, \ldots$, and can be expressed in terms of Chandrasekhar's X and Y functions (see Section IV.B), the solution of which is discussed in Chapters VIII and IX of Chandrasekhar's book (1960).

V.F Invariant Imbedding

The invariant imbedding equations are integro-differential equations giving the change in the scattering and the transmission functions when an optically thin layer is added to the principal layer. The initial analysis and the derivation of the integro-differential equations are exactly the same as in the principles of invariance (Section V.E). The name invariant imbedding thus refers to the direct numerical solution of the integro-differential equations.

If a thin layer of optical thickness $\Delta\tau \ll 1$ is added at the top of a principal layer of thickness τ, the resulting expression for the scattering function for the combined layer is simply Eq. 203 (under the same set of assumptions). Note that it is always possible to define the optical depth from the ground up. Then Eq. 203 can be used for numerical calculations to build a canopy from the ground up. Hence, from an analytical point of view, the characteristic of the invariant imbedding is to reduce a two-point boundary value problem into an initial value problem. Extensive work on the invariant imbedding method has been done by Bellman and co-workers, mostly in planetary radiative transfer (e.g., Bellman et al., 1963; also see Wing, 1962).

If one is also interested in the transmission function, Eq. 205 should be used. This equation involves the scattering function as well; hence, one should compute Eq. 203 along with Eq. 205. From a numerical point of view, Eq. 205 is less stable than Eq. 203. Hence, $\Delta\tau$ must be very small in the solution of Eq. 205. The major disadvantage of this method is the relatively large computation times, a remedy for which is discussed by Hansen and Travis (1974).

V.G The Method of Spherical Harmonics

The spherical-harmonics method, also called the P_n approximation, is very similar to the discrete-ordinates method. In fact, it is possible to rigorously prove that the discrete-ordinates equations with Gaussian quadrature of order N are equivalent to P_{N-1} equations with Mark boundary conditions (see Duderstadt and Martin, 1979, p. 476).

To illustrate the method, we consider the problem described in Section II.D. The transport equation after separation of uncollided and collided intensities and expansion in Legendre polynomials is of the form (the analogue of Eq. 64; refer to that section for the notation)

$$\mu \frac{\partial I}{\partial \tau}(\tau, \mu) + I(\tau, \mu) = \frac{\omega}{2} \sum_{n=0}^{N} b_n P_n(\mu) \int_{-1}^{1} d\mu' \, P_n(\mu') I(\tau, \mu') + Q(\tau, \mu)$$

(214)

where $Q(\tau, \mu)$ is the azimuthally averaged first-collision source.

The essence of the spherical-harmonics method consists of expanding the intensity and the first-collision source term in a complete set of Legendre polynomials:

$$I(\tau, \mu) = \sum_{n=0}^{\infty} \frac{2n+1}{4} I_n(\tau) P_n(\mu)$$

(215)

$$Q(\tau, \mu) = \sum_{n=0}^{\infty} \frac{2n+1}{4} Q_n(\tau) P_n(\mu)$$

(216)

and using the orthogonality

$$\int_{-1}^{1} d\mu \, P_n(\mu) P_{n'}(\mu) = \frac{2}{2n+1} \delta_{nn'}$$

(217)

to find

$$Q_n(\tau) = 2\pi \int_{-1}^{1} d\mu \, Q(\tau, \mu) P_n(\mu)$$

(218)

If we now substitute the expansions (Eqs. 215 and 216) into transport Equation 214, multiply by $P_{n'}(\mu)$, integrate over μ, and use the orthogonality (Eq. 217) and the following identity:

$$(n+1) P_{n+1} + n P_{n-1} = (2n+1) \mu P_n$$

(219)

we obtain an infinite set of coupled ordinary differential equations for the expansion coefficients $I_n(\tau)$:

$$\frac{n+1}{2n+1} \frac{\partial I_{n+1}}{\partial \tau} + \frac{n}{2n+1} \frac{\partial I_{n-1}}{\partial \tau} - b_n I_n(\tau) = Q_n(\tau) \qquad n = 0, 1, \ldots$$

(220)

This infinite set must be truncated and the common truncation scheme (P_n approximation) consists of demanding that $\partial I_n / \partial \tau \equiv 0$, for $n > N$. This results in a finite set of $N + 1$ equations for the $N + 1$ unknowns, $I_0, \ldots, I_N$. This system must then be solved subject to the boundary conditions on the transport equation and, for finite N, the boundary conditions cannot be satisfied for all μ. Usually, one employs Marshak's or Mark's boundary conditions, and their relative merits are discussed by Davison (1958). The main advantage of this method over the discrete-ordinates method is that by increasing the number of directions, the computation times are unchanged. However, the computation time increases with the number of terms in the expansions.

V.H Miscellaneous Methods

Several other methods such as the adjoint method (Lewins, 1965; Bell and Glasstone, 1970; Gerstl, 1979), the F_N method (Siewart, 1978; Devaux and Siewart, 1980), and the matrix-operator method (Grant and Hunt, 1969a,b,c) have been used primarily in neutron-transport theory and planetary radiative transfer. These methods will not be discussed here and the interested reader is referred to the previous citations. Instead, we will briefly discuss the integral-equation method of Gutschick and Weigel, and the Monte-Carlo methods.

V.H.1 The Integral-Equation Method

Gutschick and Weigel (1984) have developed an exact integral equation for the angularly integrated radiation intensity (scalar fluxes) intercepted by a leaf element in a layered medium under the assumption of isotropic scattering. This integral equation, which is an integral form of the transport equation (specifically, Eq. 19), can be readily solved accurately by iteration to obtain the depth-resolved scalar flux in the leaf canopy. This solution can then be numerically integrated to obtain angle-resolved intensities everywhere in the canopy. This exact integral-equation method has been applied to azimuthally symmetric problems in randomly oriented optically isotropic leaf canopies illuminated by an axially symmetric conical beam. Specifically, they evaluated the energy balance for the canopy by computing the canopy reflectance, $R(\mu_0)$, the energy deposition on the soil, $A_s(\mu_0)$, and the energy absorption in the canopy itself, $A_c(\mu_0)$:

$$R(\mu_0) = \frac{1}{|\mu_0| I_0} \int_0^1 d\mu\ \mu I(0, \mu) \tag{221}$$

$$A_s(\mu_0) = \frac{1 - r_s}{|\mu_0| I_0} \left[\int_{-1}^0 d\mu\ |\mu|\ I(\tau_T, \mu) + |\mu_0| I_0 \exp\left[\frac{G(\mu_0)\tau_T}{\mu_0} \right] \right] \tag{222}$$

$$A_c(\mu_0) = \left[1 - R(\mu_0) - A_s(\mu_0) \right] \tag{223}$$

Results obtained by Gutschick and Weigel and the discrete-ordinates exact-kernel method are presented in Table 1 for an isotropic transfer function $P(\mu \to \mu) = 1$.

TABLE 1 Comparison of Energy Fluxes Obtained with the Exact Integral-Equation Method (IEM) and the Discrete-Ordinates Exact-Kernel Method (DOM)[a]

		Canopy Reflection		Soil Absorption		Canopy Absorption	
ω	$-\mu_0$	IEM	DOM	IEM	DOM	IEM	DOM
1.0	0.5	0.6694	0.6713	0.3302	0.3286	0.0000	0.0000
0.35	0.5	0.0988	0.0990	0.0384	0.0383	0.8628	0.8627
0.35	1.0	0.0697	0.0702	0.1512	0.1509	0.7790	0.7789
0.20	0.5	0.0509	0.0510	0.0267	0.0267	0.9224	0.9224
0.20	1.0	0.0356	0.0360	0.1360	0.1358	0.8284	0.8282

[a]The problem parameters are optical depth (leaf-area index), $\tau_T = 4$; soil reflectivity, $r_s = 0.1$; isotropic leaf normal distribution, $g_L(\mu_L) = 1$; isotropic scattering, $P(\mu' \to \mu) = 1$; no diffuse sky radiation incident on the canopy, $\beta = 1$. The DOM calculations used a DP_8 quadrature set for μ, and, finally, $\Delta\tau = 0.0615$.

V.H.2   Monte-Carlo Methods

In Monte-Carlo methods, one develops a statistical analogue description of a photon's fate on the computer using random-sampling methods. Then, by running off a large number of such case histories, these results can be averaged to obtain estimates of the expected behavior of the photon population. Photon-transport processes are quite amenable to such a stochastic treatment, since scattering of an individual photon is essentially a stochastic event with the scattering transfer function being the probability density function for scattering at a certain angle. The major development of computer-oriented Monte-Carlo methods for transport problems was provided by Metropolis and Ulam (who also coined the name) in 1949. The application of these methods to neutron transport was developed by Spanier (1959), and the development of variance-reduction schemes was initiated with the works of von Neumann, Ulam, and Kahn (1954).

There are three different methods of Monte-Carlo simulation (Lenoble, 1985): (1) forward with bin averaging: the photons arriving at the detector are averaged over chosen ranges of solid angles, (2) forward with discrete angles: the radiance at the detector is estimated at each collision for specific predetermined angles, and (3) backward or adjoint with discrete angles: the photons start at particular angles at the detector and retrace their paths in the backward direction.

In general, most of the Monte-Carlo methods typically follow 10^6 photons to obtain reasonably accurate results. This means that the Monte-Carlo programs consume huge amounts of CPU time (an hour of CPU time is not uncommon), and hence, most of the mathematical sophistication goes, not into setting up the problem, but in finding ways of using these methods economically. Most results that are presented typically possess statistical fluctuations of at least a few percent (cf. Lenoble, 1985, several percent), and these errors decrease in magnitude only as the square root of the number of photons processed; in other words, the computer time increases fourfold for a twofold increase in the accuracy of the results.

Kimes and Kirchner (1982) pioneered Monte-Carlo methods for tracking photon fates in realistic leaf canopies and reported extensively the reflection functions of several canopies with varying architecture (see Kimes, 1984; Kimes et al., 1985, 1986). Ross and Marshak (1985) have recently developed a powerful Monte-Carlo procedure for photon transport in model canopies, where plants have vertical cylindrical stems and circular leaves. The photon-tracking procedure is especially attractive, since it is optimized with respect to several constituent parameters and resembles somewhat the methods used in statistical mechanics. Their results from this method correctly describe the physics of the Hot-Spot effect (the Heiligenschein) of plant stands and its utility in estimating mean leaf size in the canopy.

In closing, we note a particularly interesting book by Marchuk et al. (1980) that details most of the Monte-Carlo methods developed in the Soviet Union in atmospheric radiative transfer. A general discussion on the Monte-Carlo techniques can also be found in Lenoble (1985).

VI   APPROXIMATE METHODS FOR SOLVING THE TRANSPORT EQUATION

Methods that permit a large reduction in computation times or require no programming at all are called here approximate methods. They involve a rough approximation of the problem and their usefulness depends on the answers sought from such exercises. The method of calculating first-order scattered scalar fluxes developed by Lemeur and Blad (1974; also see Myneni et al. 1986) and the methods for calculating multiply scattered scalar fluxes developed by Norman and Jarvis (1975) and Goudriaan (1977) are all approximate methods and will not be discussed here. Instead, we will discuss the N-stream methods [low-order flux methods] that have been quite popular in canopy photon-transport studies, in spite of their deficiencies, which will also be detailed here.

VI.A *N*-Stream Approximations ($N \leq 4$)

The two-stream approximation has been used by many authors to achieve a quick approximate solution to the transport equation by decomposing the radiation field into two opposing streams. The Kubelka–Munk (K–M) solution, for instance, is based on a two-stream approximation, which goes back to an idea first used in astrophysics (Schuster, 1905; Schwarzschild, 1906). In these methods, instead of solving the full transport equation, one is now content to solve for the stream up and the stream down, if the layer is thought to be horizontal. This necessarily involves some hand waving in writing the first equation. In fact, the flux-propagation equations are derived heuristically, but they can also be deduced rigorously from the analytic transport equation. Although from now on everything else is solved exactly, the solution nevertheless remains an approximation. Low-order flux equations, because of their poor description of the angular distribution of the radiation field, often give incorrect estimates of moments and really have limited usefulness, and this fact should be recognized beforehand.

To get a flavor of these methods, we will illustrate a generalized two-stream method in the context of the standard problem discussed in Section II.D. The propagation equations for upward (F^+) and downward (F^-) scalar fluxes can be written as

$$\frac{dF^+}{d\tau}(\tau) = a_1 F^+(\tau) - a_2 F^-(\tau) - a_3 \omega \pi I_0 \exp\left(\frac{\tau}{\mu_0}\right) \qquad (224)$$

$$\frac{dF^-}{d\tau}(\tau) = a_2 F^+(\tau) - a_1 F^-(\tau) - a_4 \omega \pi I_0 \exp\left(\frac{\tau}{\mu_0}\right) \qquad (225)$$

where the scalar fluxes are

$$F^+(\tau) = \int_0^{2\pi} d\phi \int_0^1 d\mu \, \mu I(\tau, \mu, \phi) \qquad (226)$$

$$F^-(\tau) = \int_0^{2\pi} d\phi \int_{-1}^0 d\mu \, |\mu| I(\tau, \mu, \phi) \qquad (227)$$

Similar equations can be written for the total (direct + diffuse) scalar fluxes. The above equations are approximate, as they can be verified by a proper integration of the transport equation. Various two-stream methodologies, published extensively in the literature (Meador and Weaver, 1980), differ only in the approximations done to the coefficients a_1 through a_4 and in their choice of diffuse or total scalar fluxes for application to the canopy problem (Sellers, 1985).

The solution of Eqs. 224 and 225 is straightforward. The reflectivity (or the albedo) is

$$A = \frac{E\left[1 - \exp\left(2s\tau_T/\mu_0\right)\right]}{1 - E^2 \exp\left(2s\tau_T/\mu_0\right)} \qquad \text{if } \omega \neq 1 \tag{228}$$

$$A = \frac{b\tau_T/|\mu_0|}{1 + b\tau_T/|\mu_0|} \qquad \text{if } \omega = 1 \tag{229}$$

where

$$E = (r - s)/(r + s) \tag{230}$$

$$r = 1 - \omega f + \omega b \tag{231}$$

$$s = \left[(1 - \omega f)^2 - \omega^2 b^2\right]^{0.5} \tag{232}$$

The parameters f and b are the forward and backscattered fractions, respectively; for instance,

$$f = 0.5 \int_0^1 d\mu_s\, P(\mu_s) \tag{233}$$

where $\mu_s \equiv |\mathbf{\Omega}.\mathbf{\Omega}'|$. A discussion on the accuracy of such two-stream methods for multiple-scattering problems in finite layers is discussed in some detail in a book by Van de Hulst (1980, Vol. 2).

The N-stream ($N \leq 4$) methods have been particularly popular in canopy transport studies. Allen and Richardson (1968) were the first to use the Kubelka–Munk equations for the canopy problem. As described earlier, the K–M method is a two-stream method (diffuse downward and diffuse upward streams) and the relationships between these fluxes are expressed by two linear differential equations with two coefficients. Later, Allen et al. (1970) added the monodirectional direct solar-flux component, making it a three-stream method with three linear differential equations and five coefficients (called the Duntley approach). Suits (1972) added another flux, the scalar flux associated with the radiance in the view (exit) direction, making it a four-stream method with four linear differential equations and nine coefficients. Bunnik (1978) and Verhoef (1984) have extensively modified the calculation of these coefficients (called the SAIL model). The Suits method has been very popular because of its simplicity for applications in remote sensing. The Suits method, however, suffers from some serious deficiencies, which we will detail in what follows.

The Suits method provides approximate linear differential equations for the horizontally projected upward and downward diffuse fluxes. The implicit assumption is that the angular distributions of the scattered intensity inside the canopy, $I(\tau, \mathbf{\Omega})$, the intensity reflected from the soil, $I_s(\tau, \mathbf{\Omega})$, $\mu > 0$, and the directional flux of diffuse skylight that is yet unintercepted by the leaves, $I_d(\tau, \mathbf{\Omega})$, $\mu < 0$, are all constant with depth, which is of course violated quantitatively.

The propagation equations and the coefficients of Suits are entirely heuristic. However, Bunnik (1978) gives plausibility arguments for the nature of all the coefficients, though there is not even defineable any average angle of attack for the diffuse upward and downward fluxes.

Two major inconsistencies in the Suits method are particularly important. First, in a canopy of randomly oriented leaves, in the calculation of the attenuation of direct solar flux, there is +41% relative deviation in the extinction coefficient if $\theta_0 = 45°$; hence, there is a deviation of factor of 3.2 in the horizontally projected direct solar flux at the bottom of a canopy of leaf-area index 4. Second, the implied flux-interception rate is inconsistent with the known degree of isotropy of diffuse flux, and in nonrandom canopies, it is not possible to compute photon dispositions obeying the conservation principle! Gutschick and Weigel (1984) present calculations done with their exact-integral equation method and the modified (for the first inconsistency stated here) Suits method and clearly show that if the flux is incident purely as diffuse skylight, then the Suits method can give rather dramatically incorrect results. The same results obtained with the exact-integral equation method are presented in Table 1 (Section V.H.1) in comparison with the results obtained with the discrete-ordinates exact-kernel method.

VII IN RETROSPECT

In this chapter, we presented the governing equation for the transport of monochromatic radiation in an anisotropically scattering vegetation canopy. For simple idealized leaf normal distributions and scattering transfer functions, this general equation and its azimuthal average are shown to reduce to the standard forms encountered in neutron-transport and radiative transfer problems. However, for realistic canopies, such a reduction is generally not possible, and the numerical methods used to solve the transport equations must be modified for the canopy problem.

The rotationally invariant scattering transfer functions and their associated Legendre polynomial expansions, so widely used in other transport studies, are not rigorously justified for describing anisotropic scattering from leaves in a canopy. A realistic bi-Lambertian scattering transfer function is shown to have many desirable properties for canopy studies. One of the most important features of this function is the ability to rigorously reduce the transport equation to an azimuthally symmetric form involving only the polar angle. Both the azimuthally dependent and averaged canopy transport equations are readily solved with a modification of the standard methods (cf. Section V.C).

The angularly dependent reflectance from a canopy is determined not only by the leaf-scattering transfer function, but also by the leaf normal inclination distribution. Numerical results based on the bi-Lambertian transfer function and different leaf normal orientation functions show that this scattering transfer function describes realistically the scattering behavior of vegetation canopies.

We believe that the problem of photon transport in slab geometry for leaf canopies with diffuse anisotropic scattering has been rigorously investigated. This is not to say that new methods are not needed; on the contrary, we should emphasize developing simpler methods, since such methods can be readily inverted for obtaining plant-canopy attributes from remotely sensed radiometric measurements. We hope that future research in this discipline will address the following problems:

(1) A good starting point for representing specular reflection by leaves in the transport calculations would be with the works of Vanderbilt and Grant (1985) and of Reyna and Bhadwar (1985). On a clear sunny day, a full-grown wheat or corn

canopy looks like an ocean of small mirrors from certain view directions. This effect cannot be modeled by the present methods, which account only for the diffuse scattered flux.

(2) The exponential form of the uncollided flux propagation in the canopy stems from an implicit assumption of random leaf placement in the foliage space. Analytical forms of the binomial family may provide a more realistic representation of the foliage spatial disposition and this should be tested by comparing computed values with measured reflectance data (see Nilson, 1971).

(3) Currently, there is no knowledge of the importance of penumbral irradiation on the emergent flux distributions. In tall canopies (e.g., a closed forest stand), penumbral effects are significant and affect the distribution of the leaf-intercepted direct solar flux. Penumbral effects in a completely absorbing leaf canopy are discussed in the works of Miller and Norman (1971a,b), Denholm (1981a,b) and by Myneni and Impens (1985a,b,c).

(4) The specification of the boundary conditions in most canopy investigations is ambiguous. The angular distribution of the diffuse skylight incident on the canopy is simply not available in most cases. This can be overcome by considering the coupled media of vegetation and atmosphere (see Gerstl and Zardecki, 1985b; Simmer and Gerstl, 1985; Gerstl and Simmer, 1986). For radiometric radiance data gathered from satellite platforms to be truely useful, coupled media investigations are imperative to differentiate the perturbations introduced by the atmosphere in the signal from the target.

(5) Most numerical methods today are for slab-geometry problems. The only rigorous multidimensional approach to date is the Monte-Carlo method of Kimes and Kirchner (1982). We need to develop numerical methods applicable to three-dimensional plane geometries that may avoid the disadvantages of the Monte-Carlo techniques. Toward this goal, we are currently developing a three-dimensional transport code based on discrete-ordinates exact-kernel method for photon transport in a coupled medium of atmosphere and vegetation.

(6) Finally, no matter how accurate and rigorous our transport calculations are, their quality can only be improved by inputing an accurate leaf-scattering transfer function. With the exception of data of Moldau (1965, cited in Ross, 1981), Breece and Holmes (1971), and Woolley (1971), there is virtually very little new information on the leaf-scattering diagram. The data collected by these authors is useful, but is not complete in the 4π scattering space for all possible incident directions. Transport calculations are going to be of little use if such information is not available in the future, and we urge the experimentalists to fill this gap.

ACKNOWLEDGMENTS

One of the authors (RBM) acknowledges the various discussions he had with Drs. J. K. Shultis, S. A. W. Gerstl, V. P. Gutschick, B. Powers, and L. D. Travis. Dr. A. Marshak read the entire manuscript very carefully and pointed out many errors and we are very thankful for his remarks. Dr. J. V. Martonchik kindly agreed to read the whole manuscript and gave many useful suggestions. Financial support for this study was provided partly by National Aeronautics and Space Administration grant NAG 5-389.

LIST OF PRINCIPAL SYMBOLS

Each symbol is explained in the text when it is first introduced. Only the principal symbols used in this chapter are listed. Transient notations are not listed. Symbols formed by subscription and superscription are also not listed here in a consistent fashion.

STANDARD ALPHABETICAL SYMBOLS

Symbol	Description
$A(\nu)$, $a_{\pm i}$	Expansion coefficients in the singular eigenfunction method
b_n	Coefficients of the $N+1$ term Legendre expansion
$F^+(\tau)$	Upward diffuse flux density
$F^-(\tau)$	Downward diffuse flux density
$f(\mathbf{\Omega} \rightarrow \mathbf{\Omega} \,;\, \mathbf{\Omega}_L)$	Leaf scattering transfer function
$G(z,\mathbf{\Omega})$	Geometry factor
$g_L(z,\mathbf{\Omega}_L)$	Distribution function of leaf normal orientation
$g_L(z,\mu_L)$	Distribution function of leaf normal orientation along the $\cos \theta_L$ coordinate
$h_L(z,\phi_L)$	Distribution function of leaf normal orientation along the ϕ_L coordinate
$I(z,\mathbf{\Omega})$	Specific-energy intensity
$I_0(\mathbf{\Omega}_0)$	Intensity of the monodirectional solar component
$I_d(\mathbf{\Omega})$	Intensity of the diffuse sky component
$I^0(\tau,\mathbf{\Omega})$	Intensity of uncollided photons
$I^s(\tau,\mathbf{\Omega})$	Diffuse scattered (in the leaf canopy) intensity
$J(\tau,\mathbf{\Omega})$	Sum of the distributed source term, $S(\tau,\mathbf{\Omega})$, and the first collision term, $Q(\tau,\mathbf{\Omega})$
$P_n(\mu)$; $P_n(\mu')$	Legendre polynomials
$P_n^m(\mu)$; $P_n^m(\mu')$	Associated Legendre polynomials
$P(\mathbf{\Omega}' \rightarrow \mathbf{\Omega})$	Normalized scattering transfer function
$P(\mathbf{\Omega}_{nm} \rightarrow \mathbf{\Omega}_{ij})$	Exact-kernel cross section for photon transfer from direction $\mathbf{\Omega}_{nm}$ to $\mathbf{\Omega}_{ij}$
$Q(\tau,\mathbf{\Omega})$	First-collision source term
$R(\mu,\mu_0)$	Reflection function in the one-angle problem
$R(\mathbf{\Omega},\mathbf{\Omega}_0)$	Reflection function in the two-angle problem
r_L	Leaf-reflection coefficient
r_s	Soil-reflection coefficient
$S(\tau_T;\mathbf{\Omega},\mathbf{\Omega}_0)$	Chandrasekhar's scattering function
T	Physical depth of the leaf canopy
$T(\mu,\mu_0)$	Transmission function in the one-angle problem
$T(\mathbf{\Omega},\mathbf{\Omega}_0)$	Transmission function in the two-angle problem
$T(\tau_T;\mathbf{\Omega},\mathbf{\Omega}_0)$	Chandrasekhar's transmission function
t_L	Leaf-transmission coefficient
$u_L(z)$	Leaf-area density
z axis	Denotes depth; directed downward into the canopy

GREEK SYMBOLS

Symbol	Description
δ	Dirac delta function
$\Gamma(\Omega' \rightarrow \Omega)$	Area scattering transfer function
μ	Cos θ, where θ is the polar angle measured relative to the outward normal
μ_s	Cosine of the scattering angle, $\lvert \Omega.\Omega' \rvert$
ν	Eigenvalues in the expansion of the homogeneous transport equation
$\Omega(\mu,\phi)$	Solid angle
$\Omega'(\mu',\phi')$	Superscript prime refers to incoming, or incidence, beam
$\Omega_0(\mu_0,\phi_0)$	Direction of the monodirectional solar component
$\Omega_L(\mu_L,\phi_L)$	Direction of the normal to the upper face of the leaf
ω	Single-scatter albedo of the leaves
$\Phi_n^m(\mu)$	Auxilliary function
ϕ	Azimuthal angle measured from North and through West
$\phi_t(\mu)$	Boundary angle
$\Psi(\mu)$	Auxilliary function in Sections IV.B and V.E
$\Psi(\mu,\mu',\mu_L)$	Kernel of the one-angle area scattering transfer function
$\psi(\mu,\mu_L)$	Azimuthal average of the projected leaf area
$\sigma_e(z,\Omega)$	Extinction coefficient
$\sigma_s(z,\Omega' \rightarrow \Omega)$	Differential scattering coefficient
$\tau(z)$	Optical depth
$Y^m(\mu')$	Characteristic function
$\xi_k(\tau); \xi_{k+1}(\tau)$	Finite element basis functions
$\xi(\pm\nu_i,\mu)$	Eigenfunction in the expansion of the homogeneous transport equation

REFERENCES

Allen, W. A., and A. J. Richardson (1968). Interaction of light with a plant canopy. *J. Opt. Soc. Am.* **58:**1023–1029.

Allen, W. A., T. V. Gayle, and A. J. Richardson (1970). Plant canopy irradiance specified by the Duntley equations. *J. Opt. Soc. Am.* **60:**372–376.

Ambartsumian, U. A. (1943). *Theoretical Physics* (Engl. transl.). Pergamon, New York.

Asrar, G., M. Fuchs, E. T. Kanemasu, and J. L. Hatfield (1984). Estimating absorbed photosynthetic radiation and leaf area index from spectral reflectance in wheat. *Argon J.* **76:**300–306.

Bell, G. I., and S. Glasstone (1970). *Nuclear Reactor Theory.* Van Nostrand-Reinhold, New York.

Bellman, R., R. Kalaba, and M. C. Prestrud (1963). *Invariant Imbedding and Radiative Transfer in Slabs of Finite Thickness.* Elsevier, New York.

Breece, H. T., and R. A. Holmes (1971). Bidirectional scattering characteristics of healthy green soybean and corn leaves in vivo. *Appl. Opt.* **10:**119–127.

Bunnik, N. J. J. (1978). *The Multispectral Reflectance of Shortwave Radiation by Agricultural Crops in Relation with Their Morphological and Optical Properties.* Mededelingen Landbouwhogeschool, Wageningen, The Netherlands.

Carlstedt, J. L., and T. W. Mullikin (1966). Chandrasekhar's *X*- and *Y*-functions. *Astrophys. J., Suppl.* **12**:449–459.

Case, K. M., and P. F. Zweifel (1967). *Linear Transport Theory.* Addison-Wesley, Reading, Massachusetts.

Chandrasekhar, S. (1944). On the radiative equilibrium of a stellar atmosphere. II. *Astrophys. J.* **100**:76–86.

Chandrasekhar, S. (1960). *Radiative Transfer.* Dover, New York.

Cooper, K., J. A. Smith, and D. Pitts (1982). Reflectance of a vegetative canopy using the adding method. *Appl. Opt.* **21**:4112–4118.

Dave, J. V. (1964). Meaning of the successive iteration of the auxilliary equation in the theory of radiative transfer. *Astrophys. J.* **140**:181–205.

Dave, J. V. (1970). Intensity and polarization of the radiation emerging from a plane-parallel atmosphere containing monodispersed aerosols. *Appl. Opt.* **9**:2673–2684.

Dave, J. V., and J. Gazdag (1970). A modified Fourier transform method for multiple scattering calculations in a plane parallel Mie atmosphere. *Appl. Opt.* **9**:1457–1466.

Davison, B. (1958). *Neutron Transport Theory.* Oxford University Press, London.

Denholm, J. V. (1981a). The influence of penumbra on canopy photosynthesis. I. Theoretical considerations. *Agric. Meteorol.* **25**:145–166.

Denholm, J. V. (1981b). The influence of penumbra on canopy photosynthesis. II. Canopy of horizontal circular leaves. *Agric. Meteorol.* **25**:167–194.

Devaux, C., and C. E. Siewart (1980). The F_N method for radiative transfer problems without azimuthal symmetry. *J. Appl. Math. Phys.* **31**:592–604.

de Wit, C. T. (1965). *Photosynthesis of Leaf Canopies*, Agric. Res. Rep. No. 663. Pudoc, Wageningen, The Netherlands.

Duderstadt, J. J., and W. R. Martin (1979). *Transport Theory.* Wiley, New York.

Duncan, W. G., R. S. Loomis, W. A. Williams, and R. Hanau (1967). A model for simulating photosynthesis in plant communities. *Hilgardia* **4**:181–205.

Engle, W. W. (1973). ANISN-ORNL, a one-dimensional discrete ordinates transport code. *Oak Ridge Nat. Lab. [Rep.] ORNL (U.S.)* **ORNL-CCC-254**.

Gerstl, S. A. W. (1979). *Application of the Adjoint Method in Atmospheric Radiative Transfer Calculations*, LANL Rep. LA-UR-80-17. Los Alamos National Laboratory, Los Alamos, New Mexico.

Gerstl, S. A. W., and C. Simmer (1986). Radiation physics and modelling for off-nadir satellite sensing of non-Lambertian surfaces. *Remote Sens. Environ.* **20**:1–29.

Gerstl, S. A. W., and A. A. Zardecki (1985a). Discrete-ordinates finite-element method for atmospheric radiative transfer and remote sensing. *Appl. Opt.* **24**:81–93.

Gerstl, S. A. W., and A. A. Zardecki (1985b). Coupled atmosphere/canopy model for remote sensing of plant reflectance features. *Appl. Opt.* **24**:94–103.

Goel, N. S., and D. E. Strebel (1983). Inversion of vegetation canopy reflectance models for estimating agronomic variables. I. Problem definition and initial results using the Suits model. *Remote Sens. Environ.* **13**:487–507.

Goudriaan, J. (1977). *Crop Micrometeorology: A Simulation Study.* Pudoc, Wageningen. The Netherlands.

Grant, I. P., and G. E. Hunt (1969a). Discrete space theory of radiative transfer. I. Fundamentals. *Proc. R. Soc. London. Ser. A* **313**:183–197.

Grant, I. P., and G. E. Hunt (1969b). Discrete space theory of radiative transfer. II. Nonnegativity and stability. *Proc. R. Soc. London, Ser. A* **313**:199–216.

Grant, I. P., and G. E. Hunt (1969c). Discrete space theory of radiative transfer and its applications to problems in planetary atmospheres. *J. Atmos. Sci.* **26**:963–972.

Gutschick, V. P., and F. W. Weigel (1984). Radiation transfer in vegetative canopies and other layered media: Rapidly solvable extract integral equation not requiring Fourier resolution. *J. Quant. Spectrosc. Radiat. Transfer* **31**:71–82.

Hansen, J. E. (1971a). Multiple scattering of polarized light in planetary atmospheres. I. The doubling method. *J. Atmos. Sci.* **28**: 120–125.

Hansen, J. E. (1971b). Multiple scattering of polarized light in planetary atmospheres. II. Sunlight reflected by terrestrial water clouds. *J. Atmos. Sci.* **28**:1400–1426.

Hansen, J. E., and L. D. Travis (1974). Light scattering in planetary atmospheres. *Space Sci. Rev.* **16**: 527–610.

Herman, B. M., and S. R. Browning (1965). A numerical solution to the equation of radiative transfer. *J. Atmos. Sci.* **22**:529–566.

Hill, T. R. (1975). *ONETRAN: A Discrete Ordinates Finite Element Code for the Solution of the One-Dimensional Multigroup Transport Equation*, LANL Rep. LA-5990-MS. Los Alamos National Laboratory, Los Alamos, New Mexico.

Hovenier, J. W. (1971). Multiple scattering of polarized light in planetary atmospheres. *Astron. Astrophys.* **13**:7–29.

Hunt, G. E. (1964). The generation of angular discretization coefficients for radiation scattered by a spherical particle. *J. Quant. Spectrosc. Radiat. Transfer* **10**:857–864.

Irvine, W. M. (1965). Multiple scattering by large particles. *Astrophys. J.* **142**:1563–1575.

Kahn, H. (1954). *Applications of Monte Carlo*, USAEC Rep. AECU-3259. U.S. Atomic Energy Commission, Washington, D.C.

Kaper, H. G. (1966). *One Speed Transport Theory with Anisotropic Scattering Application to the Slab Albedo Problem. Part I. Theory*, Rep. TW-37. Mathematische Institut, University of Groningen, The Netherlands.

Kaper, H. G., G. Lekkerkerker, and J. Hejtmanek (1982). *Spectral Methods in Linear Transport Theory.* Birkhauser, Basel, Switzerland.

Kimes, D. S. (1984). Modelling the directional reflectance from complete homogeneous vegetation canopies with various leaf-orientation distributions. *J. Opt. Soc. Am.* **1**:725–737.

Kimes, D. S., and J. A. Kirchner (1982). Radiative transfer model for heterogeneous 3-D scenes. *Appl. Opt.* **21**:4119–4129.

Kimes, D. S., W. W. Newcomb, R. F. Nelson, and J. B. Schutt (1986). Directional reflectance distributions of hardwood and pine forest canopy. *IEEE Trans. Geosci. Remote Sens.* **GE-24**:281–293.

Kimes, D. S., J. M. Norman, and C. L. Walthall (1985). Modelling the radiant transfers of sparse vegetation canopies. *IEEE Trans. Geosci. Remote Sens.* **GE-23**:695–704.

Lemeur, R. L., and B. L. Blad (1974). A critical review of light models for estimating the shortwave radiation regime of plant canopies. *Agric. Meteorol.* **14**:255–286.

Lenoble, J. (1985). *Radiative Transfer in Scattering and Absorbing Atmospheres: Standard Computational Procedures.* A. Deepak Publ., Hampton, Virginia.

Leonard, A., and T. W. Mullikin (1964). Spectral analysis of the anisotropic neutron transport kernel in slab geometry with applications. *J. Math. Phys.* **5**:399–407.

Lewins, J. (1965). *Importance, The Adjoint Function.* Pergamon, New York.

Marchuk, G. I., G. A. Mikhailov, M. A. Navarliev, R. A. Darbinjan, B. A. Kargin, and B. S. Elepov (1980). *The Monte Carlo Methods in Atmospheric Optics.* Springer-Verlag, Berlin.

McCormick, N. J., and I. Kuscer (1973). Singular eigenfunctions expansions in neutron transport theory. *Adv. Nucl. Sci. Technol.* **7**:181–282.

Meador, W. L., and W. R. Weaver (1980). Two-stream approximations to radiative transfer in planetary atmospheres: A unified description of existing methods and a new improvement. *J. Atmos. Sci.* **37**:630–643.

Miller, E. E., and J. M. Norman (1971a). A sunfleck theory for plant canopies. I. Length of sunlit segments along a transect. *Agron. J.* **63:**735–738.

Miller, E. E., and J. M. Norman (1971b). A sunfleck theory for plant canopies. II. Penumbra effects: Intensity distribution along sunfleck segments. *Agron. J.* **63:**739–743.

Monteith, J. L. (1965). Light distribution and photosynthesis in field crops. *Ann. Bot. (London)* [N.S.] **29:**17–37.

Myneni, R. B., and I. Impens (1985a). A procedural approach for studying the radiation regime of infinite and truncated foliage spaces. I. Theoretical considerations. *Agric. For. Meteorol.* **33:**323–337.

Myneni, R. B., and I. Impens (1985b). A procedural approach for studying the radiation regime of infinite and truncated foliage spaces. II. Experimental results and discussion. *Agric. For. Meteorol.* **34:**3–16.

Myneni, R. B., and I. Impens (1985c). A procedural approach for studying the radiation regime of infinite and truncated foliage spaces. III. Effect of leaf size and inclination distributions on non-parallel beam radiation penetration and canopy photosynthesis. *Agric. For. Meteorol.* **34:**183–194.

Myneni, R. B., R. B. Burnett, G. Asrar, and E. T. Kanemasu (1986). Single scattering of parallel direct and axially symmetric diffuse solar radiation in vegetative canopies. *Remote Sens. Environ.* **20:**165–182.

Myneni, R. B., G. Asrar, and E. T. Kanemasu (1987a). Light scattering in plant canopies: The method of Successive Orders of Scattering Approximations (SOSA). *Agric. For. Meteorol.* **39:**1–12.

Myneni, R. B., G. Asrar, and E. T. Kanemasu (1987b). Reflectance of a soybean canopy using the method of Successive Orders of Scattering Approximations (SOSA). *Agric. For. Meteorol.* **40:**71–87.

Myneni, R. B., G. Asrar, R. B. Burnett, and E. T. Kanemasu (1987c). Radiative transfer in an anisotropically scattering vegetative medium. *Agric. For. Meteorol.* **41:**97–127.

Myneni, R. B., V. P. Gutschick, J. K. Shultis, G. Asrar, and E. T. Kanemasu (1988a). Photon transport in vegetation canopies with anisotropic scattering. Part I. The scattering phase function in the one-angle problem. *Agric. For. Meteorol.* **42:**1–16.

Myneni, R. B., V. P. Gutschick, J. K. Shultis, G. Asrar, and E. T. Kanemasu (1988b). Photon transport in vegetation copies with anisotropic scattering. Part II. Discrete-ordinates finite difference exact kernel technique for photon transport in slab geometry for the one-angle problem: *Agric. For. Meteorol.* **42:**17–40.

Myneni, R. B., V. P. Gutschick, J. K. Shultis, G. Asrar, and E. T. Kanemasu (1988c). Photon transport in vegetation canopies with anisotropic scattering. Part III. The scattering phase functions in the two-angle problem. *Agric. For. Meteorol.* **42:**87–99.

Myneni, R. B., V. P. Gutschick, J. K. Shultis, G. Asrar, and E. T. Kanemasu (1988d). Photon transport in vegetation canopies with anisotropic scattering. Part IV. Discrete-ordinates finite difference Exact kernel technique for photon transport in slab geometry for the two-angle problem. *Agric. For. Meteorol.* **42:**101–120.

Nilson, T. (1971). A theoretical analysis of the frequency of gaps in plant stands. *Agric. Meteorol.* **8:**25–38.

Norman, J. M. (1980). Interfacing leaf and canopy light interception models. In *Predicting Photosynthesis for Ecosystem Models* (J. D. Hesketh and J. W. Jones, Eds.), Vol. II. CRC Press, Boca Raton, Florida, pp. 49–67.

Norman, J. M., and P. G. Jarvis (1975). Photosynthesis in sitka spruce [Picea Sitchensis (Bong.) Carr.]. Part V. Radiation penetration theory and a test case. *J. Appl. Ecol.* **12:**839–878.

Odom, J. P. (1975). Neutron transport with highly anisotropic scattering. Ph.D. Dissertation, Kansas State University, Manhattan.

Ranson, J., and L. L. Biehl (1982). *Soybean Canopy Reflectance Modelling Data Set*. Purdue University, West Lafayette, Indiana.

Ranson, J., and L. L. Biehl (1983). *Maize Canopy Reflectance Modelling Data Set*. Purdue University, West Lafayette, Indiana.

Reyna, E., and G. D. Bhadwar (1985). Inclusion of specular reflectance in vegetative canopy models. *IEEE Trans. Geosci. Remote Sens.* **GE-23**:731–736.

Risner, J. M. (1985). Semi-analytical evaluation of the scattering source term in discrete ordinates transport calculations. MS Thesis, Kansas State University, Manhattan.

Ross, J. (1981). *The Radiation Regime and Architecture of Plant Stands*. Junk Publ., The Netherlands.

Ross, J., and A. L. Marshak (1985). A Monte Carlo procedure for calculating the scattering of solar radiation by plant canopies. *Sov. J. Remote Sens. (Engl. Transl.)* **4**:783–801.

Ross, J., and T. Nilson (1968). A mathematical model of radiation regime of plant cover. In *Actinometry and Atmospheric Optics*. Valgus Publ., Tallinn, Estonia, pp. 263–281.

Schuster, A. (1905). Radiation through foggy atmospheres. *Astrophys. J.* **21**:1–22.

Schwarzchild, K. (1906). In *Goettinger Nachr.*, p. 42 (as cited in Chandrasekhar, 1960).

Sellers, P. J. (1985). Canopy reflectance, photosynthesis and transpiration. *Int. J. Remote Sens.* **6**:1335–1372.

Siewart, C. E. (1978). The F_N method for solving radiative transfer problems. *Astrophys. Space Sci.* **58**:131–137.

Simmer, C., and S. A. W. Gerstl (1985). Remote sensing of the angular characteristics of canopy reflectance. *IEEE Trans. Geosci. Remote Sens.* **GE-23**:648–675.

Sobolev, V. V. (1975). *Light Scattering in Planetary Atmospheres*. Pergamon, New York.

Spanier, J. (1959). *Monte Carlo Methods and Their Application to Neutron Transport Problems*, WAPD Rep. 195.

Suits, G. W. (1972). The calculation of the directional reflectance of a vegetative canopy. *Remote Sens. Environ.* **2**:117–125.

Uesugi, A., and W. M. Irvine (1970). Multiple light scattering in a plane parallel atmosphere. *Astrophys. J.* **159**:127–135.

Van de Hulst, H. C. (1963). *A New Look at Multiple Scattering*, Tech. Rep. Goddard Institute for Space Studies, New York.

Van de Hulst, H. C. (1980). *Multiple Light Scattering. Table, Formulas and Applications*. Vols. I and II. Academic Press, New York.

Vanderbilt, V. C., and L. Grant (1985). Plant canopy specular reflectance model. *IEEE Trans. Geosci. Remote Sens.* **GE-23**:722–730.

Verhoef, W. (1984). Light scattering by leaf layers with application to canopy reflectance modelling: The SAIL model. *Remote Sens. Environ.* **16**:125–141.

Wick, C. G. (1943). Uber Ebene Diffusions probleme. *Z. Phys.* **120**:702.

Wing, G. M. (1962). *An Introduction to Transport Theory*. Wiley, New York.

Woolley, J. T. (1971). Reflectance and transmittance of light by leaves. *Plant Physiol.* **47**:656–662.

6

INVERSION OF CANOPY REFLECTANCE MODELS FOR ESTIMATION OF BIOPHYSICAL PARAMETERS FROM REFLECTANCE DATA

NARENDRA S. GOEL

Department of Systems Science
State University of New York
Binghamton, New York

I INTRODUCTION

The optical remote sensing of an object is based on two assumptions: (1) that the solar radiation, which has interacted with the object to be sensed and has been received by a remotely located sensor, carries in it the signature of the object, and (2) that this signature can be deciphered to obtain the important characteristics of the object. For vegetation, these characteristics include: (1) the identity of the vegetation (corn, soybean, blue grama grass, black spruce, etc.), (2) the growth or development stage, (3) the extent of stress on the vegetation, (4) the amount of green vegetation, expressed as the biomass or the area indices of vegetation elements (leaves, branches, stems, barks, flowers, etc.), and (5) the architecture of the canopy (the spatial distribution of vegetated and nonvegetated areas on the ground, and the density and orientation of vegetation elements within the vegetated areas).

The problem of assessing the vegetation characteristics from the spectral signature is not trivial. It is further complicated by the fact that the signal received by a remote sensor also carries with it the signatures of such external factors as soil and atmosphere, and also by the presence of stochastic elements such as wind and dew.

There are basically two approaches—statistical and physical—to assess the vegetation characteristics from the spectral signatures.

In the *statistical approach*, one collects data on the spectral signatures of a variety of objects, tries to obtain a statistical correlation between the objects and their signatures, and then uses the statistical techniques (such as cluster analysis, Bayesian inference, regression analysis, and parametric and nonparametric statistics) to determine the characteristics of an object from its signature. Such an approach has mainly been used to classify vegetation into species, and is described in detail in Chapter 14.

205

In the *physical approach*, one tries to understand the physics of the interaction between solar radiation and the vegetation elements and then incorporates this understanding into a model that relates the vegetation characteristics to its spectral signature or reflectance. The model is first validated using measurements on canopies of known characteristics. It is then used to determine the unknown characteristics of a vegetation from the measured spectral signatures. This chapter is devoted to this approach.

In general, the spectral signatures (or reflectances) of a canopy depend on the canopy characteristics or a set of parameters denoted here as C, wavelength λ, the direction of the incident solar radiation, and the viewing direction. Further, since different vegetative species grow at different rates, one would also expect the temporal profile of the spectral signature to carry information about the identity and development stage of the vegetation. Thus, if S denotes the spectral signatures of a plant canopy, then symbolically,

$$S = R(t; \lambda; \theta_s, \psi_s; \theta_o, \psi_o; C) \tag{1}$$

where t is the emergence time of the plant; θ_s and ψ_s are the solar zenith and azimuth angles, respectively; θ_o and ψ_o are the view zenith and azimuth angles, respectively; and R is the functional dependence of S on these parameters. The physical approach to remote sensing strives to determine C from the measured S, by *inverting* the relation in Eq. 1.

This determination of C through *inversion* has been attempted in two ways:

(1) *Inversion Using Nadir Reflectances in Several Wavelength Bands.* In the early days of remote sensing, because of the practical limitations of satellite-borne sensors, nadir reflectances, in several wavelength bands, were emphasized for the purpose of estimating canopy parameters from reflectance data. Since the number of measurements (usually four to seven, depending upon the number of wavelength bands used in the sensor) is significantly smaller than the number of canopy parameters that determine the reflectance, one would expect this approach to be inaccurate. To address this fundamental problem of inaccuracy, two approaches have been used.

(a) *Use of Spectral Transform or Vegetation Indices.* Here one defines linear or nonlinear combinations of nadir reflectances in several wavelength bands that are sensitive to the canopy parameter to be estimated (e.g., leaf-area index, LAI), but not to other parameters (e.g., soil reflectance). At least four dozen transforms or indices have been proposed, but most of them are functionally related (Perry and Lautenschlager, 1984). A concise review of the use of transforms and indices for the estimation of vegetation parameters can be found in Jackson (1984). This approach is also discussed in Chapter 1. It has mainly been used to estimate LAI for planted crops, but the estimated LAI can vary by as much as 100% (Curran and Williamson, 1987).

(b) *Use of Temporal Dependence of Spectral Transforms and Vegetation Indices.* This approach exploits the relationship between the temporal variation of the canopy reflectance and vegetation characteristics, and is referred to as the temporal profile approach (Chapter 1). It has been used to classify vegetation and to determine the crop's development stage and emergence date.

(2) *Inversion Using Off-Nadir Canopy Reflectance Data.* Both the limitation of the nadir reflectance data in the estimation of canopy parameters and the additional information about canopy architecture that off-nadir reflectances contain have been recognized for some time. In spite of this awareness, only quite recently have off-nadir canopy

reflectance data been used to estimate canopy parameters. This chapter will focus on this approach to estimate the vegetation characteristics.

In most of our discussions, we will ignore the complications caused by the presence of the atmosphere, other than to discuss its effects (1) on the nature of the solar radiation incident at the vegetation canopy and (2) on the choice of the wavelength and solar/viewing directions to minimize the effects of the atmosphere. Thus, our discussion will be most relevant to the estimation of canopy parameters from the spectral measurements made on top of canopy. (A discussion of the impact of atmosphere on the remote sensing of objects is given in Chapter 9.)

The canopy reflectance data or the spectral signatures can be characterized in many forms (Nicodemus et al., 1977) such as the bidirectional reflectance factor or the bidirectional reflectance function. In this chapter, we will only use the bidirectional reflectance factor, and for brevity call it simply canopy reflectance, or CR.

Figure 1 schematically shows the relationship between the problem of estimation of canopy parameters from measured CR data, the development of a CR model (the direct problem), and the inversion of the CR model. A prerequisite for using any inversion technique to estimate vegetation parameters is a good CR model, formally expressed as Eq. 1. Therefore, any discussion of inversion techniques should first address the CR

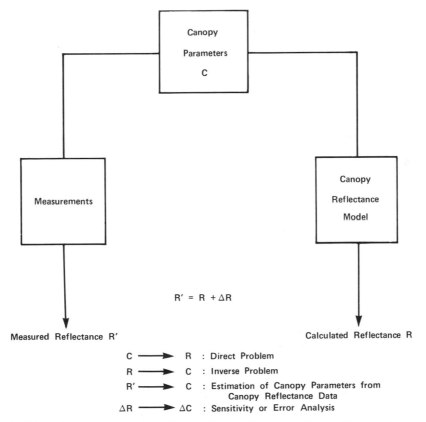

Figure 1 Schematic representation of the direct problem, the inverse problem, and the estimation of canopy parameters from reflectance data.

models. Over two dozen models have been proposed over the last 20 years. These models are discussed in considerable detail in Goel (1987) and also in Chapter 5. Therefore, in Section II, we will only provide a conceptual basis for understanding CR, delineate various factors that determine the CR, and provide a brief review of various CR models.

In Section III, we discuss the question of the inversion of CR models to estimate canopy parameters from the CR data. We review various mathematical aspects of the inversion technique. We discuss in detail the inversion of one model, the SAIL model, including the question of the accuracy of estimation of canopy parameter and how to maximize this accuracy through proper choice of the solar/view directions. The inversion of other models, including estimation of canopy parameters using measured CR data, is only summarized.

The last section is devoted to concluding remarks and some suggestions for future research in the area of inversion of CR models.

II CONCEPTUAL BASIS FOR UNDERSTANDING CANOPY REFLECTANCE

This section has two purposes: (1) to provide a conceptual basis for the physics of the interaction of solar radiations with a vegetation canopy and (2) to provide an overview of the various canopy reflectance (CR) models that incorporate this physics. An extensive review article by this author (Goel, 1987) addresses these purposes in detail and this section has been adapted from that article.

II.A Factors Affecting the Canopy Reflectance

There are many factors that determine the CR.

II.A.1 Incoming Solar Flux

The incoming solar flux consists of two parts: the direct or specular flux, which has not been scattered by the atmosphere, and the diffuse flux, which has been scattered in the downward direction. The fraction of diffuse incident radiation, SKYL, depends upon the atmospheric conditions and increases with wavelength. Sophisticated atmospheric scattering models have been developed to understand this dependence (see Chapter 9). The direction of the direct flux is characterized by the solar zenith angle θ_s and the azimuth angle ψ_s, whereas the diffuse flux is characterized by its angular distribution.

The CRs measured at orbital altitudes under cloudy skies contain less information about vegetation than those under clear skies. Therefore, it is desirable to make CR measurements when SKYL is small (about 10–15% in the near infrared, NIR, region). Most CR models decouple the atmospheric scattering, assume that the distribution of incident diffused flux is isotropic, and treat SKYL as a parameter. (There are a few exceptions where the atmosphere and canopy are treated together; see, e.g., Gerstl and Zardecki, 1985a, b).

II.A.2 Spectral Properties of the Vegetation Elements

The radiation flux incident on a vegetation element, e.g., a leaf, is subject to two processes: scattering and absorption. Scattering consists of two subprocesses: reflection from and transmission through the leaf. The reflection consists of two parts: (a) the specular reflected flux (like that from a plane mirror), where the angle of incidence is equal to the angle of reflection, and (b) the diffuse flux. Likewise, the transmitted beam consists of refracted and diffuse parts. The relative amounts of diffused and specular fluxes depend upon the characteristics of the vegetation element (surface properties, cellular structure, composition, etc.) and those of the incident flux (wavelength, angle of incidence, and polarization). The directional dependence of diffuse flux can be very complex.

Many measurements of the leaf spectral properties are available over a wide range of wavelengths (Gausman, 1985). Most of them use near-normal incidence sources and integrating hemispheres that collect all the radiation emanating from the top and bottom sides of the leaf. These measurements give what is referred to as the hemispherical reflectance ρ, the hemispherical transmittance τ, and the absorptance $a = 1 - (\rho + \tau)$. According to these measurements:

(1) ρ and τ are wavelength-dependent. In the visible (VIS) region (400–700 nm), both ρ and τ are small (and, hence, absorptance is large). They then rapidly increase with λ until they achieve their largest values (smallest absorptance) in the NIR region (800–1100 nm).

(2) ρ for the lower surface (or dorsal) is usually higher than that for the upper surface (or ventral), and this distinction is wavelength-dependent (most notable in the VIS region and negligible in the NIR region).

(3) For any of the two leaf surfaces, ρ and τ depend upon the angle between the direction of the source and the leaf normal. One can approximate these dependences by

$$\rho = a\theta_s^2 + b\theta_s + c \tag{2a}$$

$$\tau = a'\theta_s^2 + b'\theta_s + c' \tag{2b}$$

where a, b, c, a', b', and c' are constants that depend upon the wavelength, and θ_s is the source incident zenith angle in radians.

In general, the radiation scattered from a leaf also depends upon the scattering direction. Only a few measurements on corn and soybean leaves have been carried out to understand this dependence (Breece and Holmes, 1971; Norman et al., 1985; Walter-Shea, 1987). According to these measurements:

(1) The scattering of radiation is nonisotropic or non-Lambertian. The bidirectional reflectance distribution functions (BRDFs), in both the VIS and NIR regions, can generally be characterized by a shallow bowl shape, with values of reflectance increasing with the view zenith angle, for most azimuth angles. The bowl is shallower for near-normal incidence angles, and the backscattering is greater in the NIR region than in the VIS region.

(2) For a near-normal incidence angle, the scattering is approximately Lambertian and predominantly diffused. As θ_s increases, the leaf becomes a more and more non-Lambertian scatterer, and the forward scattering increases almost exponentially (due mostly to specular reflectance). The non-Lambertian component of the leaf reflectance can be as much as 40% of the total hemispherical reflectance.

(3) Bidirectional transmittance distribution functions (BTDFs) show less variability with changing incidence angle than do BRDFs. The distributions for each incidence angle are similar in geometry and magnitude.

Detailed estimates of specular reflectance, using polarization measurements at Brewster's angle, have been made by Grant et al. (1983; also see Grant, 1987). These measurements show that the specularly reflected radiation may vary from about 10% (for soybean) to more than 50% (for corn and sorghum) in the VIS region. However, for the NIR region, the specular component is much smaller. Surface characteristics, such as the extent of waxiness, affect the amount of specular reflectance.

II.A.3 Architecture of the Canopy

A vegetation element inside a canopy receives two kinds of radiation: solar radiation unintercepted by other elements and radiation intercepted and then scattered from these elements. Likewise, the sensor receives several types of fluxes: (1) flux scattered only one time from a vegetation element (single scattering), (2) flux scattered several times from many vegetation elements (multiple scattering) but without reaching the soil, and (3) flux reflected from the soil and reaching the sensor either unintercepted by a vegetation element or, if intercepted, then scattered in the direction of the sensor. Therefore, the CR depends not only on the scattering and absorbing properties of vegetation elements, but also on the architecture of the canopy (a term used for the spatial distribution of vegetated and nonvegetated areas on the ground, and the density and orientation of vegetation elements within the vegetated areas).

We will now describe how the architecture of a canopy is characterized and how it affects the canopy reflectance.

(1) *Spatial Distribution of Vegetated and Nonvegetated Areas.* As vegetation grows, more and more of the soil surface is covered with vegetation. The spatial distribution of vegetation on the ground will depend upon how the seeds are planted, the type of vegetation that is planted, and the growth stage of the plant. In CR models for sparsely covered vegetation, a canopy is assumed to be made up of subcanopies, placed on the ground either in a regular pattern (e.g., in rows or in a orchard-like lattice structure) or randomly (according to a specified distribution). The subcanopies are assumed to have simple geometrical shapes (ellipsoid, cone, cube, etc.). A canopy where the vegetation fully covers the ground is referred to as a closed canopy. If the distribution of the vegetation within a closed canopy is laterally homogeneous, it is referred to as an homogeneous canopy.

The spatial distribution of subcanopies on the ground impacts the CR because it determines which parts of the soil are lit by unintercepted solar beam and which parts are not, and which parts of the radiation reflected from the soil reach an observer (located above the canopy) unintercepted or after being intercepted by vegetation.

(2) *Leaf-Area Index (LAI) and Leaf-Angle Distribution (LAD).* For a homogeneous canopy or a subcanopy, one usually assumes that the density of all the vegetation elements is uniform. Each type of element is characterized by its area index (e.g., leaf-area index, LAI) and by its angular distribution (e.g., leaf-angle distribution, LAD).

LAI is an important canopy parameter and is related to the canopy biomass. It is also required in models that deal with plant growth and evapotranspiration. Its manual mechanical measurement, even with an electronic area meter, is quite expensive if one wants a reasonable level of accuracy (Daughtry and Hollinger, 1984).

LAD is characterized by a distribution density function $f(\theta_L, \psi_L)$, where θ_L and ψ_L are the leaf inclination and leaf azimuth angles, respectively. Thus, $f(\theta_L, \psi_L) \, d\theta_L \, d\psi_L$ is the fraction of leaf area within the leaf inclination angles of θ_L and $\theta_L + d\theta_L$ and leaf azimuth angle of ψ_L and $\psi_L + d\psi_L$. In most CR models, LAD is assumed to be azimuthally symmetric, a good approximation for many vegetations. It varies from one vegetation to another, but for most vegetations, it can be characterized (Goel and Strebel, 1984) by a two-parameter (μ and ν) beta distribution:

$$f(\theta_L) = \left[\frac{1}{(360)(90)} \right] \left[\frac{\Gamma(\mu + \nu)}{\Gamma(\mu)\Gamma(\nu)} \right] \left(1 - \frac{\theta_L}{90} \right)^{\mu - 1} \left(\frac{\theta_L}{90} \right)^{\nu - 1} \quad (3)$$

where Γ is the gamma function. The two parameters μ and ν are related to the average leaf-inclination angle (ALA) and its second moment $\langle \theta_L^2 \rangle$ by

$$\text{ALA} = \frac{90\nu}{\mu + \nu} \quad (4a)$$

$$\langle \theta_L^2 \rangle = \frac{(90)^2 \, \nu(\nu + 1)}{(\mu + \nu)(\mu + \nu + 1)} \quad (4b)$$

(A four-parameter beta distribution can also be used to represent the combined leaf-inclination angle and the leaf azimuth-angle distribution; Strebel et al., 1985).

Let us now discuss how LAI and LAD affect the CR.

In the VIS region, since most of the light incident on a leaf is absorbed as the number of leaves is increased, i.e., LAI is increased, more and more of the incident radiation is absorbed by the vegetation. Once LAI reaches a threshold value, most of the incident radiation will be absorbed by the leaves, and a further increasing of the LAI will not change the CR. Thus, in the VIS region, the CR asymptotically decreases (almost exponentially) as LAI increases and saturates at around a LAI of 2–3. In the NIR region, since absorption is minimal and scattering is maximal, the CR increases (again almost exponentially) with LAI, until it saturates at around a LAI of 6–8.

LAD impacts the CR in several different ways. One effect of the LAD has been lucidly represented by Norman et al. (1985) through a very simple example of two flat leaves in the principal plane, as shown in Figure 2. Leaf 2, which is nearly perpendicular to the direct sunlight, is well lit, whereas leaf 1, which is nearly parallel to the sunlight, is dimly lit. Observer A, with the sun behind him, will "see" the bright leaves well but not the dimly lit leaves. Thus, to him, the scene appears bright, with the brightness being influenced by the reflectances of the bottom of leaf 1 and the top of leaf 2. Observer B will "see" the dimly lit leaf 1 well, but not much of the well-lit leaf 2. Therefore, to him, the scene will appear darker. The scene's brightness in this case is determined by the transmittance of leaf 1 and the top reflectance of leaf 2. (Specular reflectance from leaf 1 and leaf 2 will not be observed by either observer.)

When the sun is directly behind the observer (or sensor), a greater proportion of the directly illuminated vegetation components will be viewed. Shadows within the canopy or on the soil surface will be hidden by the foliage (or the soil particles) that are illuminated by the sun. Therefore, the CR will tend to be highest here. This peak in the reflectance when the sun is directly behind the observer is referred to as the "hot spot" effect (see Chapters 5 and 7). A canopy is composed of many leaves with a wide range of leaf inclination and azimuth angles. Therefore, the magnitude of the hot spot effect,

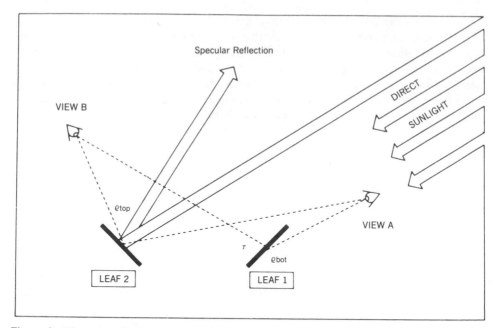

Figure 2 Illustration of canopy geometry that contributes to the observed canopy bidirectional reflectance distribution function (BDRF). [Reprinted from Norman et al. (1985). © 1985 IEEE.]

in general, will depend upon the LAD. (Since the shadowing of one leaf by another will also depend upon the leaf size, the Hot-Spot effect will also be dependent on the size of the leaves.)

Another way the LAD affects the CR is through a change in the probability of a gap through the entire canopy as a function of the solar zenith and view angle. The gap determines whether or not the incident and outgoing fluxes will be intercepted by the vegetation. For solar illuminations at a not-too-low elevation angle, a canopy with mostly vertical leaves (an erectophile canopy) will have a higher probability of a gap than a canopy with mostly horizontal leaves (a planophile canopy). Further, this probability has a stronger dependency on the view zenith angle for an erectophile canopy than for a planophile canopy.

The net effect of LAD on the CR is that canopies with the most horizontally inclined leaves have the least variability in reflectance as a function of the solar and view zenith angles, but have the highest reflectance of all the distributions. For canopies with mostly erect leaves, the reflectance decreases as the solar zenith angle increases in the VIS region, whereas the reflectance increases in the NIR region. As the view zenith angle increases, the reflectance increases (because the sensor sees more of the scattering from vegetation components in the upper layer, and sees less of the lower-layer components which scatter less), with the increase being greatest when the sun is near the horizon (there is more interception of the radiation).

(3) *Nonuniform Distribution.* When the vegetation elements are not uniformly distributed, e.g., they are clumped, one sees other effects of the architecture of the canopy on the CR. The clumping increases the probability of a gap through the entire canopy (which increases the influence of the scattering on the vegetation elements in lower layers) and the number of low-transmitting clumps of vegetation. These increases tend to increase the amount of backscattering and reduce the amount of forward scattering.

II.A.4   *Scattering from the Soil*

Like vegetation elements, soil also absorbs and reflects radiation that is incident on it. The reflected radiation consists of specular and diffused parts. The BRDF for bare soils depends upon the wavelength and the roughness of the soil. It usually is highly non-Lambertian and, in fact, it can be more non-Lambertian than many vegetated surfaces. (The nature of the soil reflectance is discussed in Chapters 3 and 4. However, in most CR models, the soil is assumed to be a Lambertian diffuse reflector, characterized by a λ dependent hemispherical reflectance ρ_s. (Typically, ρ_s in the NIR region is twice that in the VIS region. However, for snow, its value in the NIR region is about half of that in the VIS region. This may complicate the assessment of parameters for the tree canopies in the winter months.) The assumption of Lambertian soil is quite good for a dense canopy (LAI > 3), because very little radiation reaches the ground and the effect of reflectance from soil is negligible. However, for a sparse canopy, the soil reflectance can play an important role, especially for nadir and hot spot viewing directions. In general, multiple scattering within the soil layer will cause more absorption, leading to a lower CR.

II.B   Canopy Reflectance Models

Various CR models that incorporate the factors described in the preceding section can be divided into the following four categories:

(1) *Geometrical Models.* Here the canopy is assumed to consist of a ground surface (of known reflective properties) with geometrical objects or protrusions of prescribed shapes and dimensions (cylinders, spheres, cones, ellipsoids, etc.) and optical properties (reflectance, transmittance, and absorptance) placed on the ground in a defined manner. The interception of light and shadowing by the protrusions and the reflectance from the ground surface are analyzed to determine the reflectance from the canopy.

These models should adequately represent the reflectances of sparse canopies (shrubs, sparse forests, orchards, planted crops in the early stages of growth, etc.), where multiple scattering may be neglected, and the reflectance at low solar zenith angles, where the mutual shading of objects is negligible.

(2) *Turbid-Medium Models.* Here the vegetation elements are treated as small absorbing and scattering particles (plates and cylinders), with given optical properties, distributed randomly in horizontal layers and oriented in given directions. The canopy is usually treated as a horizontally uniform plane-parallel layer in which the radiation field depends only on the coordinate z (perpendicular to the canopy) and not on x or y. The canopy architecture is expressed through LAI and LAD. Quantities such as leaf dimensions, the effective distance between leaves, and the nonrandom distribution of leaves along the horizontal direction are neglected.

These models are more successful in representing the reflectance of denser and more horizontally uniform canopies in which the vegetation elements are smaller in size than the height of the canopy.

(3) *Hybrid Models.* Here, as in the geometrical models, the canopy is approximated by a distribution of geometrical-shaped plants except that the multiple scattering is not neglected. The vegetation elements are treated as absorbing and scattering particles. These models are needed for canopies that are neither sparse nor dense. They are obviously the most complex of all the CR models, but they are also quite versatile; in the limiting cases, they represent both the sparse and dense canopies.

(4) *Computer-Simulation Models.* In these models, the arrangement and orientation of vegetation elements are simulated on a computer. Each of these elements is divided into a finite number of areas. A Monte Carlo procedure, involving the selection of random numbers, is used to determine if a given beam of light will hit one of these areas. If it is hit, the direction of scattered radiation is chosen again using a Monte Carlo procedure. Thus, the interception and scattering of radiation is numerically followed, almost on a photon-by-photon basis. These models are computer intensive, but they have the advantage of allowing a more realistic simulation of a radiation regime in a canopy. Using these models, one can also investigate the statistical nature of the radiation field, i.e., one can calculate not only the average values, but also the higher moments and even the probability distribution of the reflected fluxes.

As noted in the Introduction, there are over two dozen CR models that have been proposed. They are reviewed in detail in Goel (1987). We will only briefly summarize those that have been inverted in the next section, where we discuss the inversion technique.

III INVERSION TECHNIQUES FOR CANOPY PARAMETERS ESTIMATION

A CR model relates the reflectance of a canopy for various solar and viewing directions to its parameters. Let us represent this relationship by

$$y_i = f_i(x_j) \qquad i = 1, \ldots, N; j = 1, \ldots, M, \tag{5}$$

where x_j are the M canopy parameters, and y_i are the N measurable canopy reflectances. The index i can represent wavelength or the solar and/or viewing direction. One technique to estimate canopy parameters consists of the inversion of the above relation to express x_j in terms of measured values of y_i. In this section, we will discuss this inversion technique and present some results obtained by inverting specific CR models.

III.A Mathematical Aspects of the Inversion Technique

Since the problem of estimating characteristics for a system from measurements on it is an ubiquitous one, there is a vast range of literature on this subject. The problem is referred to by many names, including systems identification, parameters estimation, diagnostics, and inversion. The fundamental concepts behind the various approaches were first laid out by the noted mathematician Karl Friedrich Gauss in 1795 (at the age of 18) through the well-known least-square error method (see Sorenson, 1970, for an excellent historical discussion). Recent advances have merely emphasized the implementation aspects of his concepts, using powerful computing techniques. (A significant development is the so-called Kalman Filter, done under the aegis of control theory and control engineering; see, e.g., Bryson and Ho, 1969.)

According to the least-square error method, one defines a merit function F by

$$F = \sum_i w_i (y_i - y_i')^2 \tag{6a}$$

where y_i and y_i' are the CRs as calculated from the model and as observed, respectively. The summation is over all sets of observations. w_i represents the weight given to the *i*th measurement; a more accurate measurement can be given a higher weight than a less accurate one. By substituting Eq. 5 into Eq. 6a, one gets

$$F = \sum_i w_i [f_i(x_j) - y_i']^2 \tag{6b}$$

The canopy parameters x_j are determined by minimizing the function F. For this purpose, one starts with a guess for the values of x_j, uses the CR model to calculate $f_i(x_j)$, and then determines F using Eq. 6b. One then iterates on the guess until F takes on its minimal value. If the model is an accurate representation of the canopy reflectance and there are no measurement errors in y_i, F should ideally approach zero, and the corresponding values of x_j are the desired estimates of the canopy parameters.

There are at least four important issues relevant to this inversion technique.

(1) *Number of Parameters vs. Number of Measurements.* For a unique determination of M canopy parameters x_j from N measurements y_i, M must be smaller than or equal to N. (In fact, it should be quite a bit smaller if f_i of Eq. 5 is a nonlinear relationship—a likely situation.) The models designed to simulate the maximum realism of a radiation regime in a canopy (such as computer-simulation models; see Section II) generally involve many canopy parameters. Therefore, such models are the least desirable candidates for inversion, and, in fact, to date no such model has been inverted.

(2) *Mathematical Invertibility of the Model.* Even with a sufficient number of (error-free) measurements y_i, it may not be possible to uniquely determine the canopy parameters through the inversion of a model. When it is possible, we will call the model mathematically invertible. Proving rigorously whether or not a model is mathematically invertible is quite difficult. One simple and practical approach (Goel and Strebel, 1983) consists of the following steps: (a) select a set of values for the canopy parameters x_j, (b) use the model, as represented by Eq. 5, to calculate the CRs y_i, and (c) treat these simulated CRs as the measured ones (this is equivalent to error-free measurements) and determine the values of x_j through the inversion of the model. If these determined values are identical to those used in generating the simulated CRs, the model is likely to be mathematically invertible. One can increase the confidence level of this inference by repeating the procedure for many sets of values for the canopy parameters x_j.

(3) *Local vs. Global Minimization of F.* In minimizing a nonlinear function such as F, one of the most common problems is that F does not change, even though there is another set of values for the canopy parameters for which it takes on a lower value. In other words, the minimization process does not lead to the global minimum of F; instead, it is trapped in a local minimum. For such a case, the estimated values of the canopy parameters will depend upon the initial guess and may be quite erroneous.

(4) *Stability of the Solution.* Even for a mathematically invertible model, one must face the fact that most measurements are not error-free. Therefore, an important aspect of the inversion technique is the stability of the solution or the sensitivity analysis of the model, i.e., the changes in the estimated values of the canopy parameters as the measured CRs change. If the changes are large, i.e., the solution is not stable, the inversion technique is not a viable approach for estimating canopy parameters from the measured CR data.

If the minimization procedure is locally trapped or the solution is not stable, one can use one of the following two approaches:

(a) Modify the merit function (Eq. 6b) so as to effectively alter the shape of the multidimensional surface between F and the unknown parameters. This can be done by: (i) changing the weight factors w_i, (ii) reducing the number of unknown parameters x_j by keeping some less important (auxiliary) canopy parameters fixed (at values either directly measured or estimated indirectly through independent measurements) in the inversion process, (iii) changing the number and/or type of CR observations, (iv) iterating on a function such as log x_j or exp x_j rather than x_j directly, and (v) using a function of y_i rather than y_i directly as the dependent variable.

(b) Use a different minimization technique and prior knowledge of the properties of the admissible solutions (regularization theory; see, e.g., Baltes, 1980). This knowledge can be incorporated by adding global bounds to the solution; using statistical properties of the solution, such as averages and standard deviations; imposing conditions like smoothness and positivity of the solution; and adding inequalities between unknown parameters (e.g., $0 \leq \rho + \tau \leq 1$, where ρ and τ are the leaf hemispherical reflectance and transmittance, respectively).

Goel and Thompson (1984a, b, 1985) addressed these four points in the specific context of the inversion of CR models. They developed a procedure that seeks a global minimum of F, defined by Eq. 6b, subject to specified constraints on the values of x_j. They also developed a general technique for quantifying the changes in the estimated values of the canopy parameters as the CRs are changed. For a given CR model, this technique can be used to determine the canopy parameters that should be kept fixed in the inversion process and to choose the optimal solar/view directions for the CRs for which the estimated values of the unknown canopy parameters are likely to be most accurate.

We will now illustrate these procedures and techniques using specific CR models.

III.B Estimation of Canopy Parameters Through Inversion of Specific CR Models

Several models have been investigated for the purpose of estimation of canopy parameters through their inversion and using measured CR data. However, due to space limitations, we will only describe in detail estimation based on the SAIL model for reflectance of a homogeneous canopy; the estimations based on other models will only be summarized.

III.B.1 SAIL Model
We will divide our discussion in several parts.

The Model. One of the earliest turbid-medium models is due to Suits (1972). In this model, the canopy is idealized as a homogeneous mixture of horizontal and vertical diffusely (Lambertian) reflecting and transmitting panels. That is, the original vegetation elements are replaced by their horizontal and vertical projections. The soil is assumed to be a Lambertian reflector, characterized by a hemispherical reflectance ρ_s. Because of the simple leaf-angle distribution, for a specific set of solar/view directions, the model uses only the following six canopy parameters:

ρ, τ, LAI, ρ_s, SKYL, and ALA

where ALA is the average leaf angle, and other parameters have been defined in Section II.A.

The Suits model was extended by Verhoef and Bunnik (1981; see also Verhoef, 1984) to allow for any distribution of leaf angles. They call their model the SAIL (*S*cattering by *A*rbitrarily *I*nclined *L*eaves) model. In the original model, the leaf-inclination distribution (LAD) of a canopy layer is approximated by a set of frequencies for 13 distinct inclination angles θ_L located at the centers of the intervals 0–10, 10–20, . . . , 70–80° and 80–82, 82–84, . . . , 88–90°, which all add up to 1. (A finer division of the interval 80–90°, nearly vertical leaves, is used because the calculated CRs for near vertical view angles, nadir view, are very sensitive to variations of LAD in this region of θ_L.)

To minimize the number of canopy parameters, desirable to simplify the task of CR model inversion, Goel and Thompson (1984b, c) used the two-parameters (μ and ν) beta distribution, defined by Eq. 3 to represent LAD. With this representation for LAD, for a specific set of solar/view directions, the SAIL model consists of the following seven canopy parameters:

$$\rho, \tau, \text{LAI}, \rho_s, \text{SKYL}, \mu, \text{ and } \nu. \tag{7}$$

Mathematical Invertibility of the Model. Using the simulated data (see Section III.A), Goel and Thompson (1984b) showed that the SAIL model is mathematically invertible. That is, one can uniquely estimate all the (seven) parameters occurring in the model using only the CR data. However, can one do so in practice? This question can be answered by applying the sensitivity analysis of the Appendix to the SAIL model.

Sensitivity Analysis and Practical Estimation of Canopy Parameters. Table 1 gives the error sensitivities for random errors ΔR_i of 1% in the CRs' R_i for the 25 solar/view directions given in Table 2. In calculating these sensitivities, the following nominal values of canopy parameters for a homogeneous soybean canopy are used:

$$\rho = 0.454; \tau = 0.518; \text{LAI} = 2.87, \mu = 1.607, \nu = 2.177;$$

$$\text{SKYL} = 0.2206, \text{ and } \rho_s = 0.2095 \tag{8}$$

TABLE 1   Error Sensitivities[a] to Random Errors of 1% in the CRs for the 25 Solar/ View Angles Given in Table 2 for a Soybean Canopy

Parameter	Cases[b]				
	1	2	3	4	5
LAI	73	27	25	1.4	1.2
μ	27	26	19	21	11
ν	36	36	25	24	12
SKYL	43	42	—	37	—
ρ	5.1	3.6	3.6	—	—
τ	5.2	3.9	3.5	—	—
ρ_s	124	—	—	—	—

[a]When the error level exceeded 10%, the actual value has been rounded to the nearest integer.
[b]In case 1, all variables are kept free in the inversion process. In case 2, ρ_s is kept fixed at its nominal value; in case 3, SKYL is kept fixed in addition. In case 4, all the spectral variables, ρ, τ, and ρ_s, are kept fixed, whereas in case 5, SKYL is also kept fixed.
Source: Goel and Thompson (1984b).

TABLE 2 Sets of Solar/View Angles Used for Sensitivity Analysis

Set	No. of Angle Combinations[a]	Solar Zenith Angle, θ_s	View Zenith Angle, θ_o	Relative Azimuth Angle, ψ
1	25	60	0	180
		30	30, 40	30, 60, 90, 120, 150, 180
		60	50, 60	30, 60, 90, 120, 150, 180
2	50	60, 30	0	180
		60, 30	30, 40, 50, 60	30, 60, 90, 120, 150, 180

[a]In each set, all the combinations of angles from each line were used. All angles are in degrees.
Source: Thompson (1984b).

Keeping in mind that typical values of random errors ($\Delta R_i / R_i$) are about 5%, from Table 1, one can see that if all the canopy parameters are kept fixed in the inversion process, the error levels in the estimation of all parameters other than ρ and τ would be several hundred percent. Thus, one cannot hope to accurately estimate important biophysical parameters like LAI and LAD while keeping all parameters free.

What then is the alternative? From Table 1, if ρ_s and SKYL are kept constant in the inversion process, the error level in the estimation of LAI is reduced by a factor of about 3. The absolute value of this error level for LAI drops to less than 10% if ρ and τ are also kept fixed in the inversion process. This suggests that SKYL and ρ_s should always be kept fixed in the inversion process (and preferably ρ and τ also) if one hopes to estimate LAI and LAD, with any reasonable level of confidence, from the CR data.

We should make two points: (1) Although, the error levels in the estimated values of the canopy parameters depend upon the nominal values of the canopy parameters, the previous conclusions generally hold. (2) One can reduce the high error levels by increasing the number of solar/view angles in the CR data. However, the decrease is not significant enough to alter the necessity of reducing the number of unknown parameters in the inversion process, as stated in the preceding paragraph.

Inversion with Measured CR Data. Goel and Thompson (1984b) inverted the SAIL model, using the measured CR data, in the NIR region (800–1000 nm), for a homogeneous soybean canopy, for 12 solar directions, all collected on August 20, 1980 (Ranson et al., 1981, 1985a). Table 3 gives the parameters for the 12 data sets.

In the inversion process, the parameters LAI, μ, and ν were kept free, while the other parameters were kept fixed at their field measured values. The estimated values of LAI, μ, and ν are given in Table 4. Keeping in mind that the measured values of LAI and ALA are 2.87 (with a standard deviation of 0.44) and 51.9°, respectively, one can see that the estimated values of LAI and ALA are quite good. However, the estimated leaf-angle distribution (LAD) appears to be wrong. From Eq. 3, when $\nu \gg 1$, $<\theta_L>^2 \approx <\theta_L^2>$, and, hence, by definition, the variance of θ_L approaches the value 0. Thus, the calculated LAD is a distribution peaked at ALA.

Figure 3 gives a comparison between the observed CRs and those calculated using the SAIL model. It shows the plots of percentage differences between the calculated and observed CRs. Each block corresponds to a pair of view zenith (θ_o) and view azimuth (ψ_o) angles. The twelve difference or error bars in each block correspond to the 12 data sets, with the leftmost bar for data set 1 and the rightmost for data set 12. The length of the smallest bar corresponds to a 2% error between the observed and calculated CRs (as

TABLE 3 Parameters for the 12 CR Data Sets Collected at Purdue University for a Soybean Canopy[a]

Data Set	Time Period Start : Stop	Solar Zenith, Min : Max	Solar Azimuth, Min : Max	ρ_s	SKYL	No. of CR Observ.
1	1555 : 1614	37 : 40	132 : 139	0.2318	0.1648	42
2	1623 : 1639	34 : 36	142 : 148	0.2361	0.1820	46
3	1707 : 1723	31 : 32	160 : 167	0.2343	0.1717	49
4	1730 : 1745	30 : 31	171 : 178	0.2411	0.1663	51
5	1817 : 1831	31 : 32	193 : 200	0.2402	0.1639	52
6	1838 : 1852	32 : 33	203 : 209	0.2400	0.1675	52
7	1906 : 1920	35 : 36	214 : 220	0.2348	0.1697	52
8	1927 : 1943	37 : 39	222 : 227	0.2354	0.1776	51
9	2010 : 2023	43 : 46	235 : 239	0.2314	0.1779	52
10	2032 : 2046	47 : 49	241 : 245	0.2254	0.1725	52
11	2108 : 2126	53 : 56	249 : 253	0.2156	0.1859	51
12	2142 : 2159	59 : 60	237 : 256	0.2095	0.2225	50

[a]The parameters shared in common are $\theta_o = 0, 7, 15, 22, 30, 45, 60°$; view azimuth angle = 0, 45, 90, 135, 180, 225, 270, 315°; LAI = 2.87, $\rho = 0.454$, $\tau = 0.518$, fraction of leaves in 10° intervals centered at 5, 15, . . . , 85° equal to 0.0349, 0.0360, 0.0750, 0.1294, 0.1674, 0.2085, 0.1387, 0.1226, and 0.0875, respectively. The values of μ and ν for the corresponding beta distribution are 1.607 and 2.177, respectively. SKYL represents the average value during the time of measurement of a data set. The time is in GMT. The values of ρ, τ, ρ_s, and SKYL are all "integrated values" for the Exotech-100 NIR band (800–1100 nm).

Source: Adapted from Goel and Thompson (1984c).

TABLE 4 Results of Inversion of Observed Canopy Reflectance Data Sets for a Soybean Canopy Using the SAIL Model When Only the Variables LAI, μ, and ν Are Kept Free

Data Set		Computed Values			RMS Errors,[b]
	LAI	μ	ν	ALA[a]	Calc. : Observ.
1	3.34	8.4	12.5	53.8	0.012805
2	3.19	5.4	8.0	53.7	0.014179
3	2.81	9.8	14.5	53.7	0.013838
4	2.79	11.1	15.8	52.9	0.014201
5	3.09	12.1	20.6	56.7	0.014958
6	3.12	20.5	32.6	55.3	0.016977
7	3.21	18.5	31.5	56.7	0.017198
8	3.14	9.8	16.3	56.2	0.015841
9	3.37	16.6	29.4	57.5	0.017935
10	3.14	10.9	18.9	57.1	0.017766
11	3.06	15.8	25.3	55.4	0.019921
12	2.87	11.1	15.4	52.3	0.022710
Avg[c]	3.09	12.5	20.1	55.1	0.016527
SD[c]	0.19	4.4	7.9	1.8	0.002842

[a]ALA is the average leaf angle.

[b]RMS error is that between the CRs calculated using the SAIL model and the observed CRs.

[c]Avg and SD, respectively, are the averages and standard deviations of the calculated values for the 12 data sets.

Source: Adapted from Goel and Thompson (1984c).

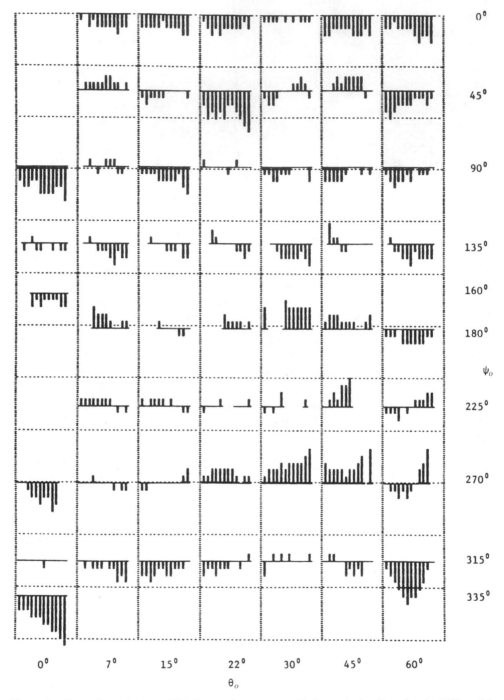

Figure 3 Comparison of observed CRs for a soybean canopy with those calculated by using the SAIL model and the estimated values of the canopy parameters. See text for fuller explanation. [Reprinted by permission of the publisher from Goel and Thompson (1984c). Copyright 1984 by Elsevier Science Publishing Co., Inc.]

a percent of the observed CR). A bar above (below) the horizontal base of the bar chart corresponds to the observed CR higher (lower) than the calculated CR. A missing bar indicates that the difference is less than 1% whereas a missing dot in the horizontal baseline indicates that that particular pair of values of the view zenith and view azimuth angles was not included in the data set.

Figure 3 shows that in most cases, the calculated CRs are generally lower than the observed values, especially for $\psi_o \leq 135°$ and $\psi_o \geq 315°$, and the differences can be as high as 10–15%. On the other hand, there are certain sets of values of θ_o and ψ_o, e.g., $7° \leq \theta_o \leq 22°$ and $180° \leq \psi_o \leq 270°$, where the values given by the SAIL model are generally in agreement with the observed values, and another set of values, e.g., $30° \leq \theta_o \leq 45°$ and $180° \leq \psi_o \leq 270°$, where the calculated values are generally higher than the observed ones.

Another measure of difference between the calculated and measured CRs is provided by the root-mean-square error (RMSE). This measure, between two sets of CRs, R_a and R_b, is defined by

$$\text{RMSE} = \left(\frac{\sum (R_a - R_b)^2}{N_{\text{obs}}} \right)^{1/2} \tag{9}$$

where N_{obs} is the number of solar/view directions in each of the two data sets, and the summation is over all of these directions. Table 4 presents the RMSEs for the 12 data sets.

In an attempt to further understand the relationship between the calculated and observed CRs and the estimated values of the canopy parameters, Goel and Thompson made systematic changes in the values of the observed CRs and the measured values of leaf hemispherical reflectance ρ and transmittance τ, and carried out the inversion process with these modified values to obtain canopy parameters. Table 5 shows the results.

In cases 1 and 3, the observed CRs, for all the solar/view directions, are reduced by 4 and 8%, respectively, whereas they are increased by the same amounts for cases 2 and 4. As to be expected, a reduction (increase) in the observed CRs corresponds to a lower (higher) value of LAI. In addition, a decrease (increase) in CRs tends to make the LAD broader (narrower). The average leaf angle, ALA, also decreases (increases) with a

TABLE 5   Results of Inversion of Modified Data Set 1[a]

Case No.	Calculated Values				RMS Errors, Calc. : Observ.
	LAI	μ	ν	ALA	
1	2.84	6.2	8.5	52.1	0.012473
2	3.84	10.4	16.2	54.8	0.013055
3	2.46	5.4	7.1	51.0	0.012081
4	4.50	17.8	30.0	56.4	0.013205
5	3.28	10.6	15.1	52.9	0.013210
6	3.31	5.5	8.2	53.9	0.012493

[a]Various cases correspond to various modifications of the measured CR. Case 1 (2) corresponds to a 4% decrease (increase), whereas case 3 (4) to a 8% decrease (increase) in the observed CRs. In case 5, ρ is decreased and τ is increased by 5% from their respective measured values, whereas in case 6, ρ is increased and τ is decreased by the same percentage. For notations see Table 4.

Source: Adapted from Goel and Thompson (1984c).

reduction (increase) in the CRs, but the effect is less pronounced than for the LAI or LAD. Although the calculated LAD tends to become closer to the measured one when the CRs are reduced, in none of the three cases of reduced CRs did the calculated LAD approach the measured LAD. A further reduction in the CRs should make the LAD closer to the measured value, but at a cost of resulting in an unacceptably low value of LAI.

To assess the effects of measurement errors in the canopy parameters, which are kept fixed in the inversion process, they changed the values of ρ and τ. In case 5, in the inversion process, the value of ρ used in 5% less than the measured value, whereas the value of τ is 5% higher. In case 6, the value of ρ used is 5% higher and the value of τ is 5% lower than the measured values. (Note that one of these parameters was increased while the other one was decreased so as not to cause unrealistic changes in the leaf absorption coefficient, which is equal to $1 - [\rho + \tau]$.) Increasing ρ and decreasing τ tends to make the LAD less peaked, whereas the LAD tends to be more peaked when ρ is decreased and τ is increased. This suggests that if one increases ρ further and simultaneously decrease τ, one can obtain a reasonable LAD. One easy way to test this conjecture is to keep the values of μ and ν fixed in the inversion process and carry out the inversion with LAI, ρ, and τ as free variables. Such an inversion gave the following values: $\rho = 0.5927$, $\tau = 0.3784$, LAI $= 2.40$, and RMSE $= 0.012850$. The calculated value of ρ is 14.4% higher than the measured value, whereas it is 14.7% lower for τ. The calculated value of LAI is toward the lower end of the range of measured values.

Keeping in mind the errors in measurement of canopy parameters and the CR, the previous results suggest that the SAIL model seems to represent the reflectance of a homogeneous soybean fairly well, and the estimated values of the important canopy parameters LAI and ALA are in good agreement with the measured ones. However, estimation of LAD is erroneous. Goel and Thompson note that the disagreement between the model and the data is due to several effects not being properly accounted for in the SAIL model, including row effects (canopy not being truly homogeneous), dependence of ρ and τ on the angle of incidence of solar radiation (leaves being non-Lambertian reflectors, see Eqs. 2), and the hot spot phenomenon (see Section II.A.3).

Optimal Solar/Viewing Directions. In connection with the previous estimation of LAI and LAD (μ and ν) from the measured CR data, there are the following two important issues:

(1) How do the accuracies of the estimations of LAI and LAD change if the ancillary parameters (which are kept fixed at their measured values in the inversion process) are only known to a certain level of accuracy?

(2) Are there optimal solar/view directions for which the estimations of LAI and LAD are most accurate?

Both of these issues have been addressed by Goel and Thompson (1985) by extending the sensitivity analysis of the Appendix. We will now describe the concept behind the analysis and give key results.

The concept behind an optimal choice of solar/view directions is as follows. The values of the three canopy parameters LAI, μ, and ν are given by the intersection of three surfaces defined by

$$R_i = R(\text{LAI}, \mu, \nu; \xi_i) \qquad i = 1, 2, 3 \tag{10}$$

where the ξ_i denote the three solar/view directions, and the R_i denote the corresponding CRs. If these surfaces (or the vectors normal to these surfaces at the point of intersection) tend to be almost parallel to one other, a small error in the observed CR (i.e., a small change in the surface) will greatly move the intersection point (i.e., the calculated values of LAI, μ, and ν). Thus, to minimize the errors in the estimation of LAI and LAD to errors in CR, it is desirable to choose those R_i for which the gradient vectors (with respect to LAI, μ, and ν) are as orthonormal to one other as possible.

In Goel and Thompson (1985), the following "error function" W was chosen (and minimized) for the overall measure of the error levels in the estimation of LAI, μ, and ν:

$$W = 0.98E\left[\left(\frac{100\Delta\text{LAI}}{\text{LAI}}\right)^2\right] + 0.01E\left[\left(\frac{100\Delta\mu}{\mu}\right)^2\right] + 0.01E\left[\left(\frac{100\Delta\nu}{\nu}\right)^2\right] \quad (11)$$

where E represents the expected value. This equation implicitly states that one wants to estimate LAI almost $(0.98/0.01)^{1/2} \approx 10$ times more accurately than the LAD parameters μ and ν.

Table 6 gives the optimal triplet (set 1) of illumination/viewing directions (out of the 25 solar/view directions of Table 2) for which W is minimum, a triplet (set 2) that is almost as desirable as set 1, and a triplet (set 3) that is one of the least desirable.

This table also gives the expected levels of error in the estimation of LAI, μ, and ν (and the value for the overall error function W) for 1% random errors in CRs under the following assumptions: (1) measurements are repeated eight times for each of the optimal solar/viewing directions (total of 24 measurements), and (2) it is assumed that the other parameters (ρ, τ, SKYL, ρ_s) of the model are exactly known. If the second assumption is not valid, the error levels are higher. For example, if the values of SKYL and ρ_s are known to within 10%, and ρ and τ to within 5%, then for the optimal triplet

TABLE 6 The Error Sensitivities for LAI, μ, and ν for Three Triplets of Solar/View Angles Selected From Those Given in Table 2[a]

Set No.	Solar/View Directions	Error Levels			
		LAI	μ	ν	W
1	60 0 180 30 40 180 60 60 30	0.94	7.6	7.5	2.01
2	30 40 150 60 50 180 60 60 30	0.96	8.4	8.1	2.28
3	30 40 60 60 50 120 60 60 180	178.6	1820.3	2109.6	1.1×10^5

[a]It is assumed that for each of the three directions in a triplet, the canopy reflectance observation is repeated eight times (for a total of 24 observations). The nominal values for the agronomic parameters are those for a soybean canopy, i.e., LAI = 2.69, μ = 1.607, and ν = 2.177. The other parameters ρ, τ, ρ_s, and SKYL are kept fixed in the inversion process at the values given in Eq. 8.

Source: Goel and Thompson (1985).

(set 1), the percentage error levels in LAI, μ, and ν are 16.3, 11.8, and 17.6%, respectively. If only one observation per solar/view direction is made, instead of eight, then W increases by a factor of 8, and the error in the estimation of LAI increases approximately by a factor of $8^{1/2} = 2.83$. On the other hand, if 16 observations per solar/viewing direction are made, then W decreases by only a factor of 2, and hence the error level in the estimation of LAI decreases only by a factor of $2^{1/2} = 1.41$.

Although the optimal triplet of solar/view directions varies somewhat with the nominal values of the canopy parameters, the analysis gives two basic strategies for selecting the optimal solar/view directions: keep the view zenith angle fixed at a high value (say 60°) and vary the solar zenith and azimuth angles or keep the solar zenith angle fixed at a similarly high value and vary the view zenith and azimuth angles. The specific recommended solar/view angles are as given in Table 7.

We should make the following observations about optimal angles:

(1) All directions given in Table 7 are essentially in the principal plane defined by the solar and view directions, and the preferred directions have sun either facing the observer or behind him.

(2) The guidelines of Table 7 are only valid for homogeneous canopies. It is obvious that for a canopy that has its planting in rows, some observations with the sun direction parallel to the rows (to minimize shadowing effects) and some perpendicular to the rows may be quite informative about the canopy parameters.

(3) If one is interested in estimating only the LAI, then one could choose an alternate solar/viewing direction. Using the SAIL model, Bunnik et al. (1983) showed that IR/red, the ratio of reflectances in the infrared and the red regions, in the "hot spot" direction for a view zenith angle of 52° is practically independent of LAD, but is strongly dependent on LAI. Thus, this ratio can be used to estimate LAI.

TABLE 7 Optimal Triplet of Solar/View Directions for a Homogeneous Canopy[a]

View Zenith Angle θ_o Fixed

(a) $\theta_o \leq 45°$		(b) $\theta_o \geq 60°$	
θ_s	ψ	θ_s	ψ
Low	High	Low	Low
High	High	High	High
High	Low	High	Low

Sun Zenith Angle θ_s Fixed

(a) $\theta_s \leq 45°$		(b) $\theta_s \geq 60°$	
θ_s	ψ	θ_s	ψ
15°	High	15°	Low
High	High	High	High
High	Low	High	Low

[a]Here low $\theta_s \leq 30°$, high $\theta_s \geq 60°$, low $\psi \approx$ 0–30°, high $\psi \approx$ 180°, and high $\theta_o \approx 60°$.

Source: Goel and Thompson (1985).

This approach has been validated with field measurements on winter-wheat canopies.

We conclude the discussion of the SAIL model by pointing out that its potential for estimation of canopy parameters was further investigated by Goel and Deering (1985) using the measurements for orchard grass and lush green soybean canopies. They found that the LAI and ALA for these canopies also can be estimated quite accurately through the inversion technique. Goel et al. (1984) showed that a one-layer SAIL model is not adequate to estimate the LAI of a homogeneous black spruce canopy, but a two-layer model is.

III.B.2 Other Models

In addition to the SAIL model, other CR models have been investigated from the point of view of estimation of canopy parameters through their inversion. Some of these models have been studied for mathematical invertibility, using simulated CR data [see Goel and Thompson (1984a) for the Suits' model, Goel and Thompson (1983a,b) for the Cupid model due to Norman (1975, 1979; see also Walthall et al., 1985; Norman et al., 1985; Walter-Shea, 1987) for homogeneous canopies, N. S. Goel and T. Grier (unpublished result) for the Bidirectional General Array (BIGAR) model by Norman and Welles (1983; Norman et al., 1985) for a canopy with three-dimensional inhomogeneities], while others have also been inverted with measured CR data. Here we will only emphasize those in the latter category.

Camillo Model. In this model (Camillo, 1987) like the SAIL model, the canopy is also treated like a turbid medium, except that the mathematical formalism is somewhat different and certain approximations for the interaction of the solar flux with the canopy elements allow an analytical expression for the CR. In this model, it is assumed that the extinction coefficient of the diffuse flux is constant throughout the canopy, the canopy consists of one layer, and the scattering from the vegetation elements is characterized by a single-scattering albedo a and the scattering phase function

$$P(\Theta) = (1 + g \cos \Theta) \tag{12}$$

where Θ is the scattering angle between the incident and outgoing radiances, and g is the asymmetry factor defined by

$$g = \tfrac{1}{2} \int_{-1}^{+1} p(\Theta) \cos \Theta \, d(\cos \Theta) \tag{13}$$

(This asymmetry factor varies between -1 and 1. It is equal to zero for isotropic scattering. The case $g = 1$ defines an infinitely narrow forward-scattered beam, and a value of g between 0 and -1 corresponds to the predominantly backscattering, as is the case for vegetation canopies.)

The leaf hemispherical reflectance and transmittance are related to a and g through the relations:

$$\rho = a \int_0^1 P(-\Theta) \cos \Theta \, d(\cos \Theta) = a \left(\frac{1}{2} - \frac{g}{3} \right) \tag{14a}$$

$$\tau = a \int_0^1 P(\Theta) \cos \Theta \, d(\cos \Theta) = a \left(\frac{1}{2} + \frac{g}{3} \right) \tag{14b}$$

That is,

$$a = \rho + \tau \tag{15a}$$

$$g = \frac{3}{2} \frac{\tau - \rho}{\tau + \rho} \tag{15b}$$

In this model, the following parameters are used:

$$a, g, \text{LAI}, \alpha_s, k(\mu_s, \psi_s), \text{ and SKYL}; \ \mu_s = \cos \theta_s \tag{16}$$

where α_s is the soil albedo, and $k(\mu_s, \psi_s)$ is the extinction coefficient of the incident direct flux that contains in it the information about LAD.

We should note that this model has one serious limitation. The solution of the radiative transfer equation for the case of direct incident flux is valid only if

$$\frac{k(\mu_s, \psi_s)}{\cos \theta_s} < 1 \tag{17}$$

Therefore, for the model to be valid, the solar zenith angle should be smaller than arccos k. For uniformly distributed leaves, $k = \frac{1}{2}$ and the solar zenith angle must be smaller than 60°. For LAD with only horizontal leaves, $k = \cos \theta_s$ and the inequality (Eq. 17) cannot be met; the model cannot be applied. For the azimuthally symmetric all-vertical-leaves case, $k = 2 \sin (\theta_s / \pi)$. Therefore, the inequality (Eq. 17) requires that $\tan \theta_s < \pi/2$, or $\theta_s < 57.5°$.

Camillo inverted his model using the same soybean data set as those used by Goel and Thompson (1984b) in the inversion of the SAIL model, except that he chose only data sets 3 and 11 of Table 3 (with $\theta_s = 31.5°$ and $\theta_s = 54.5°$). However, he inverted the CR data not only in the NIR region, but also in the VIS (500–600 nm) region. Figure 4 shows the polar plots of measured and calculated reflectances. In these figures, $u = \cos \theta_o$ (cosine of the view zenith angle) is the distance from the center, where $u = 1$ ($\theta_o = 0°$) is at the center and $u = 0.5$ ($\theta_o = 60°$) is at the edge. Circles representing $u = 0.9$ ($\theta_o = 25.8°$) and $u = 0.7$ ($\theta_o = 45.6°$) are also shown. The azimuth is measured relative to the solar azimuth (the sun is at 180° azimuth in these figures). The solid lines represent the values calculated using the model, and the dashed lines are the data. (Note that the CR values plotted are twice the values measured in the field or used in the SAIL or other CR models.)

The percent differences between the calculated and measured values of CRs are mostly 5% or less, with a few greater than 10%, and the largest value is 18.8%. Table 8 shows the estimated values of the canopy parameters obtained through inversion of the model with each of the four data sets, together with the field measured values of these parameters. In general, the agreement is quite good, with the exception of the values of g, the asymmetry parameter (see Eq. 13), and SKYL. The values of g computed from field measured values of ρ and τ are of the order of 0.1, whereas the model fits to large negative values. A proposed explanation is that the measured values of ρ and τ do not

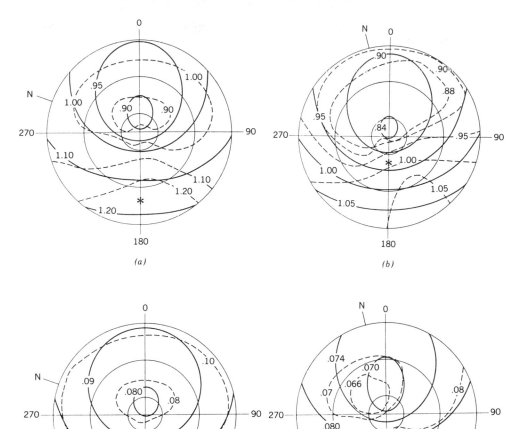

Figure 4 Polar plots of measured and calculated (solid lines) CRs. (a) and (b) in the NIR region; (c) and (d) in the VIS region. For (a) and (c), $\theta_s = 31.5°$. For (b) and (d), $\theta_s = 54.5°$. $\theta_o = 0°$ is at the center. Three concentric circles correspond to $\theta_o = 25.8$, 45.6, and $60°$ ($u = 0.9$, 0.7, and 0.5). $\psi_o - \psi_s$ is labeled around the circle. The sun position is indicated by an asterisk, and North is indicated by N. [Reprinted by permission of the publisher from Camillo (1987). Copyright 1987 by Elsevier Science Publishing Co., Inc.]

include leaf specular reflectance or reflectance from stems, which if included might account for the dominance of backscattering over forward scattering in the bidirectional reflectance data. The estimated values of SKYL are quite erroneous (this reflects the fact that the calculated reflectances are not very sensitive to changes in SKYL).

It should be noted that the Camillo model does not allow for the estimation of leaf-angle distribution (LAD). Instead, one can only compare the values of $k(\mu_s, \psi_s)$ as calculated using measured LAD against those estimated through inversion. This comparison is shown in Figure 5, where the estimated values are shown together with one sigma error bar. The agreement between two values is good, especially at $\theta_s = 54.5°$.

TABLE 8   Values of the Canopy Parameters Estimated by Fitting the Model (to Each of the Four Soybean CR Data Sets) and as Measured

Region	θ_s	a	g	α_s	LAI	k	SKYL	RMS
				Estimated				
NIR	31.5	0.93	−0.91	0.22	1.7	0.79	0.0	0.04
		±0.03	±0.16	±0.05	±1.8	±0.11	±0.09	
NIR	54.5	0.94	−1.00	0.30	1.9	0.57	0.09	0.05
		±0.04	±0.23	±0.07	±2.1	±0.12	±0.10	
VIS	31.5	0.19	−0.42	0.02	2.0	0.83	0.01	0.005
		±0.02	±0.07	±0.07	±1.6	±0.07	±0.06	
VIS	54.5	0.24	−0.20	0.03	1.6	0.57	0.00	0.007
		±0.01	±0.11	±0.07	±1.7	±0.04	±0.01	
				Measured				
NIR		0.097	0.10	0.23	2.9		—	
		±0.04	±0.05	±0.01	±0.4			
VIS	31.5	0.18	0.04	0.12	2.9		0.22	
		±0.02	±0.02	±0.01	±0.4		±0.01	
VIS	54.5	0.18	0.04	0.12	2.9		0.33	
		±0.02	±0.02	±0.01	±0.4		±0.01	

Source: Adapted from Camillo (1987).

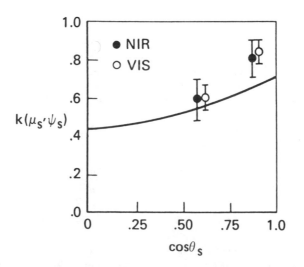

Figure 5   Estimated (together with one sigma error bar) and calculated (from measured LAD) values of $k(\mu_s, \psi_s)$ vs. cos θ_s. [Reprinted by permission of the publisher from Camillo (1987). Copyright 1987 by Elsevier Science Publishing Co., Inc.]

Goel-Grier Row Model. This hybrid model due to Goel and Grier (1986a,b) is for a row-planted canopy (two-dimensional inhomogeneous canopy). It assumes that the vegetation canopy consists of subcanopies along rows. Each subcanopy has an elliptical cross section, described by the form

$$f(\delta) = \tfrac{1}{2}\left[1 - (\delta/JP)^2\right]^{1/2} \qquad 0 \le \delta \le JP \tag{18}$$

where δ is the across-row strip displacement, P is the row period (more precisely, the ratio of the row period to the canopy's maximum height), and J ($0 \le J \le 1$) is the fraction of distance P covered by half the plant canopy (see Figure 6). In the early stages of growth, the value of J is small, and, as the plant grows, J increases and the two ellipses grow to meet each other, as shown in Figure 6.

Each subcanopy is treated like a turbid medium, and the SAIL model is used to describe the radiation regime inside each subcanopy. Thus, the model uses seven parameters for the SAIL model (see Eq. 7) plus three parameters: J, P (to characterize the shape of the subcanopies), and ROAZ, the azimuth angle of the row direction. Goel and Grier inverted their row model, using the measured CRs in the NIR regions, for a soybean canopy in early, intermediate, and full stages of growth (Ranson et al., 1981, 1985a) and a corn canopy at early and full growth stages (Ranson et al., 1985b). For each of these canopies, CR data were collected several times in a day. For all of these data sets, the estimated values of canopy parameters were in good agreement with the field measured values. Tables 9 and 10 give the estimated values of the canopy parameters as well as measures of fit between the values of CRs as calculated from the model and those measured in the field. These tables use two measures of fit: the root-mean-square error (RMSE) measure defined by Eq. (9) and the percentage root-mean-square error (PRMSE), defined by

$$\text{PRMSE} = \left(\frac{\sum (1 - R_c/R_o)^2}{N_{\text{obs}}}\right)^{1/2} \tag{19}$$

where R_o is the observed CR for a given solar/view direction, R_c is the calculated CR (as given by the model), N_{obs} is the number of solar/view directions in each of the data sets, and the summation is over all directions.

Tables 9 and 10 also use two measures for percentage of ground cover (GC). One measure, GC_1, is defined as the area of the model canopy as a fraction (or percentage) of the area ($= P$) of the rectangle formed by the row spacing and the fixed canopy height of 1. This definition is related to parameters J and P by

$$GC_1 = \begin{cases} \dfrac{2A(JP)}{P} \times 100\%, & J < 0.5 \\[2mm] \dfrac{2A(P/2)}{P} \times 100\%, & J \ge 0.5 \end{cases} \tag{20}$$

where

$$A(\delta) = \int_0^\delta f(\delta)\,d\delta \tag{21}$$

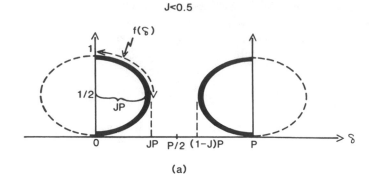

(a)

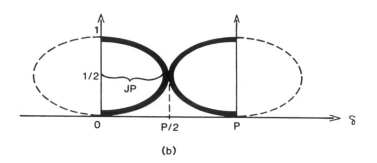

(b)

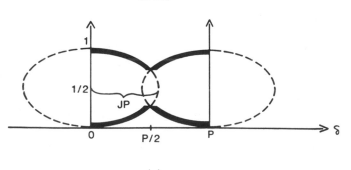

(c)

Figure 6   Idealized canopy profile for three growth stages. [From Goel and Grier (1986a).]

The other measure, GC_2, is equal to twice $J \times 100\%$ when $J \leq 0.5$ and 100% when $J \geq 0.5$. This is the measure used by the investigators who provided the data.

From Tables 9 and 10, one can see that the model fitted various data sets quite well, and the estimated values of canopy parameters, in general, are in good agreement with the measured ones. However, similar to the SAIL model, the estimation of the leaf-

TABLE 9   Estimated Values of the Canopy Parameters for Soybean Canopies[a]

Canopy Param.	July 17 Average ±St. Dev.	July 17 Measured Values	July 24 Average ±St. Dev.	July 24 Measured Values	August 28 Average ±St. Dev.	August 28 Measured Values
LAI	3.60 ± 0.23	3.0 ± 0.5	4.26 ± 0.42	3.9 ± 0.6	3.11 ± 0.22	2.9 ± 0.4
J	0.38 ± 0.01	0.38	0.43 ± 0.02	0.45	1.05 ± 0.17	0.68
μ	21.5 ± 4.1	—	$7.3 + 3.0$	2.0	8.1 ± 1.6	1.6
ν	35.5 ± 10.6	—	14.2 ± 4.4	2.4	11.8 ± 3.0	2.2
P	1.13 ± 0.24	1.10	1.13 ± 0.23	0.9	2.09 ± 1.6	0.75
ρ	0.63 ± 0.08	—	0.61 ± 0.05	—	—	0.45
τ	0.30 ± 0.10	—	0.34 ± 0.06	—	—	0.52
ALA	55.5 ± 4.1	—	60.4 ± 3.5	49.0	53.3 ± 1.2	51.8
$\%GC_1$	59.2 ± 1.1	72	67.7 ± 2.3	83	96.0 ± 0.8	99
$\%GC_2$	75.2 ± 1.1	72	86.0 ± 3.1	83	100.0 ± 0.0	99
RMSE	0.0146 ± 0.002		0.0171 ± 0.002		0.0178 ± 0.005	
PRMSE	3.5 ± 0.4		4.1 ± 0.4		3.7 ± 0.9	

[a]For July 17, the estimated values are the average and standard deviations of values obtained through inversion of the CR data sets for five solar directions. For July 24 and August 28, the number of data sets are 6 and 4, respectively. For each data set, the number of viewing directions is from 45 to 55.

Source: Goel and Grier (1986b).

TABLE 10   Estimated Values of the Canopy Parameters for Corn Canopies[a]

Canopy Param.	June 24 Average ±St. Dev.	June 24 Measured Values	July 15 Average ±St. Dev.	July 15 Measured Values
LAI	1.27 ± 0.21	1.2 ± 0.3	4.67 ± 0.03	4.0 ± 0.2
J	0.30 ± 0.04	—	1.18 ± 0.39	—
μ	3.3 ± 0.2	—	0.7 ± 0.2	1.6
ν	4.9 ± 0.6	—	1.4 ± 0.1	1.8
P	1.01 ± 0.23	1.36	1.02 ± 0.04	0.28
ρ	0.44 ± 0.04	—	0.30 ± 0.09	—
τ	0.48 ± 0.06	—	0.66 ± 0.11	—
ALA	53.7 ± 2.2	—	59.3 ± 5.0	48.0
$\%GC_1$	59.2 ± 1.1	72	95.0 ± 2.6	96
$\%GC_2$	75.2 ± 1.1	72	100.0 ± 0.0	96
RMSE	0.0135 ± 0.002		0.0205 ± 0.005	
PRMSE	5.3 ± 0.9		4.3 ± 1.0	

[a]For June 24, the estimated values are the average and standard deviations of values obtained through inversion of the CR data sets for five solar directions. For July 15, the number of data sets is 3. For each of the June 24 data sets, the number of viewing directions is about 55, whereas for each of the July 15 data sets, this number is equal to 39.

Source: Goel and Grier (1986b).

angle distribution is quite poor. But when the non-Lambertian behaviors of leaf reflectance and transmittance (i.e., their quadratic dependences on the angle of incidence; see Eqs. 2) were included, estimated values of LAD parameters μ and ν changed significantly (the estimated values of all other parameters, LAI, J, P, ρ, and τ, changed only slightly). For example, for the July 17 canopy, the average values of (μ, ν) changed from (21.5, 35.5) to (3.4, 7.0), and for the July 24 canopy, from (7.3, 14.2) to (3.8, 9.3). These changes are in the desired direction. These results indicate that non-Lambertian behavior of the leaves may be an important effect to include in the model.

In Figures 7–9 are given the measured and calculated bidirectional CR surfaces for three representative data sets. These figures further validate the model and the inversion technique. They also show that the model is capable of representing a variety of CR surfaces.

Goel-Grier TRIM Model.   Like the BIGAR model, this model, dubbed TRIM (for Three-Dimensional Radiation Interaction Model) by Goel and Grier (1988), is also for a canopy with three-dimensional inhomogeneities. It is an extension of their row model (see the last section).

They divide the ground into a rectangular grid, with one axis (denoted by η axis, which is at an angle ROAZ with the 0° azimuth) parallel to the row direction and the other axis (denoted by δ axis) perpendicular to it. The repetitive distances along the δ and η directions are, respectively, P and Q times the height of the canopy (which is taken to be equal to 1).

In a simple version of the model, referred to as the "one-ellipsoid" configuration, each cell contains one ellipsoidal subcanopy. The profile of the subcanopy is described by a continuous function, $C(\delta, \eta)$ $(0 \leq \delta \leq P, 0 \leq \eta \leq Q)$, defined by

$$C(\delta, \eta) = \begin{cases} (1 - \text{FACT})^{1/2} & \text{FACT} \leq 1 \\ 0 & \text{elsewhere in } 0 \leq \delta \leq P, 0 \leq \eta \leq Q \end{cases} \tag{22a}$$

where

$$\text{FACT} = \left(\frac{\delta - P/2}{JP}\right)^2 + \left(\frac{\eta - Q/2}{DQ}\right)^2$$

$2J$ = fraction of rectangle width P covered by one axis (δ axis)

of the ellipsoid   (22b)

$2D$ = fraction of rectangle length Q covered by the second axis

(η axis) of the ellipsoid

Thus, JP and DQ are the semiaxes of the ellipsoid in the δ and η directions, respectively. By varying the two parameters J and D, one can represent various stages of growth of a canopy (Figure 10), including a row-planted vegetation canopy (Figure 11). These parameters also determine the percentage of ground cover in any growth stage.

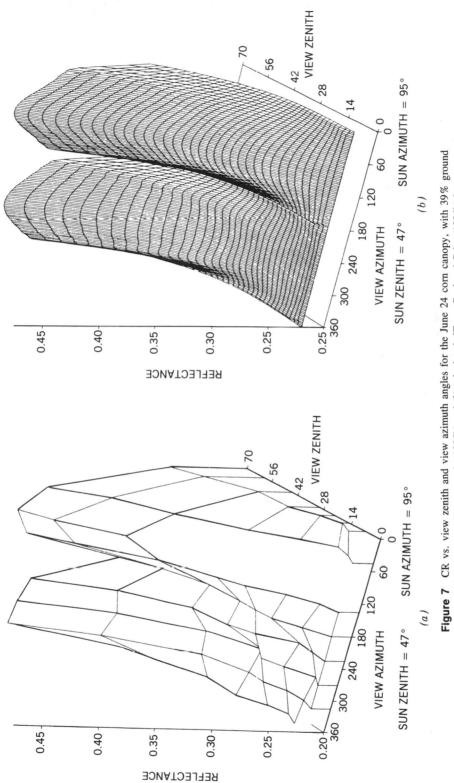

Figure 7 CR vs. view zenith and view azimuth angles for the June 24 corn canopy, with 39% ground coverage: (*a*) measured (Ranson et al., 1985b) and (*b*) calculated. [From Goel and Grier (1986b).]

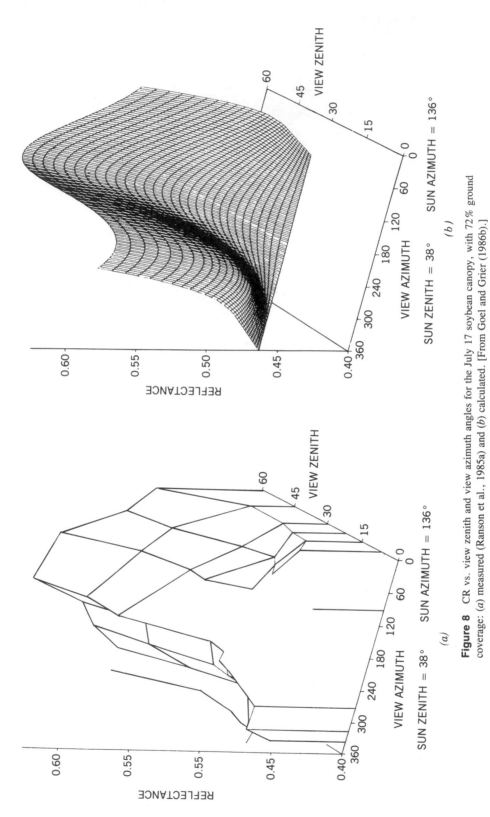

Figure 8 CR vs. view zenith and view azimuth angles for the July 17 soybean canopy, with 72% ground coverage: (*a*) measured (Ranson et al., 1985a) and (*b*) calculated. [From Goel and Grier (1986b).]

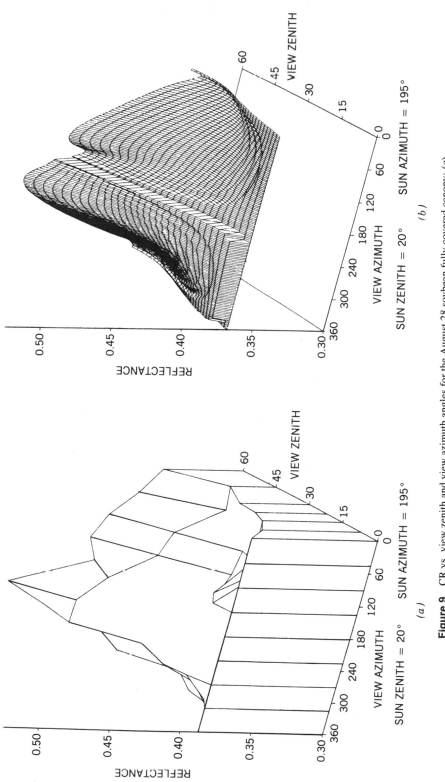

Figure 9 CR vs. view zenith and view azimuth angles for the August 28 soybean fully covered canopy: (*a*) measured (Ranson et al., 1985a) and (*b*) calculated. [From Goel and Grier (1986b).]

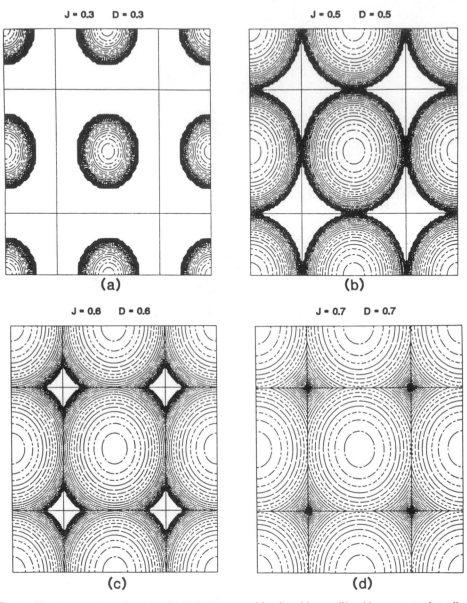

Figure 10 Cross section of rectangular lattice (at ground level), with one ellipsoid per rectangular cell and architectural parameters J and D chosen to represent the growth of the canopy. [Reprinted by permission of the publisher from Goel and Grier (1988). Copyright 1988 by Elsevier Science Publishing Co., Inc.]

For a one-layer canopy, with the ground surface divided into a rectangular grid, the model uses a total of 12 canopy parameters: 7 for the SAIL model for a homogeneous canopy; 2 parameters, P and Q, which define the grid cell size; 2 parameters, J and D, which define the two axes of the ellipsoidal subcanopy; and ROAZ, the azimuth direction of one of the axes (corresponding to P) of the grid.

Although the TRIM model is designed to represent the reflectance of canopies with three-dimensional inhomogeneities, the model has not yet been used to estimate canopy

J = 0.3   D = 0.7

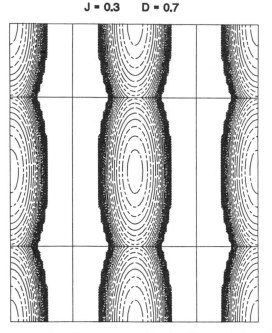

Figure 11 Schematic representation of a row canopy. [Reprinted by permission of the publisher from Goel and Grier (1988). Copyright 1988 by Elsevier Science Publishing Co., Inc.]

parameters for such canopies because of the lack of field-measured data on CRs and canopy parameters for such canopies. However, Goel and Grier (1988) inverted this model to estimate the canopy parameters for a corn canopy, in different stages of growth (it is the same corn canopy as used in conjunction with the row model in the preceding section), and for a naturally growing shinnery oak canopy. In both cases, only CR data in the NIR region have been used.

It is important to note that in the inversion process, the nature of the distribution of subcanopies is kept fixed. (For simplicity, the "one-ellipsoid" configuration previously described, with one elliptical subcanopy per rectangular cell, was used.) Thus, it is not yet established whether the distribution of subcanopies can be determined using only the CR data through inversion of the TRIM model.

The model to data has been tried only for canopies where the subcanopies are distributed in a regular pattern. The few areas that still have to be investigated include (1) extending the model to allow for random distributions of subcanopies (i.e., make the model stochastic), (2) incorporating non-Lambertian bidirectional reflectance properties of the soil, (3) including non-Lambertian behavior of the vegetation elements, and (4) treating the hot spot phenomenon more completely.

Otterman Model. In this geometrical model, Otterman (1981, 1984) assumes the protrusions to be vertical cylinders of variable heights and negligible cross sections (diameter-to-height ratio much smaller than 1), located randomly and sparsely on the plane surface (see Figure 12).

The interception of the direct solar beam by these cylinders is analyzed in terms of (1) the reflectances r_i and r_c, respectively, of the soil surface and the external vertical

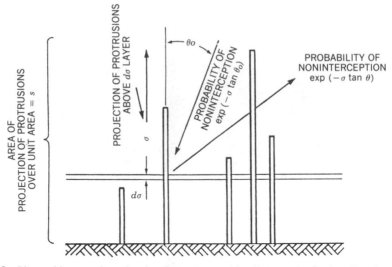

Figure 12   Plane with protrusions showing the geometry of irradiance and reflection. [Reprinted from Otterman and Weiss (1984).]

surfaces of the cylinders (both types of surfaces are assumed to be Lambertian. Further, the cylinders are assumed to be opaque, i.e., of zero transmittance); (2) the solar zenith angle θ_s; and (3) a protrusion diameter parameter s, which is the sum of the products of the height times diameter of these cylinders per unit horizontal area. [s characterizes the architecture of the canopy; large (small) s corresponds to dense (sparse) canopies.]

Only the first reflection of the direct solar beam is computed accurately. It is assumed to be made up of two parts: (1) that which is reflected from the soil surface without interception by the protrusions and (2) that which is reflected from the protrusions and then leaves the canopy without interception by other protrusions. The multiple combined reflections from the soil and protrusions are treated only approximately, and that is also for the case of $r_i = r_c$ (which may be a reasonable assumption if the ground is covered with some fallen vegetation) and $r_i = 5r_c$ (a reasonable assumption if the ground is covered with snow).

The model has been generalized (Otterman, 1985) to include the horizontal vegetation elements. These elements are characterized by a parameter u, the sum of the areas of projections of these elements on a horizontal plane per unit ground area, and are allowed to reflect from only their upper surfaces. This generalization has been extended even further by Otterman and Staenz (1985) to include mountain slopes.

The only attempt to estimate the canopy parameters through inversion of this model (which we are aware of) is by Otterman et al. (1987). They considered the question of inferring spectral reflectances of plant elements from measured BRDFs, when the architecture of the canopy is known. They assumed that the plants can be represented as a field of randomly located either vertical elements or horizontal elements (but not both; a very unrealistic assumption). Using the explicit expression for the BRDF, they showed that the spectral reflectances of soybean (assumed horizontally oriented), corn, and balsam fir (both assumed to be vertically oriented) can be accurately estimated using only the measured BRDFs.

Li-Strahler Model. The Otterman model assumes that the protrusions are small and numerous. Therefore, it is not directly applicable to the open forest canopies in which the protrusions (trees) can occur in low densities and/or are relatively large. To address these problems, Li and Strahler (1985, 1986; Li, 1981, 1985; Strahler and Li, 1981) modeled low-density timber stands as a collection of randomly spaced cones. Each cone has a fixed base/height ratio and is taken to be a flat Lambertian reflector that absorbs visible light differently (i.e., is green). Both the heights and radii of cones are assumed to be lognormally distributed, with their respective fixed means and variances, and known coefficients of variation (ratio of standard deviation to mean). Tree counts from pixel to pixel vary according to a given distribution (uniform, Poisson distribution, or Neyman) with a fixed density η of cones per unit area.

The CR is taken to be an area-weighted sum of the reflectances of four spectral scene components: sunlit area of the cone, shadowed area of the cone, shadowed area of the ground, and sunlit area of the ground.

Figure 13 shows the geometry of a cone illuminated at a solar zenith angle θ_s. The apex angle of the cone is 2α. If $\theta_s > \alpha$, a shadow will be cast. The top view in this figure shows the projection of the cone and shadow onto the horizontal plane (i.e., the cone is flattened). The angle γ describes the protion of the cone illuminated beyond the cone half and is given by

$$\sin \gamma = \tan \alpha \cot \theta_s \qquad (23)$$

The projection of the cone and shadow consists of three components: illuminated crown, shadowed crown, and shadowed background. Let A_g, A_z, A_c, and A_t, respectively, denote the areas of illuminated background, shadowed background, illuminated crown, and shadowed crown within a pixel of area A. Let G, Z, C, and T represent the corresponding reflectances per unit area in a given wavelength band. Then the reflectance S of a pixel

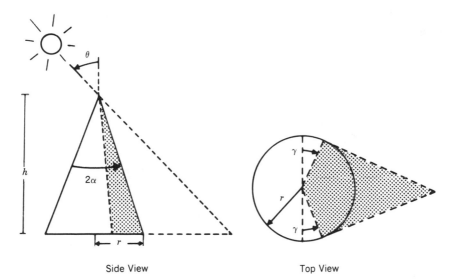

Side View   Top View

Figure 13 Geometry of an illuminated cone and its shadow. [Reprinted from Li and Strahler (1985). © 1985 IEEE.]

is given by

$$S = \frac{GA_g + ZA_z + CA_c + TA_t}{A} \tag{24}$$

All the areas occurring in this equation can be written in terms of two parameters, γ and m, the ratio of the sum of squared cone radii to the area of the pixel. The model uses the following canopy parameters:

$$\alpha, \eta, H, \text{ a spatial distribution of trees, and the four} \tag{25}$$
$$\text{reflectances } G, Z, C, \text{ and } T$$

Li and Strahler (1985) inverted their model to estimate the expected or mean value of the radii of the cones, $E(r)$; the number of cones per pixel, N; the crown height, H; and the apex angle, α, from the 10 m (pixel 10 × 10 m) simulated SPOT imagery and Landsat 80 m imagery, using a field-measured value of the coefficient of variation of radius. Tables 11 and 12 give the results of these inversions for a red fir and a mixed conifer site, together with the field-measured valued of canopy parameters (inversion of the Landsat data required the field estimation of another parameter, the variance-to-mean ratio of the number of trees in a pixel). There is a general agreement between the calculated and measured values, but there are some significant differences: the spacing parameter for the red fir stand is significantly underestimated, for example, reflecting the limitations of the model. For further discussion of the inversion procedure and the discussion of the differences between the measured and calculated value, we refer the reader to the original detailed paper (Li and Strahler, 1985).

TABLE 11   Results of Model Inversion: 10 Meter SPOT Simulation

Parameter	Red Fir Site		Mixed Conifer Site	
	Observed	Calculated	Observed	Calculated
$E(r)$	5.7 ft	4.2 ft	5.2 ft	6.9 ft
N	15.5	14.0	3.9	3.0
H	60.5 ft	37.4 ft	36.3 ft	40.8 ft
α	8.5	6.4	9.3	9.6

Source: Li and Strahler (1985).

TABLE 12   Results of Model Inversion: 80 Meter MSS Simulation

Parameter	Red Fir Site		Mixed Conifer Site	
	Observed	Calculated	Observed	Calculated
$E(r)$	5.7 ft	4.4 ft	5.2 ft	7.0 ft
N	725	493	181	192
H	60.5 ft	44.5 ft	36.3 ft	48.0 ft
α	8.5	5.6	9.3	8.3

Source: Li and Strahler (1985).

IV CONCLUDING REMARKS

In this chapter, we have discussed the potentials and limitations of the techniques of inversion of a CR model for estimating important biophysical parameters of a vegetation canopy from the measured bidirectional CR in the optical region. Most of the studies to date have been carried out with data collected using ground-based sensors. From these studies, it is fair to conclude that the technique has potential, but it has to be developed further before becoming operational for estimating canopy parameters from remotely sensed radiance data. These data differ from those collected on the ground-based sensors in two important ways: (1) the number of view angles for which CR data can be collected decreases significantly, and (2) the sensors measure the radiation reflected from the canopy after it is transmitted through the atmosphere as well as the radiation scattered by the atmospheric particles. Formally, the measured radiance can be written as

$$R' = r'IT_a + R_p \qquad (26)$$

where r' is the reflectance factor for the vegetation surface, I is the incident radiation, T_a is the atmospheric transmittance, and R_p is the path radiance. These two problems can be addressed as follows:

CR Data for a Limited Number of Viewing Directions. Here one can use a so-called reconstructive inversion technique (Goel and Grier, 1987). It uses the CR data for a limited number (3–5) of view directions to first "reconstruct" the complete bidirectional reflectance surface, which is then used to estimate the canopy parameters using the standard inversion technique.

It consists of the following three steps:

1. Fit the measured CR data to a simple three- or four-parameter CR model that represents the view-angle dependence of bidirectional CR. One such model (Walthall et al., 1985; Staenz et al., 1981) for a homogeneous canopy is

$$CR = a\theta_o^2 + b\theta_o \cos \psi + c \qquad (27a)$$

where θ_o is the view zenith angle (in radians), and ψ is the view azimuth angle relative to the sun azimuth. Constants a, b, and c implicitly depend upon the solar and canopy parameters.

The corresponding model for row-planted inhomogeneous canopies is

$$CR = a\theta_o^2 + b\theta_o \cos (\psi_s - \text{ROAZ}) \cos (\psi_o - \text{ROAZ}) +$$
$$\theta_o \sin (\psi_s - \text{ROAZ}) \sin (\psi_o - \text{ROAZ}) + c \qquad (27b)$$

where ψ_s and ψ_o are the solar and view azimuth angles, respectively, and ROAZ is the row azimuth angle. Note that when $b = d$, Eq. 27b reduces to Eq. 27a.

2. With the estimated values of parameters a, b, c, and d, use Eq. 27b (or Eq. 27a for a homogeneous canopy) to reconstruct (completely or partially) the bidirectional reflectance surface for $0° \leq \theta_o \leq 90°$ and $0° \leq \psi_o \leq 360°$. These values

of CRs are referred to as the "reconstructed" CRs, and the CR surface is referred to as a "reconstructed" CR surface.

3. Using these reconstructed CRs, carry out the "standard" inversion using a complex CR model (which explicitly uses canopy parameters as input parameters) to obtain the desired canopy parameters.

The method is schematically shown in Figure 14. Its success crucially depends upon the following:

1. How well does the parametric model defined by Eqs. 27 fit the measured CR data?
2. How sensitive are the estimated values of the parameters a, b, c, and d to errors in the measured values of CRs?
3. How well does the parametric model (Eqs. 27) fit the complex model?

An investigation into the last aspect using CR data for several vegetations shows that Eq. 27 fits two halves of the CR surface, on either side of the row direction (see Figures 6–8) better than the complete viewing hemisphere. For the part containing the sun azi-

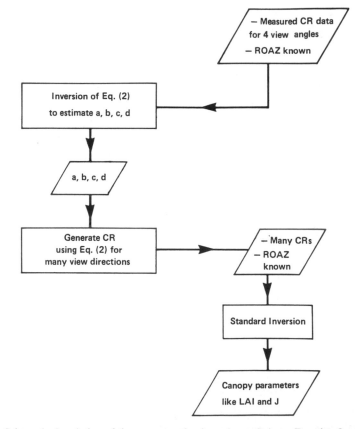

Figure 14   Schematic description of the reconstructive inversion technique. Equation 2 refers to Eq., 27b of the text. [Reprinted by permission of the publisher from Goel and Grier (1987). Copyright 1987 by Elsevier Science Publishing Co., Inc.)

muth (referred to as the primary part), an analysis of the sensitivity of the estimated values of parameters a, b, c, and d to errors in the measured values of CRs shows that the most desirable quadruplet of viewing directions (θ_o, ψ_o) is

$$(0°,0°), (45°,0°), (45°,180°) \text{ and } (45°,90°) \text{ or } (45°,270°)$$

with the last direction chosen to be in the primary part. This analysis also suggests that it is more desirable to collect CR data when the sun zenith is greater than 25° and the sun azimuth is at least 45° away from the row (preferably, perpendicular to it).

This reconstructive inversion technique has been tried on soybean and corn canopies in different stages of growth, using the Goel-Grier row model (see Section III.B.2) to carry out the standard inversion. Table 13 gives the estimated values of LAI and a parameter J, which is a measure of the growth stage of the canopy (see Section III.B.2.). This table also gives the corresponding measured values and PRMSE, the percentage root-mean-square error (see Eq. 19) between the calculated and measured CRs for all viewing directions in the principal part.

From this table, one can see that the technique indeed works quite well: it allows an accurate estimation of the important biophysical parameters, and the calculated CRs agree remarkably well with the measured values.

Effects of Atmospheric Scattering. One of the main reasons for choosing the near-infrared band for estimating canopy parameters through inversion of a CR model is to minimize the effects of atmospheric scattering. If the CR data were used in the visible region, the effects of atmospheric scattering must be addressed.

TABLE 13 Estimated Values of Canopy Parameters LAI and J Using the Reconstructive Inversion Technique with Measured CRs for Four Optimal View Directions[a]

Canopy	LAI		J		PRMSE
	Estimated	Measured	Estimated	Measured	
Soybean					
July 17	3.40	3.0	0.37	0.36	5.5
	± 0.33	± 0.5	± 0.01		± 0.4
July 24	4.10	3.9	0.40	0.42	6.9
	± 0.71	± 0.6	± 0.04		± 1.1
August 28	3.53	2.87	0.55	0.50	4.2
	± 0.58	± 0.4	± 0.08		± 0.9
Corn					
June 24	1.12	1.2	0.26	0.20	5.9
	± 0.12	± 0.3	± 0.01		± 0.0
July 23	4.92	4.8	0.48	0.48	4.6
	± 0.37	± 0.4	± 0.04		± 0.4

[a]Also given are the corresponding measured values and PRMSE, a measure of quality of fit between calculated and measured CRs for all viewing directions. See text for details.

Source: Goel and Grier (1987).

For CR data collected on ground, as discussed in the Section III.B.1, high view zenith and solar zenith angles are more desirable to maximize the accuracy of estimation of biophysical parameters such as LAI, LAD, and percentage of ground cover. For such angles, however, the atmospheric-scattering effects should also be large due to an increased radiation path between the canopy and the sensor. Therefore, before one could determine the optimal strategy for CR measurements, one has to either include atmospheric-scattering effects in the model to be inverted or else "correct for," or "filter out," these effects.

Various approaches (Rollin et al., 1985) to account for the atmospheric scattering can be broadly classified into two categories:

1. *Empirical.* Two examples are: (a) Measurement of radiance at various heights or view angles (so as to achieve various path lengths) and use of a model that gives the dependence of transmittance (T_a, see Eq. 26) on atmospheric path length. These measurements enable the determination of parameters T_a and R_p for successive layers of the atmosphere by linear regression. (b) Measurements of radiance from a reference panel or standard target with known spectral signatures, and their use in filtering out the effects of atmosphere.

2. *Analytical.* These approaches are characterized by the use of models describing atmospheric attenuation and adsorption. The input to these models can either be meteorological data or measurements from a ground-based spectroradiometer.

These approaches have relative advantages and disadvantages. However, we feel that overall the analytical approach will be more advantageous if one could use an atmospheric scattering model that uses a minimum number of atmospheric parameters (this is required from the point of view of inversion). One such model is the combined atmosphere-vegetation canopy reflectance model due to Verhoef (1985). This model has other attractive features: (a) it uses the SAIL model for the vegetation CR, which has been thoroughly investigated and found to be quite useful; (b) the atmospheric model is quite similar to the SAIL model, allowing an easy matching of boundary conditions for the radiant flux; (c) it assumes that the atmosphere and the vegetation are composed of layers, a feature quite useful in analyzing radiance data collected at different heights (on ground, aircraft, space shuttles, and satellites). This combined model has to be investigated along the directions on which the SAIL model has been investigated, using radiance data collected with sensors located on ground, aircraft, space shuttles, and satellites.

There is a fundamental problem with using the inversion technique to estimate important canopy parameters. An accurate estimation requires that one has a model that accurately represents the CR. However, the more realistic the model, the more parameters it is likely to have. This puts substantial requirements on the amount of remotely sensed data and makes the inversion of the model more difficult. Therefore, it seems that one of the fruitful (perhaps even necessary) directions for future research is on the synergistic processing of spectral signatures in different spectral channels (VIS/NIR, thermal infrared, and microwave). Although this area of investigation is in its infancy, it appears to be quite promising (Goel, 1985; Cimino et al., 1987). Elsewhere (Goel, 1985), we have discussed the possibility of developing a single model that is capable of providing the CR in the VIS/NIR region as well as the backscattering coefficient (BSC) in the microwave region. There we have shown that it is possible to simulate the observed dependences of BSC on canopy parameters (such as LAI, LAD, soil-moisture content), on angle of incidence, and on the microwave frequency by using a CR model. Thus, in principle, one could use the CR data to estimate some of the canopy parameters occurring in the BSC model and keep them fixed at these estimated values while carrying out the inversion of the BSC data. This synergistic approach ought to lead to improved accuracy of estimation of canopy parameters than the approaches based on either the BSC or the CR data alone. The synergism could naturally be extended to the thermal infrared data (see Chapter 15) and the temporal data in various spectral bands. The inversion technique will serve as the "synergizer" between various spectral channels.

ACKNOWLEDGMENTS

Over the years, the author had the opportunity to collaborate with many fine colleagues and associates, including Don Deering, Toby Grier, Keith Henderson, John Norman, Dave Pitts, Jon Ransom, Don Strebel, and Richard Thompson. They and other individuals in the remote-sensing sciences community contributed to the author's understanding of the problem addressed in this chapter. The author gratefully acknowledges and thanks NASA for supporting his research over these years.

The author wishes to express his appreciation to Namni Goel for reading and editing the rough draft of this chapter, and for making many useful suggestions from a reader's perspective.

APPENDIX: SENSITIVITY ANALYSIS OF A CANOPY REFLECTANCE MODEL

In this appendix, we will summarize a general sensitivity-analysis technique, due to Goel and Thompson (1984a), for quantifying the changes in the estimated values of canopy parameters as the CRs are changed. This technique can be used to assess the expected levels of accuracy in estimating various canopy parameters from measured CR data.

Let the measured reflectance R_i', for a specific solar/view direction, differ from the calculated R_i (given by the CR model) by a small amount. That is,

$$R_i' = R_i + \Delta R_i \tag{A1}$$

Thus, ΔR_i represents measurement errors or imperfect representation of the CR by the model. It is assumed that these errors are small enough such that the reflectance function can be adequately represented in the vicinity of a correct set x of values for the canopy parameters by the linear approximation

$$R_i(x + \Delta x) = R_i(x) + \sum_k v_{ik} \Delta x_k \tag{A2}$$

where

$$v_{ik}(x) = \frac{\partial R_i(x)}{\partial x_k} \tag{A3}$$

It is further assumed that the ΔR_i are random, independent of one another, and Gaussian distributed with a mean of 0 and a standard deviation of $[E(\Delta R_i^2)]^{1/2} = \epsilon_i R_i > 0$. With these assumptions, the standard deviation σ_k of Δx_k is found to be

$$\sigma_k^2 = \sum_i \epsilon_i R_i^2 W_{ik}^2 \tag{A4}$$

where

$$W_{ik} = \sum_j U_{kj} v_{ij} \tag{A5a}$$

and U is the inverse of the matrix

$$V_{kj} = \sum_i v_{ik} v_{ij} \tag{A5b}$$

If one assumes the same random error of 1% in all CRs (i.e., $\epsilon_i = \epsilon = 0.01$), then one can define an expected error level by the relation:

$$\text{error level} = (\sigma_k / x_k) \times 100 \tag{A6}$$

One should note that the assumption of linearity (Eq. A2) requires that the reflectance errors ΔR_i are very small. If it is not so, the actual errors Δx_k will tend to be much larger than those calculated in accordance with the assumption of linearity. Therefore, the error estimates provided by the previous formalism will tend to be conservative.

LIST OF PRINCIPAL SYMBOLS

STANDARD ALPHABETICAL SYMBOLS

Symbol	Description
A_c	Areas of illuminated tree crown
A_g	Areas of illuminated soil background
A_t	Areas of shadowed tree crown
A_z	Areas of shadowed soil background
a	Single-scattering albedo, or leaf absorptance
ALA	Average leaf-inclination angle
C	Set of canopy parameters or reflectance per unit area of illuminated tree crown
$C(\delta, \eta)$	Function defining subcanopy profile
CR	Bidirectional canopy reflectance factor
D	Fraction of Q covered by half the plant canopy
F	Merit function
$f(\delta)$	Function describing canopy cross section
$f(\theta_L, \psi_L)$	Leaf-angle distribution density function
G	Reflectance per unit area of illuminated soil background
g	Scattering asymmetry factor
GC	Percentage of ground cover

Symbol	Description
H	Height of a tree
I	Incident solar radiation
J	Fraction of P covered by half the plant canopy
$k(\mu_s, \psi_s)$	Extinction coefficient of the incident direct solar flux
LAI	Leaf-area index
LAD	Leaf-angle distribution
m	Ratio of the sum of squared cone radii to the area of pixel
N	Number of cones per pixel
NIR	Near-infrared region
P	Ratio of row period to the canopy's maximum height
$P(\Theta)$	Phase function
PRMSE	Percentage root-mean-square error
Q	Ratio of row period (perpendicular to the direction of P) to the canopy's maximum height
R	Functional dependence of S on various parameters
R'	Measured radiance above the atmosphere or measured CR
R_c	Calculated canopy reflectance
R_o	Observed canopy reflectance
R_p	Atmospheric path radiance
r	Radius of a conical tree
r'	Reflectance factor for the vegetation surface
RMSE	Root-mean-square error
ROAZ	Azimuth angle of the row direction
S	Spectral signature or reflectance of a canopy
SKYL	Fraction of incident diffuse radiation
T	Reflectance per unit area of shadowed tree crown
T_a	Atmospheric transmittance
t	Emergence time of a plant
VIS	Visible region
W	Error function
Z	Reflectance per unit area of shadowed soil background

GREEK SYMBOLS

Symbol	Description
α	Apex angle of a conical tree
α_s	Soil albedo
δ	Across-row strip displacement
η	Label for the coordinate axis at an angle ROAZ with the 0° azimuth or density of conical trees per unit area
Γ	Gamma function
γ	Angle describing the portion of a conical tree
λ	Wavelength of the incident solar radiation
μ	Parameter of a beta distribution
μ_s	$\cos \theta_s$
ν	Parameter of a beta distribution

GREEK SYMBOLS

Symbol	Description
ρ	Leaf hemispherical reflectance
ρ_s	Soil hemispherical reflectance
ψ	Relative azimuth angle between solar and view directions
ψ_L	Leaf azimuth angle
ψ_o	View azimuth angle
ψ_s	Solar azimuth angle
τ	Leaf hemispherical transmittance
θ_L	Leaf inclination angle
$<\theta_L^2>$	Second moment of θ_L
θ_o	View zenith angle
θ_s	Solar zenith angle
Θ	Scattering angle

REFERENCES

Baltes, H. P. (Ed.) (1980). *Inverse Scattering Problems in Optics*. Springer-Verlag, New York.

Breece, H. T., and R. A. Holmes (1971). Bidirectional scattering characteristics of healthy green soybean and corn leaves in vivo. *Appl. Opt.* **10**:119–127.

Bryson, A. E., and Y. Ho (1969). *Applied Optimal Control*. Blaisdell, New York.

Bunnik, N. J. J., W. Verhoef, R. W. deJongh, H. W. J. van Kasteren, R. H. M. E. Geerts, D. Uenk, and T. A. de Boer (1983). Hot spot reflectance measurements applied to green biomass estimation and crop growth monitoring. *Proc. Int. Colloq. Signatures Remotely Sens. Objects*, Bordeaux, France, pp. 111–121.

Camillo, P. (1987). A canopy reflectance model based on an analytical solution to the multiple scattering equation. *Remote Sens. Environ.* **23**:453–477.

Cimino, J. B., C. Bruegge, D. Diner, J. Paris, C. Dobson, D. Gates, F. Ulaby, N. Goel, E. Kasischke, D. Kimes, R. Lang, J. Norman and V. Vanderbilt (1987). Synergism requirements and concepts for SAR and HIRIS on EOS. *Proc. Int. Geosci. Remote Sens. Symp. (IGARSS' 87)*, Ann Arbor, Michigan, pp. 955–966.

Curran, P. J. and H. D. Williamson (1987). GLAI estimation using measurements of red, near infrared, and middle infrared radiance. *Photogramm. Eng. Remote Sens.* **53**:181–186.

Daughtry, C. S. T., and S. E. Hollinger (1984). *Costs of Measuring Leaf Area Index of Corn*. LARS Tech. Rep. 030784. Purdue University, West Lafayette, Indiana.

Gausman, H. W. (1985). *Plant Leaf Optical Parameters in Visible and Near-infrared Light*. Texas Tech. Press, Lubbock.

Gerstl, S. A., and A. Zardecki (1985a). Discrete-ordinates finite-element method for atmospheric radiative transfer and remote sensing. *Appl. Opt.* **24**:81–93.

Gerstl, S. A., and A. Zardecki (1985b). Coupled atmosphere/canopy model for remote sensing of plant reflectance features. *Appl. Opt.* **24**:94–103.

Goel, N. S. (1985). Modeling canopy reflectance and microwave backscattering coefficient. *Remote Sens. Environ.* **16**:235–253.

Goel, N. S. (1987). Models of vegetation canopy reflectance and their use in estimation of biophysical parameters from reflectance data. *Remote Sens. Rev.* **3**:1–212.

Goel, N. S., and D. W. Deering (1985). Evaluation of a canopy reflectance model for LAI estimation through its inversion. *Proc. IEEE Trans. Geosci. Remote Sens.* **GE-23**:674–684.

Goel, N.S., and T. Grier (1986a). Estimation of canopy parameters for inhomogeneous vegetation canopies from reflectance data. I. Two-dimensional row canopy. *Int. J. Remote Sens.* **7**:665–681.

Goel, N. S., and T. Grier (1986b). Estimation of canopy parameters for inhomogeneous vegetation canopies from reflectance data. II. Estimation of leaf area index and percentage of ground cover for row canopies. *Int. J. Remote Sens.* **7**:1263–1286.

Goel, N. S., and T. Grier (1987a). Estimation of canopy parameters of row planted vegetation canopies using reflectance data for only four view directions. *Remote Sens. Environ.* **21**:37–51.

Goel, N. S., and T. Grier (1988). Estimation of canopy parameters for inhomogeneous vegetation canopies from reflectance data. III. TRIM: A model for radiative transfer in heterogeneous three-dimensional canopies. *Remote Sens. Environ.* **25**:255–293.

Goel, N. S., and D. E. Strebel (1983). Inversion of vegetation canopy reflectance models for estimating agronomic variables. I. Problem definition and initial results using Suits' model. *Remote Sens. Environ.* **13**:487–507.

Goel, N. S., and D. E. Strebel (1984). Simple beta distribution representation of leaf orientation in vegetation canopies. *Agron. J.* **76**:800–803.

Goel, N. S., and R. L. Thompson (1983a). Estimation of agronomic variables using spectral signatures. *Proc. Int. Colloq. Signatures Remotely Sens. Objects*, Bordeaux, France, pp. 45–53.

Goel, N. S., and R. L. Thompson (1983b). *Cupid Model and LARS Soybean Data. Quarterly Technical Interchange on Assessing Key Vegetation Characteristics from Remote Sensing.* Rep. JSC-18894, Vol. 2, NASA/Johnson Space Center, Houston, Texas, pp. 389–404.

Goel, N. S., and R. L. Thompson (1984a). Inversion of vegetation canopy reflectance models for estimating agronomic variables. III. Estimation using only canopy reflectance data as illustrated by the Suits model. *Remote Sens. Environ.* **15**:223–236.

Goel, N. S., and R. L. Thompson (1984b). Inversion of vegetation canopy reflectance models for estimating agronomic variables. IV. Total inversion of the SAIL model. *Remote Sens. Environ.* **15**:237–253.

Goel, N. S., and R. L. Thompson (1984c). Inversion of vegetation canopy reflectance models for estimating agronomic variables. V. Estimation of LAI and average leaf angle using measured canopy reflectances. *Remote Sens. Environ.* **16**:69–85.

Goel, N. S., and R. L. Thompson (1985). Optimal solar/viewing geometry for an accurate estimation of leaf area index and leaf angle distribution from bidirectional canopy reflectance data. *Int. J. Remote Sens.* **6**:1493–1520.

Goel, N. S., K. E. Henderson, and D. E. Pitts (1984). Estimation of leaf area index from bidirectional spectral reflectance data by inverting a canopy reflectance model. *Machine Processing of Remotely Sensed Data.* Purdue University, West Lafayette, Indiana, pp. 339–347.

Grant, L. (1987). Diffuse and specular characteristics of leaf reflectance. *Remote Sens. Environ.* **22**:309–322.

Grant, L., C. S. T. Daughtry, and V. C. Vanderbilt (1983). *Measurements of Specularly Reflected Radiation from Leaves.* LARS Tech. Rep. 081583. Purdue University, West Lafayette, Indiana.

Jackson, R. D. (1984). Remote sensing of vegetation characteristics for farm management. *Proc. SPIE—Int. Soc. Opt. Eng.* **475**:81–96.

Li, X. (1981). An invertible coniferous canopy reflectance model. M.A. Thesis, University of California, Santa Barbara.

Li, X. (1985). Geometric-optical modeling of a coniferous forest canopy. Ph.D. Thesis, University of California, Santa Barbara.

Li, X., and A. H. Strahler (1985). Geometrical-optical modeling of a conifer forest canopy. *IEEE Trans. Geosci. Remote Sens.* **GE-23**:705–721.

Li, X., and A. H. Strahler (1986). Geometrical-optical bidirectional reflectance modeling of a conifer forest canopy. *IEEE Trans. Geosci. Remote Sens.* **GE-24**:906–919.

Nicodemus, F. E., J. C. Richmond, J. J. Hsia, I. W. Ginsburg, and T. Limperis (1977). *Geometrical Consideration and Nomenclature for Reflectance*. National Bureau of Standards, Washington, D.C.

Norman, J. M. (1975). Radiative transfer in vegetation. In *Heat and Mass Transfer in the Biosphere* (D. A. deVries and N. H. Afgan, Eds.), Scripta, Washington, D.C., pp. 187–206.

Norman, J. M. (1979). Modeling of complete crop canopy. In *Modification of the Aerial Environment of Plants* (B. G. Barfield and J. F. Gerber, Eds.), ASAE Monogr. No. 2. American Society of Agricultural Engineers, St. Joseph, Michigan, pp. 249–277.

Norman, J. M., and J. M. Welles (1983). Radiative transfer in an array of canopies. *Agron. J.* **75**:481–488.

Norman, J. M., J. M. Welles, and E. A. Walter (1985). Contrasts among bidirectional reflectances of leaves, canopies, and soils. *IEEE Trans. Geosci. Remote Sens.* **GE-23**:659–668.

Otterman, J. (1981). Reflection from soil with sparse vegetation. *Adv. Space Res.* **1**:115–119.

Otterman, J. (1984). Albedo of a forest modeled as a plane dense protrusions. *J. Clim. Appl. Meteorol.* **23**:297–307.

Otterman, J. (1985). Bidirectional and hemispheric reflectivities of a bright soil plane and a sparse dark canopy. *Int. J. Remote Sens.* **6**:897–902.

Otterman, J., and K. Staenz (1985). Reflectivity contrast of forested slopes with snow-covered ground. *Proc. 3rd Int. Colloq. Spectral Signatures Objects Remote Sens.* Les Arcs, France, pp. 199–203.

Otterman, J., and G. H. Weiss (1984). Reflection from a field of randomly located vertical protrusions. *Appl. Opt.* **23**:1931–1936.

Otterman, J., D. E. Strebel, and K. J. Ranson (1987). Inferring spectral reflectances of plant elements by simple inversion of bidirectional reflectance measurements. *Remote Sens. Environ.* **21**:215–228.

Perry, C. R., Jr., and L. F. Lautenschlager (1984). Functional equivalence of spectral vegetation indices. *Remote Sens. Environ.* **14**:169–182.

Ranson, K. J., V. C. Vanderbilt, L. L. Biehl, B. F. Robinson and M. E. Bauer (1981). Soybean canopy reflectance as a function of view and illumination geometry. *Proc. 15th Int. Symp. Remote Sens. Environ.*, University of Michigan, Ann Arbor, pp. 853–865.

Ranson, K. J., L. L. Biehl, and M. E. Bauer (1985a). Variation in spectral response of soybeans with respect to illumination, view and canopy geometry. *Int. J. Remote Sens.* **6**:1827–1842.

Ranson, K. J., C. S. T. Daughtry, L. L. Biehl, and M. E. Bauer (1985b). Sun-view angle effects on reflectance factors of corn canopies. *Remote Sens. Environ.* **18**:147–161.

Rollin, E. M., M. D. Steven, and P. M. Mather. (Eds.) (1985). *Atmospheric Corrections for Remote Sensing*. Department of Geography, University of Nottingham.

Sorenson, H. W. (1970). Least square estimation from Gauss to Kalman. *IEEE Spectrum*, July, 63–68.

Staenz, K., F. J. Ahern, and R. J. Brown (1981). The influence of illumination and viewing geometry on the reflectance factor of agricultural targets. *Proc. 15th Int. Symp. Remote Sens. Environ.*, University of Michigan, Ann Arbor, pp. 867–881.

Strahler, A. H., and X. Li (1981). An invertible coniferous forest canopy reflectance model. *Proc. 15th Int. Symp. Remote Sens. Environ.*, University of Michigan, Ann Arbor, pp. 1237–1244.

Strebel, D. E., N. S. Goel, and K. J. Ranson (1985). Two-dimensional leaf orientation distributions. *IEEE Trans. Geosci. Remote Sens.* **GE-23:**640–647.

Suits, G. H. (1972). The calculation of the directional reflectance of vegetative canopy. *Remote Sens. Environ.* **2:**117–125.

Verhoef, W. (1984). Light scattering by leaf layers with application to canopy reflectance modeling: The SAIL model. *Remote Sens. Environ.* **16:**125–141.

Verhoef, W. (1985). A scene radiation model based on four stream radiative transfer theory. *Proc. 3rd Int. Colloq. Spectral Signatures Objects Remote Sens.*, Les Arcs, France, pp. 143–150.

Verhoef, W., and N. J. J. Bunnik (1981). Influence of crop geometry on multispectral reflectance determined by the use of canopy reflectance models. *Proc. Int. Colloq. Signatures Remotely Sens. Objects*, Avignon, France, pp. 273–290.

Walter-Shea, E. A. (1987). Laboratory and field measurements of leaf spectral properties and canopy architecture and their effects on canopy reflectance. Ph.D. Thesis, University of Nebraska, Lincoln.

Walthall, C. L., J. M. Norman, J. M. Welles, G. Campbell, and B. L. Blad (1985). Simple equation to approximate the bidirectional reflectance from vegetative canopies and bare soil surfaces. *Appl. Opt.* **24:**383–387.

7

ESTIMATION OF PLANT-CANOPY ATTRIBUTES FROM SPECTRAL REFLECTANCE MEASUREMENTS

GHASSEM ASRAR

Headquarters
National Aeronautics and Space Administration
Washington, D.C.

RANGA B. MYNENI

Institüt für Bioklimatologie
Georg-August-Universität Göttingen
Göttingen, Federal Republic of Germany

and

EDWARD T. KANEMASU

Evapotranspiration Laboratory
Kansas State University
Manhattan, Kansas

I INTRODUCTION

Plant-canopy attributes such as green-leaf area, leaf normal distribution, green phytomass, chlorophyll density, and plant water status are usually employed to assess canopy conditions and development. These parameters are either used separately or in combination in process-based models to simulate the plant-canopy response to its surrounding environment. The techniques used commonly in measuring these parameters are laborious. The spatial and temporal trends of these parameters further complicate the measurement procedures, especially under field conditions.

Remotely sensed radiometric measurements of visible and infrared solar energy, which is reflected or emitted from plant canopies, can be used to obtain rapid and nondestructive estimates of some of these canopy attributes. Most published studies have used an empirical approach in which plant-canopy attributes were correlated directly with measured spectral reflectance. This approach has its merits because it is simple and it does not require detailed knowledge of the underlying processes. It is assumed that the remotely sensed signal represents only the parameter under study, independent of other parameters and processes. In most cases, this assumption is generally not valid, due to the complex nature of the interaction of solar energy with plant canopies. For example, water stress and nutrient deficiency both may impair plant biological activities and usually result in reduced green-phytomass production. This, in turn, affects plant-canopy architecture and its radiation regime, which may result in a reduction in absorption of photosynthetically active radiation (PAR) and, hence, an increase in scattered energy in this region of the electromagnetic spectrum. This may also result in a decrease in near-infrared reflectance from the plant canopy due to its reduced green phytomass. In this example, a change in biological activities of the plants results in a change in the remotely sensed signal; however, since there is more than one factor that can result in the same biological deficiency, the remotely sensed signal may or may not uniquely represent the condition(s) being observed. Therefore, it is imperative to understand the underlying processes that contribute to spectral characteristics of plant canopies in different regions of the electromagnetic spectrum.

The objective of this chapter is to present relatively simple methods, based on physical and/or physiological principles, for applications of remotely sensed visible and infrared reflectance in estimating plant-canopy attributes such as absorbed photosynthetically active radiation, green leaf-area index, above-ground phytomass, and leaf normal distribution. The approaches we describe in the following sections should help overcome some of the limitations of empirical methods.

II ESTIMATION OF ABSORBED PAR

A detailed discussion on the fate of solar energy as it traverses through the atmosphere is given in Chapter 1, and the theory of photon transport in leaf canopies is discussed in Chapter 5. In this section, we will discuss the relationship between the fraction of radiation that is absorbed by plant canopies and used in biological processes such as photosynthesis, photorespiration, and transpiration, and the fraction that is reflected from the canopy and recorded by a remote-sensing system. Therefore, the objective here is to investigate the relationship between canopy reflectance and absorbed photosynthetically active radiation (APAR).

The interaction between incoming solar radiation and a plant canopy determines the quantity of radiation that is absorbed, reflected, and transmitted by the plant elements such as leaves, stems, and branches. Figure 1 illustrates such interactions for the PAR region (0.40–0.70 μm) of the spectrum. From measurements of the components described in this figure, one can construct a radiation balance for the canopy to compute the absorbed PAR energy by the plant canopy:

$$APAR = (PAR_o + PAR_s) - (PAR_c + PAR_t) \qquad (1)$$

where PAR_o is the incident flux; PAR_s and PAR_c are the reflected energy from the soil surface and plant canopy, respectively; and PAR_t is the flux that is transmitted through the canopy.

The term intercepted PAR (IPAR) is commonly used interchangeably with APAR in the literature. A distinction should be made between these two terms. IPAR is purely a statistical quantity and represents only the probability of photons that are intercepted by plant elements as compared with those passing through the gaps inside the canopy with-

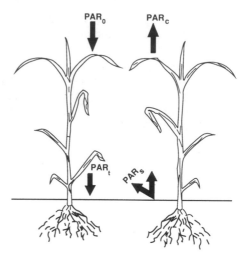

Figure 1 Components of photosynthetically active radiation (PAR) resulting from interaction of radiation with plant canopies.

out interaction with the plant elements. If all photons are intercepted by the elements, as in the case of a complete canopy cover, then IPAR = 1. If all of the photons pass through canopy gaps without being intercepted, then IPAR = 0. This definition does not include the process of energy absorption, whereas it is implicitly considered in the definition of APAR. A photon has to be intercepted by a plant element before it can be absorbed; therefore, absorption includes interception. This distinction is very important, especially in the near- and shortwave infrared regions of the spectrum, but in the PAR region, APAR and IPAR can be used interchangeably due to negligible scattering and strong absorption of photons by green-plant elements. The quantity of APAR or IPAR can be also expressed as a fraction of incident PAR flux p:

$$p = \text{APAR}/\text{PAR}_o = \text{IPAR}/\text{PAR}_o \tag{2}$$

Figure 2 illustrates the change in p as a function of solar angle during an early stage of a winter-wheat (*Triticum aestirum L.*) canopy. During the early morning and late afternoon hours, p values are maximum, since the incident PAR reaches the canopy at a large zenith angle and encounters a larger number of plant elements before reaching the soil surface. The incident flux in this case also travels through a thick atmosphere, which results in an increase in the diffuse portion of the total incident flux. During midday hours (around solar noon), the path length for incident flux through both the atmosphere and plant canopy is shorter; therefore, the p values are smaller. This diurnal trend disappears under overcast sky conditions (Figure 3), since the incident flux is primarily in diffuse form, and also with the development of plant cover through the growing season (Figure 4).

The general practice is to compute a mean daily value of p at different stages of plant development and express this as a function of green leaf-area index (L), which is the one-side green-leaf area per unit ground area. Such measurements have been conducted for a large number of species such as winter wheat (Hipps et al., 1983), corn (Gallo et al., 1985), sorghum and millet (Lapitan et al., 1988a,b), and grassland (Weiser et al., 1986). The results from these studies suggest that the seasonal behavior of p as a function of L for homogeneous canopies without a distinct row structure can be expressed by an

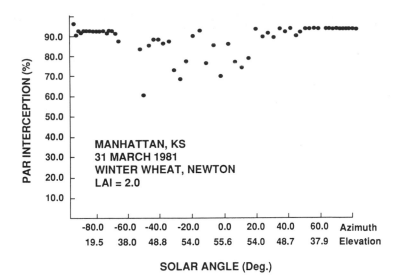

Figure 2 Interception of PAR as a function of sun angle by a winter-wheat canopy for a clear day during early stages of plant development. [Reprinted by permission of the publisher from Hipps et al. (1983). Copyright 1983 by Elsevier Scientific Publishing Co., Inc.]

exponential function based on Beer's law,

$$p = B_0 - B_1 \exp(-gL) \tag{3}$$

where B_0 and B_1 are constants, and g is a canopy geometry factor. Figure 5 shows the relationship between p and L for winter wheat (Asrar et al., 1984b), during active growth and senescent phases. The curves presented in the figure are the best exponential functions fitted to the data. It is notable that p values plateau for $L > 3$, indicating that more

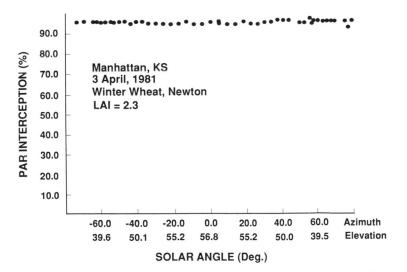

Figure 3 Same as Figure 2, except for a cloudy day. [Reprinted by permission of the publisher from Hipps et al. (1983). Copyright 1983 by Elsevier Scientific Publishing Co., Inc.]

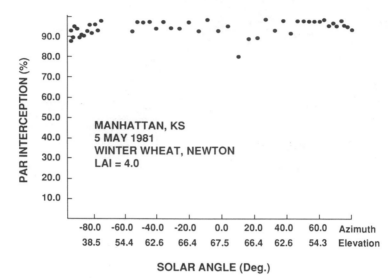

Figure 4   Interception of PAR as a function of sun angle by a winter-wheat canopy for a clear day during late stages of plant development. [Reprinted by permission of the publisher from Hipps et al. (1983). Copyright 1983 by Elsevier Scientific Publishing Co., Inc.]

than 90% of incident flux is intercepted by the canopy and does not return to the lower values observed early during the growing season. The high p values observed late in the season during the senescent phase are caused by the standing senescent vegetation that intercepts the incident flux, but it cannot absorb the energy due to loss of photosynthetically active pigments such as carotenoids and chlorophyll. This clearly illustrates the difference between interception and absorption of PAR.

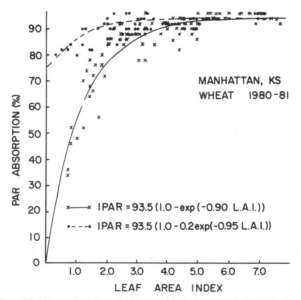

Figure 5   Interception of PAR as a function of leaf-area index for growth (solid line) and senescent (dashed line) stages of winter-wheat development. [Asrar et al. (1984). Reproduced from *Agronomy Journal*, Volume 76, 1984, pages 300–306 by permission of the American Society of Agronomy, Inc.]

II.A   Direct vs. Diffuse PAR

The p values presented in the previous figures were based on total PAR flux, which includes both direct and diffuse components. To partition these components, one has to measure the diffuse incident PAR flux using an occulting ring that blocks the direct radiation from reaching the sensor. The difference between the total and diffuse fluxes is the direct beam flux, p_0, and its interception by green plants can be expressed by (Monsi and Saeki, 1952)

$$p_0 = 1.0 - \exp\left(-gL\right) \tag{4}$$

The ratio of direct to total PAR, q, can be estimated from the equation of atmospheric transmissivity (List, 1971):

$$q = \frac{2a^m}{1 + a^m} \tag{5}$$

where m is the optical air-mass number, $m = P/P_o$; and a is the atmospheric transmission coefficient, $a = \exp\left(-\tau\right)$. P and P_0 are the atmospheric pressures at a given location and at sea level, respectively, and the ratio P/P_o allows correcting for the effects of altitude and atmospheric transmission at a given location. The scattering coefficient τ can be derived from

$$\tau = \ln\left(\frac{2\text{PAR}_o}{\text{PAR*}\cos\eta} - 1.0\right)\left(\frac{P}{P_o}\right)\cos\eta \tag{6}$$

where PAR* is the extraterrestrial flux density at normal incidence (Fuchs et al., 1984). Diffuse PAR penetration through the plant canopy, d, can be described by

$$d = \frac{1}{\pi} \int_0^{2\pi} \int_0^{\pi/2} f\text{PAR}_t \sin\eta \cos\eta \, d\eta \, d\phi \tag{7}$$

where η is the zenith angle, ϕ is the azimuth angle, and f is the relative hemispherical radiance distribution of the clear-sky condition.

Figure 6 illustrates the direct PAR fraction as a function of solar zenith angle based on the measured total incident flux and the computed diffuse component from Eq. 7. Direct PAR flux decreases gradually with an increase in solar zenith angle because of an increase in diffuse component. This is due to increased path length for the incident flux at higher solar zenith angles. Penetration of total and direct (beam) PAR flux through a wheat canopy at $L = 1.4$ is shown in Figure 7. Penetration of both components is the same when $q > 0.85$, which for this example corresponds to $\eta \leq 60°$. Beyond this limit, the penetration of the two components is distinctly different. This suggests that the contribution of the diffuse component for solar zenith angles greater than 60° should be considered in any model of biophysical processes. This is especially important in estimation of APAR or IPAR from remotely sensed spectral measurements, which will be discussed in the following section.

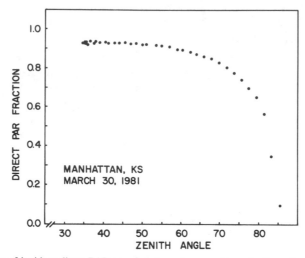

Figure 6   Fraction of incident direct PAR as a function of sun zenith angle for a clear day. [Reprinted by permission of the publisher from Fuchs et al. (1984). Copyright 1984 by Elsevier Science Publishing Co., Inc.]

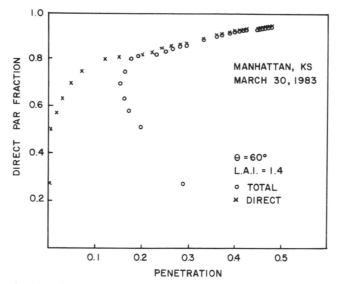

Figure 7   Penetration of total and direct PAR into a winter-wheat canopy. [Reprinted by permission of the publisher from Fuchs et al. (1984). Copyright 1984 by Elsevier Scientific Publishing Co., Inc.]

II.B   Spectral Estimates of APAR

The ability to estimate a fraction of absorbed PAR from remotely sensed spectral measurements depends on the contrast between reflectances from the soil background and green plant. In addition, remote sensors measure only one of the four components of solar energy (i.e., the reflected) that are needed for computing APAR from Eq. 1. To overcome these limitations, one can use the complementary relationships at different regions of the electromagnetic spectrum. For example, green foliage absorbs solar energy in the visible (0.40–0.70 μm) and reflects strongly in the near-infrared [0.70–1.1

μm] regions of the spectrum. Therefore, as a plant canopy develops through the growing season, the reflected solar energy diminishes in the visible region and increases in the near-infrared. This complementary relationship has been found to contain some useful information about the plant canopy and its physiological processes. Different combinations of visible and infrared reflectance measurements are used as vegetation indices (Chapter 1) to present canopy attributes (Hatfield et al., 1984) and processes (Sellers, 1986).

To illustrate some features of these indices in the estimation of APAR, we shall use a simple one-dimensional radiative transfer model (Goudrian, 1977) to simulate monochromatic reflectance from plant canopies with horizontal leaves:

$$\rho = \frac{(\rho_s \rho_h - 1) + (1 - \rho_s/\rho_h) \exp (-2g_h L)}{(\rho_s - 1/\rho_h) + (\rho_h - \rho_s) \exp (-2g_h L)} \tag{8}$$

where ρ_s is soil hemispherical reflectance; ρ_h is the reflectance from horizontal leaves; and g_h is the projected area of horizontal leaves, i.e., the extinction coefficient. ρ_h and g_h are defined by

$$\rho_h = \frac{1 - g_h}{1 + g_h} \tag{9}$$

and

$$g_h = (1 - \sigma)^{0.5} \tag{10}$$

where σ is the single-scatter albedo, which is the sum of leaf reflectance and transmittance.

In most plant canopies, leaves are inclined at an angle from a horizontal plane, and the assumption of horizontal leaves stated earlier may not hold. In this case, the radiative transfer equations must be solved numerically (Chapter 5). However, Goudrian (1977) found that Eq. 8 can be an acceptable approximation of the numerical solutions to more rigorous models if ρ_h and g_h are replaced with ρ_m and $\hat{g}$, which account for leaf normal distributions other than horizontal:

$$\rho_m = 0.206 + 1.117(\rho_f - 0.199) \tag{11}$$

where

$$\rho_f = 1 - \exp \left(\frac{-2\rho_h g'}{1 + g'}\right) \tag{12}$$

and

$$\hat{g} = 0.754 + 0.996(g_f - 0.76) \tag{13}$$

where

$$g_f = g' g_h \tag{14}$$

The parameter g' for uniform canopies with spherical leaf-angle distribution is

$$g' = \frac{0.5}{\cos \eta} \tag{15}$$

If leaves are inclined at a fixed angle from a horizontal plane, (Warren Wilson, 1967),

$$g' = \cos \theta \qquad \pi/2 - \theta \geq \eta \tag{16}$$

$$g' = \cos \theta \left[1 + \frac{2}{\pi} (\tan \omega - \omega) \right] \qquad \left(\frac{\pi}{2 - \theta} \right) < \eta < \frac{\pi}{2} \tag{17}$$

$$g' = \frac{2}{\pi} \tan \eta \qquad \eta = \frac{\pi}{2} \tag{18}$$

In the previous equations, $\omega' = \cos^{-1}(\cot \theta \cot \eta)$, and θ is the zenith angle of the leaf normal (the leaf normals are assumed to be azimuthally random).

This simple model can be used to evaluate the relationship between different vegetation indices and a fraction of absorbed PAR, p_0. If the common L term in Eqs. 4 and 8 is eliminated, we can write

$$\rho = \frac{(\rho_s \rho_h - 1) + (1 - \rho_s/\rho_h)(1 - p_0)^{-2gh}}{(\rho_s - 1/\rho_h) + (\rho_h - \rho_s)(1 - p_0)^{-2gh}} \tag{19}$$

where for $p_0 = 0$ (i.e., no vegetation), Eq. 19 yields $\rho = \rho_s$, and for $p_0 = 1$ (i.e., complete canopy cover), $\rho = \rho_h$. Equation 19 can be solved for the visible and near-infrared regions of the spectrum for different p_0 values that represent different stages of plant-canopy development.

Figure 8 illustrates the relationships between the near-infrared-to-red reflectance ratio RI and p_0 for a winter-wheat plant canopy with different leaf-angle distributions. It is clear that RI is related nonlinearly to p_0 and is very sensitive to canopy geometry and solar angle. This relationship is, however, sensitive over a wide range of p_0 values. Figure 9 illustrates the relationships between greenness index G_n and p_0 for the same conditions described in Figure 8. G_n appears to be less sensitive to canopy geometry and solar angle as compared with RI; however, the change in G_n with p_0 is not as sensitive as with RI. The relationships between NDVI and p_0 for the same conditions are given in Figure 10. The change in NDVI is linear over a wide range of p_0 values, and this relationship appears to be least sensitive to leaf-angle distribution and change in solar zenith angle; however, it is asymptotic beyond a p_0 value of 0.7. In addition, both NDVI and RI are sensitive to soil background under partial plant-canopy cover, especially during the early part of the growing season. G_n is least sensitive to a given soil background. A detailed discussion of the effects of soil background on plant-canopy reflectance and vegetation indices are discussed in Chapter 4. The effects of atmospheric conditions on vegetation indices and hence, on the relationships between p_0 and indices are discussed in Chapters 4 and 9, respectively.

Based on the previous examples given, we chose NDVI to evaluate the feasibility of estimating p_0 from measurements of spectral reflectance of some plant canopies. Figure

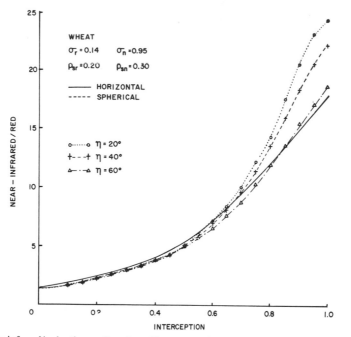

Figure 8   Near-infrared/red ratio as a function of intercepted PAR for a winter-wheat canopy with horizontal and spherical leaf-angle distribution. [Asrar et al. (1984b). Reproduced from *Agronomy Journal*, Volume 76, 1984, pages 300–306 by permission of the American Society of Agronomy, Inc.]

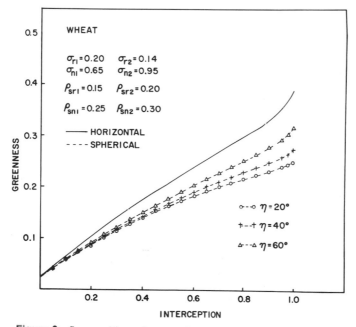

Figure 9   Same as Figure 8, except for greenness (GN) spectral index.

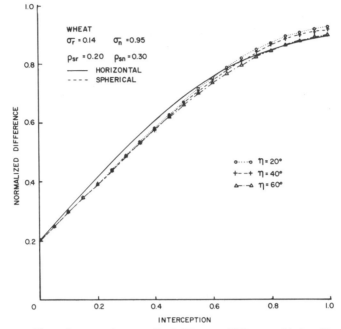

Figure 10 Same as Figure 8, except for normalized difference (ND) spectral index. [Asrar et al. (1984b). Reproduced from *Agronomy Journal*, Volume 76, 1984, pages 300–306 by permission of the American Society of Agronomy, Inc.]

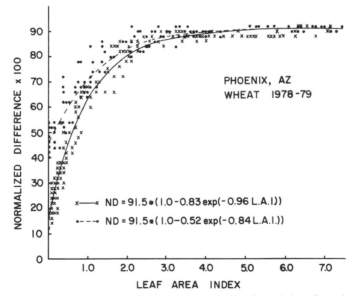

Figure 11 Normalized difference spectral index as a function of leaf-area index of a spring-wheat canopy for growth (solid line) and senescent (dashed line) periods of the growing season. [Asrar et al. (1984b). Reproduced from *Agronomy Journal*, Volume 76, 1984, pages 300–306 by permission of the American Society of Agronomy, Inc.]

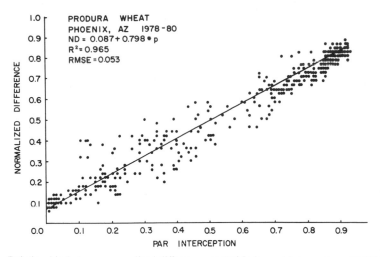

Figure 12 Relationship between normalized difference spectral index and interception of PAR for several spring-wheat canopies.

11 illustrates the change in NDVI as a function of green leaf-area index for a spring-wheat canopy. The trend is similar to that of p_0 as a function of green leaf-area index for winter wheat (Figure 5). Figure 12 illustrates the relationship between p_0 and NDVI for winter wheat (Asrar et al., 1984b). The scatter in the data during the early part of the growing season is due to the sensitivity of NDVI to soil background, which is subject to dry and wet cycles as a result of irrigation/rainfall. Also, some of this scatter could be the result of changing solar angle and atmospheric conditions. In spite of the scatter, there is a linear relationship between NDVI and p_0, which suggests NDVI can be used in estimating the fraction of absorbed PAR by uniform plant canopies. Another attractive feature of NDVI is its dynamic range, which varies between 0 and 1. This is comparable to the range of p_0 values observed under field conditions. NDVI values are typically greater than zero, since most soils (even black wet soils) reflect some of the incident solar flux.

Since p_0 can be estimated from spectral reflectance measurements, we should be able to obtain a quantitative estimate for other physical and physiological canopy attributes that are associated with p_0, from remotely sensed spectral reflectance data. In the next section, we shall discuss some simple approaches for estimating plant-stand structural variables.

III ESTIMATION OF PLANT-CANOPY STRUCTURAL VARIABLES

Spectral measurements are commonly used to estimate the leaf-area index of plant canopies with which photosynthetic capacity and transpiration rates can be evaluated. Several approaches are published in the literature for estimating L from spectral measurements. In one approach, the red- and near-infrared-canopy reflectance are used to compute different vegetation indices, which are directly correlated with measured L values. The regression equations derived from this approach in one year are used to compute L values from spectral reflectance measurements in the following years (Daughtry et al., 1983;

Hatfield et al., 1984; Gallo et al., 1985). Application of equations derived based on this approach are limited to the conditions and species for which they have been derived.

The second approach in estimating L indirectly from spectral reflectance measurements is based on models of radiation transfer through plant canopies. Several methods in this category are reported in the literature. In one method, measured canopy reflectance values are compared with simulated reflectance values from one- or two-dimensional radiative transfer models until the residual error between the two sets of reflectance values is minimum. The canopy parameters (i.e., L, mean leaf angle, leaf spectral properties, etc.) used in the simulation of canopy reflectance are then selected as the attributes for the canopy at the time of reflectance measurements. This procedure is described in detail in Chapter 6.

The other methods are based on the inversion of functions that describe the frequency of gaps inside the plant canopy. In these methods, the probability of contact between incident photons and foliage elements is represented by exact analytical functions whose inverse solutions may yield canopy attributes. The point quadrat analysis is one example in which the contact frequency function is used to obtain estimates of the leaf-area index and the leaf normal distribution function. The definitions and the theory of point quadrat analysis are discussed in what follows, and some numerical examples are presented.

III.A   Vertical Leaf-Area Density Function

Consider a volume $V(q)$ around the point $q(z; x, y)$ in the plant stand. Let s_L denote the total one-sided leaf area in this volume. Then the vertical leaf-area density function $u_L(z)$ on the assumption of horizontal homogeneity is

$$u_L(z) \equiv u_L(z; x, y) = \frac{s_L}{V(q)} \tag{20}$$

The leaf-area density function characterizes the vertical profile of leaf area inside the vegetation canopy. The downward cumulative leaf-area index $L^*(z)$ and the canopy leaf-area index L are related to $u_L(z)$ by the relations

$$L^*(z) = \int_z^{z^*} dz' \, u_L(z') \tag{21}$$

$$L = \int_0^{z^*} dz' \, u_L(z') \tag{22}$$

where z^* is the height of the canopy. The leaf-area index L of the canopy denotes the one-sided leaf area per unit ground area. Accordingly, all canopy phytometric and radiative field characteristics are given with respect to *depth L*.

III.B   Distribution Function of Leaf Normal Orientation

The spatial orientation of a leaf or a part of a leaf is described by the direction of its normal $\mathbf{\Omega}_L$ to the upper face. The unit vector $\mathbf{\Omega}_L(\mu_L, \phi_L)$, where $\mu_L = \cos \theta_L$ for $\mu_L \, \varepsilon$ 0, 1 is described by the two coordinates: θ_L, the inclination angle of the leaf normal to the vertical (also called the zenith angle), and ϕ_L, the azimuth of the normal counted off in a consistent manner from a fixed direction on the compass, usually North. In most leaf canopies, the upper surfaces of the leaves face the upper hemisphere, and, hence, the space of the leaf normals is said to be 2π. On the other hand, if leaf lamellae are twisted such that their upper surfaces face downward, the leaf normal is measured at the opposite azimuth.

Consider a horizontally homogeneous layer at height z of unit thickness in the elementary volume. Let the sum of areas of all the leaves (or parts thereof) whose normals fall within an incremental solid angle around the direction Ω_{Li} be $\hat{g}_L(z, \Omega_{Li})/\Delta\Omega_{Li}$. Upon integration over the space of leaf normal orientation, we obtain

$$\frac{1}{2\pi} \int_{2\pi} d\Omega_L \, \hat{g}_L(z, \Omega_L) = \frac{1}{2\pi} \int_0^{2\pi} d\phi_L \int_0^1 d\mu_L \, \hat{g}_L(z, \Omega_L) \equiv s_L^*(z) \qquad (23)$$

and, hence,

$$\frac{1}{2\pi} \int_{2\pi} d\Omega_L \, g_L(z, \Omega_L) = \frac{1}{2\pi} \int_0^{2\pi} d\phi_L \int_0^1 d\mu_L \, g_L(z, \Omega_L) \equiv 1 \qquad (24)$$

where $s_L^*(z)$ is the total one-sided area of all the leaves in this horizontal layer, and $g_L(z, \Omega_L) = \hat{g}_L(z, \Omega_L)/s_L^*$ is the distribution function of leaf normal orientation and is a function of height in the canopy. Numerically, $g_L(z, \Omega_L)$ denotes the fraction of total leaf area in the horizontal layer of unit thickness at height z, whose normals fall within a unit solid angle around the direction Ω_L.

Now, if in the two-dimensional distribution of leaf normal orientation, the random values of μ_L and ϕ_L are independent, then we may write

$$\frac{1}{2\pi} g_L(z, \Omega_L) = g_L(z, \mu_L) \times \frac{1}{2\pi} h_L(z, \phi_L) \qquad (25)$$

where $g_L(z, \mu_L)$ and $h_L(z, \phi_L)$ are the distribution functions of leaf normal inclination and azimuth, respectively, and

$$\int_0^1 d\mu_L \, g_L(z, \mu_L) \equiv 1 \qquad (26)$$

$$\frac{1}{2\pi} \int_0^{2\pi} d\phi_L \, h_L(z, \phi_L) \equiv 1 \qquad (27)$$

For further theoretical development, the reader is referred to the works of Ross and Nilson (1967), Nilson (1971), and Ross (1981, p. 19).

III.C Point Quadrat Analysis

Mechanical point quadrats were first developed by Levy and Madden (1933) and by Tinney et al. (1937) to study ground cover in grass canopies. Winkworth (1955), Goodall (1952), and Warren Wilson (1959a,b, 1960, 1963a,b, 1967) revised the point quadrat method for convenient investigation of the distribution of leaf-area density and the leaf normal orientation. The method involves piercing the vegetation canopy with long infinitesimally thin needles and registering the number of contacts the needles make with the canopy. From such information, it is possible to compute the vertical profile of area density and the leaf normal distribution functions described before. This method is non-destructive, but it is tedious and labor-intensive to obtain reliable data. This method has been used by researchers to assess the structure of vegetation canopies (Miller, 1967, 1969; Ford and Newbould, 1971; Knight, 1963; Turitzin, 1978; Hatley and MacMahon, 1980; amongst many others). Fiber-optic point quadrats were recently developed by Caldwell and coworkers (1983a,b) for assessing the structure of grass canopies. Because of the effective point diameter (≤ 25 mm), automatic contact detection, and enhanced sampling speed, they are much more versatile than mechanical point quadrats, and perhaps constitute the best system to date for studying the canopy architecture. More recently, Gutschick et al. (1985) proposed using a cluster of lightweight irradiance sensors as an electronic equivalent of a mechanical point quadrat.

There are two kinds of probes used in the point quadrat analysis of the plant community architecture: material and notional probes. The mechanical point quadrats of Warren Wilson (1959a), Ford and Newbould (1971), and Turitzin (1978), and the fiber-optic point quadrats of Caldwell et al. (1983a,b) are examples of material probes, in which a long thin needle is actually pushed through the vegetation canopy. Lang et al. (1985) used the collimated solar beam as a probe, albeit notionally, for assessing the structure of the canopy. MacArthur and Horn (1969) used a 35 mm camera fitted with a telephoto lens and notionally probed the canopy from beneath with the focusing optics of the system (also see Aber, 1979). The theory of point quadrat analysis is generic in the sense that one tries to solve a Fredholm integral equation of the first kind, irrespective of the kind of probe used, and this will be discussed next.

Consider an elementary volume in the horizontally homogeneous leaf canopy at height z of thickness Δz. The canopy is pierced with a point quadrat along a certain direction $\Omega_p(\mu_p, \phi_p)$ (for downward directions, $\mu_p < 0$), and the number of contacts of the point quadrat with the canopy elements (leaves) is registered. Let N_c denote the total number of contacts with the leaves in this elementary volume for a total number of n insertions of the point quadrat, all along the same direction Ω_p. The contact frequency $\nu_L(z, \Omega_p)$, i.e., the number of contacts per unit length of the probe in this elementary volume for insertions from the direction Ω_p, is

$$\nu_L(z, \Omega_p) = -\frac{\mu_p N_c}{n \, \Delta z} \tag{28}$$

It should be apparent that if the number of insertions is sufficiently great enough and if the point quadrats are indeed of infinitesimally thin radius, then $\nu_L(z, \Omega_p)$ is just the projection of the total leaf area in the elementary volume onto the plane perpendicular to the direction of insertion of the probe:

$$\nu_L(z, \Omega_p) = u_L(z) \frac{1}{2\pi} \int_0^{2\pi} d\phi_L \int_0^1 d\mu_L \, g_L(z, \Omega_L) |\Omega_L \cdot \Omega_p|$$
$$= u_L(z) \, G(z, \Omega_p) \tag{29}$$

On the assumption of azimuthal randomness of leaf normal orientation, the previous equation may be reduced to

$$\nu_L(z, \mu_p) = u_L \int_0^1 d\mu_L \, g_L(z, \mu_L) \left(\frac{1}{2\pi} \int_0^{2\pi} d\phi_L |\boldsymbol{\Omega}_L \cdot \boldsymbol{\Omega}_p| \right) \qquad (30)$$

$$= u_L \int_0^1 d\mu_L \, g_L(z, \mu_L) \, \psi(\mu_p, \mu_L) \qquad (31)$$

Equation 31 is a Fredholm integral equation (since the limits are fixed and not variable) of the first kind [since the function $g_L(z, \mu_L)$ appears only in the integrand; see Philip, 1962; Twomey, 1963]. The kernel function is based on known trignometric functions only and is explicitly

$$\psi(\mu_p, \mu_L) = |\mu_p \mu_L| \qquad \text{if } |\cot \theta_p \cot \theta_L| > 1 \qquad (32)$$

otherwise

$$\psi(\mu_p, \mu_L) = \mu_p \mu_L (2\phi_t/\pi - 1) + (2/\pi)(1 - \mu_p^2)^{1/2}(1 - \mu_L^2)^{1/2} \sin \phi_t \quad (33)$$

where

$$\phi_t = \cos^{-1}[-\cot \theta_p \cot \theta_L] \qquad (34)$$

It is now straightforward to estimate the leaf-area density function, $u_L(z)$, from the contact frequency function (also see Gates and Westcott, 1984)

$$\int_0^1 d\mu_p \, \nu_L(z, \mu_p) = u_L(z) \qquad (35)$$

The estimation of $g_L(z, \mu_L)$ from the contact frequency function involves solution of the Fredholm integral equation (Eq. 31); the theory and application of one such procedure is now described.

III.C.1   Solution of the Fredholm Integral Equation

The methods of solving Eq. 31 to obtain estimates of $g_L(z, \mu_L)$ and $u_L(z)$ from measurements of $\nu_L(z, \mu_p)$ have been subjects of numerous investigations (Miller, 1963, 1967; Philip, 1965a,b; Norman et al., 1979; Anderssen et al., 1984, 1985; Lang et al., 1985; amongst many others). Philip (1965a) warned of the unreliability of these techniques, and his conclusions were supported by recent investigations (see Lang et al., 1985). The main problem is one of instability of the derived solutions.

Consider the integral equation

$$G(\mu_p) = \int_0^1 d\mu_L \, g_L(\mu_L) \, \psi(\mu_p, \mu_L) \qquad (36)$$

and its linear transformation

$$\mathbf{G} = \mathbf{Ag} \qquad (37)$$

The idea is to invert m interpolated values of the G function to obtain n values of the $g_L(\mu_L)$ function. The method described here has been reported earlier by Twomey (1977) and more recently by Lang et al. (1985). The matrix $\mathbf{A}$ contains the quadrature coefficients, i.e., the element a_{ij} is the jth quadrature coefficient for the integral equation containing the ith kernel and is computed as

$$a_{ij} = \left(\frac{1}{\mu_{L_{j+1}} - \mu_{L_j}} \right) \left[\mu_{L_{j+1}} \int_{\mu_{L_j}}^{\mu_{L_{j+1}}} d\mu_L \, \psi(\mu_p, \mu_L) - \int_{\mu_{L_j}}^{\mu_{L_{j+1}}} d\mu_L \, \psi(\mu_p, \mu_L) \right]$$
$$- \left(\frac{1}{\mu_{L_j} - \mu_{L_{j-1}}} \right) \left[\mu_{L_{j-1}} \int_{\mu_{L_{j-1}}}^{\mu_{L_j}} d\mu_L \, \psi(\mu_p, \mu_L) - \int_{\mu_{L_{j-1}}}^{\mu_{L_j}} d\mu_L \, \psi(\mu_p, \mu_L) \right] \quad (38)$$

The integrals in the last equation are evaluated numerically using the trapezoidal rule with, say, 20 steps in μ_L. The first step is to fit a straight line between the measured G and $g_L(\mu_L)$ functions and determine the root-mean-square difference δy. Then, interpolating m values from the fitted line, the goodness of fit is determined as $\varepsilon = \delta y / (k/m)^{1/2}$, where k is the number of original measurements of the G function. Equation 37 is then inverted using a constrained least squares inversion technique that minimizes the sums of squares of the first differences (Twomey, 1977). Thus,

$$\mathbf{g} = [\mathbf{A^*A} + \alpha \mathbf{H}]^{-1} \mathbf{A^*G} \quad (39)$$

where $\mathbf{A^*}$ is the transpose of $\mathbf{A}$, and $\alpha (\leq 1)$ is a scalar whose value is chosen according to the expectation of the solution. The $\mathbf{H}$ matrix is given explicitly by Twomey (1977, p. 125), and, for any value of α, the goodness of fit of the solution is computed as

$$e^2 = |\mathbf{Ag} - \mathbf{G}|^2 \quad (40)$$

The constraint parameter α is chosen such that $e^2 > \varepsilon$. Now, let us consider a test case, in which the principles developed here can be applied.

III.C.2   Recovery of Measured $g_L(\mu_L)$

We measured the bivariate leaf normal distribution in a fully grown maize canopy during its vegetative growth phase using a custom-designed protractor-compass assembly. The assembly consisted of two separate pieces, which were inexpensive and easy to build. A thin straight iron rod of 0.15 m was fitted normal to a flat wooden block (0.16 × 0.03 × 0.02 m). A plexiglass plate (0.15 × 0.15 m) was marked into 36 equivalent sectors and a plumb bob was attached to the end of a string that was knotted through a small hole in the center of the plexiglass plate. A small compass was mounted on the plate so that all azimuthal observations could be made relative to the cardinal direction North, in a clockwise fashion. The wooden block was held parallel to the plane of a leaf, so that the iron rod simulated the normal to the upper face of the leaf. The inclination angle of the iron rod from the zenith was then measured with a protracti that had a movable pointer. Concurrently, another person held the plexiglass plate above so that the plumb bob was pointing to the origin of the iron rod (the leaf normal) on the wooden block. Then, the azimuth of the iron rod was read. The angular measurements of the leaf or its sections was made in one session on a wind-free day. The azimuthal distribution of leaf normals in this maize canopy was random (Figure 13).

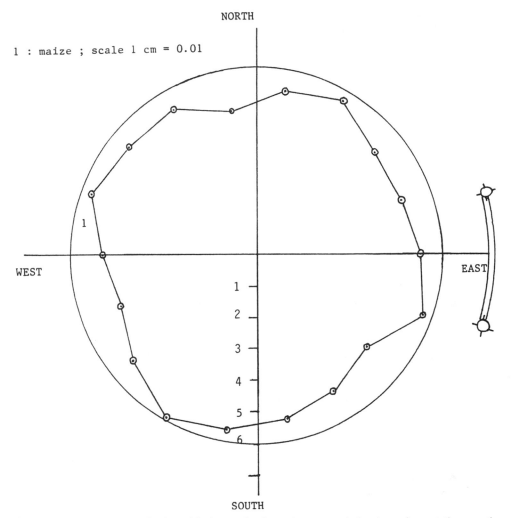

Figure 13 Azimuthal distribution of leaf normals of a maize canopy during its peak vegetative growth phase. The solar track during the period of data collection is also plotted.

The $G(\theta_p)$ function was computed using the measured leaf normal inclination distribution function by simply evaluating the integral in Eq. 31. The resulting G values were then used in the inversion procedure described before to see if we could recover the measured distribution function $g_L(\theta_L)$. The solutions are plotted in Figures 14 through 17 for different values of n (17, 36, 45, and 60) and γ. The following conclusions can be made:

1. The solutions did not appreciably change with increasing m, although $m < 5$ is too few.
2. For $\gamma \approx 1$, the solutions for all values of n were extremely constrained, i.e., g_L is invariant of θ_L.
3. The value of γ at which the inversion matrix became singular depended on the

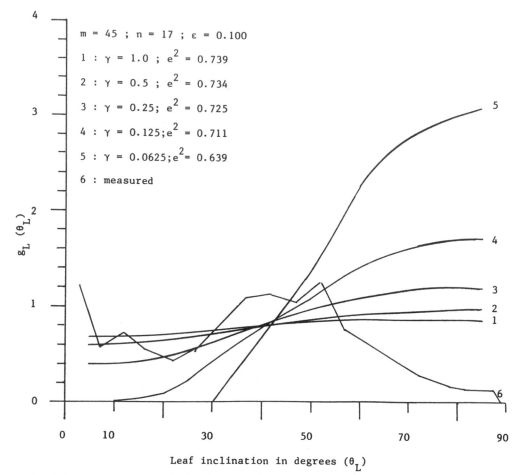

Figure 14   The leaf normal inclination distribution function obtained from the inversion of the calculated G function using measured leaf normal inclination distribution. Here $n = 17$ (for notation, see text).

value of n. For instance, when n was 17, γ could be as low as 0.0625, whereas for $n = 60$, the inversion matrix became singular when $\gamma = 1$.

4. The quality of the solution depended on the value of n. When n was 17, the unconstrained solutions were physically nonsensical. The values of g_L were negative (they were set to zero in plotting). When n was incremented to 36, the solutions were at least meaningful, but were far from the measured distribution function. When n was further incremented to 45, the solution obtained with $\gamma = 0.3$ matched exactly with the planophile distribution function, a point we will discuss next. Finally, when $n = 60$, only constrained solutions were obtained.

An interesting point of the analysis was the solution obtained when $n = 45$ and $\gamma = 0.3$ [Figure 16; curve 2]. The cumulative distribution function (CDF) of a planophile distribution and the measured CDF of the maize canopy are plotted along with the respective probability density functions (PDFs) in Figure 18. The CDF of the maize can-

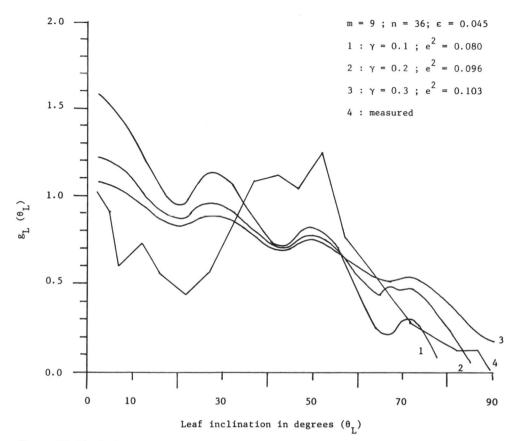

$$m = 9 \; ; \; n = 36; \; \epsilon = 0.045$$

$$1 : \gamma = 0.1 \; ; \; e^2 = 0.080$$

$$2 : \gamma = 0.2 \; ; \; e^2 = 0.096$$

$$3 : \gamma = 0.3 \; ; \; e^2 = 0.103$$

$$4 : \text{measured}$$

Figure 15 The leaf normal inclination distribution function obtained from the inversion of the calculated G function using measured leaf normal inclination distribution. Here $n = 36$ (for notation, see text).

opy resembles the planophile CDF, although the probability of leaf normal inclination between 20–50° was less in the maize canopy. The PDFs, however, show remarkable differences. When we compared the solution plotted in Figure 16, curve 2, with the planophile PDF in Figure 18, we find that they are identical. To understand this, we tested for the domain of applicability for this inversion procedure using theoretical PDFs. This is discussed in detail in the next section.

III.C.3 Recovery of Theoretical $g_L(\mu_L)$

Consider the following theoretical distribution functions of leaf normal inclination (deWit, 1965; Bunnik, 1978):

Planophile:

$$g_L(\theta_L) = 2/\pi(1 + \cos 2\theta_L) \tag{41}$$

Erectophile:

$$g_L(\theta_L) = 2/\pi(1 - \cos 2\theta_L) \tag{42}$$

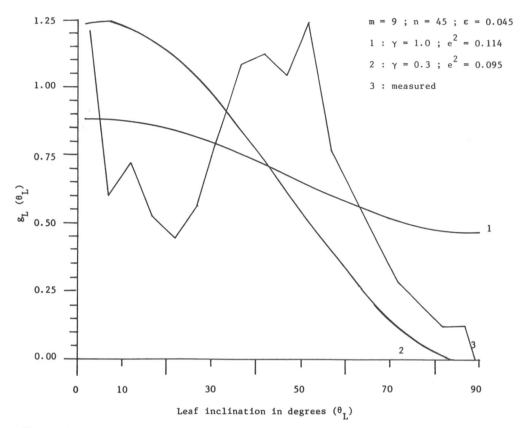

Figure 16   The leaf normal inclination distribution function obtained from the inversion of the calculated *G* function using measured leaf normal inclination distribution. Here $n = 45$ (for notation, see text).

Plagiophile:

$$g_L(\theta_L) = 2/\pi(1 - \cos 4\theta_L) \tag{43}$$

These distribution functions were used in the forward solution to compute the *G* function. Then a straight line was fit to these *G* values, and *m* (9) values were interpolated and used in the inversion ($n = 45$). The theoretical and inversion-derived distribution functions for these three theoretical cases are given in Figure 19.

In case of planophile and erectophile distributions, the inversion-derived solutions match well with the actual theoretical distribution functions (Eqs. 41 and 42); however, for plagiophile distributions, the correspondence is very poor. Clearly, the plagiophile distribution function is distinctly different, but the result from the inversion procedure is somewhat similar to the planophile distribution function. In order to understand this ambiguity, we have to look at the *G* functions for the three theoretical distributions (Figure 20). Although the distribution functions of leaf normal inclination of planophile and plagiophile types are remarkably different, the computed *G* functions are not. Hence, if one were to measure the integral entity and hope to invert it for the underlying distribution, the information is simply not contained in the measurement. Worse, if one has no knowledge of the actual distribution function (the solution), then there is no objective

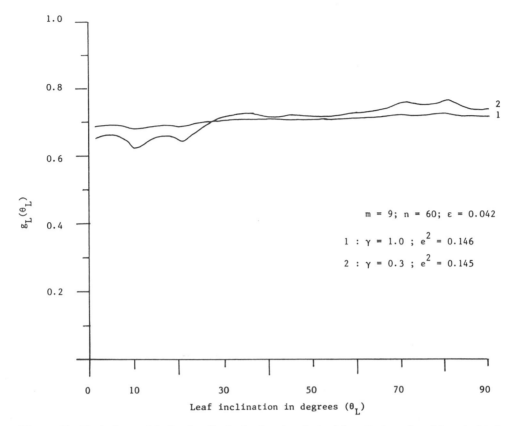

Figure 17 The leaf normal inclination distribution function obtained from the inversion of the calculated G function using measured leaf normal inclination distribution. Here $n = 60$ (for notation, see text).

criterion in choosing any one particular solution from a finite but very large set of valid solutions. This instability in the solutions is a critical drawback of the point quadrat analysis (also see Ledent, 1974, for some relevant discussion).

III.D Indirect Estimation of Green-Leaf-Area Index

In this case, we assume that the frequency of contacts can be described by the Poisson distribution (Monsi and Saeki, 1952) as

$$f_c = 1 - \exp\left(-g'L\right) \qquad (44)$$

where f_c and p_0 are considered to be equivalent, since both represent the probability for photons encountering the foliage elements. The exponential term on the right-hand side of Eq. 44 is the probability of photons penetrating through plant-canopy gaps and reaching the soil surface. The gap frequency can be also described based on binominal distribution functions (Myneni et al., 1986). Equation 44 does not account for scattering of photons by foliage elements; however, in the PAR region, the scattering is minimal due to strong absorption, and this equation is a reasonable approximation. If f_c and g'

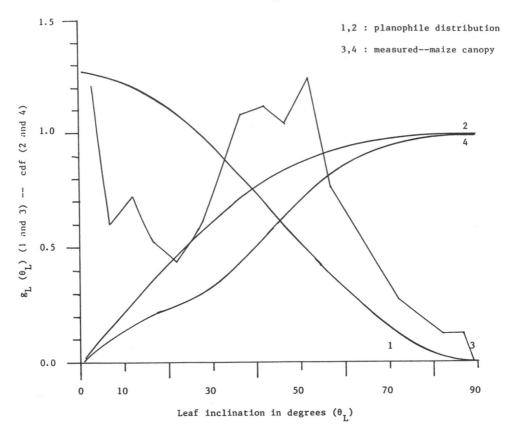

Figure 18   The cumulative and the probability distribution functions of a planophile and a maize canopy.

are known, Eq. 44 can be solved to obtain L. This procedure was tested by Fuchs et al. (1984) to estimate a mean value of L from diurnal measurement of p_0 in wheat canopies. They found that spherical leaf-angle distribution (Eq. 15) is a good approximation, and g' can be estimated from the solar elevation angle. Figure 21 illustrates the distribution of measured and computed L values based on this approach. There is good agreement between measured and computed L values, but, the computed values tend to be smaller than the measured ones for the data set obtained in Logan, Utah, for a spring-wheat canopy (Fuchs et al., 1984). This is due to the exponential term in Eq. 44, which suggests the probability of contact increases significantly at $L > 4.0$.

Asrar et al. (1984b, 1985a) used this approach to compute L values for several wheat canopies based on the measurements of red- and near-infrared-canopy reflectance under a wide range of management practices and solar elevation angles. In this approach, NDVI was used to estimate a mean p_0 value, and then L was computed as

$$L = \frac{\overline{-\ln (1 - p_0)}}{\overline{g}} \qquad (45)$$

where $\overline{g}$ and $\overline{1 - p_0}$ are the arithmetic means of the extinction coefficient and gap frequency, respectively. Figure 22 shows the results of such computations for wheat can-

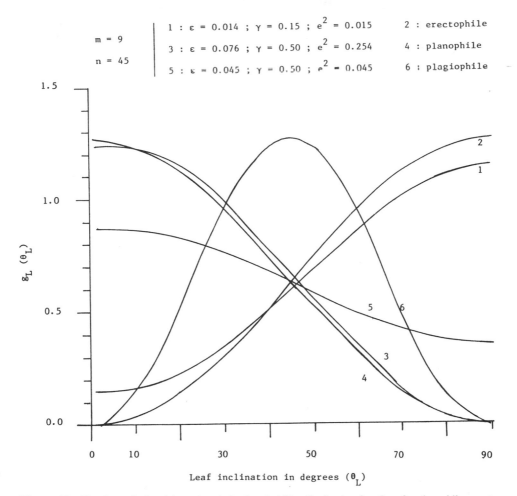

Figure 19 The theoretical and inversion-derived probability distribution functions for planophile, erectophile, and plagiophile distributions.

opies under optimum growth conditions. The agreement between measured and computed L values is generally good, except during early periods in the growing season, when the computed L values are consistently greater than the measured values due to the effect of soil background. Table 1 contains some measured and computed L values for two wheat canopies that were planted in East–West and North–South row orientations. The computed L values are based on several measurements during given days throughout the growing season, and they show the effects of different row orientations and diurnal changes in spectral characteristics of the wheat canopies.

Figure 23 illustrates the measured vs. computed L values for two spring-wheat canopies that were managed differently. When plants received an adequate supply of water, Figure 23(a), they produced the largest amount of green leaf area and maintained it over a longer period of time, as compared with the case when water was a limiting factor, Figure 23(b). The spectral characteristics of both canopies contained enough information to depict these changes, and they are represented in the computed L values for the two

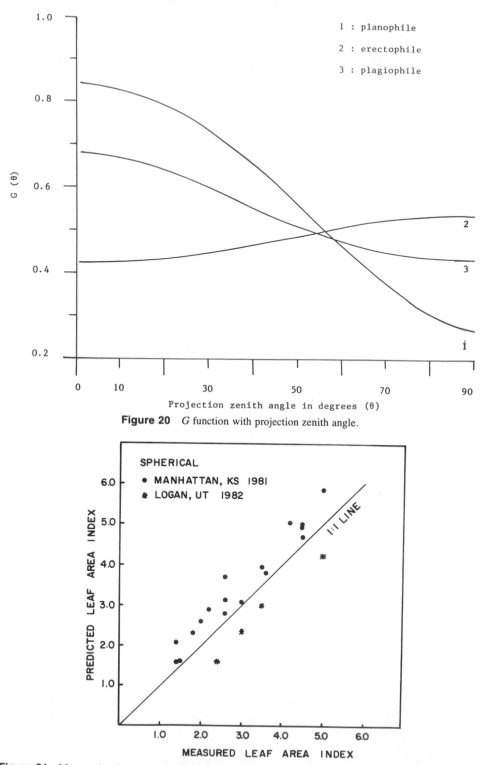

Figure 20 *G* function with projection zenith angle.

Figure 21 Measured and computed green-leaf-area indices of several wheat canopies. [Reprinted by permission of the publisher from Fuchs et al. (1984). Copyright 1984 by Elsevier Scientific Publishing Co., Inc.]

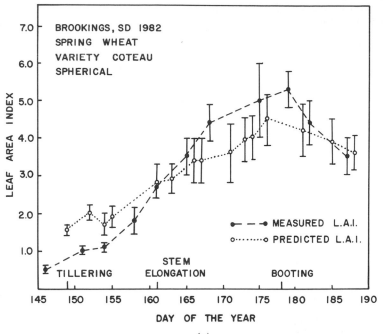

(a)

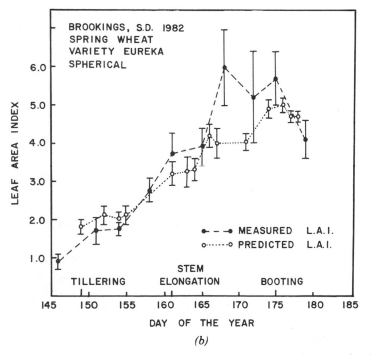

(b)

Figure 22 Measured and computed green-leaf-area indices of two spring-wheat canopies. [Asrar et al. (1984b). Reproduced from *Agronomy Journal*, Volume 76, 1984, pages 300–306 by permission of the American Society of Agronomy, Inc.]

TABLE 1   Measured and Computed Green-Leaf-Area Indices of Winter-Wheat Canopies Planted in Two Different Row Orientations and Several Sum Angles

Date	Row Orientation	Measured LAI	Solar Angle (deg)		Estimated LAI
			Azimuth	Zenith	
April 24, 1983	East–West	2.4(0.20)[a]	34.4	27.9	2.4(0.33)
			−4.1	25.4	1.8(0.22)
			−49.3	34.2	3.0(0.85)
	North–South	2.5(0.25)	28.0	29.3	2.1(0.55)
			−9.6	25.6	2.0(0.57)
			−52.8	35.7	3.3(0.33)
May 3, 1983	East–West	5.1(0.40)	51.2	31.6	6.1(1.05)
			25.5	24.5	3.0(0.92)
	North–South	3.4(0.30)	48.3	30.4	3.9(0.17)
			18.3	23.6	2.7(0.49)
May 6, 1983	East–West	5.4(0.40)	69.3	41.0	5.0(0.68)
			56.1	32.8	4.7(0.66)
			−55.6	33.6	3.8(0.42)
	North–South	3.4(0.30)	67.2	39.3	4.4(0.65)
			54.3	31.9	2.8(0.77)
			−57.9	33.9	2.7(0.55)

[a]Numbers in parenthesis are standard deviations.

Source: Reprinted by permission of the publisher from Asrar et al. (1985a). Copyright 1985 by the Elsevier Scientific Publishing Co., Inc.

canopies. Other management practices, such as different time of sowing and plant densities, also affect the development of plant canopies, and, hence, should affect their spectral characteristics. The questions frequently asked are: Do these changes significantly affect the spectral behavior of the plant canopies? Can these changes be quantified in terms of plant-canopy attributes, based on spectral reflectance measurements?

Figure 24 shows measured and computed L values for spring-wheat canopies planted at two different times and irrigated at different frequencies throughout the growing season. The treatment that was planted earliest and received the largest amount of water produced the highest amount of green-leaf area, which lasted for a longer period, Figures 24(a) and (b). The computed L values for these two canopies adequately represent the differences between the two treatments. This suggests that the effects of management and cultural practices can be depicted in spectral reflectance characteristics of the canopies. In Chapter 8, the application of such information in assessing plant-canopy processes such as photosynthesis, respiration, and transpiration is discussed in detail.

The indirect approach just described was also used to estimate green-leaf-area index in natural grassland ecosystems (Asrar et al., 1986a; Major et al., 1988; Weiser et al., 1986). In this case, the diversity in species composition and adopted management practices further increases the number of parameters that affect spectral characteristics of plant canopies. Another complicating factor in natural ecosystems is the lack of control on water status of the vegetation, since these ecosystems are primarily rain-fed. Therefore, there could be potentially a large number of variables affecting the plant-canopy condition and development, which, in turn, may affect its spectral characteristics.

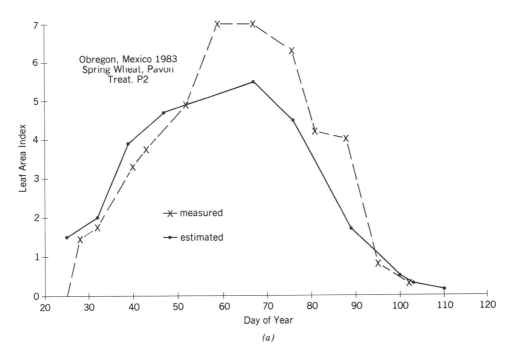

(a)

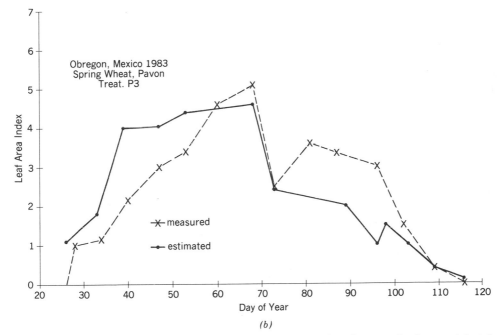

(b)

Figure 23 Measured and computed green-leaf-area indices of two spring-wheat canopies that were irrigated differently (i.e., I2, I3, I4).

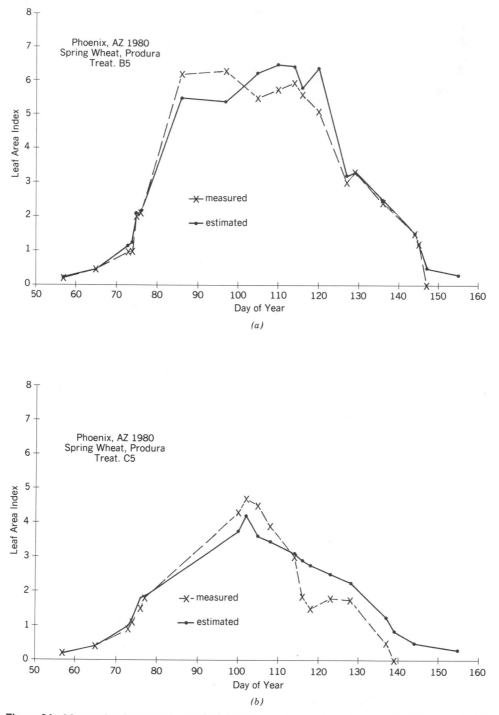

Figure 24   Measured and computed green-leaf-area indices of two spring-wheat canopies that were planted and irrigated differently.

Figure 25 illustrates a comparison between measured and computed L values for two treatments (burned vs. unburned) of a natural tallgrass prairie. These two treatments are the results of commonly adopted management practices in the tallgrass prairie region of the United States. Ecological studies conducted on these two treatments indicate that removal of senescent vegetation from the soil surface by means of burning (i.e., burned treatment) results in increased productivity and maintains a favorable balance in succession of grass vs. nongrass species (Hulbert, 1969; Knapp, 1984). Based on these findings, most land owners follow a common practice and remove the remainder of senescent vegetation from the soil surface by means of burning in early spring, prior to resumption of above-ground biological activities. We conducted several studies to assess the spectral reflectance characteristics of the burned and unburned areas of the tallgrass prairie. The two plant canopies distinctly differ in their spectral behavior in the visible and near- to shortwave infrared regions of the spectrum during selected days as well as during the growing season until the time of maximum greenness cover (Asrar et al., 1986b, 1988). The observed differences were primarily caused by the presence of senescent vegetation on the soil surface in unburned treatment. The difference in spectral characteristics of the two canopies could be used to distinguish between burned and unburned areas (Asrar et al., 1986b) or it could be used to obtain an estimate for plant-canopy attributes such as green-leaf area (Asrar et al., 1986a; Weiser et al., 1986).

In this section, we presented a simple indirect method of estimating green-leaf area using measurements of the red- and near-infrared-canopy reflectance. This method is based on the assumption that the frequency of contact between photons and canopy foilage elements follows a Poisson distribution and can be estimated from canopy re-

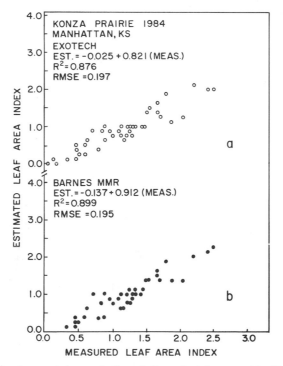

Figure 25 Measured and computed green-leaf-area indices of a tallgrass prairie. [Reprinted from Asrar et al. (1986a). © 1986 IEEE.]

flectance measurements. We further assumed that angular distribution of foliage elements in horizontally homogeneous canopies can be described by a spherical distribution and can be estimated from the solar elevation angle. This information was used in computing L values from a one-dimensional light-interception function. In the following section, we discuss methods of estimating above-ground phytomass based on the measurements of red- and near-infrared-canopy reflectance.

IV ESTIMATION OF ABOVE-GROUND PHYTOMASS

Estimates of total above-ground green phytomass may be preferred to green-leaf area in productivity studies. Spectral indices can be correlated empirically with above-ground phytomass component (Aase and Siddoway, 1981). Such relationships are site-specific and their application is limited. The alternative approach is to use the spectral indices as estimators of green-leaf area and intercepted PAR in soil–plant–atmospheric processes-based model (Kanemasu et al., 1985). This approach allows evaluation of the performance of a plant canopy in terms of its phytomass production and grain yield. Figure 26 is a schematic diagram for such a model, showing where remotely sensed data can be used as a surrogate for traditional measurements. Results from simulation studies based on this model are compared with actual measurements in Table 2. These data suggest a general improvement in simulated grain yields with remotely sensed inputs. The main limitation of such an approach is the complexity of the models, the relatively

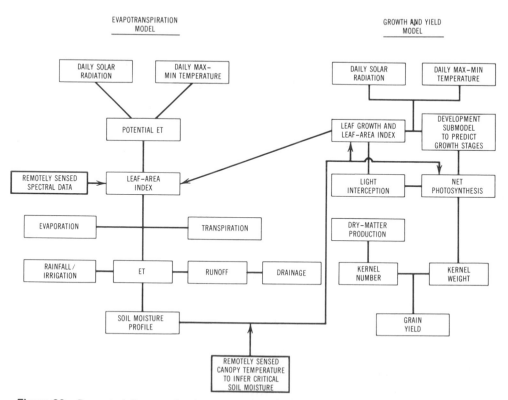

Figure 26   Conceptual diagram of a plant-processes model illustrating the use of remotely sensed data.

TABLE 2   Simulated and Observed Grain Yields of Spring Wheat Based on Combinations of Simulated and Observed Values of Leaf-Area Index (LAI) and Soil Moisture (SM)

Model Run No.	Year	LAI[a]	SM[b]	Observed Mean Yield (kg ha⁻¹)	Model Mean Yield (kg ha⁻¹)
A	78–79[c]	SIM	SIM	2372	1532
B		REF	SIM	2466	2863
C		REF	TC	2466	2486
D	79–80	SIM	SIM	3588	2379
E		REF	SIM	3588	4529
F		REF	TC	3588	3434
G	78–80	SIM	SIM	2930	1920
H		REF	SIM	2997	3420
I		REF	TC	2997	3125

[a]LAI is the leaf-area index.

[b]SM is the soil moisture; *SIM is the simulated soil moisture*; REF is the LAI estimated from reflectance; and TC is the soil moisture estimated from canopy temperature.

[c]Differences in observed mean yields within a year are due to different data availability for alternate model runs.

Source: Kanemasu et al. (1985).

large number of inputs required, and the large number of semiempirical relationships used to represent different processes.

Since net photosynthesis and phytomass production are dependent on the PAR absorption during the active growth period, it should be feasible to estimate above-ground phytomass component from red and near-infrared reflectance. In Section II, we evaluated the feasibility of estimating APAR from red and near-infrared spectral reflectance measurements. In this section, we describe a simple model for the estimation of total above-ground green phytomass that uses remotely sensed data inputs.

The total above-ground phytomass (P_m) production can be related to absorbed photosynthetically active radiation as (Asrar et al., 1985b)

$$P_m = \int_{t_1}^{t_2} \varepsilon_c \varepsilon_i \varepsilon_s I_s W_i \, dt \tag{46}$$

where ε_c and ε_i are the photochemical and interception efficiency factors, respectively; ε_s is the fraction of energy in the PAR region of the spectrum; I_s is the total incident solar energy (MJ/d); and W_i is a stress index, ε_c is defined as the ratio of chemical energy equivalent of absorbed PAR energy (g/MJ). Montieth (1970, 1977) found that ε_c was relatively constant for given species throughout the growing season. Asrar et al. (1984a) found that ε_c values computed based on the above-ground fraction of phytomass for winter wheat were affected by stages of physiological development and some management practices. Theoretical analysis and experimental measurements by Szeicz (1974) shows that the PAR fraction of the solar energy, ε_s, was nearly constant for total (direct + diffuse) radiation and nearly independent of atmospheric conditions. Diurnal and seasonal variations of ε_i were evaluated by Hipps et al. (1983) using direct measurements of components of PAR (Eq. 1) in winter-wheat canopies.

Index W_i should include the effects of nutrient deficiencies, temperature, and water stress. In the following discussion, we do not consider the effects of mineral deficiencies on plant-canopy spectral reflectance. Water stress and plant-canopy temperature, however, can be related through a ratio of actual (E_a) to potential (E_p) evapotranspiration:

$$W_i = \frac{E_a}{E_p} = \frac{\Delta + \gamma^*}{\Delta + \gamma_s(1 + r_c/r_a)} \tag{47}$$

where

$$\gamma^* = \gamma_s(1 + r_c^*/r_a) \tag{48}$$

and

$$\Delta = \frac{e_c^* - e_a^*}{T_c - T_a} \tag{49}$$

where Δ is the slope of saturation vapor pressures e_c^* and e_a^* at canopy T_c and air T_a temperatures, respectively; γ is the psychrometric constant; and r_c^* is crop resistance to vapor transfer under potential (ample water) conditions. To evaluate E_a/E_p from Eq. 47, a value for the ratio of crop resistance to water vapor transport, r_c, and one for aerodynamic resistance, r_a, are required. Jackson (1982) derived the following r_c/r_a relation based on the energy balance of a plant canopy:

$$\frac{r_c}{r_a} = \frac{(\gamma r_a R_n/\rho C_a) - (T_c - T_a)(\Delta + \gamma_s)(e_a^* - e_a)}{(T_c - T_a) - (r_a R_n/\rho C_a)} \tag{50}$$

where R_n is the net radiation; ρ and C_a are the density and heat capacity of the air, respectively; and e_a is the actual air vapor pressure. The aerodynamic resistance r_a is affected mainly by canopy architecture and wind velocity. Its computations under non-stable conditions are discussed in detail by Hatfield et al. (1983).

Remotely sensed data can be used to obtain an estimate for ε_i, R_n, T_c, and I_s in the previous equations (Asrar et al., 1985b). The other parameters, such as T_a, e_a, e_a^*, and r_c^*, should be measured directly in conjunction with remotely sensed data. The methods of estimating components of surface energy balance from remotely sensed data are discussed in Chapter 16, and application of remotely sensed data in energy-balance models is given in Chapter 17.

Figure 27 illustrates the measured and computed total above-ground phytomass values based on Eq. 46 for a spring-wheat canopy. A good correspondence among measured and computed phytomass values is observed, especially when the measured surface radiative temperatures were used to adjust the estimated phytomass values for suboptimal conditions. The effects of limited water and increased temperature on estimated phytomass values can be seen in Figure 28. These data suggest that the index proposed here is only effective under moderately stressful conditions. The main reason for disagreement between measured and estimated P_m values was poor plant-stand establishment and dominance of soil background due to late planting date (i.e., 100 days late), which coincided with a period of very warm air temperatures and high evaporative demand.

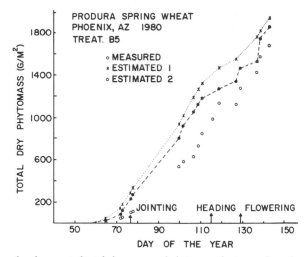

Figure 27 Measured and computed total above-ground phytomass for two spring-wheat canopies. The computed values were adjusted for water stress in method 2. [Reprinted from Asrar et al. (1986a). © 1986 IEEE.]

Under these conditions, plants were adversely affected by the environment, and the result was a substantial reduction in plant growth. It is encouraging that the combined effects of these limitations can be depicted by remotely sensed measurements; however, there is a need for minimizing the contribution of soil background. This topic is addressed in detail in Chapter 4.

The simple procedure described in this section was used successfully to evaluate the phytomass yields of cereals and native rangelands in Alberta, Canada (Major et al., 1986, 1988), and of a tallgrass prairie in the Flint Hills region of the United States (Weiser et al., 1986). In the next section, we present an overview of the studies that are being conducted on the subject of the plant-canopy Hot Spot and its utility in estimating plant-canopy structural variables.

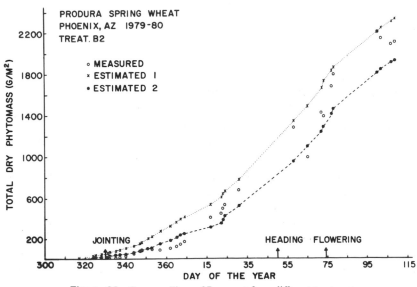

Figure 28 Same as Figure 27, except for a different treatment.

V THE CANOPY HOT SPOT

All rough and structured surfaces illuminated by a directional radiative source with a wavelength considerably smaller than the size of the constituents of the surface show a local maximum of reflected radiation within a cone around the direction of retroreflection. If the sun is considered as the only source of radiation that illuminates a vegetated surface and the angular distribution of the reflected radiation around the shadow of the observing instrument is measured, a narrow intensity peak should be detected in the reverse solar direction, when the observer's shadow can be eliminated. This effect is called the canopy hot spot in optical remote sensing, the Heiligenschein in atmospheric optics, and the opposition effect in planetary physics. When observed precisely in the direction of incident radiation, only the illuminated (sunlit) parts of the canopy are seen, and, thus, the peak in the reflected intensity. Since this effect is based only on mutual shadowing of the constituent foliage elements of the canopy, no colors are produced, which is a definite distinction from the glory effect that also appears in the retrosolar direction.

Suits (1972) was the first to introduce a correction for the joint probability of achieving a line of sight to a given layer for incident radiation and achieving the line of sight through this canopy for this layer, but this was done intuitively by analogy. Bunnik (1978) further developed the concept of the canopy hot spot and, using a chain of ingenuous arguments, concluded that the hot spot can be measured and is useful in information extraction if an active conical system is used. Later, he and his co-workers (Bunnik et al., 1984) developed a ground-based hot spot reflectance Meter (HSM) and showed that the HSM data can be used to estimate crop biomass and percent ground cover. Kuusk (1985) developed an analytical formulation for calculating the hot spot of homogeneous leaf canopies with the approximation of single scattering.

Kanevski and Ross (1982, 1985) developed a Monte-Carlo method for calculating the radiation regime of a coniferous tree. Crown structural information, such as the height, base angle, number of whorls, size of whorl, needle-area density, number of shoots, is input to the Monte-Carlo code. The results from these calculations show the brightness and the shape of the hot spot and its dependence on the structure of the crown. More recently, Ross and Marshak (1985, 1988) developed a versatile Monte-Carlo method for photon counting in realistic heterogeneous plant canopies. Canopy structural information, such as the plant density, height, number of leaves per plant, distance between leaves, dimensions and orientation of leaves, can be input to their photon-tracing routines. These tracing procedures are especially attractive, since they are optimized with respect to several constituent parameters. Based on the simulations, they concluded that the change of the bidirectional reflection function in the region of the hot spot characterizes leaf dimensions. Further, the width of the hot spot can be increased by increasing the individual leaf area or by decreasing the distance between the leaves. Recently, Gerstl and Simmer (1986) and Powers and Gerstl (1987) have been investigating the atmospheric effects on the canopy hot spot using Monte-Carlo techniques and report that the hot spot radiances are not appreciably perturbed by the atmosphere, and, hence, should be detectable by space-borne sensor systems.

The angular distribution of the canopy reflectance factors in the backscattered part of the solar plane is plotted in Figure 29 (wavelength of 0.63–0.69 μm). The data were collected on fully grown dense wheat ($L = 6.1$), maize ($L = 5.3$), soybean ($L = 2.3$), and native tallgrass prairie ($L = 1.7$) canopies on clear sunny days. A multiwavelength

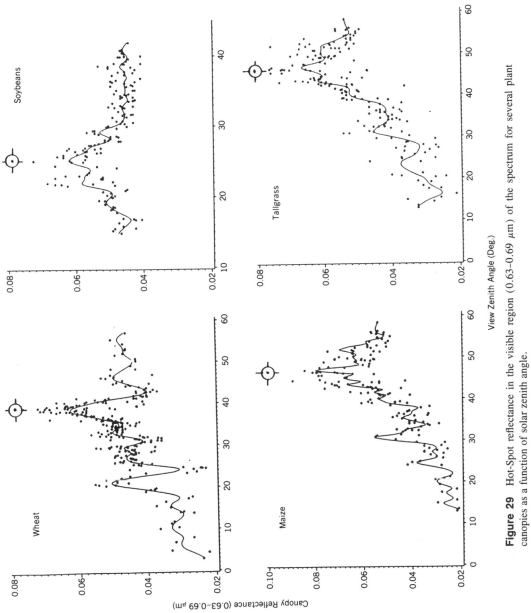

Figure 29 Hot-Spot reflectance in the visible region (0.63–0.69 μm) of the spectrum for several plant canopies as a function of solar zenith angle.

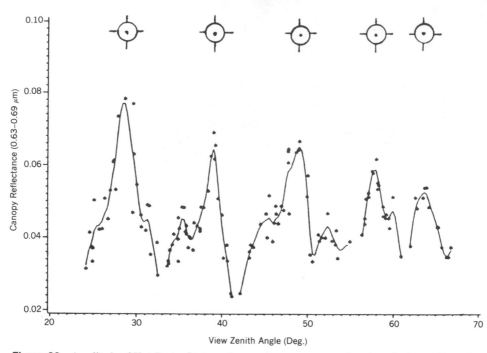

Figure 30   Amplitude of Hot-Spot reflectance for a soybean canopy as a function of solar zenith angle.

band radiometer (Exotech Model 100-A) was attached to a boom assembly mounted on a truck and was elevated to about 9 m above the canopy. The radiometer was equipped with a 1° (full planar angle) field of view (FOV) to collect the target radiances around the directions selected. The view zenith angle could be changed remotely by manipulating a linear actuator. The integration time for recording the analog signal was increased by a factor of 5 in comparison with the integration time when the radiometer had the standard 15° FOV to stabilize the signal. The position of the sun during the period of data collection (200–300 observations is about 2–3 minutes) is also plotted in Figure 29. The hot spot can be clearly seen as an amplitude spike in the signal for all four different plant canopies. Figure 30 illustrates the change in amplitude of the hot-spot signal with solar zenith angle. These figures suggest that the hot spot in plant canopies can be measured with ordinary multiband passive radiometers and the expensive active sensor systems, such as the one proposed by Bunnik et al. (1984), are not required. It is hoped that future research activity will concentrate on testing the implied utility of hot spot radiances in estimating plant-canopy structural variables.

VI IN RETROSPECT

In this chapter, we described some procedures for the estimation of plant-canopy attributes such as photosynthetically active radiation, green-leaf-area index, leaf normal distribution, and above-ground phytomass from remotely sensed radiometric measurements of visible and near-infrared plant-canopy reflectance. The objective was to avoid using empirical approaches by considering the chemical, physical, and physiological processes that lead to a particular spectral reflectance response over a given region of the electromagnetic spectrum. This was accomplished by using simple models that allow a realistic representation of the processes, yet do not require a large number of input parameters. We propose this approach as a first step toward understanding the remotely sensed spectral measurements of plant canopies.

The proposed methods were then used in estimating plant-canopy attributes from measurements of spectral reflectance for several plant species under a wide range of conditions. We found that the soil background and atmospheric conditions had major effects on estimated plant-canopy attributes. The problem associated with the effects of soil background can be avoided to a large extent by using off-nadir measurements of plant-canopy reflectance. The application of off-nadir reflectance measurements in estimation of plant-canopy attributes based on rigorous radiative transfer models is discussed in Chapter 6. The effects of atmospheric conditions on plant-canopy spectral reflectance can be minimized by conducting the measurements on clear days and around solar noon. Removal of the atmospheric effects from satellite- and aircraft-based measurements is discussed in Chapter 9.

VII   FUTURE DIRECTIONS

Estimation of plant-canopy attributes from remotely sensed radiometric measurements is gaining support by scientists from a large number of disciplines, such as agronomy, biology, ecology, forestry, and hydrology. The major accomplishments have occurred through a balanced approach between computer modeling and field measurements of the process of interaction of solar energy with plant canopy. Major accomplishments on the modeling aspect have resulted in development of two- and three-dimensional photon-transport models. The major shortcoming in this area is the lack of necessary input information for proper evaluation of such models. We presented some field data demonstrating the magnitude of the hot spot in the visible region of the spectrum for several plant canopies. The canopy hot spot is the subject of current investigations. A challenge remains to determine if the hot spot signal contains some information about the size of foliage elements and the structure of plant canopies, and if, indeed, one can quantify this information in terms of plant-canopy attributes.

The qualitative and semiquantitative use of broad wavelength band, coarse and fine spatial resolution radiometric measurements from sensor systems such as Multispectral Scanner System (MSS), Thematic Mapper (TM), Advanced Very High Resolution Radiometer (AVHRR), and Systeme Probatoire pour l'Observation de la Terre (SPOT) on board Earth-observing satellites is continuing. The potential for deriving quantitative parameters from these data is a goal that can be achieved. The research activities sponsored by the National Aeronautics and Space Administration under International Satellite Land Surface Climatology Project (ISLSCP), First ISLSCP Field Experiment (FIFE), were designed to provide algorithms for obtaining quantitative information from the data acquired by these sensor systems. The next generation of narrow wavelength band multispectral radiometers such as Airborne Visible-Infrared Imaging Spectrometer (AVIRIS) and High-Resolution Imaging Spectrometer (HIRIS) will provide the long-awaited high-resolution data and with them the challenge for proper interpretation and quantification of such information.

ACKNOWLEDGMENTS

The research results reported in this chapter was funded in part by the National Aeronautics and Space Administration grant NAG 5-389. Contribution 88-296-B from Kansas Agricultural Experiment Station.

LIST OF PRINCIPAL SYMBOLS

Each symbol is explained in the text when it is first introduced. Only the principal symbols are listed. Transient notations are not listed.

STANDARD ALPHABETICAL SYMBOLS

Symbol	Description
$\mathbf{A}$	Matrix of quadrature coefficients
$\mathbf{A}*$	Transpose of $\mathbf{A}$
$A_{s'}$	Area of the apparent solar disk
A_d	Area of the obscuring disk
$A(\mathbf{\Omega}_p)$	Probability of contact between the needle and the leaves
a	Constant parameter
a_{ij}	Elements of matrix $\mathbf{A}$
$B(\mathbf{\Omega}_p)$	Probability of no contact between the needle and the leaves
B_0, B_1	Constant parameters
b_n	Coefficients of the $N + 1$ term Legendre expansion
C_a	Heat capacity of air [J m^{-3} K^{-1}]
$D^{pj}(L, \mathbf{\Omega}_p)$	Penetration function for direct beam radiation: j is the type of leaf dispersion
$D^{ij}(L, \mathbf{\Omega}_p)$	Interception function for direct beam radiation
$D^{aj}(L + \Delta L, L; \mathbf{\Omega}_p)$	Absorption function for direct beam radiation
d	Penetration of diffuse PAR in a plant canopy
d_t	Time interval, day
E^j	Projection of the jth sky zone on the horizontal
e_a	Actual vapor pressure of air [kPa]
e_a^*	Saturated vapor pressure at a given air temperature [kPa]
e_c^*	Saturated vapor pressure at a given canopy temperature [kPa]
dE_L'	Incident energy on a leaf
$dE_{L\omega}$	Scattered energy from a leaf
E_a	Actual evapotranspiration [mm]
E_p	Potential evapotranspiration [mm]
e^2	Goodness of fit in the solution of the Fredholm integral equation
$F^+(\tau)$	Upward diffuse flux density
$F^-(\tau)$	Downward diffuse flux density
$F*(L, \mathbf{\Omega}_p)$	Sunlit-leaf-area index
f	Hemispherical radiance distribution under clear sky conditions
f_c	Frequency of contact between photons and foliage elements

Symbol	Description
$f(\Omega' \to \Omega; \Omega_L)$	Leaf scattering transfer function
$G(z, \Omega)$	Geometry factor
G_n	Greenness index
g, g'	Canopy geometry factors
g_h	Canopy geometry factor for horizontal leaves
$g_L(z, \Omega_L)$	Distribution function of leaf normal orientation
$g_L(z, \mu_L)$	Distribution function of leaf normal orientation along the $\cos \theta_L$ coordinate
$\hat{g}, g_f$	Canopy geometry factor for nonhorizontal leaves
$\mathbf{H}$	Similarity matrix
$H^{pj}(L)$	Penetration function for diffuse sky radiation: j is the type of leaf dispersion
$H^{ij}(L)$	Interception function for diffuse sky radiation
$H^{aj}(L + \Delta L)$	Absorption function for diffuse sky radiation
$h_L(z, \phi_L)$	Distribution function of leaf normal orientation along the ϕ_L coordinate
$I(z, \Omega)$	Specific energy intensity
$I_0(\Omega_0)$	Intensity of the monodirectional solar component
$I_d(\Omega)$	Intensity of the diffuse sky component
$I^0(\tau, \Omega)$	Intensity of uncollided photons
$I^s(\tau, \Omega)$	Diffuse scattered (in the leaf canopy) intensity
I_s	Incident solar energy $[\mathrm{Jd}^{-1}]$
$J(\tau, \Omega)$	Sum of the distributed source term, $S(\tau, \Omega)$, and the first collision term, $Q(\tau, \Omega)$
$K(z, \Omega_p)$	Extinction coefficient for direct beam radiation
$K'(z, \Omega_d)$	Extinction coefficient for diffuse sky radiation
L	Leaf area index
m	Optical airmass number
n	Constant parameter
NDVI	Normalized difference vegetation index
P	Fraction of intercepted or absorbed PAR
P_0	Frequency of gaps inside the plant canopy
P	Atmospheric pressure at a given altitude $[\mathrm{kPa}]$
P_0	Atmospheric pressure at sea level $[\mathrm{kPa}]$
P_m	Total above-ground phytomass
q	Ratio of direct to total PAR
RI	Near-infrared to red reflectance ratio index
R_n	Total net radiation $[\mathrm{Jd}^{-1}]$
r_a	Aerodynamic resistance $[\mathrm{sm}^{-1}]$
r_c	Canopy resistance $[\mathrm{sm}^{-1}]$
r_c^*	Canopy resistance for conditions when water is not a limiting factor
T_a	Air temperature $[\mathrm{K}]$
T_c	Canopy temperature $[\mathrm{K}]$
t	Time, day
W_i	Water stress index
$Y^m(\mu')$	Characteristic function

GREEK SYMBOLS

Symbol	Description
α'	Angular radius of the solar disk
$\Gamma(\Omega' \rightarrow \Omega)$	Area scattering transfer function
γ	Constant parameter
γ_s	Psychrometric constant
γ^*	Adjusted psychrometric constant
Δ	Slope of saturation vapor pressure and temperature relationship
δ	Dirac delta function
ε_c	Efficiency of conversion of PAR energy to dry matter [gJ^{-1}]
ε_i	Fraction of intercepted PAR
ε_s	Fraction of solar energy in the PAR region
η	Solar zenith angle
μ	Cos θ
μ_s	Cosine of the scattering angle, $\|\Omega \cdot \Omega'\|$
ν	Eigenvalues in the expansion of the homogeneous transport equation
$\xi_k(\tau), \xi_{k+1}(\tau)$	Finite element basis functions
π	Constant, 3.14. . . .
ρ_s	Hemispherical reflectance of soil
ρ_h	Hemispherical reflectance of a canopy with horizontal leaves
ρ_m	Hemispherical reflectance of a canopy with non-horizontal leaves
ρ_f	Hemispherical reflectance of a canopy with spherical leaf angle distribution
$\sigma_e(z, \Omega)$	Extinction coefficient
$\sigma_s(z, \Omega \rightarrow \Omega)$	Differential scattering coefficient
σ_ρ	Single-scatter albedo of leaves
$\tau(z)$	Optical depth
$\Phi_n^m(\mu)$	Auxilliary function
ϕ	Azimuthal angle measured from North through West; solar azimuth angle
$\phi_t(\mu)$	Boundary angle
$\Psi(\mu)$	Auxilliary function
$\Psi(\mu, \mu', \mu_L)$	Kernel of the one-angle area scattering transfer function
$\Psi(\mu, \mu_L)$	Azimuthal average of the projected leaf area
$\Omega(\mu, \phi)$	Solid angle
$\Omega'(\mu', \phi')$	Superscript prime refers to incoming, or incidence, beam
$\Omega_0(\mu_0, \phi_0)$	Direction of the monodirectional solar component
$\Omega_L(\mu_L, \phi_L)$	Direction of the normal to the upper face of the leaf
ω	Single-scatter albedo of the leaves;
ω'	fraction of backside of leaves exposed to incident photons

REFERENCES

Aase, J. K., and F. H. Siddoway (1981). Assessing winter wheat dry matter production via spectral reflectance measurements. *Remote Sens. Environ.* **11**:267–277.

Aber, J. D. (1979). A method for estimating foliage-height profiles in broad leaved forests. *J. Appl. Ecol.* **67**:35–40.

Anderssen, R. S., D. R. Jackett, and J. L. B. Jupp (1984). Linear functionals of the foliage angle distribution as tools to study the structure of plant canopies. *Aust. J. Bot.* **32**:147–156.

Anderssen, R. S., D. R. Jackett, J. L. B. Jupp, and J. M. Norman (1985). Interpretation and simple formulas for some key linear functionals of the foriage angle distribution. *Agric. For. Meteorol.*, **36**:165–188.

Asrar, G., L. E. Hipps, and E. T. Kanemasu (1984a). Assessing solar energy and water use efficiencies in winter wheat: A case study. *Agric. For. Meteorol.* **31**:47–58.

Asrar, G., M. Fuchs, E. T. Kanemasu, and J. L. Hatfield (1984b). Estimating absorbed photosynthetic radiation and leaf area index from spectral reflectance in wheat. *Agron. J.* **76**:300–306.

Asrar, G., E. T. Kanemasu, and M. Yoshida (1985a). Estimates of leaf area index from spectral reflectance of wheat under different cultural practices and solar angles. *Remote Sens. Environ.* **17**:1–11.

Asrar, G., E. T. Kanemasu, R. D. Jackson, and P. J. Pinter (1985b). Estimation of total aboveground phytomass production using remotely sensed data. *Remote Sens. Environ.* **17**:211–220.

Asrar, G., E. T. Kanemasu, G. P. Miller, and R. L. Weiser (1986a). Light interception and leaf area estimates from measurements of grass canopy reflectance. *IEEE Trans. Geosci. Remote Sens.* **GE-24**:76–82.

Asrar, G., R. L. Weiser, D. E. Johnson, E. T. Kanemasu, and J. M. Killeen (1986b). Distinguishing among a tallgrass prairie surface cover types from measurements of multispectral reflectance. *Remote Sens. Environ.* **19**:159–169.

Asrar, G., R. B. Myneni, Y. Li, and E. T. Kanemasu (1989). Measuring and modelling spectral characteristics of a tallgrass prairie. *Remote Sens. Environ.* (In press.)

Bunnik, N. J. J. (1978). *The Multispectral Reflectance of Shortwave Radiation by Agricultural Crops in Relation with Their Morphological and Optical Properties.* Pudoc, Wageningen, The Netherlands.

Bunnik, N. J. J., W. Verhoef, R. W. de Jonghe, H. W. J. van Kasteren, R. H. M. R. Geerts, H. Noordman, D. Uenk, and T. A. de Boer (1984). Evaluation of ground-based hot-spot reflectance measurements for biomass determination of agricultural crops. *Proc. 18th Int. Symp. Remote Sens. Environ.*, Paris, France.

Caldwell, M. M., T. J. Dean, R. S. Nowak, R. S. Dzurec, and J. H. Richards (1983a). Bunchgrass architecture, light interception and water-use efficiency: Assessment by fiber optic point quadrats and gas exchange. *Oecologia* **59**:178–184.

Caldwell, M. M., G. W. Harris, and R. S. Dzurec (1983b). A fiber optic point quadrat system for improved accuracy in vegetation sampling. *Oecologia* **59**:417–418.

Daughtry, C. S. T., K. P. Gallo, and M. E. Bauer (1983). Spectral estimates of solar radiation intercepted by corn canopies. *Agron. J.* **75**:527–531.

de Wit, C. T. (1965). *Photosynthesis of Leaf Canopies*, Agric. Res. Rep. No. 663. Pudoc, Wageningen, The Netherlands.

Ford, E. D., and D. J. Newbould (1971). The leaf canopy of a coppiced deciduous woodland. *J. Ecol.* **59**:843–862.

Fuchs, M., G. Asrar, E. T. Kanemasu, and L. E. Hipps (1984). Leaf area estimates from photosynthetically active radiation in wheat canopies. *Agric. For. Meteorol.* **32**:13–22.

Gallo, K. P., C. S. T. Daughtry, and M. E. Bauer (1985). Spectral estimates of absorbed photosynthetically active radiation in corn canopies. *Remote Sens. Environ.* **17**:221–232.

Gates, D. J., and M. Westcott (1984). A direct derivation of Miller's formula for average foliage density. *Aust. J. Bot.* **32**:117–119.

Gerstl, S. A. W., and C. Simmer (1986). Radiation physics and modeling for offnadir satellite-sensing of non-Lambertian surfaces. *Remote Sens. Environ.* **20**:1–29.

Goodall, D. W. (1952). Some considerations in the use of point quadrats for the analysis of vegetation. *Aust. J. Sci. Res.*, *Ser. B* **5**:1–41.

Goudrian, J. (1977). *Crop Micrometeorology: A Simulation Study.* Center for Agricultural Publishing and Documentation. Wageningen, The Netherlands.

Gutschick, V. P., M. H. Barron, D. A. Waechter, and M. A. Wolf (1985). Portable monitor for solar radiation that accumulates irradiances histograms from 32 leaf-mounted sensors. *Agric. For. Meteorol.* **33**:281–290.

Hatfield, J. L., A. Perrier, and R. D. Jackson (1983). Estimation of evapotranspiration at one time-of-day using remotely sensed surface temperature. *Agric. Water Manage.* **7**:341–350.

Hatfield, J. L., E. T. Kanemasu, G. Asrar, R. D. Jackson, P. J. Pinter, Jr., R. J. Reginato, and S. B. Idso (1984). Leaf area estimates from spectral reflectance measurements over various planting dates of wheat. *Int. J. Remote Sens.* **46**:651–656.

Hatley, C. L., and J. A. MacMahon (1980). Spider community organisation: Seasonal variation and the role of vegetation architecture. *Environ. Entomol.* **9**:632–639.

Hipps, L. E., G. Asrar, and E. T. Kanemasu (1983). Assessing the interception of photosynthetically active radiation in winter wheat. *Agric. For. Meteorol.* **28**:253–259.

Hulbert, L. C. (1969). Fire and litter effects in undisturbed bluestem prairie in Kansas. *Ecology* **50**:874–877.

Jackson, R. D. (1982). Canopy temperature and crop water stress. In *Advances in Irrigation* (D. Hillel, Ed.). Academic Press, New York, pp. 43–85.

Kanemasu, E. T., G. Asrar, and M. Fuchs (1985). Application of remotely sensed data in wheat growth modelling. In *Wheat Growth and Modelling* (D. W. Day and R. K. Atkin, Eds.), NATO Adv. Study Inst. Ser., Ser. A. Plenum, New York, p. 407.

Kanevski, V. A., and J. Ross (1982). *A Monte Carlo Simulation Model for Radiation Conditions in a Coniferous Tree*, ESSR Prepr. Tartu, USSR (with an English summary).

Kanevski, V. A., and J. Ross (1985). Effect of the architecture of a conifer on directional distribution of its reflectance: A Monte Carlo simulation. *Sov. J. Remote Sens. (Engl. Transl.)* **3**:659–663.

Knapp, A. K. (1984). Post-burn differences in solar radiation, leaf temperature and water stress influencing production in lowland tallgrass Prairie. *Am. J. Bot.* **71**:220–227.

Knight, D. H. (1963). Leaf area dynamics of a shortgrass prairie in Colorado. *Ecology* **54**:892–896.

Kuusk, A. (1985). The Hot Spot effect of a uniform vegetative cover. *Sov. J. Remote Sens. (Engl. Transl.)* **3**:645–652.

Lang, A. R. G., Xiang Yueqin, and J. M. Norman (1985). Crop structure and the penetration of direct sunlight. *Agric. For. Meteorol.* **35**:83–101.

Lapitan, R. L., G. Asrar, E. T. Kanemasu, and G. W. Wall (1989a). Spectral estimates of absorbed light and leaf area index. I. Effects of Canopy geometry. *Agron. J.* (in press).

Lapitan, R. L., G. Asrar, and E. T. Kanemasu (1989b). Spectral estimates of absorbed light and leaf area index: II. Effects of water stress. *Agron J.* (in press).

Ledent, J. F. (1974). Une formule générale pour le calcul de la surface foliare soumise au rayonnement solaire direct. Biometrie-proximetrie **14**(3):69–75.

Levy, E. B., and E. A. Madden (1933). The point quadrat method of pasture analysis. *N. Z. J. Agric.* **46**:267–279.

List, R. J. (1971). *Smithsonian Meteorological Tables*, 6th ed. Smithsonian Institute Press, Washington, D.C.

MacArthur, R. H., and H. S. Horn (1969). Foliage profile by vertical measurements. *Ecology* **50**:802–804.

Major, D. J., G. B. Schaalije, G. Asrar, and E. T. Kanemasu (1986). Estimation of whole-plant biomass and grain yield from spectral reflectance of cereals. *Can. J. Remote Sens.* **12**:47–54.

Major, D. J., S. Smoliak, G. Asrar, and E. T. Kanemasu (1988). Use of spectral reflectance to study the effects of various range improvement treatments in Southern Alberta. *Can. J. Plant Sci.* **68**:1017–1023.

Miller, J. B. (1963). An integral equation from phyology. *J. Aust. Math. Soc.* **4**:397–402.

Miller, J. B. (1967). A formula for average foliage density. *Aust. J. Bot.* **15**:141–144.

Miller, J. B. (1969). Tests of solar radiation models in three forest canopies. *Ecology* **50**:878–885.

Monsi, M., and T. Saeki (1952). Uber dem lichtfaktor in den pflanzengesellschaften und seine bedutung fur die stoffproduktion. *Jpn. J. Bot.* **52**:22 52.

Montieth, J. L. (1970). Solar radiation and productivity, in tropical ecosystems. *J. Appl. Ecol.* **9**:747–766.

Montieth, J. L. (1977). Climate and the efficiency of crop production in Britain. *Philos. Trans. R. Soc. London, Ser. B* **281**:277–294.

Myneni, R. B., R. B. Burnett, G. Asrar, and E. T. Kanemasu (1986). Single scattering of parallel direct and axially symmetric diffuse solar radiation in vegetative canopies. *Remote Sens. Environ.* **20**:165–182.

Nilson, T. (1971). A theoretical analysis of the frequency of gaps in plant stands. *Agric. Meteorol.* **8**:25–38.

Norman, J. M., S. G. Perry, A. B. Fraser, and W. Mach (1979). *Remote Sensing of Canopy Structure*. Am. Meteorol. Soc. Proc., 14th Conf. Agric. For. Meteorol., and 4th Conf. Biometeorol., 1979, Minneapolis, Minnesota. American Meteorological Society, Boston, Massachusetts, pp. 184–185.

Philip, D. L. (1962). A technique for numerical solution of certain integral equations of the first kind. *J. Assoc. Comput. Mach.* **9**:84–97.

Philip, J. R. (1965a). The distribution of foliage density with foliage angle estimated from inclined point quadrat observations. *Aust. J. Bot.* **13**:357–366.

Philip, J. R. (1965b). The distribution of foliage density of single plants. *Aust. J. Bot.* **13**:411–418.

Powers, B. J., and S. A. W. Gerstl (1987). *Modelling of Atmospheric Effects on the Angular Distribution of a Backscattering Peak*, LANL Rep. LA-UR-87-572. Los Alamos National Laboratory, Los Alamos, New Mexico.

Ross, J. (1981). *The Radiation Regime and the Architecture of Plant Stands*. Junk Publ., The Netherlands.

Ross, J., and A. L. Marshak (1985). A Monte Carlo procedure for calculating the scattering of solar radiation by plant canopies. *Sov. J. Remote Sens. (Engl. Transl.)* **4**:783–801.

Ross, J., and A. L. Marshak (1988). Calculation of the canopy bidirectional reflectance using the Monte Carlo method. *Remote Sens. Environ.* **24**:213–225.

Ross, J., and T. Nilson (1967). The spatial orientation of leaves in crop stands and its determination. In *Photosynthesis of Productive Systems* (A. A. Nichiprovich, Ed.), Israel Program of Scientific Translations, Jerusalem, pp. 86–99.

Sellers, P. J. (1986). Canopy reflectance, photosynthesis and transpiration. *Int. J. Remote Sens.* **6**:1335–1372.

Suits, G. H. (1972). The calculation of the directional reflectance of a vegetative canopy. *Remote Sens. Environ.* **2**:117–125.

Szeicz, G. (1974). Solar radiation for plant growth. *J. Appl. Ecol.* **11**:617–636.

Tinney, F. W., O. S. Aamodt, and H. L. Ahlgren (1937). Preliminary report on a study of methods used in botanical analysis of pasture swards. *J. Am. Soc. Agron.* **29**:835–840.

Turitzin, S. N. (1978). Canopy structure and potential light competition in two adjacent annual plant communities. *Ecology* **59**:161–167.

Twomey, S. (1963). On the numerical solution of Fredholm integral equations of the first kind by the inversion of the linear system produced by quadrature. *J. Assoc. Comput. Mach.* **10**:97–101.

Twomey, S. (1977). *Introduction to the Mathematics of Inversion in Remote Sensing and Indirect Measurements.* Elsevier, Amsterdam.

Warren Wilson, J. (1959a). Analysis of the spatial distribution of foliage by means of two-dimensional point quadrats. *New Phytol.* **58**:92–101.

Warren Wilson, J. (1959b). Analysis of the distribution of foliage area in grassland. In *The Measurement of Grassland Productivity* (J. D. Ivins, Ed.) Academic Press, London, pp. 51–61.

Warren Wilson, J. (1960). Inclined point quadrats. *New Phytol.* **59**:1–8.

Warren Wilson, J. (1963a). Estimation of foliage denseness and foliage angle by inclined point quadrats. *Aust. J. Bot.* **11**:95–105.

Warren Wilson, J. (1963b). Errors resulting from thickness of point quadrats. *Aust. J. Bot.* **11**:178–188.

Warren Wilson, J. (1965a). Stand structure and light penetration. I. Analysis by point quadrats. *J. Appl. Ecol.* **2**:383–390.

Warren Wilson, J. (1965b). Point quadrat analysis of foliage distribution for growing single or in rows. *Aust. J. Bot.* **13**:405–409.

Warren Wilson, J. (1967). Stand structure and light penetration. III. Sunlit foliage area. *J. Appl. Ecol.* **4**:159–165.

Weiser, R. L., G. Asrar, G. P. Miller, and E. T. Kanemasu (1986). Assessing grassland biophysical characteristics from spectral measurements. *Remote Sens. Environ.* **20**:141–152.

Winkworth, R. E. (1955). The use of point quadrats for the analysis of heathland. *Aust. J. Bot.* **3**:68–81.

8

VEGETATION-CANOPY SPECTRAL REFLECTANCE AND BIOPHYSICAL PROCESSES

Piers J. Sellers

Center for Ocean-Land-Atmosphere Interactions
Department of Meteorology
University of Maryland
College Park, Maryland

I INTRODUCTION

There have been a large number of empirical studies reported in the literature that have focused on the relationships between vegetation properties and the upwelling spectral radiances observed above the land surface.

Most of this work has made use of spectral vegetation indices (SVI), which consist of various combinations of the observed upwelling radiances in the visible (0.4–0.7 μm) and near-infrared (0.7–1.1 μm) wavelength intervals. The most common SVIs in use are the normalized difference vegetation index (NDVI) and the simple ratio (SR), which are defined as follows:

$$\text{NDVI} = \frac{L_N - L_V}{L_N + L_V} \tag{1a}$$

$$\text{SR} = \frac{L_N}{L_V} \tag{1b}$$

where L_N and L_V are the upwelling radiances as observed in the NIR and visible wavelength intervals, respectively.

The NDVI and SR have been calculated from data collected with radiometers operating over the complete range of surface, airborne, and satellite platforms. The results from these studies can be broadly grouped under three headings: biophysical states, radiation-absorption characteristics, and biophysical rates, all of which we might expect to be closely interrelated. The findings of these empirical studies are now briefly discussed.

1. *Biophysical States.* Some studies have examined the relationship between spectral radiances or reflectances and the leaf-area index or standing biomass of the vegetation. Curran (1980), Tucker et al. (1981) and Asrar et al. (1984) all refer to measurements above field crops, and Badhwar et al. (1986) reported results gathered over a Boreal forest site in the United States. Generally, the data over the arable sites exhibit a non-linear relationship between the various vegetation indices and leaf-area index (LAI) and/or biomass. All the data show a steep initial response, or good sensitivity, at low values of the LAI, with saturation occurring at an LAI of between 3 and 5. The observations above the forest canopies also show an increase in the vegetation indices with increasing LAI, but the relationship is more nearly linear. It is thought that the difference between the two sets of results is due to the structure of the different types of canopy: herbaceous vegetation tends to have near-random or slightly regular spatial distributions of leaves in the canopy, whereas trees generally have ''clumped'' or spatially aggregated foliage (see Baldocchi et al., 1985). The importance of aggregated canopy architectures is discussed later in this chapter, but in the theoretical analysis presented here, we will concentrate on the near-random case. All of the data show a fair amount of scatter for high LAI values (see, for example, Asrar et al., 1984), which suggests that the spectral vegetation indices are not uniformly reliable indicators of biophysical states.

2. *Radiation-Absorption Characteristics.* A number of studies have related the fraction of intercepted photosynthetically active radiation (IPAR) or the fraction of absorbed photosynthetically active radiation (APAR) to the SVIs. Kumar and Monteith (1982), Asrar et al. (1984), Hatfield et al. (1984), Steven et al. (1983), and others have reported on the near-linear relationships observed between IPAR or APAR and the SVIs for a variety of field crops or herbaceous covers. In all cases, the relationships are near-linear and the data exhibit surprising little scatter.

3. *Biophysical Rates.* Following directly from these results, a few researchers have related biophysical rates, notably photosynthesis and evapotranspiration rates, to the SVI. Monteith (1977), Steven et al. (1983), and Kumar and Monteith (1982) described experiments where the time integral of IPAR or APAR was used to calculate the productivity of field crops. As one would expect from the previous results, these studies show a near-linear relationship between crop production, or growth indicators, and the integrated IPAR or APAR. On a different scale, Tucker et al. (1986) and Fung et al. (1986) have correlated atmospheric CO_2 variations (on the global scale) with time series of NDVI as obtained from satellite observations. Goward et al. (1985) correlated observed net primary productivity data for several North American ecosystems with the time integral of the NDVI over the growing season as observed by the Advanced Very High Resolution Radiometer (AVHRR) on board the NOAA series of meteorological satellites. Once again, a near-linear relationship was obtained in accordance with the field studies discussed above.

The weight of empirical and experimental evidence suggests that the SVIs are rather more uniform indicators of rates or processes (photosynthesis, transpiration) associated with the vegetation, rather than particular states (leaf-area index or biomass). This chapter focuses on a discussion of the biophysical and physical reasons for this circumstance and evaluates the suitability of various satellite systems for the monitoring of global biophysical processes. The discussion is mainly theoretical, as the empirical studies have been reviewed in this introduction and elsewhere in this book, and centers on the use of simple but consistent models that describe the processes of radiative transfer, photosynthesis, and transpiration in plant canopies. The results of simulations made with these models indicate that the near-linear relationship between IPAR or APAR and the SVIs is due to a fortunate coincidence in the values of the leaf scattering coefficients in the visible and near-infrared wavelength intervals.

II CANOPY RADIATIVE TRANSFER, PHOTOSYNTHESIS AND TRANSPIRATION MODELS

A wide range of models may be found in the literature that describe the processes of radiative transfer, photosynthesis, and transpiration in plant canopies. In this discussion, we are necessarily constrained to the consideration of models that are mutually consistent, i.e., as far as possible, they should share the same input parameters, and should be relatively simple, so that the important mechanisms giving rising to the relationships operating between spectral and biophysical properties of the vegetation can be easily explored. We can split our review of the available models into two sections: radiative transfer models and photosynthesis/resistance models.

II.A Radiative Transfer Models

The description of radiative transfer within plant canopies can be tackled at a number of levels of rigor and sophistication. At the most realistic and complex end of the range, numerical ray-tracing or Monte Carlo methods may be used to describe the three-dimensional radiation field within and above a vegetated surface. These models have the particular advantages of being able to cope with spatially heterogeneous canopies and can calculate the bidirectional reflectance distribution function (BRDF) of the surface, an essential requirement for remote-sensing applications. The models of Szwarcbaum and Shaviv (1976). Kimes and Kirchner (1982), Kimes (1984), Kimes et al. (1986a), and Norman and Welles (1983) are examples of these comprehensive techniques. Although realistic, these models are computationally expensive and unless used carefully are fairly opaque to the user as to the exact causality of specific results. Additionally, these models require a very large number of input parameters, the specification of which may be difficult to provide. Clearly, for the present purpose of understanding the links between radiative transfer properties and biophysical processes, something simpler, but still realistic, is required.

The two-stream approximation model of radiative transfer has been used extensively to describe radiative flux divergence in the atmosphere (see Meador and Weaver, 1980). Ross (1975), who cites earlier references, describes the application of the model to vegetation canopies and Dickinson (1983) made use of the analysis of Norman and Jarvis (1975) to calculate hemispherical albedos of densely vegetated surfaces. Sellers (1985, 1987) extended the work of Dickinson (1983) to describe radiative transfer for sparse canopies and related these results to canopy photosynthesis and transpiration properties. [This chapter is drawn largely from the papers of Sellers (1985, 1987).] As opposed to defining a large possible number of vectors for the scattered radiation flux, as in the numerical models described before, two-stream approximation models define only two directions: a hemispherically integrated upward direction and a hemispherically integrated downward direction. This simplification allows one to write a pair of ordinary differential equations describing the divergence of the two fluxes, which may then be solved analytically. The costs of this simplification are that the model cannot be used to describe horizontally heterogeneous scattering media, for example, row crops or coniferous forests, and that only hemispherically integrated albedos and transmissivities are defined. There is considerable evidence, however, that the model may yield an acceptable description of albedos for short vegetation covers, which implies that the scattering process is described reasonably well (Sellers, 1985).

The two-stream approximation model is to be used as the basis for the theoretical discussion in this chapter. The equations describing the divergence of the scattered flux within the canopy as a result of an impinging direct flux above the canopy are written as follows:

$$\bar{\mu}\frac{dI\!\uparrow}{dL} + \left[1 - (1 - \beta)\omega\right]I\!\uparrow - \omega\beta I\!\downarrow = \omega\bar{\mu}K\beta_0 e^{-KL} \qquad (2a)$$

$$\bar{\mu}\frac{dI\!\downarrow}{dL} + \left[1 - (1 - \beta)\omega\right]I\!\downarrow - \omega\beta I\!\uparrow = \omega\bar{\mu}K(1 - \beta_0)e^{-KL} \qquad (2b)$$

where $G(\mu)$ = relative projected area of leaf elements in direction $\cos^{-1}\mu$

$I\!\uparrow, I\!\downarrow$ = upward and downward diffuse radiative fluxes, normalized by the incident flux

$K = G(\mu)/\mu$, optical depth of direct beam per unit leaf area

L = cumulative leaf-area index

α = leaf-element reflectance

β, β_0 = upscatter parameters for diffuse and direct beams, respectively

μ = cosine of the zenith angle of the incident beam

$\bar{\mu} = \int_0^1 [\mu'/G(\mu')]\,d\mu'$, average inverse diffuse optical depth per unit leaf area

μ' = direction of flux

τ = leaf-element transmittance

μ' = direction of scattered flux

$\omega = \alpha + \tau$, scattering coefficient

In Eq. 2, the individual phytoelements are treated as isotropic scatterers. Strictly speaking, this is not the case, but the use of this assumption does not seem to be responsible for serious errors.

Physical processes can be attributed to each of the terms in Eq. 2; see Figure 1. Equation 2a describes the vertical profile of the upward diffuse radiative flux $I\!\uparrow$ within the canopy: it will be remembered from the previous discussion that both the upward and downward diffuse fluxes are described as being completely isotropic. The first term in Equation 2a describes the attenuation of the upward diffuse flux as a function of the other three terms. The second term defines that fraction of $I\!\uparrow$ that is rescattered in an upward direction following interaction with leaf elements. The third term refers to the fraction of the downward diffuse flux $I\!\downarrow$ that is converted into upward diffuse flux by backscattering. The last term, on the right-hand side of Equation 2a, refers to the contribution to the upward diffuse flux by the scattering of direct incident flux penetrating

(a) $[1-(1-\beta)\omega]I\uparrow$

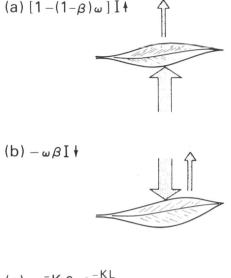

(b) $-\omega\beta I\downarrow$

(c) $\omega\bar{\mu}K\beta_o e^{-KL}$

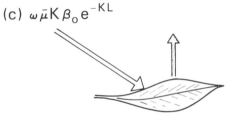

Figure 1 Radiative transfer processes occurring within a vegetation canopy, as defined by Eq. 2a, that describe the vertical divergence of the upward diffuse flux. The relevant term from Eq. 2a appears next to each diagram. (*a*) Rescattering of an upward diffuse flux into an upward direction. (*b*) Scattering of a downward diffuse flux into an upward direction. (*c*) Scattering of a downward direct beam radiation flux into an upward direction.

to the specified depth L in the canopy. Corresponding descriptions may be assigned to the four terms in Equation 2b, which describes the profile of the downward diffuse flux.

In Eq. 2, β and β_0 are the upscatter parameters for diffuse and direct beam radiation, respectively. The value of β may be inferred from the analysis of Norman and Jarvis (1975):

$$\omega\beta = \tfrac{1}{2}\left[\alpha + \tau + (\alpha - \tau)\cos^2\bar{\theta}\right] \tag{3}$$

where $\bar{\theta}$ is the mean leaf-inclination angle.

In Eq. 3, the leaf is treated as a plane with isotropic forward and backscattering properties. When the leaf is horizontal ($\cos\theta = 1$), downward fluxes will be reflected and transmitted only into the upward and downward directions, respectively. As the leaf-angle inclination increases, more of the downward diffuse flux may be reflected into the lower hemisphere and transmitted into the upper hemisphere.

Equation 2 may be solved with the conditions $\omega \to 0$ and $L \to \infty$ (single-scatter approximation and semiinfinite canopy) to estimate the canopy single-scattering albedo, which is now equal to the upward diffuse flux at the canopy top ($L = 0$). From Eq. 2, this gives

$$\beta_0 = \frac{1 + \bar{\mu}K}{\bar{\omega}\mu K} a_s(\mu) \tag{4}$$

where $a_s(\mu)$, the single-scattering albedo, is given by

$$a_s(\mu) = \omega \int_0^1 \frac{\mu' \Gamma(\mu, \mu')}{[\mu G(\mu') + \mu' G(\mu)]} d\mu' \qquad (5)$$

where $\Gamma(\mu, \mu') = G(\mu) G(\mu') P(\mu, \mu')$, and $P(\mu, \mu')$ is the scattering phase function.

If we assume that the leaves act as isotropic scatterers, we find that $P(\mu, \mu')$ is equal to $2G(\mu')^{-1}$. When this is inserted into Eq. 5, $a_s(\mu)$ may be easily obtained for a number of leaf-angle distributions; the solutions for some common cases are listed in Sellers (1985). We see from Eq. 5 that the parameters β and β_0 are functions of the leaf optical properties, τ and α, and the leaf-angle distribution function, $O(\xi, \theta)$, where ξ is the leaf azimuth angle, and θ is the angle of the leaf normal to the local vertical.

Equation 2 may be solved quite simply for the case of an isotropically reflecting soil underlying the vegetation canopy. The appropriate boundary conditions are then

$$I\!\downarrow = 0, \quad L = 0 \qquad (6a)$$

$$I\!\uparrow = \rho_s (I\!\downarrow + e^{-KL_T}), \quad L = L_T \qquad (6b)$$

where ρ_s is the soil reflectance, and L_T is the total leaf area index.

The first boundary condition states that all of the incoming flux above the canopy is made up of direct beam radiation. The second boundary condition states that the downward radiation at the soil surface, $I\!\downarrow + e^{-KL_T}$, is reflected isotropically to generate the upward diffuse flux beneath the canopy.

The solution of Eq. 2 using 6 yields

$$I\!\uparrow = \frac{h_1 e^{-KL}}{\sigma} + h_2 e^{-hL} + h_3 e^{hL} \qquad (7a)$$

$$I\!\downarrow = \frac{h_4 e^{-KL}}{\sigma} + h_5 e^{-hL} + h_6 e^{hL} \qquad (7b)$$

The values of the constants σ, h, and h_1 to h_6 are obtained from manipulation of Eq. 2 and are given in Table 1.

The canopy hemispherical reflectance $a(\mu)$ for this case is given by the upward diffuse flux leaving the top of the canopy, i.e. $I\!\uparrow$ when $L = 0$.

$$a(\mu) = I\!\uparrow(0) = \frac{h_1}{\sigma} + h_2 + h_3 \qquad (8)$$

where $a(\mu)$ is the surface reflectance for an incident beam of direction μ.

Suitable boundary conditions may be used in place of Eq. 6 and the direct radiation terms on the right-hand sides of Equation 2 dropped from the basic equation set to solve for incident diffuse radiative fluxes; see Table 1.

Of particular importance to the following discussion are the extinction terms in Eq. 7, K and h. K is defined as the mean projected area of the leaf elements in the direction of the incident flux divided by the cosine of that direction, i.e., $G(\mu)/\mu$. The diffuse

TABLE 1 Solution to Two-Stream Approximation Equations as Described in the Text

Direct Beam Radiation

$$I\uparrow = \frac{h_1 \exp(-KL)}{\sigma} + h_2 \exp(-hL) + h_3 \exp(hL)$$

$$I\downarrow = \frac{h_4 \exp(-KL)}{\sigma} + h_5 \exp(-hL) + h_6 \exp(hL)$$

$$b = 1 - (1 - \beta)\omega$$

$$c = \omega\beta$$

$$d = \omega\bar{\mu}K\beta_0$$

$$f = \omega\bar{\mu}K(1 - \beta_0)$$

$$h = \frac{(b^2 - c^2)^{1/2}}{\bar{\mu}}$$

$$\sigma = (\bar{\mu}K)^2 + c^2 - b^2$$

$$h_1 = -dp_4 - cf$$

$$h_2 = \frac{1}{D_1}\left\{\left(d - \frac{h_1}{\sigma}p_3\right)(u_1 - \bar{\mu}h)\frac{1}{S_1} - p_2\left[d - c - \frac{h_1}{\sigma}(u_1 + \bar{\mu}K)\right]S_2\right\}$$

$$h_3 = -\frac{1}{D_1}\left\{\left(d - \frac{h_1}{\sigma}p_3\right)(u_1 + \bar{\mu}h)S_1 - p_1\left[d - c - \frac{h_1}{\sigma}(u_1 + \bar{\mu}K)\right]S_2\right\}$$

$$h_4 = -fp_3 - cd$$

$$h_5 = -\frac{1}{D_2}\left\{\frac{h_4}{\sigma}(u_2 + \bar{\mu}h)\frac{1}{S_1} + \left[u_3 - \frac{h_4}{\sigma}(u_2 - \bar{\mu}K)\right]S_2\right\}$$

$$h_6 = \frac{1}{D_2}\left\{\frac{h_4}{\sigma}(u_2 - \bar{\mu}h)S_1 + \left[u_3 - \frac{h_4}{\sigma}(u_2 - \bar{\mu}K)\right]S_2\right\}$$

$$u_1 = b - c/\rho_s$$

$$u_2 = b - c\rho_s$$

$$u_3 = f + c\rho_s$$

$$S_1 = \exp(-hL_T)$$

$$S_2 = \exp(-KL_T)$$

$$p_1 = b + \bar{\mu}h$$

$$p_2 = b - \bar{\mu}h$$

$$p_3 = b + \bar{\mu}K$$

$$p_4 = b - \bar{\mu}K$$

$$D_1 = p_1(u_1 - \bar{\mu}h)\frac{1}{S_1} - p_2(u_1 + \bar{\mu}h)S_1$$

$$D_2 = (u_2 + \bar{\mu}h)\frac{1}{S_1} - (u_2 - \bar{\mu}h)S_1$$

Diffuse Radiation

$$I\uparrow = h_7 \exp(-hL) + h_8 \exp(hL)$$

$$I\downarrow = h_9 \exp(-hL) + h_{10} \exp(hL)$$

$$h_7 = \frac{c}{D_1}(u_1 - \bar{\mu}h)\frac{1}{S_1}$$

$$h_8 = \frac{-c}{D_1}(u_1 + \bar{\mu}h)S_1$$

$$h_9 = \frac{1}{D_2}(u_2 + \bar{\mu}h)\frac{1}{S_1}$$

$$h_{10} = \frac{-1}{D_2}(u_2 - \bar{\mu}h)S_1$$

attenuation term h is obtained from the solution of Eq. 2, so that

$$h = \frac{1}{\overline{\mu}} \left[(1 - \omega + 2\beta\omega) (1 - \omega) \right]^{1/2} \tag{9}$$

The average inverse diffuse optical depth per unit leaf area, $\overline{\mu}$, is equal to unity for canopies with horizontal or spherically distributed leaves, and is close to unity for most other leaf-angle distributions. In the case of horizontal leaves, or when $\tau = \alpha$, and when $\overline{\mu} = 1$, Eq. 9 simplifies to

$$h = (1 - \omega)^{1/2} \tag{10}$$

Sellers (1985) and Kimes et al. (1986b) have presented results that compare the predictions of the two-stream model with field data. For illustrative purposes, Figure 2 shows the variation of surface reflectance in the visible, a_V, and near-infrared, a_N, wavelength intervals as a function of leaf area index and soil reflectance. In these and sub-

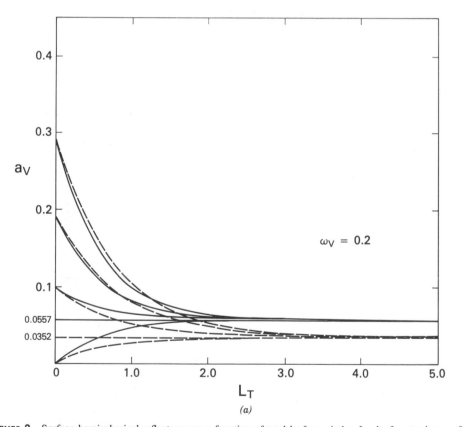

Figure 2 Surface hemispherical reflectance as a function of total leaf area index L_T, leaf scattering coefficient ω, and soil reflectance ρ_s. In all cases, solid lines refer to canopies with horizontal leaves and dashed lines refer to canopies with spherically distributed leaves. The start point of each line on the vertical (reflectance) axis indicates the background soil reflectance. (a) Reflectances in the visible region, $\omega_V = 0.2$, (b) reflectances in the near-infrared region, $\omega_N = 0.8$, and (c) reflectances in the near-infrared region, $\omega_N = 0.95$. The horizontal lines on each figure refer to the asymptotic (semiinfinite) canopy reflectance. [Reprinted by permission of the publisher from Sellers (1987). Copyright 1987 by Elsevier Science Publishing Co., Inc.]

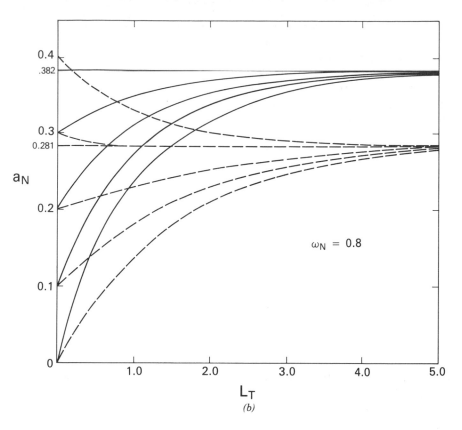

$\omega_N = 0.8$

(b)

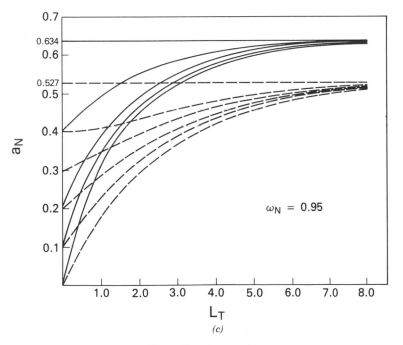

$\omega_N = 0.95$

(c)

Figure 2 *(Continued)*

sequent figures, the worked examples use the vegetation properties listed in Table 2. Two leaf-angle distributions are considered: horizontal and spherical. The latter case, which is a fair approximation for much natural vegetation, is equivalent to the leaves having an angular distribution function like the surface facets of a sphere; it is also referred to as a random distribution. The leaf-scattering coefficient in the visible region, ω_V, is taken as 0.2. Two leaf-scattering coefficients are considered in the near-infrared region: $\omega_N = 0.80$ (representative of green leaves in the 1.6–1.8 μm interval), and $\omega_N = 0.95$ (representative of green leaves in the 0.7–1.1 μm interval). The reason for considering two values of ω_N will become apparent in the discussion. In all cases, it is assumed that $\tau = \alpha$, so that Eq. 10 holds true. It is also assumed, rather less realistically, that the soil reflectance ρ_s is the same in all spectral regions. (Normally, ρ_s increases gradually with wavelength, a complication that has been neglected to keep the analysis simple.) Additionally, the incoming radiative fluxes are assumed to consist of vertical beams, close to conditions representative of a clear day with a near-overhead sun, which corresponds to good satellite remote-sensing weather.

Under normal field conditions, when the soil is relatively dark ($\rho_s < 0.1$), we see that the visible reflectance a_V does not vary much with increasing leaf area index: this is because the rescattering of light in this wavelength interval is so small ($\omega_V = 0.2$); see Figure 2(*a*). By contrast, the near-infrared surface reflectance a_N shows a much higher sensitivity to leaf area index due to the relatively high scattering coefficient of green leaves ($\omega_N = 0.8$ or 0.95) in this spectral region; see Figure 2(*b*).

Because of this, we expect the reflectance indicators SR and NDVI to be mainly functional on the near-infrared surface reflectance a_N. Figure 3 shows the variation of SR and NDVI with total leaf-area index L_T. Reference to Figure 2 and a comparison of the absolute values of a_N and a_V will confirm that for low values of ρ_s, SR and NDVI are dominated by the a_N term.

The proportion of PAR absorbed by the vegetation canopy, APAR, may be calculated by

$$APAR = 1 - a_\pi - (1 - \rho_s)\left[e^{-K_\pi L_T} + I\!\downarrow(L_T)\right] \qquad (11)$$

where the subscript π refers to the PAR wavelength interval, 0.4–0.7 μm.

TABLE 2   Optical and Physiological Properties Used for Model Canopy

Leaf-Angle Distribution	Horizontal, Spherical
Visible PAR scattering coefficients, ω_V, ω_π	0.2
Near-infrared scattering coefficient, ω_N	0.8, 0.95
Leaf photosynthesis constants	
$\quad a_1$ [mg CO_2 dm^{-1} h^{-1}]	82.6
$\quad b_1$ [W m^{-2}]	278.4
Leaf stomatal resistance constants	
$\quad a_2$ [J m^{-3}]	8750.0
$\quad b_2$ [W m^{-2}]	6.0
$\quad c_2$ [s m^{-1}]	55.0

Sources: The values of ω_V and ω_N may be compared with equivalent values of 0.175 and 0.825, respectively, reported by Dickinson (1983) for the wavelength intervals 0.4–0.7 μm (visible) and 0.7–1.1 μm (near-infrared), respectively, or with the spectra shown in Fig. 16. Values of a_1 and b_1 were obtained from curve fits to data of Hesketh and Baker (1967) and a_2, b_2, and c_2 from data of Turner (1974).

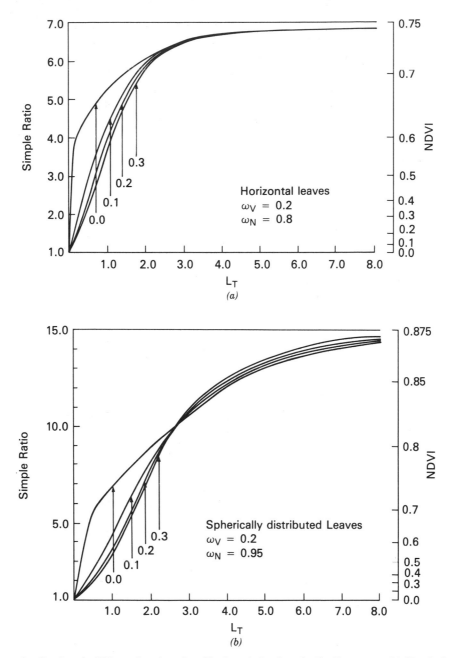

Figure 3 Simple ratio (SR) as a function of total leaf area index L_T and soil reflectance ρ_s. (*a*) SR calculated for a canopy of horizontal leaves: $\omega_V = 0.2$ and $\omega_N = 0.8$. (*b*) SR calculated for a canopy of spherically distributed leaves: $\omega_V = 0.2$ and $\omega_N = 0.95$. Values of background soil reflectance are marked against each line and are assumed to be constant over the spectrum. The equivalent normalized difference vegetation index (NDVI) is shown on the right-hand vertical axis of each figure. [Reprinted by permission of the publisher from Sellers (1987). Copyright 1987 by Elsevier Science Publishing Co., Inc.]

The first functional term in Eq. 11, a_π, refers to the "loss" of PAR that is reflected away from the surface, where a_π is given by Eq. 8. Clearly, derivation of a value for this PAR surface reflectance, a_π requires the specification of a value of ω_π, the mean leaf scattering coefficient for the effectively used PAR. The exact calculation of ω_π is set out later in this chapter; from data, ω_π is calculated to be 0.207 for a maize leaf under clear-sky conditions. Here it is taken to be equal to the specified value of $\omega_V (= 0.2)$ for now. For the bulk of the analysis presented here, therefore, the subscripts V and π are interchangeable. The second term on the left-hand side of Eq. 11 refers to the proportion of PAR reflected by the surface (vegetation and soil), and the third term refers to the proportion of PAR absorbed by the soil. Figure 4 shows the variation of APAR with total leaf-area index L_T.

II.B   Canopy Photosynthesis and Resistance Models

Canopy photosynthesis and resistance are described by extensions of simple leaf photosynthesis and resistance models, as described in Sellers (1985). These basic models are integrated over the depth of the canopy to give the area-averaged canopy photosynthesis and transpiration rates. The leaf photosynthesis model is based on that of Charles-Edwards and Ludwig (1974).

$$P = \left(\frac{a_1 \, F_\pi \cdot n}{b_1 + F_\pi \cdot n} - R_d \right) f(\psi_l) \cdot f(T_c) \cdot f(\delta e) \qquad (12)$$

where a_1, b_1 = constants [mg CO_2 dm^{-1} h^{-1}, W m^{-2}]

F_π = flux density vector of incident visible radiation (PAR) [W m^{-2}]

n = vector of leaf normal

P = leaf photosynthetic rate [mg CO_2 dm^{-1} h^{-1}]

R_d = dark respiration rate [mg CO_2 dm^{-1} h^{-1}]

In Eq. 12, the term $F_\pi \cdot n$ may be interpreted as the PAR intensity on the leaf surface. The leaf stomatal resistance model is a modification of the expression of Jarvis (1976).

$$r_s = \left(\frac{a_2}{b_2 + F_\pi \cdot n} + c_2 \right) \cdot \left[f(\psi_l) \cdot f(T_c) \cdot f(e_a) \right]^{-1} \qquad (13)$$

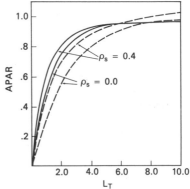

Figure 4   Variation of the fraction of absorbed photosynthetically active radiation (APAR) with leaf area index. It is assumed that $\omega_\pi = \omega_V = 0.2$ is representative for the PAR region. Horizontal and spherical leaf angle distributions are indicated by solid and dashed lines, respectively, and background soil reflectances are marked against the curves. [Reprinted by permission of the publisher from Sellers (1987). Copyright 1987 by Elsevier Science Publishing Co., Inc.]

where $\qquad a_2, b_2, c_2 =$ constants determined from data or from a_1, b_1, and R_d [J m^{-3}, W m^{-2}, s m^{-1}]

$f(\psi_l), f(T_c), f(\delta e) =$ adjustment factors for the effects of leaf water potential ψ_l, temperature T_c, and vapor pressure deficit δe.

$r_s =$ leaf stomatal resistance [s m^{-1}]

The dark respiration term, R_d, in Eq. 12 has been dropped from the ensuing analysis partly because there is some controversy about how to specify its variation with increasing leaf-area index. R_d is assumed to be a fairly small proportion of P for reasonable values of leaf-area index (< 6).

The adjustment factors, $f(\psi_l)$, $f(T_c)$, and $f(\delta e)$, are equal to unity under optimal conditions and zero when photosynthesis and transpiration are reduced by adverse environmental conditions, i.e., limiting soil moisture, extremes of temperature, and an excessive vapor pressure deficit, respectively. Jarvis (1976) presents results showing the variation of these factors with environmental conditions. It is assumed in all the following discussion that light is the only limiting factor, so that $f(\psi_l) = f(T_c) = f(\delta e) = 1$.

The quantities a_1, b_1, a_2, b_2, and c_2 are species-dependent. If one accepts the arguments favoring a constant ratio between P and the transpiration rate (Farquhar and von Caemmerer, 1982; Farquhar and Sharkey, 1982), it is relatively easy to derive values of a_2, b_2, and c_2 from a consideration of the photosynthetic equation, Eq. 12, the constant of water-use efficiency and an assumed climatic mean evaporative demand. For the present, however, the values of the constants in Eqs. 12 and 13 are determined from fitting to data for maize leaves, as given in Table 2.

Equations 12 and 13 are taken as good for processes associated with a single leaf exposed to a normal incident PAR flux $F_\pi \cdot n$. In order to obtain the area-averaged canopy photosynthetic rate P_c and resistance r_c, Eqs. 12 and 13 must be integrated over the depth of the canopy, while taking into account the fact that the leaves may have a range of orientations relative to the incoming flux.

To start with, we assume that PAR is attenuated on its passage down through the canopy according to the Goudriaan (1977) semiempirical expression

$$F_L = F_0 e^{-kL} \qquad (14)$$

$$k = \frac{G(\mu)}{\mu} (1 - \omega_\pi)^{1/2}$$

where $\qquad F_L, F_0 =$ PAR fluxes below a leaf-area index L and at the top of the canopy, respectively [W m^{-2}]

$\omega_\pi =$ effective scattering coefficient for PAR $\approx \omega_V$

Then, using the leaf-angle distribution function $O(\xi, \theta)$, we may combine Eqs. 12 and 13 with 14 to give the canopy photosynthetic rate and resistance.

$$P_c = \int_0^{L_T} \int_0^{\pi/2} \int_0^{2\pi} P(F_\pi, \xi, \theta) O(\xi, \theta) \sin \theta \, d\xi \, d\theta \, dL \qquad (15)$$

$$\frac{1}{r_c} = \int_0^{L_T} \int_0^{\pi/2} \int_0^{2\pi} \frac{O(\xi, \theta)}{r_s(F_\pi, \xi, \theta)} \sin \theta \, d\xi \, d\theta \, dL \qquad (16)$$

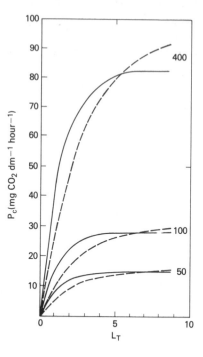

Figure 5   Canopy photosynthesis P_c as a function of leaf area index L_T, PAR flux density (marked against the curves in W m^{-2}), and leaf angle distributions (solid lines for horizontal leaves and dashed lines for spherical leaf angle distributions). [Reprinted by permission of the publisher from Sellers (1987). Copyright 1987 by Elsevier Science Publishing Co., Inc.]

[The leaf-angle distribution function is discussed in some detail in Ross (1975); basically, it describes the range of leaf azimuth, ξ, and inclination angles, θ, for a given canopy.]

Solutions to Eqs. 15 and 16 for a range of canopy types are reported in Sellers (1985). Figure 5 shows how P_c and $1/r_c$ vary with leaf-area index and incident PAR flux density for the leaf properties listed in Table 1. A point of contention in the solution of Eqs. 15 and 16 is the use of the semiempirical description of PAR attenuation, Eq. 14, in place of the more exact two-stream description of the downward flux of PAR, which may be obtained from Eq. 7(b). However, for small values of ω, the semiempirical formulation and the exact two-stream solution give very similar profiles of PAR flux density.

II.C   Theoretical Discussion

By using the expression for canopy reflectance, given by Eq. 5, and those for canopy photosynthesis and resistance, Eqs. 12 and 13, the simple ratio and normalized difference vegetation indices, SR and NDVI, respectively, can be compared to the equivalent estimates of P_c and $1/r_c$. For the sake of simplicity, we can redefine SR and NDVI in terms of surface reflectances, so that

$$\text{SR} = \frac{a_N}{a_V} \tag{17a}$$

$$\text{NDVI} = \frac{a_N - a_V}{a_N + a_V} \tag{17b}$$

where a_N and a_V are the hemispherical canopy reflectances for the near-infrared and visible wavelength intervals, respectively.

The relationship between indices based on reflectances and radiances is discussed in detail in the last section of this chapter.

For the case of our model canopy with horizontal or "spherical" leaves, where $\omega_V = 0.2$ and $\omega_N = 0.8$ or 0.95, as described in Table 2, SR and NDVI both exhibit a nonlinear increase with leaf-area index, as shown in Figure 3, with the rate of increase partly dependent upon the soil optical properties. It may be seen that the "signal" represented by the SR or NDVI becomes saturated, for all practical purposes, somewhere around a leaf area index of 4. Figure 2 may be compared to Figures 5 and 6, where similar nonlinear behaviors for APAR, P_c, and $1/r_c$ versus L_T may also be noted. Figures 7 to 9 compare the SR and NDVI with APAR, P_c, and $1/r_c$ directly, showing that for these cases at least, the value of SR is near-linearly related to APAR, P_c and, $1/r_c$. Figures 7 to 9 and other similar results indicate strongly that

$$SR \propto APAR \qquad (18a)$$

$$SR \propto P_c \qquad (18b)$$

$$SR \propto 1/r_c \qquad (18c)$$

Obviously, Eq. 18 is a very useful result from a practical viewpoint. The phenomena encapsulated by Eq. 18 do not have immediately obvious explanations: most of the rest of this chapter is devoted to discussing the biophysical processes that give rise to these relationships.

Given the relationships in Eq. 18, we should look through our equation set for a common biophysical property linking the quantities SR, P_c and $1/r_c$: clearly, it can only

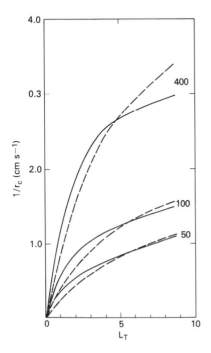

Figure 6 Inverse canopy resistance $1/r_c$ as a function of leaf area index L_T, PAR flux density (marked against each curve in W m^{-2}), and leaf angle distribution (solid lines for horizontal leaves and dashed lines for spherical leaf angle distributions). [Reprinted by permission of the publisher from Sellers (1987). Copyright 1987 by Elsevier Science Publishing Co., Inc.]

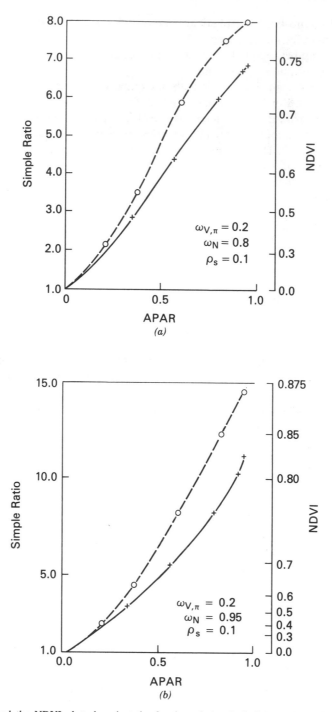

Figure 7 SR and the NDVI plotted against the fraction of absorbed photosynthetically active radiation (APAR). Background soil reflectance is 0.1. Solid lines denote horizontal leaves and dashed lines denote spherical leaf angle distributions. Circles ($\bigcirc$) and crosses ($+$) refer to increasing values of the total leaf area index L_T: O (origin), 0.5, 1.0, 2.0, 4.0, and 8.0. (a) SR calculated using $\omega_N = 0.8$. (b) SR calculated using $\omega_N = 0.95$. [Reprinted by permission of the publisher from Sellers (1987). Copyright 1987 by Elsevier Science Publishing Co., Inc.]

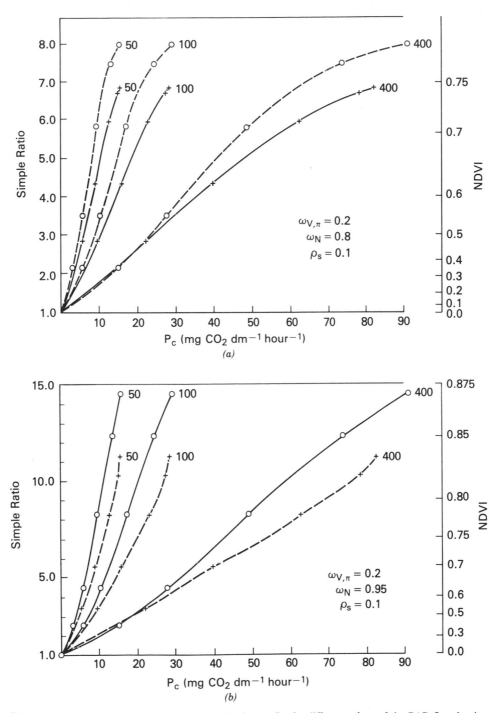

Figure 8 SR plotted against canopy photosynthetic rate P_c, for different values of the PAR flux density,. marked against each curve in W m^{-2}. Background soil reflectance is 0.1. Solid lines denote horizontal leaves and dashed lines denote a spherical leaf angle distributions. Circles ($\bigcirc$) and crosses ($+$) refer to increasing values of the total leaf area index L_T: O (origin), 0.5, 1.0, 2.0, 4.0, and 8.0. (a) SR calculated using $\omega_N = 0.8$. (b) SR calculated using $\omega_V = 0.95$. [Reprinted by permission of the publisher from Sellers (1987). Copyright 1987 by Elsevier Science Publishing Co., Inc.]

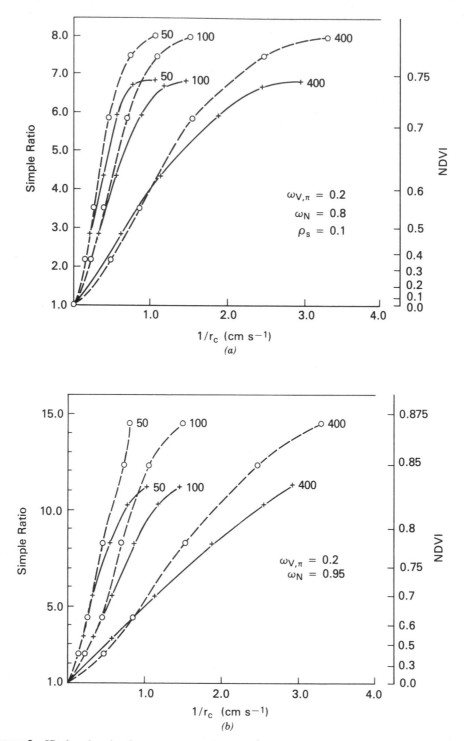

Figure 9 SR plotted against inverse canopy resistance $1/r_c$, for different values of the PAR flux density, marked against each curve in W m^{-2}. Other symbols and conditions are the same as for Figure 8. (a) SR calculated using $\omega_N = 0.8$. (b) SR calculated using $\omega_N = 0.95$. [Reprinted by permission of the publisher from Sellers (1987). Copyright 1987 by Elsevier Science Publishing Co., Inc.]

be the total leaf area index L_T. It follows from this and Eq. 18 that

$$\frac{\partial(\text{SR})}{\partial L_T} \propto \frac{\partial(\text{APAR})}{\partial L_T} \tag{19a}$$

$$\frac{\partial(\text{SR})}{\partial L_T} \propto \frac{\partial P_c}{\partial L_T} \tag{19b}$$

$$\frac{\partial(\text{SR})}{\partial L_T} \propto \frac{\partial(1/r_c)}{\partial L_T} \tag{19c}$$

The reason why Eq. 19 should hold is now to be discussed. To maintain organization in the discussion, we shall explore the dependence of the previous derivatives in two following sections: the first dealing with the left-hand side of Eq. 19, $\partial(\text{SR})/\partial L_T$, and the second with the right-hand side, $\partial(\text{APAR})/\partial L_T$, $\partial P_c/\partial L_T$, and $\partial(1/r_c)/\partial L_T$.

II.D The Dependence of the Simple Ratio Upon Total Leaf Area Index

In this section, we discuss the left-hand side of Eq. 19. Equation 17a for SR may be differentiated with respect to L_T to give

$$\frac{\partial(\text{SR})}{\partial L_T} = \frac{1}{a_V^2}\left(a_V \frac{\partial a_N}{\partial L_T} - a_N \frac{\partial a_V}{\partial L_T}\right) \tag{20}$$

For now, we are limiting the discussion to the consideration of canopies with horizontal and spherical leaf angle distributions. In this section, we shall illustrate the expansion of the general case, Eq. 20, with reference to the horizontal leaf angle distribution; an equivalent expansion for the spectral leaf angle distribution is given in the Appendix of Sellers (1987).

Canopy reflectance is given by Eq. 8. For a canopy of horizontal leaves, this reduces to

$$a = \frac{\omega}{2}\left(\frac{1-A}{P_1 - P_2 A}\right) \tag{21}$$

where $A = \left(\dfrac{P_1 - \gamma}{P_2 - \gamma}\right)e^{-2hL_T}$

a = canopy reflectance

ω = leaf scattering coefficient

$P_1 = 1 - \dfrac{\omega}{2} + h$

$P_2 = 1 - \dfrac{\omega}{2} - h$

$h = (1 - \omega)^{1/2}$

$\gamma = \dfrac{\omega}{2\rho_s}$

It is clear from Eq. 21 that the canopy reflectance depends upon the leaf scattering coefficient divided by 2, adjusted by a term that includes a dependence on the soil scattering properties and the negative exponent of *twice* the total diffuse optical-path length for a given value of L_T. This last term, e^{-2hL_T}, is of fundamental importance to the rest of the discussion.

The derivative of Eq. 21 with respect to L_T, which is required in Eq. 20, is

$$\frac{\partial a}{\partial L_T} = \omega h A \frac{P_1 - P_2}{(P_1 - P_2 A)^2} \tag{22}$$

Some experimentation with typical values of leaf and soil scattering properties reveals that the near-infrared terms in Eq. 20 are the dominant ones in controlling the variation of SR with L_T. This point may be illustrated by considering the conditions that make $\partial a_N / \partial L_T$ or $\partial a_V / \partial L_T$ approach zero. This occurs when $\gamma = P_1$ (see the definition of A in Eq. 21), which leads to

$$\frac{\omega}{2\rho_s} = 1 - \frac{\omega}{2} + h$$

or

$$\rho_s' = \frac{\omega}{2 - \omega + 2(1 - \omega)^{1/2}} \tag{23}$$

where ρ_s' is the value of soil reflectance that makes $\partial a / \partial L_T = 0$.

The value of ρ_s' given by Eq. 23 is equal to the reflectance of an infinitely thick canopy, so that when $\rho_s = \rho_s'$, the surface reflectance will show no change whether the soil is bare or is covered by the densest of canopies. This condition can be seen most clearly by inspecting Figure 2.

It is interesting to note that for the visible wavelength interval, $\omega_V = 0.2$ and $\rho_s' = 0.0557$ for horizontal leaves, whereas corresponding values for the near-infrared wavelength interval yield $\omega_N = 0.8$ and $\rho_s' = 0.38$ or $\omega_N = 0.95$ and $\rho_s' = 0.63$. Most soil reflectances lie within the range of 0.05 to 0.15 in the visible and 0.1 to 0.2 in the near-infrared. We can see then that normally $\partial a_V / \partial L_T$ is small, a_V itself is small, and that $\partial(\text{SR}) / \partial L_T$ will depend primarily on $\partial a_N / \partial L_T$.

From this discussion, we conclude that the first term in Eq. 20, which contains $\partial a_N / \partial L_T$, tends to dominate the SR response to increases in L_T. The "interference" produced by the second term being nonzero will increase as the soil reflectance increases.

It has been stated that usually $\rho_s \simeq \rho_s'$ in the visible region, so that Eq. 20 degenerates toward the approximation

$$\frac{\partial(\text{SR})}{\partial L_T} \rightarrow \frac{1}{a_V} \cdot \frac{\partial a_N}{\partial L_T} = \left(\frac{2P_1}{\omega}\right)_V \cdot \frac{\partial a_N}{\partial L_T}, \quad (\rho_s \rightarrow \rho_s')_V \tag{24}$$

The important functional term in Eq. 24 is the $e^{-2h_N L_T}$ embedded in $\partial a_N / \partial L_T$.

From this discussion, we conclude the following:

1. The simple ratio and, by extension, the normalized difference vegetation indices are primarily functions of the near-infrared reflectance under normal field conditions, that is, when $(\rho_s \to \rho_s')_V$.

2. The derivative of the simple ratio with respect to L_T is roughly proportional to $e^{-2h_N L_T}$ when $\rho_{s_N} \to 0$. The physical reason for this is that the reflectance is a function of twice the diffuse optical path length of the scattering medium: once for the attenuation of the radiation going into the medium (interception) and once for its attenuation going out of the medium (backscatter or reflectance). In the visible region, this scattering effect has little effect on the reflected signal as dark soil and a vegetation canopy have similar reflectances. In the near-infrared region, the reflectance is scaled by $e^{-2h_N L_T}$ as the canopy and soil reflectance are normally dissimilar.

We may summarize these results by

$$\frac{\partial (\text{SR})}{\partial L_T} \propto \frac{\partial a_N}{\partial L_T} \propto e^{-2h_N L_T}, \quad (\rho_s \to \rho_s')_V, \ (\rho_s \to 0)_N \quad (25)$$

where h_N = total diffuse extinction coefficient for near-infrared radiation

$= (1 - \omega_N)^{1/2}$, assuming $\bar{\mu} = 1$

$= 0.4472$ when $\omega_N = 0.8$

$= 0.2236$ when $\omega_N = 0.95$

II.E The Dependence of APAR, Canopy Photosynthetic Rate, and Canopy Stomatal Resistance Upon Total Leaf Area Index

We now turn to a discussion of the terms on the right-hand side of Eq. 19.

The derivative of APAR with respect to total leaf area index is obtained from Eq. 11 to give

$$\frac{\partial (\text{APAR})}{\partial L_T} = -\frac{\partial a_\pi}{\partial L_T} - (1 - \rho_{s_\pi}) \frac{\partial}{\partial L_T} \left[e^{-K_\pi L_T} + I\!\!\downarrow(L_T) \right] \quad (26)$$

If we use the semiempirical expression for PAR flux attenuation, Eq. 11, in place of the two-stream description, Eq. 26, we obtain

$$\frac{\partial (\text{APAR})}{\partial L_T} \propto e^{-k L_T}, \ (\rho_s \to \rho_s' \to 0)_{V, \pi} \quad (27)$$

Figure 10 shows the APAR derivative, calculated from the complete expression of Eq. 23, plotted against $e^{-k L_T}$ for the two leaf angle distributions and two values of the soil reflectance. The relationships in Eqs. 26 and 27 are clearly best satisfied when $(\rho_s \to \rho_s' \to 0)_{V, \pi}$. However, even fairly high values of ρ_{sv} do not degrade these relationships too severely. This implication is that Eq. 27 is a satisfactory description of the variation of APAR with L_T.

For the simple case of horizontal leaves, the derivatives of P_c and $1/r_c$ with respect

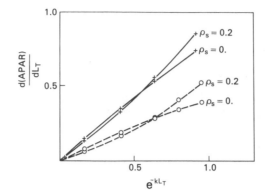

Figure 10 Derivative of APAR with respect to total leaf area index L_T, plotted against the PAR penetration function e^{-kL_T}, where k = PAR attenuation coefficient, $[G(\mu)/\mu](1 - \omega_\pi)^{1/2}$. The relationships are shown to be near-linear for low soil reflectances ($\rho_s = 0$) and become increasingly nonlinear for increasing soil reflectance ($\rho_s = 0.2$). Solid lines denote horizontal leaves and dashed lines denote spherical leaf angle distributions. Circles ($\circ$) and crosses ($+$) refer to values of L_T, increasing from *right to left;* 0.1, 0.5, 1.0, 2.0, 4.0, and 8.0. [Reprinted by permission of the publisher from Sellers (1987). Copyright 1987 by Elsevier Science Publishing Co., Inc.]

to L_T are simply

$$\frac{\partial P_c}{\partial L_T} = \frac{a_1 F_0 e^{-kL_T}}{b_1 + F_0 e^{-kL_T}} \tag{28}$$

$$\frac{\partial(1/r_c)}{\partial L_T} = \frac{b_2 + F_0 e^{-kL_T}}{a_2 + c_2(b_2 + F_0 e^{-kL_T})} \tag{29}$$

Figures 11 and 12 show the variation of the derivative terms in Eqs. 28 and 29 with e^{-kL_T}. The relationship is almost linear as PAR $\rightarrow 0$, which reduces the exponential terms in the denominators of Eqs. 28 and 29 to insignificance. As the PAR flux density increases ($F_0 = 400$ W m^{-2} represents the near-maximum value one could expect to observe in nature), the relationship becomes increasingly nonlinear as the plants' photosynthetic capacity approaches saturation. In practice, however, most indigenous species in a region reach PAR saturation at around the maximum local PAR flux density (Farquhar and von Caemmerer, 1982).

Figure 11 Derivative of canopy photosynthetic rate P_c with respect to total leaf area index L_T against PAR penetration function e^{-kL_T}. PAR flux densities are marked against each curve in W m^{-2}. Symbols and conditions are the same as for Fig. 10. [Reprinted by permission of the publisher from Sellers (1987). Copyright 1987 by Elsevier Science Publishing Co., Inc.]

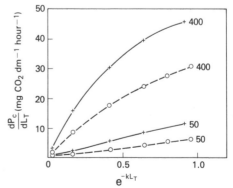

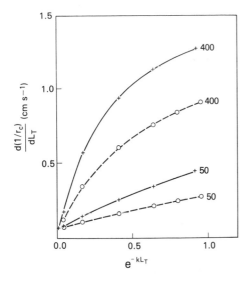

Figure 12 Derivative of inverse stomatal resistance $1/r_c$ with respect to total leaf area index L_T against PAR penetration function e^{-kL_T}. PAR flux densities are marked against each curve in W m^{-2}. Symbols and conditions are the same as for Figs. 10 and 11. [Reprinted by permission of the publisher from Sellers (1987). Copyright 1987 by Elsevier Science Publishing Co., Inc.]

The relationships represented by Eqs. 28 and 29 and Figures 10–12 are entirely self-consistent. At the risk of laboring the obvious, we should expect APAR, P_c, and $1/r_c$ to be closely linked as the plants' demand for CO_2 and the associated rate of water vapor loss are functional on the supply side of the electron-transfer process of photosynthesis, which is directly related to chlorophyll density and hence APAR.

We may summarize all the above by

$$\frac{\partial(\text{APAR})}{\partial L_T}, \frac{\partial P_c}{\partial L_T}, \frac{\partial(1/r_c)}{\partial L_T} \propto e^{-kL_T}, \qquad (F_0 \to 0, \rho_s \to \rho_s' \to 0)_V \qquad (30)$$

where k = semiempirical extinction coefficient for PAR
$= [G(\mu)/\mu](1 - \omega_\pi)^{1/2}$
$= 0.8944$ for horizontal leaves
$= 0.4472$ for spherically distributed leaves

The physical explanation for Eq. 30 is that PAR is utilized by the canopy as it is attenuated down through it. P_c and $1/r_c$ are therefore related to the *one-way* passage of PAR down through the canopy, as described by the e^{-kL_T} term.

II.F The Relationship Between the SR, APAR, Canopy Photosynthetic Rate, and Canopy Stomatal Resistance

In the introduction to this chapter, it was pointed out that data and modeling had indicated a near-linear relationship, Eq. 18, between SR, APAR, P_c, and $1/r_c$ for the normal field situation. The necessary condition for this phenomenon was summarized by Eq. 19. This equation can only hold when the ratio of the right-hand sides of equations (25) and (28) is constant. As a reminder, Eqs. 25 and 30 are

$$\frac{\partial(\text{SR})}{\partial L_T}, \frac{\partial a_N}{\partial L_T} \propto e^{-2h_N L_T} \qquad (25)$$

$$\frac{\partial(\text{APAR})}{\partial L_T}, \frac{\partial P_c}{\partial L_T}, \frac{\partial(1/r_c)}{\partial L_T} \propto e^{-kL_T} \tag{30}$$

For their ratio to be constant for all values of L_T, we must have

$$e^{-2h_N L_T} = e^{-kL_T} \tag{31a}$$

which leads to

$$2h_N = k \tag{31b}$$

$$2(1 - \omega_N)^{1/2} = \frac{G(\mu)}{\mu}(1 - \omega_\pi)^{1/2} \tag{31c}$$

and

$$\omega_N = 1 - \left[\frac{G(\mu)}{2\mu}\right]^2 (1 - \omega_\pi) \tag{31d}$$

For the worked examples discussed in this chapter, we have the following relationships.

Horizontal Leaves:

$$\omega_\pi = 0.2$$

$$G(\mu)/\mu = 1$$

so that to satisfy Eq. 31, $\omega_N = 0.8$.

Spherically Distributed Leaves:

$$\omega_\pi = 0.2$$

$$G(\mu)/\mu = 0.5 \quad \text{for an overhead sun}$$

so that to satisfy Eq. 31, $\omega_N = 0.95$.

The significance of this is as follows: if we wish to infer APAR, P_c, or $1/r_c$ from surface reflectance measurements, we should choose spectral bands in the near-infrared that come close to satisfying Eq. 31. It so happens that for the common cases of spherical and horizontal leaf-angle distributions, the values of ω_N produced by Eq. 31 are close to those observed in the 0.7–1.1 μm and 1.6–1.8 μm regions, respectively.

Kumar and Monteith (1982) presented an analysis that produced a similar result to that just given for the case of horizontal leaves. The analysis presented here shows that a perfect result, i.e. perfect linearity between APAR and SR, can only be achieved with a canopy of entirely horizontal leaves and the soil conditions $(\rho_s \to \rho_s' \to 0)_{V,\pi}$ and $(\rho_s \to 0)_N$. As the leaf inclination angle increases, the direct and diffuse beam contributions to the total flux at any level in the canopy begin to behave in increasingly different ways.

This phenomenon can be best understood by reference to the full two-stream approximation expressions for the downward and upward fluxes of radiation in the canopy.

When the near-ideal condition of $(\rho_s \rightarrow \rho'_s \rightarrow 0)_{V,\pi}$ is satisfied in the full expression for APAR, Eq. 11, the only term that varies with changes in leaf-area index is the transmission term given by

$$F_{L_T} = F_0 \left[e^{-KL_T} \left(1 + \frac{h_4}{\sigma} \right) + h_5 e^{-hL_T} + h_6 e^{hL_T} \right] \tag{32}$$

where F_{L_T} is the transmitted flux below the total leaf-area index L_T, and F_0 is the incident direct beam flux above the canopy.

For a canopy of horizontal leaves and under the conditions just specified, h_4 approaches -1, h_6 approaches zero, and h_5 approaches 1, so that Eq. 32 simply becomes

$$F_{L_T} = F_0 e^{-hL_T} \tag{33}$$

The derivative of APAR with respect to total leaf-area index is thus solely functional on the e^{-hL_T} term. (In the visible region, $h_V = k$ for horizontal leaves.) A similar phenomenon occurs with reflectance. The spectral reflectance is defined by Eqs. 8 and 21 and the derivative of the reflectance with respect to leaf area index is given by Eq. 22. It can be seen from Eq. 22 that the derivative of the near-infrared reflectance with respect to total leaf area index, $\partial a_N / \partial L_T$, will become functional solely on the $e^{-2h_N L_T}$ term as P_2 tends to zero. For this to occur, the leaf scattering coefficient must approach unity, see Eq. 22, which is a fair approximation in the near-infrared wavelength interval. From this, our condition of almost perfect linearity is met for a canopy of horizontal leaves overlying a dark soil.

As the leaf inclination increases, two effects come into play that start to degrade this ideal condition. First, h_4, h_5, and h_6 in Eqs. 7b and 32 move away from their "ideal" values of unity or zero. This is because the direct beam and diffuse beam components of the radiative flux are attenuated at different rates for nonhorizontal leaves. This being the case, the derivative of APAR with respect to L_T becomes dependent on both the K and h power terms, and K and h have increasingly different values. Similarly, the expression for the surface reflectance includes various direct beam terms (see Table 1) that increase in importance as the mean leaf inclination angle increases.

Figures 13 to 15 show the derivatives of APAR, P_c, and $1/r_c$ with respect to L_T plotted against the derivative of the near-infrared reflectance a_N. A close inspection of Figure 13 will show that when $\omega_N = 0.8$, the linearity between APAR and a_N is best for horizontal leaves, and that when $\omega_N = 0.95$, the linearity is best for spherically distributed leaves. The same trend is apparent in Figure 7, where SR is plotted against APAR. In Figures 13 and 14, the derivatives of P_c and $1/r_c$ are plotted against the "optimal" values of $\partial a_N / \partial L_T$, where $\omega_N = 0.8$ for horizontal leaves, and $\omega_N = 0.95$ for the spherical leaf-angle distribution. An inspection of Figures 8 and 9 will show, however, that Eq. 31 holds here as well; the relationships are more linear when Eq. 31 is exactly satisfied.

To sum up, the preceding analysis and Eq. 31 state that if:

(i) the spectral properties of green leaves were to conform to the condition specified by Eq. 31,

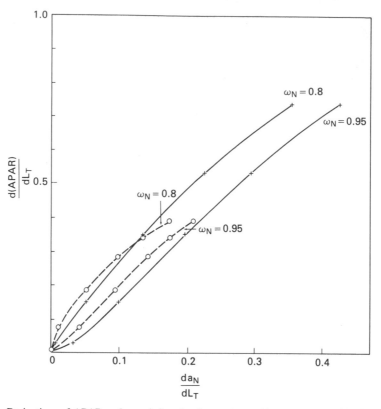

Figure 13   Derivatives of APAR and near-infrared reflectance a_N with respect to total leaf area index L_T plotted against each other. Solid lines denote horizontal leaves and dashed lines denote spherical leaf angle distributions. Circles ($\circ$) and crosses ($+$) refer to values of L_T, increasing from *right to left;* 0.1, 0.5, 1.0, 2.0, 4.0, and 8.0. $\partial a_N / \partial L_T$ was calculated for the two values of ω_N, 0.8 and 0.95, as marked on the figure. Soil reflectance is 0.1. [Reprinted by permission of the publisher from Sellers (1987). Copyright 1987 by Elsevier Science Publishing Co., Inc.]

(ii) the canopy consisted of randomly positioned leaves, and

(iii) the soil was dark so that $(\rho_s \rightarrow \rho_s')_V$, $(\rho_s \rightarrow 0)_N$,

spectral observations of the canopy reflectance would exhibit the following characteristics:

(i) the near-infrared reflectance would be a near-linear indicator of APAR, P_c, and $1/r_c$ under stress-free conditions,

(ii) the SR would be functional on the near-infrared reflectance only,

(iii) surface reflectance in the visible region would be invariant with leaf area index.

All these statements may be expressed more prosaically. If an experimenter were somehow to launch an equal number of visible/PAR and near-infrared photons at a vegetated surface, he could estimate the number of visible/PAR photons absorbed by the vegetation (as opposed to the soil) by counting the reflected near-infrared photons rather

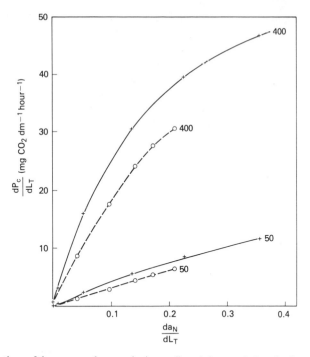

Figure 14   Derivatives of the canopy photosynthetic rate P_c and the near-infrared reflectance a_N with respect to total leaf area index L_T plotted against each other. All other symbols and conditions are the same as in Figure 13. [Reprinted by permission of the publisher from Sellers (1987). Copyright 1987 by Elsevier Science Publishing Co., Inc.]

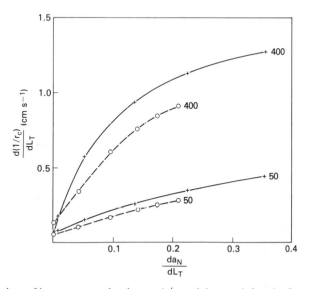

Figure 15   Derivatives of inverse stomatal resistance $1/r_c$ and the near-infrared reflectance a_N with respect to total leaf area index L_T plotted against each other. All other symbols and conditions are the same as in Figure 13. [Reprinted by permission of the publisher from Sellers (1987). Copyright 1987 by Elsevier Science Publishing Co., Inc.]

than the reflected visible/PAR photons. The relationship between the two would be most linear under the ideal conditions we have specified in Eq. 31.

We now go on to discuss the feasibility of using reflectance data to estimate APAR, P_c, and $1/r_c$. In particular, the limitations imposed by the architecture of real plant canopies and the configuration of operational satellite sensor systems are examined in the context of satisfying the condition of Eq. 31.

III APPLICATIONS

It has been shown that the near-linear relationship between canopy photosynthesis P_c, canopy inverse resistance $1/r_c$, and the simple ratio SR, as observed in one form or another by Monteith (1977), Goward et al. (1985), and Sellers (1985) is due to the near conformism of the visible and near-infrared scattering coefficients for green foliage to the rule specified in Eq. 31. It is suggested that the biophysical explanation for the near-linear relationships between SR and P_c and $1/r_c$ and APAR is that the reflected near-infrared radiation is largely dependent on *twice* the optical-path length for diffuse radiation in this wavelength interval, whereas P_c and $1/r_c$ are related to the *one-way* attenuation of PAR on its passage down through the canopy, i.e., once times the direct beam path length for PAR flux. The linearity for the relationship arrives from the fortunate proportion of ω_V to ω_N for green leaves, which more or less corresponds to the ideal relation expressed in Eq. 31.

It is important to ask at this point whether the real world is really comparable to the idealized case discussed up to now. Figure 16 shows the reflectance and transmittance spectra of green maize leaves together with a surface solar-intensity spectrum, S_λ, for a typical midlatitude clear day were taken from Miller (1972) together with the mean action efficiency of photosynthesis (energy basis), C_λ, of a number of crop plants as reported by McCree (1972). From these data, we may calculate an "effective" PAR

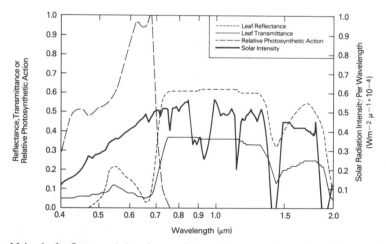

Figure 16 Maize leaf reflectance (α) and transmittance (τ) spectra for the wavelength interval 0.4–2.0 μm. Also shown are the mean photosynthetic action efficiency spectrum for a number of species' leaves, C, and the incident solar flux at the surface for a clear day, S, in W m^{-2} μm^{-1}. α, τ, and S are from Miller (1972); C is from McCree (1972). [Reprinted by permission of the publisher from Sellers (1987). Copyright 1987 by Elsevier Science Publishing Co., Inc.]

scattering coefficient for leaves ω_π, by integrating the product of the action efficiency, C_λ, and the spectral scattering coefficient (reflectance plus transmission) over all wavelengths with a suitable weighting for the intensity of the incident solar radiation, S_λ.

$$\omega_\pi = \frac{\int_0^\infty (\tau_\lambda + \alpha_\lambda)\, S_\lambda C_\lambda\, d\lambda}{\int_0^\infty S_\lambda C_\lambda\, d\lambda} \tag{34}$$

where C_λ = action efficiency of photosynthesis

λ = wavelength [μm]

S_λ = incident solar flux intensity per wavelength [W m^{-2} μm^{-1}]

τ_λ, α_λ = leaf spectral transmittance and reflectance

$\tau_\lambda + \alpha_\lambda = \omega_\lambda$

The equivalent broadband near-infrared scattering coefficient $\overline{\omega}_N$ is given by removing the C_λ terms and integrating Eq. 34 between limits of 0.68–2.9 μm. By using these spectra, a numerical calculation of the quantities ω_π and $\overline{\omega}_N$ yielded values of 0.207 and 0.848, respectively: this combination is not too far from the ideal of Eq. 31 for a planophile canopy; see the square marked ω_{PAR}, ω_{NIR} in Figure 18. However, remotely sensed observations are seldom made across such broad bands. The relative spectral responses, R_λ, of the Advanced Very High Resolution Radiometer (AVHRR) mounted on the National Oceanic and Atmospheric Administration (NOAA) series of polar orbiting satellites and the R_λ curves for the Multispectral Scanner System (MSS) and Thematic Mapper (TM) sensors mounted on the Landsat series of satellites are published in the papers of Kidwell (1981) and Markham and Barker (1983, 1985), respectively. Figure 17 shows the spectral response functions for the AVHRR sensor; similar figures

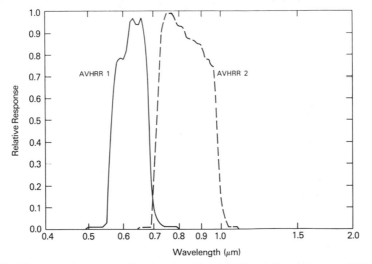

Figure 17 Relative spectral response of the Advanced Very High Resolution Radiometer (AVHRR), bands 1 and 2. The data are from Kidwell (1981). [Reprinted by permission of the publisher from Sellers (1987). Copyright 1987 by Elsevier Science Publishing Co., Inc.]

pertaining to the MSS and TM may be found in Sellers (1987). The mean leaf scattering coefficients, as observed using any one of these sensors, may be calculated by

$$\bar{\omega} = \frac{\int_0^\infty (\tau_\lambda + \rho_\lambda) S_\lambda R_\lambda \, d\lambda}{\int_0^\infty S_\lambda R_\lambda \, d\lambda} \tag{35}$$

where R_λ is the relative spectral response of a satellite sensor as a function of wavelength.

Table 3 lists the values of $\bar{\omega}$ as calculated for each sensor using Eq. 35 and numerically integrating the functions over 0.01 μm intervals.

We may now approach the question of which combination of two sensors is likely to give the best estimate of P_c and $1/r_c$, assuming that these are the vegetation quantities of most concern to terrestrial ecologists, agronomists, climatologists, etc. The criteria that we should use to choose appropriate response functions for our two sensors are simply:

1. *Visible or PAR Region Sensor,* $(R_\lambda)_V$. The apparent visible mean leaf scattering coefficient $\bar{\omega}_V$, as given by the sensor properties $(R_\lambda)_V$, and the spectral solar flux S_λ in Eq. 35 should come close to satisfying Eq. 23. This means that the resulting semiinfinite canopy reflectance observed in this spectral region would match the soil reflectance $(a \rightarrow \rho'_s)_V$ so that the total surface reflectance would be invariant

TABLE 3   (a) Effective PAR Scattering Coefficient ω_π and Broad-Band Near-Infrared Scattering Coefficient $\bar{\omega}_{NIR}$ for Maize Leaves, Eq. 35. (b) Apparent Visible, $\bar{\omega}_V$, and Near-Infrared, $\bar{\omega}_N$, Scattering Coefficients for Maize Leaves, as Would Be Measured By Satellite Sensors

(a)	ω_π	$\bar{\omega}_{NIR}$
Maize leaves	0.207	0.848
(b)	$\bar{\omega}_V$	$\bar{\omega}_N$
AVHRR 1	0.187	—
AVHRR 2	—	0.956
MSS1	0.254	—
MSS 2	0.212	—
MSS 3	—	0.880
MSS 4	—	0.980
TM 1	0.094	—
TM 2	0.278	—
TM 3	0.141	—
TM 4	—	0.980
TM 5	—	0.756

Sources: Relative spectral response data of sensors taken from data: originally from Kidwell (1981) for AVHRR, Markham and Barker (1983) for MSS, and Markham and Barker (1985) for TM.

with vegetation density. In practical terms, it is impossible to select a single sensor for this task, as the background soil reflectance varies widely over the globe. Probably the next best thing is to choose a value of $\overline{\omega}_V$ that approaches ω_π, which would ensure a good contrast between $\overline{\omega}_V$ and $\overline{\omega}_N$. This can be achieved by having the sensor response function $(R_\lambda)_V$ match the photosynthetic action efficiency function C_λ, thus making the sensor electronically equivalent to a green leaf.

2. *NIR Region Sensor $(R_\lambda)_N$.* The apparent near-infrared mean leaf scattering coefficient $\overline{\omega}_N$, as given by the sensor properties $(R_\lambda)_N$, should fall as close as possible to the value defined by Eq. 31.

It should be noted that these two criteria relate *only* to the objective of obtaining estimates of the area-averaged canopy photosynthetic capacity and inverse resistance via a simple transform applied to surface reflectance data. It is likely that these two criteria will predict "optimal" band mixes that are some way from being ideal with regard to other factors, for example, atmospheric effects, spectral contrast between sparse vegetation and bare soil, etc.

Figure 18 shows the previous two criteria as lines in the $\overline{\omega}_V : \overline{\omega}_N$ domain. The optimum point, i.e., the best combination of $\overline{\omega}_V$ and $\overline{\omega}_N$, may be found at their intersections. Various combinations of each satellite sensor configuration are plotted in the same domain and their proximity to the optimum points may be taken as a rough indication as to the linearity of the resultant SR vs. APAR, P_c, and $1/r_c$ relationship.

Figure 18 clearly illustrates that the choice of an "optimal" near-infrared sensor is dependent upon the leaf-angle distribution as well as ω_π. It is desirable that the visible, or PAR region, sensor is close to the ω_π line, marked on Figure 18 by (i). This criterion is satisfied by the AVHRR-1, MSS-1, and MSS-2 bands very well, and by the TM-1, TM-2, and TM-3 bands moderately well. (Combinations of these TM bands would achieve the desired result.) Of the near and middle-infrared bands, TM-5 and MSS-3 satisfy Eq. 31 for horizontal leaves, line (iia); and AVHRR-2 satisfies Eq. 31 for spherically distributed leaves, line (iic). Since, in reality, we are constrained to the use of sensor combinations mounted on the same satellite (the problems associated with matching different viewing times and atmospheric conditions associated with data from two or more satellites are fairly severe), we should look for the single-satellite sensor combination that best satisfies the criteria mentioned before for the most common canopy geometries found in nature, i.e., spherical or near-spherical leaf-angle distributions. Two very strong candidates immediately emerge.

The first is the combination of AVHRR-1 and AVHRR-2. These bands comprise the best possible two-band mix for spherically distributed leaves. The second is made up of MSS-2, in the visible region, and a mix of MSS-3 and MSS-4, in the near-infrared region, which can be combined in various ways to give an optimal system for a range of leaf-angle distributions. Combinations of TM-1, TM-2, and TM-3 can be used to synthesize a band that falls on the ω_π line, i.e., criterion 1. The TM-4 and TM-5 bands, according to Eq. 31, are placed at the two extremes of possible canopy geometries: planophile and erectophile. In view of this, it is possible that a mix of TM-4 and TM-5 bands may be used to infer some information about canopy geometry.

The obvious suitability of the AVHRR sensor combination is a surprising and encouraging result. Surprisingly, because the sensor was not designed with any biophysical tasks in mind, and encouraging in that it strongly supports the global biophysical analyses performed by Fung et al. (1986) and Goward et al. (1985), and the interpretative

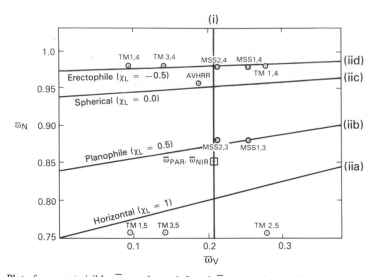

Figure 18 Plot of apparent visible, $\overline{\omega}_V$, and near-infrared, $\overline{\omega}_N$, scattering coefficients of green maize leaves as would be measured by different mixes of AVHRR, MSS, and TM bands using Eq. 35. The letters next to each point on the figure refer to an instrument (e.g., TM = Thematic Mapper) and the two numbers separated by commas refer to individual sensors on the instrument, the first number relating to the sensor operating in the visible region. If orthogonal lines are drawn linking each point to the figure axes, the $\overline{\omega}$ values for each sensor as calculated by Eq. 35 may be read. For example, the point labeled TM1, 4 refers to the Thematic Mapper where the visible-region sensor (TM-1) is projected to give a value of 0.094 for $\overline{\omega}_V$ and the NIR-region sensor (TM-4) is projected to give a value of 0.980 for $\overline{\omega}_N$ (see Table 3). Also shown, as a straight vertical line labeled (i), is the value of ω_π as calculated from Eq. 34. Proximity of the sensor points to this vertical line indicates that the visible-region sensor has a response function similar to the photosynthetic action efficiency function of McCree (1972). The lines running from left to right on the figure, (iia) through (iid), represent solutions to Eq. 31 for four different leaf angle distributions: (iia) horizontal leaves ($\chi_L = 1$), (iib) planophile leaves ($\chi_L = 0.5$), (iic) spherically distributed leaves ($\chi_L = 0$), and (iid), erectophile leaves ($\chi_L = -0.5$). Broadly speaking, the leaf angle distribution tends from horizontal to near-vertical from (iia) to (iid). Each line in (ii) represents the optimal relationship between $\overline{\omega}_N$ and $\overline{\omega}_V$ for the stated leaf angle distribution. Proximity of a sensor combination point to one of these lines indicates the suitability of the near-infrared sensor for biophysical analyses as defined by Eq. 31. Clearly, an optimal sensor combination would be close to lines (i) and (iic), representative of a spherical leaf angle distribution. It can be seen that the AVHRR combination falls very close to this optimal point. [Reprinted by permission of the publisher from Sellers (1987). Copyright 1987 by Elsevier Science Publishing Co., Inc.]

work of Tucker et al. (1981). If, on the basis of Eq. 31, we were to design a satellite sensor system to carry out inventories of global-scale photosynthetic capacity and minimum stomatal resistance, we would probably build a system with sensor-response characteristics similar to those of the AVHRR instrument.

IV SUMMARY AND CONCLUSIONS

The near-linear dependence of canopy photosynthesis (P_c) on contemporaneous simple ratio (SR) or normalized difference vegetation index (NDVI) data has been remarked on and used by a number of researchers working with remote-sensing data (e.g., Goward et al. 1985). Using simple models of canopy radiative transfer, photosynthesis, and stomatal resistance, we have seen that the spectral properties of plant leaves conform approximately to the rule $2h_N = k$, or

$$2(1 - \omega_N)^{1/2} = \frac{G(\mu)}{\mu} (1 - \omega_\pi)^{1/2} \tag{31b}$$

When Eq. 31 is satisfied, we obtain a near-linear relationship between canopy near-infrared reflectance a_N and APAR, P_c, and $1/r_c$. In the visible region, soil reflectance is typically of the same magnitude as the asymptotic canopy reflectance, so that the SR and NDVI are mainly functional on a_N and the same linear relationship more or less holds good for them, too.

All of this discussion indicates that the near-infrared surface reflectance a_N is the best indicator of biophysical properties and that the visible reflectance a_V is relatively invariant with green leaf area index. The reasons for incorporating the a_V terms in the SVIs, rather than just a_N alone, are straightforward. The main reason is that remote-sensing systems only measure radiances, not reflectances and that since these radiances are the product of an incident flux and a directional reflectance, the remotely sensed measurement is really providing the observer with the product of at least two unknowns. However, under normal terrestrial illumination conditions, it can be assumed that the ratio of incident visible and near-infrared fluxes is more or less constant, so that

$$\int_0^\infty S_\lambda (R_\lambda)_V \, d\lambda = B \int_0^\infty S_\lambda (R_\lambda)_N \, d\lambda \tag{36}$$

where B is a constant, $(R_\lambda)_V$ is the relative spectral response function of a satellite sensor operating in the visible wavelength region, $(R_\lambda)_N$ is the same as $(R_\lambda)_V$ except for the near-infrared region.

The hemispherically integrated upward diffuse fluxes $I\!\uparrow$ above the canopy as measured by the sensors would then be given by

$$I\!\uparrow_V = \bar{a}_V \int_0^\infty S_\lambda (R_\lambda)_V \, d\lambda \tag{37a}$$

$$I\!\uparrow_N = \bar{a}_N \int_0^\infty S_\lambda (R_\lambda)_N \, d\lambda \tag{37b}$$

where the broadband canopy reflectances are given by

$$\bar{a} = \frac{\displaystyle\int_0^\infty S_\lambda (R_\lambda) a_\lambda \, d\lambda}{\displaystyle\int_0^\infty S_\lambda (R_\lambda) \, d\lambda} \tag{38}$$

Combining Eqs. 36–38 yields

$$SR = \frac{\bar{a}_N}{\bar{a}_V} = \frac{I\!\uparrow_N}{I\!\uparrow_V} \cdot \frac{1}{B} \tag{39}$$

Since B should be relatively easy to estimate from theory or observations, the combination of radiances in Eq. 39 allows one to estimate the simple ratio of the surface reflectances without having to know the values of the incoming radiances. The use of the simple ratio rather than the estimated near-infrared reflectance by itself should therefore provide the observer with some greater certainty about the spectral properties of the surface. In this way, a_V may be used as a kind of standard background to normalize the a_N observations, a practice that would be perfectly justified when $(\rho_s \rightarrow \rho'_s)_V$. We may illustrate this point by going back to the example of our photon-throwing experimenter: provided the experimenter is launching an equal number, or a constant ratio $(1/B)$, of near-infrared and visible photons (i.e., the incoming spectral radiation at the surface), a count of either is not needed as long as the number of photons reflected back can be counted. When $(\rho_s - \rho'_s)_V$, the number of reflected visible/PAR photons will be a constant fraction of the number launched, regardless of the vegetation density so that the experimenter has a means of quantifying the incident radiation flux density. Since this means that the reflected visible/PAR photons are directly related to the incoming flux density, the ratio of reflected near-infrared to visible photons will be independent of their incoming flux densities and only dependent on the amount of vegetation present.

A second reason for the use of SR, rather than a_N alone, is important in marginal or desert areas, where the underlying soil surface is relatively light-colored ($\rho_s > 0.25$). Here the near-infrared signal is not necessarily a sensitive indicator of the presence or absence of vegetation. To take the extreme case, when the soil reflectance in the near-infrared region approaches the asymptotic canopy reflectance, changes in the simple ratio or normalized difference are due solely to variations in the visible reflectance. Under these conditions, the relationship between SR and P_c, $1/r_c$, and APAR becomes curvilinear.

All of the discussion covered so far in this chapter has been concerned with illustrating the biophysical reasons for the apparently near-linear relationship between SR and APAR, P_c, and/or $1/r_c$. However, a number of theoretical limitations in the analysis should be addressed prior to applying these lessons to the practical business of interpreting satellite imagery in a quantitative fashion. These limitations follow the form of the specific assumptions used throughout the analysis. Each of these will be discussed briefly in turn.

1. *Complex Leaf-Angle Distributions:* Canopies with more complex leaf angle distribution functions exhibit more complicated responses of SR with solar angle, unequal leaf reflectance, and transmittance values, etc. Similarly, the P_c and $1/r_c$ responses become dependent on solar angle. However, these effects are not too alarming for normal leaf angle distribution functions (Sellers, 1985).

2. *Nonisotropic Canopy Reflectance:* This and the question of satellite sensor narrow-field-of-view radiances cannot be addressed within the constraints of the two-stream approximation model. More sophisticated models (e.g., that of Kimes, 1984) must be used to explore the fine details of narrow-angle spectral response vs. leaf area index, leaf angle distribution function, etc. Similarly, when canopies have regular or clumped (i.e., nonrandom) leaf distributions (see, for example, Baldocchi et al., 1985), the simple analysis presented here will not translate directly. The same principle of near-linearity between SR and APAR, P_c, or $1/r_c$ should still hold as the processes both of reflectance and biophysical functioning depend on interactions between radiation and the leaf elements. The effect of non-

random leaf distributions will be to alter the dependence of SR, APAR, P_c, and $1/r_c$ upon L_T without greatly affecting their interdependence.

3. *Estimation of Mean Extinction Coefficients:* The calculation of mean scattering and reflectance coefficients (see Eqs. 35 and 38) via integration over a given wavelength interval is a valid procedure. Application of these mean quantities to estimate equivalent mean extinction coefficients and then to use these in radiative transfer calculations is a different matter. To be exact, such a procedure is mathematically and physically in error, but in this particular study of leaf optical properties, where the leaf spectral response is relatively invariant on either side of the 0.68 μm discontinuity, the resultant errors should not be too serious.

4. *Other Environmental Factors:* As mentioned previously, even if the condition as stated by Eq. 31 were perfectly met in nature (which it is not), then the SR would only provide the functional forms of P_c and $1/r_c$. Additional information, in the form of incident PAR fluxes, leaf physiological properties, and the degree of environmental stress, is required in order to determine the actual values of P_c and $1/r_c$.

In spite of these limitations, it is clear that the spectral vegetation-index data as provided by satellite remote-sensing systems should yield near-linear estimates of the area-averaged canopy photosynthetic capacities and minimum resistances, provided other quantities are available or can be estimated. This finding supports the view that the reflectance data provide indications of instantaneous rates associated with the vegetation canopies: gross primary productivity and evapotranspiration, rather than reliable estimates of any state associated with the vegetation, such as leaf area index or biomass. On a purely practical level, the AVHRR sensor mounted on the NOAA series of polar orbiting meteorological satellites appears to be a near-optimal system for carrying out global biophysical surveys. Clearly, however, more experimental work has to be done to determine the quantitative limits to the application of these remote-sensing techniques to the measurement of biophysical processes over large areas.

LIST OF SYMBOLS

STANDARD ALPHABETICAL SYMBOLS

Symbol	Description
a_A	Canopy spectral reflectance for spectral component A
a_1	Constant relating leaf photosynthetic rate to PAR flux density [mg CO_2 dm^{-1} h^{-1}]
a_2	Constant relating leaf stomatal resistance to PAR flux density [J m^{-3}]
a_s	Single-scattering albedo
APAR	(Fractional) Absorbed photosynthetically active radiation
B	Constant relating ratio of near-infrared to visible upwelling diffuse fluxes to the ratio of spectral reflectances as observed by a radiometer
b_1	Constant relating leaf photosynthetic rate to PAR flux density [W m^{-2}]

Symbol	Description
b_2	Constant relating leaf stomatal resistance to PAR flux density [W m^{-2}]
C_λ	Action efficiency of photosynthesis
c_2	Minimum leaf stomatal resistance [s m^{-1}]
F	Radiative flux [W m^{-2}]
F_0	Radiative flux above the canopy [W m^{-2}]
$f(T), f(\delta e), f(\psi_l)$	Adjustment factors for the effects of leaf temperature T, vapor pressure deficit δe, and leaf water potential ψ_l
$G(\mu)$	Projected leaf area in direction μ
h	Diffuse radiation extinction coefficient
$I\uparrow, I\downarrow$	Upward and downward diffuse radiative fluxes within the canopy, respectively (normalized by incident radiation)
K	Extinction coefficient for direct beam radiation
k_π	Approximate extinction coefficient for PAR
L	Leaf-area index [m^2 m^{-2}]
L_A	Upwelling radiance as observed by a satellite sensor for spectral component A [W m^{-2} sr]
L_T	Total leaf-area index [m^2 m^{-2}]
N	Subscript denoting near-infrared (0.7–1.1 μm) radiation
n	Vector normal to leaf surface
NDVI	Normalized difference vegetation index
$O(\xi, \theta)$	Leaf-angle distribution function as a function of azimuth ξ and inclination θ
P	Leaf photosynthetic rate [mg CO$_2$ dm^{-1} h^{-1}]
P_c	Canopy photosynthetic rate [mg CO$_2$ dm^{-1} h^{-1}]
$P(\mu, \mu')$	Scattering phase function
PAR	Photosynthetically active radiation
R_d	Leaf respiration [mg CO$_2$ dm^{-1} h^{-1}]
R_λ	Relative spectral response function (satellite sensor)
r_c	Canopy stomatal resistance [s m^{-1}]
r_s	Leaf stomatal resistance [s m^{-1}]
S_λ	Incident solar flux intensity per wavelength [W m^{-2} μm^{-1}]
SR	Simple ratio vegetation index
V	Subscript denoting visible (0.4–0.7 μm) radiation

GREEK SYMBOLS

Symbol	Description
α	Leaf reflectance
β	Diffuse upscatter parameter
β_o	Direct beam upscatter parameter
γ	$\omega/2\rho_s$
δe	Vapor pressure deficit [mb]
$\Gamma(\mu, \mu')$	$G(\mu) G(\mu') P(\mu, \mu')$

Symbol	Description
θ	Inclination angle
λ	Subscript denoting wavelength
μ	Cosine of zenith angle (incident radiation direction)
$\bar{\mu}$	Average inverse diffuse optical depth per unit leaf area
μ'	Direction of scattered radiation
ξ	Azimuth angle
π	Subscript denoting PAR
ρ_s	Soil reflectance
ρ_s'	Soil reflectance equivalent to reflectance for an infinitely thick canopy
τ	Leaf transmittance
ψ_l	Leaf water potential [m]
ω	Scattering coefficient
$\bar{\omega}$	Weighted-average scattering coefficient over a wavelength interval

REFERENCES

Asrar, G., M. Fuchs, E. T. Kanemasu, and J. L. Hatfield (1984). Estimating absorbed photosynthetic radiation and leaf area index from spectral reflectance in wheat. *Agron. J.* **76**:300–306.

Badhwar, G. D., R. B. MacDonald, and N. C. Mehta (1986). Satellite-derived leaf-area-index and vegetation maps as input to global carbon cycle models—a hierarchical approach. *Int. J. Remote Sens.* **7**(2):265–281.

Baldocchi, D. D., B. A. Hutchison, D. R. Matt, and R. T. McMillen (1985). Canopy radiative transfer models for spherical and known leaf inclination angle distributions: A test in an oak-hickory forest. *J. Appl. Ecol.* **22**:539–555.

Charles-Edwards, D. A., and L. J. Ludwig (1974). A model for photosynthesis by C_3 species. *Ann. Bot. (London)* [N.S.] **38**:921–930.

Curran, P. J. (1980). Multispectral photographic remote sensing of vegetation amount and productivity. *Proc. 14th Int. Symp. Remote Sens.*, University of Michigan, Ann Arbor, pp. 623–637.

Dickinson, R. E. (1983). Land surface processes and climate-surface albedos and energy balance. *Adv. Geophys.* **25**:305–353.

Farquhar, G. D., and T. D. Sharkey (1982). Stomatal conductance and photosynthesis. *Ann. Rev. Plant Physiol.* **33**:317–345.

Farquhar, G. D., and S. von Caemerrer (1982). "Modeling of photosynthetic response to environmental conditions." In *Encyclopedia of Plant Physiology, New Series* (O. L. Lange et al., Eds.), Vol. 12B. Springer-Verlag, Berlin, pp. 549–487.

Fung, I. Y., C. J. Tucker, and K. C. Prentice (1986). On the application of the AVHRR vegetation index to study the atmosphere–biosphere exchange of CO_2. *J. Geophys. Res.* **92**(3):2999–3015.

Goudriaan, J. (1977). *Crop Micrometeorology: A Simulation Study.* Center for Agricultural Publishing and Documentation, Wageningen, The Netherlands, p. 249.

Goward, S. N., C. J. Tucker, and D. G. Dye (1985). North American vegetation patterns observed with the Nimbus-7 advanced very high resolution radiometer. *Vegetation* **64**:3–14.

Hatfield, J. L., G. Asrar, and E. T. Kanemasu (1984). Intercepted photosynthetically active radiation estimated by spectral reflectance. *Remote Sens. Environ.* **14**:65–75.

Hesketh, J. D., and D. Baker (1967). Light and carbon assimilation by plant communities. *Crop Sci.* **7**:285–293.

Jarvis, P. G. (1976). The interpretation of the variations in leaf water potential and stomatal conductance found in canopies in the field. *Philos. Trans. R. Soc. London, Ser. B* **273**:593–610.

Kidwell, K. B. (1981). *NOAA Polar Orbiter Data (Tiros-N, NOAA-6 and NOAA-7) Users Guide.* Department of Commerce, Washington, D.C.

Kimes, D. S. (1984). Modeling the directional reflectance from complete homogeneous vegetation canopies with various leaf orientation distributions. *J. Opt. Soc. Am.* **1**:725–737.

Kimes, D. S., and J. A. Kirchner (1982). Radiative transfer model for heterogeneous 3-D scenes. *Appl. Opt.* **21**:4119–4129.

Kimes, D. S., W. W. Newcomb, R. F. Nelson, and J. B. Schutt (1986a). Directional reflectance distributions of a hardwood and pine forest canopy. *IEEE Trans. Geosci. Remote Sens.* **GE-24**:281–293.

Kimes, D. S., P. J. Sellers, and W. W. Newcomb (1986b). Hemispherical reflectance (albedo) dynamics of vegetation canopies for global and regional energy budget studies. *J. Clim. Appl. Meteorol.* **26**(8):959–972.

Kumar, M., and J. L. Monteith (1982). Remote sensing of crop growth. In *Plants and Daylight Spectrum* (H. Smith, Ed.). Academic Press, New York, pp. 133–144.

Markham, B. L., and J. L. Barker (1983). Spectral characterization of the Landsat-4 MSS sensors. *Photogramm. Eng. Remote Sens.* **6**:811–833.

Markham, B. L., and J. L. Barker (1985). Spectral characterization of the Landsat Thematic Mapper sensors. *Int. J. Remote Sens.* **5**(6):697–716.

McCree, K. J. (1972). The action spectrum, absorptance and quantum yield of photosynthesis in crop plants. *Agric. Meteorol.* **9**:191–216.

Meador, W. E., and W. R. Weaver (1980). Two-stream approximations to radiative transfer in planetary atmospheres: A unified description of existing methods and a new improvement. *J. Atmos. Sci.* **37**:630–643.

Miller, L. D. (1972). *Passive Remote Sensing of Natural Resources.* Department of Watershed Science, Colorado State University, Fort Collins, Colorado.

Monteith, J. L. (1977). Climate and the efficiency of crop production in Britain. *Philos. Trans. R. Soc. London Ser. B* **281**:277–294.

Norman, J. M., and P. G. Jarvis (1975). Photosynthesis in Sitka spruce (Picea sitchensis (Bong.) Carr). V. Radiation penetration theory and a test case. *J. Appl. Ecol.* **12**:839–878.

Norman, J. M., and J. M. Welles (1983). Radiative transfer in an array of canopies. *Agron. J.* **75**:481–488.

Ross, J. (1975). Radiative transfer in plant communities. In *Vegetation and the Atmosphere* (J. L. Monteith, Ed.), Vol. 1. Academic Press, London, pp. 13–52.

Sellers, P. J. (1985). Canopy reflectance, photosynthesis and transpiration. *Int. J. Remote Sens.* **8**(6):1335–1372.

Sellers, P. J. (1987). Canopy reflectance, photosynthesis and transpiration II. The role of biophysics in the linearity of their interdependence. *Remote Sens. Environ.* **21**:143–183.

Steven, M. D., P. V. Biscoe, and K. W. Jaggard (1983). Estimation of sugar beet productivity from reflection in the red- and infrared spectral bands. *Int. J. Remote Sens.* **4**(2):325–334.

Szwarcbaum, I., and G. Shaviv (1976). Monte-Carlo model for the radiation field in plant canopies. *Agric. Meteorol.* **17**:333–352.

Tucker, C. J., B. N. Holben, J. H. Elgin, and E. McMurtrey (1981). Remote sensing of total dry matter accumulation in winter wheat. *Remote Sens. Environ.* **11:**171–190.

Tucker, C. J., I. Y. Fung, C. D. Keeling, and R. H. Gammon (1986). Relationship between atmospheric CO_2 variations and a satellite-derived vegetation index. *Nature* (*London*) **319:**195–199.

Turner, N. C. (1974). Stomatal response to light and water under field conditions. *Bull. R. Soc. N.Z.* **12:**423–432.

9

THE ATMOSPHERIC EFFECT ON REMOTE SENSING AND ITS CORRECTION

YORAM J. KAUFMAN

University of Maryland
College Park, Maryland

and

National Aeronautics and Space Administration
Goddard Space Flight Center
Greenbelt, Maryland

I INTRODUCTION

This chapter discusses the physics of the atmospheric effect on remotely sensed signals. The emphasis is on the solar spectrum (0.4–2.5 μm) for passive remote sensing of reflected and scattered sunlight. The atmospheric effect on thermal infrared remote sensing is discussed in Chapter 16. The discussion of the atmospheric effect is followed by a discussion of the possible correction algorithms for different remote-sensing applications.

I.A The Atmospheric Effect and Its Correction

Interactions between electromagnetic radiation and the target (absorption and scattering inside the target, and reflection by its surface) modulate the characteristics of the emerging radiation (Horvath et al., 1970; Herman and Browning, 1975). The modulation may result in changes in the brightness, polarization, and direction of the radiation, as well as its wavelength dependence. This modulation serves as the signal that is used for remote sensing of the target's characteristics. For example, the ratio of the reflected radiation in the near-IR and in the visible (the vegetation index) is a measure of the photosynthetic capacity of vegetation (Tucker and Sellers, 1986).

Interactions of the direct solar radiation and of the radiation reflected from the target with the atmospheric constituents interfere with this process of remote sensing and is called the "atmospheric effect." The atmospheric gases and aerosols (airborne particulate matter), as well as clouds, scatter and absorb solar radiation and can, therefore, modulate the radiation reflected from the target by attenuating it, changing its spatial distribution, and introducing into the field of view radiation from sunlight scattered in the atmosphere. As a result, the atmosphere can affect the apparent image of the target (the image that is observed by the sensor) in several possible ways, depending on the target, the sensor characteristics, and the remote-sensing application:

- The wavelength dependence of the atmospheric effect can modulate the brightness differently for each spectral band. Therefore, it will affect target classification (Fraser et al., 1977) if the classification is based on the "color" of the target (the wavelength dependence of its reflection coefficient). It can affect the discrimination between stressed and unstressed vegetation. For example, Slater and Jackson (1982) showed that atmospheric effects can cause a delay of 3–7 days in detection of the presence of drought stress on wheat.
- Atmospheric scattering may alter the spatial distribution of reflected radiation from the target (Pearce, 1977; Kaufman, 1982). As a result, the spatial resolution of the remote-sensing system may be affected by the atmosphere (Kaufman, 1984b).
- The atmosphere may change the apparent brightness of the target, affecting its reflectance and albedo.
- Subpixel clouds may generate spatial variations in the apparent surface reflectance.

It is, therefore, natural to expect that correction of the atmospheric effect can be useful for improving the quality of remotely sensed data. Atmospheric correction algorithms basically require two steps. First, the optical characteristics of the atmosphere are estimated by using special features of the target and the atmosphere. Alternatively, the atmospheric characteristics may be measured or modeled, and the atmospheric effect can be computed by radiative-transfer algorithms. Second, the image can be corrected by inversion techniques that derive the target optical characteristics (e.g., reflectance) from the measured radiance.

I.B Satellite Systems

The particular way the atmosphere affects the satellite imagery (e.g., wavelength dependence or spatial effects) depends on the characteristics of the satellite systems. From this point of view, we should mainly be concerned with the following characteristics:

- spatial resolution,
- spectral bands, and
- polarization characteristics.

Satellite systems have a wide range of spatial resolutions. The high-resolution sensors are the European SPOT with 20 m resolution; Landsat Thematic Mapper (TM), 30 m; and Landsat Multispectral Sensor System (MSS), 80 m. The low-resolution satellites are meteorological satellites NOAA AVHRR with 1 km resolution, but data archived as 4 km resolution; the Geostationary Operational Environmental Satellite (GOES) Visible Infrared Spin-Scan Radiometer (VISSR), 1 km resolution, but archived as 8 km resolution; and the European Meteostat with a similar resolution. The spectral bands used in the satellite systems are usually located in "atmospheric windows" (spectral regions) with minimal gaseous absorption. Some of the systems feature narrow bands located in the center of atmospheric windows (e.g., the Landsat TM); others have one or two wider bands in the visible and near IR (e.g., the NOAA AVHRR). These sensors are designed to detect all polarization directions with the same sensitivity.

I.C   Surface Reflectance

The Earth's surface is usually the main target for remote sensing from space. The characteristics of the particular type of surface under investigation are important in order to determine possible atmospheric effects and methods for their removal. For example, remote sensing of a dark surface (reflectance less than 0.05) will be affected mainly by atmospheric backscattering of the direct sunlight to the sensor, whereas remote sensing of a bright surface (reflectance larger than 0.2) will be affected by atmospheric absorption (Fraser and Kaufman, 1985). For remote sensing with a high-resolution sensor (e.g., SPOT and Landsat TM of 20–30 m resolution) of a spatially variable surface (e.g., agricultural or urban area), the atmospheric forward-scattering characteristics are important (Mekler and Kaufman, 1980; Kaufman, 1982). They cause blurring of the image and reduction of contrast (Kaufman, 1984a; Kaufman and Fraser, 1984). In general, the characteristics of surface reflectance that are of importance for predicting the possible atmospheric effects and for designing methods for their removal are

- the range of values of the surface reflectance,
- the surface spatial variability (for high-resolution sensors),
- the degree of departure from a Lambertian surface, and
- the surface polarization (if the sensor is sensitive to the polarization of radiation).

I.D   Applications

The atmospheric effect and the need for its correction depends also on the application of the remotely sensed signal. For example, remote sensing of the ocean color (Gordon et al., 1983) is affected mainly by atmospheric backscattering, since the water is dark and does not have high spatial variability. Remote sensing of the water color or of the water quality of lakes, in the vicinity of bright land, would be affected by the presence of the nearby land, an effect that depends on the atmospheric forward-scattering characteristics. A different aspect can be a requirement for quantitative information such as the estimation of the surface albedo, which requires the knowledge of the absolute radiance and vegetation index, which is affected mainly by the differences between the atmospheric effect in the two wavelength bands used in the computation of the index (Holben and Fraser, 1984; Holben, 1986). In the case of classification of surface features based on a training set located in the same image, the remote-sensing procedure is affected only by variation of the atmospheric effects across the image. Therefore, different applications have different requirements and may need different atmospheric correction algorithms.

I.E   The Concept of Atmospheric Correction

In order to perform a correction of the sensed image, we have to quantize the process that contaminates the image. In the case of atmospheric effects, this process is scattering and absorption by atmospheric gases, aerosol, and clouds. Knowledge of these atmospheric optical characteristics can be obtained from several sources:

- **Climatology of the Area.** Molecular (gaseous) scattering is a stable and well-studied phenomenon (Rayleigh, 1871). The effect of aerosols, clouds, and gaseous absorption can, however, be variable. Published research results on the atmospheric characteristics and their variations can be used to estimate the expected atmospheric effect for a specific part of the world, during a given season and time of day. Such information can be obtained from the analysis of aerosol measurements (Shettle and Fenn, 1979), models of gaseous absorption (Kneizys et al., 1983), and from the analysis of satellite images (Fraser et al., 1984).

- **Remotely Sensed Images.** Some features in the image with known characteristics can be used to infer the optical characteristics of the atmosphere or, directly, the components of the atmospheric effect. The very low surface reflectance of the oceans in the red part of the spectrum is the basis of the correction algorithm for the ocean color products that are based on satellite imageries (Gordon et al., 1983). The small reflectance of trees in the blue and red parts of the spectrum is the basis of a correction algorithm of images over vegetated regions of the land surface (Kaufman and Sendra, 1988).

- **Ground Measurements.** Some aerosol characteristics can be measured from the ground (King et al., 1978; Kaufman and Fraser, 1983). These measurements together with gaseous-absorption and other aerosol characteristics obtained from climatology (Kneizys et al., 1983; Shettle and Fenn, 1979) or in situ aircraft samplings can be used to estimate the atmospheric effect and to correct for it.

These atmospheric-correction methods are discussed in depth in the following sections, after a brief review of atmospheric constituents and atmospheric optics.

II ATMOSPHERIC CONSTITUENTS

Atmospheric gases and aerosols contribute to absorption and scattering of direct sunlight and sunlight reflected form the Earth's surface. Absorption reduces the amount of energy available in a given wavelength, whereas scattering redistributes the energy by changing its direction. Although scattering does not change the properties of this radiation other than its direction, it results in a reduction in contrasts of the observed objects and causes blurring (decrease in the magnitude of the sharp edges) of the image.

II.A Gases

The main components of atmospheric gases are nitrogen, N_2 (78%); oxygen, O_2 (21%); and small amounts of water vapor, H_2O; carbon dioxide, CO_2; and ozone, O_3 (Valley, 1965). Gaseous absorption is caused by specific absorption bands that depend on the molecular structure of each absorbing gas species. Scattering is caused by molecular-density fluctuation, therefore, all atmospheric gases contribute to scattering and their combined effect is computed as a function of atmospheric density and pressure. The effect of atmospheric gases on the transmittance of sunlight to the surface is shown in Figure 1. The computations are performed by the Lowtran code (Kneizys et al., 1983) for the sun at zenith, using the U.S. 1962 Standard Atmosphere.

In the visible part of the spectrum, transmission (see Figure 1) is affected mainly by ozone absorption and by molecular scattering. Molecular scattering was first formulated by Rayleigh (1871) and is known as "Rayleigh scattering." Its contribution to the vertical path transmittance (T_r) to space is expressed by the optical thickness τ_r:

$$\tau_r = -\ln T_r \qquad (1)$$

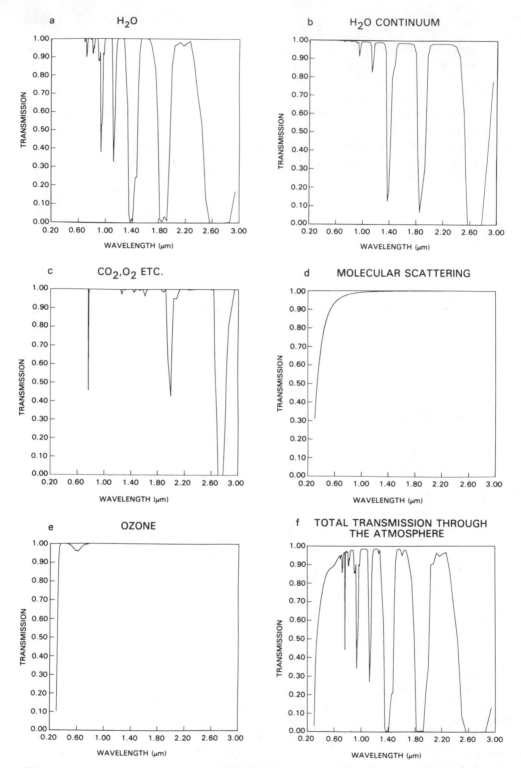

Figure 1 Atmospheric transmission for the standard United States 1962 atmosphere computed by Lowtran code (Kneizys et al., 1983) for sun at zenith. Attenuation due to absorption: (*a*) water bands, (*b*) water continuum, (*c*) CO_2, O_2, and other minor absorbers, (*d*) molecular scattering, (*e*) ozone absorption, and (*f*) total transmission.

The Rayleigh optical thickness τ_r decreases as a function of the wavelength (λ) roughly as λ^{-4}. An expression for τ_r is given by Hansen and Travis (1974):

$$\tau_r = 0.008569\lambda^{-4}(1 + 0.0113\lambda^{-2} + 0.00013\lambda^{-4}) \tag{2}$$

where λ is in μm, for a standard surface pressure of $P_o = 1013.25$ mbar. This expression differs from the theoretical λ^{-4} law due to the wavelength dependence of the index of refraction. Since air density is proportional to pressure P, the vertical optical thickness between the top of the atmosphere and a level with pressure P can be expressed by

$$\tau_r(P) = \tau_r(P_o)P/P_o \tag{3}$$

In the near-IR region, we can see a strong and narrow oxygen band at 0.76 μm, Figure 1(c), and a few water vapor and carbon dioxide bands, see Figure 1, that restrict the absorption-free wavelengths that can be used for remote sensing.

The density of dry atmospheric gas, $n(z)$, decreases approximately in an exponential rate with respect to height z from the sea level, $z = 0$:

$$n_a(z) = n_a(0)\exp(-z/H) \tag{4}$$

where H is the scale height of 8.0 km (Fraser and Curran, 1976). Note that only 1% of the atmospheric mass lies above 32 km (four scale heights). The relative deviation of this density from that of the U.S. Standard Atmosphere for the same height is less than 20% for $z < 20$ km (Fraser et al., 1975). The chief reason for this difference is that Eq. 4 is strictly valid only for an isothermal atmosphere. Atmospheric pressure also decreases exponentially with respect to height, since it results from the integration of Eq. 4 from the top of the atmosphere ($z = \infty$) to height z. Atmospheric pressure may vary slightly in time and from place to place, but it is relatively constant (standard deviation of 1%).

Water-vapor concentration is highly variable and can be found mainly in the boundary layer (lowest is 1-2 km). The variability in water-vapor concentration results from variability in water-vapor sources (evaporation) and sinks (clouds). Carbon dioxide is more stable in the atmosphere, and, therefore, is usually well mixed with other dry gases and has the same scale height, except near sources, e.g., large cities (Fraser et al., 1975). Ozone is concentrated in the stratosphere (20–50 km above the surface).

Figure 2 shows the spectral distribution of the direct solar radiation that reaches the Earth's surface after being absorbed or scattered by atmospheric gases. The solar irradiance was computed as the product of the atmospheric transmittance (Figure 1f) and the extraterrestrial irradiance reaching the Earth's atmosphere. Extraterrestrial flux was obtained from Neckel and Labs (1984) and is also plotted on the figure. The actual radiation flux that reaches the surface is reduced due to aerosol scattering and absorption (see the next section). Part of the sunlight scattered by the atmospheric gases still reaches the surface and contributes to the diffuse part of the total solar irradiance at the surface.

II.B Atmospheric Aerosols

The term "atmospheric aerosol" refers to the liquid and solid matter suspended in air. Liquid particles of size greater than 1 μm are usually called "cloud drops." In order to estimate the effect of aerosols on propagation of electromagnetic radiation through the

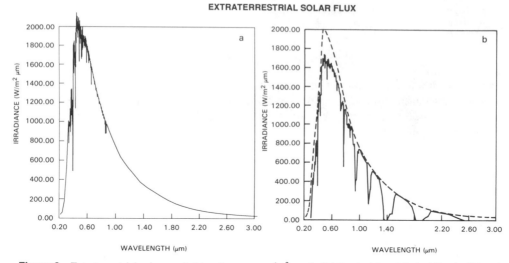

Figure 2   Extraterrestrial solar spectral irradiance, watt/m^2 μm [solid line in (*a*) and dashed line in (*b*)] and the irradiance on the Earth surface [solid line in (*b*)] for sun at zenith and no aerosol.

atmosphere, and thus on the atmospheric effect on remote sensing, their chemical and physical characteristics have to be understood. These characteristics depend on the origin of the aerosol, the process of its formation, atmospheric effects on the aerosol, and processes for its removal.

The major sources of formation of atmospheric aerosols are (Twomey, 1977)

- **Formation of Particles from Solid Surfaces.** In this process, mechanical forces due to weathering and removal by wind detach small (micron-size) solid particulates from the surface (Twomey, 1977). The lifetime of these suspended particles in the atmosphere depends on their size. As a result, the size distribution of suspended dust particles is limited by mechanical processes and by the floating time in the atmosphere.

- **Formation of Particles from the Ocean.** The major mechanism is the bursting of bubbles at the ocean surface (Twomey, 1977). Bursting of a thin film of water on the upper face of the bubble generates, typically, 100 submicron drops; relaxing of hydrostatic pressure after the bubble collapses generates, typically, a jet of 4–6 drops that subsequently evaporate and leave behind micron-size sea salt particles.

- **Formation of Particles from Gas.** Chemically active gases (e.g., sulphates, nitrates, etc.) emitted from natural (decay of plants, forest fires, and volcanoes) or man-made sources (industrial processes and agricultural burning) are transferred by chemical reactions into a liquid form in the atmosphere. The gas-to-liquid conversion may result from dry processes (aggregation of molecules), from condensation on existing particles, or by liquid-phase conversion (absorption of the gases by cloud drops that evaporate later). The dry conversion generates much smaller particles than the wet conversion (Hoppel et al., 1987). In the dry process, Whitby (1978) estimated that 95% of the mass is produced by condensation on already existing particles and only 5% is used to form new particles. These processes generate submicron particles. The evaporation of cloud droplets leaves a residue material that forms giant aerosol agglomerates (Twomey, 1977).

The optical effects of the aerosol depend on the physical characteristics of the particles:

- size distribution,
- refractive index, a function of the chemical composition (does not have to be homogeneous within the particle and may vary as a function of the particle size),
- shape of the particle (mainly whether the particle is spherical or nonspherical), and
- spatial distribution of the particle concentration, and of the characteristics mentioned before.

Particles produced by mechanical processes from solid surfaces are nonspherical (dust storms); conversely, particles produced by gas-to-liquid conversion are usually spherical, although the presence of graphitic carbon in the air, due to car exhaust emissions and other burning processes, may complicate the sphericity and the homogeneity of the liquid particles. The graphic carbon may be a center of condensation of the liquid particle or it can stick to the particle's surface.

These characteristics of the atmospheric aerosol complicate the quantitative analysis of the aerosol effect on a remotely sensed signal and its correction. One method to simplify the analysis is to group the available information into an optical model of the atmospheric aerosol. This model may describe part or all of aerosol characteristics as a function of the geographic location and the season. Such modeling efforts have been initiated by the Air Force Geophysical Laboratory (AFGL), resulting in the model of Shettle and Fenn (1979). In their report, Shettle and Fenn discussed five types of aerosol: rural, urban, maritime, tropospheric (a background aerosol that is not affected by local sources), and fog. A model for dust, discussed later by Shettle (1984), was not included. For each type of aerosol, the size distribution was modeled as a function of the relative humidity. To simplify the analysis, it was assumed that the aerosol consists of uniform spherical particles, and the refractive index was tabulated as a function of the relative humidity and wavelength. The size distributions are given by log-normal functions, as suggested by Whitby (1978):

$$\frac{dn}{dr} = \frac{N_o}{\ln (10) \cdot r \cdot \sigma \sqrt{2\pi}} \exp \left| \frac{(\log r - \log r_o)^2}{2\sigma^2} \right| \tag{5}$$

where N_o is the aerosol particle density, r is the particle radius, σ is the standard deviation of $\ln r$, and r_o is the geometrical mean radius of the size.

For each aerosol type, two log-normal distributions were used, one representing submicron particles and a second representing micron size particles. Each log-normal distribution is assumed to consist of spherical homogeneous particles with a given refractive index. A different value of refractive index was assumed for particles in each log-normal function.

A different representation of the size distribution is given by a power law (Junge, 1963):

$$\frac{dn}{d(\ln r)} = \begin{cases} 0 & \text{for } r < r_1 \text{ or } r > r_3 \\ C & \text{for } r_1 < r < r_2 \\ C(r/r_2)^{-\nu} & \text{for } r_2 < r < r_3 \end{cases} \tag{6}$$

This size distribution is a smooth envelop of two or three log-normal distributions. Typical values of the parameters are

$$r_1 = 0.01 \ \mu\text{m} \qquad r_2 = 0.1 \ \mu\text{m} \qquad r_3 = 10 \ \mu\text{m}$$

and ν is in the range $\nu = 2$–4. Log-normal distributions are a better representation of the aerosol physical characteristics (Whitby, 1978). In studying the aerosol effect on atmospheric optics, however, a power law is useful, since it is based on one major parameter, ν, which can be related to the slope of the relationship between aerosol optical thickness and wavelength (see Section III.A).

The model of Shettle and Fenn (1979) describes the size distribution and refractive index of the aerosol as a function of geographic location, season, and relative humidity, and, therefore, describes the optical effect of a "representative particle." Total aerosol loading is specified independently. Although loading can vary significantly from day to day (Peterson et al., 1981; Kaufman and Fraser, 1983), regional statistics can be used to estimate the probable range of variations (Husar and Holloway, 1984). The major deviation from the Shettle and Fenn (1979) model is the effect of aerosol nonhomogeneity and nonsphericity on the aerosol optical characteristics (Bohren, 1986). Although a general theory to model these effects does not exist, calculations for concentric spheres with different refractive indexes (Ackerman and Toon, 1981) show that the characteristics of a volume containing a given fraction of graphitic carbon and water-soluble aerosol depend on the physical structure of the aerosol (details in Section III).

II.C Clouds

The interaction of sunlight with clouds affects remote sensing of the Earth's surface. For example, it affects the remotely sensed value of the surface reflectance and products such as vegetation index, surface albedo, and field classification. Whereas large clouds (large relative to the pixel size) can be accounted for due to their high reflectance in the visible and near-IR part of the spectrum (and usually low temperature observed in the IR), subpixel clouds are difficult to account for, and, therefore, in the following, we discuss their characteristics. The optical characteristics of a cloud layer can be described by the cloud size distribution, the reflection and transmission characteristics as a function of the cloud size, and the cloud fraction, defined as the fraction of the area of the surface covered by the cloud projections.

II.C.1 Cloud Size Distribution

A summary of cumulus cloud size distribution, $n_c(d)$, where d is the cloud diameter, was given by Joseph (1985). Based on these data, Joseph (1985) suggested the use of an exponential law size distribution. An exponential distribution was also used by Plank (1969) for cumulus clouds measured from an aircraft above Florida. Joseph's data indicate that cloud size increases with cloud fraction (χ_C):

$$n_c(d) = C \exp(-\alpha d) \begin{cases} \alpha = 10\text{–}20 \ \chi_c [\text{km}^{-1}] & \text{for } \chi_c < 0.5 \\ \alpha = 0 & \text{for } \chi_c > 0.5 \end{cases} \qquad (7)$$

This size distribution is for $0.05 \text{ km} < d < 1.0 \text{ km}$. The dependence of α on χ_C was found as a linear fit to the results summarized by Joseph (1985). Schmetz (1984) suggested a statistical clustering technique as a way to explain the increase in the cloud's size with the increase in cloud fraction. An estimation of the cumulus cloud distribution from Landsat images, published recently by Wielicki and Welch (1986), shows a power law form.

II.C.2 Cloud Reflection and Transmission

Pochop et al. (1968) found an empirical relation between the transmission of a cloudy sky (T_{cs}) and the cloud fraction χ_c measured by an observer on the surface:

$$T_{cs} = 1 - (1 - C_i)\chi_c^2 \tag{8}$$

where C_i is the transmission for an overcast sky for cloud type i and for a given solar zenith angle (θ_o). Empirical expressions for C_i are given by Haurwitz (1948):

$$C_i = (A_i/94.4) \exp\left[-(B_i - 0.059)/\cos \theta_o\right] \tag{9}$$

where A_i and B_i are given in Table 1. Haurwitz analyzed ground observations of the total solar insolation on the ground as a function of the solar zenith angle θ_o and cloud type for an overcast sky. From Eq. 8, we can derive the average transmission of the clouds themselves, since T_{cs} is a weighted average transmittance of clouded and cloud-free areas (taken as 1):

$$T_{cs} = T_c\chi_c + (1 - \chi_c) \tag{10}$$

Therefore,

$$T_c = 1 - (1 - C_i)\chi_c \tag{11}$$

A comparison between the transmission model of Eq. 8, for alto-cumulus and strato-cumulus, to aircraft measurements analyzed by Schmetz (1984), for broken strato-cumulus, is shown in Figure 3. The model fits the measurements for an overcast sky, but

TABLE I   Coefficients A_i and B_i and Cloud Transmittance C_i for an Overcast Sky for Eight Cloud Types

Cloud Type	A_i	B_i	$C_i(\theta_o = 45°)$
Cirrus	82.2	0.079	0.85
Cirro-stratus	87.1	0.148	0.81
Alto-cumulus	52.5	0.112	0.51
Alto-stratus	39.0	0.063	0.41
Atrato-cumulus	34.7	0.104	0.35
Stratus	23.8	0.159	0.22
Nimbo-stratus	11.2	−0.167	0.16
Fog	15.4	0.028	0.17

Source: After Haurwitz (1948).

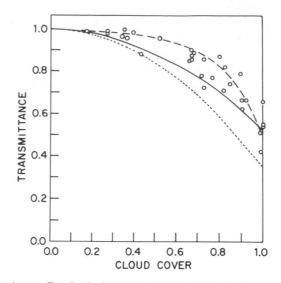

Figure 3   Solar transmittance T_{cs} of a broken strato-cumulus deck, as measured with an aircraft (after Schmetz, 1984) versus cloud fraction, and the transmittance curves (Eq. 8) for the empirical model of Pochop et al. (1968) and Haurwitz (1948) for strato-cumulus (dotted line) and alto-cumulus (solid line). Schmetz (1984) model M is also shown (dashed line). The solar zenith angle is 50° and the optical depth of the cloud layer was 10.

predicts lower transmission than measured for $\chi_c < 1$. This figure emphasizes the problems in applying the empirical cloud model to a specific experiment.

III   ATMOSPHERIC OPTICS

The variability in optical characteristics of the atmosphere results mainly from variability of gaseous absorption, discussed before, and from variability of the aerosol characteristics and loading. Physical aerosol characteristics can be described by the size distribution of the aerosol, by the refractive index of the chemical material that composes the aerosol (assuming that aerosol particles are homogeneous), and by the aerosol density. Note that these properties may vary in time and space. The refractive index may also vary as a function of particle size. In particular, smaller aerosol particles (less than 1 micron) are known to originate from a different source than larger particles, and thus have a different refractive index (Shettle and Fenn, 1979). A representation of the aerosol optical characteristics is given by its interaction with sunlight (Figure 4). A small volume of aerosol can absorb a fraction of the radiation and scatter a different fraction. Thus, the optical characteristics of a small volume of aerosol can be described by

- K_e, the extinction coefficient (km^{-1}), which describes the fraction of radiation taken from the direct beam by the aerosol,
- ω_o, the single-scattering albedo, which is the fraction of scattering from the total extinction (ω_o = scattering coefficient/extinction coefficient), and
- $P(\theta)$, the scattering phase function, which describes the angular distribution of scattered radiation (θ is the scattering angle).

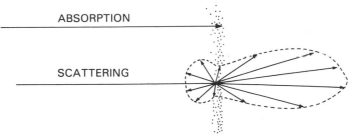

Figure 4   Schematic presentation of scattering and absorption processes in a volume of air.

The relation between the physical characteristics of the aerosol (size distribution and refractive index) and the optical characteristics (extinction coefficient, single-scattering albedo, and scattering phase function) can be computed for homogeneous spherical particles by the Mie theory (Mie, 1908; Hansen and Travis, 1974). According to this theory, the extinction cross section $\beta(\mathrm{cm}^2)$ can be computed as a function of the particle radius r (β is the effective cross section of the particle interacting with radiation). For particles much smaller than the radiation wavelength, ($r \ll \lambda$), β is smaller than the geometric cross section of the particle (see Table 2). Therefore, the extinction efficiency factor Q (the ratio of scattering cross section to total cross section) is $Q(r/\lambda) = \beta(r)/(\pi r^2) < 1$. For a particle much larger than the wavelength, $r \gg \lambda$, the extinction cross section is about twice the geometrical cross section. In between, $Q(r/\lambda)$ oscillates as a function of the ratio of the particle size to the wavelength, reaching as high as $Q = 4$ for $r/\lambda \approx 1$.

The extinction coefficient of a small volume of air (K_e) is computed as the contribution of all the particles in the volume, each with radius r, cross section β, and density $n(r)$ per radius interval dr:

$$K_e = \int \beta(r)\,n(r)\,dr \tag{12a}$$

TABLE 2   Extinction Efficiency $Q(r/\lambda) = \beta(r)/\pi r^2$ and the Single-Scattering Albedo ω_0 as a Function of the Ratio of Particle Size r to Radiation Wavelength λ (Computed by the Dave and Gazdag Code (1970) for Refractive Index 1.43–0.0035i)

r/λ	$Q(r/\lambda)$	ω_o
10.00	0.218E+01	0.743
3.167	0.255E+01	0.895
1.000	0.338E+01	0.969
0.317	0.129E+01	0.977
0.100	0.328E−01	0.834
0.032	0.177E−02	0.158

with $n(r)$ normalized as

$$\int n(r) \, dr = N_o \tag{12b}$$

where N_o is the density of the particles (cm^{-3}).

Each particle also has a different scattering phase function $P(r, \theta)$. The average phase function of the volume of aerosol is given by

$$P_a(\theta) = (1/N_o) \int P(r, \theta) \, n(r) \, dr \tag{13}$$

The scattering properties of particles vary with particle size in a similar way to the extinction coefficient. In Figure 5, the scattering phase functions are plotted for several particle sizes. In order to obtain a smooth phase function, a log-normal size distribution was used (Eq. 5) with $\sigma = 0.4$ and values of the mean number radius as indicated in the figure. An increase in the particle size (relative to the wavelength) results in a stronger forward scattering. For $r/\lambda \leq 0.1$, the phase function is close to the molecular-scattering phase function, whereas for $r/\lambda = 10$, the phase function for small scattering angles is three orders of magnitude stronger.

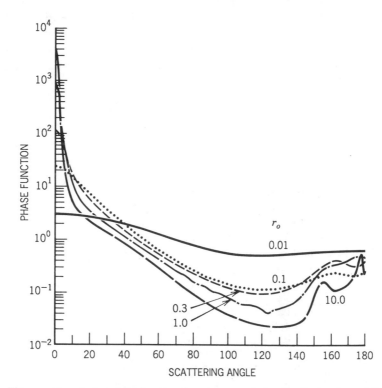

Figure 5 The aerosol scattering phase function for a log-normal size distribution ($\sigma = 0.4$) and several values of r_o (μm-indicated for each curve).

III.A Optical Thickness

The optical characteristics of a small volume of aerosol can be used to compute the characteristics of a whole aerosol layer, or the entire atmosphere. We defined previously the relation between optical thickness for gaseous scattering and the vertical transmittance of the direct sunbeam (Eq. 1). Therefore, from the relation between the extinction coefficient K_e and transmission from the top of the atmosphere ($z = \infty$) to height z – $T(z)$:

$$T(z) = \exp\left[-\int_z^\infty K_e(z)\,dz\right] \tag{14}$$

the optical thickness can be related to the extinction coefficient:

$$\tau(z) = \int_z^\infty K_e(z)\,dz \quad \text{or} \quad K_e(z) = d\tau/dz \tag{15}$$

where $z = 0$ at the Earth's surface, and $z = \infty$ at the top of the atmosphere.

The aerosol optical thickness depends on aerosol characteristics (size distribution and refractive index) and on aerosol total loading. It represents the attenuation of radiation propagating through the atmosphere (through its relation to transmission), and, therefore, can be used as a representative figure of the aerosol loading for the purpose of evaluating the aerosol effects on remote sensing. Although climatology of aerosol characteristics is available (Shettle and Fenn, 1979; Kneizys et al., 1983), it does not include statistics of the natural variability of aerosol optical thickness. One reason may be the strong spatial and temporal variations of the optical thickness. Statistics of this sort were performed for the United States for the period 1961–1969 (Flowers et al., 1969) and more recently for the eastern United States (Peterson et al., 1981). To generate such statistics, systematic measurements of the optical thickness by sun photometers are needed. Standard ground meteorological reports contain a measure of the visibility (V) on the surface. Since the optical thickness (τ) depends on the vertical extent of the aerosol, the correlation between τ and V is not very high (Kaufman and Fraser, 1983). For summer haze prevailing in the eastern United States, it was found that ground visibility can be used only to distinguish between three levels of haziness, or three groups of values of the optical thickness. Thus, it can be used as some indication of the haziness, but not a very accurate one. A small amount of optical-thickness data for the entire world exists. Roughly 100 stations world wide make once-a-day measurements of the optical thickness (R. S. Fraser, private communication). NOAA's National Climate Center in Asheville, North Carolina, disseminates the data. Presently, an effort to generate aerosol climatology is carried by D'Almeida and Koepke (1987).

The wavelength dependence of the aerosol optical thickness (τ_a) can be used to derive some information about the aerosol size distribution (averaged on a vertical column). The relation between the optical thickness and the size distribution is given in Eq. 12a for $K_e(z)$ and Eq. 15, which relates $K_e(z)$ to τ_a. The relation $\tau_a(\lambda)$ results from the fact that scattering is most efficient for particles where $r \sim \lambda$ (see Table 2). Therefore, the shape of the curve of $K_e(\lambda)$ or $\tau_a(\lambda)$ contains information on $n(r)$. For an ideal power law size distribution ($r_1 = r_2 = 0$ and $r_3 = \infty$), there is an analytical relation between ν and $\tau_a(\lambda)$ (Junge, 1963):

$$dn/d(\ln r) = Cr^{-\nu} \rightarrow \tau_a(\lambda) = \tau'\lambda^{-\nu+2} \qquad (16)$$

This relation can be used to get a first approximation on the size distribution (King et al., 1978), which often is sufficient for problems related to remote sensing (Kaufman and Fraser, 1983).

III.B Single-Scattering Albedo

The single-scattering albedo is the ratio between the aerosol scattering coefficient and the total extinction coefficient (scattering and absorption). It is a measure of aerosol absorption. A totally absorbing aerosol will have a zero value of single-scattering albedo, whereas a perfectly scattering aerosol will have a single-scattering albedo of 1.0. For a homogeneous aerosol, the single-scattering albedo is defined by a particle refractive index (mainly the imaginary part) and by a particle radius. Particles much smaller or much larger than the wavelength of radiation are more effective absorbers than particles in the intermediate range (see Table 2). For a nonhomogeneous aerosol, such as an external mixture of particles with different optical characteristics or an aerosol with composite particles (graphitic carbon core with a liquid shell), the single-scattering albedo depends on the characteristics of the composition (Ackerman and Toon, 1981). For example, for a 10% mixture of soot (originating mainly from automobiles) and 90% sulphuric compounds (generated by industry) results in a single-scattering albedo of 0.88 for an external mixture (separate particles) and 0.78 for an internal mixture (soot core and sulphuric shell). The phase function is also different in these two cases. In the backscattering direction, it can vary by a factor of 2 (Ackerman and Toon, 1981).

Waggoner et al. (1981) collected extensive measurements of the single-scattering albedo at ground level over urban and remote areas. Although these measurements do not represent necessarily the single-scattering albedo of the aerosol in the entire atmospheric column, they may be close to representing air in the boundary layer (lowest is 1–2 km), where most of the aerosol is present. They found ω_o in the range $\omega_o = 0.50$–0.65 for urban-industrial areas (such as Seattle, Denver, and Los Angeles), $\omega_o = 0.73$–0.87 for urban-residential areas (such as Seattle; St. Louis; Tyston, Missouri; Hall Mountain, Arkansas) and $\omega_o = 0.89$–0.95 in remote areas (Mauna Loa Observatory, Hawaii; Anderson Mesa, Arizona; Mesa Verde, Colorado; and the Abastumani Observatory, USSR).

III.C Scattering Phase Function

The scattering phase function is the third aerosol optical characteristic that has an effect on remote sensing. The few examples in Figure 5 show the dependence of the scattering phase function on the particle size for spherical aerosol particles. A mixture of molecular scattering (phase function similar to very small particles) with an increasing amount of aerosol will result in a phase function with increasing degree of asymmetry. Measurements of this phenomenon were performed by Barteneva (1960). The value of the phase function in the forward direction (scattering angle $\theta < 30°$) increased with the increase of optical thickness or decrease in the visibility. For $30° < \theta < 60°$, there was a mixed behavior of the phase function, and for $\theta > 60°$, the value of the phase function decreased with the increase of the optical thickness. A scattering angle of around 40° was often suggested as the angle where the variability of the scattering phase function is minimal.

III.D Radiative Transfer

In the previous sections, we discussed the characteristics of solar radiation and its interaction (absorption and scattering) with atmospheric gases and aerosol. This interaction was discussed using the microscopic approach, where a small volume of air was described by its extinction coefficient $K_e(\text{cm}^{-1})$, the single-scattering albedo ω_o, and the single-scattering phase function $P(\theta)$. Description of the radiation pattern in the Earth-atmosphere system (the macroscopic approach) requires integration of the radiative characteristics of a small volume of air through the entire atmosphere. This integration is performed with the help of the radiative-transfer (RT) equation (Chandrasekhar, 1960), which is reviewed in what follows.

By assuming a horizontally uniform atmosphere and uniform surface reflectance, the atmospheric characteristics are given by the extinction coefficient $K_e(z)$, the single-scattering albedo $\omega_o(z)$, and the phase function $P(\theta, z)$, where z is the vertical coordinate (see Figure 6), and θ is the scattering angle. The vertical variability of ω_o and P may result from variation in the mixing between aerosol and gases, variation in absorbing gas density, or in aerosol characteristics. The variability of $K_e(z)$ results mainly from variability of aerosol density (see the results of measurements given in Section V.C), although it may vary due to a variation of relative humidity and a resultant variation of particle size.

It is convenient to replace the vertical coordinate z with the dimensionless vertical optical thickness τ (Eq. 15). The RT equation traces the change in radiance $L'(\theta')$ during its transfer through a thin layer of air with vertical optical thickness $d\tau = K_e(z)\,dz$ located at τ (Hansen and Travis, 1974):

$$dL' = -L'\,d\tau/u + (J + J_o)\,d\tau/u \qquad (17)$$

where $u = \cos\theta'$, and θ' is the zenith angle of $L(\theta')$. θ' is between $0°$ and $90°$ for the downward radiation ($u > 0$) and $90° - 180°$ for the upward radiation ($u < 0$). With this definition, the ratio $d\tau/u$ is always positive. The first term on the right-hand side

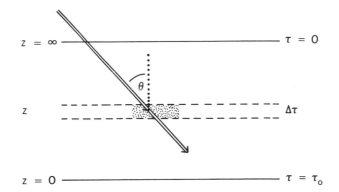

Figure 6 Schematic presentation of the radiative transfer of radiance $L'(\theta')$ through a layer of optical thickness $\Delta\tau$.

of Eq. 17 represents the attenuation of radiation by the layer. Transmittance through the layer in direction θ' is

$$T = \exp\left(-d\tau/u\right) \sim 1 - d\tau/u \tag{18}$$

Therefore, the change in the radiance $L'(\theta')$ due to attenuation is $L'\,d\tau/u$. The second term on the right-hand side of Eq. 17 represents the increase in radiance L' due to scattering of direct sunlight and diffuse light in the direction of radiation. J_o is the source term due to scattering of the direct sunlight:

$$J_o(u, \phi, \tau) = (\omega_o/4\pi) \exp\left(-\tau/\mu_o\right) P(u, \mu_o, \phi, \phi_o) E_o \tag{19}$$

where $\exp\left(-\tau/\mu_o\right)$ is the attenuation of solar radiation from the top of the atmosphere to the layer at τ, μ_o is cosine of the solar zenith angle (θ_o), P is the scattering phase function from the direction of the sun to (θ', ϕ), and E_o is the extraterrestrial solar irradiance. Since a spherical integral of P is defined here as equal to 4π, the phase function is divided by 4π in the source term (Eq. 19).

The source term due to scattering [in the direction (θ', ϕ)] of diffuse skylight and radiation reflected from the surface is termed J. It is given by

$$J(u, \phi, \tau) = \int \int P(u, u', \phi, \phi') \cdot L'(\tau, u', \phi') \, du' \, d\phi' \tag{20}$$

J is computed using sky and surface radiances, multiplied by the scattering phase function from the direction of the incoming radiation to (θ', ϕ). Dividing Eq. 17 by $d\tau/u$ results in

$$u\,dL'/d\tau = -L' + J + J_o \tag{21}$$

Equation 21 cannot be solved analytically due to the presence of the diffuse-light source term J. Radiance L' appears in a differential on the left side of the equation and in an integral on the right side. Analytical approximations to solve this equation have been suggested by using several ways to eliminate the integral in the equation. One simple way is to solve the equation for a nonreflecting surface and by assuming that there is no multiple scattering in the atmosphere. This is the single-scattering approximation. If light can be scattered only once, there is no contribution from the sky light scattered again in the atmosphere and $J = 0$. In this case, we can solve Eq. 21 (Hansen and Travis, 1974) to get the upward radiance:

$$L'_{\text{up}} = \frac{E_o \mu_o \omega_o}{4\pi(\mu + \mu_o)} \left\{1 - \exp\left[-\tau_o(1/\mu + 1/\mu_o)\right]\right\} \cdot P(-\mu, \mu_o, \phi) \tag{22}$$

and the downward radiance:

$$L'_{\text{down}} = \frac{E_o \mu_o \omega_o}{4\pi(\mu - \mu_o)} \left[\exp\left(-\tau_o/\mu\right) - \exp\left(\tau_o/\mu_o\right)\right] \cdot P(\mu, \mu_o, \phi) \tag{23}$$

where τ_o is the total atmospheric optical thickness, and $\mu = |u|$.

The single-scattering approximation is only valid for a small optical thickness; i.e., $\tau_o(1/\mu + 1/\mu_o) \ll 1$. In this case, we can also approximate the exponents in Eqs. 22 and 23 by the first two terms in a Taylor expansion, resulting in

$$L'_{\text{up}} = \frac{E_o\mu_o\omega_o\tau_o P(-\mu, \mu_o, \phi)}{4\pi\mu\mu_o} \quad \text{and} \quad L'_{\text{down}} = \frac{E_o\mu_o\omega_o\tau_o P(\mu, \mu_o, \phi)}{4\pi\mu\mu_o} \quad (24)$$

These two equations, although not accurate for most applications, present the basic characteristics of the atmospheric effect and can be used for a clear atmosphere. It shows that the upward path radiance depends on the product of the atmospheric optical thickness and the single-scattering albedo. This product is the contribution of scattering to the optical thickness. The angular distribution of the radiance for a clear atmosphere is a direct result of the scattering phase function. This explains the bright aureole around the sun and the dark sky far from the sun. Other approximate solutions that try to account for multiple scattering eliminate the integral in J (to make the equation solvable) by assuming that the radiance is independent of direction. One value is assumed for the upward radiation and a different value for the downward radiation (Meador and Weaver, 1980; Joseph et al., 1976). In this case, Eq. 21 can be solved analytically for a uniform surface as the boundary condition (Kaufman, 1978, 1979).

IV ATMOSPHERIC EFFECTS

Remote sensing of surface characteristics is based on measurements of upward radiance emerging out of the atmosphere. The atmospheric effect on upward radiance for a cloud-less sky can be computed as a solution to the radiative-transfer (RT) equation. The most general solution of the RT equation is for the case of a nonuniform non-Lambertian surface as the boundary condition. This is a complex solution and can be performed by approximate methods, such as the single-scattering approximation (Tanre et al., 1981) or by a Fourier-series representation of the radiance (Diner and Martonchik, 1985a) for several spatial frequencies. Usually, for estimation and correction of atmospheric effects on remote sensing, the general solution is not required and solutions for simpler cases, such as a uniform surface or a nonuniform but Lambertian surface, are sufficient. An alternative approach to model the coupled system of surface and atmosphere is to use a single radiative-transfer model that includes radiative transfer in the atmosphere as well as the Earth's surface. Such a model was developed by Gerstl and Zardecki (1985a,b) for the coupled atmosphere and canopy. In the model, the atmosphere was represented by several layers and the canopy by other layers, each with its specific radiative-transfer characteristics. If a fast and less accurate estimation of the atmospheric scattering and absorption on the performance of satellite and airborne platforms is needed, the newly developed LOWTRAN model (Isaacs et al., 1987) can be used. This model uses line-by-line computations of the atmospheric absorption and scattering, approximating multiple scattering by the two-stream approximation. In this approximation, the upward radiance can be computed with an error up to 20% for optical thickness of 1.0 (Isaacs et al., 1987). The error is expected to be linear with the optical thickness for $\tau_a \leq 1.0$.

In order to describe the relation between the different cases, let us consider the three main components of the upward radiance (Figure 7) as detected by a sensor with a narrow field of view.

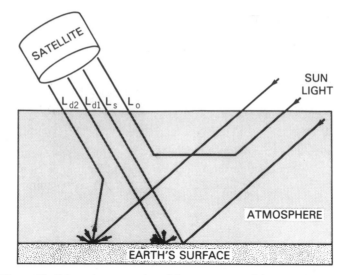

Figure 7   Schematic presentation of the components of the upward radiance.

$$L' = L'_o + L'_s + L'_{d1} + L'_{d2} \tag{25}$$

where

L'_o is the path radiance, the radiance of light scattered from the direct sunbeam by the atmosphere into the sensor's field of view without being reflected from the surface. This component is independent of the surface reflectance and, therefore, has the same value for a uniform, nonuniform, or non-Lambertian surface. This component causes a loss of contrast in the image by uniformly brightening the image.

L'_s is the attenuated signal, the radiance of light transmitted directly downward through the atmosphere, reflected from the surface, and directly transmitted through the atmosphere to the sensor. This component depends only on the reflectance of the surface in the field of view. It provides information about the surface.

L'_{d1} is the radiance of light scattered by the atmosphere before reaching the surface, reflected by the surface, and directly transmitted through the atmosphere to the sensor. This component results from reflection in different directions of the diffuse radiation and, therefore, is affected by non-Lambertian source properties.

L'_{d2} is the radiance of light reflected by the surface and then transmitted to the sensor with at least one scattering in the atmosphere. For a high-resolution sensor, this component depends on the reflectance of the surface areas located out of the field of view and, therefore, will be affected by a nonuniform surface. It also results from reflection in different directions of the diffuse radiation and, therefore, is affected by the surface non-Lambertian reflection. Light reflected from the surface more than once is also included in L'_{d2}.

The simplest solution of the RT equation is for the case of a uniform Lambertian surface. Although the surface is simplified in this case, this is the most useful solution, since in many applications, there is not enough information about the state of the atmosphere or the surface to gain from a detailed solution. For low-resolution satellite data (pixel size > 0.5 km), the effects of the surface nonuniform properties are negligible. Therefore, the solution for a nonuniform surface is important only for high-resolution imagery (e.g., SPOT and Landsat).

IV.A   Uniform Lambertian Surface

In this case, the solution to the RT equation can be represented as an analytical function of the surface reflectance (Chandrasekhar, 1960; Fraser and Kaufman, 1985):

$$L'(\rho, \mu, \phi) = L'_o(\mu, \phi) + \rho E'_d(\theta_o) T(\mu)/[\pi(1 - s\rho)] \qquad (26)$$

where L' is the upward radiance at the top of the atmosphere for surface reflectance ρ; L'_o is the radiance for $\rho = 0$; E'_d is the total irradiance of light at the surface for $\rho = 0$; ρ is the hemispherical reflectance of the surface, which reflects light according to Lambert's law; T is the transmittance from the ground to the top of the atmosphere; and s is the reflectance of the atmosphere for isotropic light entering the base of the atmosphere. The first term on the right-hand side of Eq. 26 is just the radiance of the atmosphere if the surface were nonreflecting ($\rho = 0$). The second term gives the direct and diffuse transmissions of light reflected from the ground to a satellite.

Throughout the remainder of this chapter, radiances and fluxes will be normalized to reflectance units. As a result, the normalized radiances L and L_o are related to the absolute spectral radiances L' and L'_o by

$$L = \pi L'/E_o \cos\theta_o \quad \text{and} \quad L_o = \pi L'_o/E_o \cos\theta_o \qquad (27)$$

where $E_o \cos\theta_o$ is the solar spectral irradiance incident on a horizontal surface unit area at the top of the atmosphere. The normalized irradiance E_d is related to the absolute irradiance E'_d by

$$E_d = E'_d/E_o \cos\theta_o \qquad (28)$$

Equation (26) written in terms of the normalized units becomes

$$L(\rho, \mu, \phi) = L_o(\mu, \phi) + \rho E_d(\theta_o) T(\mu)/[(1 - s\rho)] \qquad (29)$$

As an example, if the optical thickness of the atmosphere was $\tau = 0$, then $L_o = 0$, $E_d = 1$, $T = 1$, and the radiance is equal to the surface reflectance $L(\rho, \mu, \phi) = \rho$. The advantage of using reflectance units is that the difference between L and ρ clearly shows the net atmospheric effect:

$$L - \rho = L_o - \rho[1 - E_d T/(1 - s\rho)] \qquad (30)$$

In this sense, L is the apparent surface reflectance observed from space.

The net atmospheric effect ($L - \rho$) may be positive or negative. The direction and magnitude of the effect depends clearly on the value of the surface reflectance, the aerosol characteristics, and the directions of observation and solar radiation. The aerosol characteristics that affect this difference are mainly the total aerosol loading (given by the optical thickness), the single-scattering albedo, and the fraction of light scattered backward by the aerosol. The second term in Eq. 30 is the contribution from L_s and L_d ($L_d = L_{d1} + L_{d2}$).

Aerosol and molecular optical thicknesses are wavelength-dependent (see Sections II.A and II.B). As a result, the upward radiance and its three components are wavelength-dependent as well. An example of the wavelength variation is shown in Figure 8 for dark ($\rho = 0.05$) and bright ($\rho = 0.40$) surfaces. The radiance was calculated in "atmospheric windows" with minimal gaseous absorption. The atmospheric effect, shown in the magnitude of L_o and L_d relative to L_s, generally decreases with wavelength as a result of the decrease of the optical thickness. The net atmospheric effect is positive for short wavelengths (the atmospheric effect increases the upward radiance) and negative for longer wavelengths. This is due to the dominant role of atmospheric scattering in the short wavelengths and of aerosol and gaseous absorption in the long wavelengths. The wavelength dependence of the atmospheric effect for a real surface cover is shown in Figure 9. In this figure, the reflectance and the upward radiance at the top of the atmosphere, for two values of ω_o, are plotted for a dense alfalfa canopy. As in Figure 8, the atmospheric effect tends to brighten the surface in the visible part of the spectrum and to darken it in the near IR. The darkening and brightening by the atmospheric effect is a result of a delicate balance between aerosol and molecular scattering and absorption by aerosol and atmospheric gases. This balance depends also on the value of the surface reflectance and observer-sun orientations, as demonstrated in Figure 10.

Several examples of the upward radiance as a function of the aerosol optical thickness are shown in Figure 10. The computations are performed by an RT code (Dave and Gazdag, 1970) for $\theta_o = 40°$ and $\lambda = 610$ nm. The four curves on each panel are labeled with the value of surface reflectance (ρ). The difference between this value and the ordinate of the corresponding curve is the change in apparent reflectance caused by the

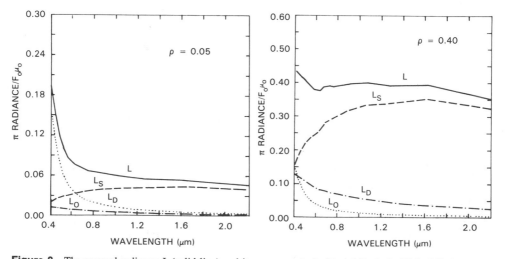

Figure 8 The upward radiance L (solid line) and its components L_o (dotted line), L_s (dashed line), and L_d (dashed-dotted line) plotted as a function of the wavelength for dark ($\rho = 0.05$) and bright ($\rho = 0.40$) surfaces. The radiance is normalized to reflectance units ($L = \pi L'/E_o\mu_o$) and is computed for nadir observations. The solar zenith angle is 30°. The aerosol optical thickness is $\tau_a = 0.36$ at $\lambda = 0.4$ μm and $\tau_a = 0.07$ at $\lambda = 2.2$ μm. (After Kaufman, 1984a.)

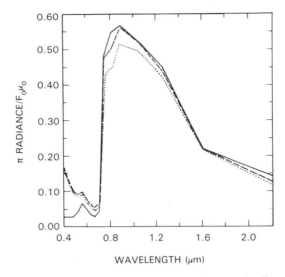

Figure 9 The surface reflectance for dense alfalfa (solid line), the upward radiance (dashed line) at nadir for relatively clear conditions (aerosol optical thickness is $\tau_a = 0.25$ at $\lambda = 0.56$ μm) and low absorption, $\omega_o = 0.96$; and the radiance for the same optical thicknesses, but moderate absorption, $\omega_o = 0.81$ (dotted line). The solar zenith angle is $30°$.

atmosphere. Figures 10(a) and (b) show the radiances for an extreme path length through the atmosphere encountered in current satellite observations ($\theta = 60°$). If the aerosol albedo of single scattering is large, Figure 10(a), the radiance increases linearly with respect to aerosol optical thickness, but the rate of increase becomes smaller as the surface reflectance becomes larger. For a smaller value of ω_o, or larger aerosol absorption, Figure 10(b), the slopes of the curves decrease and become negative for large surface reflectance. In the latter case, the apparent reflectance decreases markedly with more aerosol. Such an effect would be important in estimating the surface albedo of a desert ecosystem, for example, from satellite observations. The atmospheric effects with a minimum path length through the atmosphere (nadir viewing) are shown in Figures 10(c) and (d). For the nadir view, light reflected from the surface becomes more important relative to the L_o term of Eq. 29. The apparent reflectance of bright surfaces still decreases noticeably with additional amounts of moderately absorbing aerosol ($\omega_o = 0.81$).

The net atmospheric effect, the difference between the radiance above the atmosphere and the surface reflectance, decreases almost linearly with increasing surface reflectance (Eq. 30 and Figure 11). For any one of the two values of ω_o in Figure 11, the curves for the four values of aerosol optical thickness intersect within the neighborhood of a point. The abscissa of this point will be called the critical surface reflectance (ρ_c). The significance of ρ_c is that for $\rho = \rho_c$, the radiance stays essentially constant as the optical thickness changes. At the critical surface reflectance, an increase in aerosol amount results in more backward scattering of direct sunlight and thus larger L_o; but there is also a greater attenuation of light, resulting in smaller irradiance on the surface (E_d) and smaller atmospheric transmittance (T). Hence, the two terms on the right-hand side of Eq. 30 change by the same magnitude, but in opposite direction. Since the change in radiant energy by attenuation is weighted by the surface reflectance, the attenuation

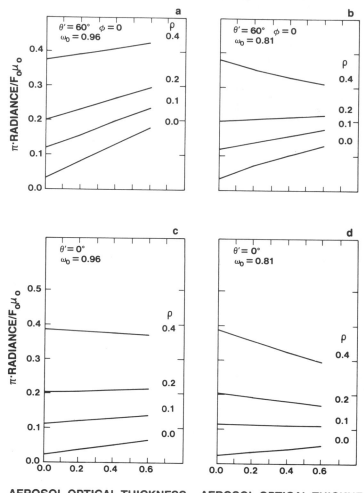

Figure 10   The radiance (in reflectance units) of light (0.61 μm) reflected from the Earth-atmosphere system as a function of the total aerosol optical thickness and surface reflectance ρ. The graphs are for different combinations of the aerosol single-scattering albedo ω_o and the observation zenith angle θ'. The solar zenith angle is $\theta_o = 40°$ (After Fraser and Kaufman, 1985.)

becomes stronger (weaker) when the surface reflectance is greater (smaller) than ρ_c. Therefore, for $\rho < \rho_c$, the radiance increases with increasing optical thickness (positive atmospheric effect), and for $\rho > \rho_c$, the radiance decreases with increasing optical thickness. The critical reflectance depends on the aerosol single-scattering albedo ω_o, the scattering phase function $P(\theta)$, the length of the path through the atmosphere, and the relative position of the line of sight and the sun (Fraser and Kaufman, 1985).

An example of the dependence of the measured atmospheric effect on the surface reflectance is given in Figure 12. Here the normalized upward radiance for a hazy day (August 2, 1982) was plotted as a function of the upward radiance for a clear day (August 20, 1982). The radiances are derived from the 700–800 nm band of a Landsat image over Washington, DC. The critical reflectance is $\rho_c = 0.18$.

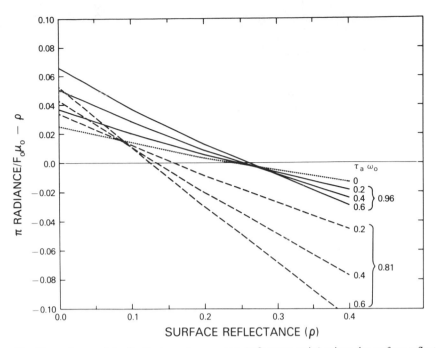

Figure 11 The radiance of the Earth-atmosphere system (reflectance units) minus the surface reflectance for nadir observations as a function of the surface reflectance. The aerosol optical thickness τ_a and the single-scattering albedo ω_o are indicated for each line. $\theta_o = 40°$, $\lambda = 610$ nm, and $\nu = 3$ (After Fraser and Kaufman, 1985.)

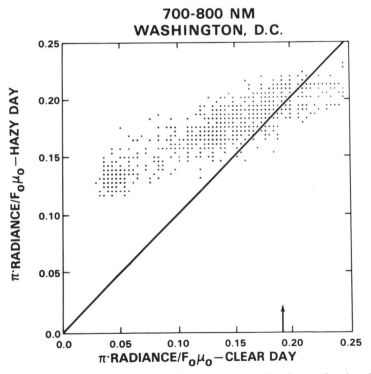

Figure 12 Scatter diagram of the radiance from a Landsat image on a hazy day as a function of the radiance on a clear day taken over Washington, D.C. The arrow indicates the radiance for which no charge occurs due to the haziness difference between the clear and hazy day. The waveband is 700–900 nm; solar zenith angle is 33° (After Fraser and Kaufman, 1985.)

IV.B   Non-Lambertian Surface

Water and vegetated surfaces have non-Lambertian reflectance characteristics, due to specular reflection, backscattering, and the effects of shadows and background soil (Kriebel, 1977; Kimes, 1983; Deering and Eck, 1987). As a result, it is interesting to test the effect of these characteristics on remote sensing under different amounts of haze. Gerstl and Simmer (1986) and Lee and Kaufman (1986) indicated that the non-Lambertian characteristics of the surface may affect remote-sensing measurements of vegetation mainly in the directions of specular reflection and backward scattering. In the following, these effects are demonstrated by computations performed by a RT model for a plane-parallel atmosphere with a horizontally uniform, non-Lambertian surface. This model was developed for rough ocean by Ahmad and Fraser (1982) and modified for vegetated surfaces by Lee and Kaufman (1986).

The components of the upward radiance above a non-Lambertian surface are illustrated in Figure 13. Here the angular distributions of the upward radiances at the top of a hazy atmosphere are compared to that just above a non-Lambertian canopy cover and to the surface reflectance ($\theta_o = 30°$). Pasture land is used as the lower boundary. At λ

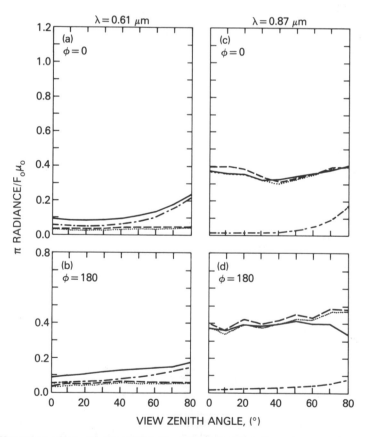

Figure 13   Upward emerging radiance (solid line), atmospheric path radiance (— · — · —), upward radiance at the base of the atmosphere (dotted line), and surface reflectance (dashed line) for (a) λ = 0.61 μm and φ = 0°, (b) λ = 0.61 μm and φ = 180°, (c) λ = 0.87 μm and φ = 0°, (d) λ = 0.87 μm and φ = 180°. The plots are for a hazy atmosphere over pasture land and solar zenith angle of $\theta_o = 30°$ (After Lee and Kaufman 1986.)

= 0.61 μm, Figures 13(a) and (b), where the reflectance is small and the atmospheric optical depth is large, the surface features are completely obscured by the atmospheric effect. In this case, the upward radiance above the atmosphere is primarily generated by the atmospheric path radiance (L_o). As a result, the upward radiance has a strong angular dependence (for view zenith angle $\theta' > 40°$) that is unrelated to the surface reflectance. At $\lambda = 0.87$ μm, Figures 13(c) and (d), where the surface reflectance is large and the optical depth is small, the angular distribution of the upward radiance above the atmosphere closely follows the surface features except for large view zenith angles at the 180° azimuth (sun at the back of the observer). The major contribution to the upward radiance is from the surface reflection, and the contribution by the atmospheric path radiance is relatively minor in this wavelength. A large deficit of the upward radiance for view angle $\theta' > 60°$ and $\phi = 180°$ is due to the combination of the relatively weak atmospheric backscattering and the extinction of the upward radiance by the larger path length.

For a low sun ($\theta_o = 60°$, Figure 14), the atmosphere exerts a stronger perturbation to the upward radiance than for a high sun ($\theta_o = 30°$). At both wavelengths, the upward radiance is very much enhanced in the forward-scattering direction, mostly due to the large atmospheric path radiances. Some significant differences between the features for low and high sun angles are found at $\lambda = 0.87$ μm, where surface reflectances for the low sun are highly anisotropic, being much stronger for backscattering. The upward

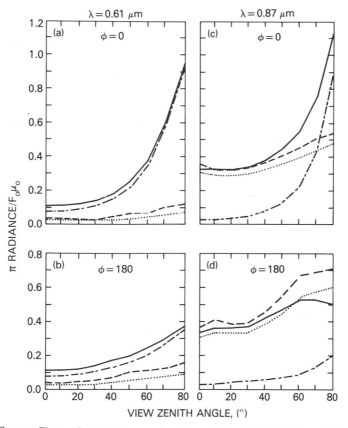

Figure 14 Same as Figure 13, but for the solar zenith angle $\theta_o = 60°$ (After Lee and Kaufman, 1986.)

radiance above the atmosphere is about equal to the surface reflectance for view zenith angle less than $50°$. But for $\theta' > 50°$, it is much larger in the forward direction ($\phi = 0°$) and much smaller in the backscattering region ($\phi = 180°$) than the surface reflectance. In this case, the angular dependence of the upward radiance just above the surface is also different than the surface reflectance. This is due to the non-Lambertian reflection of the diffuse skylight in addition to the direct sunlight (Lee and Kaufman, 1986; Deering and Eck, 1987). Deering and Eck (1987) measured the surface reflectance of orchard grass and soybeans under different haze conditions. They found that the higher fraction of diffuse skylight in a hazy day, resulted in significant differences in the surface reflectance measured from the ground (the measurements were carried for $\theta_o > 55°$). These effects are much stronger in the visible than in the near-IR, resulting in a significant change in the NDVI of ΔNDVI $= 0.05$.

In summary, these results suggest that for strong reflection (e.g., for $0.87\ \mu$m), radiance measured by a satellite sensor is similar to the canopy reflectance at most view angles for a high sun angle; however, for a low sun and large view zenith angles ($\theta' > 50°$), it is very different from the surface reflectance. In the case of weak reflection (as for $\lambda = 0.61\ \mu$m), surface features are completely masked off. Simmer and Gerstl (1985) and Gerstl and Simmer (1986) also obtained a similar result for a solar zenith angle of $29.3°$. Their results show that, for a surface reflectance larger than 0.1, local extremes in the angular distribution of surface reflectance were still detectable above the atmosphere.

It is interesting to compute the error in upward radiance obtained when using a Lambertian surface rather than the exact non-Lambertian profile. The most accurate approach is to use a different value of the Lambertian surface reflectance in each direction of observation that corresponds to the appropriate value of the non-Lambertian reflectance in this direction. A comparison between the upward radiance computed by these two approaches is shown in Figure 15. Here the upward radiance above Savannah (Kriebel, 1977) was computed (Lee and Kaufman, 1986) for a non-Lambertian surface (solid lines in Figure 15) and for a Lambertian surface with a different reflection coefficient for each view angle (dashed lines in Figure 15). For comparison, the surface reflectance is also plotted (dotted lines in Figure 15). For a high sun ($\theta_o = 30°$), Figures 15(a) and (b), the difference between the radiances resulting from the two models is generally small. Relatively large differences occur at the backscattering region ($\phi = 180°$) in both wavelengths. At a solar zenith angle of $60°$, Figures 15(c) and (d), large discrepancies are found in the backscattering plane for a view zenith angle larger than $40°$ in both wavelengths. For example, for $\lambda = 0.87\ \mu$m, the upward radiance for a Lambertian surface at $\theta' = 60°$ and $\phi = 180°$ is 0.07 larger than that of a non-Lambertian surface. This may be attributed to the anisotropic surface reflectance (dotted line). The reflectance ρ at $\theta' = 60°$ and $\phi = 180°$ is 0.43, whereas it is much smaller for most surrounding angles, except for $\theta' = 70°$ and $\phi = 180°$ (the upward radiance is affected by surface reflections from all angles). Consequently, it becomes smaller for the non-Lambertian surface than for a Lambertian surface that is represented by $\rho = 0.43$.

It is concluded that for the three types of vegetations considered, the assumption of a Lambertian surface can be used satisfactorily for computations of the upward radiance and for the derivation of surface reflectance from remotely measured radiances for view angles outside the backscattering region. Within the backscattering region, however, the use of the assumption can result in a considerable error in the upward radiance and in the derived reflectance even with a small amount of atmospheric aerosol.

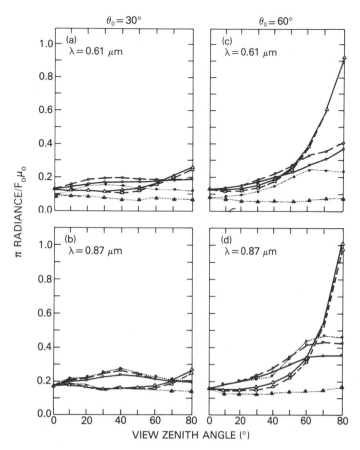

Figure 15 Upward emerging radiances for Lambertian (dashed) and non-Lambertian (solid) surfaces and the corresponding surface reflectance (dotted) in the principle plane: $\phi = 0°$ (Δ, Δ) and $\phi = 180°$ (o, o) for (a) $\theta_o = 30°$ and $\lambda = 0.61$ μm, (b) $\theta_o = 30°$ and $\lambda = 0.87$ μm, (c) $\theta_o = 60°$ and $\lambda = 0.61$ μm, (d) $\theta_o = 60°$ and $\lambda = 0.87$ μm. Hazy atmosphere over savannah. (After Lee and Kaufman, 1986.)

The strong effect of a non-Lambertian surface in the backscattering region can be understood by comparison of the relative contributions of L_s and L_d to the upward radiance. While the RT model for a non-Lambertian surface computes both L_s and L_d correctly, the model for a Lambertian surface assumes the same value of the surface reflectance in the computation of L_s and L_d. In the backscattering region, surface reflectance varies sharply with the direction of observation. As a result, there is a large difference between the surface reflectance used in L_s and an effective surface reflectance needed to compute L_d. This difference is not accounted for in the Lambertian RT model.

IV.C Nonuniform Lambertian Surface

The difference between remote sensing of a uniform and a nonuniform surface is in the values of the diffuse radiance term, L_{d1} (see Eq. 25). This radiance is affected not only by the radiation reflected from the observed area (as in L_s), but also by radiation reflected from the surrounding areas (Figure 7), out of the field of view, and scattered by the atmosphere into the field of view. This process (termed the adjacency effect; Otterman and Fraser, 1979) introduces information from areas that are not observed, and therefore contaminates the remote-sensing signal (Tanre et al., 1979, 1981; Mekler and Kaufman, 1980; Kaufman, 1982). A comparison between the atmospheric effect on the detected radiance for uniform and nonuniform surfaces is shown in Figure 16 (Kaufman and Fraser, 1984). This figure represents the changes in the three components of the radiance for an infinitesimally small field, for three different cases of background reflectance relative to the field reflectance:

(a) background much brighter than the field,
(b) uniform surface (same brightness of the field and the background), and
(c) darker background.

For a Lambertian surface, there is no way to distinguish between L_s and L_{d1}; therefore, we shall regard L_{d1} as part of L_s in the following discussion (and in Figure 16).

The effect of the atmosphere on L_o is the same for all the cases, and nearly the same for L_s, for specified optical thickness. The diffuse radiance L_{d2} decreases as the background reflectance decreases. The bright background in Figure 16(a) causes a strong increase in the total radiance L with optical thickness. This increase is very small for a uniform surface, Figure 16(b), and the dark background in Figure 16(c) causes a decrease in the radiance with an increase in the optical thickness. We conclude from this discussion that for the same atmosphere and for the same observed field, the radiance may increase or decrease, depending on the brightness of the field's background. This phenomenon (the adjacency effect) cannot be predicted by an atmospheric model that is based on the assumption of a uniform surface reflectance, and it can be computed only using multidimensional radiative-transfer computations. The adjacency effect depends on the field size; see Figure 17. This figure shows the results of Pearce's (1977) Monte-Carlo computations of the nadir radiance above the center of a square field surrounded by a different background. The reflectance of the field is $\rho_f = 0.4$ for the upper three curves and the reflectance of the background is $\rho_b = 0.2$. For the lower three curves, $\rho_f = 0.2$ and $\rho_b = 0.4$. In this figure, the radiance is shown as a function of edge length of the square field. The atmosphere is nonabsorbing and is modeled with three different aerosol optical thicknesses (0.0, 0.21, and 0.64). The asymptotic values of the radiances for infinitely large fields are given on the right side for each case. It is seen that when $\rho_f = 0.4$, a change in the aerosol optical thickness from 0 to 0.64 causes an increase in the radiance by 4% for an infinite field (uniform surface), whereas, for a finite field of edge length of 0.3 km, the effect is to reduce the radiance by 15%. This comparison emphasizes that the atmospheric effect is different for finite fields than for infinite fields. The adjacency effect is significant for field sizes up to 2 km with some remaining effect up to 10 km. Dana (1982) found from analysis of Landsat imagery, that the adjacency effect on a given pixel is due to the background reflectance that extends no more than 1 km from the pixel. This value agrees with the field size of 2km discussed before.

The results of the Monte-Carlo computations are also compared to exact numerical results (Dave and Gazdag, 1970) for a uniform surface. The latter radiances are labeled "Dave Results" on Figure 17. The comparison was performed also for an infinitesimal field on a large background. For very small fields, the Dave code was applied by the following procedure. For a uniform surface, the radiance is explicitly dependent on the surface reflectance ρ and given by Eq. 29. For an infinitesimal field, the radiance L_{in}

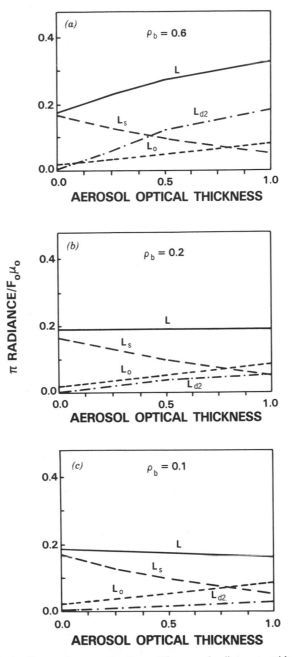

Figure 16 Atmospheric effect on the upward radiance. The upward radiance at zenith normalized to reflectance units is plotted as a function of the aerosol optical thickness. The radiance is for an infinitesimally small field of reflectance $\rho_f = 0.2$. The background reflectances ρ_b are indicated in each figure. The calculations were performed for a solar zenith angle of $30°$, wavelength of 550 nm, and aerosol refractive index of 1.43 $- 0.0035i$. (After Kaufman and Fraser, 1984.)

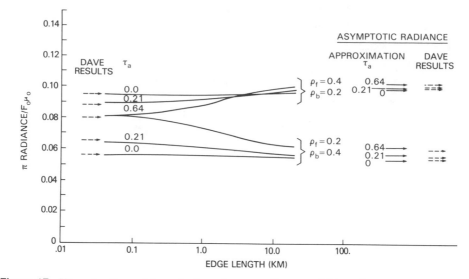

Figure 17 Upward radiance at the nadir above the center of a square field as a function of the edge length. (After Pearce, 1977.) The aerosol optical thickness τ_a is given for each curve. $\lambda = 550$ nm, $\theta_o = 40°$, and the refractive index is $1.43 - 0.0035i$. The field (ρ_f) and the background (ρ_b) reflectances are given for each group of curves. The asymptotic radiance (∞) and the radiances for an infinitesimal field are compared with exact computations according to Dave's code ($- - \rightarrow$). (Kaufman and Fraser, 1984.)

can be expressed as a function of the reflectances of the field (ρ_f) and the background (ρ_b):

$$L_{in} = L_o + E_d(\rho_f T_s + \rho_b T_d)/[(1 - s\rho_b)] \tag{31}$$

Here the diffuse radiance originates from fields with reflectance ρ_b, whereas the direct radiance originates from the field with reflectance ρ_f (Kaufman, 1979). The upward transmittance term (T) in Eq. 29 was separated here into the direct transmittance T_s of light reflected from the field with reflectance ρ_f and the diffuse transmittance T_d of light reflected from surrounding areas of reflectance ρ_b. The radiance L_{in} in Eq. (31) can be calculated by substituting L_o, E_d, and s from the Dave computations for a uniform surface and computing T_d from Eq. 29 for any uniform surface with reflectance ρ.

The magnitude of the adjacency effect is illustrated in Figure 18 for several field sizes. In this figure, the upward radiance above the center of a square field is plotted as a function of aerosol optical thickness (τ_a). The results are calculated by rescaling the Monte-Carlo radiances (Pearce, 1977) to lower contrasts in the surface reflectance. Results for a field of reflectance of $\rho_f = 0.2$ surrounded by a black area ($\rho_b = 0.0$) are shown in Figure 18(a) for four different field sizes viewed. As the aerosol optical thickness increases, the upward radiance increases for large fields ($X = \infty$ and $X = 5.2$ km). For small fields ($X = 0.11$ km and $X = 1.23$ km), the radiance decreases as a function of the optical thickness. Figure 18(b) shows the radiance above a black field surrounded by fields of reflectance 0.2 (e.g., a lake surrounded by soil or vegetation). Here the adjacency effect increases the observed radiance above the field much more than it is increased by an atmosphere without the adjacency effect ($X = \infty$).

The influence of a simple surface nonuniformity on the upward radiance (a field surrounded by a uniform surface) was demonstrated before. For applications where there

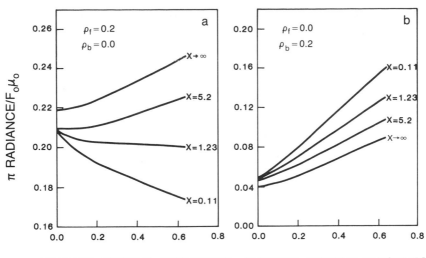

Figure 18 Reflectance of the Earth-atmosphere system in the direction toward the center of a square field located at the nadir. The length of a side is given by X (km). The reflectances of the target and of the background are ρ_f and ρ_b, respectively. $\lambda = 550$ nm, $\theta_o = 40°$, and the aerosol index of refraction is $1.5 - 0.0i$. The computations are based on a Monte-Carlo computation of Pearce (1977).

is a need to simulate the upward radiance for complex surface patterns and to design correction algorithms, there is a need to develop relations between the upward radiance and the surface reflectance pattern, such as the relations that were developed for a uniform surface (see Eq. 29). This relation is developed by the help of Fourier transforms (Mekler and Kaufman, 1980; Kaufman, 1982; Diner and Martonchik, 1985a). In order to be able to use the Fourier-transform approach, we need to assume that the upward radiance is linearly dependent on the surface reflectance. Strictly speaking, this is not correct since light reflected once from the surface may turn to it after being scattered backward in the atmosphere, which may be rereflected again by the surface. The contribution of this term to the upward radiance for a uniform surface is accounted for by the $(1 - s\rho)$ term in Eq. 29. Fortunately, since $\rho < 1$ and $s < 1$, the product $s\rho$ is usually smaller than 0.1; therefore, the effect of surface nonuniformity on this term may be approximated separately (Kaufman, 1982). By ignoring (for now) the multiple reflection of light from the surface, the atmosphere is a linear medium that transfers the signal (surface reflectance) to the observer.

In order to develop an expression for the relationship between the radiance and the nonuniform surface reflectance, $\rho(X, Y)$, imagine a black surface with one reflecting point on it, located at (X_0, Y_0) (a delta function):

$$\rho(X, Y) = \delta(X - X_0, Y - Y_0) \tag{32}$$

With no atmospheric effect, the upward radiance in any direction would be also described by a delta function. The atmospheric scattering broadens the radiance distribution to a bell-shape function called the atmospheric point-spread function:

$$L(X, Y) = m(X - X_0, Y - Y_0) + L_0 \tag{33}$$

For a surface with a varying reflectance, Fourier analysis describes the upward radiance $L(X, Y)$ as a convolution of the surface reflectance $\rho(X, Y)$ and the atmospheric point-spread function:

$$L(X, Y) = \int \rho(X', Y')\, m(X - X', Y - Y')\, dX'\, dY' + L_0 \qquad (34)$$

This equation can be written using Fourier transforms (Kaufman, 1984c):

$$L(X, Y) = \mathfrak{F}^{-1}\left\{\mathfrak{F}[\rho(X, Y)]\, M(k_x, k_y)\right\} + L_0 + L_\beta \qquad (35)$$

where $\mathfrak{F}$ and $\mathfrak{F}^{-1}$ are the Fourier transform and the inverse Fourier transform, respectively; L_o is the upward radiance for zero surface reflectance; M is the atmospheric modulation transfer function (MTF), a Fourier transform of the point-spread function; K_x and K_y are the spatial frequencies; X and Y are horizontal dimensions; and L_β is a residue radiance of light reflected by the surface and scattered backward within the atmosphere at least twice. Similar relations were developed by Kozoderov (1985) and Diner and Martonchik (1985a).

The two-dimensional MTF, $M(k_x, k_y)$, can be related to a one-dimensional MTF, $M(k)$, for nadir view (Kaufman, 1982):

$$M(k_x, k_y) = M(k) \qquad \text{where } k^2 = k_x^2 + k_y^2 \qquad (36)$$

This simplifies the computation of MTF, which is the key for simulation of the upward radiance for a two dimensional nonuniform surface, and it is used later for atmospheric corrections. For off nadir, the two-dimensional MTF have to be computed for each observation direction using a two-dimensional modulated surface by the methods of Diner and Martonchik (1985a), or Monte-Carlo computations (Pearce, 1977).

The computation of the upward radiance is performed by applying the Fourier transform of Eq. (35) to a two-dimensional array representing the surface reflectance at each point. Therefore, the algorithm has the ability to specify an independent surface reflectance at each point on the surface. The radiances L_o and L_β and the MTF M can be calculated by an approximation to the two-dimensional radiative-transfer equation (Kaufman, 1982).

The Fourier method can be applied to any two-dimensional surface reflectance pattern such as in Figure 19. In this example, the surface consists of a black triangular field ($\rho_f = 0$) on a bright background ($\rho_b = 0.4$). For the fourier transform calculations, the surface is composed of 128×128 square pixels, each with a length of 100 m. To each pixel, a reflectance of 0.0 or 0.4 is assigned. A hazy atmosphere (aerosol optical thickness of 0.63) is applied. The aerosol characteristics are the same as the aerosol used in the calculations of Figure 17. The atmospheric effect decreases the radiance over the bright area in Figure 19 and increases the radiance over the dark area close to the boundary between them. Since "close" means within the scale height of the atmospheric scattering, the effect is substantial almost all over the dark field. The degradation of the radiance of the bright background is strongest close to the center of the triangle edges and minimal close to its corners, where the "sink" of radiation is smaller. Over the black field, the atmospheric effect is strongest close to the corners of the field, where the bright background occupies a larger fraction of the total background area.

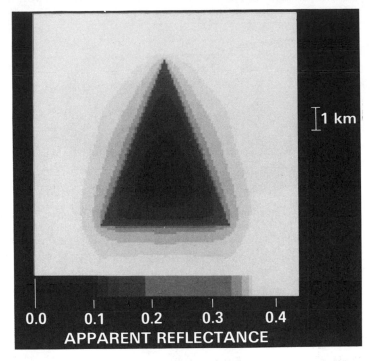

Figure 19 Atmospheric effect on the upward radiance above a nonuniform surface reflectance. The surface is composed of a triangular field of reflectance 0.0 and background of reflectance 0.4. The atmospheric effect diffuses the boundaries between the field and its background. The calculations are for $\tau_a = 0.63$, $\lambda = 550$ nm, $\theta_o = 40°$.

The atmospheric MTF $M(k)$ is the modulation of the spatial frequencies in the surface pattern by the atmosphere. Zero frequency represents infinitely large areas and the MTF is the contrast transmission through the atmosphere between two large uniform areas (Pearce, 1977):

$$M(0) = [L(\rho) - L(\rho = 0)]/\rho \qquad (37)$$

The normalized MTF $M_n(k_x, k_y) = M(k_x, k_y)/M(0)$ represents the net effect of the atmosphere coupled with the surface nonuniform reflectance. Figure 20 shows an example of the normalized MTF as a function of the spatial frequency k. Results of Monte-Carlo computations (Pearce, 1977), an analytical approximation (Kaufman, 1982), and an empirical fit (Kaufman, 1984b) are given in this figure. The empirical fit, although less accurate and limited in its validity to the range for which it was developed, can be a simple tool to simulate the atmospheric effect above a nonuniform surface. The empirical MTF is given by

$$M_n(k) = 1 - 0.5\tau_r\left[1 - \exp\left(-2.5kH_r\right)\right] - 0.7\lambda^{-0.2}\tau_a\left[1 - \exp\left(-1.3kH_a\right)\right]$$

$$(38)$$

where wavelength λ is expressed in μm. Here the MTF is given as a function of the spatial frequency for a given molecular optical thickness, aerosol optical thickness (τ_a),

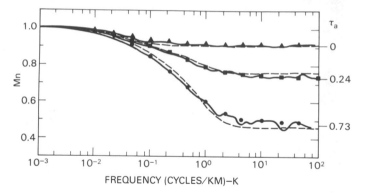

Figure 20 Normalized MTF as a function of the spatial frequency for three levels of atmospheric haziness and a wavelength of 470 nm. The solid lines are Monte-Carlo calculations; the dashed lines are empirical approximations Eq. 38; and the filled-in circles, squares, and triangles are physical approximations.

the scale height of aerosol H_a (~ 1km), and of molecules H_r(~ 8km), and wavelength. Equation 38 was tested and found valid for a wide range of wavelengths (0.47–1.65 μm) and optical thicknesses (0–0.64). The empirical formula fits the original curves within an error $\Delta M = \pm 0.04$ ($\pm 10\%$).

IV.D Effect of Subpixel Clouds

In the previous sections, the atmospheric effect on remote sensing was discussed for a cloudless sky. The presence of clouds has an additional effect on the quality of remotely sensed data. Due to the optical and thermal properties of clouds, we can distinguish between four cloud types:

- Clouds that are much larger than the field of view of the sensor with a large optical thickness. These clouds can be easily distinguished in the visible region due to their high wavelength-independent reflectance.
- Thin high clouds (cirrus), which may not be seen in the visible channels, but can be distinguished in the IR due to their low temperature.
- Low stratified clouds with a reflectance that is not high enough to be detected easily in the visible channels and with a temperature close to the surface temperature. These clouds may be considered as part of the aerosol layer (only much larger particles) and can be taken into account as such.
- Broken cloudiness with many of the clouds much smaller than the field of view of the sensor. These clouds do not belong to any of the above categories and therefore deserve a special treatment (Kaufman, 1987b).

In this section, we shall, therefore, discuss the radiative effects of low subpixel clouds on the upward radiance and on remote sensing of the surface. Expressions for the effect of the subpixel clouds on the upward radiance are given, and a method to minimize their effect is discussed.

IV.D.1 Theory (Kaufman, 1987b)

The algorithm presented here is based on radiative-transfer theory. In order to develop simple, yet sufficiently accurate, expressions to model the effect of broken cloudiness on remote sensing, it is assumed that the clouds form one thin layer of randomly distributed clouds of rectangular shape. It is further assumed that there is no multiple scattering between the clouds themselves, although an individual photon can be affected by one cloud on its way down and by a second cloud on its way up; however, this multiple-scattering effect is less than 25% (Schmetz, 1984; Aida, 1977).

The clouds are characterized by the cloud fraction χ_c, dimensions of the individual clouds X_c and Y_c, distribution of the cloud width $n(X_c)$, and the cloud's reflectance and transmittance (can be size-dependent). For computation of the diffused light, it is assumed that cloud reflectance (ρ_{cl}) and transmittance (T_c) are Lambertian. Also, the probability of cloud fraction $P(\chi_c)$ has to be specified as well as the cloud base altitude (H_c).

Since the atmosphere can contain absorbing and scattering gases and aerosols in addition to clouds, the following atmospheric optical characteristics are used for cloud-free conditions: L_o, path radiance; E_{dd}, the downward diffuse irradiance due to aerosol and molecular scattering for $\rho = 0$ (in reflectance units; see Eq. 28); E_{dd}^*, the downward diffuse irradiance, including cloud scattering for cloud fraction χ_c:

$$E_{dd}^* = E_{dd}(1 - \chi_c) + T_c\chi_c \tag{39}$$

The molecules and aerosol are assumed to be located under the cloud layer. The surface is assumed to be Lambertian with reflectance ρ. Computation of the upward radiance in the presence of clouds is based on four different surface observation conditions (see Figure 21):

(a) no cloud effect—sunny surface observed directly from space,
(b) the surface is in a sunny area but observed through a cloud,
(c) the surface is in a shade but observed directly from space (not through a cloud), or
(d) the surface is in a shade and observed through a cloud.

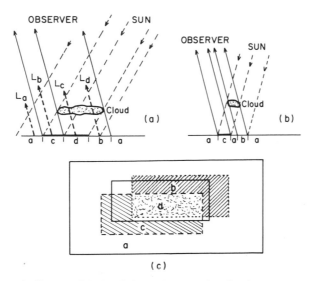

Figure 21 A schematic diagram of the irradiation and observation directions: (a) with an overlap between the cloud and its shade from the direction of the observer; (b) no overlap; and (c) the cloud (enclosed in the solid line), its shadow (dashed line), and the area covered by the cloud from the observer (dotted line).

For each of the previous areas, a separate computation of the upward radiance and its relative contribution to the average radiance over the area is performed.

(a) *A Sunny Area Observed Directly from Space.* The upward radiance L_a is given by (similar to Eq. 29)

$$L_a = L_o + (E_s + E_{dd}^*)\rho T/(1 - s'\rho) \tag{40}$$

Here the upward radiance results from path radiance L_o and total solar irradiance $E_s + E_{dd}^*$ reaching the ground, reflected from it with reflectance ρ and transmitted upward through the atmosphere, to the observer, with transmittance T. E_s is the direct solar irradiance (reflectance units). The denominator, $1 - s'\rho$, represents the enhancement of the upward radiance due to the multiple reflectance of light between the surface and the atmosphere. The only cloud effect presented in this expression is in the atmospheric backscattering s':

$$s' = s + \chi_c \rho_{cl} \tag{41}$$

It is assumed that the cloud layer is above the aerosol layer, and thus, in addition to the reflectance from the aerosols and molecules s, we find the cloud reflectance ρ_{cl} weighted by the cloud fraction χ_c.

(b) *Above a Sunny Area Viewed Through a Cloud.* The radiance is composed of the cloud reflectance ρ_{cl} and transmittance through the cloud of the upward radiance from under the cloud. Since it is assumed that the cloud transmittance T_c is completely diffused, the transmitted radiance is a weighted sum of the radiance above a shady area L_c (see below) and above a sunny area L_a:

$$L_b = \rho_{cl} + T_c[L_a(1 - \chi_c) + L_c \chi_c] \tag{42}$$

(c) *Above a Shady Area Observed from Space.* The radiance results from reflection of the downward irradiance E_{dd}^* by the surface and transmission T through the atmosphere to the observer:

$$L_c = E_{dd}^* T\rho/(1 - s'\rho) \tag{43}$$

The cloud shadow on part of the atmosphere decreases the path radiance L_o and generates a complex dependence of path radiance on the location, observation, and solar irradiation directions. Therefore, the path radiance L_o is not included in this expression.

(d) *Over a Shady Area Observed Through a Cloud.* The radiance is given by the same expression as L_b, due to the diffuse nature of the cloud transmittance ($L_d = L_b$)

The total upward radiance is computed as the weighted sum of all these terms:

$$L = W_a L_a + W_b L_b + W_c L_c + W_d L_d \tag{44}$$

The weighting functions W_i are proportional to the fraction of the surface that belongs to each of the categories.

To compute the weighting functions W_i, we should distinguish between two situations:

(A) From the observer's point of view, there is an overlap between the cloud shadow and the cloud itself; see Figure 21(a).
(B) There is no overlap between the two; see Figure 21(b).

The overlap area S_d, see Figure 21(c), is

$$S_d = [X_c - H_c(\tan \theta' \cos \phi + \tan \theta_o \cos \phi_o)] \\ \cdot [Y_c - H_c(\tan \theta' \sin \phi + \tan \theta' \sin \phi_o)] \qquad S_d \geq 0 \qquad (45)$$

where θ and ϕ are the observer zenith and azimuth angles, respectively, and ϕ_o is the azimuth of the sun.

In the case of an overlap (case A), $S_d > 0$; therefore, the areas S_b and S_c in Figure 21(c) are

$$S_b = S_c = X_c Y_c - S_d \qquad (46)$$

If there is no overlap (case B), Eq. 45 will result in a negative value of S_d. In this case, S_b and S_c are equal to the cloud area and the overlap area S_d is zero.

For a single cloud, with no other clouds around, the weighting functions W_a, W_b, W_c, and W_d are expressed by the ratio of the areas S_b, S_c, and S_d to the cloud area $(X_c Y_c)$ and weighted by the cloud fraction χ_c:

$$W_i = \chi_c S_i/(X_c Y_c) \qquad i = b, c, d \qquad \text{and} \qquad W_a = 1 - (W_b + W_c + W_d) \quad (47)$$

For randomly distributed clouds, the weigting functions should allow for observation through one cloud of an area shaded by a second cloud. The resulting weighting functions are (Kaufman, 1987b)

$$W_b' = W_c' = \chi_c(1 - \chi_c)(X_c Y_c - S_d)/(X_c Y_c) \qquad (48a)$$
$$W_d' = \chi_c[S_d + (X_c Y_c - S_d)\chi_c]/(X_c Y_c) \qquad (48b)$$

and

$$W_a' = 1 - (W_b' + W_c' + W_d') \qquad (48c)$$

Once the upward radiance is found as a function of cloud size, the radiance for an ensemble of clouds with size distribution $n(X_c)$ is obtained from the integral

$$L' = \int L(X_c) n(X_c) dX_c \qquad (49)$$

Here $L(X_c)$ is given by Eq. 44, the individual radiances by Eqs. 40–43, and the weighting functions by Eqs. 48a, 48b and 48c.

IV.D.2 Radiances

The effect of cloudiness on the upward radiance, as given by Eq. 49, is shown in Figure 22. The radiance was computed for an exponential cloud-size distribution. The observer and the sun are in the principal plane ($\theta' = \theta_o = 45°$), $\phi_o = 0$, with azimuth $\phi = 0°$ and $\phi = 180°$. Two cloud transmission models were used in the computations. In Figure 22(a), the cloud transmittance was independent of the cloud fraction and the cloud size ($T_c = 0.45$, $\rho_{cl} = 0.5$). In Figure 22(b), the cloud transmittance varied according to Eq. 11 for three different cloud types. Note that the cloud reflectance should be expressed as a function of the cloud size. But, since such measurements are not available, we preferred to use the empirical relations of the cloud transmissivity as a function of the cloud fraction instead. The cloud base was $H_c = 1$ km. In both cases, the cloud effect on the upward radiance is stronger for $\phi = 180°$. For this azimuth, the sun is behind the observer and, therefore, there are no shadows that decrease the brightness and the cloud effect. For $\phi = 0°$, the shadows are maximal and the radiance does not increase as much with the increase of cloud fraction. The effect is smaller for clouds with a transmittance that is proportional to the cloud fraction (Eq. 11).

IV.D.3 Minimizing the Cloud Effect

In order to minimize the cloud effect on remotely sensed surface reflectance or vegetation index, the surface can be sensed on several days and a day with minimal reflectance or maximal vegetation index should be used (Tucker et al., 1983). This procedure is based on the assumption that the surface does not change substantially during a period of a few days. This procedure reduces the cloud effect, but some residual effect may be still left. In order to model the residual cloud effect, the following statistical question should be answered: What is the effect of cloudiness on the upward radiance when the best case

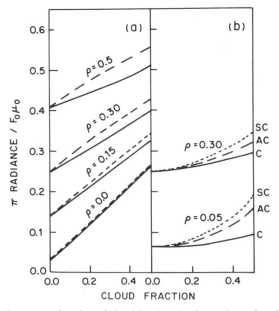

Figure 22 Upward radiance as a function of cloud fraction for four values of surface reflectance ρ. The atmospheric parameters are $L_o = 0.03$, $E_s = 0.7$, $E_{dd} = 0.2$, $T = 0.8$, and $s = 0.1$. (a) Constant cloud transmittance of $T_c = 0.45$, and $\rho_{cl} = 0.50$ (solid lines for $\phi = 0°$ and dashed lines for $\phi = 180°$. (b) Varying cloud reflectance based on Eq. 11.

(minimal cloud amount) out of N independent remote-sensing cases is chosen? The answer depends on the probability of occurrence of a cloud fraction $P(\chi_c)$ and the accumulated probability function $A(\chi_c)$ defined by

$$A(\chi_c) = \int_0^{\chi_c} P(\chi_c') \, d\chi_c' \tag{50}$$

In this case, the average residual cloud fraction χ_c^* is (Kaufman, 1987b)

$$\chi_c^* = \int_0^1 [1 - A(\chi_c)]^N \, d\chi_c \tag{51}$$

For this particular cloud fraction χ_c^*, the radiance can be found from one of the curves in Figure 22. This radiance will correspond to the best one out of N independent cases chosen. In Figure 23, the average cloud fraction χ_c^* is plotted as a function of N for two cloud fraction probability distributions $P(\chi_c)$: a log-normal cloud-fraction probability and a step-function probability. The log-normal probability is spread around $\chi_c = 0.2$, and the step-function probability average is $\chi_c = 0.2$. The relative cloud effect on the radiance that corresponds to these cloud fractions is plotted in Figure 23(b) for two values of surface reflectance ρ. The cloud characteristics are the same as in Figure 22(a). Despite the similarity between the two probability functions, there is a big difference between the effect on radiance of these two cloud-fraction probability functions. Therefore, the structure of the probability function (and not only the median) substantially affects the residual cloud effect on remote sensing. The effect of cloudiness on the vegetation index is shown in Figure 23(c). The vegetation index is defined here as

$$\text{NDVI} = (L_{\text{NIR}} - L_{\text{VIS}})/(L_{\text{NIR}} + L_{\text{VIS}}) \tag{52}$$

where L_{VIS} is the radiance in a visible band (taken here as the radiance for $\rho = 0.05$), and L_{NIR} is the radiance in the near-IR (taken here as the radiance for $\rho = 0.3$). Note that the number of independent remote-sensing cases is usually smaller than the number of days for which data were collected. Since weather systems have a duration of 2–4 days, usually, 2–4 consecutive days can be considered as one independent case.

IV.D.4   Subpixel Cloud Detection

Methods for the detection of subpixel clouds from satellite images were reported by Coakley and Bretherton (1982), Joseph (1985), Cahalan (1983), and Arking and Childs (1985). These methods are based on a contrast between the cold high-level clouds and the warmer ground in the thermal region, as well as on the increase in the brightness and standard deviation of the histogram of detected radiances. The presence of clouds will increase the radiance detected above pixels that are contaminated with clouds. The contrast between contaminated and cloud-free pixels is strong for a dark uniform surface. Therefore, this method was applied for images of the oceans (Coakley and Bretherton, 1982; Joseph, 1985; Cahalan, 1983; Arking and Childs, 1985). In Figure 24, the upward radiances for three values of cloud fraction and two surface reflectances were simulated. Clouds of random size were generated and randomly located in an image (containing 100 pixels 4 × 4 kilometers each) until the desired cloud fraction was reached. It was

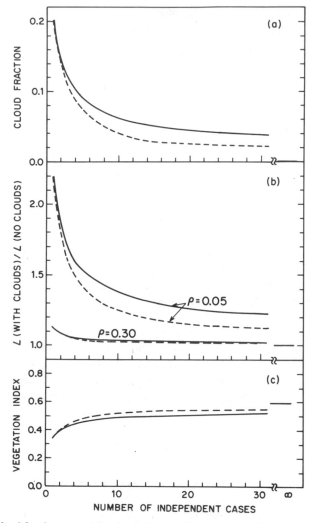

Figure 23   The cloud fraction χ_c, and the cloud effect on (b) the upward radiance and on (c) the vegetation index NDVI as a function of the number of independent days from which the lowest radiance or highest NDVI are chosen. The radiance is plotted for two values of surface reflectance ρ. The solid lines are for a log-normal cloud fraction probabilities with a maximum at $\chi_c = 0.2$ and a standard deviation $\sigma = 2.0$. The dashed lines are for constant probabilities between $\chi_c = 0.0$ and $\chi_c = 0.4$. The cloud transmittance was a constant of $T_c = 0.45$, and $\rho_{cl} = 0.50$.

assumed that the cloud and its shade are located in the same pixel. The increase in the standard deviation with an increase of cloud fraction is evident in the figure.

IV.D.5   Application

In this section, we shall apply the algorithm described in the previous section to the cloud conditions prevailing in three climatic areas in Israel: Mediterranean, Tel-Aviv (540 mm rain/year); desert, Eilat (25 mm rain/year); and mountainous, Mount Kenaan (720 mm rain/year). In Figure 25, the cloud fraction frequencies in January, April, and

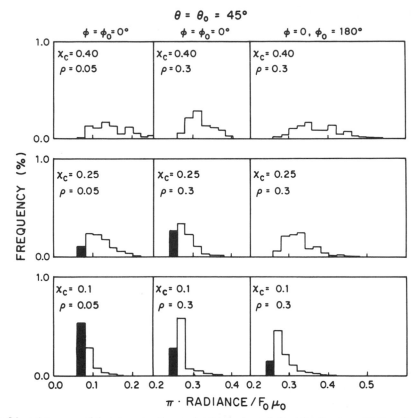

Figure 24 Histograms of the upward radiances for three values of cloud fraction χ_c and two surface reflectances ρ. In the simulation, clouds with random size were generated and randomly located in an image (containing 100 pixels, 4×4 kilometers each) until the desired cloud fraction was achieved. The pixels that were not affected by the clouds are shaded. $\theta_o = \theta' = 45°$; the cloud transmittance was constant of $T_c = 0.45$, and $\rho_{cl} = 0.50$.

July are shown for these three locations. The frequencies are averages of the last 15 years (obtained from the Israel Meteorological Service). For these locations, choosing the best one out of N independent remote-sensing cases will result in the cloud fractions shown in Figure 26(a) as a function of N. In Figures 26(b) and 26(c), the cloud effect on the upward radiance and the vegetation index is shown for these locations. It was assumed that the cloud transmittance is a function of the cloud fraction, as in Eq. 11, and the alto-cumulus cloudiness was chosen for the computations. The Eilat July curve represents the driest (almost cloud-free) conditions. The Mount Kenaan January curve represents wet conditions. For the meteorological conditions and cloud characteristics shown in Figure 26, the use of $N = 4$ will remove most of the cloud effect. This corresponds to 8–16 measurements in consecutive days. This number of measurements can be compared with the number of measurements used in the generation of the maximum-value composite (see Holben, 1986). The number of acquisitions to process a single composite image varies from 7 in the standard NOAA global vegetation index (National Oceanic and Atmosphere Administration, 1983) up to 30 for monthly representations of the vegetation index (Tucker et al., 1985; Justice et al., 1985).

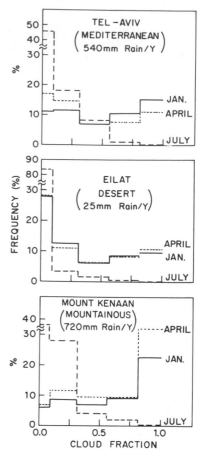

Figure 25 Cloud-fraction frequencies prevailing in three climate areas in Israel: Mediterranean, Tel-Aviv (540 mm rain/year); desert, Eilat (25 mm rain/year); and mountainous, Mount Kenaan (720 mm rain/year) for January, April, and July. The frequencies are averages for the last 15 years. The data were obtained from the Israeli Meteorological Service.

IV.E   Effects on Classification

Atmospheric scattering and absorption by gases and particulates may affect the classification of surface features (Fraser et al., 1977). For a uniform surface, or for low-resolution imagery, the wavelength dependence of the atmospheric effect results from the wavelength dependence of aerosol and molecular scattering (see Sections II.A and III.A). For high-resolution imagery of a nonuniform surface, the spectral characteristics of one field may affect the spectral dependence of the radiance detected above a second nearby field (the adjacency effect).

An example of the atmospheric effect on classification for low-resolution imagery is shown in Figure 27. The simulation is performed for five surface covers (water; bare soil; and alfalfa with low, medium, and high biomass) and four atmospheric models (Fraser and Kaufman, 1985). The atmospheric models differ by the values of the optical thickness and the single-scattering albedo. Models 1 and 3 are for low aerosol absorption ($\omega_o = 0.96$) and models 2 and 4 for moderate aerosol absorption ($\omega_o = 0.88$). Models 1 and 2 are for $\tau_a = 0.29$ and models 3 and 4 are for $\tau_a = 0.86$ for $\lambda = 660$ nm. The resulting radiances for $\lambda = 660$ nm vs. $\lambda = 820$ nm are plotted in Figure 27. The spread between the values associated with the four atmospheric models indicates the atmospheric effect on classification of surface objects by means of clustering. For a given

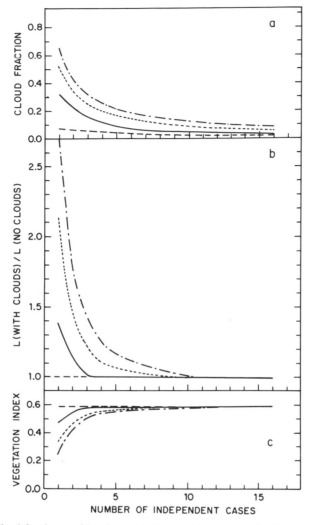

Figure 26 (a) Cloud fraction resulting from choosing the best one out of N independent remote-sensing cases. Results are shown for Tel-Aviv in January (dotted line), Eilat in January (solid lines), Eilat in July (dashed lines), and Mount Kenaan in January (dashed-dotted lines). (b) The ratio of the radiance with and without clouds that corresponds to the cloud fraction in (a). The cloud transmittance is a function of the cloud fraction (Eq. 11), and the alto-cumulus cloudiness was chosen for the computations. (c) The vegetation index as a function of N for the cloud fraction in (a).

atmospheric model, the classes are well separated. Once all the atmospheric models are introduced, as in Figure 27, the classes are less separable. This simulates the conditions of an image with spatially variable atmospheric characteristics. For example, bare soil for high absorption and both small and large amounts of haze (models 2 and 4) is located very close to alfalfa with low biomass for a hazy atmosphere with low absorption (model 3). Therefore, in this simplified two-band classification, uncertainty in the atmospheric absorption brings the two classes to a separation of only $\Delta L = 0.02$.

As a result, the atmospheric effect may reduce significantly the separation between classes if the characteristics of the atmosphere vary across the image. In the case of

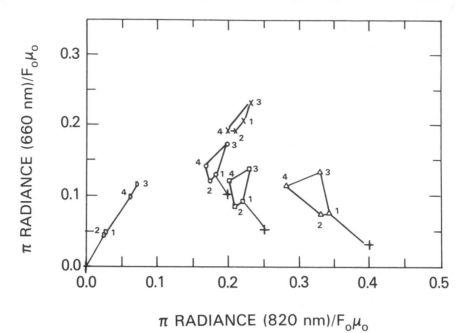

Figure 27   The radiances in two spectral bands of the Earth-atmosphere for five surface covers and four atmospheric models (see text). $\theta = 0°$, $\theta_o = 50°$. + is the surface reflectance; x is for bare soil; and the circles, square, and triangle are for low, medium, and high biomass alfalfa, respectively; D is the water.

supervised classification, bassed on classes determined from a different satellite image or from ground measurements, the classification may not be accurate even if the separation between the classes is good, due to a difference between the atmospheric characteristics in the classified and the supervised data set. For high-resolution imagery, even for a horizontally homogeneous atmosphere, atmospheric effect on the upward radiance may vary across the image due to the adjacency effect (Kauffman, 1984c). A different reflectance of the surrounding area of a given field may result in different values of the L_{d2} term in the upward radiance detected above a particular field, similar to the effect shown in Figure 27. In order to simulate this effect, a nonuniform surface was constructed (Figures 28 and 29) and the upward radiance computed by the Fourier algorithm (Eq. 35).

The surface fields are arranged in a checkerboard form, Figure 28(*a*), so that one-half of the darkest fields are surrounded by the brightest fields and the other half are surrounded by the medium-bright fields. Each image consists of 128×128 pixels. In order to simulate realistic field reflectances, a random deviation from the original reflectances ρ of each field was applied according to

$$\rho' = \rho(1 + r_1) + r_2 \tag{53}$$

where r_1 and r_2 are random numbers with zero averages and standard deviations of 0.1 and 0.01, respectively. The resultant probability distribution is shown in Figure 28(*b*). In this case (no atmospheric effect), the field classes are separated. In Figure 28(*c*), the atmospheric effect is applied to the reflectances of Figure 28(*b*) for field sizes of $X = 0.05$ km. For these small fields, all fields are equally affected by the adjacency effect,

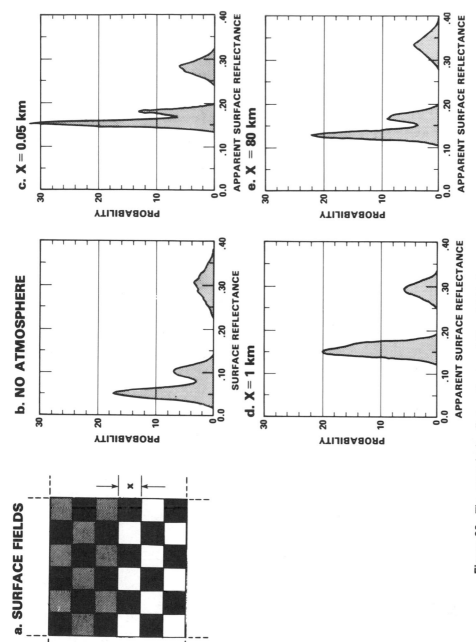

Figure 28 The atmospheric effect on the separability of field classes as a function of the field sizes for "organized" fields. The calculations were performed for a hazy atmophere (aerosol optical thickness of 0.50). (*a*) The field organization. The reflectances were 0.05 for the black fields, 0.10 for the gray fields, and 0.30 for the white fields. A random variation in the reflectance was added, as indicated in the text. (*b*) The resultant probability distribution (for no atmosphere). Graphs (*c*), (*d*) and (*e*) show the probability distribution for field sizes X of 0.50, 1, and 80 km, respectively.

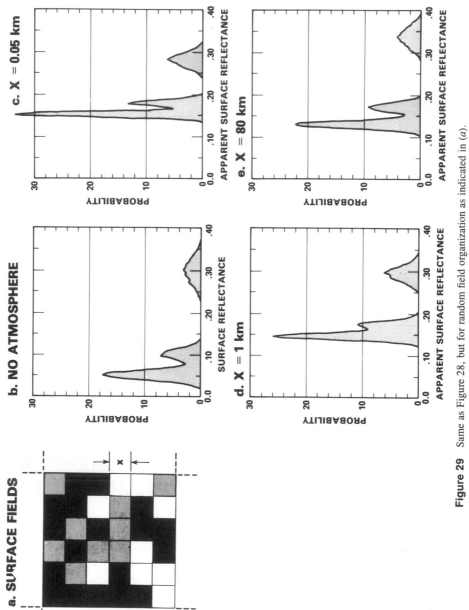

Figure 29 Same as Figure 28, but for random field organization as indicated in (*a*).

382

since in this case, the diffused upward radiance L_{d2} is almost uniform over the surface. The radiance is lower than the surface reflectance for the bright fields and higher than the surface reflectance for the dark fields. As a result, the range of radiances is narrower and the differences between radiances of the classes is smaller. The separability is not changed, since a uniform atmospheric effect was applied all over the area.

In the opposite extreme case, for very big fields ($X = 80$ km) shown in Figure 28(e), the fields are so big that the adjacency effect on the boundaries of these fields affects only a small fraction of the total number of pixels. Except near the boundaries, the atmospheric effect over each field is given by Eq. 29, where ρ is the reflectance of the field, ρ_f. The effect is, therefore, similar to the effect for small fields, and the change in the separability of the large fields is negligible. For intermediate field sizes (X of the order of the scale height of the aerosol layer), the atmospheric effect is nonuniform across the area and is affected mainly by the immediately surrounding fields. As a result, the width of each radiance distribution increases relative to the total radiance range. The two distributions that were separated by a reflectance difference of $\Delta\rho = 0.05$ in Figure 28(b) are totally merged in Figure 28(d). Since half of the dark fields have much brighter background than the remaining dark fields, the adjacency effect caused a bigger change in the radiance above these fields, "shifting" this part of the distribution to fill the gap between the two distributions. In Figure 29, a similar simulation was performed, but in this case, a random distribution of the fields was chosen, as indicated in Figure 29(a). The distribution for the intermediate field size is affected by the different locations of the fields. The separability is higher than in the case of Figure 28, although still very small.

Since atmospheric effect in the presence of surface nonuniformity affects the separability between field classes, it should affect classification as well. The influence of atmospheric effect in the presence of a nonuniform surface on the classification of agricultural fields was studied by Kaufman and Fraser (1984). In this study, 15 different fields, including soil, six different densities of alfalfa, water, savannah, bog, pasture land, forest, corn, soybean, and wheat stubble, were used. The amospheric effect was simulated for three different haze conditions and several sizes of fields. The fields were chosen as squares and surrounded by a uniform background composed of one of the other fields in the simulation. The calculations were based on the numerical results of Pearce (1977) and were performed for the four Landsat MSS spectral bands. To test the atmospheric effects on classification, 15 reference fields were determined. Each one was selected from the previously mentioned list, with an edge length of 1.2 km, surrounded by soil. The upward radiances above the reference fields were calculated for average atmospheric haziness [$\tau_a(550$ nm$) = 0.21$]. The classification of an unknown (test) field is performed by comparing its radiance with that of each of the reference fields. A class is determined by a minimum value of s_k:

$$s_k^2 = \tfrac{1}{4} \sum_{i=1}^{4} \left(L_{ti} - L_{ri} \right)^2 \qquad k = 1, 2, \ldots, 15 \tag{54}$$

where L_{ti} and L_{ri} are the radiances above the center of the test and reference fields, respectively. The subscript i refers to one of the four spectral bands. The superscript k ranges through the complete set of integers from 1 to 15, corresponding to the 15 classes.

The classification results are shown in Figures 30(a)–(d), with tables atop each figure giving details about the fields, their backgrounds and aerosol optical thicknesses τ_a. The

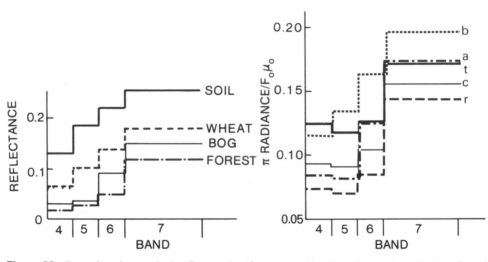

a   **THE EFFECT OF OPTICAL THICKNESS**

	FIELD	BACK-GROUND	X (KM)	τ_a BAND 4	C	RADIANCE FOR LANDSAT DATA				CLASSIFIED AS
						4	5	6	7	
t	CONIFEROUS FOREST	SOIL	1.23	0.64		.125	.118	.126	.171	WHEAT STUBBLE
r	CONIFEROUS FOREST	SOIL	1.23	0.21		.073	.070	.085	.144	–
c	CONIFEROUS FOREST	SOIL	1.23	0.21	YES	.092	.091	.104	.156	BOG
a	BOG	SOIL	1.23	0.21		.084	.082	.125	.172	–
b	WHEAT STUBBLE	SOIL	1.23	0.21		.115	.135	.163	.197	–

Figure 30 Examples of atmospheric effect on classification: (*a*) the effect of haze alone, (*b*) the effect of size alone, (*c*) combined effect, and (*d*) classification of water. Each example gives the radiances of the fields both in tabular and graphical forms. In addition, the reflectance of each surface cover is given.

word "yes" in column C indicates data after an atmospheric correction is applied. In order to test the importance of inclusion of the adjacency effect in a correction scheme, the correction is made for a surface of uniform reflectance, ignoring the adjacency effect (in this simulation, a correction that would include the adjacency effect would result in the correct surface reflectance). The fields are designated at the left by a letter: *r* for reference field, *t* for test field, *c* for the test field after the atmospheric correction has been applied, and *a* and *b* for one or two fields that have the closest spectral characteristics to fields *t* and *c*. The reflectance of the fields at the surface are shown in the lower left parts of the figures. The radiances above the atmosphere are taken from the table and reproduced in the lower right parts of the figures. Only the aerosol optical thickness τ_a changes between the reference ($\tau_a = 0.21$ for $\lambda = 0.55$ μm) and test ($\tau_a = 0.63$ for $\lambda = 0.55$ μm) fields in Figure 30(*a*). The edge length of all fields in the simulation is $X = 1.2$ km. The reference and test fields are coniferous forests surrounded by a background of soil. Because the radiances above the test field are higher than those of the reference field, the forest is misclassified as wheat stubble *b*. The corrected data *c* is

b ## THE EFFECT OF SIZE

	FIELD	BACKGROUND	X (KM)	τ_a BAND 4	RADIANCE FOR LANDSAT BAND				CLASSIFIED AS
					4	5	6	7	
t	ALFALFA 2280	CONIFEROUS FOREST	0.111	0.21	.097	.066	.297	.411	ALFALFA 1650
r	ALFALFA 2280	CONIFEROUS FOREST	1.23	0.21	.101	.066	.309	.423	–
a	ALFALFA 1650	CONIFEROUS FOREST	1.23	0.21	.101	.075	.305	.402	–

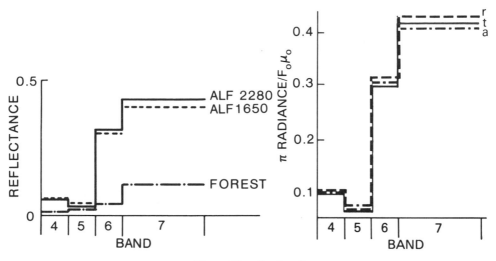

Figure 30 (*Continued*)

classified as bog *a*. The atmospheric turbidity does not usually vary so strongly between reference and test fields, but when it does, serious classification errors arise.

Field size is the only parameter that changes between the reference ($X_r = 1.2$ km) and the test ($X_t = 0.11$ km) fields for Figure 30(*b*). The fields are alfalfa of density 2280 kg/ha with backgrounds of coniferous forest. The test field is still classified as alfalfa, but of lower density (1650 kg/ha). The size difference results in a misclassification, even though the atmospheric characteristics and the surface reflectances are the same for both the reference and test regions.

In Figure 30(*c*), the combined effect of haze, background reflectance, and size is examined. The test field is wheat stubble with a background of coniferous forest for $X_t = 0.11$ km and $\tau_a = 0.63$ for $\lambda = 0.55$ μm. A comparison of radiance from wheat stubble shows that the atmospheric effect is weak because of compensating effects: the denser haze results in an increased radiance, whereas a darker background lowers the radiance. Therefore, the field is classified correctly as wheat stubble. The atmospheric effect would be much stronger for a uniform field of what stubble; thus, the resultant uniform atmospheric correction applied to radiances of *t* results in a decrease of radiances shown in *c*. This decrease is caused by the uniform atmospheric correction, which does

c

COMBINED EFFECT

	FIELD	BACKGROUND	X (KM)	τ_a BAND 4	C	RADIANCE FOR LANDSAT BAND				CLASSIFIED AS
						4	5	6	7	
t	WHEAT STUBBLE	CONIFEROUS FOREST	0.111	0.64		.122	.130	.153	.194	WHEAT STUBBLE
r	WHEAT STUBBLE	SOIL	1.23	0.21		.115	.135	.163	.197	–
c	WHEAT STUBBLE	CONIFEROUS FOREST	0.111	0.64	YES	.088	.104	.133	.180	BOG
a	BOG	SOIL	1.23	0.21		.088	.082	.125	.172	–

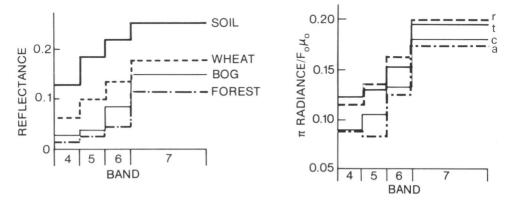

Figure 30 (*Continued*)

not account for the adjacency effect. The correction procedure results in the misclassification of the field as bog *a*. This example shows how uniform atmospheric correction can cause a misclassification of data that were classified correctly before.

An example of classification of water basins is shown in Figure 30(*d*). A square water pond of 0.11 km on a side with a background of soybeans for $\tau_a = 0.21$ ($\lambda = 0.55 \ \mu$m) is misclassified as wheat stubble due to the large increase of radiances from line *r* to line *t*. The uniform atmospheric correction causes here only a slight correction to line *c*, and the water pond is still misclassified, this time as bog *a*. Therefore, small water bodies on bright backgrounds can be misclassified from space, and a correction based upon uniform surface reflectance is not adequate.

The simulation study presented shows that nonuniform reflectance of the Earth's surface may be an important factor in determining the effect of atmosphere on remote sensing of the surface and, thus, on classification of surface features for a high-resolution satellite imagery. The adjacency effect, which increases the radiances detected above dark fields surrounded by adjacent bright areas (and vice versa), creates an atmospheric effect that cannot be corrected by methods that assume a uniform surface. Therefore, atmospheric correction algorithms of high-resolution imagery should account for the nonuniform nature of the Earth's surface.

IV.E.1 Effect on the Vegetation Index
A special case of classification of surface cover is the use of the vegetation index (NDVI) defined in Eq. 52. This index represents the ratio of the radiance reflected by the surface

d

CLASSIFICATION OF WATER

	FIELD	BACKGROUND	X (KM)	τ_a BAND 4	C	RADIANCE FOR LANDSAT BAND 4	5	6	7	CLASSIFIED AS
t	WATER	SOYBEAN	0.111	0.64		.147	.127	.164	.159	WHEAT STUBBLE
r	WATER	SOIL	1.23	0.21		.070	.058	.050	.044	-
c	WATER	SOYBEAN	0.111	0.64	YES	.115	.101	.145	.144	BOG
a	BOG	SOIL	1.23	0.21		.084	.082	.125	.172	-
b	WHEAT STUBBLE	SOIL	1.23	0.21		.115	.135	.163	.197	-

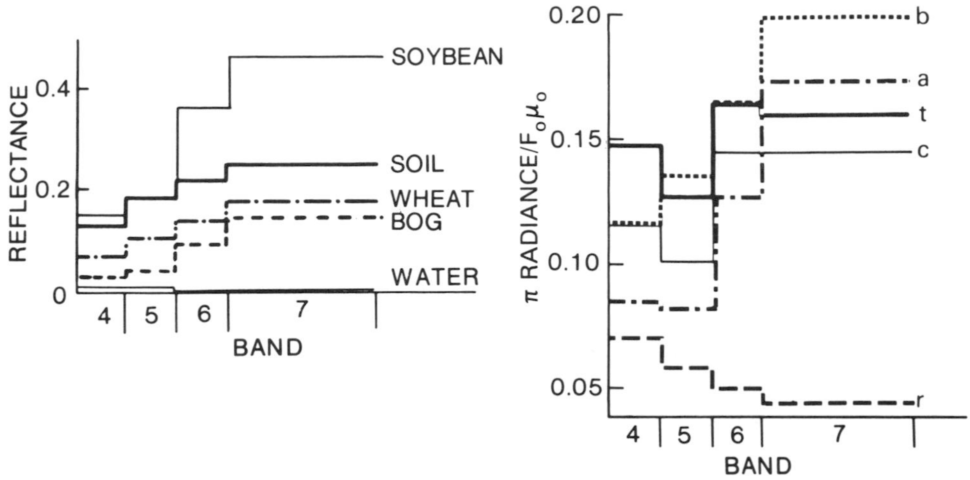

Figure 30 *(Continued)*

in the near IR to the red part of the spectrum. A review of application of this index for monitoring surface vegetation by the NOAA AVHRR sensor is given by Holben (1986). Green vegetation has a strong absorption in the red part of the spectrum and a weak absorption in the near IR, whereas other surface cover types have a continuously varying characteristic as a function of the wavelength in the solar spectrum. As a result, the NDVI may be as high as 0.8 for dense vegetation, close to zero for bare soil and clouds, and negative for water (as detected by the AVHRR in clear conditions; Holben, 1986). The atmospheric effect on remote sensing of vegetation will increase the radiance in the red part of the spectrum (due to the low vegetation reflection, Figures 8 and 9) and will decrease the radiance in the near-IR part of the spectrum (due to the high vegetation reflection, Figures 8 and 9). For example, and increase in the aerosol optical thickness by 0.3 will decrease the vegetation index by ΔNDVI $= 0.05$ for the nadir view and 0.10 for a 45° direction of observation (Holben, 1986).

IV.F Atmospheric Effect on the Apparent Spatial Resolution

The apparent spatial resolution of the satellite imagery may be affected by the adjacency effect. To calculate this effect, it is convenient to use the modulation transfer function (MTF), see Eq. 35, since both the satellite field of view and the atmospheric effect can be described by the MTF of each of these systems (Kaufman, 1984b). The two-dimensional MTF is the Fourier transform of the point spread function. The signal detected by the sensor from a surface pattern in the form of a delta function (black surface with one bright point) is expressed as a convolution of the atmospheric point spread function with the sensor point spread function. To understand this statement, we may consider the radiance out of the atmosphere as an input to the satellite sensor's optical and electronic system that generates an output that is a convolution of the sensor's point spread function with the input that is in this case the atmospheric spread function. The Fourier transfer of each of the spread functions is the MTF of the appropriate system (sensor or atmosphere). The Fourier transform of the convolution of the point spread functions of the sensor and the atmosphere is the product of the two MTFs. Therefore, the MTF is an easy tool to represent the net effect on resolution of more than one system (e.g., sensor and atmosphere).

A high value of the MTF (close to unity) represents a small effect of the transferring system (the atmosphere and the sensor) on the signal. Note that in this section, the absolute (nonnormalized) MTF is discussed (the normalized MTF is shown in Figure 20). Figure 31 shows the MTF of the Landsat Thematic Mapper (solid line). The product of the Thematic Mapper's MTF and the atmospheric MTF is also plotted. The atmospheric MTF was derived for $\theta_o = 20°$, $\lambda = 550$ nm, and aerosol optical thickness $\tau_a = 0.5$. For a given value of the MTF (e.g., 0.35 or 0.50 in Figure 31), the spatial resolution (given by the field-edge length) is substantially smaller than the resolution without the atmospheric effect. An additional MTF is plotted in Figure 31, representing the product of the atmospheric and Thematic Mapper MTFs, but after applying an atmospheric correction that assumes a uniform surface reflectance. The corrected MTF curve was obtained from the uncorrected one by dividing by its value at $k = 0$, which

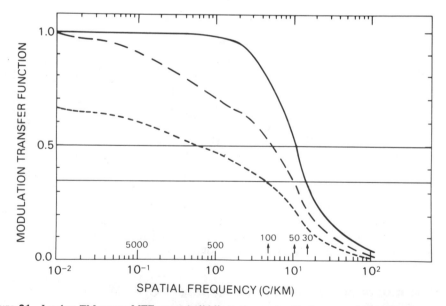

Figure 31 Landsat TM sensor MTF curve (solid line); the curve with the atmospheric effect for $\tau_a = 0.5$ and $\lambda = 550$ nm (short dashed line); with the atmospheric effect after applying a correction for a uniform surface (long dashed line). The thin horizontal lines show the atmospheric effect for MTF values 0.35 and 0.50.

is the contrast attenuation for big fields (Eq. 37). This correction reconstructs the contrasts between very large fields, but only partially corrects for contrast of fields with more common sizes of 100 to 2000 m in diameter. Figure 31 shows that the sensor resolution (for an MTF value of 0.35) is reduced from 30 to 100 m because of the atmospheric effect. The adjacency effects alone would decrease it to 50 m (dashed line).

This analysis of the atmospheric effect on the spatial resolution of remotely sensed imagery from space shows that the atmospheric scattering decreases the apparent resolution considerably. This decrease depends on the shape of the sensor's MTF as well as the atmospheric optical characteristics.

V MEASUREMENTS OF THE ATMOSPHERIC EFFECT

In the previous sections, the atmospheric optical characteristics were reviewed as well as the effect of these characteristics on radiation transfer in the atmosphere and on remote sensing of the Earth's surface. In this section, we shall test the theoretical models against measurements of the atmospheric radiation and the atmospheric characteristics. These measurements are conducted for the natural atmosphere (Kaufman et al., 1986; Kaufman, 1988) as well as in the laboratory (Mekler et al., 1984), where one can control the "atmospheric" characteristics.

V.A Measurements of the Spectral Dependence

Atmospheric scattering and absorption depend substantially on the wavelength of the radiation. Scattering is usually much stronger for short wavelengths than for long wavelengths. It varies as λ^{-4} for molecular scattering and as λ^0 to λ^{-3} for aerosols. This wavelength dependence of scattering causes a similar wavelength dependence of the upward and downward diffuse radiations.

In order to measure the atmospheric effect and its wavelength dependence, a field experiment was conducted (Kaufman et al., 1986). The atmospheric effect on the upward radiance was measured by an airborne scattering radiometer, with simultaneous measurements of the atmospheric characteristics. The aircraft flew close to the ground (300 m) and measured the radiance entering the atmosphere from below (this radiance is affected only by the aerosol in the lowest 300 m), and also flew above the haze (at 3800 m), where the measured upward radiance is affected by the atmosphere.

The experiment included three measuring systems (Kaufman et al., 1986):

Airborne Scanning Radiometer. The Ocean Color Scanner (OCS) was mounted on the NASA Skyvan aircraft to collect digital images of the upward radiance. The radiance was detected at nine narrow spectral bands in the visible and near-IR regions of the spectrum (465–773 nm).

Aerosol Sampler. During the OCS flights, profiles of the aerosol scattering coefficient $[K_s(z) = \omega_o K_e(z)]$ were measured by an airborne nephelometer. The air temperature and dew point were also measured.

Ground Measurements. The wavelength dependence of the aerosol optical thickness $\tau_a(\lambda)$ was measured from the ground by four sun photometers; two had eight spectral bands from 440 to 1030 nm, and two had four bands from 370 to 945 nm. The data were collected on September 10, 1982, over the eastern shore of the

Chesapeake Bay, around 1 p.m. Eastern Standard Time. On this day, a dense haze developed during a few days of meteorological stagnation conditions over the eastern United States (Kaufman et al., 1986). These conditions are rather common during the summer months in the eastern United States (Kaufman and Fraser, 1983; Peterson et al., 1981. The measured aerosol optical thickness is shown in Figure 32. It decreased sharply with wavelength.

These measurements are used in this section to study the spectral dependence of the atmospheric effect on remotely sensed signals. The upward radiances detected over a forest, a corn field, and water are shown in Figures 33–35 for the low (dashed lines) and the high (solid lines) OCS flights. The plotted radiances are averages of the measured radiances over each site (forest, corn field, etc.). There are differences between the direction of the radiances derived from the low and the high flight data, as well as in the solar orientation (the flights were taken about 1 hour apart). In Figures 33–35, the path radiance is dominant for the dark vegetation (in the visible region) and for water, and it is similar to the computations of Figure 9. For $\lambda = 780$ nm, for the bright vegetation, the atmospheric effect results usually in decreasing the upward radiance (Figures 33 and 34). The magnitude and the direction of the atmospheric effect on a bright surface depends on the balance between atmospheric scattering and absorption (Fraser and Kaufman, 1985). In the case of a small lake surrounded by vegetation, Figure 35(*b*), the high reflectance of the vegetation in the near IR (773 nm) causes the radiance above the lake to be brighter in the near IR than in the visible due to the adjacency effect. In this case, when the detector observes the lake, the radiance reflected from the bright vegetation surrounding the lake is scattered by the atmosphere into the field of view of the detector.

To test our theoretical knowledge of the atmospheric effect, the upward radiance in

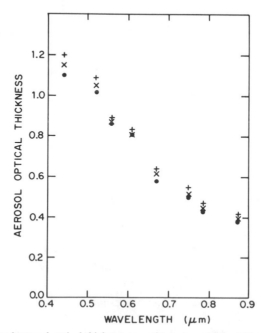

Figure 32   The measured aerosol optical thickness over the eastern shore of Chesapeake Bay as a function of the radiation wavelength: • for 12E.S.T., + for 13E.S.T., and x for 14E.S.T.

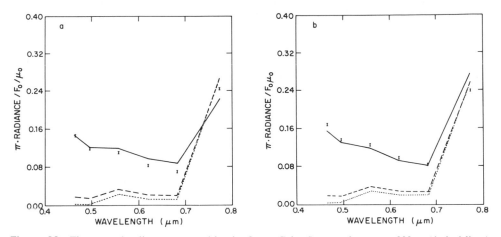

Figure 33 The upward radiance measured by the Ocean Color Scanner detector at 300 m (dashed lines) and at 3800 m (solid lines). The radiance of 300 m is corrected for the lowest 300 m of the atmosphere (dotted lines). The bars represent the theoretical simulation for the upward radiance using the optical thickness measured from the ground. (*a*) Forest nadir observation. (*b*) Forest, $\theta = 32°$, and $\phi = 120°$.

high flight is simulated by radiative-transfer computations and compared with measurements. The input data to the simulation are the radiances in low flight (transformed into the surface reflectance) and the aerosol spectral optical thicknesses. The computations are performed by the Dave and Gazdag (1970) radiative-transfer model. To convert the low-flight radiance into surface reflectance, the radiance was corrected for the atmospheric effect in the lowest 300 meters of the atmosphere (under the low flight, dotted lines in Figures 33–35). The simulation is performed for an aerosol with refractive index $n = 1.43 - 0.0035i$ and a power law size distribution (Junge, 1963) with a power ($\nu = 3.7$) that corresponds to the average aerosol characteristics in the Washington, D.C., area during the summer (Kaufman and Fraser, 1983).

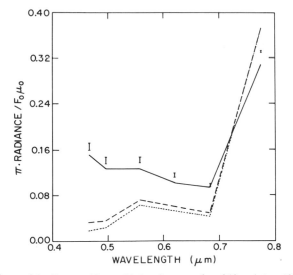

Figure 34 Same as Figure 33, but for corn, $\theta = 24°$ and $\phi = 120°$.

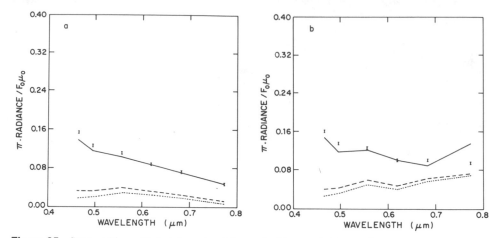

Figure 35 Same as Figure 33, but for water: (*a*) over the Chesapeake Bay, nadir observation, and (*b*) over the lake, $\theta = 10°$ and $\phi = 120°$.

The simulated radiances as a function of wavelength are shown in Figures 33–35 (bars). The average of the absolute deviations between the simulated radiance and the measured radiance is 8%. In short wavelengths, the simulated radiances are higher than the measured radiances by 3–20%. This systematic error can be due to error in the assumed phase function or error in the measured optical thickness. As expected, the theoretical model that assumes a uniform surface reflectance fails to predict the adjacency effect over the lake in the near IR, Figure 35(*b*), due to light reflected from the bright land surrounding the lake.

The effect of observation direction is enhanced in Figure 36 by displaying the atmospheric effect measured in all the cases for each wavelength. The atmospheric effect is calculated here as the difference between the radiance at the high flight and the radiance at the low flight. This difference approximates the path radiance from the visible part of the spectrum, since the surface reflectance here is much weaker than the atmospheric backscattering. Path radiance L_o increases with the observation zenith angle for zenith angles above 20°, due to the longer optical path μ^{-1} and different value of the scattering phase function. The simulated radiance difference is also shown in the figure (solid line).

This simulation shows that radiative-transfer computations, based on the measured and assumed atmospheric and surface characteristics, fit the measured radiances. The surface bidirectional reflectance characteristics, the difference in the sampling procedure between the low and high flights, and the time difference between flights are the main sources of errors. In addition, uncertainties in the aerosol scattering phase function also contribute to the errors.

V.B Atmospheric Effect Above a Nonuniform Surface: Laboratory Experiment

In this section, results are given of a laboratory experiment to measure the adjacency effect and to test the theoretical approach against experimental data (Mekler et al., 1984). In the experiment, a solar simulator provided a stable radiation source. The collimated beam from the simulator was reflected into the measured media by a turnable mirror (Figure 37). The Earth's surface was simulated by a surface painted with an almost

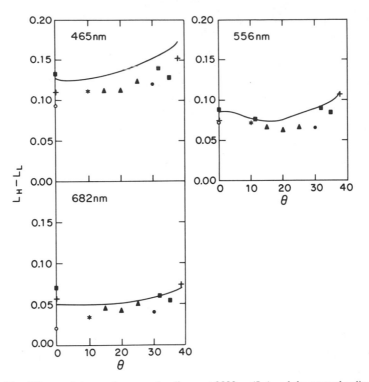

Figure 36 The difference between the upward radiance at 3800 m (L_H) and the upward radiance at 300 m (L_L). The symbols represent the experimental points and the dashed line represents the theoretical simulation. The radiance difference is plotted as a function of the observation zenith angle for all the surface covers for each wavelength. Δ is for corn, • is for pasture, ■ is for forest, + is for water in the bay, * is for water in the lake. The radiance difference is normalized to reflectance units.

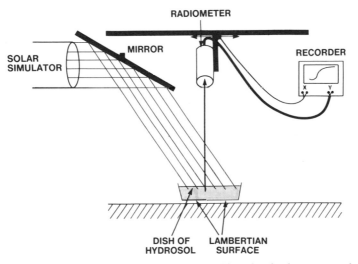

Figure 37 Schematic diagram of the laboratory experiment. The solar simulator generated a steady solar beam that was reflected by the mirror into the measured media. The measured media consisted of a Lambertian surface composed out of two field halves, black and white, and a suspended hydrosol. The measured radiance was recorded as a function of the position of the radiometer.

Lambertian paint, half white and half black, with a sharp boundary between the two
field halves. The atmosphere was simulated by a hydrosol, water with suspended latex
hydrosols 1.09 ± 0.08 μm in diameter. Since the optical characteristics of the latex
spheres are well known, as well as the absorption characteristics of water, the simulated
atmosphere can be modeled easily. The upward radiance of the reflected and scattered
light at $\lambda = 660 \pm 30$ nm was detected by a calibrated radiometer. The output of the
radiometer was plotted as a function of its position above the dish. An example of the
results is shown in Figure 38, where the detected radiance is plotted as a function of the
position of the radiometer for several optical thicknesses of the latex hydrosol. The
adjacency effect is evident from the figure when the latex spheres are added to the water,
increasing the optical thickness τ of the hydrosol.

The main purpose of the experiment was to measure the amplitude and the range of
the adjacency effect. The amplitude of the adjacency effect is the absolute difference
between the radiance at a given point and the radiance far from the boundary. Maximum
amplitude is reached at the boundary of the two field halves. The range of the adjacency
effect is measured by the distance from the boundary at which the amplitude of the
adjacency effect is $1/e$ of its value at the boundary. The linearity of the logarithm of
the radiance difference as a function of the distance from the two field halves boundary
permits an analysis of three radiance quantities: the radiance at "infinity" (far from the
boundary), the radiance near the boundary, and the range of the effect. The radiance at
infinity from the boundary of the two halves and close to the boundary (over the dark
field) is plotted as a function of the optical thickness in Figure 39 for the nadir observation direction. The radiance at infinity increases as a function of the optical thickness
due to a partially compensating effect of the water absorption and the scattering by the

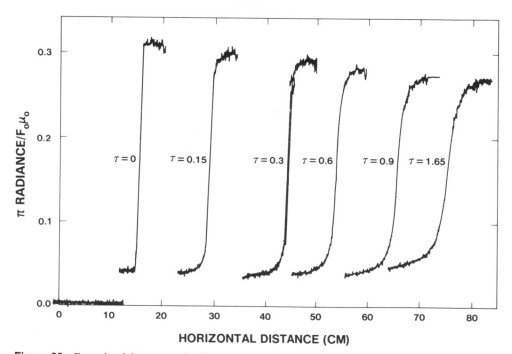

Figure 38 Example of the measured radiance as a function of the horizontal distance across the two field
halves boundary. The hydrosol optical thickness is indicated for each curve.

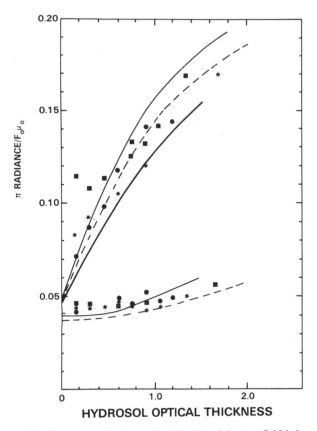

Figure 39 The upward radiance at nadir above the dark side of the two field halves, very far from the boundary (lower part of the figure) and very close to the boundary ($X = 5$ mm; upper part of the figure) as obtained from the measurements. Each symbol corresponds to one measurement sequence. The thin solid lines represent the approximate solutions above the two field halves (Kaufman, 1982) for the radiance close to ($X = 5$ mm) and far from the boundary. The thick solid line represents the approximate solution at the boundary. The dashed lines represent the solutions obtained from Dave's code at the boundary and far from the boundary (Dave and Gazdag, 1970). The source zenith angle is 30°.

hydrosols with a relatively small backward scattering phase function. The radiance close to the boundary increases strongly as a function of the optical thickness, as predicted theoretically (Mekler and Kaufman, 1980).

The range of the adjacency effect depends linearly on the scale height of the atmospheric scatterers (Mekler and Kaufman, 1980; Kaufman, 1982) or, in the present simulation, on the height of the water in the measured media. Therefore, a range factor is defined as a normalized reciprocal of the range of the adjacency effect. The range factor is defined so that it is independent of the scale height of the scatterers and, therefore, can be directly applied from results of the laboratory simulation to the atmosphere:

$$\eta = H/X_e \tag{55}$$

where H is the height of the water in the measured media (7–10 cm in the experiment) or the height of a homogeneous aerosol layer in the atmosphere, and X_e is the horizontal distance at which the adjacency effect is reduced to $1/e$ of its value at the boundary, the range of the adjacency effect:

$$\frac{1}{e} = \frac{L(X_e) - L(\infty)}{L(0) - L(\infty)} \qquad (56)$$

The range factor was calculated from each measured radiance $L(X)$ by a linear least-square fit to the logarithm of the radiance. In Figure 40, the range factor is plotted as a function of the hydrosol optical thickness for nadir observations above the dark surface. The range factor is plotted here for four independent measurement sequences; the variations among these sequences represent the uncertainty in the measured values. The uncertainty in the range factor is 10–20%.

The measurements can be compared with theoretical predictions described in Section IV.C. The optical characteristics of the water are specified here by the water absorption $(0.008\ \mathrm{cm}^{-1})$ and by negligible scattering (Mekler et al., 1984). The scattering phase function of the hydrosol was calculated by a Mie scattering program, under the assumption that there are no coagulations between the latex particles.

The radiances for nadir observation are compared with theoretical predictions in Figure 39. The solid thin lines represent the calculations for the two field halves model

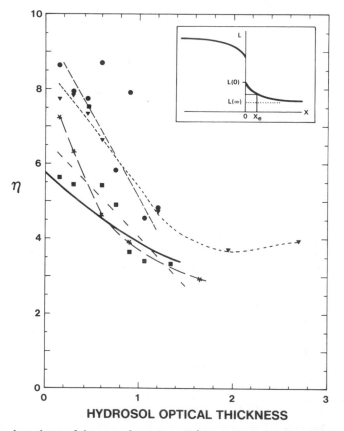

Figure 40 The dependence of the range factor ($\eta = H/X_e$) on the hydrosol optical thickness for nadir observations. Each symbol represents one measurement sequence. The source zenith angle is 30°. The dashed lines represent visual averages of each measurement sequence. The solid line represents the theoretical results.

(Section IV.C) far from the boundary and at the boundary. The measured points were not taken at the boundary, but 5 mm from it; thus, the thick solid line represents the approximation at the 5 mm point. The dashed lines represent radiances calculated from the Dave code (Dave and Gazdag, 1970) at the boundary and far from it. Far from the boundary, Dave radiances for uniform surface were used directly. At the boundary, the Dave radiances were transformed from the far field to the close field by (Kaufman, 1982):

$$L(0) = L_o + (T_s\rho + T_d\rho')F_d/(1 - s\rho') \tag{57}$$

Equation 57 is similar to Eq. 31 except that the background reflectance ρ' is the average of the reflectances of the two field halves, and ρ is the reflectance of the point on the surface for which $L(0)$ is calculated. The comparison in Figure 39 shows that the far-field radiances fit the theoretical predictions by both models within $\Delta L = 0.005$. The measured radiances close to the boundary are higher by $\Delta L = 0.02$ than the curve 5 mm from the boundary.

In Figure 40, the theoretical predictions of the two-dimensional radiative-transfer approximation (Kaufman, 1982, and Eq. 35) for the range factor are compared with the measured results over the dark area. It is seen that the theoretical results have the same dependence on the hydrosol optical thickness as the measured results in both cases. The measured range factor is higher than the theoretical predictions by 0–30%; this may be due to an incomplete treatment of multiple scattering in the theory. It is concluded that the theoretical approximation to the three-dimensional radiative-transfer equation (Kaufman, 1982), as well as the Monte-Carlo method (Pearce, 1977), can be used to calculate the atmospheric effect on the upward radiance in the presence of nonuniform surfaces. The adjacency effect for the off-nadir view was found to be similar to the effect for nadir observation (Mekler et al., 1984).

V.C Atmospheric Effect Above a Nonuniform Surface: Field Experiment

In the laboratory experiment described in the previous section, the characteristics of the "surface" and the "atmosphere" are well controlled. In a field experiment, there are many uncertainties affecting the surface properties and the atmospheric characteristics. These uncertainties and even more do exist also in remote-sensing applications. It is interesting, therefore, to test the theory for such conditions. The same experiment described in Section V.A is used here to measure the effects of the atmosphere above a surface with variable reflectance. These measurements are used to quantify the atmospheric effect on remote sensing (from space) of surface reflectance, including quantifying the adjacency effect, and to test theoretical radiative-transfer models.

V.C.1 Measurements

The wavelength dependence of the aerosol optical thickness was presented in Figure 32. Except for small variations, partially due to uncertainty in the gaseous absorption and calibration, the logarithm of the aerosol optical thickness τ_a is linearly dependent on the logarithm of the wavelength $\tau_a = C\lambda^{-\nu}$. This implies that $\tau_a(\lambda)$ can be used to derive the aerosol size distribution by assuming a power law distribution (see Eq. 6). The derived size distribution will be used in the radiative-transfer computations.

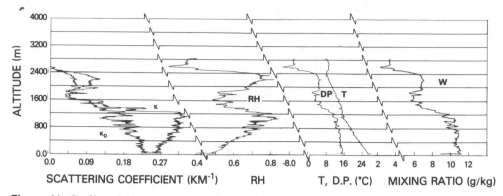

Figure 41 Profiles of the measured scattering coefficient (k_s), calculated dry scattering coefficient (k_D), relative humidity (RH), dew point (DP), temperature (T), and mixing ratio (W). The profiles were taken over Chesapeake Bay from 0843 to 0923 E.S.T.

Two of the aerosol profiles taken during the experiment are shown in Figures 41 and 42 for morning and afternoon, respectively. The scattering coefficient K_s (km^{-1}), the temperature T, and the dew point DP were measured. The mixing ratio W and relative humidity RH were calculated from the temperature and dew point. Two main groups of layers can be distinguished in the profile: 0–1700 and 1700–2800 m. Similarity between the scattering coefficient K_s and the relative humidity can be seen in Figure 41. There is a high correlation between relative humidity and the scattering coefficient mainly in the higher layer, which suggests an effect of the relative humidity on the aerosol characteristics. The dry scattering coefficient k_d, shown in Figures 41 and 42, lacks the vertical variations presented in k_s (k_d was computed based on an expression given by Hanel, 1981).

V.C.2 Upward Radiance

The aircraft sensor measured the upward radiance in low and high flights in the range of view angles $0° \pm 45°$. The radiances plotted in Figure 43 were averaged over the direction parallel to the seashore for the low flight and the high flight separately (for the low flight, view angles of $0° \pm 9°$; and for the high flight, view angles of $0° \pm 4°$). The difference between the upward radiance $L(X)$ (for the high flight) at distance X from the boundary and the radiance very far from the boundary $L(\infty)$ are plotted. The

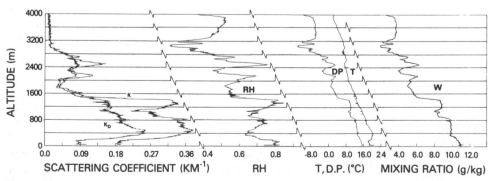

Figure 42 Same as Figure 41, except for the afternoon profile, which was taken from 0120 to 0141 E.S.T.

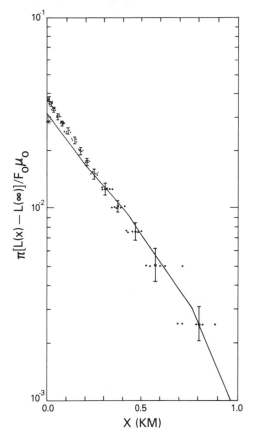

Figure 43 The measured difference (solid circles) between the upward radiance at nadir at distance X from the seashore and the radiance 2-km from the seashore (above the water). The error bars represent the uncertainty due to the digital counts. (The solid lines represents the theoretical predictions based on an approximate model (Kaufman, 1982). The star is the numerical "exact" computation by Dave and Gazdag (1970) code.

radiances from six different segments of data taken over water are presented (as dots) and compared with theoretical predictions. This figure demonstrates the magnitude and range of the adjacency effect. The apparent reflectance of the water close to the seashore is increased by $\Delta L = \pm 0.038$ due to the presence of an adjacent bright land surface. Within a distance of $X = 270$ m from the seashore, the adjacency effect is reduced to e^{-1} of its value at the seashore. For off-nadir view, the radiance difference due to the adjacency effect is plotted for four observation directions in Figure 44(a). The aircraft flew perpendicular to the seashore, scanning in a direction parallel to the seashore.

V.C.3 Comparison with Theory

Calculation of adjacency effects is based on multidimensional solutions of the equation of radiative transfer (see Section IV.C). For the off-nadir direction, Monte-Carlo computations (Pearce, 1977) are applied. The theoretical prediction of the upward radiance is based on the following aerosol characteristics: vertical distribution, total vertical optical thickness, size distribution, and refractive index. It is assumed that except for concentration, all other aerosol characteristics are independent of height and that the atmosphere is horizontally uniform. It is also assumed that the aerosol concentration is proportional to the wet scattering coefficient. The scattering phase function was not measured; therefore, the dependence of the aerosol optical thickness on wavelength was used to derive the size distribution. The aerosol optical thickness for $\lambda = 773$ nm interpolated

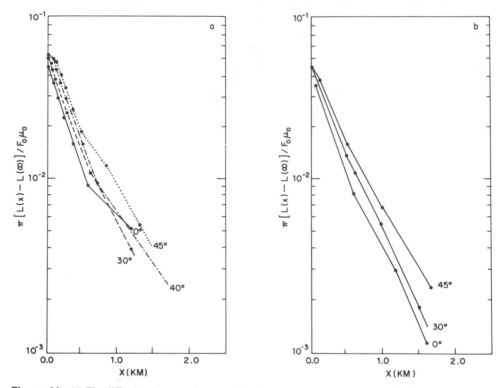

Figure 44 (a) The difference between the upward radiance at distance X from the seashore and the upward radiance far from the seashore (about 2 km). The data were collected in the afternoon (1405–1415 E.S.T.) for view angles of 0°, 30°, 40° and 45°. (b) Monte-Carlo approximation for the data in (a) for view angles 0°, 30°, and 45°.

from the measured data is $\tau_a = 0.5$ and the single-scattering albedo was assumed to be $\omega_o = 0.97$ (Fraser et al. 1984). A Lambertian surface reflectance was assumed.

The theoretical predictions of the nadir radiances are shown in Figure 43. The theory differs from the experimental results by 0 to 20%. The radiances for a uniform surface calculated by Dave's program (Dave and Gazdag, 1970) were also used to derive the magnitude of the adjacency effect (see Eq. 57). The error in the magnitude of the adjacency effect (Figure 43) can be due to two sources of error: uncertainty in the land reflectance, mainly through its non-Lambertian characteristics, or an underestimation of the number of large aerosols and, therefore, an overestimation of angular width of the scattering phase function. It also can be due to approximations in the theoretical model.

For off-nadir observations, the experimental results are compared with Monte-Carlo computations (Kaufman et al., 1986) in Figure 44(b). It is seen that the measured radiances have a similar dependence on the distance from the seashore as the computed radiances. Also the view angle dependence is similar. As for the nadir view, the measured radiances are 10–20% higher than the computed values.

V.D Conclusions

The simultaneous measurements of upward radiance below and above a haze layer, and of the relevant atmospheric characteristics, demonstrated clearly the adjacency effect (the effect of a bright area on the upward radiance measured above a dark adjacent area) both in the laboratory and in the field experiments. The measurements of aerosol profiles and optical thicknesses enabled a quantitative comparison between the measured radiance and theoretical predictions. Good agreements were obtained. The difference between the theory and the measurements was larger for the radiance close to the seashore (or the black-white boundary) than for the radiances far from the seashore. As a result, it was found that the magnitude of the measured adjacency effect (the difference between the radiance at the seashore and far from the seashore, due to the presence of the bright land) is larger than those predicted based on the theory by 10–20%. This difference is suspected to be due mainly to effects of non-Lambertian and nonuniform properties of the land surface, and to the effect of large aerosols on the forward-scattering phase function, that were not included in the model. Similar results were obtained from the laboratory experiment as they were suggested from Landsat data by Kaufman (1982).

For off-nadir view directions up to 45°, the adjacency effect is similar to the effect at nadir but with a larger magnitude. The theory also predicts that the magnitude should increase with the angle of observation, due to effectively thicker atmosphere (and thus more scattered light) and due to an increase in the surface reflectance with the observation direction.

VI ATMOSPHERIC CORRECTIONS

In the previous sections, it was shown that images of the Earth's surface in the solar spectrum that are viewed by satellites are contaminated by sunlight scattered toward the sensor due to atmospheric molecules, aerosol, and clouds (path radiance, L_o). Therefore, solar energy that is reflected from the Earth's surface and serves as the remote-sensing signal is attenuated by the atmosphere. This atmosphere effect is wavelength-dependent, varies in time and space, and depends on the surface reflectance and its spatial variation. Correction for the atmospheric effect can produce remote-sensing signals that are better related to the surface characteristics. In this section, we shall discuss the basis of the atmospheric-correction mechanism and a few specific correction methods.

VI.A Basic Concept

Molecular scattering and absorption can be accounted for satisfactorily. It is practically invariant in time and space. Since gaseous absorption is usually minimized by choosing bands in the atmospheric windows, aerosol scattering and absorption are the main variables in the atmospheric effect on satellite imagery. Aerosol scattering is the major variable component of the atmospheric effect for dark surfaces, whereas absorption is important for bright surfaces (Fraser and Kaufman, 1985).

In order to perform atmospheric correction of remotely sensed data, the optical characteristics of the atmosphere must be known. These characteristics may be known in their microscopic form (profile of the extinction coefficient, the single-scattering albedo, and the scattering phase function) or in the macroscopic form (the vertical optical thickness, the average aerosol scattering phase function, and the single-scattering albedo). Alternatively, if it is impossible to measure or derive the aerosol characteristics and the gaseous absorption, a measure of path radiance L_o and atmospheric transmission functions T and F_d can be used for the correction.

The correction procedure depends to some extent on the type of atmospheric effect and on the remote-sensing application. We can distinguish between three major types of atmospheric effects on remote sensing. These depend on the sensor spatial resolution and the specific remote-sensing application. These effects were discussed in Section IV and are summarized below:

1. For a uniform surface (or low-resolution imagery, e.g., AVHRR), the effect of a cloudless atmosphere can be darkening if the surface is bright (sand or vegetation in the near IR) and brightening if the surface is dark (water or vegetation in the visible region). Brightening is caused by the atmospheric path radiance L_o and darkening by absorption and backscattering of the downward irradiance and upward radiance $L_s + L_D$.

2. For a nonuniform surface, sensed by a high-resolution satellite (e.g., Landsat MSS and TM bands and SPOT), the atmosphere reduces the spatial variations of the upward radiance. This reduction is caused by atmospheric diffusion of photons that were reflected over bright areas to the field of view over dark areas.

3. In cloudy conditions the subpixel (unresolved) clouds increase the apparent brightness of the Earth-atmosphere system. This increase is proportional to the square of the cloud cover, due to the increase of the cloud thickness with the increase of cloud fraction.

VI.B   Atmospheric Correction

VI.B.1   Determination of Atmospheric Optical Characteristics

The basic philosophy of the atmospheric correction is to obtain information about the atmospheric optical characteristics and to apply this information in a correction scheme. One way to describe this information is by the aerosol optical thickness, phase function, the single-scattering albedo, and the gaseous absorption. For high-resolution imagery, some information about the aerosol vertical profile is also required. The problems in the atmospheric correction are due to the difficulty in determining these characteristics. As a result, the only operational use of atmospheric corrections today is that of the ocean color (Gordon et al., 1983), reviewed in what follows, where the very low reflectance of the water in the red can be used for the correction. In principle, information on the atmospheric optical characteristics can be obtained from three different sources:

Climatology. Documented information on the atmospheric characteristics and their variation can be used to estimate the expected atmospheric effect for a specific part of the world and a specific season. Such documentation can be obtained from the analysis of measurements taken from the ground and partially from the analysis of satellite data (Fraser et al., 1984). This source of information will be used for optical characteristics that cannot be determined otherwise for the particular image being corrected.

Measurements from the Ground: The aerosol optical thickness can be obtained from sun photometer measurements (King et al., 1978; Kaufman and Fraser, 1983). The phase function can be determined from inversion of solar almucantar measurements (Nakajima et al., 1986), and the single-scattering albedo can be determined from the collection of particles on filters, preferably by aircraft sampling of the entire atmospheric boundary layer. The application of such measurements for atmospheric corrections is useful for intense field measurements or for establishing the climatology of a given area.

Determination from Satellite Imagery: Satellite imagery has been used to determine the aerosol optical thickness. Using a single image for its determination, the method required a dark surface and, therefore, was applied only over water (Griggs, 1975; Mekler et al., 1977; Carlson, 1979; Koepke and Quenzel, 1979; Takayama and Takashima, 1986). For the correction of the ocean color, instead of the optical thickness, the path radiance was determined (Gordon et al., 1983). By using the difference in the brightness between a clear and hazy day, the difference in the optical thickness can be computed over land also (Fraser et al., 1984). In order to determine the optical thickness from a single image, some assumptions about the surface reflectance should be made. A few methods to derive the aerosol optical thickness and the required assumptions are reviewed in Sections VI.C–VI.F. The methods reviewed in Section VI.C use measurements in a few slant directions to derive τ_a. For this purpose, the surface reflectance is assumed to be lambertian (Turner, 1979) or some smooth dependence of the surface reflectance on the observation angle is assumed (Diner and Martonchik, 1985b). A different method that assumes the existence of a step function in the surface reflectance is reviewed in Section VI.D (Kaufman and Joseph, 1982). A suggestion to detect patches of dense dark vegetation (e.g., forests) and to assume a given value of its reflectance in the blue and red parts of the spectrum is in Section VI.F (Kaufman, 1988; Kaufman and Sendra, 1988). Determination of the aerosol single-scattering albedo and particle size (which can be used to compute the scattering phase function) was suggested by Kaufman et al. (1985); Kaufman, (1987a). This method is useful to determine the aerosol characteristics (from imagery that includes water-land interfaces) in areas that suffer from substantial aerosol outbreaks.

VI.B.2 Derivation of the Surface Reflectance

Once the aerosol characteristics are determined by the use of these methods, the correction is staightforward:

For a uniform surface, it can be corrected by the inverse of the analytical relation, Eq. 29:

$$\rho = f/(1 + sf) \quad \text{where} \quad f = (L - L_o)/TE_d \tag{58}$$

for each pixel separately, where functions s, L_o, T, and E_d are found from the atmospheric characteristics, determined by the methods previously described, and from a theoretical look-up table of these functions for several values of the solar zenith angle θ_o, observation zenith angle θ, azimuth ϕ, and atmospheric characteristics. The look-up table is computed by radiative-transfer methods based on Mie computations of the scattering from uniform spheres (Kaufman and Sendra, 1988).

For a nonuniform surface, the surface reflectance $\rho(x, y)$ can be determined from the inverse of Eq. 35:

$$\rho(x, y) = \mathcal{F}^{-1}\left\{\mathcal{F}[L(x, y) - L_o - L_\beta]/M(k_x, k_y)\right\} \tag{59}$$

This correction can be applied to part of the image, a "window" (Kozoderov, 1985) of dimensions that should be of the order of the atmospheric height or larger (2–10 km; Kaufman, 1982).

VI.C   Correction Based on Slant-Direction Measurements

The method suggests to perform the correction by a derivation of the aerosol optical thickness τ_a from the satellite imagery using slant-direction observations. The method is based on a slant observation in a few directions with a high-resolution sensor of the upward radiance above a contrast in the surface reflectance (Turner, 1979). Assuming that the surface reflectance is spatially variable, but that the average surface reflectance ρ_b over distances of the order of 1 km can be assumed as spatially uniform, we can use Eq. 31 to compute the contrast between two pixels on the surface with lambertian reflectances ρ_1 and ρ_2:

$$L_1 = L_o + E_d(\rho_1 T_s + \rho_b T_d)/(1 - s\rho_b)$$

$$L_2 = L_o + E_d(\rho_2 T_s + \rho_b T_d)/(1 - s\rho_b) \tag{60}$$

$T_s = \exp(-\tau/\mu)$. Therefore,

$$\Delta L = L_2 - L_1 = (\rho_2 - \rho_1) E_d/(1 - s\rho_b) \cdot \exp(-\tau/\mu) \tag{61}$$

Since the only dependence on μ is in the exponent, the aerosol optical thickness can be derived as the absolute value of the slope of $\ln(\Delta L)$ as a function of μ. For a non-Lambertian surface, the difference $\rho_2 - \rho_1$ depends on μ and, therefore, τ_a is no longer the absolute value of the slope (Steven and Rollins, 1986). This method utilizes imagery with high spatial resolution.

Diner and Martonchik (1985b) suggested a modification to this method. Instead of the contrast between two adjacent pixels, they suggested to use the radiances from an image with variable surface reflectance. The power spectrum of the image (a product of the Fourier transform of the radiance and the complex conjugate of the Fourier transform) is averaged over the spatial frequencies except for the zero frequency. The average power spectrum represents the average contrasts of pixels located in different distances one from another. The dependence of the average power spectrum on scan angle θ' is used to derive the aerosol optical thickness. The advantage of this method over Turner's method is that the whole image is used in the analysis, not only one or few pixels. Furthermore, they suggested to reduce the effect of surface non-Lambertian property by averaging radiances for a given value of μ for two opposite azimuths between the direction of observation and the direction of the sun. Limiting the observation to a narrow range of observation angles (e.g., 51°–60°) results in a smaller sensitivity of the resultant optical thickness on the surface non-Lambertian property. Diner and Martonchik (1985b) tested their method analytically for a non-Lambertian surface that varies smoothly with the observation angle. They suggested that the method may result in accuracy of the derived vertical transmittance of better than 7%, or an error in τ_a of $\Delta\tau_a = 0.07$. This method requires imagery with high spatial resolution for a pointable system such as SPOT.

In order to correct the satellite imagery, in addition to the aerosol optical thickness, the phase function and the single-scattering albedo have to be known. If the optical thickness is measured as a function of wavelength, then it may be used to estimate the aerosol size distribution (King et al., 1978) and from it, the phase function. If the error in τ_a is $\Delta\tau_a = 0.07$, the derivation may require well-separated spectral bands and be used only for hazy conditions.

VI.D Correction Based on a Sharp Contrast in the Surface Reflectance

According to this method, the derivation of the aerosol optical thickness and correction of satellite imagery is performed using a step function in the surface reflectance (Kaufman and Joseph, 1982). In this method, the radiances from only the nadir view are used; therefore, the assumption of Lambertian surface property is not as important as in the slant-path methods. On the other hand, the surface is assumed to be composed of a step function in the surface reflectance (e.g., a seashore), which limits its applicability. It is important that the reflectance over the dark part of the step function be spatially uniform. The bright side (the land) does not have to be as uniform. An example of the computed upward radiance above a step function in the surface reflectance is shown in Figure 45. The derivation of the aerosol optical thickness is based on the radiances far from steps $L(\infty)$ and $L(-\infty)$ and the radiances close to the steps $L(0)$ and $L(-0)$.

$$L(\pm\infty) = L_o + E_d(T_s\rho^\pm + T_d\rho^\pm)/(1 - s\rho^\pm) \tag{62}$$

$$L(\pm 0) = L_o + E_d(T_s\rho^\pm + T_d\rho_b)/(1 - s\rho_b) \tag{63}$$

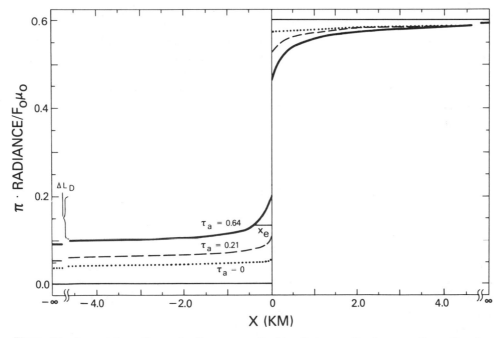

Figure 45 Computations of upward radiance, normalized to reflectance units, above a surface with a step-function reflectance. Thin solid line for no atmosphere, dotted line for $\tau_a = 0.0$, dashed line for $\tau_a = 0.21$, and thick solid line for $\tau_a = 0.64$. The calculations are based on the model of Kaufman (1982) for $\lambda = 550$ nm power law size distribution ($\nu = 3$), solar zenith angle of $40°$, nadir observation, aerosol refractive index of $1.5 - 0.01i$, and the standard Elterman's vertical profile.

Defining a function A_c, which is a ratio of radiance differences,

$$A_c^{\pm} = [L(\pm\infty) - L(\pm 0)]/[L(\pm 0) - L^*] \quad \text{where} \quad L^* = 0.5\{L(0) - L(-0)\}$$

$$(64)$$

we get

$$A_c^{\pm} = [T_d + (T_s + T_d)s\rho^{\pm}]/[T_s(1 - s\rho^{\pm})] \quad\quad (65)$$

Since $s\rho$ is usually small, A_c is approximately:

$$A_c^{\pm} \approx T_d/T_s \quad\quad (66)$$

This ratio is independent of the surface reflectance and can be used to find one piece of information on the aerosol characteristics (e.g., the optical thickness). Since A_c depends, to some degree, on the surface reflectance, an iterative method is used to derive the aerosol optical thickness and the surface reflectances. This method was applied to a Landsat image of the African coast during optical-thickness measurements from a nearby ship (Fraser, 1976; Kaufman and Joseph, 1982). The derived optical thickness $\tau_a = 0.60 \pm 0.08$ was close to the measured value from the ship, i.e., $\tau_a = 0.56$. A sensitivity study showed that an error of $\Delta\omega_o = \pm 0.05$ results in an error in the derived value of τ_a of $\Delta\tau_a = \pm 0.03$. No substantial sensitivity of the method to the size distribution was found, since $A_c^{\pm}$ does not depend on L_o. The main disadvantage of this method is that the derivation of the aerosol optical thickness and of the surface reflectance is based on the presence of a sharp discontinuity in the surface reflectance. Such discontinuities may be found usually close to large bodies of water. An atmospheric correction that is based on the presence of contrasts in the image, but without a necessary step-function contrast, was suggested by Ueno et al. (1978). They corrected the image for a varying value of the aerosol optical thickness and choose a value that resulted in apparent right contrasts in the image. A too small value of the aerosol optical thickness would result in undercorrection, and in a still blurred image, whereas a too high value of the aerosol optical thickness would result in overcorrection, an image with bright lines enhancing the contrasts.

VI.E   Corrections of the Effect Over the Oceans

Remote sensing of the chlorophyll concentration in ocean waters (Clarke et al., 1970; Gordon et al., 1983; 1985) is the only present example of a routine application of atmospheric correction in remote sensing (Gordon, 1978; Gordon and Clark, 1981). This correction is possible due to the unique characteristics of the ocean properties that allow the derivation of the atmospheric characteristics directly from the satellite image and the application of a correction to the same image. The basis of the correction is that ocean water is almost completely black in the red region of the spectrum for an observation direction far from the solar-observer specular direction. Thus, the upward radiance detected in this region of the spectrum results mainly due to molecular and aerosol scattering, and it can be used to estimate the amount of haze or aerosol optical thickness in this region. This estimation is performed for each pixel separately, and, therefore, there is no need to assume a uniform aerosol loading across the image. The wavelength dependence of the atmospheric effect is found in a particular point in the image where the chlorophyll concentration is low (less than 0.25 mg/m^3), and as a result, the ocean contribution to the upward radiance can be assumed to be a priori known for wavelengths $\lambda \geq 0.52$ μm. Table 3 gives some typical values of the ocean and atmospheric contributions to the upward radiance for two extreme chlorophyll (C) concentrations.

The algorithm for atmospheric corrections can be summarized as follows (Gordon et al., 1983). The upward radiance at the top of the atmosphere over the ocean is expressed by

$$L(\lambda) = L_{mo}(\lambda) + L_{ae}(\lambda) + T(\lambda)L_w(\lambda) \tag{67}$$

where the path radiance L_o is separated to the molecular (L_{mo}) and aerosol (L_{ae}) contributions. This separation is exact only for single scattering, and it is only an approximation for multiple scattering. T is the transmission through the atmosphere of radiance emerging from the water (L_w). For a very small optical thickness, radiance L_{ae} can be approximated by Eq. 24. In the correction algorithm, it is assumed that the actual upward radiance is proportional to the product $\omega_o \tau_a P_a$, where ω_o is the aerosol single-scattering albedo, τ_a is the aerosol optical thickness, and P_a is the aerosol scattering phase function. Therefore, if the ratio ϵ_i defined as

$$\epsilon_{i,4} = L_{ae}(\lambda_i)/L_{ae}(\lambda_4) \qquad i = 1, 2, 3 \tag{68}$$

is replaced with

$$\epsilon_{i,4} = \left[\omega_o(\lambda_i)\,\tau_a(\lambda_i)\,P_a(\lambda_i)\right]/\left[\omega_o(\lambda_4)\,\tau_a(\lambda_4)\,P_a(\lambda_4)\right] \qquad i = 1, 2, 3 \tag{69}$$

where $\epsilon_{i,4}$ can be established in a given point in the image and it can be assumed to be independent of position (which means that while the aerosol amount may vary from pixel to pixel, its characteristics are assumed to be invariant), then the radiance emerging from the ocean L_w can be derived from

$$L(\lambda_i) = L_{mo}(\lambda_i) + L_{ae}(\lambda_i) + T(\lambda_i)L_w(\lambda_i) \qquad i = 1, 2, 3$$
$$L(\lambda_4) = L_{mo}(\lambda_4) + L_{ae}(\lambda_4) + T(\lambda_4)L_w(\lambda_4) \tag{70}$$

resulting in

$$L_w(\lambda_i) = T(\lambda_i)^{-1}\left\{L(\lambda_i) - L_{mo}(\lambda_i) - \epsilon_{i,4}\left[L(\lambda_4) - L_{mo}(\lambda_4) - T(\lambda_4)L_w(\lambda_4)\right]\right\}$$
$$\tag{71}$$

TABLE 3   Typical Values of the Ocean and Atmospheric Contribution to the Upward Radiance

Channel	Wavelength (μm)	L_{mo}, Molecular Radiance	L_{ae}, Aerosol Radiance	L_w, Ocean $C = 0.03$ mg/m^3	$C = 10$ mg/m^3
1	443	0.10	0.015	0.035	0.0008
2	520	0.05	0.013	0.008	0.010
3	550	0.04	0.012	0.005	0.015
4	670	0.02	0.010	0.0001	0.002

This equation can be used to derive $L_w(\lambda_i)$ from $L(\lambda_i)$, since $L_{mo}(\lambda_i)$ can be computed (it is the upward radiance for a Rayleigh atmosphere for a black surface) and $T(\lambda_i)$ is computed as the gaseous attenuation in the atmosphere. The radiances $L_w(\lambda_i)$ are further related to the chlorophyll concentration. This atmospheric correction was extensively applied to the Coastal Zone Color Scanner (CZCS) imagery and compared successfully to measurements of chlorophyll from ships.

Over inland water bodies, this correction cannot be applied as such due to the presence of the adjacency effect from the surrounding land (Kaufman and Fraser, 1984). Tanre et al. (1987) found from analysis of Landsat TM imagery that the adjacency effect can increase the apparent inland water reflectance by 0.01 in the red region of the spectrum and 0.02 in the near-IR for lakes surrounded by vegetation. For lakes surrounded by soil, sand, or an urban area, the effect should be stronger. A similar effect was observed from aircraft data (see Figure 35). The correction procedure, in this case, has to be based on Fourier transform inversion algorithm (Eq. 59). Another problem concerning inland water bodies and mainly rivers is the presence of turbidity, which may increase the water reflectance. Thus, the assumption of zero reflectance in the red and a known reflectance in the blue parts of the spectrum (Gordon et al., 1983) is not valid any more. In this case, the only possibility for correction is the method applied over the land, discussed in the next section.

VI.F Corrections Over the Land Surface

In this section, a correction procedure is discussed in which the atmospheric optical thickness is estimated for two wavelengths or more from the upward radiance detected above dark objects. Dense dark vegetation was chosen as a controllable dark object. The derived aerosol optical thickness and other aerosol characteristics derived from a suitable aerosol model are used in the correction. This method requires only a single image for its correction, can be applied to images acquired from any direction of observation, and, basically, for any resolution and any surface that contains dense dark vegetation (Kaufman and Sendra, 1988).

Dense dark vegetation has a low reflectance in the visible part of the spectrum. In Table 4, measured reflectances of pasture and forests are summarized. The visible reflectances of pasture are in the range of 2.5–4%. This reflectance can be decreased by the effect of shadows present in tall vegetation, such as trees and bushes, to 1–3% (see

TABLE 4 Vegetation Reflectance (Nadir View)

Vegetation/λ (μm)	Reflectances (Solar Zenith Angle 40°)			
	0.4–0.5	0.5–0.6	0.6–0.7	Source
Pasture	0.026	0.030	0.039	Kriebel (1977)[a]
Pasture	0.027	0.039	0.032	Kaufman (1988)[b]
Coniferous forest	—	0.011	0.017	Kriebel (1977)
Decideous forest	0.010	0.022	0.012	Kaufman (1988)
Hardwood forest	—	—	0.025	Kimes et al. (1986)[c]
Pine forest	—	—	0.028	Kimes et al. (1986)

[a]The exact wavelengths used are 0.43 μm, 0.52 μm, and 0.61 μm.
[b]The exact wavelengths used are 0.46 μm, 0.56 μm, and 0.62 μm.
[c]The exact wavelengths used are 0.58–0.68 μm.

Table 5). The visible reflectance of forests is in the range of 1–2% in the blue (0.4–0.5 μm) and red (0.6–0.7 μm) regions, and 2–3% in the green (0.5–0.6 μm) region. For atmospheric corrections of satellite imagery, it is suggested to use the upper limit of these reflectances due to the possible presence of bare soil. Forested areas can be found in many locations of the world and their optical characteristics are rather stable (except for deciduous trees in the winter). Therefore, dense dark vegetation is suggested as the controlled dark surface used to derive the aerosol optical thickness and to correct for the atmospheric effects on remotely sensed images of the Earth surface. In the following, we discuss the proposed correction procedure, its sensitivity to sources of errors, and some results of applications.

VI.F.1 Algorithm
The following algorithm is developed assuming that the image being corrected contains some patches of dense dark vegetation (e.g., trees and shrubs) larger than the pixel size.

TABLE 5 Results of the Correction Algorithm

(a) Original Radiances (L) and Standard Deviations ($\pm\Delta L$)

| Subarea | Date | $\pi \cdot$ Radiances$/F_o\mu_o$ | | | Vegetation Index |
		Band 1 (L_1)	Band 2 (L_2)	Band 3 (L_3)	NDVI $= (L_3 - L_2)/(L_3 + L_2)$
Northern	8/20/82	0.109 ± 0.019	0.081 ± 0.028	0.177 ± 0.029	0.386 ± 0.115
Northern	8/02/82	0.175 ± 0.015	0.134 ± 0.018	0.203 ± 0.016	0.208 ± 0.016
Southern	8/20/82	0.100 ± 0.011	0.070 ± 0.013	0.155 ± 0.055	0.473 ± 0.090
Southern	8/02/82	0.147 ± 0.012	0.105 ± 0.012	0.175 ± 0.046	0.293 ± 0.069

(b) Aerosol Optical Thickness Measured by Sun Photometer and the Slope (ν)

| Subarea | Date | Aerosol Optical Thickness | | | Slope (ν) |
		Band 1 (τ_1)	Band 2 (τ_2)	Band 3 (τ_3)	
Northern	8/20/82	0.34 ± 0.08	0.23 ± 0.02	0.17 ± 0.01	2.9 ± 0.3
Northern	8/02/82	1.20 ± 0.05	0.96 ± 0.02	0.77 ± 0.01	1.7 ± 0.2

(c) Results of the Algorithm

| Subarea | Date | Derived Reflectance | | | Corrected Vegetation Index |
		Band 1 (ρ_1)	Band 2 (ρ_2)	Band 3 (ρ_3)	
Northern	8/20/82	0.051 ± 0.026	0.049 ± 0.033	0.170 ± 0.032	0.58 ± 0.18
Northern	8/02/82	$0.049 + 0.025$	0.050 ± 0.027	0.176 ± 0.022	0.57 ± 0.16
Southern	8/20/82	0.043 ± 0.013	0.033 ± 0.016	0.145 ± 0.063	0.66 ± 0.14
Southern	8/02/82	0.049 ± 0.016	0.038 ± 0.015	0.151 ± 0.058	0.60 ± 0.19

| Subarea | Date | Aerosol Optical Thickness | | | Slope (ν) |
		Band 1 (τ_1)	Band 2 (τ_2)	Band 3 (τ_3)	
Northern	8/20/82	0.496 ± 0.084	0.279 ± 0.053	0.177 ± 0.033	3.41 ± 0.60
Northern	8/02/82	1.400 ± 0.065	0.977 ± 0.036	0.747 ± 0.083	2.18 ± 0.15
Southern	8/20/82	0.440 ± 0.051	0.275 ± 0.044	0.197 ± 0.036	2.72 ± 0.36
Southern	8/02/82	1.039 ± 0.043	0.716 ± 0.036	0.573 ± 0.086	2.17 ± 0.11

These patches are used as a controlled dark surface for the estimation of the aerosol optical thickness. Although this requirement limits the applicability of the algorithm, it results in an automatic correction in which the atmospheric haziness is estimated from the image itself. The algorithm is based on the following assumptions:

1. There is at least a fraction f_1 of the pixels in the image that are covered fully by dense dark vegetation. The algorithm will determine in the image a fraction f_1 of the pixels that best fit the criterion of dense dark vegetation.

2. The surface reflectance of these patches of dense dark vegetation in the spectral bands used in the algorithm is known. For example, for the blue and the red parts of the spectrum, we can assume $\rho = 0.02 \pm 0.01$ and for the green part of the spectrum, $\rho = 0.03 \pm 0.01$ (see Table 5).

3. The aerosol scattering phase function P and the single-scattering albedo ω_o are known (an error in P and ω_o affects the derived optical thickness, but cancels out partially in the correction of the surface reflectance; see the section on "evaluation").

4. The surface can be considered as Lambertian for the purpose of the correction (for a discussion of this assumption, see Section IV.B). This assumption is valid for most irradiation and viewing directions. Large errors can be expected for high solar zenith angles and for viewing directions that are close to specular reflection and backscattering.

The sensitivity of the algorithm to these assumptions will be tested in the latter part of this section. The following are the steps executed in the algorithm.

A. Estimate the gaseous absorption based on climatology of the area by the Lowtran 6 code.

B. Remove cloudy areas by masking out pixels with high wavelength-independent reflectance or low IR radiative temperature. The remaining subpixel clouds that cannot be removed by this method are considered as part of the aerosol layer and will be treated as such.

C. Estimate the aerosol single-scattering phase function P, single-scattering albedo ω_o, and the height of the aerosol above the surface from climatology of the area, or any other souce of information such as ground measurements, if they are available.

D. Optional. If the area has a substantial aerosol loading (e.g., location where Saharan dust or forest-fire smoke is present or anthropogenic air pollution, as in the summer in the eastern United States), update step C by the derivation of aerosol characteristics from the difference in the brightness between satellite images during clear and hazy conditions (Kaufman et al., 1988).

E. Generate a theoretical look-up table of the upward radiance L as a function of τ_a and ρ, $L_\lambda(\tau_a, \rho)$, for the solar direction and the observation direction, and for the chosen aerosol characteristics (phase function P and single-scattering albedo ω_o).

F. For the satellite image being corrected, identify pixels covered by dense dark vegetation. This determination is performed by choosing a fraction f_1 (e.g., $f_1 = 0.2$) from the image of the pixels with the highest vegetation index, and from this group, choosing a fraction f_2 (e.g., $f_2 = 0.5$) of pixels with the lowest radiance in the near IR (thus choosing pixels covered with dark vegetation from the vegetated pixels). The fractions f_1 and f_2 are determined a priori based on an estimate of the occurrence of such vegetation.

G. The aerosol optical thickness τ_a is computed for each pixel identified as covered by dense dark vegetation. The optical thickness is found by interpolation of τ_a from the library $L_\lambda(\tau_a, \rho)$ for the radiance L_m measured for each pixel, and for the assumed reflectance ρ in each band. The determination of the optical thickness is performed for the red and, if available, also for the blue parts of the spectrum, where vegetation is dark.

H. Generate a map of $\tau_a(\lambda, x, y)$ by interpolating and smoothing the derived values. Interpolate and extrapolate τ_a as a function of the wavelength by using the wavelength dependence of τ_a found in step G or by the climatological dependence of τ_a on the wavelength. The wavelength dependence of τ_a can also be used to test and possibly change the assumed phase function (Kaufman et al., 1988).

I. Perform the atmospheric correction by using Eq. 58 for a uniform surface (or low sensor spatial resolution) or Eq. 59 for a nonuniform surface (or high spatial resolution).

VI.F.2 Application and Validation

The suggested algorithm was applied to Landsat MSS images taken over the Washington, D.C., and Chesapeake bay areas. Cloud-free images during relatively clear (August 20, 1982, referred to here as a ''clear day'') and hazy (August 2, 1982, referred to here as a ''hazy day'') conditions were used. For each day, two subsets of the image were used, each containing 50×100 pixels. Each pixel in the subset is an average of 10×10 pixels in the original Landsat image, and thus each subset represents an area of 30×70 km, with a resolution of ~ 700 m. The two subsets are aligned in the North–South direction.

In Figure 46, an example is shown of the relation between the brightness in band 2 (600–700 nm) and the vegetation index [NDVI is the ratio of band 3 (700–800 nm), L_3, to band 2, L_2] as well as the normalized difference NDVI for the northern subset. This relation is plotted for the clear day, Figure 46(a), and for the hazy day, Figure 46(b). For 20% of the pixels ($f_1 = 0.2$) with the highest NDVI (= 0.46 for the clear day and 0.2 for the hazy day), radiance L_2 each day is within $\Delta L_2 = \pm 0.01$, which corresponds to an uncertainty in the reflectance ρ of $\Delta\rho = \pm 0.01$. Note that the value of NDVI is different in the hazy day from that in the clear day. The radiance is more disperse in band 1 (500–600 nm) and its value is also higher, due to the higher reflectance of vegetation in the green part of the spectrum. Unfortunately, no blue band is present in the MSS, and, therefore, band 1 (in addition to band 2) is used here in the atmospheric correction. These two bands (1 and 2) will be used here to derive the aerosol optical thickness in order to perform atmospheric corrections of bands 1–3 of the Landsat MSS image. In Figure 47, a comparison is shown between the histograms of the radiances in band 2 of the original northern subset (solid lines) and histograms of these dark pixels selected by the procedure (step F). The application of the procedure results in a much narrower histogram, shown as the shaded area in the figure. These histograms of the selected pixels correspond to the lowest radiances in the image.,

The histograms of the aerosol optical thickness computed from the chosen pixels in bands 1 and 2 for the northern subset (step G) are shown in Figure 48. The average and standard deviations of the histograms are given in Table 5c. Comparison between the

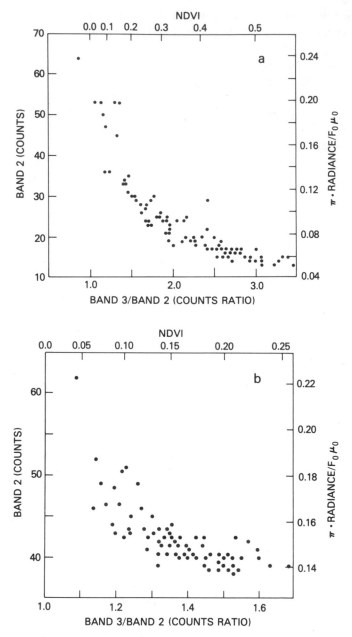

Figure 46   A scatter diagram of the radiances from Landsat, MSS band 2 (600–700 nm) and the vegetation index (ratio between the counts in bands 2 and 3). Each point represents an average of 10×10 pixels. (*a*) August 20, 1982, a clear day, and (*b*) August 2, 1982, hazy day.

satellite-derived optical thickness and the optical thickness simultaneously measured from the ground (arrows in Figure 48) shows a good agreement between them within $\Delta \tau = \pm 0.20$ in band 1 and $\Delta \tau = \pm 0.05$ in band 2 (see also Table 5b). In this case, it was assumed that the surface reflectances are $\rho_1 = 0.03$ and $\rho_2 = 0.02$ in the two MSS bands. We can see the superiority of band 2 (red) over band 1 (green).

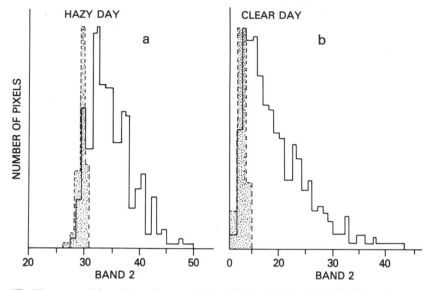

Figure 47 Histogram of the radiance (in counts) from Landsat MSS band 2 (600–700 nm) for a clear day (August 20, 1982) and a hazy day (August 2, 1982). The solid line represents the original histogram and the dashed line (also shaded) represents the histogram after the selection algorithm was applied (see text).

The clear day represents here a background aerosol that is present in the eastern United States during the summer (Kaufman and Fraser, 1983). The hazy day represents the effect of forest-fire smoke originating in northwest Canada and transported to the Washington, D.C., area (Kaufman et al., 1988). A plot of the spatial variation of the optical thickness derived in the two bands on the clear and hazy days is shown in Figure 49. In this plot, the values of the optical thickness derived in each East–West line was averaged to yield a single value representing the line. The plots show the variation of

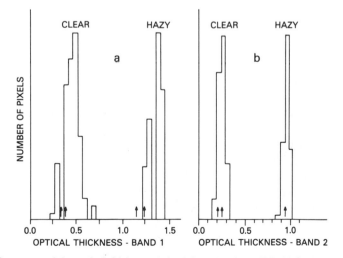

Figure 48 Histograms of the optical thickness derived from Landsat MSS: (a) band 1 (500–600 nm) and (b) band 2 (600–700 nm) for a clear day (August 20, 1982) and a hazy day (August 2, 1982). The arrows show the corresponding measurements from sun photometers (see also Table 6).

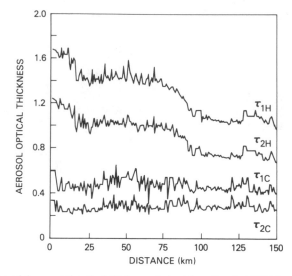

Figure 49   Variation of the aerosol optical thickness derived from Landsat MSS bands 1 and 2 for August 2, 1982 (upper two plots, τ_{1H} and τ_{2H}), and for August 20, 1982 (lower two plots, τ_{1C} and τ_{2C}).

the optical thickness along 150 km. For a clear day, the optical thickness is very stable across the area represented here. For a hazy day, a gradient in the smoke density can be seen in the plot. In the case of such a gradient, the procedure tends to select pixels with a lower optical thickness, because the vegetation index, used as a criterion in the selection, is reduced with an increase in the optical thickness.

As a result, application of the procedure to the two subsets of the image on a hazy day resulted in areas where the optical thickness was not computed at all, due to the higher values of the optical thickness in these areas. This higher optical thickness reduced the value of the NDVI in these areas under the threshold value. In order to add the missing information, the values of the optical thickness were recomputed for these portions separately. The procedure can compute the values of the optical thickness with a spatial nonhomogeneity of $\Delta\tau_a = 0.2$. Beyond this range, there is a need to apply the procedure to each section of the image that has a semiuniform τ_a. The computations in the different sections of the subareas (resulted in an optical thickness that varies smoothly across the area (Figure 49).

In order to correct the images and remove the atmospheric effect, the values of the optical thickness are interpolated (and, if necessary, extrapolated) across the whole scene (step H). From the derived optical thickness in each band, the exponent ν is computed (see Eq. 16). Average values of ν are tabulated in Table 6c for each subarea. For the northern subarea, the derived value of ν can be compared with the value of ν derived from optical thicknesses measured by sun photometers from the ground (Kaufman et al., 1986). The values of ν derived from satellite-measured radiances are larger by $\Delta\nu = 0.5$ from values measured from the ground. This difference is due mainly to the higher optical thickness derived in band 1 from the satellite imagery. The value of ν is used to extrapolate the optical thickness to the near-IR (band 3, τ_3) in order to perform atmospheric corrections in that band. A comparison between the extrapolated value of τ_3 (Table 6c) and the value measured by the sun photometer from the ground (Table 6b) shows an excellent agreement.

The correction algorithm (Eq. 58) was applied to the images, resulting in the histograms of the vegetation index in Figure 50. Values of the derived average reflectance and standard deviation are given in Table 6c. It is seen that the atmospheric effect generated a large difference between the radiances and the corresponding vegetation indexes in the two days. The correction scheme, transformed the histograms in the clear and hazy days, generating an almost complete overlap between them. This shift in the histogram is also shown in two dimensions in Figure 51. Each ellipse in the figure is centered on the average value of the radiances in the image (in bands 2 and 3). The axis of the ellipse is twice the standard deviation of the histogram. A plot of the original and corrected radiances is shown in Figure 52 for bands 2 and 3. In this plot, the average of 10 columns (on the eastern part of the subarea) is plotted as a function of the location before and after the correction was applied, on clear and hazy days. The large difference in the upward radiance between the clear and hazy days was drastically reduced by the atmospheric correction. The major difference in the corrected radiance in the two days is the high spatial frequency of the radiance variation. The hazy-day corrected radiances are less variable than the clear-day radiances (in high spatial frequencies). This difference may arise from the non-Lambertian properties of the surface. On the clear day, most of the downward irradiance is the direct solar radiation, whereas on the hazy day, most of the downward irradiance is diffuse. As a result, specular reflection (of slant surfaces) is more pronounced on the clear day than on the hazy day. The assumption, used in the correction, that the surface is Lambertian prohibits the regeneration of the specular reflection in the hazy day. In this particular case (August 2, 1982), the haze was so dense that there is hardly any specular reflection left to be restored.

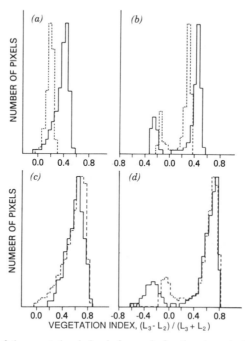

Figure 50 Histograms of the vegetation index before and after the atmospheric correction. (*a*) Northern subarea, original radiances; (*b*) southern subarea, original radiances; (*c*) northern subarea, corrected radiances; and (*d*) southern subarea, corrected radiances. Solid lines represent vegetation index for a hazy day (August 2, 1982) and dashed lines for a clear day (Aug. 20, 1982).

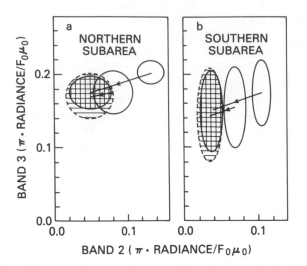

Figure 51 The effect of the correction on the location of the radiances in the band 2 by band 3 radiance space (*a*) for the northern subarea and (*b*) for the sourthern subarea. The ellipses are centered at the average values and their dimensions represent the standard deviation of the radiances. The arrows show the transformation caused by the atmospheric correction. Empty ellipses show the original radiances, ellipses with a vertical pattern show the correction of the hazy-day data (August 2, 1982), and ellipses with a horizontal pattern show the correction of the clear-day data (August 20, 1982).

VI.F.3 Evaluation

In this section, we shall discuss the applicability of the algorithm to different parts of the world and the sensitivity of the correction method to the accuracy of the satellite calibration and to the three major assumptions in the procedure: the fraction of dense

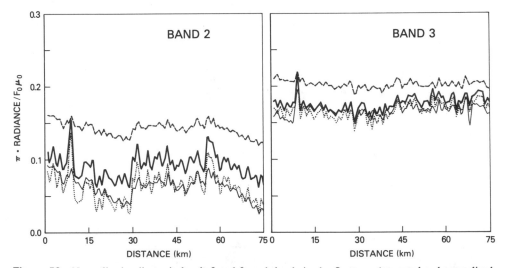

Figure 52 Normalized radiance in bands 2 and 3, and the derived reflectance (corrected and normalized radiance) for the northern subset. Line average of the first 10 columns is plotted as a function of the line location in the subarea. Radiances from the original image: thick dashed lines for August 20, 1982; dashed lines for August, 2, 1982; corrected radiances; dotted lines for August 20, 1982; solid lines for August 2, 1982.

dark vegetation in the image (f_1), the value of the surface reflectance of the dense dark vegetation, and the aerosol characteristics.

The major uncertainty in the suggested algorithm results from the uncertainty in the assumed reflectance of the dense dark vegetation. Assuming an error of $\delta\rho = 0.01$, the expected error in the optical thickness would be $\delta\tau_a = 0.13$ for optical thickness in the range $\tau_a = 0.20-0.40$ and $\delta\tau_a = 0.09$ for $\tau_a = 0.8-1.0$. This generates an error in the corrected surface reflectance of $\delta\rho = 0.01$. For comparison in the clear day, the atmospheric correction reduced the apparent surface reflectance in the red band by $\Delta\rho_2 = 0.032$ for the nothern subset and $\Delta\rho_2 = 0.084$ for the hazy day. The histograms in Figure 50 and the values in Table 2 show that the actual difference between the corrected reflectance between the clear and the hazy days was usually smaller than the theoretical predicted error of $\delta\rho = 0.01$.

For the northern subset, the atmospheric correction decreased the difference between the average vegetation index on the hazy and clear days from $\Delta\text{NDVI} = 0.18$ to $\Delta\text{NDVI} = 0.01$. For the southern subset, the correction reduced the difference in the average vegetation index between the hazy and clear days from $\Delta\text{VI} = 0.14$ to $\Delta\text{VI} = 0.06$. This larger difference is not due to the correction algorithm, since the histograms of the corrected images are not shifted relatively one to another (see Figure 51), but the "tail" of the histogram for small and for negative vegetation indexes differs. These values of vegetation indexes correspond to the water in the Bay area.

The derived surface reflectance depends on the satellite calibration in two ways:

1. A degradation in the sensor gain will decrease the radiances detected by the satellite, and if the preflight calibration is used, the derived apparent surface reflectance will be smaller than the true one.

2. The degradation will also reduce the values of the optical thickness derived from the radiances above dense dark vegetation. The reduced values of the optical thickness will result in a smaller atmospheric correction. For a bright surface ($\rho > 0.3$), the smaller correction will result in further smaller values of the corrected reflectances. For a dark surface ($\rho < 0.1$), the smaller correction will result in larger reflectances, thus, compensating partially for the initial reduction in the reflectance.

The sensitivity of the derived surface reflectance and optical thickness to calibration error was tested for two kind of errors: (a) error in all the bands (degradation of 10%), and (b) degradation of 10% in band 1 only. Degradation in all the bands (case a) decreased all the optical thicknesses by 30% for the clear day and by 15% for the hazy day, resulting in the same slope ν. In bands 1 and 2, the dark surface became darker only by $\delta\rho = -0.003$. This is due to the partial compensation discussed before. For band 3, the brighter surface was darkened by $\delta\rho = -0.014$ (8%). Calibration error in one band only (case b, error in band 1) resulted in a change in ν and in a change in the surface reflectance in band 1 only.

The sensitivity of the correction algorithm to fraction f_1 shows that if fraction f_1 is larger than the actual fraction of dense dark vegetation ($f_1 = 0.5-1.0$), the resultant optical thickness is too high, resulting in overcorrection of the radiances and the corresponding vegetation index. For $0.1 < f_1 < 0.5$, the sensitivity to f_1 is not substantial, resulting in an error in the optical thickness of $\delta\tau \leq 0.04$. This sensitivity is not a general result and depends on the actual fraction of dense dark vegetation in a given image. Therefore, if the fractional cover of dense vegetation is unknown, a small value of f_1 will be safe. A very small value of f_1 (< 0.05) produced poor statistics and noisy data.

Uncertainty in the aerosol scattering phase function, due to uncertainty in its size distribution or composition, may result in large errors in the derived optical thickness. Rather large uncertainty in the size distribution was used in the sensitivity test. In addition to the power law distribution with exponent of $\nu = 3.7$, representative to this climate (Kaufman and Fraser, 1983), a log-normal distribution was used with mass mode radius of $r_o = 0.20$ and $r_o = 0.60$ μm. Although the error in the derived optical thickness is large (e.g., for a clear day, τ_1 increases from $\tau_1 = 0.50$ for the power law to $\tau_1 = 0.76$ for $r_o = 0.20$ μm and $\tau_1 = 0.89$ for $r_o = 0.60$ μm), the corrected surface reflectance is not very sensitive to particle size and the resulting scattering phase function (e.g., for a clear day, ρ_1 varies from 0.051 to 0.056 for these three size distributions, and ρ_2 varies from 0.049 to 0.050). The small sensitivity of the derived reflectance results because the correction is applied using the same phase function that was used to compute the optical thickness, thus resulting in cancellation of most of the errors involved. The vegetation index varies for a clear day between NDVI = 0.57 and 0.58 and for a hazy day between 0.54 to 0.57 due to the uncertainty in the size distribution. Therefore, the correction increased the vegetation index from the uncorrected values of NDVI = 0.386 for a clear day and NDVI = 0.208 for a hazy day to NDVI = 0.56 $\pm$ 0.02 for both days.

The sensitivity of the algorithm to the single-scattering albedo ω_o was tested by a relatively large error in $\omega_o - \Delta\omega_o = 0.1$. This difference may result from a change from a rural to an urban aerosol. The "true" value in the simulation was $\omega_o = 0.95$, and the value used in the correction was $\omega_o = 0.85$. This underestimation of ω_o causes an overestimation in τ_a by 50%. On a clear day, the error in the derived surface reflectance is only $\Delta\rho = 0.005$ for bands 1 and 2 and $\Delta\rho = 0.01$ for band 3. The small error in the surface reflectance for the large error in the derived value of the optical thickness is a result (as we already have seen for an error in the phase function) of a compensating effect of the error in $\Delta\omega_o$ in computation of τ_a and in the correction (recomputation of the path radiance L_o). On a hazy day, the error is much larger due to the relatively small contribution of the surface reflectance to the upward radiance.

It is concluded that the major limitation to the application of the algorithm is the presence of pixels fully covered by dense dark vegetation. Figure 53 shows the areas around the world where

1. the algorithm is expected to be successful,
2. there is an uncertainty in its application, and
3. the successful application is very unlikely.

The figure is based on a map of natural vegetation (Kühler, 1979). In the first area, forests with evergreen and deciduous trees, as well as tall shrubs, are included. This area seems to be suitable for the application of the procedure. The second area includes the same type of vegetation as in first area, but growing in patches or singly; thus, a full cover of a pixel is not certain. The applicability of the algorithm to this type of area depends also on the spatial resolution of the sensor. In the third area, the trees or bushes are far apart or completely not present.

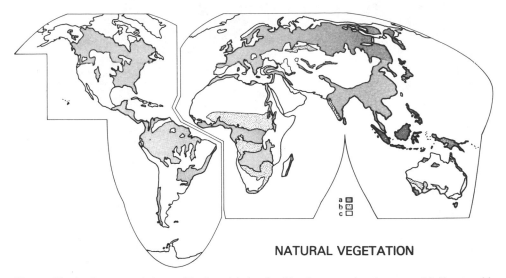

NATURAL VEGETATION

Figure 53 Regions around the world where (*a*) the algorithm is expected to be successful (forests with evergreen and deciduous trees as well as high shrubs) and (*b*) there is uncertainty in the application [same vegetation as in (*a*), but growing in patches or singly]. (*c*) The success of the algorithm is very unlikely (trees or high shrubs do not grow or grow singly). (Adopted from Kühler, 1979.)

VII CONCLUSIONS

The atmospheric effect on the upward radiance depends on the optical characteristics of the aerosol, the absorbing gases in the atmosphere, and on the reflection characteristics of the surface. The atmospheric effect on remote sensing depends also on the surface parameters, the surface features being sensed, and on the resolution of the satellite sensor. In order to correct for the atmospheric effect, the aerosol characteristics have to be known. Usually, the aerosol optical thickness is the major parameter, since it is used as a representative of the aerosol total loading. The aerosol phase function has to be known mainly in order to estimate the atmospheric-path radiance that makes up the major component of the atmospheric effect above dark surfaces. Above bright surfaces, knowledge of the single-scattering albedo is very important, and for high-resolution remote sensing, the vertical distribution of the aerosol has to be estimated.

Correction of the atmospheric effect is based on the estimation of these aerosol parameters from climatology, measurements from the ground, or from the satellite imagery itself. The determination of the aerosol loading or the optical thickness from the satellite imagery can be done by assuming some characteristics of the surface. It is performed in conditions where the effect of the atmosphere is maximal (slant view direction or dark surface is used to derive the aerosol optical thickness). Once the atmospheric parameters are estimated, the correction can use an algebraic relation between the detected radiance and the surface reflectance for a uniform surface or for imagery with low spatial resolution (1 km and up) or by the Fourier-transform method if the atmospheric effects on the nonuniform surface have to be preserved.

In Figure 54, an example is given of a correction of Landsat imagery of the Washington, D.C., area taken during relatively clear conditions and on a very hazy day. On the left side, the original images are shown for bands 1 ($\lambda = 0.55$ μm) and 3 ($\lambda = 0.75$

Figure 54 Images of the Washington, D.C., area taken from Landsat 3 MSS bands 1 (0.55 μm) and 3 (0.75 μm) on a clear day (August 20, 1982) and a hazy day (August, 2, 1982). The corrected images are shown to the right of the original images. The correction was performed by the Fourier method (taking into account the adjacency effect).

μm). The atmosphere masks the features of the surface, especially in band 1. The Fourier correction of the image, based on a measured optical thickness and assumed phase function and single-scattering albedo, is shown on the right side of the image. The contrast of the image was restored and the corrected images on the clear and hazy days look very similar.

A combination of methods for remote sensing of the aerosol characteristics from space together with climatology of the atmospheric aerosol (yet to be derived) can be used for an efficient atmospheric correction of satellite imagery.

LIST OF SYMBOLS

STANDARD ALPHABETICAL SYMBOLS

Symbol	Description
A_c	Accumulated probability function of cloud fraction
$A_c^{\pm}$	Function of the radiances
A_i	Coefficient in the cloud transmission equation for cloud type i
B_i	Coefficient in the cloud transmission equation for cloud type i
C	Coefficient
C_i	Transmittance of an overcast sky for cloud type i
d	Diameter
DP	Dew-point temperature
E_o	Extraterrestrial solar irradiance
E_d	Total irradiance of the surface (normalized to reflectance units)
E_d'	Total irradiance of the surface
E_d	Total irradiance of the surface (normalized to reflectance units)
E_{dd}	Diffuse irradiance on the surface (normalized to reflectance units)
E_{dd}^*	Diffuse irradiance on the surface, including the cloud effect (normalized to reflectance units)
E_s	Direct solar irradiance on the surface (normalized to reflectance units)
$\mathcal{F}$	Fourier transform
$\mathcal{F}^{-1}$	Inverse Fourier transform
H	Atmospheric density scale height
H_a	Aerosol density scale height
H_c	Cloud top height
H_r	Rayleigh scattering scale height
J_o	Source term due to scattering of direct sunlight
K_d	Dry scattering coefficient
K_e	Extinction coefficient
K_s	Scattering coefficient
k_x	Spatial frequency in the x direction
k_y	Spatial frequency in the y direction
L	Radiance normalized to reflectance units
$L(\pm 0)$	Upward radiances close to the boundary of the "two field halves"
$L(\pm \infty)$	Upward radiance far from the boundary of the "two field halves"
L'	Radiance

Symbol	Description
L_a	Upward radiance with no cloud effect (reflectance units)
L_{ae}	Aerosol part of the path radiance
L_b	Upward radiance for a sunny area observed through a cloud (reflectance units)
L_c	Upward radiance for a shaded area observed directly from space (reflectance units)
L_d	Upward radiance for a shaded area observed through a cloud (reflectance units)
L_d	$L_{d1} + L_{d2}$
L'_{d1}	Diffuse radiance
L_{d1}	Diffuse radiance (normalized to reflectance units)
L'_{d2}	Diffuse radiance
L_{d2}	Diffuse radiance (normalized to reflectance units)
L_{in}	Radiance above an infinitesimally small field
L_m	Measured radiance
L_{mo}	Molecular part of the path radiance
L_{NIR}	Radiance in the near-IR region
L_o	Path radiance (reflectance units)
L'_o	Path radiance
L_{ri}	Radiance above a reference field (normalized to reflectance units)
L'_s	Radiance of the attenuated signal
L_s	Radiance of the attenuated signal (normalized to reflectance units)
L_{ti}	Radiance above a test field (normalized to reflectance units)
L_{VIS}	Radiance in the visible region
L_w	Radiance emerging from the water
L_β	Residual radiance (reflectance units)
$m(x, y)$	Point spread function
M	Modulation transfer function (MTF)
M_n	Normalized modulation transfer function
n	Aerosol particle density
n_a	Atmospheric density
n_c	Cloud size distribution
N	Number of remote-sensing observations
N_o	Density of aerosol particles
P	Atmospheric pressure
$P(\theta)$	Scattering phase function
$P_a(\theta)$	Average scattering phase function of a volume of aerosol
$P(\chi_c)$	Probability of cloud fraction χ_c
P_o	Standard atmospheric pressure at sea level.
Q	Extinction efficiency factor
r	Particle radius
r_o	Geometric mean radius of the log-normal size distribution
r_1	Particle radius in the power law size distribution
r_2	Particle radius in the power law size distribution
RH	Relative humidity
s	Atmospheric backscattering of the upward flux
s'	Atmospheric backscattering in the presence of clouds

Symbol	Description
S_b	Sunny area observed through a cloud
S_c	Shaded area observed directly from space
S_d	Overlap area between clouds
S_k	Classification measure
T	Temperature
T_c	Cloud transmittance
T_r	Molecular transmittance
T_d	Diffuse transmittance
T_s	Direct transmittance
T_{sc}	Transmittance of a cloudy sky
u	Cos θ
V	Visibility
W	Mixing ratio
W_a	Coefficient for the upward radiance with no cloud effect
W_b	Coefficient for the upward radiance for a sunny area observed through a cloud
W_c	Coefficient for the upward radiance for a shaded area observed directly from space
W_d	Coefficient for the upward radiance for a shaded area observed through a cloud
X	Horizontal coordinate
X_c	Cloud width
X_e	Range of the adjacency effect
Y	Vertical coordinate
Y_c	Cloud length
z	Height

STANDARD GREEK SYMBOLS

Symbol	Description
α	Coefficient
β	Extinction cross section
δ	Delta function
ε	Ratio
θ	Scattering angle
θ_o	Solar zenith angle
θ'	Observer zenith angle
λ	Wavelength
μ	$|u|$
μ_o	Cos θ_o
ν	Power in a power law function
ρ	Hemispherical reflectance of the surface
ρ_b	Background reflectance
ρ_c	Critical surface reflectance
ρ_{cl}	Cloud reflectance
ρ_f	Reflectance of the observed field
σ	Standard deviation of ln r in the particle log-normal size distribution

Symbol	Description
τ	Optical thickness
τ_a	Aerosol optical thickness
τ_r	Rayleigh optical thickness
τ_o	Total atmospheric optical thickness
ϕ	Azimuth angle
ϕ_o	Solar azimuth angle
χ_c	Cloud fraction
ω_o	Single-scattering albedo
η	Range factor of the adjacency effect

REFERENCES

Ackerman, T. P., and O. B. Toon (1981). "Absorption of visible radiation in atmosphere containing mixtures of absorbing and nonabsorbing particles. *Appl. Opt.* **20**:3661–3668.

Ahmad, Z., and R. S. Fraser (1982). An iterative radiative transfer code for ocean-atmosphere system. *J. Atmos. Sci.* **39**:656–665.

Aida N. (1977). Scattering of solar radiation as a function of cloud dimensions and orientation. *J. Quant. Spectrosc. Radiat. Transfer* **17**:303–310.

Arking, A., and J. Childs (1985). Retrieval of cloud cover parameters from multispectral satellite images. *J. Clim. Appl. Meterol.* **24**:322–333.

Barteneva, O. D. (1960). Scattering functions of light ion the atmospheric boundary layer. *Izv. Akad. Nauk SSSR*, Ser. *Geofiz.*, pp. 1237–1244.

Bohren, C. F. (1986). Absorption and scattering of light by nonspherical particles. *Conf. Atmos. Radiat.*, Williamsburg, Virginia, pp. 1–7.

Cahalan, R. F. (1983), Climatological statistics of cloudiness. *Proc. 5th Conf. Atmos. Radiat.*, American Meteorological Society, Boston, Massachusetts, pp. 206–213.

Carlson, T. N. (1979). Atmospheric turbidities in Saharan dust outbreaks as determined by analysis of satellite brightness data. *Mon. Weather Rev.* **107**:322–335.

Chandrasekhar, S. (1960). *Radiative Transfer*. Dover, New York.

Clarke, G. K., G. C. Ewing, and C. J. Lorenzen (1979). Spectra of backscattered light from the sea obtained from aircraft as a measure of chlorophyll concentration. *Science* **167**:1119–1121.

Coakley, J. A., Jr., and F. P. Bretherton (1982). Cloud cover from high resolution scanner data: Detecting and allowing for partially filled fields of view. *J. Geophys. Res.* **87**:4017–4932.

D'Almeida, G. A., and P. Koepke (1987). An approach to global aerosol climatology. *IUGG-IAMAP 19th General Assembly*, Vancouver, Canada.

Dana, R. W. (1982). Background reflectance effects in Landsat data. *Appl. Opt.* **21**:4106–4111.

Dave, J. V., and J. Gazdag (1970). A modified Fourier transform method for multiple scattering calculations in a plane parallel Mie atmosphere. *Appl. Opt.* **9**:1457–1466.

Deering, D. W., and T. F. Eck (1987). Atmospheric optical depth effects on angular anisotropy of plant canopy reflectance. *Int. J. Remote Sens.* **8**:893–916.

Diner, D. J., and J. V. Martonchik (1985a). Influence of aerosol scattering on atmospheric bluring of surface features. *IEEE Trans. Geophys. Remote Sens.* **GE-23**:618–624.

Diner, D. J., and J. V. Martonchik (1985b). Atmospheric transmittance from spacecraft using multiple view angle imagery. *Appl. Opt.* **24**:3503–3511.

Flowers, E. C., R. A. McCormic, and K. R. Kurfis (1969). Atmospheric turbidity over the United States. *J. Appl. Meteor.* **8**:955–962.

Fraser, R. S. (1976). Satellite measurements of mass of sahara dust in the atmosphere. *Appl. Opt.* **15**:2471–2479.

Fraser, R. S., and R. J. Curran (1976). Effects of the atmosphere on remote sensing. In *Remote Sensing of Environment* (J. Lintz and D. S. Simonett, Eds.) Addison-Wesley, London, pp. 34–84.

Fraser, R. S., and Y. J. Kaufman (1985). The relative importance of scattering and absorption in remote sensing. *IEEE Trans. Geosci. Remote Sens.* **23**:625–633.

Fraser, R. S., N. E. Gaut, E. C. Reifenstein, and H. Sievering (1975). Interaction mechanism—within the atmosphere. In *Manual of Remote Sensing*. (R. G. Reeves, A. Anson, and D. Landen, Eds. American Society of Photogrammetry, Falls Church, Virginia.

Fraser, R. S., O. P., Bahethi, and A. H. Al-Abbas (1977). The effect of the atmosphere on classification of satellite observations to identify surface features. *Remote Sens. Environ.* **6**:229.

Fraser, R. S., Y. J. Kaufman, and R. L. Mahoney (1984). Satellite measurements of aerosol mass and transport. *Atmos. Environ.* **18**:2577–2584.

Gerstl, S. A. W., and C. Simmer (1986). Radiation physics and modelling and off nadir satellite sensing of non-Lambertian surfaces. *Remote Sens. Environ.* **20**:1.

Gerstl, S. A. W., and A. Zardecki (1985a). Discrete-ordinate finite element method for atmospheric radiative transfer and remote sensing. *Appl. Opt.* **24**:81–93.

Gerstl, S. A. W., and A. Zardecki (1985b). Coupled atmosphere/canopy model for remote sensing of plant reflectance features. *Appl. Opt.* **24**:94–103.

Gordon, H. R. (1978). Removal of the atmospheric effects from satellite imagery of the oceans. *Appl. Opt.* **17**:1631–1636.

Gordon, H. R., and D. K. Clark (1981). Clear water radiances for atmospheric corrections of CZCS imagery. *Appl. Opt.* **20**:4175–4180.

Gordon, H. R., D. K. Clark, J. W. Brown, O. B. Brown, R. H. Evans, and W. W. Broenkow (1983). Phytoplankton pigment concentration in the middle Atlantic bight: Comparison of ship determination and CZCS estimates. *Appl. Opt.* **22**:20–36.

Gordon, H. R., R. W. Austin, D. K. Clark, W. A. Hovis, and C. S. Yentsch (1985). Ocean color measurements. *Adv. Geophys.* **27**:297–333.

Griggs, M. (1975). Measurements of atmospheric aerosol optical thickness over water using ERTS-1 data. *J. Air Pollut. Control Assoc.* **25**:622–626.

Hanel, G. (1981). An attempt to intercept the humidity dependencies of the aerosol extinction and scattering coefficient. *Atmos. Environ.* **15**:403–406.

Hansen, J. E., and L. D. Travis (1974). Light scattering in planetary atmosphere. *Space Sci. Rev.* **16**:527–610.

Haurwitz, B. (1948). Insolation in relation to cloud type. *J. Meteorol.* **5**:110–113.

Herman, B. M., and S. R. Browning (1975). The effect of aerosols on the Earth—Atmosphere albedo. *J. Atmos. Sci.* **32**:1430–1445.

Holben, B. N. (1986). Characteristics of maximum value composite images for temporal AVHRR data. *Int. J. Remote Sens.* **7**:1417–1437.

Holben, B. N., and R. S. Fraser (1984). Red and near IR sensor response to off-nadir viewing. *Int. J. Remote Sens.* **5**:145–160.

Hoppel, W. A., J. W. Fitzgerald, G. M. Frick, R. E. Larson, and B. J. Wattle (1987). *Preliminary Investigation of the Role that DMS and Cloud Cycles Play in the Formation of the Aerosol Size Distribution*, NRL 9032. Naval Research Laboratory, Arlington, Virginia.

Horvath, R. B., J. G. Polcyn, and C. Fabian (1970). Effect of atmospheric path on airborne multispectral sensors. *Remote. Sens. Environ.* **1**:203.

Husar, R. B. and J. M. Holloway (1984). The properties and climate of atmospheric haze. In *Hygroscopic Aerosol* (L. H. Ruhnke and A. Deepak, Eds.). A. Deepak Pub., Hampton, Virginia, pp. 129–170.

Isaacs, R. G., W. C. Wang, R. D. Worsham, and S. Goldenberg (1987). Multiple scattering Lowtran and Fascode models. *Appl. Opt.* **26:**1272–1281.

Joseph, J. H. (1985). The morphology of Fair weather cumulus cloud fields as remotely sensed from satellites and some applications. *Adv. Space Res.* **5:**213–216.

Joseph, J. H., W. Wiscombe, and J. Weinman (1976). The delta Eddington approximation for radiative flux transfer. *J. Atmos. Sci.* **33:**2452.

Junge, C. E. (1963). *Air Chemistry and Radiochemistry.* Academic Press, New York.

Justice, C. O., J. R. Townshed, B. N. Holben, and C. J. Tucker (1985). Phenology of global vegetation using meteorological satellite data. *Int. J. Remote Sens.* **6:** 1271.

Kaufman, Y. J. (1978). Influence of the atmosphere on the contrast of the Landsat images. *Space Res.* **17:**65–68.

Kaufman, Y. J. (1979). Effect of the Earth atmosphere on contrast for zenith observation. *J. Geophys. Res.* **84:**3165–3172.

Kaufman, Y. J. (1982). Solution of the equation of radiative transfer for remote sensing over non-uniform surface reflectivity. *J. Geophys. Res.* **81:**4137–4147.

Kaufman, Y. J. (1984a). Atmospheric effects on remote sensing of surface reflectance. *Proc. SPIE—Int. Soc. Opt. Eng.* **475:**20–33.

Kaufman, Y. J. (1984b). Atmospheric effect on spatial resolution of surface imagery. *Appl. Opt.* **23:**3400–3408.

Kaufman, Y. J. (1984c). The atmospheric effect on separability of field classes measured from satellite. *Remote Sens. Environ.* **18:**21–34.

Kaufman, Y. J. (1987a). Satellite sensing of aerosol absorption. *J. Geophys. Res.* **92:**4307–4317.

Kaufman, Y. J. (1987b). The effect of subpixel clouds on remote sensing. *Int. J. Remote Sens.* **8:**839–857.

Kaufman, Y. J. (1988). Atmospheric effect on spectral signature—measurements and corrections. *IEEE Trans. Geosci. Remote Sens.* **26:**441–450.

Kaufman, Y. J., and R. S. Fraser (1983). Light extinction by aerosols during summer air pollution. *J. Appl. Meteorol.* **22:**1694–1706.

Kaufman, Y. J., and R. S. Fraser (1984). The effect of finite field size on classification and atmospheric correction. *Remote. Sens. Environ.* **15:**95–118.

Kaufman Y. J., and J. H. Joseph (1982). Determination of surface albedos and aerosol extinction characteristics from satellite imagery. *J. Geophys. Res.* **20:**1287–1299.

Kaufman, Y. J., and C. Sendra (1988). Algorithm for automatic atmospheric corrections to visible and near-IR satellite imagery. *Int. J. Remote Sens.* **9:**1357–1381.

Kaufman, Y. J., T. W. Brakke, and E. Eloranta (1986). Field experiment to measure the radiative characteristics of a hazy atmosphere. *J. Atmos. Sci.* **43:**1135–1151.

Kaufman, Y. J., R. S. Fraser, and R. A. Ferrare (1988). Remote sensing of aerosol over the land—method. *J. Geophys. Res.* (submitted for publication).

Kimes, D. S. (1983). Dynamics of directional reflectance factor distributions for vegetation canopies. *Appl. Opt.* **22:**1364–1372.

Kimes, D. S., W. W. Newcomb, R. F. Nelson, and J. B. Schutt (1986). Directional reflectance distributions of a hardwood and pine forest canopy. *IEEE Trans. Geosci. Remote Sens.* **GE-24:**281–297.

King, M. D., D. M. Byrne, B. M. Herman, and J. A. Reagan (1978). Aerosol size distribution obtained by inversion of optical depth measurements. *J. Atmos. Sci.* **35:**2153–2167.

Kneizys, F. X., E. P. Shettle, W. O. Gallexy, J. H. Chetwynd, L. W. Abreu, J. E. A. Selby, S. A. Clough, and R. W. Fenn (1983). *Atmospheric Transmittance/Radiance: Computer Code LOWTRAN 6.* Air Force Geophysics Laboratory, Hanscom Air Force Base, Massachusetts.

Koepke, P., and H. Quenzel (1979). Turbidity of the atmosphere determined from satellite calculation of optimum viewing geometry. *J. Geophys. Res.* **84:**7847–7855.

Kozoderov, V. V. (1985). Correction of space images for atmospheric effects. *Sov. Int. J. Remote Sens.* **3(2): 255–271.**

Kriebel, K. T. (1977). Reflection properties of vegetated surfaces: Tables of measured spectral biconical reflectance factors. *Muench. Univ., Meteorol. Inst., Wiss. Mitt.* **29.**

Kühler, A. W. (1979). Map of natural vegetation. In *The World Book Atlas* (E. B. Espenshade and J. L. Morrison, Eds.). World Book Encyclopedia, Chicago, Illinois, pp. 16–17.

Lee, T., and Y. J. Kaufman (1986). The effect of surface nonlambertianity on remote sensing. *IEEE Trans. Geophys. Remote Sens.* **GE-24:**699–708.

Meador, W. E., and W. R. Weaver (1980). Two stream approximation to radiative transfer in planetary atmospheres: A unified description of existing methods and new improvements. *J. Atmos. Sci.* **37:**630.

Mekler, Y., and Y. J. Kaufman (1980). The effect of the Earth atmosphere on contrast reduction for a non-uniform surface albedo and two halves field. *J. Geophys. Res.* **85:**4067–4083.

Mekler, Y., H. Quenzel, G. Ohring, and I. Marcus (1977). Relative atmospheric aerosol content from ERTS observations. *J. Geophys. Res.* **83:**967–972.

Mekler, Y., Y. J. Kaufman, and R. S. Fraser (1984). Reflectivity of the atmosphere-inhomogeneous surface system: Laboratory simulation. *J. Atmos. Sci.* **41:**2595–2604.

Mie, G. (1908). A contribution to the optics of turbid media, especially colloidal metallic suspensions. *Ann. Phys. (Leipzig)* [4] **25:**377 (in German).

Nakajima, T., T. Takamura, M. Yamano, and M. Shiobara, (1986). Consistency of aerosol size distributions inferred from measurements of solar radiation and aurole. *J. Meteor. Soc. Japan*, **64:**765–776.

Neckel, H., and D. Labs (1984). The solar radiation between 3300 and 12500 A. *Sol. Phys.* **90:**205–258.

(NOAA) National Oceanic and Atmospheric Administration (1983). *Global Vegetation Index User's Guide SDS/NESDIS.* National Climate Data Center, Washington, D.C.

Otterman, J., and R. S. Fraser (1979). Adjacency effects on imaging by surface reflection and atmospheric scattering: Cross radiance to zenith. *Appl. Opt.* **18:**2852.

Pearce, W. A. (1977). *A Study of the Effects of the Atmosphere on Thematic Mapper Observations*, Rep. 004-77. EG&G/Was. Anal. Serv. Center, Riverdale, Maryland.

Peterson, J. T., E. C. Flowers, G. J. Berri, C. L. Reynolds, and J. H. Rudisil (1981). Atmospheric turbidity over central north Carolina. *J. Appl. Meteorol.* **20:**229–241.

Plank, V. G. (1969). The size distribution of cumulus clouds in representative Florida populations, *J. Appl. Meteorol.* **8:**46–67.

Pochop, L. O., M. D. Shanklin, and D. A. Horner (1968). Sky cover influence on total hemispheric radiation during daylight hours. *J. Appl. Meteorol.* **7:**484–489.

Rayleigh, Lord (J. W. Strutt) (1871). *Philos. Mag.* [4] **41:**107–274.

Schmetz, J. (1984). On the parameterization of radiative properties of broken clouds. *Tellus* **36a:**417–432.

Shettle, E. P. (1984). Optical and radiative properties of a desert aerosol model. *Proc. Int. Radiat. Symp.*, pp. 71–74.

Shettle, E. P., and R. W. Fenn (1979). *Models for the Aerosol of the Lower Atmosphere and the Effect of Humidity Variations on Their Optical Properties*, AFGL-TR 790214. Opt. Phys. Div., Air Force Geophysics Laboratory, Hanscom Air Force Base, Massachusetts.

Simmer, C., and S. A. W. Gerstle (1985). Remote sensing of angular characteristics of canopy reflectances. *IEEE Trans. Geosci. Remote Sens.* **GE-23:**648–658.

Slater, P. N., and R. D. Jackson (1982). Atmospheric effects on radiation reflected from soil and vegetation as measured by orbital sensors using various scanning directions. *Appl. Opt.* **21**:3923–3931.

Steven, M. D., and E. M. Rollins (1986). Estimation of atmospheric correction from multiple aircraft imagery. *Int. J. Remote Sens.* **7**:481–497.

Takayama, Y., and T. Takashima (1986). Aerosol optical thickness of yellow sand over the yellow sea derived from NOAA satellite data. *Atmos. Environ.* **20**:631–638.

Tanre D., M. Herman, P. Y. Deschamps, and A. De Leffe (1979). Atmospheric modelling for space measurements of ground reflectance including bidirectional properties. *Appl. Opt.* **18**:3587–3594.

Tanre D., M. Herman, and P. Y. Deschamps (1981). Influence of the background contribution upon space measurements of ground reflectance. *Appl. Opt.* **20**:3676–3684.

Tanre, D., P. Y. Deschamps, P. Duhaut, and M. Herman (1987). Adjacency effect produced by the atmospheric scattering in TM data. *J. Geophys. Res.* **92**:12000–12006.

Tucker, C. J., and P. J. Sellers (1986). Satellite remote sensing of primary productivity. *Int. J. Remote Sens.* **7**:1395–1416.

Tucker, C. J., C. Vanpraet, E. Boerwinkel, and A. Gaston (1983). Satellite remote sensing of total dry matter production in the Senegalese Sahel. *Remote Sens. Enviorn.* **13**:461–474.

Tucker, C. J., J. R. G. Townshend, and T. E. Goff (1985). African land cover classification using satellite data. *Science* **227**:369.

Turner, R. E. (1979). *Determination of Atmospheric Optical Parameters Using the Multispectral Resource Sampler*, MRS proof of concept study on atmospheric corrections. Goddard Space Flight Center, Greenbelt, Maryland.

Twomey, S. (1977). *Atmospheric Aerosols*. Elsevier, New York.

Ueno, S., Y. Haba, Y. Kawata, T. Kusaka, and Y. Terashita (1978). The atmospheric blurring effect on remotely sensed Earth imagery. In *Remote Sensing of the Atmosphere: Inversion Methods and Applications* (A. L. Fymat and V. E. Zuev, Eds.). Elsevier, Amsterdam, pp. 305–319.

Valley, S. L. (Ed.) (1965), *Handbook of Geophysics and Space Environments*. AFCRL.

Waggoner, A. P., R. E. Weiss, N. C. Ahlquist, D. S. Covert, S. Will, and R. J. Charlson (1981). Optical characteristics of atmospheric aerosol. *Atmos. Environ.* **15**:1891–1909.

Whitby, K. T. (1978). The physical characteristics of sulfur aerosols. *Atmos. Environ.* **12**:135–159.

Wielicki, B. A., and R. M. Welch (1986). Cumulus cloud properties derived using Landsat satellite data. *J. Clim. Appl. Meteorol.* **25**:261–276.

10

APPLICATIONS IN FOREST SCIENCE AND MANAGEMENT

DAVID L. PETERSON

Ames Research Center
National Aeronautics and Space Administration
Moffett Field, California

and

STEVEN W. RUNNING

School of Forestry
University of Montana
Missoula, Montana

I INTRODUCTION

Forests and woodlands cover approximately 40% of the global land surface and annually produce 70% of the net global terrestrial carbon accumulation (Lieth, 1975). Forest products are a major natural resource commodity, valued at billions of dollars annually worldwide. These impressive statistics imply geographic and economic advantages in applying remote-sensing methods to forest science and management. Such general facts fail to capture the incredible diversity of forests, which produces many difficulties in the development of general remote-sensing procedures. Boreal spruce forests, tropical rain forests, redwood and sequoia forests, desert pinyon-juniper forests, and eucalyptus forests have this much in common: they all contain a large number of trees.

The frontiers of a new Earth system science are beginning to shift attention to global concerns. The importance of forests in all of Earth's cycles and the widespread disruption of forestlands has focused attention on forest ecosystems. A diverse new sensing capability has opened up opportunities for quantitative analysis of forests and other ecosystems at all geographic and spatial scales. Sensors are now, or will soon be, available, collecting data with broad to fine spectral resolution, with large to small spatial and temporal resolution, and with various radiometric and angular characteristics. And all of these sensor attributes are obtained by optical means, i.e., measuring either reflected or emitted radiation. There are also good reasons to expect data from optical sensors to be more effective when combined with data from active sensors, such as radar. This chapter will emphasize the analysis of data from optical sensors, data for which substantial potential continues to expand.

What are these potentials? Can quantitative properties of forest canopies be derived from the reflectance of solar radiation? Properties being identified by ecologists and forest scientists include biophysical and structural, biochemical, and thermal information. What radiometric factors influence these estimates, and can radiative models of the interactions of canopies and associated environment with the solar beam be used to help sort out the role of these factors? Remote-sensing data have been used since 1972 to derive categorical information with mixed success. These categorical data have been incorporated into geographical information systems with other physical, socioeconomic, and institutional information to conduct spatial modeling analyses for many forestry problems. Coupling remote-sensing data with collateral data and using the frequent acquisition capabilities of remote-sensing platforms permit both improved information extraction as well as statistical estimations of forest conditions and changes in forest conditions over time.

Perhaps the most exciting new use of information derived from remote-sensing data is to provide driving data for ecosystem models. These simulation models are explicitly formulated for those state variables to be derived from remote-sensing data. Various mechanistic models predict ecosystem processes such as productivity, nutrient pools and turnover, energy and matter exchange, and so on, processes that cannot be directly sensed. Changes in landscape pattern, measured by remote-sensing means, are now being used in succession models. This chapter will concentrate on ways in which each of these potential applications of such importance to forest science and management can be accomplished through the use of data derived from optical remote-sensing data.

II UNIQUE BIOPHYSICAL PROBLEMS IN REMOTE SENSING OF FORESTS

Forests make particularly difficult remote-sensing targets. Tree leaves occur in all different sizes, shapes, and colors. From dark green fir needles, thin and cylindrical in shape, leaves of other species can be as expansive as the light green broad fronds of palms. Broadleaf deciduous forests have a marked seasonal cycle of leaf display, whereas evergreen conifer forests present a virtually static remote-sensing target, with visible change in conifers remaining undetectable for many years. Remote-sensing problems occur where deciduous and coniferous species form mixed stands. This common occurrence produces mixed spectral signals that are hard to characterize. Across the long lifetime of trees, forests can develop very deep canopies with leaves distributed vertically across as much as 20 m. The geometric orientation of leaves on trees can take all possible distributions, often displayed in clumps and layers, as yet defying simple mathematical description. The ground surface beneath the forest overstory canopy may vary from bare rock, soil, and snow to grasses, annual forbs, perennial shrubs, or small trees. Each of these understory features may contribute to the reflected signal measured by a sensor.

Understanding the factors regulating forest development and the interaction of forest canopies with solar radiation is complicated by the wide geographic distribution and the range of sites and climates occupied by forests. This range is perhaps greatest of any biome type. Forests prosper and grow on virtually the hottest and wettest to almost the coldest and driest places on Earth, from sea level to 4000 m elevation. Many forests grow in uneven mountainous terrain. The terrain relief produces large variations in how solar radiation reaches the forest and produces landform shadows. Terrain relief also generates large microclimate differences in temperature, precipitation, and soil properties that produce large differences in forest composition and activity over relatively small geographic areas. Shadowing and local variations in elevation, slope, and aspect make corrections to remotely sensed images for atmospheric and terrain difficult because so much collateral information is required. These corrections are especially important for coniferous forests. The small reflectance signal generated by conifer canopies is strongly influenced by terrain and atmospheric effects.

Studies of forests are never routine. The large size and long life of forests make one cope with logistical experimental difficulties. Measurements of seasonal reflectance with truck-mounted spectrometers are fairly routine and invaluable procedures for remote sensing of crops and grasslands. Despite the value of such measurements, few such procedures have been attempted for mature forests. Collecting the corresponding seasonal changes in biophysical variables would be labor intensive as one might imagine in measuring the needle area of a hectare of Douglas fir forest.

Forests may have longevities of 50–5000 years, which means they are exposed to a nearly endless variety of disturbances and stresses. Insect and disease epidemics, wildfires, windstorms, snow damage, and drought periodically disrupt forests. The activities and interference by humans can be added to these natural events, powerful effects such as harvesting by clear-cutting, land conversion to agricultural and cultural uses, and chronic air pollution. This panoply of disruptions to a forested landscape makes one realize the structural and functional heterogeneity and nearly continuous state of successional change typical of most forests.

Finally, the most important processes occurring in forests, carbon exchange (photosynthesis and respiration), transpiration, biogenic trace gas emission, and even growth are not directly sensible by optical sensors. Scientists must evaluate what forest attributes are best observable, how they can be measured by current optical sensors, and how these attributes relate to important functional processes.

III THEORETICAL MODELING AND LEAF SPECTRAL PROPERTIES

III.A Canopy Radiation Models

The interaction between incident solar radiation and forest canopies has been analyzed and modeled in a number of different ways. Smith (1983) has reviewed various approaches to canopy radiation models, most of which were formulated for agricultural canopies. This book also provides detailed discussion of plant radiation models. Although many of the existing models were formulated for agricultural canopies, they often can be applied to forest canopies. This book also provides detailed discussion of plant radiation models (see Chapters 5 and 6). Most of the existing models use turbid-medium theory for radiation penetration and abstract the canopy as horizontally homogeneous, of infinite extent, usually containing isotropic scattering elements, leaves, oriented according to some distribution function. Examples of such models include the Suits model (1972), which treats directional effects as arising from the vertical component of the leaf elements, and Sellers' Simple Biosphere model (1985). Kimes, et al. (1980) used a Monte-Carlo model considering the probability of gaps for radiation penetration into various canopy geometries to treat radiation reflectance in *Pinus contorta* and grass canopies. They examined the behavior of vegetation-canopy reflectance as a function of solar zenith angle and showed a great amount of variability in the simulation of highly absorbed radiation (0.68 nm). The Suits model, in a modified form, has been used to study the reflectance from aspen canopies and is discussed briefly in the next section. Fox (1978) used the Suits model to study the bidirectional reflectance properties of conifer canopies in Michigan. He found that the model tended to consistently overestimate reflectance. Norman and Jarvis (1974, 1975) developed a model for *Picea sitchensis* plantations by treating the shoots as the scattering medium rather than individual leaves. Their study was primarily directed at the radiation environment within the canopy rather then relating the simulated values to remotely sensed data. They did find that the ratio of visible to near-infrared canopy transmittance did not prove to be a good index to leaf-area index, but simulation of transmittance distribution did agree well with measurements. This model does not require radiance measurements within the canopy, but only the canopy structural and spectral properties and incident radiation above the canopy. Caldwell et al. (1986) have modeled in detail the radiation interception and photosynthetic light harvesting in a diverse array of *Quercus coccifera* L. canopies of different leaf-area indices with nonrandom leaf azimuth orientations. They have used these simulations to calculate microclimate and gas exchange of all leaf layers for four seasons and compared their calculations to measured values with good results.

Research has been directed since 1980 at increasing multidimensionality of the models to enable study of natural and artificial spacing effects on canopy reflectance. In some models, the morphological characteristics of the individual canopies are treated as either rectangular or ellipsoids of revolution, considering the radiation passage through the individual canopy as a turbid medium (Mohren et al., 1984). Kimes (1984) and Kimes and Kirchner (1982) have simulated the bidirectional properties of agricultural and forest canopies and found that both canopies have similar bidirectional properties. Strahler and Li (1981) have developed a model of forest-canopy reflectance using a geometric optics approach. Each tree is abstracted as a cone (or other specified shape) of height and base diameter from a statistical distribution. The trees are distributed according to a spatial function determined from measurement, over a known background reflectance. The model can be inverted, once having calibrated the coefficients in the model, to predict tree size and height from remotely sensed data. Li and Strahler (1986) used their model to simulate the bidirectional reflectance properties and to predict the occurrence of the canopy "hot spot" or retrodirection increase in reflected radiance. Franklin et al. (1986b) have extended this model with some success to the ellipsoidal canopies of savanna forest in West-Central Africa. Gerstl and Zardecki (1985) and Shultis and Myneni (1988) have used the methods of discrete ordinates to model the radiation fields using the theory of multidimensional photon transport. Most of their studies to date have been concerned with agricultural fields where the plant spacing and structural conditions can be simulated as two-dimensional problems. The extension of these models to the three-dimensional heterogeneous canopies of forest ecosystems is warranted.

Jarvis and Leverenz (1984) stated that the interception of radiation by canopies is determined by structural and optical properties of a canopy. They give the order of importance as leaf-area index, vertical foliage distribution, leaf inclination-angle distribution, leaf reflectance and transmittance, grouping or clumping of the foliage, and leaf azimuth-angle distribution. Most of these factors have been incorporated in the models already discussed and are also important in the reflection of radiation. For the study of remotely sensed data, other characteristics of forests may be important. Most of the existing models treat the foliar component of the plants, and this treatment should be sufficient for forest canopy foliage. Forest canopies can be highly heterogeneous, consisting of mixed species with variations in leaf optical and structural properties. Fewer models can treat this heterogeneity. Forests are commonly multistoried, containing the highly variable optical properties of both the overstory and understory plants. In addition to the optical properties of the leaves, the reflectance and absorption by bark and the background elements (bare rock, soils, snow, litter layer, and mineral substrates) may be required to completely understand and model the reflectance patterns of forests. Existing models generally fail to simulate this variability.

Spatial effects on a larger scale may be caused by variable mountainous terrain, affecting both the illumination conditions as well as the atmospheric path length. Orographic effects also can perturb local atmospheric properties that influence scattering and absorption. Temporal effects must be considered in forest canopies that exhibit a seasonal phenology of spring leaf growth, autumn and drought-induced leaf litterfall, and leaf flushes and elongation several times during a growing season. This phenology includes total canopy replacement in broadleaf deciduous forests and 5–100% turnover in evergreen conifer forests. Seasonal changes in pigmentation and biochemical condition are also important temporal characteristics. Aperiodic and periodic phenology may be seen in tropical forests, where leaf phenology is controlled by seasonal rainfall rather than the more predictable solar seasons. The dynamic characteristics of canopy optical properties are important factors in ecosystem processes and make remote-sensing investigations complex.

III.B Leaf Optical Properties

Pioneering work by Gates et al. (1965), Myers (1970), Gausman et al. (1969), and others have established the clear importance of leaf pigments, leaf water, and internal scattering as explanations for the reflectance and transmittance properties of leaves. Gates showed that chlorophyll a and b and the associated pigments were primarily responsible for the strong leaf absorption characteristics at visible wavelengths (400–700 nm). The absorption properties of water are responsible for the general characteristics of leaf reflectance in the shortwave-infrared region (SWIR, taken in this discussion to be 1200–2500 nm) (Knipling, 1970; Gausman et al., 1969; Woolley, 1971). Liquid water possesses four absorption peaks of increasing strength at 970, 1190, 1450, and 1940 nm (Curcio and Petty, 1951). Water absorbs between the peaks, with increasing strength at longer wavelengths, although the spectral variation is smooth and gradual. The high diffuse reflectance of the near-infrared (NIR, 700–1200 nm) reflectance plateau of plant leaves was shown to be due to cell wall-air interfaces within the leaf (Gausman et al., 1969). Vanderbilt et al. (1985) have shown that the cuticle of broadleaf plants is responsible for specular reflectance from the surface and for polarization of the reflected radiation. The stochastic compartment model of leaf reflectance of Tucker and Garratt (1977) considers a transition matrix to pass radiation from one cellular compartment (e.g., the palisade cells) to other compartments, and the radiation is either scattered or absorbed within cells according to turbid-media theory. This model simulates the combined effects of internal scattering and absorption by pigments and water by empirical selection of coefficients. Allen et al. (1973) and Kumar and Silva (1973) employed optical ray tracing techniques, treating the cell wall-air interface as smooth reflecting-refracting surfaces, to simulate the forward and backscattering properties of leaves. Myers (1970) studied the influence of water content on the reflectance of leaves progressively dried and found the dry-leaf reflectance spectrum was not flat, but resembled the reflectance spectrum of cellulose or cotton. Except for the observation of Myers, none of the other studies considered absorbing media other than leaf pigments and water.

Based on spectroscopic research in the near- and shortwave-infrared regions dating back to the 1950s, Karl Norris of the USDA and his associates have demonstrated that the absorption spectra of leaf material contain absorption properties attributable to other biochemical constituents. They have shown that the protein, lignin, starch, sugar, and other biochemical constituents of forage and other organic samples can be accurately estimated from their reflectance spectra in the near- and shortwave-infrared regions (Martin et al., 1985; Norris and Barnes, 1976; Norris et al., 1976). The basis for these results is the occurrence of spectral absorption features associated with organic constituents throughout the visible and infrared regions. Organic compounds absorb radiation in the middle-infrared and in the ultraviolet regions (<400 nm) at fundamental stretching and bending vibrations of strong molecular bonds between light atoms, e.g., hydrogen

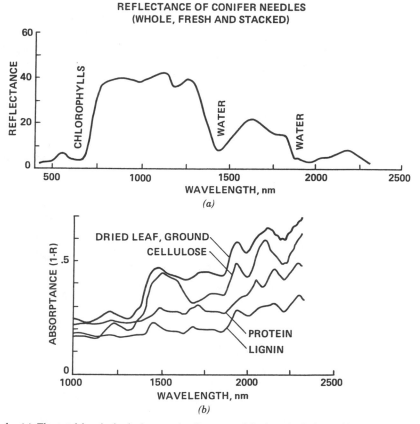

Figure 1 (*a*) The total hemispherical spectral reflectance of fresh and whole conifer needles arranged in contiguous rows and stacked five deep, measured in a scanning spectrophotometer with integrating sphere. (*b*) Comparison of the absorptance (plotted as 1 − reflectance) spectrum between 1000 and 2400 nm for deciduous leaves that were oven-dried and ground to a uniform particle size with the absorptance characteristics of three biochemical constituents measured in powdered pure form: cellulose (surgical cotton), protein (adapted from Rotolo, 1979), and lignin (extracted from the wood of loblolly pine by the University of Wisconsin, Forest Products Lab, John Obst, personal communication) (from Peterson et al., 1988). (Note: Curves have been shifted vertically a slight amount to improve clarity; absolute values are not exact.)

bonds associated with carbon, oxygen, and nitrogen. The absorption bands observed in the near- and shortwave-infrared largely originate as harmonics and overtones of the fundamental stretching frequencies of C–H, N–H, and O–H bonds together with various combination bands (Hergert, 1971; Hirschfeld, 1985). The nitrogen-containing compounds include proteins, chlorophyll, and nucleic and amino acids. Important carbon compounds include lignin, cellulose, and carbohydrates. Chlorophyll absorption in the visible region is by electronic transitions (Danks et al.,1984).

The reflectance spectrum of conifer needles, as shown in Figure 1(a), is characteristic of the major absorption and multiple-scattering phenomena. The visible region is dominated by the leaf pigments, including chlorophyll. The water-absorption peaks are the major features in the shortwave infrared. The low absorption and efficient diffuse scattering are responsible for the near-infrared "plateau." The reflectance spectrum and the transmittance spectrum are generally very similar in shape. In Figure 1(b), the absorption spectrum of a deciduous leaf is plotted for the infrared region. This spectrum is for the total hemispherical reflectance (R) transformed to absorptance ($1 - R$) of dried (no free water) and uniformly ground foliage. The absorption spectra of protein, cellulose, and lignin are compared to the leaf spectrum. Protein displays five major absorption peaks that are overlapped with the absorption characteristics of cellulose and lignin. Other compounds typically present in leaves are starch (very similar to cellulose) and glucose. Many of the biochemical constituents contain O–H bonds that possess absorption characteristics of water and thus contribute to the "water"-absorption peaks of leaves. This comparison suggests that the absorption properties of the biochemical fractions are related to the absorption characteristics of the intact leaves. The leaves of deciduous and coniferous forests from Wisconsin, Alaska, and California have been analyzed using the techniques of near-infrared spectroscopy. Wessman et al. (1987b) have shown that the nitrogen content, Figure 2(a), and the lignin content, Figure 2(b), of the Wisconsin foliage can be predicted using spectral reflectance of dried and ground foliar samples. Card et al. (1988) has shown similar results for protein, Figure 2(c), and other biochemical fractions of samples pooled from all sites in these three states, comprising both deciduous and coniferous foliage. The standard errors of prediction have been shown to be comparable to wet chemical laboratory techniques.

IV REMOTE SENSING OF FOREST-CANOPY CHARACTERISTICS

IV.A Leaf-Area Index and Other Structural Properties

Leaf-area index is central to the estimation of canopy processes such as evapotranspiration, photosynthesis, leaf litterfall, nitrogen capitals, and gaseous emissions, and it structurally quantifies the primary energy exchange surface of a forest. The total surface area of all leaves above a specified ground area is the all-sided leaf-area index, a dimensionless quantity. The projected surface area of individual leaves, without considering leaf-angular distribution, is the projected (one-sided) leaf-area index (LAI) to be used in this discussion. Leaf-angle distributions to estimate the actual projected surfaces in the direction of the solar beam are considered only in canopy radiation-transfer models.

Numerous studies of crop canopies and of low stature seasonal natural communities have found that increases in LAI produce an increase in reflectance in the near-infrared (NIR) and a rapid decrease in the red and shortwave-infrared (SWIR) reflectance. The absorption in the visible region by leaf pigments and in the SWIR by leaf water resulted

PERCENT NITROGEN
WISCONSIN

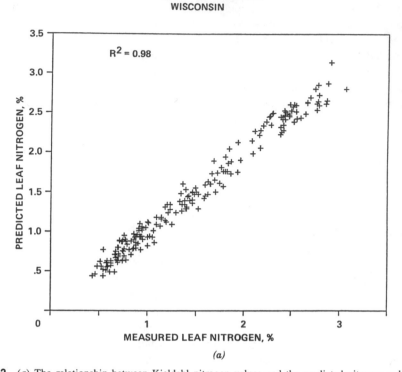

(a)

Figure 2 (*a*) The relationship between Kjeldahl nitrogen values and the predicted nitrogen values using spectroscopic analysis of the infrared reflectance of dried ground leaves of various species. Foliage samples include 163 samples from 18 deciduous and 2 coniferous tree species, and 40 from native prairie grasses of Wisconsin. The total hemispherical spectral reflectance was obtained from a Neotec 51A Scanning Filter Instrument (spectral range: 1590–2357 nm). The prediction equation included five wavelengths, selected by modified multiple stepwise regression using 138 samples for calibration and based on the first derivative of the log $1/R$ data, to predict the remaining 65 samples with $R^2 = 0.98$; the standard error of calibration and of prediction was 0.11%. (From Wessman et al., 1987b.) (*b*) The relationship between leaf lignin concentration, determined by the sulphuric acid digest method, and the predicted lignin concentration using spectroscopic analysis of the infrared reflectance of dried ground leaves of 18 deciduous and coniferous trees ($n = 173$). Spectra obtained as in (*a*). The prediction equation included six wavelengths based on the second derivative of log $1/R$ data with 132 samples for calibration to predict the remaining 41 samples with $R2 = 0.78$; the standard error of calibration was 2.90% and of prediction, 3.14%. (From Wessman et al., 1987b.) (*c*) The relationship between lead protein content and the predicted protein content using spectroscopic analysis of total hemispherical reflectance of the visible and near-infrared reflectance of dried ground leaves of various tree species from Alaska, California, and Wisconsin. Spectra were obtained from a Perkin-Elmer model 330 spectrophotometer (range: 400–2400 nm). The prediction equation included six wavelengths based on log $1/R$ data using 52 samples (circles) for calibration to predict the remaining 51 samples (squares) with $R^2 = 0.77$; the standard error of prediction was 1.28%. (From Card et al., 1988.)

in the asymptotic nature of canopy reflectance (Tucker, 1977; Ripple, 1985). While this asymptote is acheived at LAI values of only 2–3, an asymptote in the near-infrared is approached more slowly to LAI values of 6–7. The direct positive response in the near-infrared has been attributed to scattering and weak absorption at cell-wall interfaces within leaves, so that canopy reflectance continues to increase with more leaf area. To minimize the effects of variations in solar irradiance, various normalizing transforms

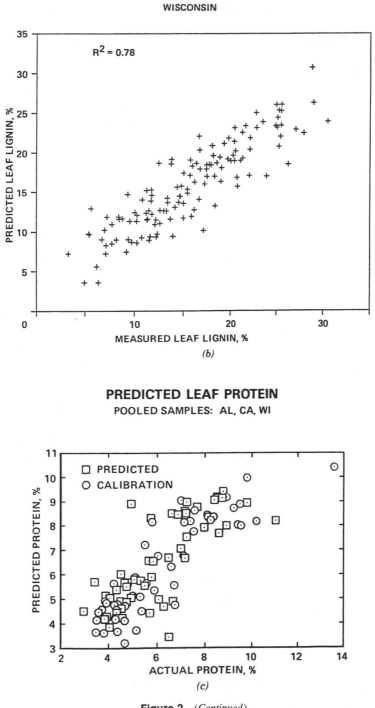

Figure 2   (*Continued*)

have been investigated. The two most common indices are the simple ratio of near-infrared to either the visible or shortwave-infrared reflectances, NIR/RED and NIR/SWIR, and the normalized difference vegetation index, NDVI and (NIR − Red)/(NIR + Red). Steven (1985) and Perry and Lautenschlager (1984) have discussed the functional equivalence of these indices. Asrar et al. (1984) and Sellers (1985, 1987) have related NDVI to absorbed photosynthetically active radiation in cultivated crops.

In conifer forests of Earth's temperate zone, leaf-area development is strongly controlled by environmental factors. Grier and Running (1977) showed that LAI was inversely related to a site water balance for Oregon forests. In a more comprehensive analysis, the three-dimensional relationship between climate, soil-water holding capacity, and leaf-area index has been described for northern Rocky Mountain forests, Figure 3 (Nemani and Running, 1989b). The relationship couples the causal climate, the system storage, and soil water in an ecosystem model predicting the ecosystem response, LAI. The relations shown in Figure 3 have general applicability for temperate forests in the absence of control exerted by nutrient limitations or other factors.

Although the leaf area of an evergreen tree is very difficult to measure directly, allometric relations between more easily measured variables and LAI have been developed by destructive sampling (Gholz et al., 1979; Westman, 1987). Three allometric variables are commonly used in forest studies: the diameter at breast height (DBH), the cross-sectional area of conducting sapwood tissue, and the annual increment of tree rings, the latter two determined by increment coring the tree at breast height. The LAI of 73 study plots was estimated using these relationships. These plots were located throughout the western United States, covering the entire regional range in LAI. The LAI plot data were related to the 0.1 ha data of the Landsat Thematic Mapper (TM) satellite and to an airborne simulator of the TM sensor.

The relationship of LAI to red and SWIR reflected radiation measured by the TM was inverse and became asymptotic at LAI values of about 4–5 (for red, Figure 4; for SWIR and other visible data, the relationships are very similar). For the near-infrared reflectance, the relation to LAI was relatively flat and quite variable. D.L. Williams (personal communication) and Kleman (1984) have found similar results for conifer for-

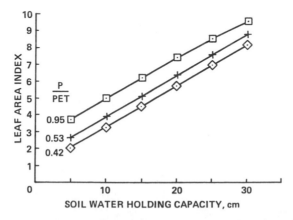

Figure 3   The interrelationship between climate, soil water capacity, and maximum LAI found for conifer stands in Montana. Climate, defined here as annual precipitation/potential evaporation, represents ecosystem water availability. Soil water capacity represents ecosystem response to site water limitations. The relation between LAI and soil water capacity is derived from simulations using the FOREST-BGC model in Figure 15. (From Nemani and Running, 1989b.)

LAI vs. CORRECTED RED RADIANCE
OREGON, SEQUOIA, MONTANA TM

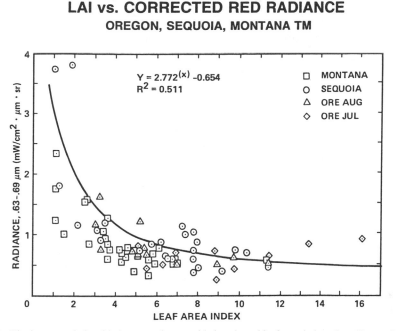

Figure 4   The inverse relationship between the one-sided projected leaf-area index of coniferous forests and the reflected radiance in the red region as measured by the Landsat Thematic Mapper satellite, corrected for atmospheric, solar, and topographic differences between stands and sampling regions. Seventy-three conifer stands from West-Central Oregon, northern Montana, and Sequoia National Park, California, are included and two Thematic Mapper scenes in Oregon from July and August 1984. (From Peterson et al., 1986, 1987; Spanner et al., 1988.)

ests of New England and Sweden, respectively. When considering only crown closures greater than 89%, a direct linear relation, Figure 5(a), was found between LAI and NIR reflectance. This result was consistent with findings for other canopies, although the dynamic reflectance range of conifers was much lower than for other vegetation. The relation of LAI and NIR reflectance of more open stands (25–89% canopy closure) differed from the results of Figure 5(a), depending on background reflectance. For a dark background, such as forest floor litter beneath deep canopies casting shadows, the NIR response is reduced below the closed-canopy line. Most of the points with 50–74% canopy closure and higher values of LAI are represented by stands from Sequoia National Park. At lower values of LAI but higher canopy closures (74–89%), the NIR values approach the closed-canopy line. A contrasting situation occurs for stands with either a bright granite and/or a broadleaf vegetation understory. As canopy closure increases, these points approach the closed-canopy line from above, i.e., higher radiance values at lower LAI and canopy closure. In both cases, the pronounced effect of the background is diminished as canopy closure increases or shows no effect if the background and overstory reflectances are about the same. An interesting problem occurs for broadleaf-conifer mixes in the overstory. The broadleaf component has on outsized effect on reflectance in the near-infrared but not in the red or shortwave-infrared regions.

Although spectral indices generally have a direct positive relationship with LAI, at Thematic Mapper scales, the local variations in LAI and radiance lead to high variances. At low LAI, the ratio tends to compensate for the effects of a variable vegetated back-

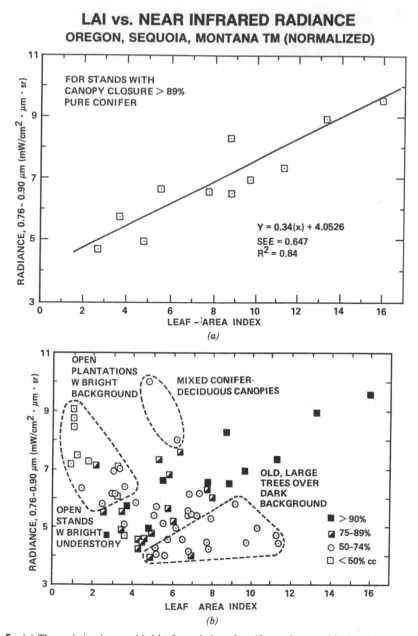

Figure 5 (*a*) The variation in one-sided leaf area index of coniferous forests with closed canopies (greater than 89% as determined from photointerpretation of aerial photography) is directly related to the reflected radiance in the near-infrared (0.76–0.90 μm) as measured by the Thematic Mapper satellite, corrected for atmospheric, solar, and topographic differences between regions (Montana, California, and Oregon). (From Spanner et al., 1988.) (*b*) The influence of background variations on the spectral response of a coniferous forest of varying canopy closure. The central direct trend for closed canopies is as in (*a*), shown here as solid squares. The cluster of points encircled below this trend consists of stands, mostly from Sequoia National Park, of old growth and very large red and white fir trees having a dark shadowed background of forest floor litter and rock. Open stands (50–89%) just above the central trend have vegetated backgrounds or understory, or bare granite surfaces. Plantations that were sampled generally had open canopies over bright exposed soils or broadleaf understory plants. Two stands above the trend had canopies codominated by conifers and deciduous trees, the latter having an outsized effect on the combined signal. Thus, these open stands have trends opposite to the closed-canopy stands due to variations in both crown closure and background reflectance. (From Spanner et al., 1989.)

ground, but less so for a dark background. The relationship for Oregon sites alone was based on averages of four plots per stand rather than single points, and tended to reduce effects of local variance (Peterson et al., 1987; Running et al., 1986). Reduction in variance by using larger sample frames was tested in two ways. Firstly, individual plots in close geographic proximity, thus in similar environments, were averaged. The resulting relationship between LAI and the NIR/RED ratio is shown in Figure 6. The relationship for LAI versus NDVI was asymptotic at LAI values of about 8, saturating at NDVI = 0.8. Secondly, Spanner et al. (1987) have studied the relation of LAI to daily Local Area Coverage AVHRR spectral data for large pure conifer stands of over 10 km^2 from throughout the West. Direct relationships between LAI and the NIR/RED ratio and NDVI prevailed, similar to Figure 6. The AVHRR indices varied only slightly throughout the growing season and declined up to 30% in early spring and late fall. The decline has been suggested to be due to leaf phenology, shadow changes, and possibly snow (M.A. Spanner, personal communication.)

The weak reflectance signal from conifer canopies is particularly affected by atmospheric scattering. Additive path radiance for even clear skies (23 km visibility for the Oregon data set) can contribute most of the signal in visible channels, up to 50–60% in red wavelengths and 20% of the NIR signal (Spanner et al., 1984a). The additive radiance suppressed the simple ratio relation threefold, with a 30% reduction in sensitivity for the simple ratio. The additive radiance reduced the NDVI relation by 40% to 0.5.

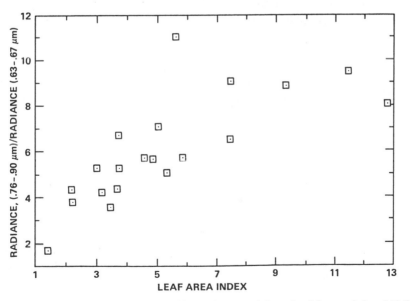

Figure 6 The relationship between LAI of coniferous forests and the ratio of the near-infrared (0.76–0.90 μm) to red (0.63–0.69 μm) radiance when stands in close geographic proximity or in similar environments are averaged ($R2 = 0.89$). The geographic proximity represents a larger sampling frame applied to the data of Figures 4 and 5, and tends to reduce the high variance encountered in conifer forest at the scale of the Thematic Mapper. The variation in LAI between vegetation zones is significantly greater than that within. The variation between is controlled mainly by climate, so that the close geographic proximity corresponds to stands having similar environment.

Although additive effects are dependent on elevation, once these effects are corrected, the values found for NDVI (0.8 maximum) are very similar to those calculated for broad-leaf forests (Goward et al., 1985).

A study of LAI and radiance relationships was reported for both deciduous (aspen, birch) and coniferous (black spruce) stands from the boreal forests of Minnesota (Shen et al., 1985; Badhwar et al., 1986a,b). These investigations showed increased red and SWIR absorption as LAI increased early in the growing season, and a decrease in the fall, whereas the trends for NIR varied in the opposite direction. They used measurements of leaf optical properties with angular and height distributions and bark optical properties in a modified Suits model to relate radiance observed to overstory LAI. A direct linear relationship for the aspen stands was observed between the NIR/RED ratio and LAI for June reflectance data, but a flat relationship for August data. For an overstory LAI variation up to 3.5, the presence of an understory LAI of only 1.0 significantly reduced the simulated relationship of LAI to the NIR/RED ratio. In contrast, a grey background without vegetation led to simulated reflectance response that is direct and positive with LAI. The trends for open stands of black spruce over a bright moss background were inverse with LAI for both the NIR and the red, even though the black spruce canopy was modeled as a homogeneous layer. A similar response has been demonstrated on other open canopy stands having bright substrates: as LAI and canopy closure increase, overall reflectance decreases (Butera, 1986; Mead et al., 1979; see also Figure 5). Franklin (1986) showed that visible reflectance was inversely related to conifer basal area and foliar biomass for fir stands. The ratio of the SWIR to NIR was inversely related to stand basal area in white fir forests of Sequoia National Park (Peterson et al., 1986). Basal area and foliar biomass are strongly related to LAI variations. Franklin attributed part of this behavior to variations in canopy closure, shown to be the major structural variable in conifer stands related to scene brightness (e.g., see the section on classification; Butera, 1986; Mead et al., 1979). Ranson et al. (1986) and Ranson and Daughtry (1987) have demonstrated the sensitivity of spectral measurements and ratios to variations in sun angle, view angle, and background in studies of simulated balsam fir canopies. The varying effects of shadows on a grass background showed that neither the NDVI nor greenness index were sensitive to foliar biomass of the firs, but that strong relationships were found for a background with spectral characteristics similar to snow.

Variance in canopy reflectances with LAI are also due to increases in radiation absorption with leaf-water content. Everitt and Nixon (1986) attributed higher SWIR reflectance of two drought-stressed shrubs to less water in the foliage of the stressed plants. Ripple (1986) came to similar conclusions for laboratory studies of snapbean. Westman and Price (1987) followed the changes in reflectance of conifer leaves of Jeffery pine and white fir as the leaves were air-dried and found that SWIR reflectance progressively increased with loss of water content, whereas NIR reflectance initially increased and then decreased, the latter attributed to changes in leaf anatomy and morphology. Hunt et al. (1987) defined a leaf-water content index from middle and near-infrared TM bands and found a strong relationship to measured leaf relative water content of three desert shrubs.

For remote sensing of other canopy structural attributes, the most successful approach to the prediction of tree height and crown diameter has been by inversion of the geometric model of Strahler and Li (1981). Several investigators have used airborne laser profiling to estimate canopy height. The beam of a small optical laser is directed vertically downward and the time trace of the reflected signal is related to both canopy height and the underlying surface (Arp et al., 1982). Stand volume and biomass have been correlated with canopy closure and used in multistage sampling designs to estimate total volume across large regions (e.g., Peterson and Card, 1980; Peterson et al., 1981, 1983; Brass et al., 1981). Larger-scale variations in forest versus nonforest area determined by Landsat MSS data have been used to characterize the spectral response of the 1 km AVHRR data by Logan (1983). Since the MSS data had been previously calibrated relative to timber volume, these investigators could relate aggregated volume to the AVHRR data as a series of volume isolines.

IV.B Canopy Chemistry

The principal advancement in technology that permits the investigation of remote sensing of canopy biochemical content is high spectral resolution imaging spectrometers (Goetz et al., 1985; Vane et al., 1984). The first such instrument available was the Airborne Imaging Spectrometer (AIS) flown on NASA aircraft. This instrument acquired imagery of 128 contiguous spectral bands, each 9.6 nm in width, forming a narrow swath width image (300 m) of 32 pixels, with a spectral range of 800–2400 nm. Due to data-quality problems associated with this experimental sensor, the analyses to be discussed were limited to the spectral range of 1200–1600 required various techniques to reduce noise effects (see Wessman et al., 1987a, 1988; Peterson et al., 1988; Hlavka, 1986; Swanberg and Matson, 1987).

In our research (with J. D. Aber, P. A. Matson and P. M. Vitousek), we acquired AIS imagery over several sites that were characterized by natural ecological gradients in soil fertility or nutrient availability, and climate. These gradients show variations in total canopy biochemical contents because of differences in concentration and total foliar biomass. Two gradients will be discussed here: a nitrogen mineralization gradient across Blackhawk Island in Wisconsin (Pastor et al., 1983) and the temperature-moisture gradient across West-Central Oregon (Gholz, 1982). The variation in biochemical content in the Oregon transect was principally a function of differences in foliar biomass, although concentration differences existed between vegetation zones. In Wisconsin, the nitrogen content varied much less than expected, but a large gradient was found in lignin content (Wessman et al., 1988). Sites from the University of Wisconsin Arboretum in Madison were also included in the analyses.

Based on a limited deciduous sample set from Blackhawk Island alone ($n = 6$) and simple correlation against untransformed AIS data, we found broad-scale changes in brightness in the infrared associated with differences in total canopy water content, whereas lignin and starch were associated with narrower spectral regions or features of the spectra in the 1500 nm region (Peterson et al., 1988). The sample set was increased to 18 plots by including both conifer and deciduous sites from Blackhawk Island and the Arboretum. A multiple stepwise linear regression ($R^2 = 0.85$) was found between lignin content and AIS spectral response taken as the first difference, Figure 7 (Wessman et al., 1988; Wessman, 1987). The estimation equation was based on three wavelengths: 1256, 1555, and 1311 nm (alternatively, 1592 nm) in descending order of importance. We have speculated that the selection of 1256 nm is related to the C–H bonds in organic molecules having a harmonic stretching frequency in this region (see Figure 1). The 1555 nm band is comparable to the 1560 nm band found by Norris and Barnes (1976) for lignin in forage samples and by Card et al. (1988) for laboratory spectral analysis of lignin in Wisconsin deciduous leaves. Direct physical or chemical reasons for the selec-

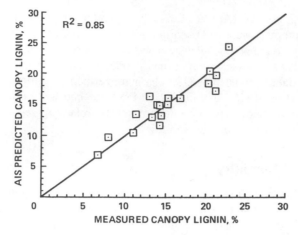

Figure 7   Percent canopy lignin observed in the field for 18 forest (mixed northern hardwood and coniferous) stands in Wisconsin (Blackhawk Island and the University of Wisconsin Arboretum) versus the percent lignin values predicted from stepwise multiple-regression analysis of reflectance spectra measured by the NASA Airborne Imaging Spectrometer (AIS). The spectra have been treated as the first difference and the prediction equation is based on three bands centered at 1256, 1555, and 1311 nm ($R2 = 0.85$, standard error = 2.0%, $p < 0.05$). (From Wessman, 1987; Wessman et al., 1987a, 1988; Aber et al., 1989.)

tion of 1555 nm will require future spectroscopic research to characterize the specific absorptivity of leaf lignin in its correct physical condition, within the scattering matrix of leaves. Although the 1500–1600 nm region has been related to differences in water content, absorption by water varies slowly in the spectral direction and the first difference largely reduces such slowly varying characteristics. However, 1311 nm is located along the rapidly varying wing of the water-absorption peak centered at 1450 nm, and the first difference transformation emphasizes these kind of features. We speculate that 1311 nm was selected, though contributing only 3%, because of the moderate covariance of water content and lignin content. The relationship between nitrogen and AIS first difference or second difference was low, likely due to the small variation in canopy nitrogen. We have analyzed fresh conifer foliage for nitrogen and lignin using laboratory spectrophotometers and found that the wavelength region near 1240 nm was predictive of nitrogen content.

Wessman et al. (1988) mosaicked the three AIS flight line data sets over Blackhawk Island to create a continuous image and applied the lignin equation to generate the spatial distribution of lignin concentration across the entire island, about 2 km^2, using the estimation equation (Figure 8). The spatial distribution is highly consistent with the pattern of lignin, lower values toward the eastern side, corresponding to regions of high nitrogen availability and rapid litter decomposition; moderate values in the interior, where nitrogen availability is reduced; and high values toward the west and around the periphery of the island occupied generally by conifer species on low-fertility sandy soils. Three sites not included in the regression were available to check the predicted lignin content, and predictions were 18.2, 17.0 and 14.6% compared to 15.5, 15.1 and 14.4%, respectively, measured in the field the year before (field estimates tend to vary by about 10%, including year-to-year and sampling/measurement errors).

Swanberg and Peterson (1987) conducted a comparable analysis for the Oregon sites for which AIS data were available. Nitrogen content was estimated (Figure 9) using a four-term equation based on the spectral bands centered at 1215, 1522, 1559, and 1587

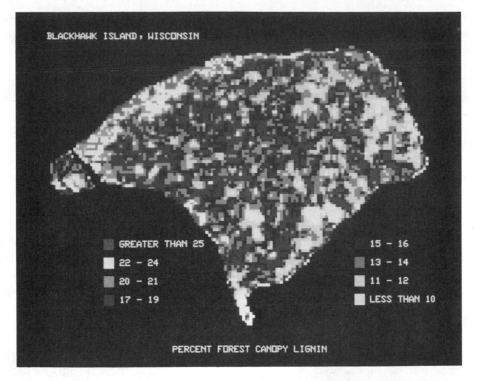

Figure 8 Spatial distribution of canopy lignin concentration across Blackhawk Island, Wisconsin. The image is a mosaic of data from six AIS flight lines, having two different spectral operating modes. Only data from the spectral range of 1200–1600 nm are used. Changes in percent lignin from left to right across the island reflect a continuous change in soil texture, resulting from sediment sorting when the island was an early postglacial floodplain. (From Wessman et al., 1988; Wessman, 1987.)

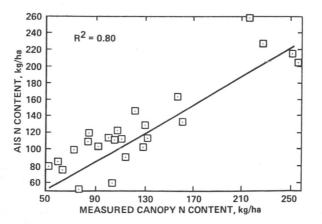

Figure 9 Canopy nitrogen content measured in the field for 24 coniferous forest stands across West-Central Oregon versus the canopy nitrogen content predicted from stepwise multiple-regression analysis of reflectance spectra measured by the AIS. The spectra have been treated as the first difference and the prediction equation is based on four bands centered approximately at 1215, 1522, 1559, and 1587 nm. Only the spectral range from 1200–1600 nm was used in this analysis. (From Swanberg and Peterson, 1987; Peterson et al., 1988.)

nm. Some of these bands correspond roughly to those selected for lignin in the Wisconsin analyses. This correspondence might be due to cross correlation between nitrogen and lignin and with foliar biomass and 1522 nm is close to the absorbing peaks of N–H bonds.

The next generation of such imaging spectrometers is the Airborne Visible-Infrared Image Spectrometer (AVIRIS) (Vane et al., 1984) that will also be flown on NASA aircraft prior to the deployment of the High-Resolution Imaging Spectrometer (HIRIS) being planned for the Earth Observing System in the mid-1990s. AVIRIS began operation in 1987, has 228 spectral bands covering the spectral range from 400–2400 nm, and an 11 km swath width. Another prototype instrument, called the Advanced Solid-state Array Spectroradiometer (ASAS), has 30 spectral bands between 455 and 873 nm, each about 14 nm wide, with the capability to acquire bidirectional reflectance data over specific ground targets (Barnes et al., 1985). Laboratory fertilization studies of coniferous seedlings such as those by Tsay et al. (1982) indicate that the visible region should be effective for the estimation of nitrogen content. Their work and our own of a similar nature (Peterson et al., 1985) showed that increases in nitrogen concentration were strongly correlated with inverse reflectance in the green and red portions of the visible spectrum. Off-nadir observations may help to reduce background effects in forest stands.

IV.C Canopy Thermal Properties

The thermal emissions of broadleaf and needleleaf forests have been studied using the Thermal Infrared Multispectral Scanner (TIMS), an airborne instrument with four bands in the region of 8–12 μm. Sader (1984) found the effective radiant temperature of eight conifer stands in Oregon to average 24.4 °C, whereas ground measured temperatures averaged 25.8 °C. Older forests with deeper canopies were 2.5 °C cooler than 25–23-year-old stands. Nemani and Running (1989a) found surface temperatures ranging from 26–41 °C from a 20 × 25 km AVHRR scene of a Montana forested watershed on August 6, 1985. The variation in temperatures was produced by variable contributions of thermal radiance from surfaces exposed by changes in LAI and canopy closure throughout the watershed. Nemani and Running (1989a) calculated an effective watershed canopy resistance from the AVHRR data for eight days during the summer of 1985 using energy budget and Penman-Monteith techniques (Running, 1984). Canopy resistance increased from 80 to 280 m/sec as the forest developed seasonal water stress through the summer, with canopy energy partitioning progressively, favoring sensible heat (estimated from surface temperatures from the AVHRR) over latent heat flux. For closed conifer stands, the small characteristic dimension of needles and the aerodynamic roughness of canopies produce canopy and needle temperatures very close to ambient air temperature, a feature that makes sensing of these forests useful for surface climate studies. However, as the canopy opens, the much more diurnally variable ground surface temperature progressively contributes to the overall scene temperature.

Much higher thermal conditions characterize wildland fires. Temperatures as high as 1300 °C have been measured in some chaparral fires (Brass et al., 1987) for short time durations. The flaming front, smoldering and cooldown temperatures, their duration, the fuels being burned, and the rate of spread of the fire front are critical determinants of postfire and combustive loss of nutrients and of soil erosion. Anomalously high radiant emission rates of the active fire front measured by TMS bands 5 and 7 cannot be satisfactorily explained by including variations in optical depth in theoretical models. Brass et al. found that the TMS sensors are nonlinear at these high-temperature values. Matson and Dozier (1981) have devised an area-weighting algorithm based on Planck's function to estimate surface temperatures of fires using the 1 km resolution of the AVHRR at two thermal channels, the 3.5 and 10 μm bands. This algorithm cannot account for possible differences in emissivity of surfaces throughout the temperature field being sensed or for the elevated temperatures of postfire soils and ash layers observed by Brass and his colleagues.

V SPATIAL CHARACTERIZATION OF FOREST ATTRIBUTES

V.A Classification of Single-Date Images

Multivariate statistics have been the most common choice in the analysis of multispectral data for forestry applications. The method generally consists of, first, clustering algorithms to establish a set of multivariate class statistics against which each pixel measurement vector in the scene can be compared. Then, a classification decision rule, such as the probability of maximum likelihood that the pixel belongs to a class from among the statistics set, is calculated and the pixel is assigned to the particular class (see Haralick and Fu, 1983). The information classes most often considered include both cover type or community type descriptors as well as limited structural categories, such as crown closure and size class of the trees.

Although two different approaches to the development of the multivariate statistics are used, so-called unsupervised and supervised, their combination is generally superior. In the unsupervised method, the radiance values of the image data set are submitted to clustering algorithms that generate statistics until some stopping rule is reached, such as minimum distance between clusters, separability measures, and minimum number of points per cluster. Another approach is to "seed" spectral space with starting points, either artificially or by using the data themselves to establish candidate mean values for clusters, and then iterate the clustering procedure until a minimization criterion is achieved. In the supervised method, training sites with known properties are used to extract spectral statistics from the image data by interactively identifying the sites in the imagery. In the unsupervised method, identification of the cluster is done after the classification is complete by comparing the spatial distribution of the mapped classes with reference data (sometimes called "ground truth," often obtained from aerial photographs).

Many studies have used a combination of these two techniques, for example, "guided" clustering, in which the cluster statistics are edited after the clusters are pooled (Gaydos and Newland, 1978). An excellent example of analysis of Landsat MSS data by guided clustering techniques is given by the analysis of the McCloud Ranger District in northeastern California (Fox and Mayer, 1980). They found that plots of similar spectral shape but different albedo or brightness represented a family of one community type at different cover densities (Mayer and Fox, 1981). In this mixed conifer region, cover types as detailed as species type (Ponderosa pine, white fir, lodgepole pine, and mixed conifer) were successfully mapped and, in addition, cover density (good or poor stocking) and average tree size (large or small trees) could be discriminated. The accuracy of the classification for tree species/density/size varied from 68 to 86%. The procedures were repeated to identify wildlife habitat types with comparable success.

Similar results were achieved using simulated Thematic Mapper data for a region of the Clearwater National Forest in Idaho (Spanner et al., 1984b). Three size classes (pole timber, saw timber, and saplings/seedlings) were discriminated using four principal components, involving a band from each spectral region (visible, near-infrared, short-wave infrared, thermal), indicating that TM data has increased information content as opposed to two spectral dimensionality of MSS data. Studies of ponderosa pine in Colorado (Butera, 1986; Mead et al., 1979) and pinyon pine/Utah juniper stands in Nevada (Brass et al., 1981) illustrate that cover density or crown closure can be successfully discriminated. The color changes occurring in late season, during leaf senescence, have been studied to identify and map different hardwood communities in West Virginia (Rock, 1982) using simulated TM data.

Studies of the spatial, spectral, and radiometric properties of TM data versus MSS data have shown that greater detail can be achieved in information classes by TM, although often at a loss of accuracy due to increased number of "edge" or mixed signatures (Williams et al., 1984; Sadowski et al., 1978; Markham and Townshend, 1981). Woodcock and Strahler (1987) have analyzed the factor of scale in remotely sensed images for a variety of cover types including forested areas. Their use of semivariograms indicate the conditions under which different clustering and classification techniques may be most effective, depending on the scale of objects in the scene. Tropical forests, in contrast to temperate forests, are characterized by high species diversity, frequent disturbance, and multistoried canopies. Discrimination of species assemblages have been less successful, but identification of disturbance and unique communities like bamboo forests have been successful (Singh, 1984). Discrimination of seasonally flooded versus upland forest has been successful in some instances (Hauck et al., 1986). Baltaxe (1980) and Sader and Joyce (1985) have reviewed the application of Landsat data to tropical forest surveys.

V.B   Multiple-Date Imagery and Collateral Data

The use of spectral data alone generally results in ambiguities caused by the limited dynamic range and similarity in spectral response of important information classes. Many studies have used either multiple dates of imagery or have incorporated collateral information into the classification process. Strahler (1980) has discussed the use of prior probabilities to modify the normally distributed cluster statistics by adjusting the maximum likelihood calculation based on collateral conditioning variables. For example, topographic information (slope gradient, aspect and elevation) was used with known ecological trends in species distribution associated with environmental parameters related to topography to improve classification performance for a coniferous region in northern California (Strahler, 1981; Franklin et al., 1986a; Strahler et al., 1979, 1980). Community types having similar spectral signatures but different ecological affinities could be separated by suppressing the probabilities associated with the incorrect class in favor of the *a priori* class.

Maynard and Strahler (1981) extended these multivariate techniques to the use of the logit, the natural logarithm of the probability that a pixel belongs to a class versus the probability that it does not belong. The logit classifier can be used to assign a pixel to a class (as in the maximum likelihood case), to estimate the proportions of different cover types or species types within each pixel (i.e., the predicted community structure), or to use multiple dates to estimate change. The logit classifier was used to map the presence of permafrost based on its association with vegetation type and surface temperature derived from Thematic Mapper data for a boreal forest region (Morrissey et al., 1986).

Collateral information can also be used to develop a classification decision tree (as was done for a study in Colorado by Hoffer et al., 1975; Fleming and Hoffer, 1979; Swain et al., 1975). Ancillary variables such as soil type and terrain parameters are used to limit decisions in a ''layered'' classifier to a subset of spectral clusters associated with each collateral distinction being drawn. Price (1985) made similar associations for the ridge-and-valley province of Pennsylvania to map different hardwood communities by topography. In all of these cases, a great deal of *a priori* information must be known.

Structural variables such as cover density or size class corresponding to spectral variations associated with plant spacing, tree form, shadows, and background conditions produce texture in the imagery. This can cause problems with classifiers that treat each pixel without reference to its neighbors. Texture measures, such as local spectral variance, have been used as an additional variable in clustering analyses to discriminate cover types (Strahler et al., 1979). These local sources of variance often lead to misclassification of pixels, causing a ''salt and pepper'' appearance in map products. A variety of techniques may be tested to reduce this effect by considering the spatial context of a pixel. For example, region-growing algorithms examine contiguous pixels for similarity so as to group them according to some decision rule, such as majority membership (Hoffer et al., 1975). These smoothing algorithms, which can emphasize or reduce uniqueness such as linear features, produce a product having the appearance of manually photointerpreted stand maps.

V.C Mosaics

A clearly attractive feature of satellite imagery is that it provides large areas repetitively imaged in a fairly uniform way. Repetitive coverage presents the possibilities of overcoming loss of data due to cloud cover, using seasonality to improve discrimination, and monitoring change. Also, adjacent and overlapping scenes can be mosaicked to cover very large areas. Digital mosaics include simultaneous map projection and adjustment of frame edges to eliminate geometric and radiometric seams (Zobrist et al., 1983). The projection to a specified mapping system, e.g., Lambert conic conformal or Universal Transverse Mercator, is accomplished with control points, a rotation of images to North, and ''rubber sheeting'' of the imagery. Radiometric differences between scenes are adjusted using the regions of overlap between scenes and fitting a surface to the brightness values. Mosaics have been prepared by the Jet Propulsion Laboratory for Arizona, Pennsylvania, Los Angeles, and Bolivia, and the largest mosaic, involving 36 scenes, the entire state of California.

For the California mosaic, the state was subdivided by irregular ecological regions of similar potential vegetation based on the map of Küchler. Each ecological region (a total of 53) was independently analyzed by unsupervised clustering to produce up to 64 spectral classes per region, resulting in over 1200 spectral clusters. Each cluster was identified from interpretation of classification maps of a particular ecoregion by experienced foresters from each region using an interactive image display and high-altitude aerial photographs. The clusters were labeled as one of sixteen information classes. In a few regions, digital terrain data was used to reduce spectral ambiguities. A color-coded statewide thematic map (Figure 10) was produced (Peterson et al., 1980; Peterson, 1983).

The statewide map of California and its associated spectral statistics were incorporated into many detailed smaller-area studies and usually coupled with guided clustering techniques to improve local specificity. These analyses addressed questions of prime timberland (Fox et al., 1985), reforestation and fire-hazard prediction (Brass and Peter-

Figure 10   Color-coded (or black-and-white) mapping of the resources of California based on the classification of Landsat Multispectral Scanner data taken in August 1976. The 36 scenes of Landsat data were digitally mosaicked and resampled into 80 × 80 meter pixels, and combined with digital topographic data. Over 1200 spectral clusters from 53 ecological regions were generated using unsupervised techniques and labels assigned through photointerpretation of high-altitude aerial photography. (From Peterson, 1983; Peterson et al., 1980.)

son, 1983); urban greenbelt and fire-corridor selection (Likens and Maw, 1981; Hodson and Likens, 1982), and regional inventories of biogenic hydrocarbon emissions (Moreland and Fosnight, 1981) among others. These studies demonstrated one of the strengths of such large data sets: their reuse for many unexpected environmental applications usually involving geographic information systems.

V.D Change Detection and Multidate Analysis

The decline in the forest land base worldwide is well known. In the tropics, forests are felled, burned, and either converted to pasture or used for short-term crops (swidden agriculture). Timber harvesting in the temperate zone by clear-cutting methods has also been a very common practice. Other changes are less the direct result of anthropogenic causes, such as insect defoliation and fire and storm damage. Forests are also in decline because of pollution, either through direct foliar injury or through indirect effects on the entire ecosystem. Woodwell et al. (1984) has argued that change detection of forest boundaries globally is the most important first step in following dynamics of terrestrial vegetation in the global carbon cycle. Nelson (1983) has itemized many of the change-detection methods in use, including image differences, multiple-image pair ratioing, vegetation-index differences, change vector, chi-square analysis, changes in classification, and thresholds. Likens and Maw (1981) developed a hierarchical classification scheme combining multiple Landsat MSS imagery with existing land-use maps to identify and map urban and nearby fire-prone chaparral forestlands in San Bernardino County, California.

Clear-cutting practices produce an abrupt and spatially well-defined alteration of albedo. Year-to-year and even monthly analyses of simple brightness changes in MSS data of the Pacific Northwest have been used to monitor compliance with logging plans and taxation reporting timetables (Roger A. Harding, Washington Dept. of Natural Resources, personal communication). Johnson et al. (1979) and Lee (1976) have shown how Landsat MSS data acquired during the summer months can be analyzed to estimate areas of clear-cutting operations in coniferous forests of western Washington and Canada, respectively. Hafker and Philipson (1982) have shown that clear-cutting patterns in hardwood forests of Pennsylvania are best detected using winter imagery by recognition of snow-covered openings based on high reflectance in the near-infrared bands of Landsat MSS data. Extensive forest clearing in Rondonia, Brazil, was observed using the AVHRR 1 km data by Tucker et al. (1984). The expansion of forest conversions could be determined along with the pattern associated with the development of paved roads. Mayer et al. (1980) demonstrated that the regrowth of brush in clear-cut areas could be accurately estimated in 10% increments of cover density from 10–50% in a coniferous region with natural regeneration in northern California. They found that the brush density was not well correlated with the age since the clear-cutting operation.

Williams and Nelson (1986), using MSS, found that areas of heavy defoliation by gypsy moths (>60%) could be discriminated from areas of little or no defoliation (<30%), but that areas of moderate defoliation (30–60%) were spectrally variable and confused with both classes (Nelson, 1983). Vogelmann and Rock (1986) analyzed Thematic Mapper data for spruce-fir forests in Vermont showing decline related to acid precipitation. Areas of heavy damage were clearly discriminated and an index of the near-infrared and shortwave (1.5 μm band) infrared data was related to intermediate levels of damage along transects following environmental gradients (Rock et al., 1986).

Changing atmospheric conditions between different dates can produce apparent shifts when none exists or obscure real change. Malila (1980) has proposed a change-vector procedure to predict both the kind of change (direction) and magnitude, and applied this to a study of Kershaw County, Georgia, to monitor forest practices. Logical decision rules have been developed to monitor change in tropical regions of Costa Rica (Joyce et al., 1980) with multidate imagery.

VI GEOGRAPHIC AGGREGATIONS AND ESTIMATION

Statistical estimation of vegetation properties such as acreage of specific cover types, timber volume or biomass, and ecosystem parameters is an important use of remote sensing. Estimation may be based on correlative models between a parameter directly sensed by the satellite correlated with a second, not directly sensible, variable. For example, our research in the deciduous forests of Wisconsin showed that the content of lignin in the canopies was inversely related ($r^2 = 0.96$) to the annual mineralization of nitrogen from litter decomposition (Aber et al., 1988). Thus, the spatial distribution of lignin given before (Figure 8) from the AIS data for Blackhawk Island can be transformed into a spatial distribution of annual nitrogen mineralization (Figure 11, Wessman, 1987; Wessman et al., 1988; Aber et al., 1988). Similarly, Myrold et al. (1986) have shown that nitrogen content in canopies of Oregon forests is strongly related to anaerobic nitrogen mineralization, CO_2 evolution from soils, and net primary productivity.

More commonly, remotely sensed data have been used as a first stage in various multistage inventory designs to estimate forest properties across large landscapes (Peterson et al., 1981). Stratification of forested regions into "stands," or small irregular areas of relative homogeneity, has been the basis of many existing inventory systems.

The classification map defines the population size and geographic distribution for each strata to serve as a base for sample design, to allocate and select samples by strata, and

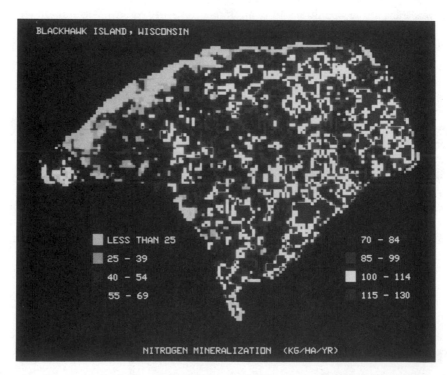

Figure 11 The spatial distribution of annual nitrogen mineralization for Blackhawk Island, Wisconsin, as estimated from its inverse relationship with canopy lignin concentrations applied to the spatial distribution of canopy lignin shown in Figure 8. (From Aber et al., 1989; Wessman et al., 1987b, 1988; mineralization data from Pastor et al., 1983.)

to act as expansion factors in the estimation model. The classification of Fox and Mayer (1980) for the McCloud Ranger District discussed before was used to restratify an existing set of ground plots for one forest compartment. They found that when stand stocking levels were high, the average coefficient of variation in timber volume for the classification strata was 11.1% compared to 9.7–15.7% by conventional aerial photomapping means. When stocking levels were low and tree size was small, the coefficient of variation of classified strata was higher (28.4–92.9%) and generally greater than air photo techniques (9.9–100%), depending on the strata being compared.

A comprehensive forest classification and inventory system (FOCIS) was developed by Strahler and his associates for application to timber volume estimates of the USDA/Forest Service in California (Strahler, 1981; Strahler et al., 1981). The relationship between species occurrence and terrain variables, coupled with a natural region concept, was used to delineate the probable distribution of the species type. The Landsat data was divided into shaded and illuminated hillslopes using the terrain data, and each spectral cluster was assigned to one of several stocking and size classes. For the western portion of the forest, the stratified estimate for total volume using existing plots was 4.46 billion cubic feet at a standard error of 6.3%.

The previous examples illustrate the use of Landsat classifications to reassign samples originally selected by stratified simple random sampling. Such poststratification can be effective when the original sample framework was not already biased by the selection process. Poststratification using independent data has been recommended for Forest Service surveys (Cunia, 1978). Combining a Landsat classification with a systematic sample set was done for Douglas County, Nevada (Ambrosia et al., 1983). The classification of the pinyon pine/Utah juniper woodlands and Jeffrey pine forest discriminated four cover-density classes (<30%, 30–50%, 50–75%, and >75%) at accuracies exceeding 90% (Brass et al., 1981). The Landsat acreage totals were used to expand directly the average strata volumes per class for a countywide estimate. The conventional method estimated 2.07 million cubic feet with 11.2% error, whereas the Landsat stratified estimate was 3.08 million cubic feet at 6.5% error.

Multistage designs based on sample selection with unequal probability have also been used. The variance of timber volume for the primary or first-stage sampling units (to be estimated from Landsat data) is assumed to vary directly or exponentially with actual volume, as shown in Figure 12(a). The Landsat data is partitioned into primary sampling units (PSUs) within which secondary subsamples will be selected (aerial photo plots and often double-sampled ground plots, cluster sampling). Some characteristic of the PSU is derived from the satellite data related to the variable of interest, such as volume, and used to calculate the PSU selection probabilities (Wi) among the sum of all of the PSUs (Langley, 1971). These selection probabilities in turn are used to expand each PSU volume (Vi), estimated by conventional methods, by dividing by the selection probability, thus generating an independent estimate of the total volume of the population. The mean of the expanded PSU volumes is the estimated total volume, since the single PSU volume is proportional to the total population volume as the PSU selection weight is to the total selection probability. Langley (1971) showed that acreage of forest versus nonforest in PSUs could act to reduce variance, and this was used in a probability proportional to size design, as these designs are also called, for a study in Oregon (Peterson and Card, 1980). Kirby and Van Eck (1977) used an area of softwoods derived from a summer and winter Landsat MSS classification of the boreal region of Canada for 10 × 10 km primary sample units. Six different categories (old-growth and second-growth conifers, hardwoods, reproduction, nonforest, and nonstocked forest) were differentiated

UNEQUAL PROBABILITY SAMPLING

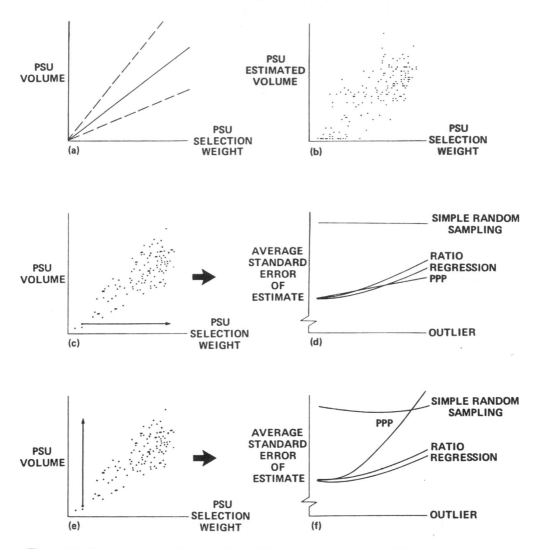

Figure 12   The assumed model for unequal probability sampling designs (probability proportional to size, PPS or prediction, PPP), actual data from a study in Idaho forests, and a comparison of the performance of such designs versus other sampling methods given sample outliers from the assumed models. (*a*) The assumed model is that the probability of selection of a particular primary sample unit (PSU) as derived from remote-sensing data (see text) varies directly as the actual PSU property, timber volume in this case, and that the variance increases as the volume or volume squared (dashed lines). (*b*) Actual distribution of volume and PSU selection weights of 400 acre PSUs sampled from throughout central Idaho forests. By using simulated data corresponding to the unequal probability model, various sampling methods are tested for their sensitivity to points falling outside the model: (*c*) horizontal outliers (overestimating the PSU selection weight), ratio, regression, and PPP designs have comparable sensitivity (*d*). For vertical outliers (*e*), cases where actual volume is greater than expected from the selection weight, the ratio and regression methods behave as before, but the PPP design is very sensitive. These cases can easily occur through misclassification of poorly stocked stands in which the estimation model inflates the contribution of such PSUs because of the underestimation of the selection weight. (From Peterson et al., 1983.)

from 14 MSS scenes of western Washington state (Harding and Scott, 1978), the non-forest to forest acreage per township was used to select PSUs, and the acreage of the different categories was used as expansion factors for basal area and volume estimates by ownership (Peterson et al., 1983). A probability proportional to prediction design was used in Idaho by assigning spectral classes to three volume classes determined by analysis of variance of existing ground data to stratify by volume. Thus, the selection probabilities and expansion factors were based on predicted volume per sample PSU (Peterson et al., 1983).

Des Raj (1968) has shown that a probability proportional to size (PPS) sampling is more precise than a simple random sampling (SRS) if the correlation of Wi to Vi^2/Wi is positive. The use of volume classes in Idaho showed that Wi and Vi were correlated between 0.72 and 0.80, and that the more sensitive Des Raj correlation ranged from 0.38 to 0.45. In Oregon, this correlation was only 0.13; Washington, 0.21. Although PPS sampling is theoretically superior to SRS, its added complexity makes equal probability sampling with regression or ratio estimation often more attractive. The sensitivity to outliers from the model relationship was tested for each estimator. Figure 12 shows the effect on the standard error as the severity of the outlier increases. Of 34 PSUs selected in Oregon, two samples were serious vertical outliers and the estimate of total volume was substantially overestimated. These two PSUs increased the total volume estimate by 1.8 million cubic feet and the relative standard error from 7.45 to 13.54%.

VII INTEGRATION OF SATELLITE IMAGERY WITH GEOGRAPHIC INFORMATION SYSTEMS

A geographic information system (GIS) is a collection of digital software and hardware to acquire, sort, and manipulate georeferenced data. The GIS should be able to accept data in any of four formats: cellular or raster, polygonal/line or vector, point, and tabular or textual. Once independent data planes are captured and registered to a common geo-reference system, they can be merged to perform analysis, including spatial modeling, spatial interactions and overlays, and data aggregation and estimation. Each data type (raster, polygon/line, point) will usually include pointers to the textual data base. This data base will be either hierarchical or relational in structure and include descriptions and attribute lists of measured properties.

An example of a GIS-based analysis, including all of the characteristics just described, is provided by a study of Santa Cruz County in central coastal California (Figure 13, Maw and Brass, 1981). The data base was constructed to conform to the 1:100,000 scale map series of the U.S. Geological Survey; the cell size was 100 × 100 meters. Landsat MSS data from the statewide mosaic described before was intensively studied using guided clustering techniques to produce a regional cover type map. An accurate classification map was produced to the level of species type and species mixes, including redwood, Douglas fir, mixed fir/redwood, madrone, tan oak, and live oak. Ancillary data, including soil series, land use, land ownership, roads, digital terrain, and fire-danger weather, were added.

The Santa Cruz data base was used for several model analyses: reforestation potential, fire-hazard mapping (Brass and Peterson, 1983), and soil-loss estimation (Spanner et al., 1983). In California, legislation was passed to provide financial support to medium-sized landowners (20–200 acres) for brush removal and replanting of softwoods into areas

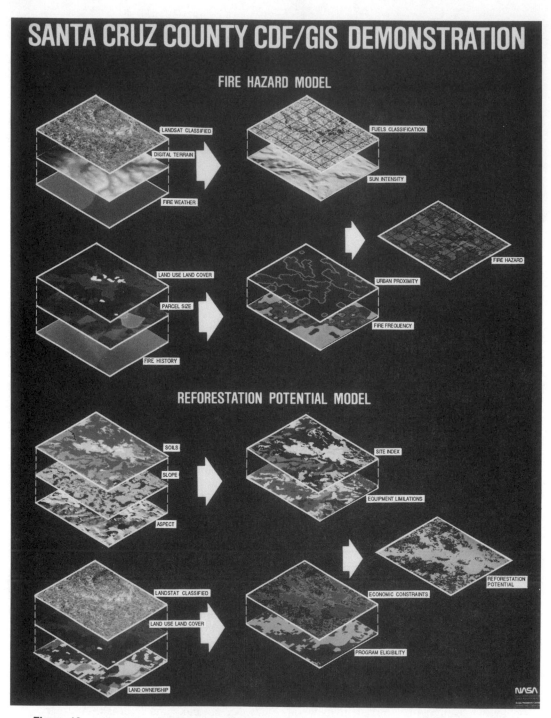

Figure 13 A black-and-white representation of the application of multiple geographic data layers to forest management and policy analyses for Santa Cruz County, California. The left-hand column presents coregistered data for on 7.5 minute quad of four different data types: cellular, Landsat MSS vegetation classification and digital elevation and slope data; vector or polygonal, soils, land ownership, USGS LUDA land use, fire weather; and, point, fire-start history. The middle column represents intermediate results of spatial analysis derived from these data layers for (top of figure) fire-hazard assessment and (bottom) potential for reforestation. The right-hand column presents the final maps of the spatial analyses, incorporating all four data types. (See text and Brass and Peterson, 1983; Maw and Brass, 1981.)

456

having good softwood site growth potential. The areas had to be currently covered with brush resulting from past harvest or other conversion practices. By using a sieve or masking model, the soil data modified by terrain variables to improve specificity were used to assess site growth potential from published textual information. This was overlaid by ownership class to isolate qualified areas and by the vegetation classification to identify the current condition of the qualified areas. A map with acreage estimates, Figure 13, indicated that whereas a high potential existed in the county to increase redwood/Douglas fir production on the highest site classes, most of the current potential was occupied by hardwood species such as madrone, tan oak, and other large trees, eliminating these candidate lands, since only brush removal was being qualified. This GIS analysis, examining the validity of policy decisions, led to a change in policy by the state.

Modeling of fire hazard represented an example of forest-management applications. Four factors were considered: (1) fuel loading, (2) fire-spread characteristics, (3) risks to structure and resources, and (4) ignition potential.

The Landsat vegetation classification was relabeled by county foresters into fuel-loading classes described by the National Fire Danger Rating System. The context of vegetation in relation to terrain slope was characterized by calculating the steepest local slope using the digital terrain data ($<30\%$, $30–60\%$, $>60\%$) according to recommendations of the National Wildfire Hazard Classification System. The vegetation/fuels and slope classes were multiplied to rate each pixel for fuel loading.

Risks to structures and proximity to areas of expected higher-density population were defined. The perimeters in 100 m increments were calculated around boundaries of low-acreage landowners and residential land use. The risk to structures was rated by their proximity to high-fuel cells: risk ratings decreased with distance (by 100 m increments) from the ''structure'' map, whereas risk ratings increased when resources were in close proximity to high-fuel cells. A similar analysis was carried out for high-value timber resources.

The point data layer representing the history of past fire starts was converted to a continuous surface using a proximal algorithm that rated each cell on the basis of cumulative past fire starts nearby. This surface-reflected ignition potential concentrated the risk near highways and urban areas, where most fires start. The final two maps, for risks to structure and resources are valuable to allocate fire-fighting resources, to concentrate risk-reduction programs, and, with inclusion of the road network and timing access records, to plan for fire extinction (Figure 13).

In the Santa Cruz study, some operations led to poor model results, which were addressed in an ad hoc way by rescaling results at intermediate stages of the analysis. Robinson and Strahler (1984) have discussed the concerns of inexactness in data layers and of the use of fuzzy set theory to manage data. The Santa Cruz study is an example of their special case, which assumes nonfuzzy data and a nonfuzzy modeling scheme.

A promising approach to problems of inexactness is to partition a forested landscape according to landscape patterns that control variance of a desired property. If the boundaries of these patterns can be identified (such as field boundaries), then an analysis can proceed between fields rather than pixels and the field values treated as distribution functions. For example, topographic variables such as slope and aspect often exert a strong influence upon the development of LAI and other biophysical parameters in many forest landscapes, such that opposing hillslopes can carry much different canopy and biomass properties. Within a hillslope, high local heterogeneity due to factors other than the desired one to be estimated can complicate the use of the data at the pixel level. Band (1986a,b) has devised an analytical system to process digital terrain data into hydrologically meaningful segments. This system can identify automatically the entire stream network to various orders and also the drainage divides associated with each link in the stream network (Figure 14). Each combination of stream link and drainage divide on each side of the stream represents a "field." Topological operations make this system into an effective GIS based on facets or fields maintaining all the geographic fidelity of the data base. Each slope facet is a topographically homogeneous landscape unit and becomes an ideal template for building a GIS of other ecosystem attributes. Soil factors,

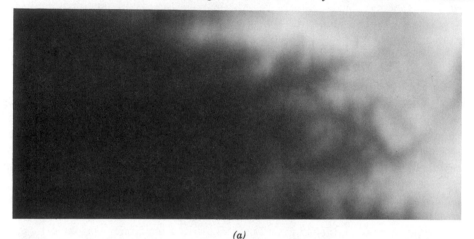

(a)

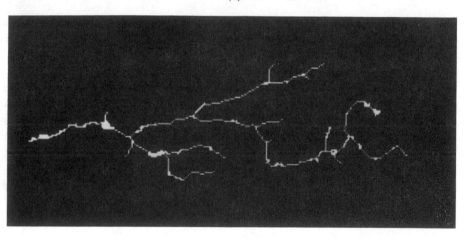

(b)

Figure 14 Results of the automated derivation of hydrologic information from digital terrain data (Digital Elevation Model data from USGS). (*a*) A grey-scale representation of the digital elevation model data for the North Fork of Elk Creek, Montana. (*b*) The drainage network defined by transforming each pixel into the number of pixels defined to drain through them (a drainage-area transform) and creating a discrete form of the Elk Creek channel network by choosing a threshold drainage area defined to support a stream channel. (*c*) The drainage divides associated with each link of the channel network, forming a watershed partition into subcatchment areas. The combination of (*b*) and (*c*) provides a topographically meaningful segmentation and logical data base structure for managing various data layers for spatial data analyses such as estimates of forest evapotranspiration. (From Band, 1986a,b.)

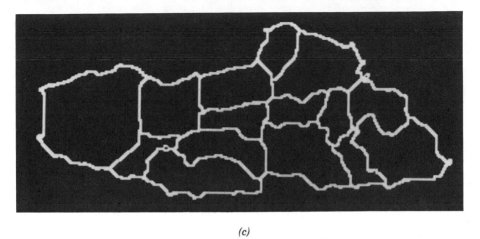

(c)

Figure 14   *(Continued)*

forest types, microclimate, and other attributes frequently can be represented homogeneously within each of the landscape facets. This system is being used to help model forest evapotranspiration and net carbon assimilation and productivity in a 2600 square kilometer region in Montana.

VIII   COUPLING SATELLITE DATA AND FOREST ECOSYSTEM MODELS

The preceding discussion illustrates that forest remote sensing is evolving beyond basic classification of cover types to a range of other capabilities, many incorporating ancillary data and GIS. In the introduction, we stated that many of the processes of greatest scientific interest in forestry (evapotranspiration, photosynthesis, respiration, decomposition) are not directly sensible by optical remote-sensing techniques. A new application of remote sensing is to use satellite data to drive ecosystem-process models that calculate those indirect variables. A model designed for this purpose (Figure 15, Running and Coughlan, 1988) requires canopy meteorology, leaf-area index, and leaf nitrogen and lignin concentrations to be measured and input from satellites. From these key structural and chemical ecosystem measures, the simulation model calculates the important carbon, hydrologic, and nitrogen cycle processes of interest. Because the critical canopy variables are measured by satellite, this type of model then has the potential to be executed over large areas in a GIS framework of topographically homogeneous landscape units, as suggested before. The result will be process-level calculations of key ecosystem processes that can be mapped across the landscape. The first example of this capability is given in Figure 16, where LAI estimated from NDVI values obtained from AVHRR satellite data was combined with ground climatological data to drive the ecosystem-process model (Figure 15) to produce maps of annual evapotranspiration and photosynthesis for a 1200 km^2 region of western Montana forest land (Running et al., 1988).

We have also used the ecosystem simulation model in Figure 15 to "calibrate" the AVHRR NDVI product that has been used to map North America (Goward et al., 1985), Africa (Tucker et al., 1985), world vegetation (Justice et al., 1985), and global CO_2

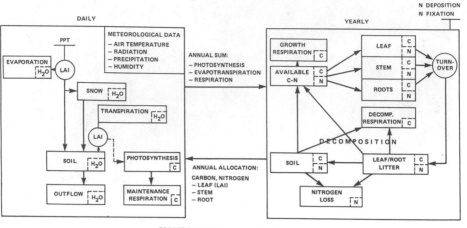

Figure 15 A compartment flow diagram for FOREST-BGC, a forest ecosystem-process model designed to be driven by remotely sensed inputs of daily surface climate, canopy LAI, canopy nitrogen, and canopy lignin. The model then predicts daily rates of forest evapotranspiration, photosynthesis and respiration, and annual rates of primary production, litterfall, decomposition and nitrogen mineralization. (From Running and Coughlan, 1988.)

exchange activity (Fung et al., 1987). Daily meteorological data and weekly NDVI were acquired for seven sites of diverse climate around the United States. The ecosystem model was run in 1983 and 1984 at the seven sites and the predictions of photosynthesis, transpiration, and net primary production compared to the activity implied by the weekly NDVI for each site. Integrated annual NDVI proved to be highly correlated with the annual photosynthesis and transpiration simulated for forests at these sites, Figure 17, but correlation of NDVI with seasonal trends was less successful, Figure 18 (Running and Nemani, 1988). Errors were greatest where substantial canopy water stress during midsummer caused low canopy gas exchange to be associated with periods of high NDVI values. Clearly, the NDVI time series provides a good indication of the active growing season length, but is not responsive to dynamic changes in canopy physiology.

Other forest ecosystem models also have the potential to utilize information derived from satellite data. The modelers must build a model sensitive to the type and resolution of data that satellite sensors can provide. For example, the widely used JABOWA derivatives (Botkin et al., 1972; Shugart and West, 1980) simulate individual tree growth processes as an annual diameter increment, another meaningful ecosystem variable not directly sensible from satellite data. However, the birth and death subroutines could be defined to be controlled by remotely sensed measures of developing leaf area and mortality, but at the stand rather than at the individual tree level. A different use of remote-sensing data is possible with these models. They can simulate changes in the proportions of landscape processes over long time frames. The patterns resulting from these long-term processes may produce community and structural reflectance changes potentially sensible by remote-sensing techniques. Hall et al. (1987) compared the classifications of two Landsat multispectral scanner scenes of northern Minnesota spaced about 10 years apart (Figure 19). Land use and fairly rapid successional change in communities

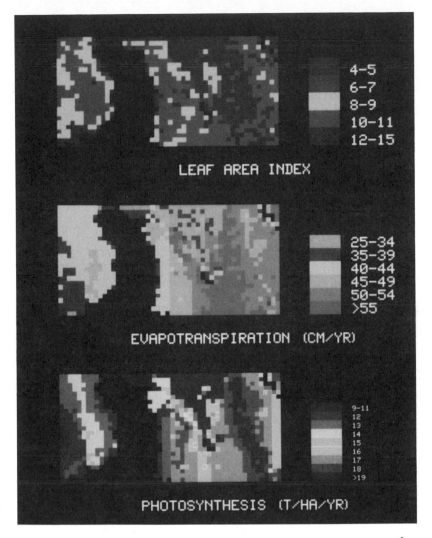

Figure 16 A map of LAI, annual evapotranspiration, and net photosynthesis for 1200 km^2 of forested landscape in Montana. LAI was estimated with AVHRR data and the NDVI algorithm; evapotranspiration and net photosynthesis were calculated with the FOREST-BGC model (Figure 15) for each 1.1 km^2 pixel, using daily meteorological data generated by the MT-CLIM mountain microclimate simulator (Running et al., 1987). (From Running et al., 1989.)

could occur across this time period. The proportions of acreage in each discriminable community was used to estimate the matrix of transition and retention probabilities among the various recognizable stages.

Global-scale models, such as the global carbon models (Emmanuel et al., 1984; Moore et al., 1981) currently define terrestrial vegetation into aggregated compartments of carbon that are not geographically referenced. Newer versions of these models, built upon worldwide- and regional-scale GIS systems incorporate satellite-defined measures of terrestrial vegetation. The global distribution and borders between major vegetation types could provide useful data for these models, particularly for a potentially changing en-

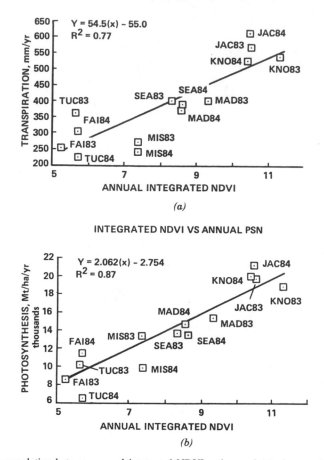

Figure 17 The correlation between annual integrated NDVI and annual (*a*) photosynthesis and (*b*) transpiration calculated with the FOREST-BGC model (Figure 15) for seven sites across a climatic gradient of North America for 1983 and 1984. The sites are Fairbanks, Alaska (FAI); Seattle, Washington (SEA); Missoula, Montana (MIS); Madison, Wisconsin (MAD); Knoxville, Tennessee (KNO); Jacksonville, Florida (JAC); and Tucson, Arizona (TUC). The NDVI data are weekly composites from the AVHRR sensor, and each site was simulated with average LAI = 2.7. (From Running and Nemani, 1988b.)

vironment. The repetitive measurement capability of satellites would then make change detection, deforestation–reforestation, desertification, and land conversion measurable inputs into a global carbon cycle rather than the current best estimates.

Two models are being developed to define terrestrial vegetation processes in a general way and incorporate them into General Circulation Models (GCMs). Sellers (1985, 1987) and Sellers and Dorman (1987) have developed a Simple Biosphere (SiB) model that requires, among other things, LAI, canopy height, canopy reflectance, and surface roughness that can be measured by satellite sensors. Dickinson (1984) and Wilson et al. (1987) have developed a model, called the Biosphere–Atmosphere Transfer Scheme (BATS), that also requires LAI, canopy albedo, soil albedo, and other variables and vegetation cover as given by the classification system and maps of Matthews (1983),

NDVI, RELATIVE PHOTOSYNTHESIS, TRANSPIRATION

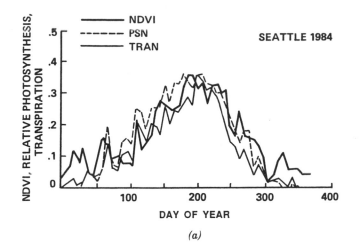

(a)

NDVI, RELATIVE PHOTOSYNTHESIS, TRANSPIRATION

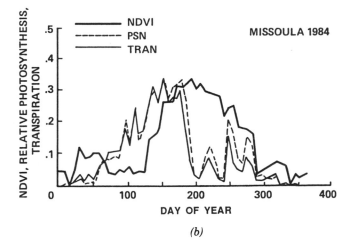

(b)

Figure 18 The seasonal trend of weekly composite NDVI compared to scaled weekly sums of photosynthesis (PSN) and transpiration (TRAN) for (*a*) Seattle, Washington, and (*b*) Missoula, Montana. (*a*) For Seattle, the correlation of weekly NDVI was, with PSN, $R^2 = 0.81$, scaling of 1.69 T carbon/ha/week/NDVI; and for TRAN, $R^2 = 0.82$, scale of 58.5 mm/week/NDVI. (*b*) For Missoula, correlations were, with PSN, $R^2 = 0.24$, scale of 1.78 T carbon/ha/week/NDVI, and for TRAN, $R^2 = 0.13$, scale of 48.1 mm/week/NDVI. Weekly NDVI can successfully estimate canopy processes in a mesic site like Seattle, but becomes very inaccurate for water-limited forests like Missoula. (From Running and Nemani, 1988.)

which also may be obtained from satellite sensor data. Although neither model was built explicitly for satellite data inputs, both define vegetation at the appropriate scale of complexity for remote-sensing inputs. When coupled to GCMs, both will require satellite inputs to provide any dynamic evaluation of terrestrial vegetation effects on global climate.

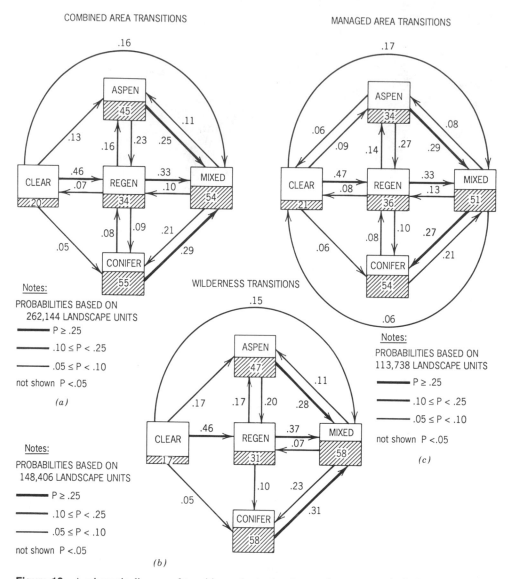

Figure 19 A schematic diagram of transition and retention frequencies among ecological states as derived from classification of Landsat MSS data for 1973 and Thematic Mapper data from 1983 for a large area of boreal forest in Superior National Forest, Minnesota. Spectrally distinguishable cover conditions included areas of sparse or low-density vegetation (clearings), young stands of aspen and jack pine (regeneration), areas dominated by aspen and broadleaf species (aspen), by mixed conifers and broadleaf trees (mixed), and by conifer species (conifer), the oldest successional stage. The frequencies were observed by comparing the classification results of the coregistered classified scenes, although possible errors of misclassification and bias in the process have yet to be examined. The different results obtained for two distinct subdivisions of the forest are shown in (*b*) a wilderness area and in (*c*) an area managed for production. Such data might be effectively employed in population-statistics models of forest dynamics over long time periods and in studying the effects of patchiness in the landscape. (Adapted from Hall et al., 1987.)

ACKNOWLEDGMENTS

We wish to gratefully acknowledge the cooperation and stimulation afforded us by many of our colleagues and associates, most of whom are cited within the text. They have made substantial contributions to the research covered in this chapter. So many topics are covered spanning 11 years of research that it is not possible to acknowledge everyone by name. The editorial review and comments of Judith Cody were specifically helpful in arriving at a final draft of the chapter. We wish to thank her sincerely.

REFERENCES

Aber, J. D., C. A. Wessman, D. L. Peterson, and J. H. Fownes (1989). Estimation of canopy chemistry, nitrogen mineralization and net primary productivity in forest ecosystems using remote sensing. In: H. Mooney and R. Hobbs (eds.). Remote sensing of biosphere functioning. (forthcoming)

Allen, W. A., H. W. Gausman, and A. J. Richardson (1973). Willstatter–Stoll theory of leaf reflectance evaluated by ray tracing. *Appl. Opt.* **12**(10):2448–2453.

Ambrosia, V. G., D. L. Peterson, and J. A. Brass (1983). Volume estimation techniques for Pinyon pine/Utah juniper woodlands using Landsat data and ground information. In *Renewable Resource Inventories for Monitoring Changes and Trends*. Society of American Foresters, Corvallis, Oregon, pp. 88–192.

Arp, H., J. C. Greisbach, and J. P. Burns (1982). Mapping in tropical forests: A new approach using the Laser APR. *Photogramm. Eng. Remote Sens.*, 48(1):91–100.

Asrar, G., M. Fuchs, E. T. Kanemasu, and J. L. Hatfield (1984). Estimating absorbed photosynthetic radiation and leaf area index from spectral reflectance in wheat. *Agronomy J.*, 76:300–306.

Badhwar, G. D., R. B. MacDonald, and N. C. Mehta (1986a). Satellite-derived leaf-area-index and vegetation maps as input to global carbon cycle models—a hierarchical approach. *Int. J. Remote Sens.*, 7(2):265–281.

Badhwar, G. E., R. B. MacDonald, F. G. Hall, and J. G. Carnes (1986b). Spectral characterization of biophysical characteristics in a boreal forest: Relationship between Thematic Mapper band reflectance and leaf area index for aspen. *IEEE Trans. Geosci. Remote Sens.*, **GE-24**(3):322–326.

Baltaxe, R. (1980). *The Application of Landstat Data to Tropical Forest Surveys*. Forest Resources Division, Forestry Department, FAO, Rome.

Band, L. E. (1986a). Topographic partition of watersheds with digital elevation models. *Water Resour. Res.*, **22**:15–24.

Band, L.E. (1986b). Analysis and representation of drainage basin structure with digital elevation data. *Proc. 2nd Int. Conf. Spatial Data Handling*, Seattle, Washington.

Barnes, W. L., F. G. Huegel, and J. R. Irons (1985). An airborne imaging system for measuring anisotropic reflectance patterns. *Proc. PECORA 10th Symp. Remote Sens. For. Range Resour. Manage.*, Fort Collins, Colorado, pp. 508–516.

Botkin, D. B., J. F. Janak, and J. R. Wallis (1972). Some ecological consequences of a computer model of forest growth. *J. Ecol.* **60**:849–872.

Brass, J. A., and D. L. Peterson (1983). Reforestation and fire hazard mapping in Santa Cruz County, California. In *The Application of Remote Sensing to Resource Management*, Renewable Resouces Foundation, Seattle, Washington.

Brass, J. A., W. Likens, and R. Thornhill (1981). Wildland inventory and resource modeling for Douglas and Carson City Counties, Nevada, using Landstat and digital terrain data. *NASA Tech. Pap.* **TP-2137:**–43.

Brass, J. A., V. G. Ambrosia, P. J. Riggan, J. S. Myers, and J. C. Arvesen (1987). Aircraft and satellite thermographic systems for wildfire mapping and assessment. *25th Aerosp. Sci. Meet.*, Reno, Nevada, Paper AIAA-87-0187.

Butera, C. (1986). A correlation and regression analysis of per cent canopy closure versus TMS spectral response for selected forest sites in the San Juan National Forest, Colorado. *IEEE Trans. Geosci. Remote Sens.* **GE-24**(1):122–129.

Caldwell, M. M., H.-P. Meister, J. D. Tenhunen, and O. L. Lange (1986). Canopy structure, light microclimate and leaf gas exchange of *Quercus coccifera* L. in a Portuguese macchina: Measurements in different canopy layers and simulations with a canopy model. *Trees* **1:**25–41.

Card, D. H., D. L. Peterson, P. A. Matson, and J. D. Aber (1988). Prediction of leaf chemistry by the use of visible and near infrared reflectance spectroscopy. *Remote Sens. Environ.* **26:**123–147.

Cunia, T. (1978). On the objectives and methodology of national forest inventory systems. *IUFRO Meeting of Subject Group S4.02 on Forest Resouce Inventory*, Bucharest, Romania.

Curcio, J. A., and C. C. Petty (1951). The near infrared absorption spectrum of liquid water. *J. Opt. Soc. Am.* **41**(5):302–304.

Danks, S. M., E. H. Evans, and P. A. Whittaker (1984). *Photosynthetic Systems: Structure, Function and Assembly*. Wiley, New York, p. 37.

Dickinson, R. E. (1984). Modeling evapotranspiration for three-dimensional global climate models. *Geophys. Monogr., Am. Geophys. Union* **29:**58–72.

Emmanuel, W. R., G. G. Killough, W. M. Post, and H. H. Shugart (1984). Modelling terrestrial ecosystems in the global carbon cycle with shifts in carbon storage capacity by land-use change. *Ecology* **65:**970–983.

Everitt, J. H., and P. R. Nixon (1986). Canopy reflectance of two drought-stressed shrubs. *Photogramm. Eng. Remote Sens.* **52**(8):1189–1192.

Fleming, M. D., and R. M. Hoffer (1979). Machine processing of Landsat MSS data and DMA topographic data for forest cover type mapping. In *Machine Processing of Remotely Sensed Data*. Purdue University, West Lafayette, Indiana, pp. 377–390.

Fox, L. III (1978). The effect of canopy composition on the measured and calculated reflectance of conifer forests in Michigan. In *Remote Sensing for Vegetation Damage Assessment*, Seattle, Washington, pp. 89–114.

Fox, L., III, and K. E. Mayer (1980). *Forest Resource Classification of the McCloud Ranger District, Mt. Shasta, California, Using Landsat Digital Data*, Final Rep. EM-7145-2. USDA/ Forest Service, Washington, D.C.

Fox, L. III, J. A. Brockhaus, and N. D. Tosta (1985). Classification of timberland productivitiy in northwestern California using Landsat, topographic and ecological data. *Photogramm. Eng. Remote Sens.* **51**(11):1745–1752.

Franklin, J. (1986). Thematic Mapper analysis of coniferous forest structure and composition. *Int. J. Remote Sens.* **7**(10):1287–1301.

Franklin, J., T. L. Logan, C. E. Woodcock, and A. H. Strahler (1986a). Coniferous forest classification and inventory using Landsat and digital terrain data. *IEEE Trans. Geosci. Remote Sens.* **GE-24**(1):139–149.

Franklin, J., X. Li, and A. H. Strahler (1986b). Canopy reflectance modeling in Sahelian and Sudanian woodland and savanna. *Proc. 20th Int. Symp. Remote Sens. Environ.*, Nairobi, Kenya (in press).

Fung, I. Y., C. J. Tucker, and K. C. Prentice (1987). Application of AVHRR vegetation index to study atmosphere-biosphere exchange of CO_2. *J. Geophys. Res.* **92**:2299–3015.

Gates, D. M., H. J. Keegan, J. C. Schleter, and V. R. Weidner (1965). Spectral properties of plants. *Appl. Opt.* **4**(1):11–20.

Gausman, H. W., W. A. Allen, and R. Cardenas (1969). Reflectance of cotton leaves and their structure. *Remote Sens. Environ.* **1**:19–22.

Gaydos, L. and W. L. Newland (1978). Inventory of land use and land cover of the Puget Sound Region using Landsat digital data. *J. Res. U.S. Geol. Surv.* **6**(6):807–814.

Gerstl, S. A. W., and A. Zardecki (1985). Coupled atmosphere/canopy model for remote sensing of plant reflectance features. *Appl. Opt.* **24**(1):94–103.

Gholz, H. L. (1982). Environmental limits on aboveground net primary production, leaf area, and biomass in vegetation zones of the Pacific Northwest. *Ecology* **63**(2):469–481.

Gholz, H. L., C. C. Grier, A. G. Campbell, and A. T. Brown (1979). *Equations for Estimating Biomass and Leaf Area of Plants in the Pacific Northwest*, Res. Pap. 41. Forest Research Laboratory, Oregon State University, Corvallis.

Goetz, A. F. H., G. Vane, J. E. Solomon, and B. N. Rock (1985). Imaging spectrometry for earth remote sensing. *Science* **228**;1147–1153.

Goward, S. N., C. J. Tucker, and D. G. Dye (1985). North American vegetation patterns observed with the NOAA-7 Advanced Very High Resolution Radiometer. *Vegetatio* **64**:3–14.

Grier, C. C., and S. W. Running (1977). Leaf area of mature Northwestern coniferous forests: Relation to site water balance. *Ecology* **58**(4):893–899.

Hafker, W. R., and W. R. Philipson (1982). Landsat detection of hardwood forest clearcuts. *Photogramm. Eng. Remote Sens.* **48**(5):779–780.

Hall, F. G., and D. E. Strebel, S. J. Goetz, K. D. Woods, and D. B. Botkin (1987). Landscape pattern and successional dynamics in the boreal forest. *Proc. Int. Geosci. Remote Sens. Symp. (IGARSS'87)*, Ann Arbor, Michigan, pp. 473–482.

Haralick, R. M., and K.-S. Fu (1983). Pattern recognition and classification. In *Manual of Remote Sensing, Vol. I* (R. N. Colwell, Ed.), American Society of Photogrammetry, Falls Church, Virginia, pp. 793–805.

Harding, R. A., and R. B. Scott (1978). *Forest Inventory with Landsat: Phase II*. Washington Forest Productivity Study, Department of Natural Resources, State of Washington, Olympia.

Hauck, M., K. Hiller, and H. Kux (1986). Southern Pantanal Matogrossense/interpretation of MOMS-data preliminary results. *Proc. Int. Geosci. Remote Sens. Symp. (IGARSS'86)*, Zurich, Switzerland, pp. 1451–1456.

Hergert, H. L. (1971). Infrared spectra. In *Lignins: Occurrence, Formation Structure and Reactions* (K. V. Sarkanen and C. H. Ludwig, Eds.), Wiley (Interscience), New York.

Hirschfeld, T. (1985). Instrumentation in the next decade. *Science* **230**:286–291.

Hlavka, C. (1986). Destriping AIS data using Fourier filtering techniques. *JPL Publ.* **86-35**: 74–80.

Hodson, W., and W. Likens (1982). Data base modeling to determine development scenarios in the San Bernardino County: A project summary. In *Auto-Carto V*. Crystal City, Virginia, and American Society of Photogrammetry, Falls Church, Virginia.

Hoffer, R. M. and Staff (1975). *Computer-aided Analysis of SKYLAB Multispectral Scanner Data in Mountainous Terrain for Land Use, Forestry, Water Resources and Geologic Applications*, LARS Inf. Note 121275. Purdue University, West Lafayette, Indiana.

Hunt, R. E., Jr., B. N. Rock, and P. S. Nobel (1987). Measurement of leaf relative water content by infrared reflectance. *Remote Sens. Environ.* **22**:429–435.

Jarvis, P. G., and J. W. Leverenz (1984). Productivity of temperate, deciduous and evergreen forests. In *Physiological Plant Ecology* (O. L. Lange, P. S. Nobel, C. B. Osmond, and H. Zeigler, Eds.), Vol. IV. Springer-Verlag, New York, pp. 233–280.

Johnson, G. R., E. W. Barthmaier, T. W. Gregg, and R. E. Aulds (1979). Forest stand classification in western Washington using Landsat and computer-based resource data. *Proc. 13th Int. Symp. Remote Sens. Environ.*, University of Michigan, Ann Arbor, pp. 1681–1696.

Joyce, A. T., J. H. Ivey, and G. S. Burns (1980). The use of Landsat MSS data for detecting land use changes in forestland. *Proc. 14th Int. Symp. Remote Sens. Environ.*, San Jose, Costa Rica, pp. 979–988.

Justice, C.O., J. R. G. Townshend, B. N. Holben, and C. J. Tucker (1985). Analysis of the phenology of global vegetation using meteorological satellite data. *Int. J. Remote Sens.* **6**:1271–1318.

Kimes, D. S. (1984). Modeling the directional reflectance from complete homogeneous vegetation canopies with various leaf orientation distributions. *J. Opt. Soc. Am.* **1**:725–737.

Kimes, D. S., and J. A. Kirchner (1982). Radiative transfer model for heterogeneous 3-D scenes. *Appl. Opt.* **21**:4119.

Kimes, D. S., J. A. Smith, and K. J. Ranson (1980). Vegetation reflectance measurements as a function of solar zenith angle. *Photogramm. Eng. Remote Sens.* **46**(12):1563–1573.

Kirby, C. L., and P. I. Van Eck (1977). A basis for multistage forest inventory in the boreal forest region. *Proc. 4th Can. Symp. Remote Sens.*, Quebec, pp. 72–94.

Kleman, J. (1984). The spectral signatures of coniferous tree stands measured from a helicopter. *Proc. 8th Can. Symp. Remote Sens.*, pp. 575–580.

Knipling, E. B. (1970). Physical and physiological basis for the reflectance of visible and near-infrared radiation from vegetation. *Remote Sens. Environ.* **1**:155–159.

Kumar, R., and L. F. Silva (1973). Light ray tracing through leaf cross section. *Appl. Opt.* **12**(2):2950–2954.

Langley, P. G. (1971). Multistage sampling of earth resources with aerial and space photography. *NASA [Spec. Publ.] SP* **NASA SP-275**:129–141.

Lee, Y. J. (1976). Computer-assisted forest land classification in British Columbia and Yukon Territory: A case study. *Proc. Fall Conv. Am. Soc. Photogramm.*, Falls Church, Virginia, pp. 240–250.

Li, X., and A. H. Strahler (1986). Geometric-optical bidirectional reflectance modeling of a conifer forest canopy. *IEEE Trans. Geosci. Remote Sens.* GE-**24**:906–918.

Lieth, H. (1975). Primary production of the major vegetation units of the world. *Ecol. Stud.* **14**:203–215.

Likens, W., and K. Maw (1981). Hierarchical modeling for image classification. *Proc. PECORA 7th Symp. Remote Sens. For. Range Resour.*, Sioux Falls, South Dakota.

Logan, T. L. (1983). Regional forest biomass modeling from combined AVHRR and Landsat data. *Proc. CERMA Energy Resour. Manage. Conf.*, San Francisco, California.

Malila, W. A. (1980). Change vector analysis: An approach for detecting forest change with Landsat. In *Machine Processing of Remotely Sensed Data.*, Purdue University, West Lafayette, Indiana, pp. 326–335.

Markham, B. L., and J. R. G. Townshend (1981). Land cover classification accuracy as a function of sensor spatial resolution. *Proc. 15th Int. Symp. Remote Sens. Environ.*, University of Michigan, Ann Arbor, pp. 1079–1090.

Martin, G. C., J. S. Shenk, and F. E. Barton, II (Eds.) (1985). Near infrared reflectance spectroscopy (NIRS): Analysis of forage quality. *U.S., Dep. Agric., Agric. Handb.* **643**:1–96.

Matson, M., and J. Dozier (1981). Information of subresolution high temperature sources using a thermal IR sensor. *Photogramm. Eng. Remote Sens.* **47**:1311–1318.

Matthews, E. (1983). Global vegetation and land use: New high-resolution data bases for climate studies. *J. Clim. Appl. Meteorol.* **22**:474–487.

Maw, K. D., and J. A. Brass (1981). Forest management applications of Landsat data in a geographic information system. *Proc. PECORA 7th Symp. Remote Sens. For. Range Resour.*, Sioux Falls, South Dakota.

Mayer, K. E., and L. Fox, III (1981). Identification of conifer species groupings from Landsat digital classifications. *Photogramm. Eng. Remote Sens.* **48**(1):1607–1614.

Mayer, K. E., L. Fox, III, and J. L. Webster (1980). Forest condition mapping of the Hoopa Valley Indian Reservation using Landsat data. *Proc. 1st Int. Symp. Remote Sens. Nat. Resour.*, University of Idaho, Moscow, pp. 217–242.

Maynard, P. F., and A. H. Strahler (1981). The logit classifier for multi-image data. *Proc. IEEE Workshop Picture Data Descr. Manage.*, pp. 18–26.

Mead, R. A., R. S. Driscoll, and J. A. Smith (1979). Effects of tree distribution and canopy cover on classification of Ponderosa pine forest from Landsat-1 data. *U.S., For. Serv., Res. Note RM* **RM-375**:1–3.

Mohren, G. M. J., C. P. Van Gerwen, and C. J. T. Spitters (1984). Simulation of primary production in even-aged stands of Douglas fir. *For. Ecol. Manage.* **9**:27–49.

Moore, B., R. D. Boone, J. E. Hobbie, R. A. Houghton, J. M. Melillo, B. Peterson, J. R. Shaver, C. J. Vorosmarty, and G. M. Woodwell (1981). A simple model for analysis of the role of terrestrial ecosystems in the global carbon budget. In *Carbon Cycle Modeling—Scope Report 16*, (B. Bolin, Ed.), Wiley, New York, vol. 16, pp. 365–385.

Moreland, R. M., and E. A. Fosnight (1981). Remote sensing data integration into a geographic information system for the creation of a biogenic hydrocarbon emissions inventory of the San Francisco Bay Area. *Proc. PECORA 7th Symp. Remote Sens. For. Range Resour.*, Sioux Falls, South Dakota.

Morrissey, L. A., L. L. Strong, and D. H. Card (1986). Mapping permafrost in the boreal forest with Thematic Mapper satellite data. *Photogramm. Eng. Remote Sens.* **52**(9):1513–1520.

Myers, V. I. (1970). Soil, water and plant relations. In *Remote Sensing with Special Reference to Agriculture and Forestry*. National Academy of Sciences, Washington, D.C., pp. 253–297.

Myrold, D. D., P. A. Matson, and D. L. Peterson (1986). Relationships between N cycle processes and remotely-sensed data across a transect of vegetation zones in Oregon. *Symp. Bull. 4th Int. Congr. Ecol.*, Syracuse, New York.

Nelson, R. F. (1983). Detecting forest canopy change due to insect activity using Landsat MSS. *Photogramm. Eng. Remote Sens.* **49**(4):1303–1314.

Nemani, R. R., and S. W. Running (1989a). Estimating regional surface resistance to evapotranspiration from NDVI and thermal-IR AVHRR data. *J. Appl. Meteor.* (in press).

Nemani, R. R., and S. W. Running (1989b). Testing a theoretical climate-soil-leaf area hydrologic equilibrium of forests using satellite data and ecosystem simulation. *Agric. For. Meteorol. (in press)*.

Norman, J. M., and P. G. Jarvis (1974). Photosynthesis in Sitka spruce. III. Measurements of canopy structure and radiation environment. *J. Appl. Ecol.* **11**:375–398.

Norman, J. M., and P. G. Jarvis (1975). Photosynthesis in Sitka spruce. V. Radiation penetration theory and a test case. *J. Appl. Ecol.* **12**(3):839–878.

Norris, K. H., and R. F. Barnes (1976). Infrared reflectance analysis of nutritive value of feedstuffs. In *Feed Composition, Animal Nutrient Requirements, and Computerization of Diets*. Utah State University, Logan, pp. 237–241.

Norris, K. H., R. F. Barnes, J. E. Moore, and J. S. Shenk (1976). Predicting forage quality by infrared reflectance spectroscopy. *J. Anim. Sci.* **43**(4):839–878.

Pastor, J., J. D. Aber, C. A. McClaugherty, and J. M. Melillo (1983). Aboveground production and N and P cycling along a nitrogen mineralization gradient on Blackhawk Island, Wisconsin. *Ecology* **65**(1):256–268.

Perry, C. R., and L. F. Lautenschlager (1984). Functional equivalence of spectral vegetation indices. *Remote Sens. Environ.* **14**:169–182.

Peterson, D. L. (1983). Remote sensing technology applied to forest assessment in California. *Adv. Space Res.* **3**(2):193–197.

Peterson, D. L., and D. H. Card (1980). Issues arising from the demonstration of Landsat-based technologies to inventories and mapping of the forest resources of the Pacific Northwest states. In *Remote Sensing of Earth Resources*, (F. Shahrokhi, Ed.), Vol. VII. University of Tennessee Space Institute, Tullahoma, pp. 65–100.

Peterson, D. L., N. Tosta-Miller, S. D. Norman, D. Wierman, and W. Newland (1980). Land cover classification of California using mosaicking and high speed processing. *Proc. 14th Int. Symp. Remote Sens. Environ.*, San Jose, Costa Rica, pp. 279–305.

Peterson, D. L., J. A. Brass, and D. H. Card (1981). Incorporating remote sensing in inventory designs. *Proc. In-place Resour. Inventory Workshop*, Orono, Maine.

Peterson, D. L., D. Noren, and G. Gnauck (1983). Methods and results of three unequal probability multistage sampling designs for timber volume in the Pacific Northwest. In *Renewable Resource Inventories for Monitoring Changes and Trends*. Society of American Foresters, Corvallis, Oregon, pp. 326–329.

Peterson, D. L., P. A. Matson, J. G. Lawless, J. D. Aber, P. M. Vitousek, and S. W. Running (1985). Biogeochemical cycling in terrestrial ecosystems: Modeling, measurement and remote sensing. *Proc. 36th Int. Astronaut. Fed. Congr.*, Stockholm, Sweden. (Available: NASA/Ames Research Center, Moffett Field, California.

Peterson, D. L., W. E. Westman, N. J. Stephenson, V. G. Ambrosia, J. A. Brass, and M. A. Spanner (1986). Analysis of forest structure using Thematic Mapper simulator data. *IEEE Trans. Geosci. Remote Sens.* **GE-24**(1):113–121.

Peterson, D. L., M. A. Spanner, S. W. Running, and K. B. Teuber (1987). Relationship of Thematic Mapper simulator data to leaf area index of temperate coniferous forests. *Remote Sens. Environ.* **22**:323–341.

Peterson, D. L., J. D. Aber, P. A. Matson, D. H. Card, N. Swanberg, C. Wessman, and M. A. Spanner (1988). Remote sensing of forest canopy and leaf biochemical contents. *Remote Sens. Environ.* **24**:85–108.

Price, C. V. (1985). Discrimination of lithologic units on the basis of botanical data and Landsat Thematic Mapper spectral data from the Ridge and Valley Province, Pennsylvania. MS Thesis, Dartmouth Collge, Hanover, New Hampshire.

Raj, Des (1968). *Sampling Theory*. McGraw-Hill, New York, pp. 50, 100.

Ranson, K. J., and C. S. T. Daughtry (1987). Scene shadow effects on multispectral response. *IEEE Trans. Geosci. Remote Sens.* **GE-25**(4):502–509.

Ranson, K. J., C. S. T. Daughtry, and L. L. Biehl (1986). Sun angle, view angle, and background effects on spectral response of simulated balsam fir canopies. *Photogramm. Eng. Remote Sens.* **52**(5):649–658.

Ripple, W. J. (1985). Landsat Thematic Mapper bands for characterizing fescue grass vegetation. *Int. J. Remote Sens.* **6**(8):1373–1384.

Ripple, W. J. (1986). Spectral reflectance relationships to leaf water stress. *Photogramm. Eng. Remote Sens.* **52**(10):1669–1675.

Robinson, V. B., and A. H. Strahler (1984). Issues in designing geographic information systems under conditions of inexactness. In *Machine Processing of Remotely Sensed Data*. Purdue University, West Lafayette, Indiana, pp. 389–394.

Rock, B. N. (1982). Mapping deciduous forest cover using simulated Landsat D TM data. *Proc. Int. Geosc. Remote Sens. Symp. (IGARSS'82)*, Munich, Germany, **WP5**:3.1–3.5.

Rock, B. N., J. E. Vogelmann, D. L. Williams, A. F. Vogelmann, and T. Hoshizaki (1986). Remote detection of forest damage. *BioScience* **36**:439–445.

Rotolo, P. (1979). Near infrared reflectance instrumentation. *Cereal Foods World* **24**(3):94–98.

Running, S. W. (1984). Microclimate control of forest productivity: Analysis by computer simulation of annual photosynthesis/transpiration balance. *Agric. For. Meteror.* **32**:267–288.

Running, S. W., and J. C. Coughlan (1988). A general model of forest ecosystem processes for regional applications. I. Hydrologic balance, canopy gas exchange and primary production processes. *Ecol. Model.* **42**:125–154.

Running, S. W., and R. R. Nemani (1988). Relating the seasonal pattern of the AVHRR normalized difference vegetation index to simulated photosynthesis and transpiration of forests in different climates. *Remote Sens. Environ.* **17**:472–483.

Running, S. W., D. L. Peterson, M. A. Spanner, and K. B. Teuber (1986). Remote sensing of coniferous forest leaf area. *Ecology* **67**:273–276.

Running, S. W., R. R. Nemani, and R. D. Hungerford (1987). Extrapolation of synoptic meterological data in mountainous terrain and its use for simulating forest evapotranspiration and photosynthesis. *Can. J. For. Res.* **17**:472–483.

Running, S. W., R. R. Nemani, D. L. Peterson, L. E. Band, D. F. Potts, L. L. Pierce, and M. A. Spanner (1989). Mapping regional forest evapotranspiration and photosynthesis by coupling satellite data with ecosystem simulation. *Ecology* (in press).

Sader, S. A. (1984). *Thermal Differences in Reforested Clearcuts and Old Growth Forest on Mountainous Terrain Using Thermal Infrared Multispectral Scanner data*, Earth Resour. Lab. Rep. No. 229. National Space Technology Laboratory,

Sader, S. A., and A. T. Joyce (1985). Global tropical forest monitoring. *Proc. Int. Conf. Adv. Technol. Monit. Process. Global Environ. Inf.*, London, England.

Sadowski, F. G., W. A. Malila, and R. F. Nalepka (1978). Applications of MSS systems to natural resource inventories. *U.S., For. Serv., Res. Note RM* **RM-55**:248–256.

Sellers, P. J. (1985). Canopy reflectance, photosynthesis and transpiration. *Int. J. Remote Sens.* **6**(8):1335–1372.

Sellers, P. J. (1987). Canopy reflectance, photosynthesis and transpiration. II. The role of biophysics in the linearity of their interdependence. *Remote Sens. Environ.* **21**:143–183.

Sellers, P. J., and J. L. Dorman (1987). Testing the Simple Biosphere Model (SiB) using point micrometeorological and biophysical data. *J. Clim. Appl. Meteorol.* **26**:622–651.

Shen, S. S., G. D. Badhwar, and J. G. Carnes (1985). Separability of boreal forest species in the Lake Jennette area, Minnesota. *Photogramm. Eng. Remote Sens.* **51**(11):1775–1783.

Shugart, H. H., and D. C. West (1980). Forest succession models. *BioScience* **30**:308–313.

Shultis, J. K., and R. B. Myneni (1988). Radiative transfer in vegetation canopies with anisotropic scattering. *J. Quant. Spectrosc. Radiat. Transfer* **39**:115–129.

Singh, A. (1984). Discrimination of tropical forest cover types using Landsat MSS data. In *Machine Processing of Remotely Sensed Data*. Purdue University, West Lafayette, Indiana, pp. 395–404.

Smith, J. A. (1983). Matter-energy interactions in the optical region. In *Manual of Remote Sensing* (R. N. Colwell, Ed.). American Society of Photogrammetry, Falls Church, Virginia, pp. 62–114.

Spanner, M. A., J. A. Brass, and D. L. Peterson (1983). Use of a preexisting digital data base for soil erosion prediction. *Proc. CERMA Energy Resour. Manag. Conf.*, San Francisco, California.

Spanner, M. A., D. L. Peterson, M. J. Hall, R. C. Wrigley, D. H. Card, and S. W. Running (1984a). Atmospheric effects on the remote sensing estimation of forest leaf area index. *Proc. 18th Int. Symp. Remote Sens. Environ.*, Paris, France, pp. 1295–1308.

Spanner, M.A., J. A. Brass, and D. L. Peterson (1984b). Feature selection and the information content of Thematic Mapper simulator data for forest structural assessment. *IEEE Trans. Geosci. Remote Sens.* **GE-22**(6):482–489.

Spanner, M. A., D. L. Peterson, S. W. Running, and L. Pierce (1987). The relationship of AVHRR data of the leaf area index of western coniferous forests. *Proc. Space Life Sci. Symp.* NASA, Office of Space Science and Applications. Washington, D.C., pp. 358–359.

Spanner, M. A., L. Pierce, D. L. Peterson, and S. W. Running (1989). Remote sensing of temperate coniferous forest leaf area index: Influence of canopy closure, understory vegetation and background reflectance. *Int. J. Remote Sens.* (in press).

Steven, M. D. (1985). The physical and physiological interpretation of vegetation spectral signatures. *Proc. 3rd Int. Colloq. Spectral Signatures Objects Remote Sens.*, Les Arcs, France, ESA SP-247, pp. 205–208.

Strahler, A. H. (1980). The use of prior probabilities in maximum likelihood classification of remotely sensed data. *Remote Sens. Environ.* **10**:135–163.

Strahler, A. H. (1981). Stratification of natural vegetation for forest and rangeland and inventory using Landsat digital imagery and collateral data. *Int. J. Remote Sens.* **2**(1):15–41.

Strahler, A. H., and X. Li (1981). An invertible coniferous forest canopy reflectance model. *Proc. 15th Int. Symp. Remote Sens. Environ.*, University of Michigan, Ann Arbor, p. 1151.

Strahler, A. H., T. L. Logan, and C. E. Woodcock (1979). Forest classification and inventory system using Landsat, digital terrain and ground sample data. *Proc. 13th Int. Symp. Remote Sens. Environ.*, University of Michigan, Ann Arbor, p. 1541–1557.

Strahler, A. H., J. E. Estes, P. F. Maynard, F. C. Mertz, and D. A. Stow (1980). Incorporating collateral data in Landsat classification and modeling procedures. *Proc. 14th Int. Symp. Remote Sens. Environ.*, San Jose, Costa Rica, pp. 1009–1026.

Strahler, A. H., J. Franklin, C. E. Woodcock, and T. L. Logan (1981). *FOCIS: A Forest Classification and Inventory System Using Landsat and Digital Terrain Data*, Rep. No. NFAP-255. Nationwide Forestry Applications Program, USDA/Forest Service, Salt Lake City, Utah.

Suits, G. H. (1972). The cause of azimuthal variations in directional reflectance of vegetative canopies. *Remote Sens. Environ.* **2**:175–182.

Swain, P. H., C. L. Wu, D. A. Landgrebe, and H. Hauska (1975). Layered classification techniques for remote sensing applications. *Proc. 1st NASA Earth Resour. Surv. Symp.*, Houston, Texas, Vol. I-A, pp. 1087–1097.

Swanberg, N. A., and P. A. Matson (1987). The use of Airborne Imaging Spectrometer data to determine experimentally induced variation in canopy chemistry. *JPL Publ.* **87-30**:70–74.

Swanberg, N. A., and D. L. Peterson (1987). Using the Airborne Imaging Spectrometer to determine nitrogen content in coniferous forest canopies. *Proc. Int. Geosci. Remote Sens. Symp. (IGARSS'87)*, Ann Arbor, Michigan, p. 981.

Tsay, M.-L., D. H. Gjerstad, and G. R. Glover (1982). Tree leaf reflectance: A promising technique to rapidly determine nitrogen and chlorophyll content. *Can. J. For. Res.* **12**:788–792.

Tucker, C. J. (1977). Asymptotic nature of grass canopy reflectance. *Appl. Opt.* **16**(5):1151–1157.

Tucker, C. J., and M. W. Garratt (1977). Leaf optical system modeled as a stochastic process. *Appl. Opt.* **16**(3):635–642.

Tucker, C. J., B. N. Holben, and T. E. Goff (1984). Intensive forest clearing in Rondonia, Brazil, as detected by satellite remote sensing. *Remote Sens. Environ.* **15**:255–261.

Tucker, C. J., J. R. G. Townshend, and T. E. Goff (1985). African landcover classification using satellite data. *Science* **277**:369–375.

Vanderbilt, V. C., L. Grant, L. L. Biehl, and B. F. Robinson (1985). Specular, diffuse and polarized light scattered by two wheat canopies. *Appl. Opt.* **24**(5):2408–2418.

Vane, G., A. F. H. Goetz, and J. B. Wellman (1984). Airborne Imaging Spectrometer: A new tool for remote sensing. *IEEE Trans. Geosci. Remote Sens.* **GE-22**(6):546–549.

Vogelmann, J. E., and B. N. Rock (1986). Assessing forest decline in coniferous forests of Vermont using NS-001 Thematic Mapper simulator data. *Int. J. Remote Sens.* **7**(10):1303–1321.

Wessman, C. A. (1987). An evaluation of imaging spectroscopy for estimating forest canopy chemistry. Ph.D. Dissertation, University of Wisconsin, Madison, pp. 54–84.

Wessman, C. A., J. D. Aber, and D. L. Peterson (1987a). Estimating key forest parameters through remote sensing. *Proc. Int. Geosci. Remote Sens. Symp. (IGARSS'87)*, Ann Arbor, Michigan, pp. 1189–1194.

Wessman, C. A., J. D. Aber, D. L. Peterson, and J. M. Melillo (1987b). Foliar analysis using near infrared reflectance spectroscopy. *Can. J. For. Res.* **18**:6–11.

Wessman, C. A., J. D. Aber, D. L. Peterson, and J. M. Melillo (1988). Remote sensing of canopy chemistry and nitrogen cycling in temperate forest ecosystems. *Nature,* **335**:154–156.

Westman, W. E. (1987). Aboveground biomass, surface area, and production relations of red fir (*Abies magnifica*) and white fir (*A. concolor*). *Can. J. For. Res.* **17**:311–319.

Westman, W. E., and C. V. Price (1987). Remote detection of air pollution stress to vegetation: Laboratory-level studies. *Proc. Int. Geosci. Remote Sens. Symp. (IGARSS'87)*, Ann Arbor, Michigan, pp. 451–456.

Willaims, D. L., and R. F. Nelson (1986). Use of remotely sensed data for assessing forest stand conditions in the eastern United States. *IEEE Trans. Geosci. Remote Sens.* **GE-24**(1):130–138.

Williams, D. L., J. R. Irons, B. L. Markham, R. F. Nelson, D. L. Toll, R. S. Latty, and M. L. Stauffer (1984). A statistical evaluation of the advantages of Landsat Thematic Mapper data in comparison of Multispectral Scanner data. *IEEE Trans. Geosci. Remote Sens.* **GE-22**(3):294–301.

Wilson, M. F., A. Henderson-Sellers, R. E. Dickinson, and P. J. Kennedy (1987). Sensitivity of the Biosphere–Atmosphere Transfer Scheme (BATS) to the inclusion of variable soil characteristics. *J. Clim. Appl. Meteorol.* **26**:341–362.

Woodcock, C. E., and A. H. Strahler (1987). The factor of scale in remote sensing. *Remote Sens. Environ.* **21**:311–332.

Woodwell, G. M., J. E. Hobbie, R. A. Houghton, J. M. Melillo, B. Moore, A. Park, B. J. Peterson, and G. R. Shaver (1984). Measurement of changes in the vegetation of the earth by satellite imagery. In *The Role of Terrestrial Vegetation in the Global Carbon Cycle: Measurement by Remote Sensing (SCOPE 23)* (G. M. Woodwell, Ed.). Wiley, New York, pp. 221–240.

Woolley, J. T. (1971). Reflectance and transmittance of light by leaves. *Plant Physiol.* **47**:656–662.

Zobrist, A. L., N. A. Bryant, and R. G. McLeod (1983). Technology for large digital mosaics of Landsat data. *Photogramm. Eng. Remote Sens.* **49**(9):1325–1335.

11

APPLICATIONS TO COASTAL WETLANDS VEGETATION

MICHAEL F. GROSS

College of Marine Studies
University of Delaware
Newark, Delaware

MICHAEL A. HARDISKY

Department of Biology
University of Scranton
Scranton, Pennsylvania

VYTAUTAS KLEMAS

College of Marine Studies
University of Delaware
Newark, Delaware

I INTRODUCTION

Coastal wetlands form a highly productive buffer zone between the sea and the upland. Much of the productivity is derived from vascular plants. The inherent values of wetlands include carbon reduction (energy fixation), nursery habitat for fish and many invertebrates, nutrient assimilation, geochemical cycling, water storage, and sediment stabilization (Odum, 1983). These physical contributions of wetlands to coastal systems, as well as the less tangible values related to aesthetics, education, and recreation (Reimold and Hardisky, 1979), suggest that coastal wetlands are extremely important for the maintenance of the quality of coastal environments. Most of these wetland values are directly related to the quality and quantity of the vascular vegetation. Remote-sensing techniques are useful for scientific investigations directed toward identification, classification, and understanding functions of coastal wetlands vegetation because of the difficulties involved in gaining access to and maneuvering in wetland environments.

Coastal wetlands may be inhabited by herbaceous or woody plants or both. Most remote-sensing research of wetlands has involved those dominated by herbaceous vegetation, which will be the focus of this chapter. Coastal wetlands vegetation communities possess several characteristics that differentiate them from other herbaceous vegetation communities that are typically the subject of remote-sensing investigations. For example, in contrast to agricultural systems, wetland soils are inundated regularly by tides, the vegetation does not grow in rows, there is frequently a large dead biomass component throughout the growing season, and in saline environments, the plants may be coated

474

with salt much of the time. Because of these differences, observed interactions between electromagnetic radiation and nonwetland canopies could not necessarily be assumed to be valid for wetland canopies. Thus, separate research involving wetland vegetation was conducted.

Optical remote-sensing applications to wetlands can be broken into two broad categories: mapping and measuring biophysical parameters. Mapping wetlands vegetation involves delineation of wetlands from uplands and quantifying the distribution of various types of vegetation within the wetlands. Measuring biophysical parameters involves using spectral data to predict one or more characteristics of a vegetation type. These characteristics include height, density, percent cover, and biomass or productivity. Of these, biomass has been the most frequently studied because of its central role as an indicator of wetland value.

II MAPPING COASTAL WETLANDS VEGETATION

II.A Aerial Photography

In the 1970s, legislative activity at the state and national levels (Haueisen, 1973) mandated governmental agencies to inventory and regulate land use within coastal wetlands. The need for a rapid cost-effective method for mapping large tracts of wetlands and assessing changes in their acreage necessitated the use of remote sensing. Aerial photographic surveys were already in widespread use for evaluating ecological conditions in natural terrestrial (Colwell, 1967) and agricultural (Shay, 1967) vegetation. Aerial survey techniques were modified for coastal mapping. Photography is most useful when acquired within two hours of low tide, and coastal weather tends to be less predictable than at inland sites (Thompson et al., 1973). Color infrared photography is the preferred photographic medium because it tends to increase the contrast between wetland and upland areas, making boundary decisions much easier (Reimold et al., 1972; Anderson and Wobber, 1973). Determining wetland/upland boundaries is important because many states have laws restricting development of coastal wetlands. The high spatial resolution possible with photography makes it superior to other optical remote-sensing techniques for boundary delineations. Reports describing state inventory techniques include Egan and Hair (1971) for Maryland, Gallagher et al. (1972a,b) for Georgia, Anderson and Wobber (1973) for New Jersey, Klemas et al. (1974) for Delaware, and Penny and Gordon (1975) for Virginia. Albeit discrimination of wetlands from upland vegetation is often of primary importance, the differences in primary productivity and habitat value among wetland species encourage the classification of the major vegetation communities within wetlands. Although more costly and more time consuming, comparisons of multispectral photographs (Pestrong, 1969; Russell and Wobber, 1972; Howland, 1980, among others) can be very effective in improving species discrimination.

The availability of aerial photographs dating back several decades makes this remote-sensing medium particularly valuable for historical studies of natural and manmade changes to wetlands. Historical alterations in wetlands acreage and use were measured using aerial photography in Virginia (Niedzwiedz and Batie, 1984) and Connecticut (Civco et al., 1986). Neill and Deegan (1986) evaluated patterns of habitat changes that occur during the growth and decay of Mississippi River deltaic lobes by comparing different-aged lobes using habitat maps constructed from aerial photographs. Such studies are not possible using data from aircraft or satellite scanners, which were not in wide use before the 1970s.

II.B Aircraft and Satellite Sensors

With the launching of the Landsat series of satellites in the early 1970s, Multispectral Scanner (MSS) data became available. The coarse resolution of the data (57 × 79 m pixels) limited surveys using MSS data to relatively large tracts of wetlands, and the MSS data were usually supplemented with high spatial resolution aerial photography. Major advantages provided by the satellite data include (1) near-simultaneous coverage of wetlands over a large geographic area, eliminating the possible problem of variability in spectral tone on photographs caused by different photography overflight dates and different batches of film; and (2) digital values, which favored the development of automated techniques for classifying wetlands vegetation rapidly. Mapping and classification of marshlands using MSS data were accomplished in South Carolina and Georgia coastal marshes (Anderson et al., 1973), in Virginia coastal marshes (Carter and Schubert, 1974), in Delaware coastal marshes (Klemas et al., 1975), in Louisiana marshes (Butera, 1978), and in the Columbia River wetlands of Oregon (Lyon, 1979). A combination of MSS data and aerial photography has also been used to develop vegetation maps for forested wetlands such as the Great Dismal Swamp (Garrett and Carter, 1977; Carter et al., 1977). Although few investigators have tried to map coastal mangrove systems, Butera (1983) used aircraft and Landsat MSS data to classify large wetland tracts, achieving accuracy up to 87% for mangrove areas.

In recent years, better spatial, spectral, and radiometric resolutions have been provided by the Landsat Thematic Mapper (TM) and SPOT satellites. These improvements make more accurate and more detailed mapping of wetlands possible. Successful discrimination of vegetation and substrate types in the Wash estuary, England, was accomplished using TM simulator data (Hobbs and Shennan, 1986) and Landsat TM imagery (Donoghue and Shennan, 1987). Levasseur et al. (1987) found SPOT satellite multispectral data to be valuable for vegetation and soil mapping in the marshes of the Bay of Mont St. Michel, France. A multisensor approach involving aircraft-borne multispectral scanners similar to the Landsat MSS, aerial photography, and Landsat MSS and TM data, was used for mapping wetlands in the Savannah River swamp ecosystem (Jensen et al., 1986). Aerial photography provided information on wetland change, and airborne scanners were used to evaluate the type and spatial distribution of wetland vegetation and to measure surface temperatures of thermal effluent. Landsat MSS and TM imagery were employed for larger spatial coverage to map the wetlands of the entire watershed.

The major advantages of Landsat TM over Landsat MSS for mapping include the TM's greater spatial resolution (pixels approximately 900 m^2 in size versus about 4800 m^2 for MSS) and narrower spectral bands. SPOT offers an even better resolution than TM (400 m^2 multispectral pixels, 100 m^2 pixels for panchromatic data). The improved spatial resolution is important in coastal wetlands, where patches of vegetation may be small, and where ponds and tidal creeks are numerous and often not very large. However, the availability of only three spectral bands for SPOT multispectral data, versus four for MSS and six nonthermal bands for TM, suggests that classification of vegetation type may be less accurate using SPOT data. The best approach to mapping studies may be a multisensor one, where various sensors are used to complement each other.

Although TM, MSS, and SPOT bands are wide, they do not include the entire electromagnetic spectrum. A major recent advance in remote sensing, known as imaging spectrometry, makes it possible to collect images in many narrow contiguous spectral bands spanning the visible, near-infrared, and middle-infrared regions (Goetz et al., 1985). As a result, a complete reflectance spectrum can be derived for each pixel in the image, and imaging spectrometry can reveal reflectance features unique to a particular vegetation type, soil, or environmental condition. Such features could either be hidden within a broad spectral band or, by residing outside of the wavelengths sensed, be missed entirely by MSS, TM, or SPOT sensors.

An Airborne Imaging Spectrometer (AIS) was developed by the Jet Propulsion Laboratory in Pasadena, California. The instrument gathers data in up to 160 separate bands between 0.8 and 2.4 μm. The instrument was operational from 1984 through 1986. Although problems with data quality and calibration limited the usefulness of the AIS data, there is evidence that AIS data may be helpful in discriminating wetland vegetation types and soils on the East (Gross and Klemas, 1986) and West coasts (Wood and Beck, 1986; Gross et al., 1987a). A more advanced imaging spectrometer, AVIRIS (Airborne Visible-Infrared Imaging Spectrometer), was first operational in 1987. It acquires data in 224 bands between 0.4 and 2.4 μm. Better calibration and the availability of visible waveband data should make AVIRIS a more valuable tool for wetlands mapping, once new software is developed to process the increased number of bands and volume of data.

III SPECTRAL ESTIMATION OF MACROPHYTIC BIOMASS AND PRODUCTIVITY

III.A Handheld Radiometer Studies

Early reports describing the interaction of leaf tissue with light (performed on nonwetland species) indicated that changes in the spectral quality of reflected radiation were directly related to the quantity of leaf tissue and pigment concentrations (Allen and Richardson, 1968; Colwell, 1974; Gausman, 1974). The most significant of the spectral changes was a decrease in red radiation, resulting from strong absorption by the chlorophylls, and an increase in near-infrared (NIR) radiation resulting from intra- and interleaf scattering.

Based on this information, measurements of wetland-canopy reflectance were made using handheld radiometers. The majority of this work has been done in salt marshes. Figure 1 compares the major spectral bands (and transformations thereof) as they relate to green biomass of *Spartina alterniflora*, a common North-American salt marsh grass. Red radiance is negatively and NIR radiance is positively related to changes in green biomass. Transforming the raw data helps smooth high-frequency noise and yields variables strongly correlated to changes in green biomass [Figures 1(c) and (d); see also Tucker, 1979].

Several investigators have applied these spectral relationships to studies of wetland plant communities. Gramineous marsh plants, such as *Spartina patens*, *Distichlis spicata*, and *S. alterniflora*, show a strong correlation between NIR/red radiance, or reflectance ratios, and green biomass in Delaware salt marshes (Bartlett, 1976; Bartlett and Klemas, 1981). Drake (1976) found good correlations between green biomass and the inverse of red reflectance for grass, sedge, and shrub communities typical of brackish marshes along the East coast of the United States. Using a ratio of NIR radiance from

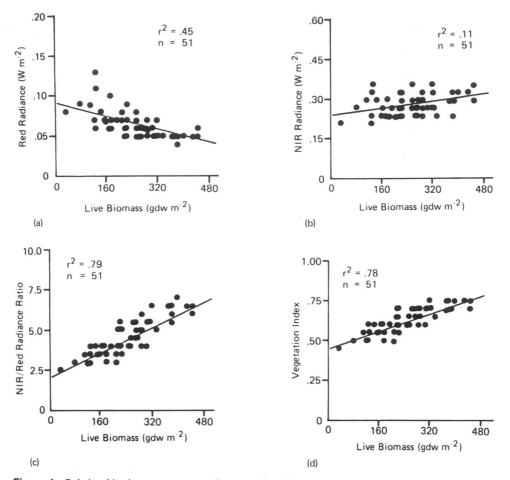

Figure 1 Relationships between canopy radiance and live biomass for *Spartina alterniflora*. (*a*) Red radiance (TM band 3) is negatively correlated with green biomass, and (*b*) near-infrared radiance (TM band 4) is weakly positively correlated with live biomass. (*c*) The ratio TM 4/TM 3 and (*d*) the vegetation index (TM 4 − TM 3)/(TM 4 + TM 3) are highly positively correlated with green biomass. Least-squares regression lines are shown. (From M. A. Hardisky, M. F. Gross, and V. Klemas, "Remote Sensing of Coastal Wetlands," *BioScience* **36**:453–460. © 1986 American Institute of Biological Sciences.)

a reference panel to NIR radiance from the target, Jensen (1980) showed a strong relationship ($r = 0.93$) between the NIR reflectance and biomass of photosynthetic tissue in a canopy of the European marsh shrub *Halimione portulacoides*. In an independent test of the relationship, he found a correlation coefficient of 0.98 for biomass estimated from harvests and predicted from NIR reflectance data. Hardisky et al. (1983c) used a normalized difference TM vegetation index [(NIR − red)/(NIR + red); hereafter referred to as vegetation index] and found very strong correlations between the spectral information and live leaf biomass of various *Spartina alterniflora* swards in a Delaware salt marsh. Gross et al. (1986, 1988), who also used a vegetation index, were able to accurately predict live aerial biomass of *S. anglica* and other French salt marsh species based on spectral data.

If an equation relating spectral reflectance to biomass can be used to predict biomass

TABLE 1   A Comparison of Remotely Sensed and Harvest-Estimated Annual Net Aerial Primary Productivity for *Spartina alterniflora* in Canary Creek Marsh, Delaware[a]

Method	All Height Forms		Short Form		Tall Form	
	Harvest[b]	VI[c]	Harvest	VI	Harvest	VI
Peak Standing Crop	540	600	372	411	826	913
Milner and Hughes (1968)	540	600	372	420	826	913
Morgan (1961)	540	600	405	411	826	913
Smalley (1958)	601	603	505	441	980	974

[a]Units on production values are gdw m^{-2} yr^{-1}.
[b]Harvest denotes NAPP computed from harvested data.
[c]VI denotes NAPP computed from vegetation-index data.

Source: Modified from Hardisky et al., 1984.

(From M. A. Hardisky, M. F. Gross, and V. Klemas, "Remote Sensing of Coastal Wetlands," *BioScience* **36:**453–460. © 1986 American Institute of Biological Sciences.)

remotely, then repetitive spectral measurements spanning the growing season should yield an estimate of annual net aerial primary productivity (NAPP). Hardisky et al. (1984) used this type of predictive model with handheld radiometric data to estimate seasonal biomass changes in a Delaware salt marsh. The spectral estimate was made concomitantly with harvesting of the same areas to test the relationship between spectrally estimated and harvested biomass and productivity for an entire growing season. In general, spectral NAPP estimates were within 10% of harvest estimates (Table 1). This study suggests that similar estimates could be made using airborne or satellite-based multispectral scanners.

Salt marshes are often characterized by large monospecific stands of vegetation. In contrast, the physiognomy of brackish marshes is usually more varied because a particular plant community often comprises many species. Different plant morphologies thus coalesce to produce canopy architectures that reflect incident radiation differently from a monospecific canopy. The differential reflectance results primarily from two factors: percentage of leaf surfaces in the horizontal plane versus the vertical plane, and how much light penetrates into the canopy. Based on general reflectance characteristics, three distinct canopy architectures can be identified: broadleaf (most reflective surfaces, leaves, in the horizontal plane), gramineous (a more random distribution of reflective surfaces between horizontal and vertical planes), and leafless (most reflective surfaces in the vertical plane).

The rationale for lumping plant communities into three canopy types is that plants with similar morphology produce a similar reflecting or absorbing surface for downwelling, or incident, radiation. Figure 2 illustrates the effects of the three canopy types on observed vegetation index (Hardisky and Klemas, 1985). Since the quality of reflected radiation (expressed as a vegetation index) differs for each canopy architecture, accurate biomass predictions must rely on separate empirically derived models describing each type.

Studies by Hardisky (1984) suggested that biomass could indeed be predicted for communities of one canopy type using a single model. For example, a model developed for gramineous canopies successfully predicted biomass in communities dominated by *Spartina alterniflora, S. cynosuroides, S. patens,* and *Typha angustifolia.* Ground-based

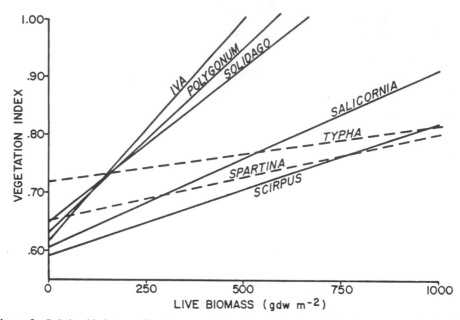

Figure 2 Relationship between live biomass and vegetation index for wetland plants representing broadleaf (*Iva, Polygonum, Solidago*), gramineous (*Spartina, Typha*), and leafless (*Salicornia, Scirpus*) canopies. (From M. A. Hardisky, M. F. Gross, and V. Klemas, "Remote Sensing of Coastal Wetlands," *BioScience* **36**:453–460. © 1986 American Institute of Biological Sciences.)

spectral measurements used in existing models generally resulted in biomass predictions differing by 0–20% from biomass means estimated by harvesting.

Remote-sensing studies for estimation of biomass in the coastal mangrove systems are limited. A positive relationship ($r = 0.79$) between the TM vegetation index and live leaf biomass for the black mangrove, *Avicennia germinans*, has been described by Hardisky (unpublished data) for a dwarf mangrove community near Puntarenas, Costa Rica. The more ubiquitous taller mangrove forms will require extensive ground comparisons before an operational biomass-estimation procedure can be developed.

III.B Factors Influencing Spectral Estimates

The two most common environmental factors influencing target radiance are the quantity and orientation of dead biomass and the amount of soil reflectance in comparison with vegetation reflectance in a given target area. The presence of dead material tends to decrease the observed vegetation index; its position and orientation within a marsh canopy, however, are controlled by decomposition and physical factors such as tidal activity and wind. This means that the same quantity of dead biomass can cause different changes in the observed vegetation index and becomes difficult to factor out of biomass-estimation procedures. Except in marshes with a very sparse canopy (<30% cover), soil reflectance is not usually a problem. Richardson and Wiegand (1977) have proposed a perpendicular-vegetation index (PVI), which factors out the influence of soil reflectance. In Chapter 4, this topic is discussed in depth.

Several other variables that can affect the observed vegetation index and, therefore, predicted biomass are presented in Table 2. Except for nitrogen enrichment, these en-

TABLE 2 Factors Affecting Spectral Reflectance from Salt Marsh Canopies

Environmental Factor	Effect Upon Red Radiance	Effect Upon NIR Radiance	Effect Upon Vegetation Index[a]	Reference
Dead biomass in canopy	Increase	Little or no change	Decrease	Hardisky et al. (1984)
Dark wet-soil background	Little or no change	Decrease	Decrease	Hardisky et al. (1984)
Salt accumulation on leaves	Increase	Little or no change	Decrease	Hardisky (1984)
Decrease in leaf moisture	Increase	Little or no change	Decrease	Hardisky et al. (1983b)
Nitrogen addition	Decrease	Increase	Increase	Hardisky et al. (1983d); Hardisky (1984)
Heavy metal contamination (copper)	Increase	Little or no change	Decrease	Hardisky et al. (1983a)

[a]A positive relationship exists between vegetation-index values and quantity of green biomass. Thus, increases or decreases in vegetation index indicate similar changes in estimated biomass.
(From M. A. Hardisky, M. F. Gross, and V. Klemas, "Remote Sensing of Coastal Wetlands," *BioScience* **36**:453–460. © 1986 American Institute of Biological Sciences.)

vironmental factors tend to decrease vegetation indices, thus resulting in underestimation of green biomass from remotely sensed data. In addition to these factors, in wetlands inundated by very turbid water, silt deposition on plants can be a problem. Budd and Milton (1982) reported poor correlations between spectral data and biomass of *Spartina anglica* when the plants were coated with silt. The effect of silt on reflectance is probably similar to that of salt (Hardisky, 1984). Standing water on the marsh is another potential problem. Because turbid estuarine water absorbs visible and NIR radiation strongly, the reflectance of inundated vegetation will be markedly different from that of canopies not partially obscured by water. Consequently, spectral data used for biomass or productivity estimation should be acquired at tidal levels not high enough to cover the substrate.

Another consideration is that at high green-biomass levels, an asymptote in the relationship between vegetation index and green biomass is reached (Gausman et al., 1976), such that increases in live biomass are not detected by the vegetation index. Although this has not generally been reported to be a problem in studies of gramineous biomass in wetlands, it may have caused poor biomass predictions for dense *Halimione portulacoides* shrub canopies in French marshes (Gross et al., 1988). Sellers (1985, 1987) reported that the relationship between the vegetation index and leaf-area index (LAI, which is usually highly correlated with biomass) of winter wheat becomes nonlinear at a certain LAI, whereas the relationship between the vegetation index and absorbed photosynthetically active radiation (APAR, an indication of photosynthetic rate), or canopy stomatal resistance, remains nearly linear. Thus, for many situations, spectral reflectance data may be better indicators of processes (such as photosynthesis or transpiration) associated with the vegetation, than of vegetation state (LAI, biomass; Sellers, 1987). Research is in progress to define the relationship between spectral data and photosynthetic rates for wetland vegetation.

Solar angle—a function of latitude, date, and time of day—may also influence observed spectral reflectance. The three contributors to reflectance from a canopy are soil, shadow, and vegetation (Colwell, 1974). The reflectance of a canopy will be determined by the spectral characteristics of these components and the proportion of each component illuminated. Changes in the proportion of each component illuminated as a function of solar angle may, therefore, cause the reflectance of the canopy to vary. Nadir reflectance from broadleaf wetland canopies, which have most of their biomass oriented horizontally and contain few if any gaps for shadows or soil, varies little with solar angle (Gross, 1987). In gramineous and leafless canopies, a large proportion of the biomass is oriented vertically. Consequently, there are gaps in the canopy and the proportion of soil, shadow, live vegetation, and dead vegetation illuminated at different sun angles varies. This causes the nadir-measured reflectance from these canopy types to change substantially as a function of solar angle.

For gramineous and leafless canopies, red reflectance tends to be positively correlated with solar elevation angle (the angle of the sun above the horizon), whereas NIR reflectance is often negatively correlated with solar elevation angle. Transformation of reflectance into a simple ratio (NIR/red) or a vegetation index accentuates the dependence on solar angle. Gross (1987) found that the nadir-measured simple ratio and vegetation index were linearly and negatively correlated with solar elevation angle. An example of this dependence of the vegetation index on solar angle is shown in Figure 3, a graph of the mean vegetation index versus time of day for eight *Spartina alterniflora* plots on several different dates. The same plots were used on each date. The magnitude of the

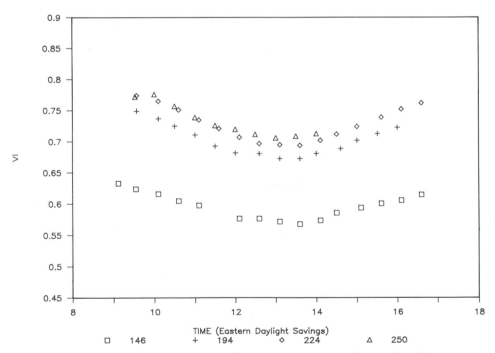

Figure 3   Mean vegetation index (VI) vs. time for eight intermediate-height *Spartina alterniflora* permanent plots. The symbols represent the mean VI from the plots at each sampling date. In the legend, the numbers beside the symbols are the Julian dates of the days sampled. *Source:* Reprinted by permission of the publisher from ''Effects of Solar Angle on Reflectance from Wetland Vegetation,'' by M. F. Gross, M. A. Hardisky, and V. Klemas, *Remote Sens. Environ.* **26:**195–212. Copyright 1988 by Elsevier Science Publishing Co., Inc.

change in reflectance per degree change in solar angle (1) may vary for the same plot in the morning as opposed to the afternoon on the same day, as well as seasonally, (2) may vary between plots of the same species, and (3) may vary between species. This variability may be a result of differences in canopy architecture and the arrangement and orientation of stems and leaves on the plants. Canopy architecture in turn may be influenced by water stress, wind, heliotropism or some other cause of directional orientation of leaves or stems, and by the quantity and distribution of dead biomass. Additionally, dew deposits on leaves and stems, which have been shown to cause diurnal variability in reflectance in wheat (Pinter, 1986), are probably also a factor in wetlands.

It might not be possible to predict the exact rate of change of reflectance per degree solar angle for a particular plot or canopy, but it is possible to approximate it and to use this information to partially compensate for differences in reflectance due to differences in solar angle at the time of measurement. This may be accomplished by converting reflectances made at various solar angles to their predicted reflectance at a common solar angle, using one or more empirically determined rates of change in reflectance per degree change in solar angle. Application of this technique to estimation of *Spartina alterniflora* biomass from spectral data demonstrated that compensation for solar angle differences does indeed result in improved estimates of biomass (Table 3; Gross, 1987).

The physical geometry of the sensor in relation to the target can also influence spectral estimates. Bartlett et al. (1986) reported an increase in observed vegetation index due to off-nadir viewing of the target. The off-nadir effect is particularly important in scanner systems operating at relatively low altitudes. In such systems, sensor view angle deviates rapidly from nadir on either side of the flight line. Fortunately, the off-nadir increase in vegetation index (or individual spectral components) can be modeled, and view angle effects can be removed, if necessary, from the data.

Wetland vegetation reflectance may also be a function of solar azimuth angle or viewing azimuth angle. Coastal wetlands are often subjected to nonrandom wind and tidal flow patterns. This may cause canopies to grow with an azimuthal orientation. This may be a particularly serious problem for canopies containing substantial dead tissue, since dead-biomass orientation is controlled by physical factors.

A latitudinal study of *Spartina alterniflora* from the Gulf of Mexico (latitude 29°N)

TABLE 3   Relationship between *Spartina alterniflora* Observed and Predicted Live Aerial Biomasses with and without Adjustment for Solar Angle[a]

Solar Angle Adjustment	O[b]	P[c]	P − O[d]
No	237	170	−67
	(92.6)[e]	(105.2)	(57.7)
Yes	237	213	−24
	(92.6)	(104.2)	(58.0)

[a] $n = 53$.
[b] O is the mean observed live aerial biomass (gdw/m^2).
[c] P is the mean predicted live aerial biomass (gdw/m^2).
[d] P − O is the mean prediction error (gdw/m^2).
[e] One standard deviation is shown in parentheses.
Source: Reprinted by permission of the publisher from "Effects of Solar Angle on Reflectance from Wetland Vegetation," by M. F. Gross, M. A. Hardisky, and V. Klemas, *Remote Sens. Environ.* **26:**195–212. Copyright 1988 by Elsevier Science Publishing Co., Inc.

to the Bay of Fundy in Canada (latitude 46°N) has shown that the relationship between reflectance and biomass is substantially different for canopies south of about 37°30'N than it is for canopies north of that latitude (Bartlett et al., 1988). Within the two latitude regimes, spectral index/biomass relationships are linear and consistent despite substantial variability in climate, soils, and hydrology. The relationship of vegetation index to horizontal LAI, however, is consistent throughout the entire range of latitude. The cause of this phenomenon is not known, but it is being investigated. One hypothesis is that there are latitudinal differences in leaf curling or morphology, with an abrupt change for some reason occurring near 37°30'N. This study demonstrates that geographically distant canopies of one species should not be assumed to be identical or to have identical reflectance for a given level of biomass, and that when correlations between reflectance and biomass are poor, one should consider relating spectral data to other parameters such as horizontal LAI.

III.C Aerial Photography and Aircraft Sensors

Studies based on handheld radiometry provide a basic understanding of the relationship between canopy characteristics and spectral reflectance, and the factors that affect this relationship. Large-area estimates of biomass require airborne platforms, due to the need for an increased instantaneous field of view and large-areal coverage in a relatively short time period. While aerial photography has primarily served mapping purposes, several studies combined aerial photography and appropriate ground measurements to quantify biomass and productivity for large wetland areas (Stroud and Cooper, 1968; Reimold et al., 1973; Curran, 1980, 1982; Dale et al., 1986). The color infrared photographs were essentially used as templates from which the areal extent of various color tones (corresponding to different vegetation communities) could be determined and the appropriate biomass and/or productivity values from ground measurements could be assigned to each color tone. The resultant combination provided an effective means of extrapolating a limited number of ground samples to a much larger marsh area. The major disadvantages of using photographs to estimate biomass are that the color tones are dependent upon the developing process and the photographs are not a digital source of data. It is therefore difficult to make temporal comparisons or obtain accurate estimates of biomass.

Hardisky (1984) recorded upwelling radiation from brackish marsh vegetation in Delaware using a handheld radiometer mounted in a low-altitude aircraft. He noted that the increased field of view provided a more representative community reflectance value because more of the natural spatial heterogeneity was integrated into the measured reflectance. Use of a vegetation index based on TM bands yielded biomass estimates within 10% of those estimated by harvesting. A vegetation index based on the MSS wavelength bands was less effective, probably because of the wider spectral bands.

D. S. Bartlett (personal communication, 1988) had an airborne multispectral scanner flown over a *Spartina alterniflora* marsh and noted close agreement between harvested biomass and spectrally estimated biomass. Butera et al. (1984) developed a remote-sensing-based model to predict wetland productive capacity, expressed as detrital export, for a Gulf Coast marsh dominated by *S. patens*. The model considers macrophytes as the primary source of carbon production, and remote-sensing determinations of wetlands vegetation are used to quantify the carbon available for export as detritus. Regression analyses of TM simulator data against vegetation parameters revealed that, for this marsh, the spectral data more successfully estimated percent total vegetative cover than biomass.

Biomass estimations based on aircraft scanner data are influenced by variability in sensor view angle on either side of the flight line. Although the off-nadir effect can be modeled (Bartlett et al., 1986), some error invariably remains, limiting the value of aircraft sensors using a scanning mechanism for biomass estimation. Aircraft sensors using the nonscanning linear array technologies like those used by the AIS or the SPOT satellite would be less subject to the off-nadir view angle problem.

III.D Satellite Sensors

A drawback of handheld and aircraft-borne sensors for estimating biomass is the small area that can be covered at one time. An advantage of using Landsat or SPOT satellite data is their provision of near-simultaneous imagery of a large geographic region. Within one scene, there are no complications caused by different solar elevation or view angles. The data are, therefore, amenable to digital analysis without much preprocessing. Perhaps the chief disadvantage of satellite data is the problem of atmospheric interference. The distance between the target and the satellite is usually on the order of several hundred kilometers. The spatial and temporal variability of the composition of the atmosphere make correcting for atmospheric effects difficult. This problem is discussed further in Chapter 9. Another limitation of satellite imagery includes the lack of scheduling flexibility to avoid unfavorable tidal conditions or cloud cover.

Satellite imagery (and also data from aircraft sensors) usually include several types of wetlands vegetation. As noted before (Figure 2), the relationship between spectral data and biomass is species-specific. Prior to estimating biomass, the various species of vegetation must be separated so that the appropriate biomass-estimation model can be used. This is accomplished by classifying the image. Classification accuracy can be verified effectively by using a combination of ground observations and high-resolution aerial photographs.

Gross et al. (1987b) used a Landsat TM image to quantify the distribution of live aerial *Spartina alterniflora* biomass throughout 580 ha in a marsh in Delaware, based on the relationship between live aerial biomass and the vegetation index. To establish this distribution, the relationship between biomass and ground-based spectral data was formulated. Using equations developed from satellite-based radiance values and ground-based reflectance values for several large, homogeneous, and spectrally distinct targets, the satellite data were converted to their predicted reflectance at ground level to avoid atmospheric effects. Using a maximum likelihood algorithm, the image was classified into isolated pixels dominated by *S. alterniflora*. Satellite-derived biomass estimates for the *S. alterniflora* parts of the marsh were obtained by substituting the predicted ground-level reflectance of the satellite data into the equation relating biomass to ground-measured reflectance. Verification of the validity of this procedure revealed that biomass estimates computed from satellite-based data were within 13 % of those derived from ground-based data. This study demonstrates that satellite data can be used to produce accurate high spatial resolution estimates of *S. alterniflora* live aerial biomass. Similar analysis of several images from one growing season could result in remote-sensing estimates of net aerial primary productivity for wetlands.

SPOT imagery could also be used for this purpose. The SPOT sensors can be aimed (look angle can be altered), providing more frequent temporal coverage of a site. For example, at midlatitudes, SPOT imagery can be acquired approximately ten times every 26 days, whereas a Landsat satellite passes over a site only once every 16 days. This is an important consideration in many coastal areas, where cloud cover limits the frequency of the collection of useful data. However, as mentioned previously, when vegetation is viewed off-nadir, it has a different vegetation index than when viewed near-nadir. This introduces a potential source of variability into SPOT data acquired at angles substantially different from nadir.

The spatial resolution of Landsat MSS and AVHRR (Advanced Very High Resolution Radiometer) satellite data is too poor for applications requiring high spatial resolution estimates of biomass. However, the AVHRR satellite offers daily coverage of every point on Earth, and the probability of obtaining a clear weather overpass over a particular site is much better. The spatial resolution of AVHRR data is on the order of 1×1 km or 4×4 km. A 1×1 km pixel is about 1000 times larger than Landsat TM pixels; consequently, the utility of AVHRR satellite data for estimation of biomass at high spatial resolution is limited. The AVHRR also has a wide scan angle, so a correction for off-nadir viewing must be made.

IV FUTURE RESEARCH AREAS

Soils of coastal wetlands are often saturated with water and are anoxic. Under anaerobic conditions, biogenic gases such as methane and hydrogen sulphide are produced and may be released into the atmosphere (Bartlett, 1984). The roots of macrophytes may influence gas production and release by serving as a source of carbon (energy) for microbes, and by serving as a conduit for the transfer of oxygen downward into the soil and of other gases upward into the atmosphere. Knowledge of the quantity of root material in wetland soils would facilitate estimates of gas flux from wetlands. Optical remote sensing cannot provide estimates of root biomass directly because short wavelengths of electromagnetic radiation do not penetrate soil. If a relationship between aboveground and belowground biomass exists, it would be possible to use optical remote sensing to estimate aboveground biomass, and then estimate belowground biomass from aboveground biomass. Refinements of remote-sensing-based models for estimating aboveground biomass, and development of models relating aboveground and belowground biomass, are necessary for accurate predictions of belowground biomass from spectral data.

The location of coastal wetlands at the interface between land and sea makes them susceptible to pollution and to changes in salinity, which may act as stresses on the vegetation. Increases in relative sea level and in population pressure along coastal areas are likely to result in increased salinity and pollution. If such stresses cause changes in reflectance (by affecting plant internal composition and physiology), it may be possible to monitor environmental conditions remotely by examining spectra from vegetation. Research is needed to determine the extent to which wetlands vegetation spectra are affected by environmental factors. The excellent spectral resolution offered by imaging spectrometers makes them ideal instruments for evaluating the spectral characteristics of plants.

The less saline coastal wetlands are usually composed of a number of species growing together. In these areas, accurate classification of vegetation type and estimation of productivity using remote sensing is complicated by the difficulty in identifying the species comprising the canopies. Species discrimination may be possible if species' spectra contain unique features. Photography and the MSS, TM, SPOT, and AVHRR sensors do not provide the high spectral resolution needed to detect such differences, but imaging spectrometry does. Additional research involving imaging spectrometers is needed to determine their utility for species discrimination.

V SUMMARY

Mapping of coastal wetlands vegetation began with the use of aerial photography. Currently, aerial photography is being supplemented by airborne and spaceborne sensor systems to improve mapping capabilities. Studies based on handheld radiometry have shown that numerous factors such as dead biomass, soil, salt or silt accumulation on plants, changes in leaf-moisture content, concentration of soil nutrients or toxins, solar zenith and azimuth angles, and viewing zenith and azimuth angles can affect spectral characteristics of wetland vegetation canopies. Despite the complications introduced by these factors, accurate estimates of biomass have been made using ground-based, aircraft-based, and satellite-based sensor systems. It is expected that the increased availability of data from relatively new sensors with improved spectral and spatial resolution will result in further advances in optical remote sensing of coastal wetlands vegetation.

REFERENCES

Allen, W. A., and A. J. Richardson (1968). Interaction of light with a plant canopy. *J. Opt. Soc. Am.* **58:**1023–1028.

Anderson, R. R., and F. J. Wobber (1973). Wetlands mapping in New Jersey. *Photogramm. Eng. Remote Sens.* **39:**353–358.

Anderson, R. R., V. Carter, and J. McGinness (1973). Mapping southern Atlantic coastal wetlands, South Carolina—Georgia, using ERTS-1 imagery. In *Remote Sensing of Earth Resources* (F. Shahrokhi, Ed.). University of Tennessee, Tullahoma, pp. 1021–1028.

Bartlett, D. S. (1976). Variability of wetland reflectance and its effect on automatic categorization of satellite imagery. M.S. Thesis, University of Delaware, Newark.

Bartlett, D. S. (1984). Global biology research program; biogeochemical processes in wetlands. *NASA Conf. Publ.* **NASA CP-2316:**1–39.

Bartlett, D. S., and V. Klemas (1981). In situ spectral reflectance studies of tidal wetland grasses. *Photogramm. Eng. Remote Sens.* **47:**1695–1703.

Bartlett, D. S., R. W. Johnson, M. A. Hardisky, and V. Klemas (1986). Assessing impacts of off-nadir observation on remote sensing of vegetation: Use of the Suits model. *Int. J. Remote Sens.* **7:**247–264.

Bartlett, D. S., M. A. Hardisky, R. W. Johnson, M. F. Gross, V. Klemas, and J. M. Hartman (1988). Continental-scale variability in vegetation reflectance and its relationship to canopy morphology. *Int. J. Remote Sens.* **9:**1223–1241.

Budd, J. T. C., and E. J. Milton (1982). Remote sensing of salt marsh vegetation in the first four proposed Thematic Mapper bands. *Int. J. Remote Sens.* **3:**147–161.

Butera, M. K. (1978). A determination of the optimum time of year for remotely classifying marsh vegetation from Landsat multispectral scanner data. *NASA Tech. Memo.* **NASA TM-58212:**1–14.

Butera, M. K. (1983). Remote sensing of wetlands. *IEEE Trans. Geosci. Remote Sens.* **GE-21:**383–392.

Butera, M. K., J. A. Browder, and A. L. Frick (1984). A preliminary report on the assessment of wetland productive capacity from a remote-sensing-based model—a NASA/NMFS joint research project. *IEEE Trans. Geosci. Remote Sens.* **GE-22:**502–511.

Carter, V. P., and J. Schubert (1974). Coastal wetlands analysis from ERTS MSS digital data and field spectral measurements. *Proc. 9th Int. Symp. Remote Sens. Environ.*, University of Michigan, Ann Arbor, pp. 1241–1260.

Carter, V. P., M. K. Garrett, L. Shima, and P. Gammon (1977). The Great Dismal Swamp: Management of a hydrologic resource with the aid of remote sensing. *Water Resour. Bull.* **13:**1–12.

Civco, D. L., W. C. Kennard, and M. W. Lefor (1986). Changes in Connecticut salt-marsh vegetation as revealed by historical aerial photographs and computer-assisted cartographics. *Environ. Manage.* **10:**229–239.

Colwell, J. E. (1974). Grass canopy bidirectional spectral reflectance. *Proc. 9th Int. Symp. Remote Sens. Environ.*, University of Michigan, Ann Arbor, pp. 1061–1085.

Colwell, R. N. (1967). Remote sensing as a means of determining ecological conditions. *BioScience* **17:**444–449.

Curran, P. J. (1980). Multispectral photographic remote sensing of vegetation amount and productivity. *Proc. 9th Int. Symp. Remote Sens. Environ.*, University of Michigan, Ann Arbor, pp. 623–637.

Curran, P. J. (1982). Multispectral photographic remote sensing of green vegetation biomass and productivity. *Photogramm. Eng. Remote Sens.* **48:**243–250.

Dale, P. E. R., K. Hulsman, and A. L. Chandica (1986). Seasonal consistency of salt marsh vegetation classes classified from large-scale color infrared aerial photographs. *Photogramm. Eng. Remote Sens.* **52:**243–250.

Donoghue, D. N. M., and I. Shennan (1987). A preliminary assessment of Landsat TM imagery for mapping vegetation and sediment distribution in the Wash estuary. *Int. J. Remote Sens.* **8:**1101–1108.

Drake, B. G. (1976). Seasonal changes in reflectance and standing crop biomass in three salt marsh communities. *Plant Physiol.* **58:**696–699.

Egan, W. G., and M. E. Hair (1971). Automated delineation of wetlands in photographic remote sensing. *Proc. 7th Int. Symp. Remote Sens. Environ.*, University of Michigan, Ann Arbor, pp. 2231–2251.

Gallagher, J. L., R. J. Reimold, and D. E. Thompson (1972a). Remote sensing and salt marsh productivity. *Proc. 38th Annu. Meet.*, *Am. Soc. Photogramm.*, Washington, D.C., pp. 338–348.

Gallagher, J. L., R. J. Reimold, and D. E. Thompson (1972b). A comparison of four remote sensing media for assessing salt marsh productivity. *Proc. 8th Int. Symp. Remote Sens. Environ.*, University of Michigan, Ann Arbor, p. 1287–1295.

Garrett, M. K., and V. Carter (1977). Contribution of remote sensing to habitat evaluation and management in a highly altered ecosystem. *Trans. North Am. Wildl. Nat. Resour. Conf.* **42:**56–65.

Gausman, H. W. (1974). Leaf reflectance of near-infrared. *Photogramm. Eng. Remote Sens.* **40:**183–191.

Gausman, H. W., R. R. Rodriguez, and A. J. Richardson (1976). Infinite reflectance of dead compared with live vegetation. *Agron. J.* **68:**295–296.

Goetz, A. F. H., G. Vane, J. E. Solomon, and B. N. Rock (1985). Imaging spectrometry for earth remote sensing. *Science* **228:**1147–1153.

Gross, M. F. (1987). Remote sensing of tidal wetlands vegetation and its biomass. Ph.D. Dissertation, University of Delaware, Newark.

Gross, M. F., and V. Klemas (1986). The use of airborne imaging spectrometer (AIS) data to differentiate marsh vegetation. *Remote Sens. Environ.* **19:**97–103.

Gross, M. F., V. Klemas, and J. E. Levasseur (1986). Remote sensing of *Spartina anglica* biomass in five French salt marshes. *Int. J. Remote Sens.* **7**:657–664.

Gross, M. F., S. L. Ustin, and V. Klemas (1987a). AIS-2 spectra of California wetland vegetation. *JPL Publ.* **87-30**:83–90.

Gross, M. F., M. A. Hardisky, V. Klemas, and P. L. Wolf (1987b). Quantification of biomass of the marsh grass *Spartina alterniflora* Loisel using Landsat Thematic Mapper imagery. *Photogramm. Eng. Remote Sens.* **53**:1577–1583.

Gross, M. F., V. Klemas, and J. E. Levasseur (1988). Remote sensing of biomass of salt marsh vegetation in France. *Int. J. Remote Sens.* **9**:397–408.

Hardisky, M. A. (1984). Remote sensing of aboveground biomass and annual net aerial primary productivity in tidal wetlands. Ph.D. Dissertation, University of Delaware, Newark.

Hardisky, M. A., and V. Klemas (1985). Remote sensing of coastal wetland biomass using Thematic Mapper wavebands. In *Landsat-4 Early Results Symposium*, Vol. IV. NASA/Goddard Space Flight Center, Greenbelt, Maryland, pp. 251–269.

Hardisky, M. A., V. Klemas, and F. C. Daiber (1983a). Remote sensing salt marsh biomass and stress detection. *Adv. Space Res.* **2**:219–229.

Hardisky, M. A., V. Klemas, and R. M. Smart (1983b). The influence of soil salinity, growth form and leaf moisture on the spectral radiance of *Spartina alterniflora* canopies. *Photogramm. Eng. Remote Sens.* **49**:77–83.

Hardisky, M. A., R. M. Smart, and V. Klemas (1983c). Seasonal spectral characteristics and aboveground biomass of the tidal marsh plant, *Spartina alterniflora. Photogramm. Eng. Remote Sens.* **49**:85–92.

Hardisky, M. A., R. M. Smart, and V. Klemas (1983d). Growth response and spectral characteristics of a short *Spartina alterniflora* salt marsh irrigated with freshwater and sewage effluent. *Remote Sens. Environ.* **13**:57–67.

Hardisky, M. A., F. C. Daiber, C. T. Roman, and V. Klemas (1984). Remote sensing of biomass and annual net aerial primary productivity of a salt marsh. *Remote Sens. Environ.* **16**:91–106.

Haueisen, A. J. (1973). An examination of legislation for the protection of the wetlands of the Atlantic and Gulf coast states. *Gulf Res. Rep.* **4**:233–263.

Hobbs, A. J., and I. Shennan (1986). Remote sensing of salt marsh reclamation in the Wash, England. *J. Coastal Res.* **2**:181–198.

Howland, W. G. (1980). Multispectral aerial photography for wetland vegetation mapping. *Photogramm. Eng. Remote Sens.* **46**:87–99.

Jensen, A. (1980). Seasonal changes in near infrared reflectance ratio and standing crop biomass in a salt marsh dominated by *Halimione portulacoides* (L.) Aellen. *New Phytol.* **86**:57–67.

Jensen, J. R., M. E. Hodgson, E. Christensen, H. E. Mackey, Jr., L. R. Tinney, and R. Sharitz (1986). Remote sensing inland wetlands: A multispectral approach. *Photogramm. Eng. Remote Sens.* **52**:87–100.

Klemas, V., F. C. Daiber, D. S. Bartlett, O. Crichton, and A. O. Fornes (1974). Inventory of Delaware's wetlands. *Photogramm. Eng. Remote Sens.* **40**:433–439.

Klemas, V., F. C. Daiber, D. S. Bartlett, and R. H. Rogers (1975). Coastal zone classification from satellite imagery. *Photogramm. Eng. Remote Sens.* **41**:499–513.

Levasseur, J. E., J. LeRhun, M. F. Gross, and V. Klemas (1988). Pertinence d'une carte infographique (données SPOT), en regard de la réalité floristique et structurale d'un couvert végétal de marais sales (Baie du Mont-Saint-Michel, France). In *SPOT 1 Image Utilization, Assessment, Results*, Cepadues-Editions, Toulouse, France, pp. 1055–1062.

Lyon, J. G. (1979). *An Analysis of Vegetation Communities in the Lower Columbia River Basin*, MICHU-SG-79-311. Michigan Sea Grant, Ann Arbor.

Milner, C., and R. E. Hughes (1968). Methods for the measurement of the primary production of grasslands. IBP Handbook No. 6, Blackwell Scientific Publ., Oxford, U.K.

Morgan, M. H. (1961). Annual angiosperm production on a salt marsh. M.S. Thesis, University of Delaware, Newark.

Neill, C., and L. A. Deegan (1986). The effect of Mississippi River delta lobe development on the habitat composition and diversity of Louisiana coastal wetlands. *Am. Midl. Nat.* **116:**296–303.

Niedzwiedz, W. R., and S. S. Batie (1984). An assessment of urban development into coastal wetlands using historical aerial photography: A case study. *Environ. Manage.* **8:**205–214.

Odum, E. P. (1983). *Basic Ecology.* Saunders, Philadelphia, Pennsylvania.

Penny, M. E., and H. H. Gordon (1975). Remote sensing of wetlands in Virginia. *Proc. 10th Int. Symp. Remote Sens. Environ.*, University of Michigan, Ann Arbor, pp. 495–503.

Pestrong, R. (1969). Multiband photos for a tidal marsh. *Photogramm. Eng.* **35:**453–470.

Pinter, P. J., Jr. (1986). Effect of dew on canopy reflectance and temperature. *Remote Sens. Environ.* **19:**187–205.

Reimold, R. J., and M. A. Hardisky (1979). Nonconsumptive use values of wetlands. In *Wetland Functions and Values* (P. E. Greeson, J. R. Clark, and J. E. Clark, Eds.), American Water Resources Association, Minneapolis, Minnesota, pp. 558–564.

Reimold, R. J., J. L. Gallagher, and D. E. Thompson (1972). Coastal mapping with remote sensors. In *Proceedings of the Coastal Mapping Symposium of the American Society of Photogrammetry.* ASP, Falls Church, Virginia, pp. 99–112.

Reimold, R. J., J. L. Gallagher, and D. E. Thompson (1973). Remote sensing of tidal marsh. *Photogramm. Eng. Remote Sens.* **39:**477–488.

Richardson, A. J., and C. L. Wiegand (1977). Distinguishing vegetation from soil background information. *Photogramm. Eng. Remote Sens.* **43:**1541–1552.

Russell, O., and F. J. Wobber (1972). Aerial multiband wetlands mapping. *Photogramm. Eng.* **38:**1188–1189.

Sellers, P. J. (1985). Canopy reflectance, photosynthesis and transpiration. *Int. J. Remote Sens.* **6:**1335–1372.

Sellers, P. J. (1987). Canopy reflectance, photosynthesis and transpiration. II. The role of biophysics in the linearity of their interdependence. *Remote Sens. Environ.* **21:**143–183.

Shay, J. R. (1967). Remote sensing for agricultural purposes. *BioScience* **17:**450–451.

Smalley, A. E. (1958). The role of two invertebrate populations, *Littorina irrorata* and *Orchelimum fidicinum*, in the energy flow of a salt marsh ecosystem. Ph.D. Dissertation, University of Georgia, Athens.

Stroud, L. M., and A. W. Cooper (1968). *Color-Infrared Aerial Photographic Interpretation and Net Primary Productivity of Regularly Flooded North Carolina Salt Marsh*, No. 14. University of North Carolina, Water Resources Research Institute, Asheville.

Thompson, D. E., J. E. Ragsdale, R. J. Reimold, and J. L. Gallagher (1973). Seasonal aspects of remote sensing coastal resources. In *Remote Sensing of Earth Resources* (F. Shahrokhi, Ed.), University of Tennessee, Tullahoma, pp. 1201–1249.

Tucker, C. J. (1979). Red and photographic infrared linear combinations for monitoring vegetation. *Remote Sens. Environ.* **8:**127–150.

Wood, B. L., and L. H. Beck (1986). Trace element-induced stress in freshwater wetland vegetation: Preliminary results. *JPL Publ.* **86-35:**171–179.

12

SPECTRAL REMOTE SENSING IN GEOLOGY

ALEXANDER F. H. GOETZ

Center for the Study of Earth from Space
Cooperative Institute for Research in Environmental Sciences
University of Colorado at Boulder
Boulder, Colorado

I   INTRODUCTION

Practical remote sensing has come a long way since the first aerial photographs were taken from a balloon over 130 years ago. The desire to attain a bird's-eye view, rising above the 1.5 m elevation view of the surroundings, has spurred the development of new and better ways to look at the Earth. As aerial photography moved from balloon and pigeon platforms to aircraft and finally spacecraft, the degree of sophistication in information extraction increased concomitantly, ultimately to today's computer image-processing techniques of enhancement and pattern recognition. Except for a few pioneering studies in spectral analysis in the 1950s and 1960s, the only information derived from remote-sensing data at that time came from the spatial context. The disciplines of photogrammetry and geomorphology dominated the field of air photo interpretation until the advent of Landsat in 1972. The multispectral scanner aboard the first Landsat, then called ERTS-1, gave most geologists their first look at images taken in a spectral region beyond that seen by the eye. The multispectral perspective spawned a new generation of interpreters bound to take advantage of the new dimension of spectral remote sensing.

This chapter establishes the context and the rationale for spectral remote sensing, beginning with an in-depth review of spectral properties of geologic materials, then a discussion of modern sensors and data-analysis techniques and a selection of key applications taken from the literature. Finally, there is a discussion of future directions and expectations to lead us into the next millennium.

II   HISTORICAL PERSPECTIVE

Unbeknownst to most of them, geologists depend heavily on remote sensing in their work. Any measurement made from afar, and that includes seeing with our eyes, can be classed as remote sensing. Therefore, remote sensing is necessary to interpret morphology, grain size, texture, spatial relationships, and color. Only when we alter the surface or sample in some way in order to acquire more information, do we move beyond remote sensing. The geologist must use all the tools at his disposal appropriate to the problem at hand, which can include optical mineralogy, x-ray diffraction, x-ray fluorescence, microprobe analysis, and age dating, to name a few. Regardless of the amount of laboratory work done, the results still must be put into the proper context and, therefore, field relationships are very important. The field geologist will never be replaced by technology.

Modern field work is not undertaken without aerial photography. Since the 1930s, air photos, in particular stereo air photos, have been used to compile and interpret information for reconnaissance-level geologic mapping; rock-type discrimination; mapping of faults, fractures, and joints; geomorphic analysis of land forms; hazard assessment; and resource exploration.

The advent of images acquired from space, first from handheld cameras on the Gemini and Apollo spacecraft during the 1960s and finally consistent, worldwide coverage from Landsat beginning in 1972, provided a regional perspective heretofore unavailable. Although Landsat did not provide useful stereo coverage because of its sun-synchronous orbit designed for efficient repetitive worldwide coverage, the same interpretation techniques used with aerial photography could be applied to the more regional perspective by altering the techniques where necessary to accommodate the change in scale. Two recent texts discuss the interpretation of the spatial aspects of space-acquired images in detail (Drury, 1987; Short and Blair, 1986). Therefore, the subject will not be pursued further.

In addition to the expanded regional perspective, the Landsat Multispectral Scanner System (MSS) data provided the new spectral dimension. The development of digital image processing made it possible to exploit the radiometric data acquired in the visible and near-infrared portion of the spectrum and to derive spectral signatures of rock types heretofore not feasible with photographic techniques. The identification of iron oxides associated with hydrothermal alteration zones, using color-ratio compositing from digitally processed data, was the first major application of multispectral data in geologic mapping (Rowan et al., 1974; 1977).

The new-found capability to acquire images in spectral regions beyond the sensitivity range of the eye required intensive ground studies, laboratory studies, and a need to verify the results acquired with the Landsat images. The mid-to-late 1970s saw the development of field spectrometers and a new respect for the complexities of understanding the interaction between electromagnetic energy and the surface, particularly averaged over the 79×79 m ground instantaneous field of view (GIFOV) of the MSS (Goetz et al., 1975).

Field spectral measurements and renewed interest in laboratory spectroradiometry made it clear that restricting the wavelength coverage of imaging instruments to the region short of 1.0 μm would eliminate the 1.6–2.5 μm region that contained diagnostic vibrational overtone features important to geologic mapping. In particular, the 2.0–2.5 μm region contains the diagnostic features for minerals containing OH and CO_3 ions. A seminal study of the Cuprite mining district using airborne multispectral scanner data with bands in the 1.6 and 2.2 μm regions (Abrams et al., 1977) convinced decision makers within NASA to add the seventh band to the Thematic Mapper (TM) then under construction. The addition of this 2.08–2.36 μm band along with the 1.55–1.75 μm, band 5, made it possible to identify clay and carbonate-bearing units independent of iron content.

During the late 1970s, it became increasingly apparent that higher spectral resolution data in the 0.4–2.5 μm region, and multispectral data from the thermal infrared, would provide a vast improvement in remote mineral identification. New airborne instruments were developed, first, a profiling spectroradiometer (Chiu and Collins, 1978) and a multispectral scanner for the thermal IR region (Kahle and Goetz, 1983). These instruments led to new insights in geobotanical mapping (Collins et al., 1983) and lithologic mapping (Kahle and Goetz, 1983).

Profiling spectroradiometry was carried out in orbit with the Shuttle Multispectral Infrared Radiometer in 1981 (Goetz et al., 1982), and positive results from this experiment gave confidence that the then nascent imaging spectrometer program at NASA/JPL would provide valuable data from both airborne and spaceborne platforms. Two imaging spectrometers now exist for airborne use, the airborne imaging spectrometer (AIS) (Vane and Goetz, 1988; Goetz et al., 1985b) and the Airborne Visible and Infrared Imaging Spectrometer (AVIRIS) (Vane, 1987). These instruments are providing image data in hundreds of contiguous spectral bands simultaneously, with band widths of 10 nm or less. Concurrently, a field spectrometer was being developed with the capability to acquire data at more than twice the spectral resolution of AVIRIS in only 2 seconds. The Portable Instant Display and Analysis Spectrometer (PIDAS) (Goetz, 1987) for the first time makes it possible to measure the spectral reflectance intrapixel variance.

As the reader may have perceived, over the last 15 years, the development of remote-sensing techniques for geological applications has been directed primarily toward better identification of surface mineralogy. In order to make use of this new technology, it is necessary to have an understanding of the reflectance and emittance properties of minerals.

III SPECTRAL PROPERTIES OF MINERALS

The key to deriving information from remote-sensing data is to understand the interaction between electromagnetic energy and the surface being observed. The following discussion is derived in part from three authors (Hunt, 1980; Nassau, 1980; Burns, 1970). The reader is referred to their papers, and those additional publications referenced therein, for further detail.

The radiance measured by a sensor is related to the irradiance on a Lambert surface by the following expression:

$$L(\lambda) = \frac{R(\lambda)E(\lambda)}{\pi} \tag{1}$$

where $L(\lambda)$ is the spectral radiance measured in $W\,m^{-2}\,sr^{-1}\,\mu m^{-1}$, $R(\lambda)$ is the absolute spectral reflectance, and $E(\lambda)$ is the irradiance from the source, which in most cases is the sun, modified by transmission and scattering in the atmosphere. The most commonly used term in remote sensing is bidirectional reflectance, $\rho(\lambda)$, which is

$$\rho(\lambda) = \frac{R_s}{R_d} \tag{2}$$

where R_s is the reflectance of the sample, and R_d is the reflectance of a standard. Common standard materials are halon (Weidner and Hsia, 1981), Fiberfrax, and barium sulfate. In the visible (0.4–0.7 μm), near-infrared (0.7–1.1 μm), and short-wavelength infrared (1.1–2.5 μm), $\rho(\lambda)$ is of primary interest. However, in the thermal infrared region (8–14 μm), where over 99% of the energy received by the sensor is emitted from

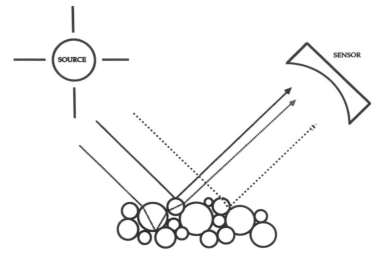

Figure 1   Paths taken by photons emitted from the source and collected by the sensor after being scattered by a solid surface.

the Earth's surface, the emittance, $\varepsilon(\lambda)$, is the parameter of interest. Kirchhoff's law states that under conditions of thermodynamic equilibrium,

$$\varepsilon(\lambda) = 1 - \rho(\lambda) \tag{3}$$

The interaction of electromagnetic radiation with materials on the macroscopic level (effects such as refraction, diffraction, and scattering) can be described by classical theory, i.e., Maxwell's equations. However, the interaction of electromagnetic energy with molecules and crystal lattices requires quantum mechanics and the particulate view of electromagnetic energy. The characteristic spectral absorption features seen in reflectance spectra of minerals are a mixture of scattering and absorption and depend not only on quantum mechanical interactions, but on particle size and the single-scatter albedo as well (Hapke, 1981, 1984, 1986). Even though light is being scattered by the surface, the absorption features created are due to transmission through the particles, as shown in Figure 1.

Two major types of interaction between photons and crystal lattices, electronic and vibrational, cause diagnostic absorption features in reflectance and emittance spectra throughout the optical region, 0.4–14 μm, discussed here. At wavelengths short of 2 μm, electronic transitions, i.e., changes in the energy state of electrons bound to atoms or molecules or lattices, create diagnostic absorption features. The electronic transitions require higher energy levels and, therefore, take place at shorter wavelengths than do vibrational transitions.

IV   ELECTRONIC TRANSITIONS

Quantum mechanics specifies that electrons in an atom must occupy specific quantized orbits, and there are four quantum numbers: a principal, an angular momentum, a magnetic, and a spin quantum number. At energies corresponding to the visible and near-infrared portion of the spectrum, the angular momentum quantum number is the most important for polyatomic compounds such as minerals. The transition rules become quite complicated and, therefore, the electronic states of polyatomic molecules are usually described in terms of their symmetry behavior, i.e., the symmetry of the electronic wave function that is compatible with the symmetry of the crystal lattice. Only a small number of symmetry groups are required to encompass hundreds of thousands of molecules, and the molecular symmetries can be described in terms of five symmetry elements: identity, plane of symmetry, center of symmetry, axis of symmetry, and rotation reflection axis (Hunt, 1980). Only a few combinations of the symmetry elements occur, and the combination of elements is called a *group*. Symmetry space groups are appropriate for crystal lattices since the entire unit cell is translated by an operation to another completely equivalent position in the crystal. Energy levels are assigned to various symmetries called *symmetry species*. Group theory allows certain states to be designated and as well provides selection rules that determine whether a transition between energy levels is allowed or forbidden. When cations such as iron are embedded in a crystal lattice, some of their electrons may end up being shared by the solid as a whole instead of with a particular atom. The energy levels in this case become more or less continuous and separated into regions called *valence* and *conduction bands* separated by a forbidden region, as shown in Figure 2. This behavior is associated with semiconductors. The spectral reflectance of semiconductors, as shown in Figure 3, exhibits very low reflectance at short wavelengths until the critical wavelength equivalent to the width of the forbidden band is reached, and the reflectance rapidly rises to a high value. The sloped edge of the reflectance curve is due to impurities and defects in the lattice and lack of order in the crystal.

Charge-transfer transitions are a special case of semiconductor behavior in which electrons do not enter into a conduction band but rather transfer from one atom to another and remain localized in the lattice. The charge transfer between iron and oxygen is the

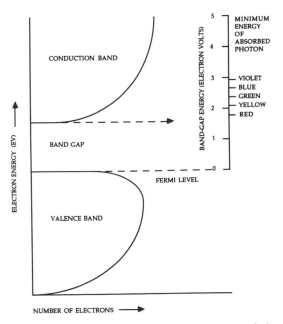

Figure 2 Semiconductor band structure. An incoming photon with energy ≥ the band-gap energy will cause an electron to be promoted into the conduction band and will cause a change in the electrical resistance of the material.

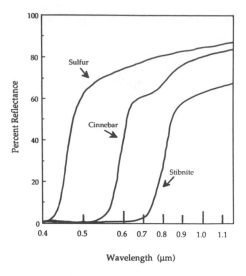

Figure 3 Spectral reflectance of three semiconducting minerals having different band-gap widths, causing the short-wavelength absorption edge to shift in wavelength.

most common and results in a strong absorption in the UV region. The wing of the band reaches into the visible portion of the spectrum, giving rise to an increasing reflectance toward longer wavelengths. The fact that almost all materials have lower reflectance in the blue portion of the spectrum than in the red attests to the fact that iron is nearly ubiquitous in minerals.

Silicon, aluminum, and oxygen, which are the major constituents of crustal rocks, do not have electronic energy levels that show features in the visible and near-infrared portions of the spectrum. However, the transition elements Fe, Cr, Ni, Ti, Co, Mn, Wo, and Sc, all have unfilled 3d shells that determine the energy levels but are under the influence of the crystal field in which they are embedded. The symmetry of the field determines the energy levels and the transitions. The excited states of these electrons have energies corresponding to the visible wavelengths and are responsible for a wide range of colors. An example is the transition element chromium in corundum (Al_2O_3). In corundum (ruby), the substitution of chromium for a few percent of the aluminum ions creates the red color. The chromium ion has three unpaired electrons, creating a complicated spectrum of excited states. The excited states form bands modified by the presence of the crystal matrix. The position of each level in the energy spectrum is determined by the electric field in which the ion is placed. The symmetry and strength of the material depends on the other ions surrounding the chromium.

In corundum, each aluminum ion is surrounded by six oxygen ions in a distorted octahedron. The crystal field of corundum is brought about because the valence electron pairs are more closely coupled with the oxygen ions than they are with the aluminum, and this gives rise to an electric field called the *crystal field* or *ligand field*. A chromium ion placed in this field has three excited states of its unpaired electrons, which have energies in the visible portion of the spectrum. Figure 4 is a schematic representation of an energy transition in ruby, showing the violet as well as green and yellow absorptions due to excitations from the ground level, $4A_2$, to the excited states, $4T_2$ and $4T_1$. The selection rules govern the energy spectrum of the reemitted light. Most of these transitions fall at lower energies at infrared wavelengths. But there is a fluorescence caused by the reemission of an energy corresponding to the difference between $2E$ and $4A_2$ states. Ruby appears red because the violet and yellow-green component have been ab-

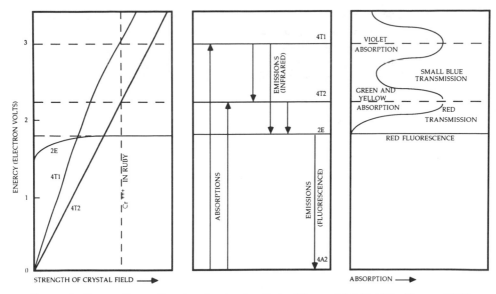

Figure 4 Energy diagram for ruby, showing the effects of the crystal field. (After Nassau, 1980.)

sorbed. Essentially, all the red is transmitted along with some blue, and this gives rise to the deep red-purple color in ruby.

The subtle differences in energy levels created by the crystal field can make a significant change in its color, as demonstrated by the change in energy states in the chromium ion when immersed in beryllium-aluminum silicate. Again, the chromium is surrounded by six oxygen ions in octahedral coordination. However, the crystal lattice has different chemical bonds and the magnitude of the electric field surrounding the chromium ion is somewhat reduced. As shown in Figure 5, emerald has a violet absorption,

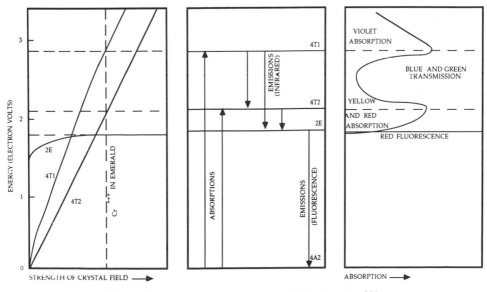

Figure 5 Energy diagram for emerald. (After Nassau, 1980.)

but the longer wave absorption has now been displaced into the yellow and red portions of the spectrum, allowing blue light and green light to be transmitted.

Crystal field colors arise when ions containing unpaired electrons are introduced into a solid. Yellow quartz, jasper, and jade, as well as garnet, owe their color to the presence of iron, whereas azurite, turquoise, and malachite obtain their blue or green color from copper.

V COLOR CENTERS

The electrons responsible for the crystal fields need not be attached to a transition metal ion. An excess electron trapped in some structural defect such as a missing ion or hole (the absence of one electron from a pair) can have similar effects. These anomalies are called *color centers* or *F centers*, where F stands for the German word for color, Farbe.

An example of a crystal lattice containing an excess electron is fluorite (CaF_2). An F center forms when a fluorine ion is missing from its usual position. It can be driven out by high-energy radiation or a strong electric field. When the fluorine atom is driven out, a negatively charged entity must take up the position in the lattice. If the charge is supplied by an electron, an F center is created. The electron is bound by the crystal field of all the surrounding ions and can occupy various energy states similar to those in transition elements. One of the energy states gives rise to the purple color seen in fluorite.

The color of smokey quartz is created by substitution of aluminium, which is trivalent, for silicon and quartz, which are quadrivalent. The aluminum itself does not create the color because there are no unpaired electrons. However, with exposure to x-rays or gamma rays, the paired electron bond of an oxygen atom adjacent to the aluminum is broken, leaving an unpaired electron in the orbital. The missing electron is called a *hole*, and the unpaired electron has a set of excited states similar to that of the excess electron in fluorite.

VI ELECTRONIC EFFECTS OF IRON

Iron is the most common constituent of minerals that creates features in the visible portion of the spectrum. Iron substitutes for magnesium and aluminum in octahedral sites. The electronic energy levels of 3d electrons are split by interaction with the surrounding ions and assume new energy values. The new values are determined by the valence state, coordination number and site symmetry, and the type of ligand formed (Fe–O, Cr–O) and the metal-ligand interatomic distances. Figure 6 shows spectra of iron-bearing minerals with Fe^{3+} in various coordinations. In summary, the positions of absorption features caused by cations are not diagnostic of the cation itself, but rather of the crystal lattice in which the ion is imbedded.

VII VIBRATIONAL PROCESSES

The vibrations of molecules in lattices result in small displacements of the atoms about their resting positions. These displacements are subject to quantum mechanical rules.

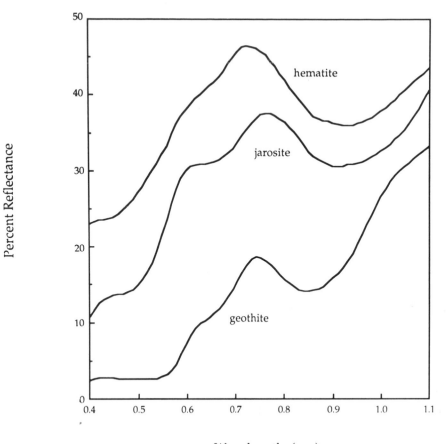

Figure 6   Reflectance spectra of ferric-iron-bearing minerals.

The energy levels in a linear harmonic oscillator are given by

$$E_v = (v_i + \tfrac{1}{2})\, h\nu_i \tag{4}$$

where v_i is an integer, h is Planck's constant, and ν_i is the oscillator frequency. There are $3N - 6$ possible degrees of freedom, where N is the number of atoms in the molecule. There are $3N - 5$ degrees of freedom for linear molecules. Absorption features in reflectance spectra arise as a result of transitions from one state to another. Fundamentals are frequencies arising from a transition from the ground state, $v_i = 0$ to $v_i = 1$, and all other $v_i = 0$. Here i represents the mode of vibration, either translation or rotation.

Overtones occur when there is a transition from the ground state to $v_i = 2$ or more, while all others remain in the ground state. Combinations occur when there is a transition from the ground state to the sum of two or more fundamental or overtone states. In the case of water vapor (see Figure 7), ν_1 is a symmetrical stretch occurring at 2.74 μm or 3651.7 cm^{-1}; ν_2, the bending mode occurring at 6.27 μm or 1595 cm^{-1}; and ν_3, the asymmetric stretch at 2.66 μm or 3755.8 cm^{-1}. Overtones and combinations of these modes make up the well-known atmospheric water-vapor absorption bands at 1.875 μm

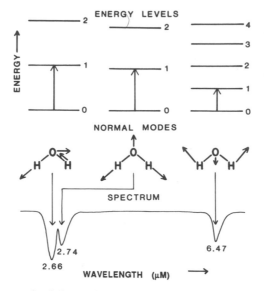

Figure 7   Vibrational-energy-level diagram for water vapor showing three fundamental vibrations along with the infrared absorption spectrum corresponding to the ν_3, ν_1, and ν_2 vibrations. (Hunt, 1980.)

($\nu_2 + \nu_3$), 1.454 μm ($2\nu_2 + \nu_3$), 1.38 μm ($\nu_1 + \nu_3$), 1.135 μm ($\nu_1 + \nu_2 + \nu_3$), and 0.942 μm ($2\nu_1 + \nu_3$).

Liquid water has features at generally longer wavelengths, with ν_1 at 3.11 μm or 3219 cm^{-1}, ν_2 at 6.08 μm or 1645 cm^{-1}, and ν_3 at 2.90 μm or 3445 μm. No rotational degrees of freedom are possible in crystal lattices and, therefore, no hyperfine structure is seen. In minerals and rocks, the presence of water is evidenced by $2\nu_3$, 1.4 μm and ($\nu_2 + \nu_3$), 1.9 μm absorption features, which together are completely diagnostic of its presence (Figure 8).

The OH$^-$ ion has one fundamental stretching vibration at or near 2.77 μm, depending on the site in the lattice. Several fundamental absorption features are possible if several different sites in the lattice are occupied. The $2\nu_1$ first overtone causes an absorption near 1.4 μm and is found in any OH-bearing mineral (Figure 9).

Features near 2.2 μm are created by Al–OH bonding coupled with fundamental OH stretching modes (Figure 10) and Mg–OH gives rise to absorptions near 2.3 μm (Figure 11). The latter can be confused with CO_3-produced features in the same region. The CO_3 feature arises from four fundamental vibrations: a C–O stretch, ν_1; an out-of-plane bend, ν_2; a doubly degenerate asymmetric stretch, ν_3; and a doubly degenerate n-plane bend, ν_4. Combination overtones produce five features in the short wavelength infrared. They occur in ascending order: ($\nu_1 + 2\nu_3$) near 2.55 μm, $3\nu_3$ near 2.35 μm, ($3\nu_1 + 2\nu_4$) near 2.16 μm, $2\nu_1 + 2\nu_3$ near 2.0 μm, and ($\nu_1 + 3\nu_3$) near 1.9 μm (Figure 12).

Sulphate (SO_4) ions have fundamental vibration frequencies at 9.06, 10.19, 16.31, and 22.17 μm. Because of their long wavelengths, no features that are a direct result of overtones or combinations appear in the short-wavelength infrared. At least a fourth overtone would be necessary and these are not in evidence. Vibrational features for sulphates are the result of OH overtones or combinations with AlO–OH as in alunite. The features in the gypsum spectrum are the result of OH and H_2O overtones and combinations, not the result of SO_4 overtones (Figure 13).

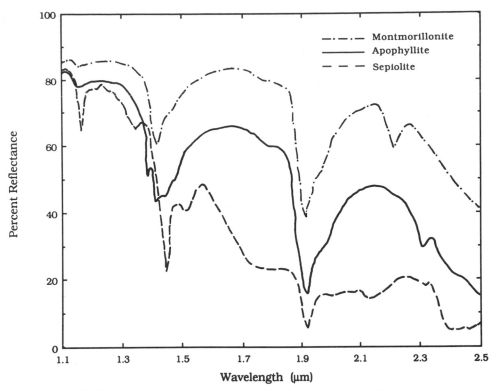

Figure 8 H₂O-bearing minerals showing diagnostic absorption features at 1.4 and 1.9 μm.

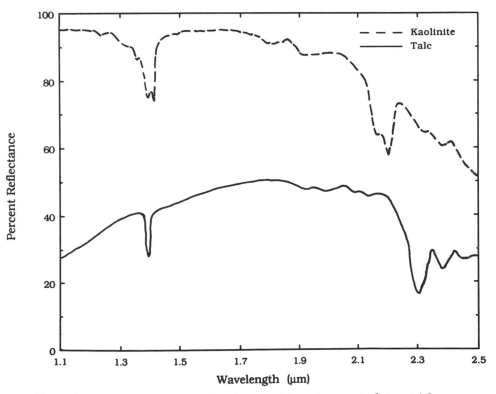

Figure 9 OH-bearing minerals showing a feature at 1.4 μm, but no water feature at 1.9 μm.

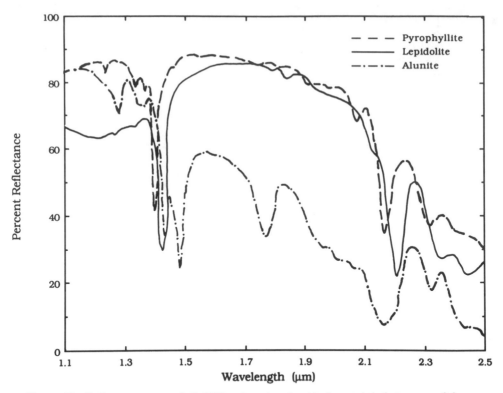

Figure 10 Reflectance spectra of Al–OH-bearing minerals with characteristic features near 2.2 μm.

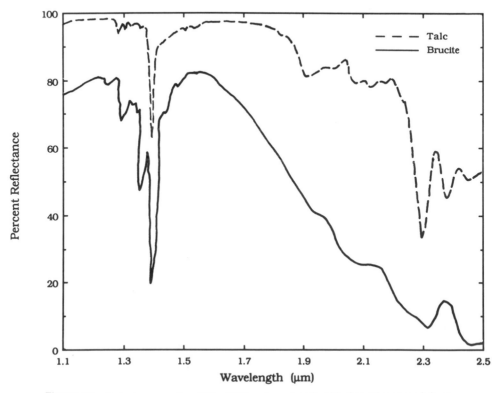

Figure 11 Reflectance spectra of Mg–OH-bearing minerals with absorptions near 2.3 μm.

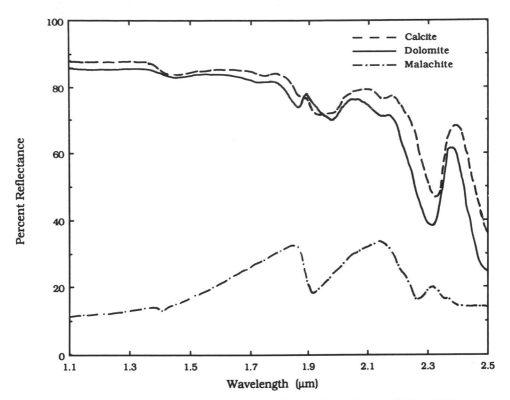

Figure 12 Reflectance spectra of carbonate minerals showing features between 2.28 and 2.34 μm.

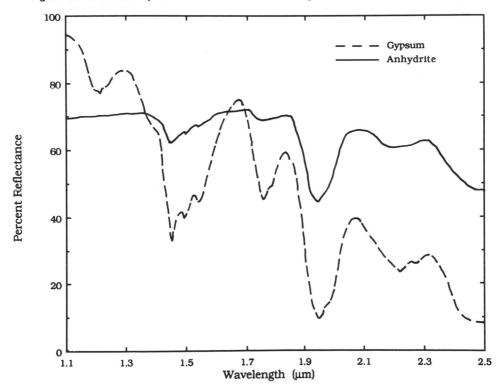

Figure 13 Reflectance spectra of sulfate minerals showing OH and H_2O overtones and combinations. SO_4 overtones are not visible.

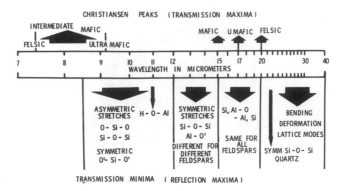

Figure 14   Location of features and the types of vibrations that produce the spectral signatures of silicates in the midinfrared region. (Hunt, 1980.)

VIII   SILICATES

Silicates, which make up the bulk of the crustal rocks, exhibit fundamental vibrational stretching modes in the 10 μm region (Figure 14). The fundamental vibrational features are conveniently located in the 8–14 μm atmospheric transmission window and are, therefore, useful for remote sensing. Remote observations here relate to the bulk composition of the material rather than the individual cations or nonsilicate anions. Reflectance and emittance measurements are complicated in this region by particle-size effects since the particles are in the range of size of the observational wavelength. Figure 15 shows the classic behavior of the index refraction n and the extinction coefficient k at and near the fundamental vibration frequency. This effect is called *anomalous dispersion* although this is not anomalous behavior. Fresnel's formula gives the reflection coefficient in terms of n and k:

$$R = \frac{(n - 1)^2 + k^2}{(n + 1)^2 + k^2} \tag{5}$$

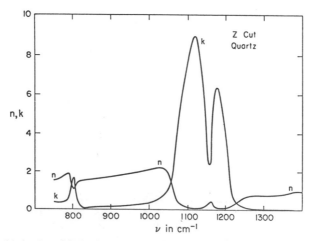

Figure 15   Spectral behavior of the real (n) and imaginary (k) indices of refraction for Z-cut quartz in the region of the fundamental Si–O stretching vibration at 1100 cm^{-1}. (From Launer, 1952.)

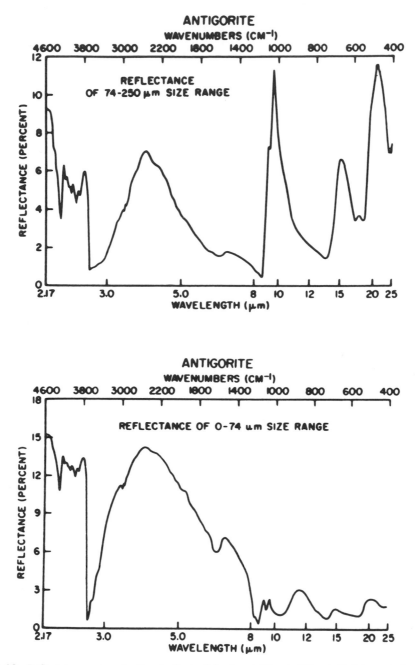

Figure 16   Reflectance spectra of antigorite for particle sizes >74 and <74 μm. (Salisbury et al., 1987.)

As n approaches 1 and k, too, is small, R approaches 0. This is the so-called *Christiansen peak*. The reflection peak at or near the fundamental vibration frequency is called the *reststrahlen* or *residual ray peak*. This peak gives rise to an emittance minimum that can be detected with multispectral sensors. There is a variation in the appearance of the reflectance and emittance peaks, depending on the sample particle size (Salisbury and

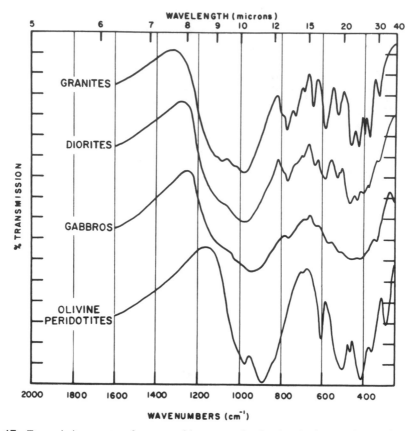

Figure 17 Transmission spectra of a range of igneous rocks showing the increase in wavelength of the absorption maximum as the rock type becomes more mafic. (Hunt, 1980.)

Hunt, 1968). The relative amounts of surface and volume scattering is determined primarily by the absorption coefficient and particle size, and the effects can be dramatic, as shown in Figure 16 (Salisbury et al., 1987). The spectral emittance curve is also complicated by thermal gradients in the surface. To some extent, transmission spectra and emittance spectra are similar, but the former is more readily measured. Figure 17 shows transmission spectra of some igneous rocks. Notice the shift of the transmission minima to longer wavelengths, moving from quartz-rich to quartz-poor rocks or as the crystal lattice moves from a completely shared oxygen structure (quartz) to an isolated SiO_4 tetrahedron structure (olivine). This shift makes possible rock identification from thermal infrared multispectral scanners.

IX SENSORS

The advances in spectral remote sensing in geology have been directly dependent on the development of new sensing systems. As discussed before, the advent of the first Landsat Multispectral Scanner System (MSS) changed the geologist's view of the Earth, on the one hand because of the synoptic view, but, perhaps more importantly, because for the first time, it was possible to observe the Earth at wavelengths not only outside the range of sensitivity of the human eye, but of infrared film as well. Computer image processing, discussed in more detail later, made it possible to extract information from the four bands of data that could not otherwise be combined directly by optical means.

It soon became apparent that, since the wavelength region covered by the MSS did not extend beyond 1.0 μm, the only major Earth surface cover materials that could be identified directly were vegetation and iron-oxide-bearing rocks and soils. Although variation in vegetation cover and the relative abundance of iron in rocks led to separation of surface materials into several categories or classes, the discrimination was not sufficient to make skeptics into true believers. The next step in sensor development was to build a multispectral scanner with spectral bands that covered the regions of the spectrum containing diagnostic vibrational features discussed in the preceding section.

The Thematic Mapper (TM), as the MSS preceding it, was designed primarily to satisfy the needs of the vegetation community. The first four bands of TM (Table 1) coincide with the chlorophyll absorption, green peak, and IR plateau features of the vegetation reflectance spectrum. The positioning of bands 3 and 4 of the TM is not the same as the last two bands of the MSS, and the detection of the 0.85–0.9 μm Fe^{3+} absorption features in minerals such as hematite and goethite (Figure 6) is often not possible. On the other hand, bands 5 and 7 in TM have proven very valuable for the discrimination of OH^-- and CO_3^--bearing minerals based on the rapidly decreasing reflectance between 1.6 and 2.5 μm. Therefore, TM data has been widely used to map areas of hydrothermal alteration in which iron minerals may not be present (Goetz et al., 1983).

The TM, like the MSS, is an optomechanical scanner. Sixteen detectors in each band are swept across an 185 km wide swath beneath the spacecraft. In order to achieve a high signal-to-noise ratio, a bidirectional scanning mirror is used. The resulting Z pattern of the scan lines is removed by a scan-line corrector in the optics (Figure 18). Data are

TABLE 1   Comparison of TM and MSS

Spectral Band		Radiometric Sensitivity		
		Thematic Mapper (TM)		Multispectral Scanner (MSS)
1	0.45–0.52 μm	0.8% (*NE* $\Delta\rho$)	0.5–0.6 μm	0.57% (*NE* $\Delta\rho$)
2	0.52–0.60	0.5%	0.6–0.7	0.57%
3	0.63–0.69	0.5%	0.7–0.8	0.65%
4	0.76–0.90	0.5%	0.8–1.1	0.70%
5	1.55–1.75	1.0%		
6	10.40–12.50	0.5K (*NE* ΔT)		
7	2.08–2.35	2.4% (*NE* $\Delta\rho$)		

Specifications	Thematic Mapper (TM)	Multispectral Scanner (MSS)
Ground IFOV	30 m (bands 1–6)	82 m (bands 1–4)
	120 m (band 7)	
Data rate	85 Mbits/s	15 Mbits/s
Quantization levels	256	64
Weight	258 kg	68 kg
Size	1.1 × 0.7 × 2.0 m	0.35 × 0.4 × 0.9 m
Power	322 W	50 W

Source: Goetz et al., 1985b.

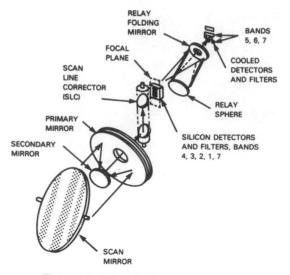

Figure 18   Thematic Mapper optical layout.

encoded to 8 bits or 256 grey levels. A detailed description of TM can be found in Slater (1980).

The Thermal Infrared Multispectral Scanner (TIMS) (Kahle and Goetz, 1983) is an airborne optomechanical scanner system covering the 8–12 μm region in six spectral bands 0.4–1.0 μm wide (Figure 19). Table 2 lists the design parameters of TIMS. TIMS has opened up the thermal infrared region for remote rock and mineral identification of silicates that do not have diagnostic vibrational or electronic features at shorter wavelengths. High sensitivity is required since contrasts in spectral emittance are small because surface microroughness creates cavities that exhibit blackbody behavior. The minimum spectral emittance for natural surfaces is approximately 0.8 and the maximum approaches 1.0. This contrast is less than one-fourth of that seen in the visible and near-IR regions. At a surface temperature of 300 K, an $NE\Delta T$ of 0.1 is equivalent to an $NE\Delta\varepsilon$

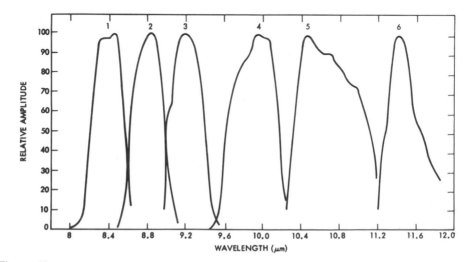

Figure 19   Bandpasses of the Thermal Infrared Multispectral Scanner (TIMS). (Palluconi and Meeks, 1985.)

TABLE 2   Design Parameters of TIMS

Parameter	Value
Spectral channels	6
Field of view	80°
Instantaneous FOV	2.5 mrad
Optics diameter	19 cm
Objects plane scanner	45° inclined flat mirror
Spectral separation	Czerny-Tuner spectrometer
Detectors	HgCdTe
Cooling	LN$_2$
NE ΔT	0.1 to 0.3 K at 300 K

Source: Palluconi and Meeks, 1985.

of 0.002 at 8 μm. Therefore, the range of possible spectral emittances is divided into approximately 100 levels. TIMS presently is flown in the NASA-NSTL* (National Space Technology Laboratory) Lear Jet at 12,000 m altitude or the NASA-AMES C-130 at 4000 m altitude, yielding a ground instantaneous field of view (GIFOV) of 30 and 10 m, respectively.

The recent development of sensors utilizing line- and area-array electronic scanning promises to revolutionize spectral remote sensing. The multiplex advantage provided by an individual detector for each GIFOV across track, in the case of line-array detectors, and additionally one detector for each wavelength in the case of area-array detectors, makes it possible to increase either the spatial or spectral resolution or both as a trade-off for the increased signal-to-noise-ratio obtained. Figure 20 shows the four main categories of spectral imaging systems.

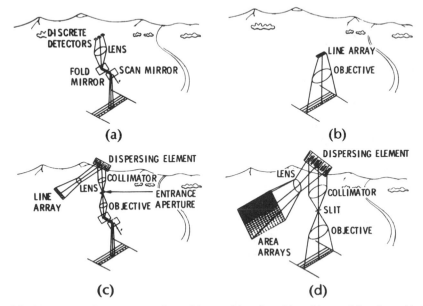

Figure 20 Four approaches to sensors for multispectral imaging: (*a*) multispectral imaging with discrete detectors, (*b*) multispectral imaging with line arrays (*c*) imaging spectrometry with line arrays and (*d*) imaging spectrometry with area arrays. (Goetz et al., 1985a.)

*Currently known as Stennis Space Center (SSC).

The Systeme Probatoire pour l'Observation de la Terre (SPOT) (Courtois and Weill, 1985), launched in 1986, utilizes line-array detectors, Figure 20(b), and, with the increased sensitivity, achieves a 20 m GIFOV for the three-band VNIR scanner and a 10 m GIFOV for the broadband panchromatic imager. While the increased spatial resolution provided by SPOT has proved useful to geologists, only recently has the increased sensitivity of electronically scanned imagers been used in imaging spectrometers to increase the spectral resolution to the point at which direct identification of minerals is possible.

Imaging spectrometers acquire images in hundreds of contiguous spectral bands simultaneously, such that for any pixel, a complete reflectance spectrum can be drawn. The spectral resolution is defined by the separation between bands, and, according to sampling theory, two bands are required to resolve a feature. For minerals, the diagnostic absorption features can be described by the wavelength of maximum absorption and the full width of the absorption band at half-maximum depth (FWHM) (Clark et al., 1988). Since solids do not exhibit rotational degrees of freedom, there is a minimum width to the absorption features found in the visible and short-wavelength infrared. Outside the 1.4 and 1.9 μm water bands, the minimum FWHM, with a few exceptions, is 20 nm or greater. Therefore, by sampling at 10 nm intervals or less, it is possible to completely describe the reflectance spectrum of the Earth's solid surface and hence derive all the information available in the reflected signal. An interesting consequence of the minimum FWHM is that, unlike in the case of spatial resolution, where there is a continual desire for more, a limit can be reached in the spectral case beyond which no more information can be derived. New sensor developments in imaging spectrometry are yielding instruments with a sampling interval of 10 nm or better in the 0.4–2.5 μm region.

The first of the high-resolution imaging spectrometers was the Airborne Imaging Spectrometer (AIS), first flown in 1983 (Vane et al., 1984; Goetz et al., 1985b). AIS first employed a 32 × 32 element hybrid planar mercury-cadmium telluride array detector, Figure 20(d), to cover the 1.2–2.4 μm region. A movable grating is used to step the grating through four positions during the period necessary to move forward one GIFOV. By this means, 128 spectral bands are acquired. The nominal spectral sampling interval is 9.6 nm. Figure 21 shows an example of direct mineral identification from AIS data from an area of hydrothermal alteration in Cuprite, Nevada. The AIS data were normalized to an area of known uniform spectral reflectance to remove the effects of spectral solar insolation, atmospheric absorption, and instrument response, including the response of each of the 1024 individual detectors. The laboratory spectra shown are for samples collected in the field at the approximate locations shown. The host rock for the kaolinite sample is an altered ash flow tuff, consisting primarily of quartz and feldspar, most of which has been altered to kaolinite. Between 10 and 20% of the rock is kaolinite, which is the only mineral constituent having diagnostic spectral absorption features in the 2.03–2.28 μm region shown. These results showed that it is possible to develop mineralogical maps from imaging spectrometer data and gave impetus to the development of a more capable airborne sensor known as the Airborne Visible and Infrared Image Spectrometer (AVIRIS) (Vane et al., 1984).

AVIRIS uses a unique "whiskbroom" approach, Figure 20(c), in which line arrays of detectors are used as exit slits in spectrometers to provide simultaneous registered images in 224 spectral bands. The optical layout is shown in Figure 22. Separation of the scanning head and the spectrometers is necessitated by the size of the instrument, which completely fills the Q bay of the NASA U-2 and ER-2 aircraft. The wavelength region covered, 0.4–2.45 μm, necessitated the development of special high-numerical-

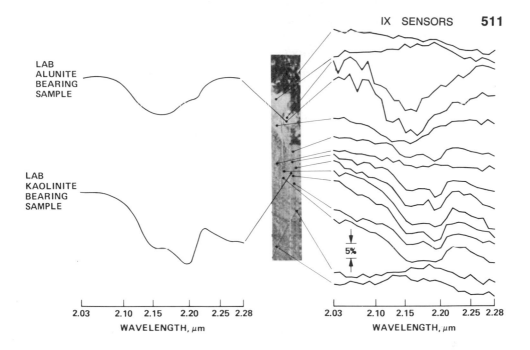

Figure 21 AIS image from Cuprite, Nevada, showing 3 × 3 pixel spectra of three representative surface units. Direct identification of the dominant mineral in each area can be made on the basis of the 2.0–2.3 μm spectral response. Laboratory spectra of field collected samples are also shown, and they verify the AIS results. (After Goetz et al., 1985a.)

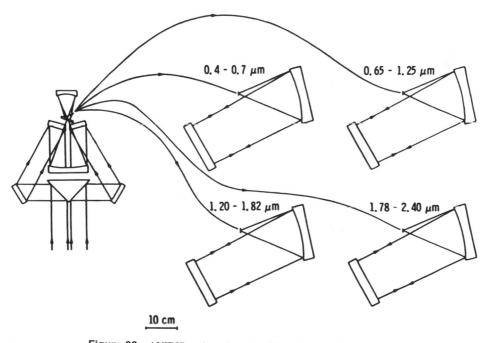

Figure 22 AVIRIS optics schematic. (Macenka and Chrisp, 1988.)

TABLE 3   Functional Parameters of AVIRIS

Parameter	Value
Instantaneous field of view	1.0 mrad
Field of View	30°
Total scan angle	33°
GIFOV (20 km altitude)	20 m
Swath width (20 km altitude)	11 km
Spatial oversampling	15%
Cross-track pixels per scan (after resampling)	550
Spectral coverage	0.4–2.4 μm
Number of spectral bands	224
Spectral sampling intervals	9.4–9.7 nm
Data encoding	10 bits per pixel
Data rate	17 Mbits/s

Source: Goetz et al., 1985b.

aperture fluoride optical fibers for wavelengths beyond 1.2 μm. From an altitude of 20 km, the swath width is approximately 11 km and the GIFOV is 20 m. Further details of the instrument parameters are given in Table 3.

X   DATA-ANALYSIS TECHNIQUES

All modern spectral sensors produce data in digital format, necessitating the use of digital image-processing techniques for analysis and interpretation. There are a number of good texts covering the basics of image processing (Castleman, 1979; Gillespie, 1980; Richards, 1986) and, therefore, only those techniques currently widely applied to multispectral data and those being developed for imaging spectrometer data will be discussed.

All data sets, whether from the Landsat MSS or a 208-band imaging spectrometer, contain more data than can be displayed effectively in one image. The role of image processing then is to reduce the dimensionality or size of the data set to fit within the dynamic range of the display medium and display it in such a fashion that the human interpreter can derive information from the data. Information can be extracted manually by interpreting the spatial relationships of various mappable features or by aiding in drawing boundaries between units on the computer-derived interpretation based on the spectral data. The processing procedures may be divided into two types, image enhancement and image classification.

The human eye–brain system is capable of distinguishing among 16–30 grey levels in a black-and-white image and over a million hues in a color image. Therefore, the challenge in image enhancement is to display as much useful data as possible while taking full advantage of the dynamic range of the display medium. For instance, in the case of TM data, each pixel is assigned one of 256 grey levels or, in other words, encoded to 8 bits. Because the instrument is configured to acquire data under a wide range of lighting conditions, any one scene normally fills only one-third of the full dynamic range. Contrast stretching is used to expand the data to fill the digital number

(DN) range from 0 to 255, or black to white. Three individual bands or band combinations, and no more, can be combined and projected in the three primary colors to produce a color image. Therefore, a major challenge to the analysis procedure is to insert as much useful data as possible into the three images that make up the color composite.

Ninety percent of the variance in any image of the Earth, having a spatial resolution equivalent to the MSS or greater, is attributable to the effects of topography. Since the shading due to topography is the same regardless of the spectral channel, there is a very high correlation between spectral channels in an image. There are four common techniques presently used to enhance spectral data by removing correlation and eliminating the influence of topography.

Ratioing individual spectral bands and making color composites of the resulting images was the first technique widely used in geologic applications of remote-sensing data (Rowan et al., 1974, 1977). Ratio images are created by ratioing individual pixels in each of two spectral band images to produce a new black-and-white image and stretching the result about a midgrey DN of 128. The effect of the ratio can be seen in Figure 23. If there are no additive offsets in the image, to the first order, ratio values for the same materials will be the same regardless of their topographic position. Therefore, materials

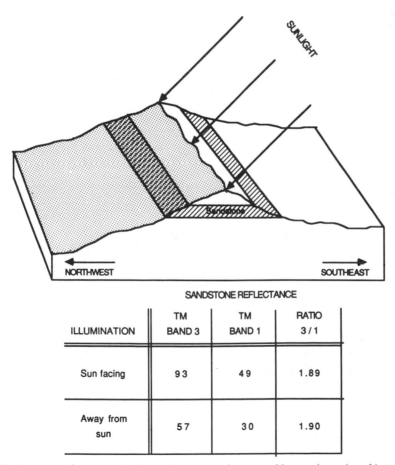

SANDSTONE REFLECTANCE

ILLUMINATION	TM BAND 3	TM BAND 1	RATIO 3 / 1
Sun facing	9 3	4 9	1.89
Away from sun	5 7	3 0	1.90

Figure 23 The effect of topographic slope and aspect can be removed by creating ratios of images in two different spectral bands. The effect of sky illumination in the shaded areas is not shown.

in an image having the same spectral reflectance characteristics will appear as the same color in a color-ratio composite (CRC) regardless of brightness differences in the original images caused by differing surface aspect and slope. The DN value in a ratio image is proportional to the slope in the spectral reflectance curve of the surface material, provided the image has been properly normalized for spectral solar insolation and sensor response. Therefore, it is possible from a CRC to work out the relative slopes of material reflectance curves and hence deduce composition.

Ratio images often appear to have low contrast and are grainy or noisy. These effects result because most of the variance in the original images has been removed, allowing the images to be stretched by a factor of ten or more, which makes low-level noise visible. The apparent lack of contrast derives from the fact that boundaries between materials are often not sharp but gradational and the separation between units in display color space is not optimal. One simple remedy is to replace one of the ratios in a CRC with one of the original bands, hence reintroducing contrast due to topographic shading. A more direct technique is to remove all correlation between spectral bands, thereby providing for the optimal separation among surface units in display space.

The principal components (PC) transformation is a means of creating a new set of images that are orthogonal to one another, or, in other words, completely decorrelated (Gonzalez and Wintz, 1977). The transformation is derived from the variance-covariance matrix and yields a new set of images in which each pixel value is a linear combination of the original data values. The number of PC images equals the number of spectral bands, and they are ranked according to the percentage of total variance of the original data contained in them. As might be expected, the first principal component contains the brightness information associated with topographic shading and gross variation in overall spectral reflectance, sometimes called albedo (Figure 24). The second and sub-

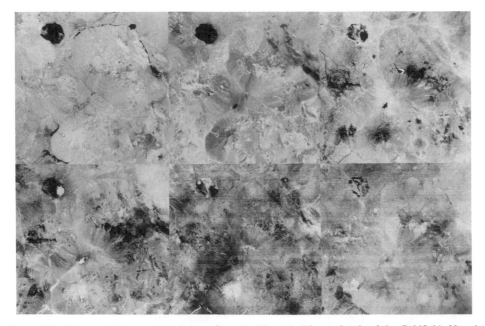

Figure 24 Six principal component images from six Thematic Mapper bands of the Goldfield, Nevada, area.

sequent principal components reflect differences in spectral reflectance among surface materials. However, because the transformation is image-data-dependent, it is not possible to derive the true or even relative spectral reflectance curves from the transformed data. Because the transformation provides the greatest separation among units in display space, color composites are extremely colorful, but caution must be exercised in interpreting the results because the colors do not have physical meaning in terms of geologically relevant parameters. The color images are most valuable for outlining units whose relevance can be field checked at a later time. There are two other disadvantages that should be recognized. One is that because the transformation is data-dependent, similar materials may have different colors in adjacent scenes, or in the same scene taken at different times. The other results from the fact that the statistics of the image and the variance-covariance matrix are weighted by the number of points in the image having similar spectral reflectance characteristics. If a surface material of interest has only a very small area of exposure, then its reflectance characteristics will have little bearing on the transformation statistics and it may not be differentiable from other units in the PC images. This would not be the case for a ratio image in which pixel values are independent of the image statistics.

There are two other methods of processing to enhance color contrast in highly correlated images while still maintaining the relative spectral reflectance relationships among surface units. One, called the *decorrelation stretch*, is based on image statistics, and the other, called the *HSI stretch*, is invariant (Gillespie et al., 1986). The decorrelation stretch is based on the PC transformation and is illustrated in Figure 25. The image data, Figure 25(a), are transformed to the principal axes, Figure 25(b), and stretched, usually linearly, to equalize the variance along the axes, Figure 25(c). The data are then transformed back to the original coordinates using the inverse of the PC transformation, Figure 25(d). The data are now much more poorly correlated and, as a result, will produce more saturated hues in a color composite. However, because of the inverse transformation, the relative colors among various units in the original image are maintained, although they are much more saturated. This property makes it possible to interpret colors in the image in terms of spectral reflectance differences, something not possible with PC color composites.

The HSI (hue, saturation, intensity) transformation is another technique to spread image values into a greater range of hues for better separation of surface material units. However, the transformation is not dependent on image statistics. The technique (Gillespie, 1980; Gillespie et al., 1986) is depicted in Figure 26, and is a transformation from cartesian to spherical coordinates. The achromatic line, consisting of equal contributions of red (R), green (G), and blue (B), is transformed to coincide with the polar axis of the spherical coordinate system (θ, ϕ, ρ). A normalization is performed to distribute the hues in a color wheel in the XZ plane (Figure 26). The intensity is taken as the scalar sum of R, G, and B. A similar effect to decorrelation stretching can be achieved by stretching the saturation and intensity images before transformation back to the original space. The HSI transformation is limited to three spectral images that are assigned to R, G, and B, regardless of their wavelength, whereas decorrelation stretches can be made from any three of the total number of the principal components.

Another application of the HSI transformation is in combining a sharpening band, an image having higher resolution, with a lower-resolution data set. For instance, replacing the intensity image of the 30 m GIFOV TM multispectral data set with 10 m GIFOV SPOT panchromatic image and making the inverse transformation results in a color im-

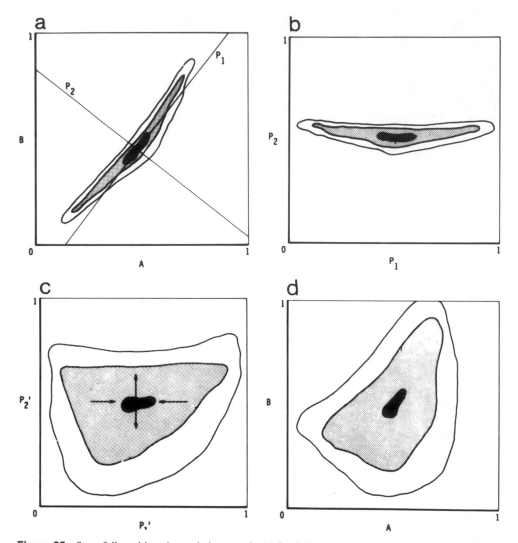

Figure 25 Steps followed in a decorrelation stretch: (*a*) for the brightness values in band A plotted against band B, the data are highly correlated; (*b*) after rotation by principal components transformation; (*c*) a stretch is applied to each principal component axis independently to fill the dynamic range of the data space; and (*d*) a transform of the eigenvector matrix is used to return the stretched principal component data to the original axis A and B. (Gillespie et al., 1986.)

age that appears to have the full resolution of SPOT. Care must be taken to register the two data sets properly prior to performing the transformation.

Image classification, either supervised or unsupervised, has not found great favor with geologists interpreting broadband images such as those from Landsat. Classification techniques (Swain and Davis, 1978) have been applied extensively to agricultural scenes with good success, partly because the boundaries between fields are linear or smoothly curved and little spatial association is necessary in the interpretation. Geologists, on the other hand, must use spatial associations since boundaries between units may be gradational and surface morphology plays a strong role in interpretation. An exception to this rule is developing in the interpretation of imaging spectrometer data.

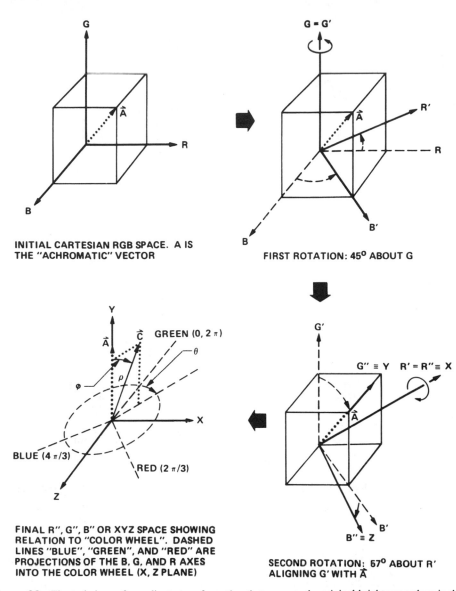

INITIAL CARTESIAN RGB SPACE. A IS
THE "ACHROMATIC" VECTOR

FIRST ROTATION: 45° ABOUT G

FINAL R", G", B" OR XYZ SPACE SHOWING
RELATION TO "COLOR WHEEL". DASHED
LINES "BLUE", "GREEN", AND "RED" ARE
PROJECTIONS OF THE B, G, AND R AXES
INTO THE COLOR WHEEL (X, Z PLANE)

SECOND ROTATION: 57° ABOUT R'
ALIGNING G' WITH Ā

Figure 26   The technique of coordinate transformation that converts the original brightness values in three spectral bands to hue, saturation, and intensity (HSI). (Gillespie et al., 1986.)

Imaging spectrometer data, unlike TM or MSS data, have sufficiently high spectral resolution to provide direct compositional information. Because of the number of spectral bands required for direct mineral identification, the image-enhancement techniques discussed before are not applicable. New pattern-recognition techniques are being developed to facilitate the extraction of mineralogical information from AIS and AVIRIS data (Goetz et al., 1985a). Deterministic rather than statistical methods of analysis are appropriate because imaging spectrometry provides spectral sampling sufficient to produce unique spectral signatures. One approach is a fast signature-matching algorithm, which involves creating a binary vector (Figure 27) from each pixel and from a library

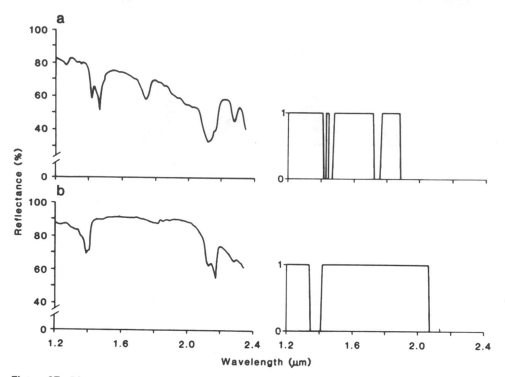

Figure 27   Binary encoding scheme for laboratory reflectance spectra of two minerals: (*a*) alunite and (*b*) kaolinite. (Goetz et al., 1985a.)

of reference spectra. Each pixel in the image is compared with each library spectrum by a rapid logical exclusive-OR operation. A Hamming distance (Tou and Gonzalez, 1974) is used to account for the natural variability in the image spectra caused by impure pixels and sensor noise. The results have been promising and a number of mineral identification successes have been reported in the literature (Goetz et al., 1985a; Goetz and Srivastava, 1985; Lang et al., 1987; Pieters and Mustard, 1988).

All surfaces are heterogeneous in composition, which means that within a pixel, there will always be several surface cover components. Adams et al. (1986) and Smith et al. (1985) have developed techniques for spectral mixture modeling. A multispectral image, whether 4 bands or 208, can be modeled as mixtures of end-member spectra and image spectra can be compared directly with laboratory spectra. Principal-components analysis is used to develop mixing lines between pairs of end members. Variables such as grain size can be modeled, as well as shade, secondary illumination, and vegetation cover. With this technique, it is possible to make quantitative determinations of surface-material abundances.

XI   APPLICATIONS

Since the launch of Landsat-1 in 1972, hundreds of papers have been written on the applications to geologic mapping and interpretation. Goetz and Rowan (1981) summarized the major results obtained during the 1970s. The following is a summary of more recent research using Landsat as well as aircraft scanner data from TIMS and AIS. This is by no means an exhaustive summary, but rather a sampling of results from the sub-disciplines of lithologic and structural mapping and interpretation, mineral and petro-leum exploration research, and, finally, mineralogical mapping using imaging spec-trometry.

It has been long recognized that lithologic mapping using spectral remote-sensing data is most advantageously carried out in arid regions. The synoptic view afforded by Land-sat means that mapping techniques can be applied consistently to large areas, much larger than would be covered by even a substantial field mapping expedition. Sultan et al. (1986) have applied Landsat TM data to mapping serpentines in the eastern desert of Egypt over a 60,000 km^2 area. They showed that serpentines could be distinguished reliably from surrounding rocks using quantitative mapping criteria based on the reflec-tance ratios. A threshold classifier based on three reflectance ratios was used. The ratio 5/7 was used to estimate the abundance of hydroxyl-bearing phases and 5/1 for mag-netite content. A third component of the classifier was the calculated value of reflectance for band 4, based on a linear interpolation between bands 3 and 5, divided by the ob-served band-4 reflectance. This component was used to distinguish mafic rocks contain-ing substantial amounts of magnetite and hydroxyl-bearing phases from serpentines. This technique makes possible the location of suture zones in poorly-mapped, arid continental regions.

The use of TM data to study the Oman ophiolite, the most complete example of ocean floor rocks outcropping on land, has shown that, even for a well-studied area mapped at a scale of 1:100,000, substantial new information can be derived (Abrams et al., 1988). Decorrelation-stretch color composites were used to recognize variations in gabbro com-position; to identify small acidic, gabbroic, and ultramafic intrusions; and to discriminate the uppermost mantle from the deeper mantle. In addition, it was possible to locate the Moho precisely and to map gossans and areas of chloritic-epidotic alteration. The con-sistency in mapping, available through the space perspective, made it possible to rec-ognize subtle changes and minor outcrops across a 100 km wide area that could not otherwise be seen in the field unless unrealistically exhaustive mapping was undertaken.

Spectral remote sensing is not only valuable for mapping in igneous terranes, but also in sedimentary basins. Lang et al. (1987) report on stratigraphic and structural analyses of the Wind River basin and Big Horn basin areas of central Wyoming. They use TM spectral images together with topographic data to map lithologic contacts, measure strike and dip, and develop a stratigraphic column that is correlated with conventional surface and subsurface sections. In addition, they used TIMS and AIS data to add mineralogic information, such as the distribution of quartz, calcite, dolomite, smectite, and gypsum, to the column derived from TM data. The combination of these data types makes it possible to develop information about areal variations in attitude, sequence, thickness, and lithology of strata exposed at the Earth's surface, and in a manner consistent with conventional mapping techniques. The latter is an important point, because only when a new technique can be related to conventional methods will it gain acceptance among the skeptics.

Respect for the value of multispectral infrared image data has been increasing ever since the publication of the first results derived from them by Kahle and Rowan (1980). Prior to the development of TIMS, this was the only study published on the analysis of multispectral thermal data from the now defunct Bendix 24-channel scanner. This study of the value of 8–14 μm multispectral data for lithologic mapping in the East Tintic

Mountains, Utah, showed that, from decorrelation-stretch images, it is possible to discriminate among several rock types primarily based on their silica content. Combined with data from the visible and short-wavelength region, it was possible to discriminate among quartzite, carbonate rocks, quartz latitic and quartz monzonitic rocks, latitic and monzonitic rocks, silicified altered rocks, argillized altered rocks, and vegetation.

This study provided the justification for the development of a multispectral scanner dedicated to the thermal infrared region. TIMS (Kahle and Goetz, 1983) showed that it was possible to map readily quartz-bearing rocks that would not otherwise be discriminable in multispectral images in the spectral region covered by TM. The silicified core of the hydrothermal system at Cuprite, Nevada is distinguishable based on its quartz content rather than its desert varnish coatings as seen at shorter wavelengths. In addition, basalts, unaltered tuffs, siltstone, and carbonates are discriminable. In Death Valley, California, TIMS has been used to map alluvial fans (Gillespie et al., 1984). The authors were able to recognize both composition and relative age differences that were generally consistent with units mapped by conventional methods. The composition of the fan gravels could be related to the source rock based on spectral emittance, and relative age mapping with TIMS was possible because the original composition of the fans was modified by differential erosion and weathering, changing the spectral emittance.

Resource exploration technique development was among the first geologic applications of spectral satellite images. Rowan et al. (1974) published the first detailed study of the identification of gossans or zones of hydrothermal alteration containing iron oxides. Color-ratio composites were used to map zones of limonite and jarosite in Goldfield, Nevada. Ratios of MSS bands 4/5 and 6/7 were key in discriminating limonite from the surrounding rocks. Based on further mapping with the MSS in Nevada and elsewhere, as well as field and laboratory spectral reflectance measurements (Rowan et al., 1977), it soon became apparent that the MSS did not have the spectral coverage to afford the detection of alteration zones devoid of iron oxides.

A study of the Cuprite, Nevada, mining district by Abrams et al. (1977) using Bendix 24-channel scanner data showed that altered rocks in the district could be distinguished from unaltered rocks with few ambiguities on a color-ratio composite. The ratios were 1.6/2.2, 1.6/0.48, and 0.6/1.0 μm, where the bands are designated by the approximate centers of the channels. The first ratio delineated hydroxyl-bearing rocks, the second was sensitive to ferric-iron content, and the third helped highlight rocks having only a moderately rising reflectance in the VNIR region. This study played a key role in the decision to place the 2.08–2.36 μm band 7 on Landsat TM.

A major joint industry-NASA study was undertaken in the late 1970s to test the value of remote-sensing data in the exploration process for the commodities copper, uranium, and oil and gas (Abrams et al., 1984). The study of three porphyry copper deposits (Abrams et al., 1983) showed that while the MSS was useful for delineating areas of iron oxide on the surface, it was not possible to distinguish among altered rocks and sedimentary red beds, volcanic rocks, and weathered alluvium. However, regional tectonic patterns were well displayed on the images. Airborne thematic-mapper simulator images showed that the ambiguities in mapping of iron-oxide-bearing surfaces could be resolved, and that hydroxyl-bearing minerals, important constituents of the alteration haloes, could be detected.

In a recent study, TM images of Spain have been used to distinguish soil developed on contact metamorphic rocks in aureoles around granitic plutons from soil formed on stratigraphically equivalent slate and metagraywacke that has been regionally metamorphosed to the greenschist facies (Rowan et al., 1987). Silver, lead, zinc, and tin deposits are associated with the plutons in the Extremadura region of western Spain. The reflectance of the contact metamorphic soil is lower, particularly around 1.6 μm, and exhibits weaker hydroxyl and Fe^{3+} absorption features than the spectra of the slate-metagraywacke soil. The spectral differences are attributed to an increase in carbonaceous material in the contact metamorphic soil around the plutons. Since the region is heavily cultivated, first a mask was made from the 4/3 ratio is identify tilled, vegetation-free fields that could be used for mapping. Field evaluation of a classification map based on TM band 5 alone and the TM ratio 3/1 shows more extensive aureoles than on published maps. Only a few areas were misclassified.

Most of the published studies in geologic remote sensing are concerned with arid and semiarid regions, since rock units are well exposed. Over 70% of the land masses have complete vegetation cover and, therefore, clues to the rock and soil substrate in these areas can only be gained by studying the vegetation. Remote sensing of vegetation is concerned with either mapping species, species association, cover density, or leaf chemistry as evidenced by spectral reflectance. The last is potentially a fertile field of study with the coming of high spectral resolution imaging spectrometers. Collins et al. (1983) have pioneered the study of the chlorophyll absorption as an indicator of plant stress associated with soil-metal anomalies.

Collins et al. (1983) used an airborne 500-channel spectroradiometer in a profiling mode to study a known forest-covered, copper-soil anomaly. The study revealed previously unknown spectral changes in the near-infrared chlorophyll-absorption spectrum. The long-wavelength, or red edge, of the 0.68 μm chlorophyll absorption was seen to shift to shorter wavelengths in trees growing in the high-copper soil. The so-called "blue shift" was reproduced in plants grown in a greenhouse under controlled conditions (Chang and Collins, 1983), and salts of copper, zinc, and nickel showed the greatest effect. A blue shift of up to 40 nm was observed. It was concluded that metal stress impairs or retards the chlorophyll production in plants and that there is no relationship between chlorophyll a/b ratios and metal stress. Collins et al. (1983) used polynomial approximation techniques to reduce the dimensionality of their 500-channel radiometer data. Fits were made to the data using Chebyshev polynomials, and ratios of coefficients were used that were sensitive to the blue shift to map out the anomalous regions in the airborne data from two overflights of the Cotter basin in Montana and one of Spirit Lake in Washington. The latter was carried out before the Mt. St. Helens eruption. The results in the Cotter basin site show good correlation with known geochemical anomalies and that the measurements are reproducible over a two-year interval. Mapping of the blue shift requires a high spectral resolution sensor with a spectral sampling interval of at least 10 nm, but in order to detect subtle shifts, a sampling interval of 2 nm is required (W. Collins, private communication). Of all the imaging sensors available or planned, only AVIRIS, with 9.6 nm continuous sampling and the Fluorescence Line Imager (Hollinger et al., 1988), have sufficient resolution to detect a blue shift.

The coming of age of imaging spectrometry brings with it exciting possibilities for geological remote sensing, particularly the prospect for direct mineral identification and mapping. The preliminary results from AIS show promise, and in the next few years, a substantial body of literature will develop as results from AVIRIS data analysis are published. The AIS has been a test bed for area-array detector development at the Jet Propulsion Laboratory, and numerous difficulties with the instrument have precluded its use as an operational data-collection system. However, enough data has been analyzed to show the potential of the technique. A recent special journal issue (Vane and Goetz, 1988) is devoted to results from AIS.

The first indications that remote mineral identification was possible with AIS were seen in data from Cuprite, Nevada (Goetz et al., 1985a). Here characteristic spectra were obtained for kaolinite, alunite, and secondary quartz containing an overtone Si–OH absorption feature. Ground spectral measurements and laboratory spectra of field-collected samples verified the AIS spectra. In the case of the kaolinite-bearing rocks, a maximum of 20% kaolinite was present in the altered tuff matrix. A repeat flight over the same area led to a serendipitous event when the aircraft overflew an area not planned for data acquisition. Goetz and Srivastava (1985) reported the discovery of buddingtonite identified by its characteristic 2.12 μm absorption feature. Ground sampling for verification was made difficult because buddingtonite cannot be identified by eye. The discovery at Cuprite was the fifth known location of the mineral in igneous rocks (M.D. Krohn, private communication). NH_4-bearing minerals are now being actively studied because of their association with precious-metal deposits in Nevada and California Krohn and Altaner, 1987).

In another study of hydrothermally altered rocks, Kruse (1988) mapped areas of quartz-sericite-pyrite alteration, areas of argillic alteration containing montmorillonite, and calcite and dolomite with AIS data. An algorithm was developed to calculate automatically band position, depth, and width and map them into IHS space. The IHS image was used to map areas of potential alteration based on the predicted relationships between the color image and the mineral absorption band. The areas mapped using the AIS data corresponded well to the areas identified by field mapping techniques.

XII   FUTURE DIRECTIONS

From the foregoing, it should be clear that geologic remote sensing is moving in the direction of acquiring data with more information content relevant to compositional mapping and ultimately to data acquisition from space platforms to facilitate access to any region of the globe. The sources of new data will be imaging spectrometers. AVIRIS will be the main source of high spectral resolution, broad spectral coverage, image data for research for most of the next decade. A new commercial imaging spectrometer service is being offered by Geophysical Environmental Research of New York City. Their instrument contains 64 channels distributed in key portions of the 0.4–2.5 μm region and creates a swath 512 pixels in width over a scan angle of 80°. Multispectral thermal images are presently only available from TIMS and there is no commercial service available in the northern hemisphere.

In the second half of the next decade, NASA intends to launch an earth-observing satellite (Eos) aboard a polar orbiting platform that will carry a number of NOAA and NASA instruments as well as instruments from the European and Japanese space agencies (Butler et al., 1984). The instrument of greatest potential interest to geologists is the High-Resolution Imaging Spectrometer (HIRIS) (Goetz and Herring, 1987; Goetz et al., 1987). HIRIS has similar capabilities to AVIRIS (Table 4) but covers a three times larger swath and will have 2–5 times greater radiometric sensitivity. The pointing capability of HIRIS will allow any part of the Earth, other than the poles, to be imaged every four days. Given the very high data rate, only specific target areas can be covered as opposed to the general coverage afforded by Landsat.

TABLE 4   Current HIRIS Parameters

Parameter	Baseline
Design altitude	824 km
Swath width	30 km
IFOV (ground footprint)	30 m
Spectral coverage	0.4–2.5 μm, 192 channels
Focal ratio	f/3.8
Pointing, down-track	+60°/−30°
cross-track	+20°/−20°
Encoding	12 bits/pixel
Raw data rate (12 bits)	512 Mbits/s
Data edit modes	Spectral select
	Encoding select
	(8 out of 12 bits)
	Serial spatial average
	(cross-track)

Source: Goetz and Herring, 1987.

Finally, the next advances in data analysis and interpretation will be in combining maps and models derived from remote sensing with other types of geological and geophysical data (Lang et al., 1987; Kowalik and Glenn, 1987). Advances in geographic information systems make it possible to combine vector (point data) and raster (image) data in models. This ability will make it practical to synthesize large quantities of diverse information and pave the way for the widespread use of spectral remote sensing in studies of continental geology.

REFERENCES

Abrams, M. J., R. Ashley, L. C. Rowan, A. F. H. Goetz, and A. B. Kahle (1977). Mapping of hydrothermal alteration in the Cuprite Mining District, Nevada, using aircraft scanner imagery for the 0.46–2.36 μm spectral region. *Geology* 5:713–718.

Abrams, M. J., D. Brown, L. Lepley, and R. Sadowski (1983). Remote sensing for porphyry copper deposits in southern Arizona. *Econ. Geol.* **78**:591–604.

Abrams, M. J., J. E. Conel, H. R. Lang, and H. N. Paley (1984). *The Joint NASA/Geosat Test Case Project*, Final Rep. American Association of Petroleum Geologists, Tulsa, Oklahoma.

Abrams, M. J., D. A. Rothery, and A. Pontual (1988). Mapping in the Oman ophiolite using enhanced Landsat Thematic Mapper images. *Tectonophysics* **151**:387–401.

Adams, J. B., M. O. Smith, and P. E. Johnson (1986). Spectral mixture modeling: A new analysis of rock and soil types at the Viking Lander 1 site. *J. Geophys. Res.* **91**:8098–8112.

Burns, R. G. (1970). *Mineralogical Application of Crystal Field Theory*. Cambridge University Press, Cambridge, Maryland.

Butler, D. M. et al. (1984). Earth observing system. *NASA Tech. Memo.* **NASA TH-X-86129**.

Castleman, K. R. (1979). *Digital Image Processing*. Prentice-Hall, Englewood Cliffs, New Jersey.

Chang, S. H., and W. Collins (1983). Confirmation of the airborne biogeophysical mineral exploration technique using laboratory methods. *Econ. Geol.* **78**:723–736.

Chiu, H. Y., and W. E. Collins (1978). A spectroradiometer for airborne remote sensing. *Photogramm. Eng. Remote Sens.* **44:**507–517.

Clark, R. N., T. King, M. Klejwa, G. A. Swayze, and N. Vergo (1989). High spectral resolution reflectance spectroscopy of minerals. *J. Geophys. Res.* (in press).

Collins, W., S. H. Chang, G. Raines, F. Canney, and R. Ashley (1983). Airborne biogeophysical mapping of hidden mineral deposits. *Econ. Geol.* **78:**737–749.

Courtois, M., and G. Weill (1985). The Spot satellite system. *Prog. Astronaut. Aeronaut. Sci.* **97:**493–523.

Drury, S. A. (1987). *Image Interpretation in Geology.* Allen & Unwin, London.

Gillespie, A. R. (1980). Digital techniques of image enhancement. In *Remote Sensing in Geology* (B. Siegal and A. R. Gillespie, Eds.). Wiley, New York, pp. 139–226.

Gillespie, A. R., A. B. Kahle, and F. D. Palluconi (1984). Mapping alluvial fans in Death Valley, California, using multichannel thermal infrared images. *Geophys. Res. Lett.* **11:**1153–1156.

Gillespie, A. R., A. B. Kahle, and R. E. Walker (1986). Color enhancement of highly correlated images. I. Decorrelation and HSI contrast stretches. *Remote Sens. Environ.* **20:**209–235.

Goetz, A. F. H. (1987). The portable instant display and analysis spectrometer. JPL Publ. 87-30:8–17.

Goetz, A. F. H., and M. Herring (1987). The High Resolution Imaging Spectrometer (HIRIS) for EOS. *Proc. Int. Geosci. Remote Sens. Symp. (IGARSS'87)*, Ann Arbor, Michigan, Vol. 1, pp. 367–372.

Goetz, A. F. H., and L. C. Rowan (1981). Geologic remote sensing. *Science* **211:**781–791.

Goetz, A. F. H., and V. Srivastava (1985). Mineralogical mapping in the Cuprite mining district, Nevada. *JPL Publ.* **85-41.**

Goetz, A. F. H., F. C. Billingsley, D. Elston, I. Lucchitta, E. M. Shoemaker, M. J. Abrams, A. R. Gillespie, and R. L. Squires (1975). Portable field reflectance spectrometer. In *Applications of ERTS Images and Image Processing to Regional Geologic Problems and Geologic Mapping in Northern Arizona*, JPL Tech. Rep. 32. Jet Propulsion Laboratory, California Institute of Technology, Pasadena, p. 1597.

Goetz, A. F. H., L. C. Rowan, and M. J. Kingston (1982). Mineral identification from orbit: Initial results from the Shuttle Multispectral Infrared Radiometer. *Science* **218:**1020–1024.

Goetz, A. F. H., B. N. Rock, and L. C. Rowan (1983). Remote sensing for exploration: An overview. *Econ. Geol.* **78:**573–590.

Goetz, A. F. H., G. Vane, J. Solomon, and B. N. Rock (1985a). Imaging spectrometry for Earth remote sensing. *Science* **228:**1147–1153.

Goetz, A. F. H., J. B. Wellman, and W. L. Barnes (1985b). Optical remote sensing of the Earth. *IEEE Proc.* **73:**950–969.

Goetz, A. F. H. et al. (1987). High resolution imaging spectrometer: Science opportunities for the 1990s. In *Earth Observing System*, Vol. IIC, Instrum. Panel Rep. NASA, Washington, D.C.

Gonzalez, R. C., and P. Wintz (1977). *Digital Image Processing.* Addison-Wesley, Reading, Massachusetts, pp. 103–112 and 309–317.

Hapke, B. (1981). Bidirectional reflectance spectroscopy. 1. Theory. *J. Geophys. Res.* **86:**3039–3054.

Hapke, B. (1984). Bidirectional reflectance spectroscopy. 3. Correction for macroscopic roughness. *Icarus* **59:**41–59.

Hapke, B. (1986). Bidirectional reflectance spectroscopy. 4. The extinction coefficient and the opposition effect. *Icarus* **67:**264–280.

Hollinger, A. B., L. H. Gray, and J. F. R. Gower (1988). The fluorescence line imager: An imaging spectrometer for ocean and land remote sensing. *Proc. SPIE—Int. Soc. Opt. Eng.* **834:**2–11.

Hunt, G. R. (1980). Electromagnetic radiation: The communication link in remote sensing. In *Remote Sensing in Geology* (B. S. Siegal and A. R. Gillespie, Eds.). Wiley, New York, pp. 5–45.

Kahle, A. B., and A. F. H. Goetz (1983). Mineralogic information from a new airborne thermal infrared multispectral scanner. *Science* **222:**24–27.

Kahle, A. B., and L. C. Rowan (1980). Evaluation of multispectral middle infrared aircraft images for lithologic mapping in the East Tintic Mountains, Utah. *Geology* **8:**234–239.

Kowalik, W. S., and W. E. Glenn (1987). Image processing of aeromagnetic data and integration with Landsat images for improved structural interpretation. *Geophysics* **52:**859–875.

Krohn, M. D., and S. P. Altaner (1987). Near–infrared detection of ammonium minerals. *Geophysics* **52:**924–930.

Kruse, F. A. (1988). Use of airborne imaging spectrometer data to map minerals associated with hydrothermally altered rocks in the northern Grapevine Mountains, Nevada and California. *Remote Sens. Environ.* **24:**31–52.

Lang, H. R., S. L. Adams, J. E. Conel, B. A. McGuffre, E. D. Paylor, and R. E. Walker (1987). Multispectral remote sensing as stratigraphic and structural tool, Wind River Basin and Big Horn Basin areas, Wyoming. *Amer. Assoc. Pet. Geol. Bull.* **71:**389–402.

Launer, P. J. (1952). Regularities in the infrared absorption spectra of silicate minerals. *Am. Mineral.* **37:**764.

Macenka, S. A., and M. P. Chrisp (1988). Airborne Visible/Infrared Imaging Spectrometer (AVIRIS) spectrometer design and performance. *Proc. SPIE—Int. Soc. Opt. Eng.* **834:**32–43.

Nassau, K. (1980). The causes of color. *Sci. Am.*, October, pp. 124–154.

Palluconi, F. D., and G. R. Meeks (1985). Thermal Infrared Multispectral Scanner (TIMS): An investigator's guide to TIMS data. *JPL Publ.* **85-32**, p. 6.

Pieters, C. M., and J. F. Mustard (1988). Exploration of crustal/mantle material for the earth and moon using reflectance spectroscopy. *Remote Sens. Environ.* **24:**151–178.

Richards, J. A. (1986). *Remote Sensing Digital Image Analysis.* Springer-Verlag, Berlin.

Rowan, L. C., P. H. Wetlaufer, A. F. H. Goetz, F. C. Billingsley, and J. H. Stewart (1974). Discrimination of rock types and detection of hydrothermally altered areas in South-Central Nevada by the use of computer-enhanced ERTS images. *Geol. Surv. Prof. Pap. (U.S.)* **883**, p. 35.

Rowan, L. C., A. F. H. Goetz, and R. Ashley (1977). Discrimination of hydrothermally altered and unaltered rocks in visible and near-infrared multispectral images. *Geophysics* **42:**522–535.

Rowan, L. C., C. Anton-Pacheco, D. W. Brickey, M. J. Kingston, A. Payas, N. Vergo, and J. K. Crowley (1987). Digital classification of contact metamorphic rocks in Extremadura, Spain using Landsat Thematic Mapper data. *Geophysics* **52:**885–897.

Salisbury, J. W., and G. R. Hunt (1968). Martian surface materials: Effect of particle size on spectral behavior. *Science* **161:**365–366.

Salisbury, J. W., L. S. Walter, and N. Vergo (1987). Mid-infrared (2.1–25 μm) spectra of minerals: First edition. *Geol. Surv. Open-File Rep. (U.S.)* **87-263**.

Short, N. M., and R. W. Blair, Jr. (Eds.) (1986). *Geomorphology from Space*, NASA SP-486. National Aeronautics and Space Administration, Washington, D.C.

Slater, P. N. (1980). *Remote Sensing Optics and Optical Systems.* Addison-Wesley, Reading, Massachusetts.

Smith, M. O., P. E. Johnson, and J. B. Adams (1985). Quantitative determination of mineral types and abundances from reflectance spectra using principal components analysis. *JGR, J. Geophys. Res.* **90** (Suppl.):C797–C804.

Sultan, M., R. E. Arvidson, and N. C. Starchio (1986). Mapping of serpentinites in the eastern desert of Egypt by using Landsat Thematic Mapper data. *Geology* **14**:995–999.

Swain, P. H., and S. M. Davis (Eds.) (1978). *The Quantitative Approach*. McGraw-Hill, New York.

Tou, J. T., and R. C. Gonzalez (1974). *Pattern Recognition Principles*. Addison-Wesley, Reading, Massachusetts, Chapter 3.

Vane, G. (Ed.) (1987). *AVIRIS: A Description of the Sensor, Ground Data Processing Facility, Laboratory Calibration and Preliminary Results*, JPL Publ. 87-38. Jet Propulsion Laboratory, Pasadena, California.

Vane, G., and A. F. H. Goetz (1988). Terrestrial imaging spectroscopy. *Remote Sens. Environ.* **24**:1–29.

Vane, G., M. Chrisp, H. Enmark, S. Macenka, and J. Solomon (1984). Airborne visible and IR imaging spectrometer: An advanced tool for Earth remote sensing. In *Proceedings of the 1984 IEEE International Geoscience and Remote Sensing Symposium*, SP215. European Space Agency, Paris, pp. 751–757.

Weidner, V. R., and J. J. Hsia (1981). Reflection properties of pressed polytetrafluoroethylene powder. *J. Opt. Soc. Am.* **71**:856–859.

13

REMOTE SENSING OF SNOW IN VISIBLE AND NEAR-INFRARED WAVELENGTHS

JEFF DOZIER

Center for Remote Sensing and Environmental Optics
University of California
Santa Barbara, California
and
Jet Propulsion Laboratory
California Institute of Technology
Pasadena, California

I INTRODUCTION

Snow, glaciers, ice in lakes and rivers, and ice in the ground comprise the frozen part of the land portion of the hydrologic cycle. Water in these frozen states accounts for more than 80% of the total fresh water on Earth and is the largest contributor to runoff in rivers and ground water over major portions of the middle and high latitudes. Snow and ice also play important interactive roles in the Earth's radiation balance, because snow has a higher albedo than any other natural surface. Over 30% of the Earth's land surface is seasonally covered by snow, and 10% is permanently covered by glaciers.

Therefore, snow cover represents a changing atmospheric output, resulting from variability in the Earth's climate, and it is also an important changing boundary condition in climate models. Thus, it is important that we monitor the temporal and spatial variability of the snow cover over land areas, from the scale of small drainage basins to continents. Over the last 10–20 years, satellite remote sensing has opened the possibility of data acquisition at regular intervals, and operational as well as research-oriented satellites have provided information on snow cover. The Landsat systems, in particular, have provided multiwavelength visible and near-infrared data for hydrological and glaciological research at the scale of drainage basins. In this chapter, we describe remote sensing of snow in the optical wavelengths.

The most important information that can be derived about the snow cover from measurements from satellite of the reflected solar radiation are

1. Maps of snow-covered areas and rates of snow-cover depletion.
2. Estimates of spectral albedo throughout the solar spectrum.

II REMOTE SENSING OF SNOW PROPERTIES FOR HYDROLOGY AND CLIMATOLOGY

Satellite remote sensing has become increasingly important to snow hydrologists, because the data provide information on the spatial distribution of parameters of hydrologic importance, snow-covered areas, surface albedo, and snow-water equivalence. For the seasonal snow cover, remote sensing has been used to improve the monitoring of existing conditions and has been incorporated into several runoff forecasting and management systems.

In the polar regions, the emphasis in remote sensing of snow and ice is primarily on frequent large-scale coverage. The albedo of snow is much greater than that of any other common natural substance, and because of its broad geographic distribution, it has an important influence on the global radiation budget and on global climate. The relevant time scale for incorporation of snow cover, albedo, and water equivalance into large-scale climate models is about two weeks. Because much of the snow cover is in remote inhospitable areas with a paucity of surface measurement stations, remote sensing from satellite offers the only reasonable hope of obtaining such data.

In addition to the snow cover in the polar regions, the alpine snow cover and alpine glaciers in the midlatitudes are important to our understanding of global and regional climates and to our use of water resources (Colbeck et al., 1979; Walsh, 1984). Much of the uncertainty and sensitivity in the global hydrologic cycle lies in these reservoirs of frozen water, and their recent melting appears to account for much of the recent rise in sea level (Meier, 1984).

The most common use of remote sensing in snow studies is to monitor snowcovered areas (Rango and Itten, 1976; Rango et al., 1977). These efforts have been carried one step further by including satellite-derived measurements of snow-covered areas as an index in a snowmelt runoff model (Rango and Martinec, 1979). The next step is to use satellite radiometric data to measure or estimate snow water equivalence and snow surface properties that are necessary for the calculation of the surface energy balance (Dozier et al., 1981).

Identification of snow during daylight hours is straightforward during clear weather, because of the high albedo of snow in the visible wavelengths (Bohren and Barkstrom, 1974; Wiscombe and Warren, 1980). Typically, the visible wavelength bands of either the NOAA AVHRR or Defense Meteorological Satellite Program (DMSP) can be used. Automatic discrimination between snow and clouds is possible with a wavelength band between 1.55–1.75 μm, where snow is dark but clouds are bright (Valovcin, 1976; Crane and Anderson, 1984; Dozier, 1984). Because persistent cloud cover over the Arctic during the snow melt period, passive microwave data are needed to supplement the observation in the visible wavelengths (Foster et al., 1984), but the principles of observation in these wavelengths is outside the scope of this chapter.

The last decade has been marked by advances in our understanding of the radiative properties of snow. The important conclusions from the considerable work in the optical wavelengths, where all radiation of importance to the energy balance is contained, is that the characteristics of snow reflectance can be modeled by the radiative-transfer equation (Warren, 1982). In the visible wavelengths, ice is highly transparent, so the albedo of snow is sensitive to small amounts of absorbing impurities (Warren and Wiscombe, 1980). In the near-infrared wavelengths, ice is more absorptive, so the albedo depends primarily on grain size (Bohren and Barkstrom, 1974; Wiscombe and Warren, 1980).

While grain size and the amount of absorbing impurities can be roughly estimated with present-day satellite sensors (Dozier, 1984; Dozier and Marks, 1987), the advent of high spectral resolution data in the 1990s should enable us to measure snow reflectance and estimate surface snow properties more accurately.

In alpine areas, the problems of spatial resolution are severe. Measurement of snow and ice properties by remote sensing requires satellite data on similar spatial scale to that of the topographic relief, i.e., a few tens of meters. Such data are available, either from the Landsat Thematic Mapper (TM) or from the French Systeme Probatoire pour l'Observation de la Terre (SPOT). The TM has the better spectral coverage, whereas SPOT has a finer spatial resolution and pointing capability. In either case, the extremely large data volumes make the analysis difficult. How we use such data to measure key areas of the alpine snow and ice cover is a challenging issue.

III OPTICAL PROPERTIES OF ICE

Snow is a collection of ice grains and air, and, when at $0°C$, it also has a significant fraction of liquid water. Snow also often includes particulate impurities—dust, soot, pollen and other plant material—and chemical impurities—small amounts of the major cations and anions. Thus, the optical properties of snow depend on the bulk optical properties and the geometry of the ice grains, the liquid water inclusions, and the solid and soluble impurities. In the visible and near-infrared wavelengths, the bulk optical properties of ice and water are very similar, so the reflectance and transmittance of the snow pack in this region of the electromagnetic spectrum depend on the wavelength variation of the refractive index of ice, the grain size distribution of the snow, the depth and density of the snow pack, and the size and amount of those impurities whose refractive indices are substantially different from those of ice and water.

In the microwave part of the spectrum, water is substantially more absorptive than ice, so in those wavelengths, minute amounts of water dramatically change the electromagnetic signature of the snow pack.

The most important optical property of ice, which causes spectral variation in the reflectance of snow in visible and near-infrared wavelengths, is that the absorption coefficient (i.e., the imaginary part of the refractive index) varies by seven orders of magnitude at wavelengths from 0.4–2.5 μm. Normally, the index of refraction is expressed as a complex number, $n + ik$. Figure 1 shows both the real and imaginary parts of the refractive index for ice and water. The important properties to note are (1) the spectral variation in the real part n is small and the difference between ice and water is not significant; (2) the absorption coefficients k of ice and water are very similar, except for the region between 1.55–1.75 μm, where ice is slightly more absorptive: (3) in the visible wavelengths, k is small and ice is transparent; and (4) in the near-infrared wavelengths, ice is moderately absorptive, and the absorption increases with wavelength.

The bottom part of Figure 1 shows the absorption in a slightly different manner. Transmission of light along distance s through a pure substance decays as $e^{-4\pi ks/\lambda}$, and the lower left graph shows spectral variation of $4\pi k/\lambda$. We also show in the lower right graph the e-folding distance for ice as a function of wavelength, i.e., the distance through which light will propagate through pure ice before its intensity is reduced to e^{-1} times its initial value.

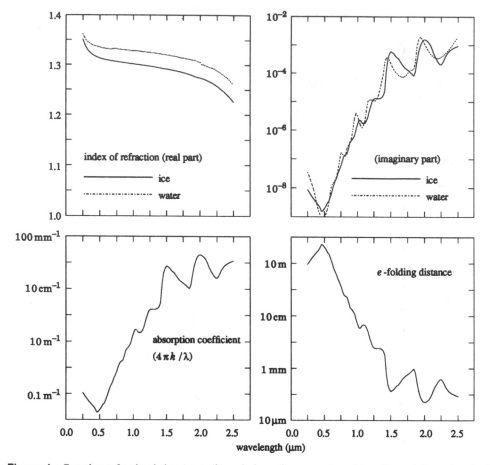

Figure 1 Complex refractive index ($n + ik$) and absorption properties of ice. Upper left: real part of refractive index (n). Upper right: imaginary part of refractive index (k). Lower left: absorption coefficient ($4\pi k/\lambda$). Lower right: e-folding distance, at which intensity is reduced to e^{-1}.

IV A MODEL FOR THE REFLECTANCE OF SNOW

We make the following assumptions in modeling the reflectance of snow, based on the equation of radiative transfer. Most of these assumptions have yet to be tested by rigorous measurements of physical properties and spectral reflectance of the same snow pack, but the model produces reflectance spectra that match those of snow. Warren (1982) has discussed these issues in his review of the optical properties of snow.

1. The reflectance of snow is modeled as a multiple-scattering problem, as first noted by Bohren and Barkstrom (1974). Scattering properties of irregularly shaped grains are mimicked by Mie calculations for an "equivalent sphere." Possible candidates for the equivalent sphere are the sphere with the same surface-to-volume ratio, the same projected area, the same volume, or the same surface area. These can be measured by stereological methods, as shown in Figure 2, and Davis et al. (1987) has used such analyses to investigate microwave properties of snow. Although snow grains are irregularly shaped, they are usually not oriented, so the assumption that their scattering prop-

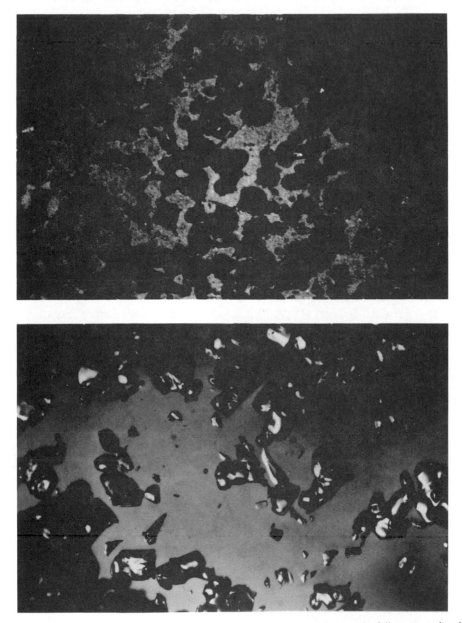

Figure 2   Stereological representation of snow microstructure (top) and photograph of disaggregated grains (bottom). By digitizing the stereological photograph, we can objectively measure geometrical properties of the ice grains. (Dozier et al., 1987.)

erties can be mimicked by some spherical radius r is reasonable, especially when we want to describe the general spectral properties. When we want details about the angular characteristics of the reflectance, the spherical assumption could become more critical.

2. Near-field effects are assumed unimportant. The fact that the ice grains in a snow-pack touch each other apparently does not affect the snow's reflectance, because the

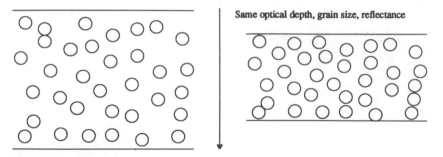

Figure 3   Near-field effects are assumed unimportant. The two representations of snow would have the same reflectance. Even though their densities are different, other properties are the same.

center-to-center spacing is still much larger than the wavelength. That is, snow reflectance is independent of density up to about 650 kg-m^{-3}. Reflectance measurements carried out under field conditions over a season and simply analyzed statistically *will* show a significant inverse relationship between density and reflectance, but the physical model shows that the explanation for changes in reflectance lies in other properties of the snow cover, namely an increase in grain size and in the amount of contaminants near the surface. Bohren and Beschta (1979) measured snow reflectance before and after artificially compacting the snow by driving a snowmobile over it, and found no change in reflectance. In Figure 3, the two representations of snow would have the same reflectance, even though their densities are different.

3. The effect of absorbing impurities (dust, soot) can be modeled either as separate spheres (smallest effect) or as concentric spheres with the impurity in the center (largest effect), as shown in Figure 4. These should bound the magnitude.

4. Because the complex indices of refraction of ice and water are similar, liquid water per se has little effect on the reflectance of snow. Instead, liquid water causes grain clusters (Colbeck, 1979), shown in Figure 5, which behave optically as large single grains.

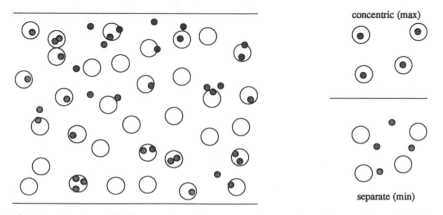

Figure 4   Absorbing impurities in snow (e.g., dust, soot, pollen) can be modeled either as separate spheres or as concentric spheres. The concentric-sphere configuration has the maximum effect, because refraction focuses light on the absorber.

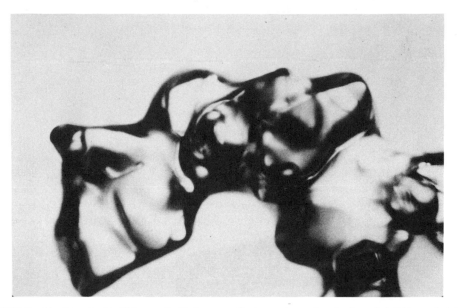

Figure 5   Liquid water itself does not affect snow reflectance at wavelengths from 0.4–2.5 μm, but the liquid water causes the snow to form grain clusters, which behave optically as larger single grains.

Given these assumptions, the scattering properties of the ice grains can be calculated by the Mie equations (Wiscombe, 1980), the complex angular-momentum approximation (Nussenzveig and Wiscombe, 1980), or, for larger grains, by geometric optics (Bohren, 1987). Then the radiative-transfer equation (Chandrasekhar, 1960) can be used to calculate the multiple scattering and absorption of the incident radiation.

$$\mu \frac{dL(\tau, \mu, \phi)}{d\tau} = -L(\tau, \mu, \phi) + J(\tau, \mu, \phi) \tag{1}$$

where L is radiance at optical depth τ in direction θ, ϕ; ϕ is the azimuth; θ is the angle from zenith, and $\mu = \cos\theta$. J is the source function; it results from scattering of both direct and diffuse radiation or, at thermal wavelengths, emission.

In our examination of the spectral properties of snow, it is computationally time consuming to calculate the angular distribution of the reflected radiation. But it is comparatively simple to examine the reflectance integrated over all angles. That is, we will restrict discussion to the spectral "distributional-hemispherical" reflectance (see Nicodemus et al., 1977). This reflectance is defined as the ratio of the reflected radiation over all angles divided by the incoming solar beam:

$$R_s(\theta_0) = \frac{\int_0^{2\pi} \int_0^1 \mu L(\mu, \phi)\, d\mu\, d\phi}{\mu_0 S_0} = \frac{F_\uparrow}{\mu_0 S_0} \tag{2}$$

where θ_0 is the incident illumination angle, $\mu_0 = \cos\theta_0$, and S_0 is the direct illumination, measured on a plane normal to the solar beam.

We can solve this kind of problem analytically with the "two-stream equations" for radiative transfer in a homogeneous medium (Meador and Weaver, 1980).

$$\frac{dF_\uparrow(\tau)}{d\tau} = \gamma_1 F\uparrow(\tau) - \gamma_2 F\downarrow(\tau) - \gamma_3 \omega_0 S_0 e^{-\tau/\mu_0} \tag{3a}$$

$$\frac{dF_\downarrow(\tau)}{d\tau} = \gamma_2 F\uparrow(\tau) - \gamma_1 F\downarrow(\tau) + \gamma_4 \omega_0 S_0 e^{-\tau/\mu_0} \tag{3b}$$

where $F_\uparrow$ and $F_\downarrow$ are upward and downward fluxes, respectively, ω_0 is the single-scattering albedo (i.e., the ratio of extinction by scattering to total extinction), and γ_i are parameters to approximate the scattering phase function. The Mie equations are used to calculate the single-scattering albedo ω_0 and the scattering asymmetry parameter g, and γ_i are functions of ω_0, g, and μ_0. To estimate the optical depth coordinate τ as a function of physical properties, we also need Q_{ext}, the extinction efficiency.

The total optical depth of a snowpack, τ_0, is a function of the extinction efficiency and the snow-water equivalence W (mass per unit area). The snow-water equivalence W is the product of mean snow density, ρ_{snow}, and depth d. Therefore,

$$\tau_0 = \frac{3W Q_{ext}}{4r\rho_{ice}} = \frac{3\rho_{snow}\, d Q_{ext}}{4r\rho_{ice}} \tag{4}$$

Wiscombe and Warren (1980) applied the delta-Eddington approximation to the γ_i parameters in the two-stream equations, i.e.,

$$\omega^* = \frac{(1 - g^2)\omega_0}{1 - g^2\omega_0}$$

$$g^* = \frac{g}{1 + g}$$

$$\gamma_1 = \frac{7 - \omega^*(4 + 3g^*)}{4}$$

$$\gamma_2 = -\frac{1 - \omega^*(4 - 3g^*)}{4}$$

$$\gamma_3 = \frac{2 - 3g^*\mu_0}{4}$$

$$\gamma_4 = 1 - \gamma_3$$

They then solved the two-stream equations for the directional-hemispherical reflectance of snow. For deep snowpacks, "semiinfinite," the underlying surface has no effect. Reflectance R_s is a function of illumination angle θ_0:

$$R_s(\theta_0) = \frac{\omega^*[\gamma_3(\xi + \gamma_1 - \gamma_2) + \gamma_2]}{(\xi + \gamma_1)(1 + \xi\mu_0)} \tag{5}$$

where

$$\xi = (\gamma_1^2 - \gamma_2^2)^{1/2}$$

For shallower snowpacks, the optical depth and the reflectance of the substrate, R_0, are needed.

$$R_s(\theta_0) = \frac{2\gamma_2 \xi e^{-\tau^*/\mu_0} \left(R_0 - \dfrac{\omega^* R^*}{1 - \xi^2 \mu_0^2} \right) + \omega^*(Q^+ P^+ - Q^- P^-)}{Q^+(\gamma_1 + \xi) - Q^-(\gamma_1 - \xi)} \qquad (6)$$

The additional variables needed for the finite-depth snowpack are

$$\tau^* = (1 - \omega_0 g^2)\tau_0$$

$$Q^{\pm} = e^{\pm \xi \tau^*}[\gamma_2 - R_0(\gamma_1 \pm \xi)]$$

$$P^{\pm} = \frac{\alpha_2 \pm \varepsilon \gamma_3}{1 \pm \xi \mu_0}$$

$$R^* = \mu_0(R_0 \alpha_1 - \alpha_2) + R_0 \gamma_4 + \gamma_3$$

$$\alpha_1 = \gamma_1 \gamma_4 + \gamma_2 \gamma_3$$

$$\alpha_2 = \gamma_2 \gamma_4 + \gamma_1 \gamma_3$$

V SPECTRAL CHARACTERISTICS OF THE REFLECTANCE OF SNOW

Figure 6 shows the spectral reflectance of pure deep snow for visible and near-infrared wavelengths, for snow grain radii from 50–1000 μm, representing a range for new snow to spring snow, although the grain clusters in coarse spring snow can exceed 0.5 cm in radius. Because ice is so transparent in the visible wavelengths, increasing the grain size does not appreciably affect the reflectance. The probability that a photon will be absorbed once it enters an ice grain is small, and that probability is not increased very much if the ice grain is larger. In the near infrared, however, ice is moderately absorptive. Therefore, the reflectance is sensitive to grain size, and the sensitivity is greatest at 1.0–1.3 μm. Because the ice grains are strongly forward-scattering in the near-infrared, reflectance increases with illumination angle (Figure 7), especially for larger grains.

The presence of liquid water in the snow does not by itself affect the reflectance. Except for meltwater ponds in depressions, where melting snow overlies an impermeable substrate, liquid-water content in snow rarely exceeds 5 or 6%. This small amount of water does not appreciably affect the bulk radiative-transfer properties, except possibly in those wavelength regions where the absorption coefficients are appreciably different. Instead, the changes in reflectance that occur in melting snow result from the increased crystal sizes and from an effective size increase caused by the two-to-four grain clusters that form in wet unsaturated snow (Colbeck, 1979, 1986). These apparently behave optically as single grains, causing decreased reflectance in near-infrared wavelengths.

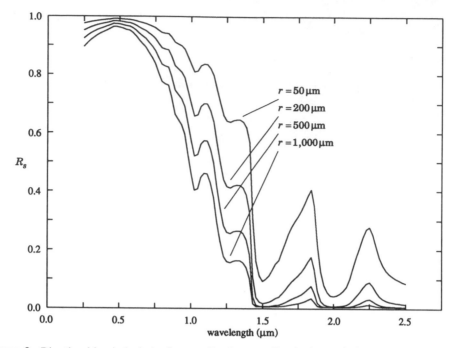

Figure 6 Directional-hemispherical reflectance R_s of snow at illumination angle $\theta_0 = 60°$ for wavelengths from 0.4–2.5 μm and grain radii r from 50–1 000 μm.

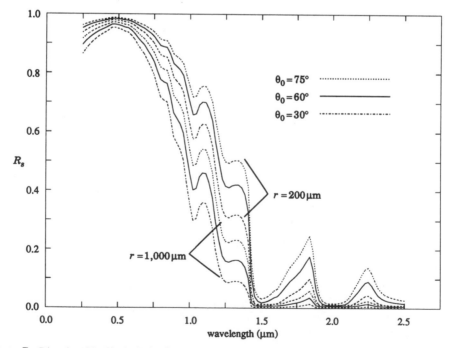

Figure 7 Directional-hemispherical reflectance R_s of snow at illumination angles $\theta_0 = 30°$, 60°, and 75°, for wavelengths from 0.4–2.5 μm and grain radii $r = 200$ and 1 000 μm.

O'Brien and Munis (1975) observed the spectral reflectance of a snow sample to be lower after warm air had been blown over it, but that the reflectance did not increase when the snow was refrozen.

There is no explicit dependence on density, at least for the semiinfinite snowpack. The natural increases in density are usually accompanied by increases in grain size;

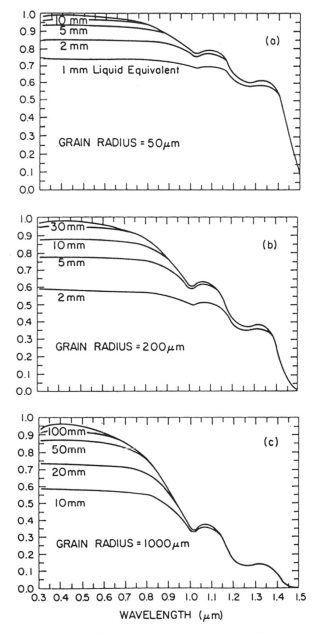

Figure 8 Directional-hemispherical reflectance R_s of finite-depth snow at illumination angle $\theta_0 = 60°$, for wavelengths from 0.4–1.5 μm, and grain radii $r = 50$, 200, and 1 000 μm. (From Wiscombe and Warren, 1980.)

hence, there will be a statistical correlation in field observations carried out over a season, but in an experiment where other variables were held constant, Bohren and Beschta (1979) measured spectrally integrated albedo before and after compacting the snow with a snowmobile and found no change.

In the visible wavelengths, reflectance is insensitive to grain size, but is affected by two variables, finite depth and the presence of absorbing impurities. The transmission of visible light through snow increases with grain size. Figure 8 shows the relationship of snow reflectance to snow-water equivalence, for a black substrate, i.e., $R_0 = 0$. For grain radius $r = 1000$ μm (1 mm), the reflectance is perceptibly reduced when the snow amount is reduced to 100 mm snow-water equivalance.

Warren and Wiscombe (1980) showed that minute amounts of absorbing impurities reduce snow reflectance in the visible wavelengths, where ice is highly transparent. Figure 9 shows that soot concentrations as low as 0.1 ppmw (parts per million by weight) are enough to reduce reflectance. The effect of the aborbing impurities is apparently enhanced when they are inside the snow grains because refraction focuses the light on the absorbers (Grenfell et al., 1981; Chýlek et al., 1983; Bohren, 1986).

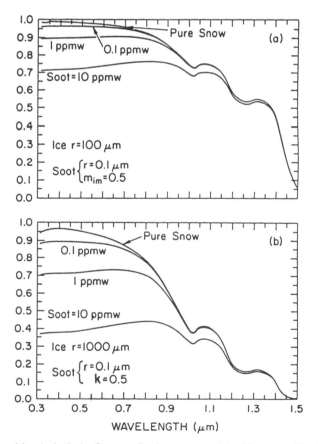

Figure 9   Directional-hemispherical reflectance R_s of snow contaminated by soot at illumination angle $\theta_0 = 60°$, for wavelengths from 0.4–1.5 μm, grain radii $r = 100$ (top) and 1 000 μm (bottom). (From Warren and Wiscombe, 1980.)

VI   MEASUREMENT OF SNOW PROPERTIES BY REMOTE SENSING

Satellite remote sensing in the visible and near-infrared wavelengths has become increasingly important to snow hydrologists because the data provide information on the spatial distribution of parameters of hydrologic importance. In snow and ice studies, remote sensing has been used to improve the monitoring of existing conditions and has been incorporated into several runoff-forecasting and -management systems. The principal operational use of remote sensing of snow properties has been to map the content of the snow cover. Throughout the world, in both small and large basins, maps of the snow cover throughout the snow season are used to forecast melt, both in areas with excellent ancillary data and in remote areas with no ancillary data (Rango et al., 1977; Andersen, 1982; Martinec and Rango, 1986).

Since the first mapping of snow cover from satellite, the spectral and spatial resolution of the available sensors has been much improved. The high spatial resolution satellites such as Landsat and SPOT and the medium-resolution sensors such as the NOAA AVHRR are widely used for mapping snow cover. The selection of the appropriate sensor depends on a trade-off between spatial and temporal resolution (Rott, 1987). The Landsat Multispectral Scanning System (MSS) has a spatial resolution of about 80 m and is suitable for snow mapping in basins larger than about 10 km^2 (Rango et al., 1983). Improved spatial resolution has been available since 1982 from the Landsat Thematic Mapper (30 m) and since 1984 from the French SPOT satellite (20 m in the multispectral mode and 10 m in the panchromatic mode). SPOT has the finest spatial resolution, but the Thematic Mapper has the best spectral coverage. Its band 4 (0.78–0.90 μm) allows estimation of the grain size from the measured reflectance, and its band 5 (1.57–1.78 μm) allows snow/cloud discrimination (Dozier, 1984; Dozier and Marks, 1987). Figure 10 shows an example of snow/cloud discrimination in the southern Sierra Nevada, California. In the left-hand image, composed from bands 2–4 of the Landsat Thematic Mapper, clouds can be discerned along the front of the mountains by their textural properties, but not their spectral characteristics. Texture, unfortunately, is difficult to analyze in image processing. The right-hand image, composed of TM bands 2, 5, and 7, clearly discriminates snow from clouds by spectral features.

Table 1, which uses data from Markham and Barker (1986) and Kidwell (1984), specifies the wavelength bands and saturation radiances for some common sensors used for remote sensing of snow: the Multispectral Scanner System (MSS) on Landsats 1–5, the Thematic Mapper (TM) on Landsats 4–5, and the Advanced Very High Resolution Radiometer (AVHRR) on NOAA 6–10. For the solar part of the electromagnetic spectrum, the values of the exoatmospheric solar irradiance at the mean Earth–Sun distance integrated over the wavelength bands are also given. The solar irradiance data used are from Neckel and Labs (1984) and Iqbal (1983). In the last column of the table, the sensor saturation radiance is expressed as a percentage of the exoatmospheric solar irradiance. If the product of the planetary reflectance and the cosine of the solar zenith angle exceeds this value, the sensor will saturate in this band.

Thus, in the visible wavelengths, we should be able to estimate the extent to which the reflectance of snow has been degraded, either by absorbing impurities or by shallow depth. However, this sensitivity would be best for the blue wavelengths, where the low saturation values of the Thematic Mapper make its use for this purpose difficult. In the near-infrared wavelengths, we should be able to estimate the grain size, and thus extend

TABLE 1   Satellite Sensor Radiometric Characteristics

Band	Wavelength Range (μm)		Radiances ($\mathrm{Wm^{-2}\,\mu m^{-1}\,sr^{-1}}$)		
			L_{max}	L_{solar}	%
LANDSAT Multispectral Scanning System (mean of LANDSATS 1–5)					
MSS1	0.49	0.61	268	590	45
MSS2	0.60	0.70	179	503	36
MSS3	0.70	0.81	159	404	39
MSS4	0.79	1.07	123	285	43
LANDSAT-4 Thematic Mapper					
TM1	0.45	0.52	158.4	623.3	25.4
TM2	0.53	0.61	308.2	581.9	53.0
TM3	0.62	0.69	234.6	496.2	47.3
TM4	0.78	0.90	224.3	332.6	67.4
TM5	1.57	1.78	32.42	69.74	46.4
TM7	2.10	2.35	17.00	23.74	71.6
LANDSAT-5 Thematic Mapper					
TM1	0.45	0.52	152.1	622.9	24.4
TM2	0.53	0.61	296.8	582.2	51.0
TM3	0.62	0.69	204.3	495.6	41.2
TM4	0.78	0.90	206.2	333.3	61.9
TM5	1.57	1.78	27.19	69.81	39.0
TM7	2.10	2.35	14.38	23.72	60.6
NOAA Advanced Very-High Resolution Radiometer					
AVHRR1	0.56	0.72	518	485	107
AVHRR2	0.71	0.98	341	364	94

SOURCES: Markham and Barker (1986); Kidwell (1984).

the estimate of the spectral albedo throughout the solar wavelength. Moreover, this information would help us interpret the spectral signature of snow at microwave frequencies.

Figure 11 shows some efforts to accomplish this interpretation, but the results are qualitative. The figure shows three pairs of images of the southern Sierra Nevada in the winter of 1985; January 24, February 25, and April 14. In the top row of the figure, the images are made from TM bands 1, 2, and 4, and in the bottom row, they are made from TM bands 5, 4, and 2. The increasingly yellowish cast to the bottom row of images as the season progresses indicates a decreasing reflectance in bands 5 and 4, and thus an increasing grain size.

Figure 10   Snow/cloud discrimination from the Landsat Thematic Mapper. The top photograph shows a false-color image made from TM bands 2–4, and the bottom photograph was made from bands 2, 5, and 7. The ability of TM band 5 to discriminate clouds from snow shows clearly.

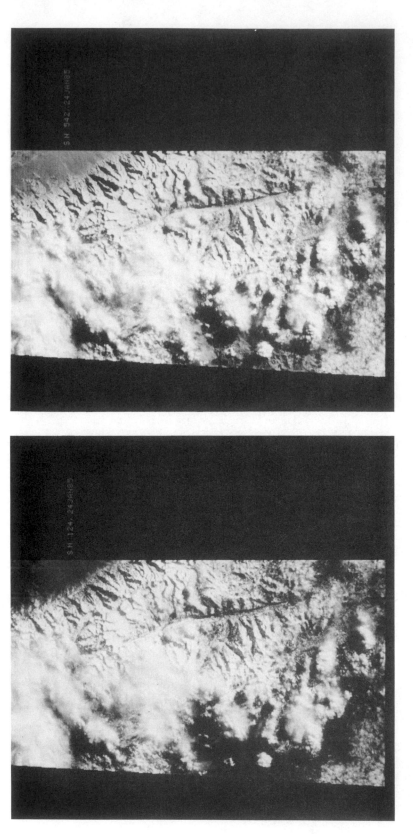

(b)

(a)

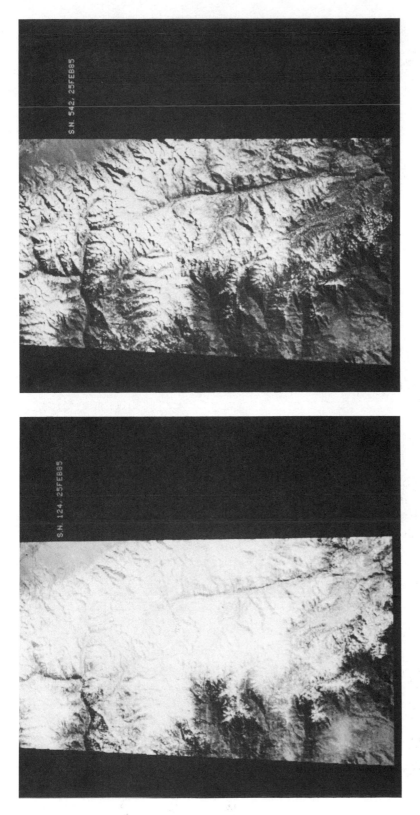

(c)

(d)

Figure 11 Snow properties from the Landsat Thematic Mapper in the southern Sierra Nevada in 1985: January 24 (*a, b*), February 25 (*c, d*), and April 14 (*e, f*). Parts *a, c,* and *e* show images in bands 1 (assigned to blue in the color rendition), 2 (assigned to green), and 4 (assigned to red). Parts *b, d,* and *f* show images masked for snow cover in bands 5 (assigned to blue), 4 (assigned to green), and 2 (assigned to red). The increasingly yellowish cast in parts *b, d,* and *f* from January to April results from a decrease in reflectance in bands 4 and, especially, 5, indicating an increase in grain size.

543

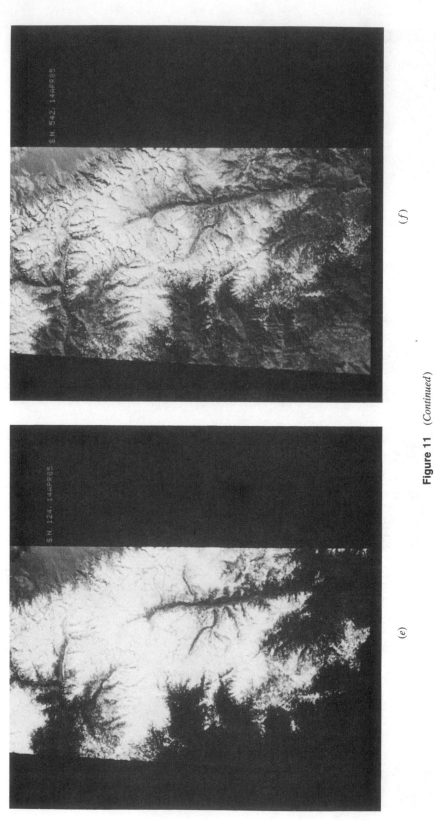

S.N. 542. 14APR85

(f)

S.N. 124. 14APR85

(e)

Figure 11 *(Continued)*

544

The largest sensitivity of snow reflectance to grain size occurs in the wavelengths from 1.0–1.3 μm, beyond the range of TM band 4 and which are currently not covered by any sensor. Future planned sensors with continuous spectral coverage, such as the High-Resolution Imaging Spectrometer (HIRIS) planned for the Earth-Observing Satellite (National Aeronautics and Space Administration, 1987), should allow much more precise interpretation of the physical properties of surface snow.

VII SUMMARY

Multispectral measurements in visible and near-infrared wavelengths have been used to map snow for more than two decades. Improved spectral coverage from the Landsat Thematic Mapper has allowed better estimation of snow properties and discrimination between snow cover and cloud cover. Future sensors with better spectral resolution should allow estimation of grain size and contamination by absorbing impurities, which in turn can be used to calculate spectral reflectance through the wavelengths of the solar spectrum.

ACKNOWLEDGMENTS

My work on snow has been supported principally by the National Aeronautics and Space Administration and the California Air Resources Board. Reference herein to any specific commercial product, process, or service by trade name, trademark, manufacturer, or otherwise, does not constitute or imply its endorsement by the United States Government, the University of California, Santa Barbara, or the Jet Propulsion Laboratory, California Institute of Technology.

REFERENCES

Andersen, T. (1982). Operational snow mapping by satellites. *IAHS Publ.* **138**:149–154.

Bohren, C. F. (1986). Applicability of effective-medium theories to problems of scattering and absorption by nonhomogeneous atmospheric particles. *J. Atmos. Sci.*, **43**:468–475.

Bohren, C. F. (1987). Multiple scattering of light and some of its observable consequences. *Amer. J. Phys.*, **55**:524–533.

Bohren, C. F., and B. R. Barkstrom (1974). Theory of the optical properties of snow. *J. Geophys. Res.* **79**:4527–4535.

Bohren, C. F., and R. L. Beschta (1979). Snowpack albedo and snow density. *Cold Reg. Sci. Technol.* **1**:47–50.

Chandrasekhar, S. (1960). *Radiative Transfer*. Dover, New York.

Chýlek, P., V. Ramaswamy, and V. Srivastava (1983). Albedo of soot-contaminated snow. *J. Geophys. Res.* **88**:10,837–10,843.

Colbeck, S. C. (1979). Grain clusters in wet snow. *J. Colloid Interface Sci.* **72**:371–384.

Colbeck. S. C. (1986). Classification of seasonal snow cover crystals. *Water Resour. Res.* **22**:59S–70S.

Colbeck, S. C., E. A. Anderson, V. C. Bissel, A. G. Crook, D. H. Male, C. W. Slaughter, and D. R. Wiesnet (1979). Snow accumulation, distribution, melt, and runoff. *Eos, Trans. Am. Geophs. Union* **60**:465–474.

Crane, R. G., and M. R. Anderson (1984). Satellite discrimination of snow/cloud surfaces. *Int. J. Remote Sens.* **5**:213–223.

Davis, R. E., J. Dozier, and A. T. C. Chang (1987). Snow property measurements correlative to microwave emission at 35 GHz. *IEEE Trans. Geosci. Remote Sens.* **GE-25**:751–757.

Dozier, J. (1984). Snow reflectance from Landsat-4 Thematic Mapper. *IEEE Trans. Geosci. Remote Sens.* **GE-22**:323–328.

Dozier, J., and D. Marks (1987). Snow mapping and classification from Landsat Thematic Mapper data. *Ann. Glaciol.* **9**:97–103.

Dozier, J., S. R. Schneider, and D. F. McGinnis, Jr. (1981). Effect of grain size and snowpack water equivalence on visible and near-infrared satellite observation of snow. *Water Resour. Res.* **17**:1213–1221.

Dozier, J., R. E. Davis, and R. Perla (1987). On the objective analysis of snow microstructure. *IAHS Publ.* **162**:49–59.

Foster, J. L., D. K. Hall, A. T. C. Chang, and A. Rango (1984). An overview of passive microwave snow research and results. *Rev. Geophys. Space Phys.* **22**:195–208.

Grenfell, T. C., D. K. Perovich, and J. A. Ogren (1981). Spectral albedoes of an alpine snowpack. *Cold Reg. Sci. Technol.* **4**:121–127.

Iqbal, M. (1983). *An Introduction to Solar Radiation*. Academic Press, Toronto.

Kidwell, K. B. (1984). *NOAA Polar Orbiter Data (TIROS-N, NOAA-6, NOAA-7, and NOAA-8)*, User Guide, NOAA/NESDIS, Washington, D.C.

Markham, B. L., and J. L. Barker (1986). Landsat MSS and TM post-calibration dynamic ranges, exoatmospheric reflectances and at-satellite temperatures. *EOSAT Landsat Tech. Notes* **1**:2–8.

Martinec, J., and A. Rango (1986). Parameter values for snowmelt runoff modelling. *J. Hydrol.* **84**:197–219.

Meador, W. E., and W. R. Weaver (1980). Two-stream approximations to radiative transfer in planetary atmospheres: A unified description of existing methods and a new improvement. *J. Atmos. Sci.* **37**:630–643.

Meier, M. F. (1984). Contribution of small glaciers to global sea level. *Science* **226**:1418–1421.

National Aeronautics and Space Administration, (1987). HIRIS Instrument Panel, *HIRIS—High-Resolution Imaging Spectrometry: Science Opportunities for the 1990s*, Earth Observ. Syst. Rep. Vol. 2c, NASA; Washington, D.C.

Neckel, H., and D. Labs (1984). The solar radiation between 3300Å and 12500Å. *Sol. Phys.* **90**:205–258.

Nicodemus, F. E., J. C. Richmond, J. J. Hsia, I. W. Ginsberg, and T. Limperis (1977). Geometrical considerations and nomenclature for reflectance. *NBS Monogr. (U.S.)* **160**:1–52.

Nussenzveig, H. M., and W. J. Wiscombe (1980). Efficiency factors in Mie scattering. *Phys. Rev. Lett.* **45**:1490–1494.

O'Brien, H. W., and R. W. Munis (1975). *Red and Near-infrared Spectral Reflectance of Snow*, Res. Rep. No. 332. U.S. Army Cold Regions Research and Engineering Lab., Hanover, New Hamsphire.

Rango, A., and K. I. Itten (1976). Satellite potentials in snowcover monitoring and runoff prediction. *Nord. Hydrol.* **7**:209–230.

Rango, A., and J. Morrinec (1979). Application of snowmelt-runoff model using Landsat data. *Nord. Hydrol.* **10**:225–238.

Rango, A., V. V. Salomonson, and J. L. Foster (1977). Seasonal streamflow estimation in the Himalayan region employing meteorological satellite snow cover observations. *Water Resour. Res.* **14**:359–373.

Rango, A., J. Martinec, J. Foster, and D. Marks (1983). Resolution in operational remote sensing of snow cover. *IAHS Publ.* **145**:371–382.

Rott, H. (1987). Remote sensing of snow. *IAHS Publ.* **166:**279–290.

Valovcin, R. F. (1976). *Snow/cloud Discrimination*, Rep. AFGL-TR-76-0174. Air Force Geophysics Laboratory, Hanscom Air Force Base, Massachusetts.

Walsh, J. E. (1984). Snow cover and atmospheric variability. *Amer. Sci.* **72:**50–57.

Warren, S. G. (1982). Optical properties of snow. *Rev. Geophys. Space Phys.* **20:**67–89.

Warren. S. G., and W. J. Wiscombe (1980). A model for the spectral albedo of snow. II. Snow containing atmospheric aerosols. *J. Atmos. Sci.* **37:**2734–2745.

Wiscombe, W. J. (1980). Improved Mie scattering algorithms. *Appl. Opt.* **19:**1505–1509.

Wiscombe, W. J., and S. G. Warren (1980). A model for the spectral albedo of snow, I. Pure snow. *J. Atmos. Sci.* **37:**2712–2733.

14

KNOWLEDGE-BASED SPECTRAL CLASSIFICATION OF REMOTELY SENSED IMAGE DATA

STEPHEN W. WHARTON

Earth Resources Branch
Goddard Space Flight Center
National Aeronautics and Space Administration
Greenbelt, Maryland

I INTRODUCTION

Parametric methods for the spectral classification of remotely sensed data are based on the assumption that the target objects of interest can be systematically discriminated on the basis of the statistical distribution of their spectral signatures, i.e., mean vector and covariance matrix. The recording and analysis of these signatures, and their correlation with the targets they represent, provide the basis for the parametric representation of spectral knowledge. The problem with this representation is that the scope of the statistics is generally limited to the image from which it was derived because of the externally induced spectral variation (i.e., the effects of atmosphere, illumination, and viewing geometry) and the intrinsic variation associated with the target (e.g., vegetation varies with maturity, canopy closure, and moisture stress; soils vary with cultivation and moisture content; and man-made materials, rocks, and minerals vary with weathering). Each image must be treated as a separate case in which the statistical relationship between the targets and the observed spectral patterns must be uniquely defined.

The purpose of this chapter is to examine alternative methods for the representation and use of spectral knowledge as a means of avoiding the need for scene-specific optimization of statistical parameters, with particular attention given to the application of expert systems. The scene-independent representation of spectral knowledge is a critical element in the development of procedures, such as expert systems, to provide accurate and consistent spectral classification of multiple images for a finite scope of application (e.g., a designated set of targets, geographic locations, and sensor characteristics). Two explicit sets of knowledge are required by an expert system to accomplish this objective: a spectral knowledge data base that may be sensor- and target-specific, but not image-specific; and the procedural knowledge to perform image processing, interact with the spectral knowledge base, and make classification decisions. The primary advantage of the expert system approach is that the explicit representation of knowledge provides the means in which the pertinent information for classification can be quantified, documented, effectively shared with other researchers, and made available to less expert users.

The remainder of this chapter is organized as follows. The background section provides an overview of conventional representations of spectral knowledge in greater detail and provides an introduction to expert systems. A case study of a prototype expert system for spectral analysis of urban land cover is given in the next section to illustrate the types of spectral and procedural knowledge that are required. The prototype system was developed to demonstrate the feasibility of the spectral expert approach under ideal conditions, i.e., moderately high-resolution Thematic Mapper Simulator data (5 m pixel size and eight spectral bands) and highly contrasting spectral categories. The conclusions section identifies promising areas for future development of expert systems for spectral land cover classification by focusing on the extensions and enhancements required to make the prototype system a more effective and practical tool for general use.

An exhaustive survey of all of the different pattern-recognition techniques that have been used in the analysis of the various types of remotely sensed data would fill an entire book, if not several volumes. The following texts are suggested for further reading: image processing—Castleman (1979), Gonzalez and Wintz (1987), Pratt (1978), and Rosenfeld and Kak (1982); syntactic pattern recognition—Fu (1982); artificial intelligence—Winston (1984) and Tanimoto (1987); computer vision—Ballard and Brown (1982). The May 1987 issue of the *IEEE Transactions on Geoscience and Remote Sensing* includes a number of papers regarding applications of artificial-intelligence techniques to the analysis of remotely sensed data. A survey of different approaches to image understanding is given by Brady (1982). Other analysis techniques include segmentation (Haralick and Shapiro, 1985; Nazif and Levine, 1984), contextual classification (Gurney and Townsend, 1983), and texture analysis (Haralick, 1979; Modestino et al., 1981).

II BACKGROUND

II.A Parametric Classification

Parametric pattern-recognition techniques are widely used to classify spectral target categories in multispectral remotely sensed data (Landgrebe, 1981). The spectral signatures, generally referred to as *training statistics*, must be defined prior to classification for each spectral category of interest. The signatures are used in the classifier as prototypes to represent the characteristic spectral response of each target. The spectral vector of each pixel to be classified is considered in turn and compared via a distance function (e.g., euclidian distance or maximum likelihood) to each of the category prototypes. The unknown pixel is classified, i.e., assigned to a prototype class, for which the distance function is minimized. A classification map can be produced to show the spatial distribution of the classes identified. The texts by Duda and Hart (1973) and Swain and Davis (1978) provide additional discussion on parametric techniques.

The training statistics can be identified by a supervised or unsupervised analysis of the data. Supervised methods require the derivation of training statistics from the analysis of pixels within areas of spectral uniformity. These training areas must be located for each category of interest. It may sometimes be difficult or impossible to specify a full list of the categories to be identified, or to define training areas for small, irregular, or sparsely distributed categories. Unsupervised methods such as cluster analysis can be used to estimate signature statistics without the use of training areas and to map the important categories in a scene without predetermining them. Clustering methods classify data by locating groups of pixels having similar spectral responses without *a priori* knowledge regarding the location of the groups. The resulting clusters are subsequently labeled by the analyst. Additional background on cluster analysis is given by Anderberg (1973) and Hartigan (1975).

Parametric techniques can generally be used to produce accurate results if the target categories are spectrally separable and if sufficient human expertise is available to define and optimize the parameters for recognition. The decision rules are relatively simple and are known to be optimal, provided that the underlying assumptions are valid regarding the statistical distribution of the spectral responses (e.g., unimodal and multivariate normal). The problem is that the training statistics cannot be applied to data from other sensors, other areas, or other dates for the same area and sensor due to external spectral variation caused by differences in slope/aspect, illumination, atmospheric conditions, viewing geometry, and sensor calibration. The external variation could be alleviated by radiometric correction, i.e., computation of the characteristic reflectance associated with each target that is invariant to topography, position of the sun, atmosphere, and position of the viewer. However, it is not obvious that such correction is possible (Woodham and Gray, 1987). The intrinsic spectral variation associated with the targets themselves also contributes to this problem.

A review of the results derived from investigations conducted as part of the Large Area Crop Inventory Experiment (LACIE) by Heydorn et al. (1979) serves to further illustrate the difficulties imposed by parametric classification. The objective of LACIE was to estimate wheat production for an entire country using Landsat Multispectral Scanner data with the 90/90 accuracy criterion, i.e., that the wheat production estimate should be within 10% of the true production at least 90% of the time. The initial approach used in LACIE required the analyst to define training areas of wheat fields and nonwheat areas. Spectral statistics from the training areas were used as category prototypes in the classifier to label all pixels as wheat or nonwheat. The classification performance was found to be erratic as a result of the subjective nature of the training process and the results did not satisfy the accuracy criterion. The requirement to process a substantial amount of data in a short period of time prevented the analyst from qualitatively optimizing the training statistics for any one data set. An attempt was made to reduce analyst dependence by signature extension, i.e., using training statistics from one data set to classify other data. For example, one approach extended signatures only to those scenes immediately adjacent to the training data set. This approach failed due to the spectral variations induced by regional effects such as soil background, cropping practices, and atmospheric effects (Minter, 1979).

II.B   Spectral Transformations

An alternative approach to the representation of spectral knowledge is to organize or synthesize information from the multispectral bands into information that can be associated with the physical characteristics of ground classes. Crist and Cicone (1984) define this process as having three parts: understanding the relationships among the spectral bands for the ground cover classes of interest, compressing information from the original spectral bands into a smaller more manageable number of features, and extracting physical scene characteristics from the spectral features. This process is based on the observation that the spectral reflectance properties of natural land areas show only a limited range of spectral band measurement combinations in the data space defined by Landsat multispectral sensors.

The relative position of these spectral measurements in the data space is related to specific phenomena on the ground. The distribution of vegetation and soil reflectance properties in the visible (0.5–0.7 μm) and near-infrared (0.7–1.1 μm) illustrates this concept. Studies have shown that the land spectral patterns vary in overall brightness because of soil properties and vary spectrally as a result of the amount of "green" vegetation present (Kauth and Thomas, 1976). This two-component explanation of land spectral reflectance produces a triangular-shaped region in the visible and near-infrared space. The base of the triangle corresponds to the line of soils (Kauth and Thomas, 1976; Richardson and Wiegand, 1977). As vegetation grows on the soil, the red radiance generally decreases due to chlorophyll absorption and the near-infrared radiance increases due to cellular reflectance (Jackson, 1983a). The apex of the triangle is defined as the point at which additional layers of leaves produce no further increase in the near-infrared reflectance, nor red absorbtance. The position of the apex is essentially unaffected by soil background because the canopy is fully developed at this point (Jackson, 1983b).

The stage of vegetation or crop development can be inferred by the position of the visible and near-infrared measurements relative to the line of soils and canopy-saturation apex. Numerous vegetation indices have been developed to quantify this relationship using ratios and linear combinations of visible and near-infrared bands (Wiegand and Richardson, 1982; Tucker, 1979). One such method, called the *Tasseled Cap Transform*, was developed by Kauth and Thomas (1976) to rotate Landsat Multispectral Scanner data into dimensions that support a "physical" interpretation. The Tasseled Cap is a linear transform of the Multispectral Scanner spectral bands that captures the majority of the variation in agricultural scenes in two orthogonal dimensions. The first dimension, called *brightness*, corresponds to spectral variation related to target albedo. The second dimension, called *greenness*, is aligned in the spectral direction of principal variations associated with the amount of photosynthetically active green vegetation in the scene. The third and fourth dimensions, called *yellowness* and *nonsuch*, respectively, are used primarily as indicators of atmospheric condition. Crist and Cicone (1984) have developed an analogous Tasseled Cap transform for the six reflective Thematic Mapper spectral bands. With TM observations, they noted that at least a third dimension is added to the data that appears to be related to the "wetness" of the soils present.

Several methods have been developed to perform automated land cover analysis for agricultural areas on the basis of information derived from vegetative indices. Lambeck et al. (1978) devised a screening procedure to automatically identify pixels corresponding to clouds, cloud shadows, snow, water, or anomalous pixels in Landsat-2 Multispectral Scanner data. The data are first corrected for differences in illumination using a cosine correction for zenith angle. The Tasseled Cap transform is applied and fixed linear thresholds are used as decision boundaries for classification. It was found that the thresholds were sensitive to striping effects, especially for low sun elevation data, thereby causing false alarms. Richardson and Wiegand (1977) developed a procedure to automatically partition Landsat-1 or 2-Multispectral Scanner data into ten decision regions on the basis of the band-5 and band-7 responses. The decision boundaries between water, bare soil, and vegetation were referenced to the line of soils. The decision boundaries for low-, medium-, and high-cover vegetation were determined arbitrarily. The percentages of ground cover identified were found to agree with ground reference data. The absolute definition of decision boundaries used in both studies introduces difficulties similar to those of spectral classification, i.e., the results cannot be easily extended to other sensors nor land cover types other than agriculture.

II.C Expert Systems

An expert system is a computer program that uses knowledge and inference procedures to solve problems that are difficult enough to require significant human expertise for their solution (Mooneyhan, 1983). These programs represent, for a restricted domain (i.e., scope of application), expert knowledge in symbolic form that can be manipulated to solve limited sets of problems (Goldberg et al., 1985). In other words, expert systems emulate the decision processes of human experts via a formal declaration of decision rules. An expert system is intended to solve complex problems by the application of solution procedures in a self-determined processing order. The expert system should also function reliably in the presence of noise without necessarily having complete information. An overview of applications of expert and knowledge-based systems are given by Davis (1986) and Duda and Shortliffe (1983).

A major advantage of expert systems is the explicit representation of knowledge. Waterman (1986) describes artificial expertise as permanent, easy to transfer, easy to document, consistent, and affordable, as opposed to human expertise, which is perishable, difficult to transfer, difficult to document, unpredictable, and expensive. Knowledge as characterized by Goodenough (1986) includes the following advantages: it captures generalizations, it can be understood by people who must provide it, it can be easily changed to correct errors or to adopt a different point of view, and it can be used in a great many situations even if it is not complete or totally accurate. Expert systems provide a means of quantifying domain knowledge as used by expert analysts and making it available to less expert users.

Expert systems generally include three major components: a knowledge base, a rules interpreter, and a working memory (Mooneyhan, 1983). The knowledge base is an explicit representation of facts and rules pertinent to the problem to be solved. In traditional programs, the knowledge is in procedural form, i.e., an algorithm that states the exact steps for solution (Goldberg et al., 1985). Facts represent basic information about the problem. Rules represent the logical procedures that a human expert would follow in solving a problem. A rule specifies the prerequisite conditions for its activation and the actions to be performed on the data base. The rules interpreter accesses the knowledge base, interacts with the working memory, and controls the order in which the rules are applied. The working memory (sometimes referred to as a blackboard) keeps track of the current status of the problem solution.

The size of the knowledge base is subject to two limitations. The first is the difficulty of defining and organizing expert knowledge (Goldberg et al., 1985). This limitation is generally acknowledged to be the hardest task in implementing an expert system. The second limitation is the computational restriction on the number of rules. These limitations necessarily restrict the scope of a given knowledge base. In the 1960s, considerable effort was devoted to the development of general-purpose problem-solving programs. However, it was observed that single programs designed to address multiple types of problems performed poorly on any individual problem (Waterman, 1986). As a result, expert systems are generally designed to perform highly specific tasks. Waterman (1986) provides an index of over 150 expert systems, including applications in agriculture, chemistry, computer systems, electronics, engineering, geology, law, manufacturing, medicine, meteorology, physics, and space technology.

Papers reviewing the applications of expert systems to remotely sensed data are given by Estes et al. (1986) and Mooneyhan (1983). Many of the expert systems developed to date were designed to recognize objects in high spatial resolution aerial photography on the basis of spatial information, i.e., size, shape, texture, or association. For examples of such systems, see McKeown (1984), Selfridge and Sloan (1982), Matsuyama (1987), and Hwang et al. (1983). Expert systems developed for analysis of multispectral image data include temporal classification of crops (Metzler et al., 1983), recognition of ridges and valleys (Wang et al., 1983), automated updating of forestry maps (Goldberg et al., 1985), and urban land-cover classification from color aerial photography (Nagao and Matsuyama, 1980).

Examples of systems that make classification decisions primarily on the basis of spectral information include the SPECTRUM system for analysis of imaging spectrometer data with 100 or more spectral bands (Borchardt, 1986; Choiu, 1985), the MSIAS system for Thematic Mapper classification (Ferrante et al., 1984), and a knowledge-based urban land-cover classification in TM Simulator data (Wharton, 1987). The SPECTRUM and MSIAS systems are briefly described in what follows. The system by Wharton is presented as a case study in the next section to illustrate a rule-based representation of spectral knowledge for moderate spectral resolution data (eight spectral bands). This system was designed to minimize the incidence of classification errors due to noise by (1) making classification decisions on the basis of convergent evidence, i.e., multiple features that support the same conclusion; (2) maximizing the amount of evidence available for each pixel by using multiple spectral and contextual features and multiple rules per feature; and (3) utilizing spectral and spatial consistency rules that are specifically tailored for uniform, border, noise, and outlier contextual interpretations.

II.C.1 SPECTRUM

The SPECTRUM system is a prototype knowledge-based for geological interpretation of imaging spectrometer data (Borchardt, 1987). The imaging spectrometer is a descendent of the Thematic Mapper and Multispectral Scanner instruments, recording image data over 100 or more spectral bands, as opposed to four and seven bands for the Thematic Mapper and Multispectral Scanner, respectively (Goetz et al., 1985). NASA is developing a series of imaging spectrometers, including the Airborne Imaging Spectrometer (AIS) with 128 spectral bands over the wavelengths 1.2–2.4 μm and the Airborne Visible and Infrared Imaging Spectrometer (AVIRS) with 224 spectral bands over the wavelengths 0.4–2.4 μm (Borchardt, 1986). The information content of this data, allowing direct identification of minerals, was not present in previous related instruments.

A primary objective in the analysis of imaging spectrometer data is the identification of surface minerals, as various minerals exhibit distinct spectral reflectance characteristics. The image data are clustered into classes of pixels based on spectral similarity, and the classes are then identified through comparison with a library of spectral reflectance characteristics for tested samples of minerals. The spectral library matching is performed on the basis of the locations of the reflectance and absorption features of the minerals as represented by the peaks and valleys of their corresponding reflectance curves (i.e., plots of reflectance versus wavelength). Classes that are not readily identified are analyzed further to determine whether or not they may correspond to mixtures of minerals. From the perspective of knowledge-based systems, the interpretation of this data is improved through the consideration of domain knowledge from the field of geology and from contextual information drawn from the image itself (Borchardt, 1986). Domain knowledge for this task includes standard classifications of minerals, their weathering properties, relative abundances, common associations, and occurrences for particular geological settings. Contextual information drawn from the image includes relative sizes and textures of the regions of similarity in the image and adjacencies between these regions.

II.C.2   MSIAS

The multispectral image analysis system (MSIAS) was developed by Ferrante et al. (1984) to perform knowledge-based spectral classification of Landsat Thematic Mapper data. Their approach was to develop rules for image classification on the basis of expert knowledge of the appearance of materials as viewed by a particular sensor under specified imaging conditions. The intent was to develop a representation that will allow materials to be recognized over a wider range of scene conditions than possible with statistical classification. Because of the human tendency to describe things in relative terms, rather than in absolute terms, they proposed a knowledge representation based on relative image measures.

Two types of relative image measures were developed to characterize spectral categories. The first measure is an analysis of trends in the per-pixel response, e.g., a peak at a particular wavelength. Examination of this spectral signature provides some indication of the appearance of the material as a function of wavelength. For example, green vegetation tends to have a maximum response in the near infrared. The second measure is the distribution of intensities for a particular spectral band, i.e., histogram analysis. The histogram summarizes the relative frequencies of intensities in the given band for the image. Knowledge of the scene contents makes it possible to relate modes in the histogram to instances of particular materials in the image. For example, because the reflectivity of water in the infrared is low, if the scene contains water, then the darkest regions in the infrared are likely to be water.

MSIAS was designed to use a hierarchical decision-tree structure to represent knowledge of spectral classes. The decision-tree structure as applied to image data recursively classifies each pixel, where each level of the tree defines a more refined classification. An application of this technique to TM data had two classification levels. The first level is divided into three spectral classes: soil-like, water, and vegetation. In the second level, the soil-like class is subdivided into concrete/silt, plowed fields, and other. The vegetation class is subdivided into crops and other. The spectral knowledge regarding signatures and histograms is represented in the form of production rules. A rule consists of an antecedent (if portion) and a consequent (then portion) (Carlutto et al., 1984). If all of the conditions specified in the antecedent are satisfied, then the rule is executed, i.e., the action(s) specified in the consequent are performed.

One set of rules is applied at the first level to recognize soil-like, water, and vegetation categories. A second set of rules is used to subdivide the vegetation category and a third set is used to subdivide the soil-like category. Examples of the rules used to identify the water, soil-like, and vegetation categories in an example of Thematic Mapper classification are given as follows:

> If: (band-4 relative intensity is dark)
> Then: (assert water)

If: (band-4 > band-3) and (band-4 > band-5)
Then: (assert vegetation)

If: (band-4 < band-5)
Then: (assert soil-like)

III SPECTRAL EXPERT SYSTEM CASE STUDY

III.A Approach

A prototype expert system was developed by Wharton (1987) to demonstrate the feasibility of using knowledge of the spectral relationships within and between categories as the basis for classification. The assumption is that these relationships are valid over a wider range of scene conditions than absolute spectral measurements. The performance objective is to correctly recognize pixels whose land-cover category can be reliably determined from visual interpretation of the reflectance data. This objective has four implications: (1) the data are assumed to be calibrated to ground reflectance; (2) the spectral categories are well illuminated, i.e., not in diffuse or dense shadow; (3) the pixels to be recognized are relatively pure, i.e., the spectral response is dominated by a single category; and (4) the categories can be recognized without the use of category-specific spatial knowledge, i.e., size, shape, and texture.

The feasibility of the prototype spectral expert was evaluated by testing its classification accuracy as applied to Thematic Mapper Simulator (TMS) and Thematic Mapper (TM) test data collected over the Washington, D.C., metropolitan area. The TMS data were collected by the NASA-Ames Research Center C130 aircraft in a series of North-South flight lines on October 27, 1982. The TMS data have seven spectral bands in the visible, near-infrared, and middle-infrared wavelengths and one band in the thermal-infrared wavelengths with 5 m spatial resolution and are assumed to satisfy the pure-pixel assumption. The TMS test images include the Goddard Space Flight Center, the Greenway Plaza shopping center, and the University of Maryland in College Park. The TM data have six spectral bands in the visible, near-infrared, and middle-infrared wavelengths with 30 m spatial resolution and one band in the thermal-infrared wavelengths with 120 m spatial resolution and are not assumed to satisfy the pure-pixel assumption. The TM test image includes Greenbelt Lake and the Greenway Plaza.

False-color images for all of the test data are shown in Figure 1. The black-and-white images corresponding to the individual spectral bands for the Greenway Plaza TMS image are shown in Figure 2. Examination of false-color imagery and aerial photography of the area shows that the dominant land cover is vegetation, consisting of forest, senescent crop fields, and grass. The second most common land cover is concrete and asphalt pavement, corresponding to the many parking lots, highways, and streets typical of urban development. Numerous buildings are represented, including single-unit houses, apartment buildings, dormitories, shopping malls, warehouses, and classrooms. Roof materials include shingle, slag, asphalt, metal, gravel, and slate. Exposed soil is evident in several construction sites and harvested cropfields. The area also includes several small lakes and ponds, primarily in parks or golf courses. Various areas are shadowed by adjacent forest or buildings.

The accuracy of visual interpretation was attributed, in part, to the ability of the analyst to integrate knowledge of the color, contrast, and tonal properties of each category and to consider the local distribution of spectral categories that serves to confirm

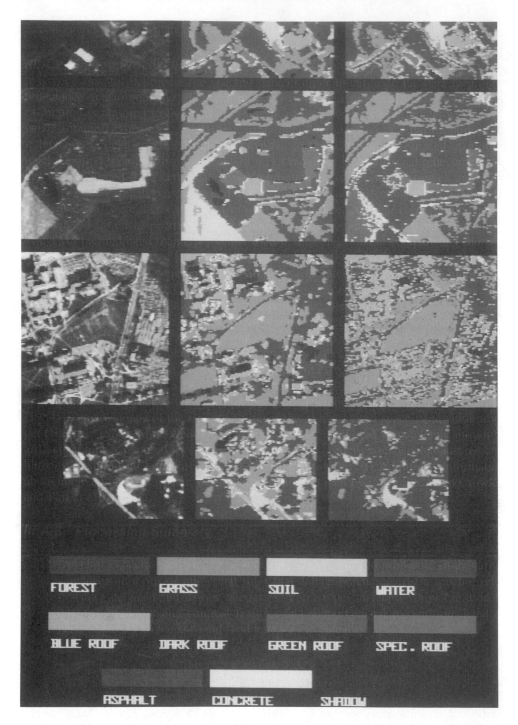

Figure 1   False-color (left column), color-coded spectral expert classifications (middle column), and color-coded KMEANS classifications (right column) for the test images: top row—TMS image of the Goddard Space Flight Center; second row—TMS image of Greenway Plaza; third row—TMS image of the University of Maryland; and bottom row—TM image of the city of Greenbelt.

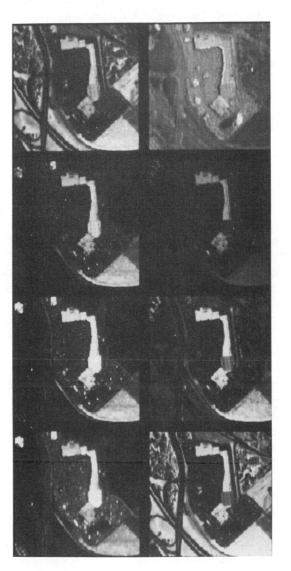

Figure 2 Black-and-white images corresponding to the eight individual spectral bands in the Greenway TMS image. The bands are numbered from left to right with bands 1–4 in the top row and bands 5–8 in the bottom row.

or contradict the label associated with a given pixel. In other words, the analyst uses multiple sources of spectral and contextual information and assigns labels on the basis of convergent evidence, i.e., two or more features that corroborate one another to support the same conclusion. To adopt a similar approach in the spectral expert, it is necessary to quantify the implicit criteria as used in the decision-making process, i.e., the spectral, contextual, and procedural knowledge must be explicitly defined. The design concepts, image processing, and overview of the processing steps follow. The spectral and procedural knowledge are described in subsequent sections.

III.A.1 Design Concepts

Convergent Evidence. The concept of convergent evidence as described by Milgram (1978) is used in making classification decisions. The intent is to base decisions upon multiple information sources to minimize dependence upon any single source. The prototype spectral expert is intended to emulate human performance in correctly recognizing pixels whose spectral categories can be determined unambiguously in manual interpretation of the TMS spectral data. The system was designed to minimize the effects of noise by making classification decisions on the basis of convergent evidence, i.e., multiple features that support the same conclusion. The advantage of using convergent evidence as opposed to stepwise decisions is that failure of individual rules due to random variation, noise, or mixed pixels does not necessarily jeopardize the final decision.

Pyramid-Image Representation. It was assumed, in the initial development of the expert system, that the use of convergent evidence would be sufficient to promote accurate recognition. The classification rule was to select the category that has the maximum amount of supporting evidence. Preliminary experimentation showed that it was not possible to achieve a reliable classification at a single spatial resolution. The problem is that the system lacks sufficient information to resolve ambiguous or incorrect spectral evidence due to noise in the data. A hypothetical example of such noise is the false recognition of roofs within a parking lot due to the presence of highly reflective car roofs. A potential solution is to smooth the data by computing a weighted average of the pixels in the local neighborhood.

The amount of smoothing increases with the size of the averaging filter, i.e., the number of pixels used in computing the weighted average. The application of the averaging filter, in effect, reduces the spatial resolution of the image. The problem with this approach is that there is generally no single optimal filter size for a given image. An increase in the size of the filter provides greater noise reduction, but also blurs the boundaries between adjacent spectral categories. A decrease in the filter size reduces edge blurring, but eliminates less noise. To avoid the limitations of a fixed spatial resolution, the prototype system was designed to process data in which each pixel is represented by a range of spatial resolutions.

A mean pyramid is used to implement the multiple-resolution image. The pyramid is an image data structure consisting of multiple levels numbered from 0 to n. Each succeeding level represents a property of the image computed at a decreasing spatial resolution. The bottom level (level 0) represents the image at full spatial resolution. Each pixel at the ith level is computed as a function of a window of pixels at the $(i - 1)$st level. For example, windows based on nonoverlapping kernels of two by two pixels yield a pyramid whose spatial resolution decreases by a factor of 2 between successive levels. The function used in the mean pyramid is simply the weighted average of pixel values in the window. Additional smoothing can be introduced by computing the average over a window that includes the two by two pixel kernel surrounded by a one-pixel border. Only the bottom three levels of the image pyramid (i.e., levels 0, 1, and 2) are used by the spectral expert.

An advantage of the image pyramid is the computational efficiency resulting from being able to replace global operations by local operations (Hartley, 1984). This allows features representative of large portions of the image to be computed in approximately the same time as local features. For example, the contrast between a pixel and its local neighborhood can be computed by subtracting values from two pyramid levels. Without the pyramid structure, it would be necessary to process the nine pixels in the local neighborhood to compute the background response. Another advantage of the pyramid is that the computations do not have to be modified to account for differences in scale, i.e., the size of the objects or neighborhoods being considered. Such differences only require that the same computations be performed at another level of the pyramid. For additional detail regarding pyramids, see Burt et al. (1981) and Tanimoto and Pavlidis (1975).

Contextual Interpretation. The multiple spatial resolutions as represented by the image pyramid are used, in effect, to reinforce or refute a given classification decision. The category having the maximum amount of supporting evidence is selected for each pyramid level. The result is a vector of decisions for each pixel. The rules used to accept or reject the classification decision at a given level depend upon the local contextual interpretation associated with the pixel under consideration. Four types of contextual interpretations were used: (1) uniform—the pixel occurs in the interior of a relatively large spectrally uniform region, (2) border—the pixel occurs within a spectrally uniform region, but is close to its border, (3) noise—the pixel occurs within or on the edge of a uniform region and has a different classification, and (4) outlier—the pixel does not occur within or next to a uniform region pixel and has a different classification. Contextual interpretation experts were developed to make classification decisions for each of the four possible interpretations. Each expert also incorporates additional rules to verify that its respective interpretation is consistent with the local pixel distribution. Spatial consistency is verified by examining the local frequency distribution of previously identified categories. The interpretation experts are described in detail in a subsequent section.

Spectral Features. The spectral features to be used to discriminate the categories were derived from an analysis of the spectral clues that support category recognition in manual interpretation. According to Estes et al. (1983) the most important information source for manual interpretation is provided by contrast in the tone/color properties of objects. The importance of contrast supports the hypothesis that knowledge of the spectral relationships within and between land-cover categories can be used to avoid the scene-specific limitations associated with fixed thresholds and parametric measures. Color is represented by the band-to-band (BB) feature. Contrast is represented by the category-to-background (CB) and category-to-category (CC) features. The band-to-band feature represents the expected relationship between spectral bands and is a generalization of color perception. The category-to-background feature describes the contrast relationships between pixels belonging to a given category and the average response of neighboring pixels belonging to other categories. The category-to-category feature represents the contrast relationships between spectral categories.

This approach takes advantage of the human tendency to describe things in relative terms and simplifies the development of a spectral knowledge base because category descriptions can be derived from visual interpretation. A similar rationale was used for feature derivation by Carlutto et al. (1984) and Ferrante et al. (1984). Multiple rules are used within each relevant feature to characterize the spectral appearance of each category. Confidence measures are also used as part of the knowledge to indicate how much weight is to be attached to the evidence provided by a given rule in supporting or contradicting a category. The flexibility afforded by multiple rules and confidence measures allows the knowledge designer to distinguish between and exploit differences in false rejection and false acceptance-error rates.

III.A.2 Image Processing

It is necessary to preprocess the spectral data in order to calibrate the data to ground reflectance and to compute the mean pyramid from the calibrated image. Reflectance calibration is required to facilitate meaningful comparisons between reflective spectral bands for the BB feature. A relative calibration is required so that the proportion of reflectance in each spectral band is equivalent to the distribution recorded at ground level. Calibration to an absolute standard is not required. Ground-based reflectance values were recorded for spectral categories representative of opposite extremes in brightness, i.e., new asphalt pavement and gravel roofs. The digital counts corresponding to the same areas were then extracted from the spectral data. A linear regression equation was derived for each spectral band to transform digital counts into ground reflectance. The TMS thermal band (band 8) was not used in the band-to-band comparisons and was not calibrated.

The mean pyramids were computed from the calibrated image data. The number of pyramid levels was selected to provide five spatial scales for computation of the spectral features. A total of six resolutions were required because the CB feature is computed between adjoining levels. Level 0 of the pyramid represents the full resolution image. The remaining five levels were computed for each spectral band by computing the weighted average for nonoverlapping two by two pixel kernels with a one-pixel wide border, i.e., each weighted average is computed over 16 pixels. Spatial resolution decreases by approximately a factor of 2 between levels. The weights were selected to approximate a Gaussian filter for the 4 by 4 pixel window as determined by Hartley (1984). The advantage of the Gaussian filter is that it has a good space-bandwidth product (Hartley, 1984). More weight on the pixels within the 2 by 2 kernel would introduce artifacts at higher pyramid levels. More weight on the border pixels would cause greater blurring, resulting in a loss of detail at higher pyramid levels. Additional information about applications of Gaussian weighted pyramids is given by Burt (1981) and Crowley and Parker (1984).

III.A.3 Processing Summary

1. Develop a spectral knowledge base to describe the spectral properties of the target categories in terms of numerical rules that define the characteristic spectral relationships between and within categories. Three types of relationships are used to quantify the color and contrast features that support visual recognition: band to band, category to background, and category to category.

2. Calibrate the image by converting each of the reflective spectral bands from digital number to ground reflectance. A relative calibration was used in which the proportion of reflectance in each band is equivalent to the distribution recorded at ground level. Calibration is required to facilitate meaningful comparisons between spectral bands for evaluation of the band-to-band feature.

3. Average the calibrated data to compute a mean pyramid. The bottom level of the pyramid represents the image at full resolution. Spatial resolution decreases by a factor of 2 for each succeeding level. The mean pyramid is used to expedite the evaluation of spectral evidence (step 5) by providing a representation in which each pixel is represented at multiple spatial resolutions.

4. Iteratively process the image by repeating steps 5–8 until the number of pixels that change categories between iterations is less than a given percentage (e.g., 5%). Multiple iterations are required to provide reliable estimates of the category distributions (updated in step 7) as used for evaluation of the category-to-category feature.

5. Compute the spectral evidence for each pixel by applying the rules in the spectral knowledge base to the mean pyramid data. An evidence array is used to record the results for all target categories at each resolution level in the pyramid.

6. Select the appropriate contextual interpretation for each pixel, i.e., uniform, border, noise, or outlier, by examining the evidence array and the local frequency distribution of classified pixels from the previous iteration (updated in step 8). Classification decisions are made on the basis of criteria associated with each interpretation.

7. Record the classification decision for each pixel and update the distribution statistics for the appropriate category for use in evaluating the category-to-category feature in the next iteration.

8. After each iteration, the local frequency distribution of classified categories is updated at each pyramid resolution level for all pixels.

III.B Spectral Knowledge

The spectral expert is designed to perform spectral classification on the basis of a spectral knowledge base that describes the spectral properties of the categories to be recognized. Two steps are required for the implementation of the knowledge base. The first is to define the spectral features to be used for category discrimination. The second is to describe each category in terms of the relevant spectral features. Numerical rules are used to represent these descriptions in the spectral knowledge base. The spectral features and the knowledge-base developments are discussed in the following sections.

III.B.1 Spectral Features

The band-to-band (BB) feature represents the relationship between spectral bands, i.e., the shape of the spectral reflectance curve. Figures 3 and 4 show the reflectance profiles of typical natural and man-made categories in the TMS data. The data must be calibrated in order for these profiles to be apparent, otherwise, they may be masked by sensor- or atmosphere-induced variations. The BB constraints are used to characterize the shape of

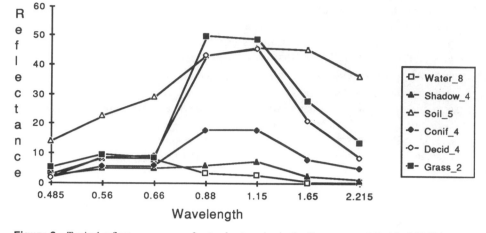

Figure 3 Typical reflectance curves of natural categories in the Greenway and Goddard TMS images.

the reflectance curve in terms of inequalities, for combinations of spectral bands, to avoid the use of absolute thresholds.

The category-to-background (CB) feature is based on the observation that the land-cover categories are often consistently brighter or darker than the surrounding pixels that belong to other categories. This feature is defined in terms of the expected sign of the contrast gradient between category pixels and the nominal background response. The sign of the contrast is defined for each spectral band. Examples of the spectral contrast patterns are given in Figure 2, which shows the eight individual spectral bands for the Greenway Plaza TMS image.

The category-to-category (CC) feature represents the contrast relationships between specific spectral categories. The CC feature is based on the observation that the distribution of a known category can be used to define limits on the distribution of other categories. The CC feature is defined for a given spectral band and comparison category in terms of the maximum overlap percentage with the comparison distribution and the designation of whether the overlap occurs on the upper or lower tail of the distribution.

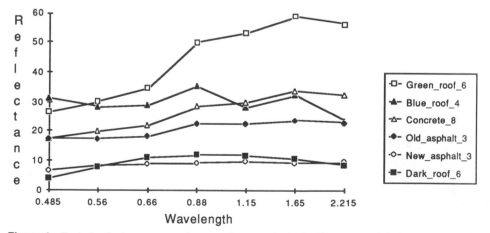

Figure 4 Typical reflectance curves of man-made categories in the Greenway and Goddard TMS images.

The tail percentages of 0, 10, and 50% are used to represent zero, nominal, and significant overlap between categories, respectively. In other words, a category can be greater than or less than 100%, greater than or less than 90%, or greater than or less than 50% of the comparison category. Because the comparison distribution is not known *a priori*, the spectral expert must be applied iteratively to obtain a reliable estimate.

III.B.2   Feature Representation

The features are represented by category-specific numerical rules that are evaluated as a function of the local distribution of reflectance values from the pyramid. Two problems must be addressed in the specification of rules. The first is to determine the number of rules to use. Type I (false rejection) errors could be introduced by having a large number of rules that must be satisfied for a category to be selected. Type II (false acceptance) errors could be introduced by having too few rules. The second problem is that all rules do not necessarily support binary decisions, i.e., accept if true and reject if false. These problems were alleviated by relaxing the implicit assumption that all rules must be satisfied for a category to be selected. The approach is to represent the knowledge-base rules in terms of three parameters that are used to direct the accumulation of evidence, i.e., a constraint, a supporting weight, and an opposing weight. The constraint is used for rule validation and is evaluated as a function of the local distribution of reflectance values for each pixel. The weights determine the relative significance of the evidence provided by a given constraint in supporting or opposing a category. If the constraint is validated, then the supporting evidence is incremented by the supporting weight. Otherwise, the opposing evidence is incremented by the opposing weight. Classification decisions are made on the basis of convergent evidence, i.e., the category having the highest ratio of supporting to opposing evidence is selected.

This method allows the analyst to exploit three types of knowledge: (1) mandatory constraints that describe spectral properties that are expected to be satisfied by all pixels belonging to a category; (2) optional constraints that describe spectral properties that are expected to be satisfied by a subset of the pixels belonging to a category; and (3) contradictory constraints that describe spectral properties that are known to contradict a category. Each type of knowledge has a different implication regarding the way in which evidence is conditionally updated. Mandatory constraints are represented by having the supporting and opposing weights greater than zero, so that evidence is added for valid constraints and subtracted otherwise. Optional constraints are represented by having only the supporting weight greater than zero, so that evidence is added if the constraint is valid; otherwise, do nothing because the failure of an optional constraint does not necessarily refute a category. Contradictory constraints are handled in a similar manner, except that evidence is subtracted if contradictory evidence is found; otherwise, do nothing because the absence of contradictory evidence does not necessarily indicate that a category is present.

III.B.3   Knowledge-Base Development

The knowledge base was developed from visual interpretation of per-pixel plots of reflectance values (e.g., Figures 3 and 4), black-and-white images of individual spectral bands (e.g., Figure 2), and three band false-color composite images (e.g., Figure 1) derived from the Goddard Space Flight Center and Greenway Plaza TMS test images. Eleven spectral categories were identified by visual interpretation: forest, grass, soil, water, asphalt, pavement, blue roofs, dark roofs, green roofs, specular roofs, and shadow. The roof categories were defined in an attempt to represent the spectral variation due to apparent differences in roof composition, i.e., asphalt, gravel, or metal. Test pixels representative of each spectral category were selected and used to develop preliminary descriptions of the relational features. Examination of the image and tabular data shows that the spectral features are not equally effective for all spectral categories. The natural categories of soil, vegetation, and water (Figure 3) were found, as expected, to have distinctive reflectance patterns. Contrast was found to be more useful for man-made categories of pavement and roof, because the reflectance patterns were highly variable and less distinctive (Figure 4). To illustrate the knowledge definition process, the spectral observations associated with the pavement and vegetation categories are described.

Pavement Categories.   Asphalt and concrete pavements have the reflectance profile typical of man-made materials, i.e., a generally increasing reflectance with increasing wavelength. The brightness of asphalt varies with the age of the pavement as a result of weathering, i.e., new asphalt is relatively dark and old asphalt is relatively bright. The shape of the spectral profile is highly variable due to the influence of cars, paint markings, oil marks, etc. An attempt was made to discriminate between asphalt and concrete by exploiting relative differences in the expected overlap with reference categories. The system was unable to reliably discriminate asphalt from concrete due to the similarity of their spectral curves; however, the pavement category as a whole could be readily discriminated from the natural and roof categories.

Examination of the black-and-white images of individual spectral bands shows that asphalt can be discriminated from roofs on the basis of differences in contrast, i.e., asphalt generally has a lower brightness (albedo) and a lower temperature than roofs. The ground base below the pavement acts as a heat sink to conduct heat away from the surface. Roofs generally have a higher surface temperature than asphalt because of the lack of such a heat sink (Goward, 1981). Knowledge of the relative distribution differences is exploited by using the expected background contrast and the expected overlap with other categories to define the limits of the asphalt distribution.

The flexibility provided by the variable evidence increments facilitated the use of the BB feature even though the reflectance profile of asphalt was known to be highly variable. This capability was exploited by including the three types of constraints mentioned earlier. Mandatory constraints were used to incorporate the consistent reflectance gradients, i.e., the average of bands 3 and 4 and the average of bands 5 and 6 are greater than the average of bands 1 and 2. An optional constraint was used to represent the ideal reflectance gradient, i.e., the average of bands 5 and 6 is greater than the average of bands 3 and 4. Contradictory constraints were used to represent the knowledge that asphalt is known to be dissimilar from vegetation and water.

The CB and CC rules were derived from visual interpretation of the black-and-white imagery. Asphalt is brighter than its background in bands 1, 2, 3, 7, and 8, and darker in bands 4, 5, and 6. The CB gradients in bands 1, 2, 6, and 7 were considered to be somewhat context-dependent and were treated as optional constraints. The remaining CB gradients were considered to be sufficiently consistent to be treated as mandatory constraints. The other categories except shadow were used as reference categories for the CC feature rules. The spectral rules for the asphalt category are given in Table 1.

Vegetation Categories.   Green or photosynthetically active vegetation can be recognized with relatively high accuracy because of its unique reflectance curve. These curves,

TABLE 1 Listing of the Band-to-Band, Category-to-Background, and Category-to-Category Spectral-Knowledge-Base Rules for the Asphalt Category

---

avg(3, 4)^a > avg(1, 2) [10, 10]^b; avg(5, 6) > avg(1, 2) [10, 10];
avg(5, 6) > avg(3, 4) [10, 0]; band(7) > avg(1, 2) [0,10];
avg(5, 6) < sum(2, 3)^c [0, 10];
avg(4, 5) < sum(1, 2, 3) {not forest vegetation} [0, 50];
avg(4, 5) < sum(6, 7) {not forest vegetation} [0, 50];
min(4, 5, 6)^d < sum(1, 2) {not soil} [0, 20];
avg(2, 3) < sum(4, 5, 6) {not water} [0, 20];

band(1) > background [10, 10]; band(2) > background [10, 0];
band(3) > background [10, 0]; band(4) < background [10, 10];
band(5) < background [10, 10]; band(6) < background [10, 10];
band(7) > background [10, 0] band(8) > background [10, 0];

band(1): <90% broof^e [0, 10]; >90% forest [10, 10]; <100% groof^f [0, 10]; <90% soil [0, 10];

band(2): <90% broof [0,10]; >90% forest [10,10]; <100% groof [0,10]; <90% soil [0, 10];

band(3): <90% broof [0, 10]; >50% forest [10, 10]; <90% groof [0, 10]; <90% soil [0, 10];

band(4): <50% forest [10, 10]; <100% grass [10, 10]; <90% groof [0, 10]; <90% soil [0, 10];

band(5): <50% forest [10, 10]; <100% grass [10, 10]; <90% groof [0, 10]; <90% soil [0, 10];

band(6): <50% grass [0, 10]; <90% groof [0, 10]; <90% soil [0, 10];

band(8): <10% broof [0, 10]; <90% droof^g [0, 10]; >90% forest [10, 10]; <50% groof [0, 10]; <90% soil [0, 10];

---

^{a}The term avg ( ) is the average of the bands in parentheses.
^{b}The numbers in [ ] are the supporting and opposing evidence weights, respectively.
^{c}The term sum ( ) is the sum of the bands in parentheses.
^{d}The term min ( ) is the minimum of the bands in parentheses.
^{e}The term broof stands for blue roof.
^{f}The term groof stands for green roof.
^{g}The term droof stands for dark roof.

with their typical peak reflectance in the near infrared bands, result from the differential reflectance and transmittance properties of green leaves (Myers, 1983). Leaves have a low reflectance and transmittance in the 0.35–0.7 μm region, due to chlorophyll and other pigment absorption at the blue and red wavelengths, with a slight rise at 0.55 μm in the green region. Transmittance and reflectance is higher in the near-infrared (0.75–1.35 μm) because of multiple scattering due to the leaf mesophyll structure. Strong water-absorption bands occur in the middle-infrared centered at 1.4, 1.9, and 2.7 μm.

The reflectance profile was considered to be such a reliable indication of vegetation that the CB and CC features were used only to discriminate forest from grass vegetation. Forest and grass have similar reflectance profiles. The difference is that grass generally has a higher reflectance in the near-infrared bands. The method used to discriminate between forest and grass was similar to that used for asphalt and concrete, i.e., the CB and CC features were used to exploit relative differences in the expected overlap with reference categories. The discrimination was more accurate, however, because of the greater difference in brightness. The spectral rules for the grass vegetation category are given in Table 2.

TABLE 2   Listing of the Band-to-Band, Category-to-Background, and Category-to-Category Spectral-Knowledge-Base Rules for the grass vegetation category[a]

$avg(4, 5) > avg(6, 7)$ [10, 10]; $band(6) > sum(2, 3)$ [10, 10];
$avg(4, 5) > sum(1, 2, 3)$ [5, 1]; $avg(4, 5) > sum(6, 7)$ [10, 0];
$avg(4, 5) > sum(2, 3)$ [5, 20];

$band(4) > background$ [10, 10]; $band(5) > background$ [10, 10];
$band(6) > background$ [10, 10]; $band(7) > background$ [10, 10];

$band(4)$: $>90\%$ forest [10, 10]; $>50\%$ soil [10, 10];
$band(5)$: $>50\%$ forest [10, 0]; $>50\%$ soil [10, 10];
$band(6)$: $>90\%$ forest [10, 10];
$band(8)$: $>50\%$ forest [10, 0];

[a]Same notation as in Table 1.

III.C   Procedural Knowledge

The following steps are used in the spectral expert to classify each pixel: (1) compute spectral evidence by applying the rules in the spectral knowledge base to the mean pyramid data, (2) select a contextual interpretation, (3) use criteria associated with each interpretation to make a classification decision, and (4) update the univariate distribution for the selected category. The input data set is the mean pyramid representation of the spectral image that has been calibrated to ground reflectance. The classification results are the output data set. An interim data set, i.e., the image process status, is computed after each iteration to summarize the results of the previous classification for use in the next iteration. The image-process status includes the univariate histograms for each category and the frequency distribution of categories in the local neighborhood surrounding each pixel. Multiple iterations are required for the category histograms to converge, i.e., provide a reliable approximation of the underlying distribution.

III.C.1   Evidence Computation

The rules in the spectral knowledge base are used to compute spectral evidence by evaluating the local reflectance properties recorded in the pyramid for each pixel. The rules specify the amount by which the supporting and opposing evidence totals are conditionally incremented as determined by the validity of the constraint. The rule constraints are validated from analysis of the series of reflectance vectors in the pyramid corresponding to each pixel. The BB constraints are validated by performing the indicated operations (average, maximum, minimum, sum) for the designated spectral bands within each reflectance vector. The CB constraints are validated by calculating the spectral gradient between the pixel and the background response. The CC constraints are validated by examining the univariate histogram for the designated comparison category and spectral band. The histograms were derived by recording the distribution of categories identified in the previous iteration. The CC rules cannot be used on the first iteration. After all of the rules have been applied, the evidence distribution is summarized by computing category scores as a function of the relative proportion of supporting and opposing evidence. These scores provide the basis for subsequent decision making.

III.C.2   Decision Expert

The evidence as derived from the multiple spatial resolutions in the pyramid are used, in effect, to reinforce or refute a given classification decision. The categories having maximal scores are selected for each pyramid level. The result is a vector of decisions for each pixel. The initial classification rule was to use a joint maximum support criteria, i.e., to select the category having maximal support at the bottom two pyramid levels, i.e., levels 0 and 1. This criteria fails when the spectral evidence is ambiguous, i.e., level 0 and level 1 classifications do not agree. Such disagreement was found to occur for edge pixels, mixed pixels, and pixels corrupted by noise. Examination of ambiguous pixels in the TMS images showed that many such cases could be resolved in visual interpretation on the basis of local context. The decision criteria depend upon the local contextual interpretation, i.e., uniform, border, noise, or outlier.

The decision expert accommodates these differences in decision methodology by using contextual experts that incorporate separate decision criteria for uniform, noise, border, and outlier contextual interpretations. The decision expert first applies the uniform expert. If the uniform interpretation is validated, then no further processing is necessary. Otherwise, it is necessary to apply the border, noise, and outlier experts. These experts cannot be used on the first iteration because the *a posteriori* contextual information, i.e., distribution of previously recognized categories, is not yet available. The following sections describe the interpretation experts in greater detail.

Uniform. The uniform expert is used to verify the contextual interpretation that the pixel under consideration is part of a relatively large spectrally uniform region. The assumption is that pixels belonging to such regions will have unambiguous spectral evidence. The joint maximum criterion is used to determine whether the spectral evidence is ambiguous. This criterion is satisfied if pyramid level 0 and level 1 classifications agree, i.e., the same category had the maximum amount of supporting spectral evidence at both levels. For example, the four by four windows for the circled forest and asphalt pixels in Figure 5(a) indicate that this criterion would be satisfied.

Border. The border expert is used to verify the contextual interpretation that the pixel under consideration occurs within a spectrally uniform region but is located adjacent to a boundary. The border interpretation is validated by examining the local frequency of previously identified categories within the three by three pixel window at pyramid level 0, e.g., Figure 5(b). The border interpretation is considered to be valid if a pixel has at least two adjacent neighbors that agree with the level 0 classification. Examination of the three by three windows shows that this criterion is satisfied for the forest and asphalt pixels shown in Figure 5(b). The classification decision is to select the pyramid level 0

(a)   (b)

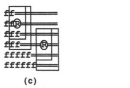

(c)   (d)

Figure 5 Hypothetical examples of pixel arrangements corresponding to (a) uniform, (b) border, (c) noise, and (d) outlier contextual interpretations, where "f" is forest, "=" is asphalt, "g" is grass, "s" is soil, and "R" is roof. The 3 by 3 (small box) and 4 by 4 (large box) windows are also shown.

classification as valid and to reject the level 1 classification. The level 1 classification is assumed to be the result of blurring within the 4 by 4 window used to compute the level 1 spectral response. For example, the 4 by 4 windows in Figure 5(*b*) show that the level 1 response for both pixels represents a mixture of asphalt and forest pixels.

Noise.   The noise expert is used to verify the interpretation that the pixel under consideration is the result of noise within an otherwise spectrally uniform region. The noise interpretation is validated by examination of the frequency distribution within the 4 by 4 windows used in computing the spectral response of successive pyramid levels. The noise interpretation is considered to be valid if the local neighborhood at pyramid levels 1 and 2 is dominated by a single category, i.e., the percentage of that category is greater than 50%. Examination of Figure 5(*c*) shows that this criterion is satisfied for both roof pixels in which the level 1 spectral response is dominated by the asphalt category. The dominance of a single category at pyramid level 1 was considered to provide sufficient contextual evidence to reject the level 0 classification and to accept the level 1 classification as correct.

Outlier.   The outlier interpretation is assigned to a pixel if all the other interpretation experts fail. Figure 5(*d*) illustrates a hypothetical case of an isolated roof pixel that occurs with a small patch of asphalt surrounded by soil, grass, and forest. The level 1 spectral response does not resemble a roof response because the 4 by 4 window contains a mixture of asphalt, forest, grass, and soil. As a result, the uniform interpretation fails. The border expert fails because the 3 by 3 window does not contain any other roof pixels. The noise expert fails because there is no single dominant category in the surrounding 4 by 4 window. The uniform, border, and noise experts fail because neither the level 0 nor the level 1 classifications are strongly supported. The classification decision was to reject the level 1 classification because it was considered to have the greatest amount of evidence against it. The level 1 classification was considered to be the weaker of the two because the level 1 spectral response is known to be highly mixed because no single category is dominant. There is insufficient contextual evidence to reject the level 0 classification and it is accepted by default.

III.C.3   Decision Recording
The classification decision is recorded for each pixel by updating the classification map and by conditionally updating the distribution statistics, i.e., univariate spectral histograms for the selected category. The statistics update is performed only for those pixels that satisfy the uniform interpretation to avoid the introduction of outliers due to edge, mixed, or noise pixels. The distribution statistics are used in the next iteration for evaluation of the CC rules. After each iteration, i.e., all pixels have been processed, the last step is to record the frequency distribution of classified categories in the neighborhood surrounding each pixel for each pyramid level. This information is used in selecting the appropriate contextual interpretation as noted before.

III.D   Testing and Results

Classification accuracies were computed for each test image on the basis of evaluating 300 ground reference pixels. The coordinates of the ground reference pixels were computed by a random-number generator. The ground reference classifications were derived by manual interpretation. The classification accuracies were computed on a per-pixel basis by comparing the spectral expert results to the ground reference data. Five iterations of the spectral expert were used for each test image to provide a uniform basis for comparison and to allow the image process status to converge. Approximately one-third of the 300 test pixels represented edges, mixtures, or shadows.

The spectral expert accuracy results were compared to those from an unsupervised parametric approach. A cluster-analysis program (KMEANS) was used in which only the number of clusters and a convergence criteria must be specified. This program was used to avoid the need for training statistics and to minimize the number of clustering parameters. The KMEANS program is similar to the CLUS program at the Laboratory for Applications of Remote Sensing System (LARSYS) at Purdue University (Spenser and Philips, 1973). The number of clusters was set to 22, i.e., two for each spectral category. The convergence criterion were defined so that the program stops if less than 5% of the pixels changed clusters between iterations. Color-coded images of the spectral expert and the KMEANS cluster-analysis classifications are shown in Figure 1.

The KMEANS analysis took approximately five minutes on a VAX 11/780 for the 120 by 120 pixel test images. The spectral expert took about two hours on a Sun 2/120 for the same data. A summary of the accuracy results from the four test images for the spectral expert and KMEANS classifications is shown in Table 3. The spectral knowledge base used in the spectral expert was derived from analysis of the Goddard and Greenway TMS images. As would be expected, the spectral expert achieved the highest classification accuracies (85 and 89% correct, respectively) in classifying these two scenes (as compared to 82 and 87% correct for KMEANS, respectively). The spectral expert was used to classify a third TMS scene of the University of Maryland without modification to the spectral knowledge base. The University of Maryland image was collected about 45 minutes later than the Goddard and Greenway images. The University of Maryland image was difficult to interpret because of the smaller size and greater density of buildings and greater diversity of roof materials. Despite these limitations, a

TABLE 3   Summary of Results for Spectral Expert and KMEANS Classifications

	Expert (left) and KMEANS (right) Classification Accuracies (%) per Test Image							
Category	Goddard (TMS)		Greenway (TMS)		University (TMS)		Greenbelt (TM)	
Forest	88	92	79	86	83	48	64	96
Grass	82	73	88	85	92	86	90	69
Soil	94	94	92	43	100	0	90	82
Water	72	75	—[a]	—	0	0	0	100
Blue Roof	100	100	92	50	63	62	88	0
Dark Roof	100	100	100	100	71	61	—	—
Green Roof	100	0	92	0	—	—	—	—
Specular Roof	100	100	—	—	—	—	—	—
Asphalt	87	85	93	97	86	86	50	65
Concrete	—	—	0	100	0	0	0	0
Shadow	25	0	50	0	31	0	0	0
Overall	85	82	89	87	80	74	63	85

[a]Category was not present.

classification accuracy of 80% was achieved (as compared to 74% correct for KMEANS). As noted before, the spectral expert accuracies were greater than the cluster-analysis results for the TMS images by margins of 3, 2, and 6% correct.

The spectral expert was used to classify a TM scene of Greenbelt, also without modification to the spectral knowledge base. The classification accuracy of 63% was lower than all of the TMS images and was less than the KMEANS accuracy of 85% correct. The following scene characteristics may have contributed to the poor accuracy observed with the TM test data: lower spatial resolution; lower spectral resolution; topographic shading; and poor contrast in the thermal band. The spatial resolution of 30 m results in the data having a higher proportion of mixed pixels as compared to the 5 m resolution of the TMS. Many of the features that were clearly visible in the TMS data could not be located or were blurred in the TM data, particularly roads and buildings. The TM scene appeared to have a higher incidence of topographic shading as compared to the TMS scenes because it was collected earlier in the morning. The thermal band has a spatial resolution of 120 m and had a minimal contrast in the Washington, D.C., TM scene.

III.E Discussion

The 80% correct or higher classification accuracies achieved for urban land-cover classification of 5 m resolution TMS suggest that the knowledge-based classification approach is feasible. The accuracy of the expert system as applied to other data, i.e., different sensors, dates, and areas, is dependent on the generality of the spectral knowledge base. The accuracy can be expected to decrease when the spectral expert is used to classify data that do not satisfy the assumptions made in the development of the knowledge base, e.g., the 63% correct accuracy achieved with Thematic Mapper data collected on the same date over the same area.

III.E.1 Advantages

The system was designed to minimize the effects of noise in the data. Classification decisions are made on the basis of convergent evidence, i.e., multiple information sources that support the same conclusion. The advantage of convergent evidence as opposed to stepwise decisions is that failure of individual rules due to random variation, noise, or mixed pixels does not necessarily jeopardize the final decision. The rules used in making the classification decisions depend upon the local spatial interpretation that is associated with the pixel under consideration. The advantage of this approach is that the spectral expert can utilize spectral and spatial consistency rules that are specifically tailored for each of four possible interpretations, uniform, border, noise, and outlier.

Three techniques are used to maximize the amount of evidence available for evaluation of spectral and spatial consistency and for subsequent decision making. The first is that multiple features and multiple rules are used to represent the spectral knowledge for each category. The second is that the spectral evidence is computed over a range of scales, i.e., multiple spatial resolutions for each pixel. The third is that the local frequency distribution of previously identified categories is used to ensure that the classification decisions are spatially consistent. Spectral knowledge is defined explicitly in terms of constraints that are used to define the expected relationships within and between categories. This tends to promote the development of a physical-based rather than a statistical understanding of the scene.

III.E.2 Limitations

The feasibility of the spectral expert has not yet been successfully demonstrated for data such as TM with relatively low spectral resolution, low spatial resolution, and low contrast, as compared to the TMS data. The classification performance was unreliable for discriminating between categories on the basis of subtle differences in brightness, i.e., asphalt/concrete and forest/grass. The system is inefficient in terms of disk space and CPU usage in comparison with most parametric approaches. Ground measurements are required to correct the data to reflectance. The spectral expert lacks the capability to modify the spectral rules to adjust for marked changes in the scene or in imaging conditions, such as differences in time of day, time of year, dominant land cover, or spectral bands.

IV CONCLUSIONS

Conventional classification techniques rely upon a parametric representation of spectral knowledge, i.e., statistics and thresholds, for assigning pixels to spectral categories. The scope of application of this knowledge is limited to the image from which it was derived. Each image must be treated as a separate case in which the statistical relationship between the targets and the observed spectral patterns must be uniquely defined. The spectral knowledge-based expert systems, discussed in the previous section, are designed to avoid the need for scene-by-scene parameter optimization by making classification decisions on the basis of knowledge of the spectral relationships within and between categories. The assumption is that these relationships are relatively stable, compared to statistics, thereby allowing the targets to be recognized over a wider range of imaging conditions within the logical scope of the knowledge base (e.g., a specified set of targets, geographic locations, dates, and sensor characteristics). A prototype spectral expert system was presented as a case study to illustrate the kinds of spectral and procedural knowledge required for urban land-cover analysis.

The advantage of the expert system approach, compared to procedural methods, is the explicit representation of knowledge. Waterman (1986) describes artificial expertise as permanent, easy to transfer, easy to document, consistent, and affordable, as opposed to human expertise, which is perishable, difficult to transfer, difficult to document, unpredictable, and expensive. The benefit to the remote-sensing community is that expert systems can be used to facilitate the extraction, documentation, and dissemination of the pertinent spectral and procedural knowledge for classification. However, this benefit is not fully realized in practice, judging from the descriptions of the spectral expert systems for land-cover classification reported therein, because the knowledge bases are sensor-specific and include only a limited number of spectral categories. Further testing is required to more fully evaluate the effectiveness of the relational representation of spectral knowledge for other targets and images.

IV.A Future Development

Additional research is necessary for spectral knowledge-based techniques to gain acceptance as practical alternatives to parametric classifiers. The functional objectives for this continuing development include the following: (1) improve the accuracy, consistency, and speed of spectral target recognition; (2) increase the number of spectral categories that can be recognized; (3) increase the number of images that can be processed by expanding the scope of recognition to include a greater number of sensors, locations, and dates; (4) develop a sensor-independent representation of the pertinent knowledge for spectral analysis; and (5) reduce the level of human expertise required to develop the knowledge base. A list of the major spectral and procedural knowledge enhancements, necessary for the prototype spectral expert to achieve the above objectives, is given in the following sections.

IV.A.1 Spectral-Knowledge Enhancements

1. Increase the number of spectral categories represented in the spectral-knowledge base for use in other applications in addition to urban land-cover mapping, e.g., agriculture, forestry, geology, and hydrology.

2. Incorporate knowledge of the spectral characteristics of targets for spectral bands in the visible, infrared, and microwave portions of the electromagnetic spectrum so that the spectral expert can be applied to data from other sensors in addition to the TMS.

3. Reference the spectral bands in terms of the specific wavelength ranges as opposed to spectral band number. This modification would simplify the development of a common spectral data base that could be used for multiple sensors without regard to the number of bands per sensor or their numbering scheme.

4. Provide the information necessary for the system to automatically select a set of potential spectral targets for a given location, time, and spatial/spectral sensor resolution. The assumption is that the spectral expert may have access to information about a very large number of spectral targets, most of which are not present in any single image. Knowledge of the geographic and temporal distribution of each target can be used to eliminate targets from consideration that are known to be absent for the area and time of a given image. In addition, knowledge of the spectral and spatial resolutions necessary for each target to be apparent, can be used to eliminate targets that are known to be present for a given area and time, but are not visible because of the insufficient spatial and spectral resolution of the sensor.

   It may also be possible to reduce the number of rules to be considered, in a similar fashion, by recording the spectral, geographical, and temporal limits associated with each rule. This information could be used to generate the list of applicable rules for a given set of targets that are likely to be effective for a given sensor and time.

5. Incorporate knowledge of the intrinsic temporal variation in the spectral properties of a target as an additional classification feature. For example, multiple images of the same area taken at different times of year could be used to discriminate vegetation types or vegetation communities on the basis of phenological differences. The lack of such variation with time could be used to reinforce the conclusion that vegetation is not present.

6. Maintain a data base of the spectral reflectance curves that were used in the development of the spectral rules for identification. This information could be used in the development of spectral rules for other sensors and could also serve to facilitate the development of a general-purpose reference library of the ''standard'' reflectance patterns associated with different spectral targets. The assumptions are

that "standard" reflectance curves can be said to exist for the spectral targets and that they can be measured in the field/laboratory, predicted from models, or derived from remotely sensed imagery.

7. As an alternative to thematic classification, provide the capability of describing the land cover in terms of measurements of physical characteristics derived from the spectral values. This would be useful if the objective is to extract properties of the image rather than to classify it. For example, rather than attempt to discriminate between forest and grass, it may be equally useful to estimate the amount of vegetation biomass, canopy closure, moisture content, percent imperviousness, etc.

IV.A.2 *Procedural-Knowledge Enhancements*

1. Provide a capability for automatic rule generation from analysis of the reflectance curves in the spectral data base. This would reduce the level of human expertise needed to initially formulate the rules for spectral recognition. The analyst could elect to subsequently edit the rules to incorporate heuristic rules or to exploit any additional information that might not be apparent from the reflectance curves. It would be useful to predict the effectiveness of the rules.

2. Provide the capability for spectral calibration. There are three alternatives. The first is to deterministically calibrate the image according to "housekeeping" information recorded by the sensor in concert with atmospheric parameters assumed for or derived from the image. The second is to calibrate the image data to the same "standard(s)" that were used in the development of the reflectance data base by matching regions in the image to the appropriate reflectance curves. The third alternative is to use either of the previous techniques in conjunction with atmospheric and sensor models to generate spectral rules that have been specifically calibrated for specific set of scene characteristics.

3. Reduce the number of decisions required by processing regions rather than individual pixels. The first step would be to segment the image into spectrally homogeneous regions by grouping spatially connected pixels having relatively similar spectral responses. To avoid analyst dependence, any thresholds used to control pixel grouping would be adaptively determined as a function of the local image contrast. Spectral classification decisions would be made on the basis of overall spectral properties of the region as a whole. In the degenerate case, a region may consist of a single pixel.

4. Provide the capability of processing spectrally mixed regions. This capability would be most useful for relatively low spatial resolution data in which the incidence of mixed pixels is relatively high. There is a need to analytically determine the probable combinations and proportions of the spectral constituents that comprise such mixed regions. This analysis of mixtures would not be necessary if the spectral constituents could be identified on the basis of local context, e.g., a mixed pixel that occurs on the boundary between grass and asphalt regions is likely to consist of a mixture of grass and asphalt.

   As part of the mixture analysis, it would also be useful to be able to identify the effects of shadowing upon the spectral response for a given region. Shadow information could be useful in discriminating between different types of vegetation communities on the basis of height or canopy closure. Shadow information could also be used to reduce the effect of topographic shading. This would be most useful for areas with high relief and/or scenes collected at low sun angles.

5. Incorporate a control strategy to reduce the number of rules and categories to be considered in making classification decisions. The intent is to build criteria into the decision-making process so that strongly contradicted targets, for a given region, are eliminated from consideration. For example, the technique of evidential reasoning (Goldberg et al., 1985) in which categories are considered in terms of support and plausibility could be applied. It would also be necessary for the system to be able to detect and backtrack from initial decisions made on the basis of preliminary evidence that are contradicted in light of additional evidence provided by subsequent processing.

6. An explanation facility is needed so that the analyst can examine the sequence of rules that were applied by the spectral expert in making a particular decision and assess its relative merit in contributing to that decision. This capability could be used in conjunction with ground reference data to automatically evaluate the performance of the decision rules in terms of the total number of applications for each rule and the proportion of those applications in which the rule supported the correct decision.

7. Develop criteria for the additional contextual interpretation of ''linear.'' It was noted in the Greenway test image that the spectral expert missed a narrow curved highway ramp because the asphalt classification was only supported at pyramid level 0. Failure to find local support (at least two neighboring asphalt pixels) and higher pyramid level support was due to the effect of mixed pixels. In addition, the error tends to propagate itself, i.e., failure to recognize one pixel of a linear feature jeopardizes the recognition of its neighbors. This type of error could be avoided by considering the spectral evidence of neighboring pixels (as opposed to their classification) when processing regions that may correspond to linear features. The contextual interpretation of ''textured'' might also prove useful, e.g., for recognition of forests at high spatial resolution.

8. Develop an interactive interface to a color-image display. The assumption is that the development of the procedural- and spectral-knowledge bases can be expedited by allowing the user to interactively manipulate the image and experiment with alternate rules for grouping pixels into regions and for making classification decisions. The display capability provides rapid visual assessment of results and allows the user to quickly detect and isolate areas that may have been incorrectly grouped or labeled.

REFERENCES

Anderberg, M. R. (1973). *Cluster Analysis for Applications*. Academic Press, New York.

Ballard, D. H., and C. M. Brown (1982). *Computer Vision*. Prentice-Hall, Englewood Cliffs, New Jersey.

Borchardt, G. C. (1986). STAR: A computer language for hybrid AI applications. In *Coupling Symbolic and Numerical Computing in Expert Systems* (J. S. Kowalik, Ed.), Elsevier, Amsterdam, pp. 169–177.

Borchardt, G. C. (1987). Hybrid organization in SPECTRUM, a system for the analysis of imaging spectrometer data. *Proc. AAAI Workshop Coupling Symbolic Numer. Comput. Knowledge-Based Syst.*, July, 1987.

Brady, M. (1982). Computational approaches to image understanding. *Comput. Surv. ACM* **14**(1): 3–71.

Burt, P. J. (1981). Fast filter transforms for image processing. *Comput. Graph. Image Process.* **16**: 20–51.

Burt, P. J., T. Hong, and A. Rosenfeld (1981). Segmentation and estimation of image region properties through cooperative hierarchical computation. *IEEE Trans. Syst. Man, Cybernet.* **11**(12): 802–809.

Carlutto, M. J., V. T. Tom, P. B. Baim, and R. A. Upton (1984). Knowledge-based multi-spectral image classification. *Proc. SPIE—Int. Soc. Opt. Eng.* **504**.

Castleman, K. R. (1979). *Digital Image Processing.* Prentice-Hall, Englewood Cliffs, New Jersey.

Choiu, W. C. (1985). NASA image-based geological expert system development project for hy-perspectral image analysis. *Appl. Opt.* **24**(14):2085–2091.

Crist, E. P., and R. C. Cicone (1984). A physically-based transformation of thematic mapper data—The TM tasseled cap. *IEEE Trans. Geosci. Remote Sens.* **22**(3):256–263.

Crowley, J., and A. Parker (1984). A representation for shape based on peaks and ridges in the difference of low-pass transform. *IEEE Trans. Pattern Anal. Mach. Intell.* **PAMI-6:**156–170.

Davis, R. (1986). Knowledge-based systems. *Science* **231:**957–962.

Duda, R. O., and P. F. Hart (1973). *Pattern Classification and Scene Analysis.* Wiley, New York.

Duda, R. O., and E. H. Shortliffe (1983). Expert systems research. *Science* **220:**261–268.

Estes, J. E., E. J. Hajic, and L. R. Tinney (1983). Manual and digital analysis in the visible and infrared regions. In *Manual of Remote Sensing* (R. N. Colwell, D. S. Simonett, and F. T. Ulaby, (Eds.), 2nd ed., Vol. I. American Society of Photogrammetry, Falls Church, Virginia, pp. 987–1123.

Estes, J. E., C. Sailer, and L. R. Tinney (1986). Applications of artificial intelligence techniques to remote sensing. *Prof. Geogra.* **38**(2):133–141.

Ferrante, R. D., M. J. Carlutto, J. Pomaraede, and P. W. Baim (1984). Multi-spectral image analysis system. *Proc. 1st. Conf. Artif. Intell. Appl.*, Denver, Colorado, pp. 357–363.

Fu, K. S. (1982). *Syntactic Pattern Recognition and Applications*, Prentice-Hall, Englewood Cliffs, New Jersey.

Goetz, A. F. H., J. B. Wellman, and W. L. Barnes (1985). Optical remote sensing of the earth. *Proc. IEEE* **73**(6):950–969.

Goldberg, M., D. G. Goodenough, M. Alvo, and G. M. Karam (1985). A hierarchical expert system for updating forestry maps with Landsat data. *Proc. IEEE* **73**(6):1054–1063.

Gonzalez, R. C., and P. Wintz (1987). *Digital Image Processing*, 2nd ed. Addison-Wesley, Reading, Massachusetts.

Goodenough, D. G. (1986). A hierarchical expert system. Presentation to NASA/GSFC, May 1, 1986.

Goward, S. N. (1981). The thermal behavior of urban landscapes and the urban heat island. *Phys. Geogr.* **2:**19–33.

Gurney, C. M., and J. R. G. Townsend (1983). The use of contextual information in the classi-fication of remotely sensed data. *Photogramm. Eng. Remote Sens.* **49**(1):55–64.

Haralick, R. M. (1979). Statistical and structural approaches to texture. *Proc. IEEE* **67**(5): 786–804.

Haralick, R. M., and L. G. Shapiro (1985). Image segmentation techniques. *Comput. Vision, Graphics, Image Process.* **29**(3):100–132.

Hartigan, J. A. (1975). *Clustering Algorithms.* Wiley, New York.

Hartley, R. (1984). Multi-scale models in image analysis. Ph.D. Thesis, University of Maryland, College Park.

Heydorn, R. P., M. C. Trichel, and J. D. Erickson (1979). Methods for segment wheat area estimation. In *Large Area Crop Inventory Equipment*, Vol. II. NASA/Johnson Space Center, Houston, Texas, pp. 621–632.

Hwang, V. S., T. Matsuyama, A. Rosenfeld, and L. S. Davis. (1983). *Evidence Accumulation for Spatial Reasoning in Aerial Photograph Understanding*, Tech. Rep. 1366. Computer Vision Laboratory, University of Maryland, College Park.

Jackson, R. D. (1983a). Spectral indices in n-space. *Remote Sens. Environ.* **13**:409–421.

Jackson, R. D. (1983b). Adjusting the tasseled-cap brightness and greenness factors for atmospheric path radiance and absorption on a pixel by pixel basis. *Int. J. Remote Sens.* **4**(2): 313–323.

Kauth, R. J., and G. S. Thomas (1976). The tasseled cap—A graphic description of the spectral-temporal development of agricultural crops as seen by Landsat. In *Machine Processing of Remotely Sensed Data*. Purdue University, West Lafayette, Indiana, pp. 4B-41–4B-51.

Lambeck, P. F., R. Kauth, and G. S. Thomas (1978). Data screening and preprocessing for Landsat MSS data. *Proc. 12th Int. Symp. Remote Sens. Environ.*, University of Michigan, Ann Arbor, Vol. II, pp. 999–1008.

Landgrebe, D. A. (1981). Analysis technology for land remote sensing. *Proc. IEEE* **69**(5):628–642.

Matsuyama, T. (1987). Knowledge-based aerial image understanding systems and expert systems for image processing, *IEEE Trans. Geosci. Remote Sens.* **25**(3):305–316.

McKeown, D. M. (1984). Knowledge-based aerial photo interpretation. *Photogrammetria* **39**:91–123.

Metzler, M. D., R. C. Cicone, and K. I. Johnson (1983). Experiments with an expert-based crop area estimation technique for corn and soybeans. *Proc. 17th Int. Symp. Remote Sens. Environ.*, University of Michigan, Ann Arbor, Vol. III, pp. 965–972.

Milgram, D. L. (1978). *Region Extraction Using Convergent Evidence*, Tech. Rep. 674 Computer Vision Laboratory, University of Maryland, College Park.

Minter, T. C. (1979). Methods of Extending Crop Signatures from One Area to Another. In *Large Area Crop Inventory Equipment*, Vol. II. NASA/Johnson Space Center, Houston, Texas, pp. 757–800.

Modestino, J. W., R. W. Fries, and A. L. Vickers (1981). Texture discrimination based upon an assumed stochastic texture model. *IEEE Trans. Pattern Anal. Mach. Intell.* **3**(5):557–580.

Mooneyhan, D. W. (1983). The potential of expert systems for remote sensing applications. *Proc. Geosci. Remote Sens. Symp. (IGARSS '83)*, San Francisco, California, pp. 4.1–4.5.3.

Myers, V. I. (Ed.) (1983). Remote sensing applications in agriculture. In *Manual of Remote Sensing* (R. N. Colwell, D. S. Simonett, and F. T. Ulaby, Eds.), 2nd ed., Vol. II. American Society of Photogrammetry, Falls Church, Virginia, Chapter 33.

Nagao, N., and T. Matsuyama (1980). *A Structural Analysis of Complex Aerial Photographs*. Plenum, New York.

Nazif, A. M., and M. D. Levine (1984). Low level image segmentation: An expert system. *IEEE Trans. Pattern Anal. Mach. Intell.*, **PAMI-6**(5):555–577.

Pratt, W. K. (1978). *Digital Image Processing*. Wiley (Interscience), New York.

Richardson, A. J., and C. L. Wiegand (1977). A table look-up procedure for rapidly mapping vegetation cover and crop development. In *Machine Processing of Remotely Sensed Data*. Purdue University, West Lafayette, Indiana, pp. 284–296.

Rosenfeld, A. and A. C. Kak (1982). *Digital Picture Processing*, 2nd ed. Academic Press, New York.

Selfridge, P. G., and K. Sloan (1982). Reasoning about success and failure in aerial image understanding. *Proc. IEEE Conf. Pattern Recognition Image Process., 1982*, Las Vegas, Nevada, pp. 44–49.

Spencer, P. W., and T. Philips (Eds.) (1973). *LARSYS Users Manual*, Vol. I. Purdue University, West Lafayette, Indiana.

Swain, P. H., and S. M. Davis (1978). *Remote Sensing—The Quantitative Approach*. McGraw-Hill, New York.

Tanimoto, S. L. (1987). *The Elements of Artificial Intelligence*. Computer Science Press, Rockville, Maryland.

Tanimoto, S. L., and T. Pavlidis (1975). A hierarchical data structure for picture processing. *Comput. Graphics, Vision Image Process.* **4:**104–119.

Tucker, C. J. (1979). Red and photographic infrared linear combinations for classification using NOAA-7 AVHRR data. *Remote Sens. Environ.* **8:**127–150.

Wang, S., D. B. Elliott, J. B. Campbell, R. W. Welch, and R. W. Haralick (1983). Spatial reasoning in remotely sensed data. *IEEE Trans. Geosci. Remote Sens.* **21**(1):94–101.

Waterman, D. A. (1986). *A Guide to Expert Systems*. Addison-Wesley, Reading, Massachusetts.

Wharton, S. W. (1987). A spectral-knowledge based approach for urban land-cover discrimination. *IEEE Trans. Geosci. Remote Sens.* **25**(3):272–282.

Wiegand, C. L., and A. J. Richardson (1982). Comparisons among a new soil index and other two- and four-dimensional vegetation indices. *Tech. Pap., 48th Annu. Meet. Am. Soc. Photogramms.*, pp. 210–227.

Winston, P. H. (1984). *Artificial Intelligence*, 2nd ed. Addison-Wesley, Reading, Massachusetts.

Woodham, R. J., and M. H. Gray (1987). An analytic method for radiometric correction of satellite multispectral scanner data. *IEEE Trans. Geosci. Remote Sens.* **25**(3):258–271.

15

QUANTITATIVE ASPECTS OF REMOTE SENSING IN THE THERMAL INFRARED

JOHN C. PRICE

Beltsville Agricultural Research Center
Agricultural Research Service
United States Department of Agriculture
Beltsville, Maryland

I INTRODUCTION

In several respects, remote sensing in the thermal-infrared spectral region is more complicated than in the visible near-infrared and the microwave. Because thermal-infrared measurements imply a value of *temperature*, which in turn pertains to the *heat* stored in objects, observations of the Earth's surface are complicated by factors that are unique to this spectral interval:

1. Objects are capable of storing and then releasing heat, so that the temperature of an object depends on its past history as well as present conditions.
2. At the Earth's surface, heat energy may be gained or lost not only by radiation, but also through exchange of sensible and latent heat (evaporation) with the atmosphere.

The need to describe the energy exchanges of each area viewed by a remote-sensing instrument implies a mathematical formulation of energy fluxes and the time evolution of surface temperature. This formulation must include sufficient detail to permit calculation of temperature values of the required accuracy. In this chapter, the mathematical techniques for analyzing remote measurements in the thermal infrared are developed and limitations to existing procedures are described. The exception to the rule of mathematical development, multispectral sensing in the thermal infrared, is discussed at the end of the chapter. This later topic is in a very early stage of development, and conventional image-analysis/photointerpretation techniques are appropriate without regard to theoretical formulations.

Radiation measurements in the thermal infrared provide information about the radiating temperature of the object being viewed, which in turn relates to the physical properties of the object and the energy fluxes that impinge on it. Typically, the land and ocean surfaces absorb energy from the sun during the day, radiating a fraction of it back to space and transfering the balance to the atmosphere as sensible heating and latent heat of evaporation. At night, stored energy is released and the surface cools until dawn, when the heating cycle begins again. In practice, remote thermal-infrared measurements provide information chiefly about evaporation, which represents a net cooling effect, and about the heat-storing capacity of the near surface layer. Numerous micrometeorological and radiation parameters must be included in a full description. Figure 1 illustrates a midmorning thermal infrared image as obtained by the Landsat-4 Thematic Mapper. In the image, darker areas are cooler, whereas lighter areas are warmer. The black areas are water bodies, which are relatively cool during the day. The goal of remote sensing in this spectral range is to derive a quantitative description of conditions or properties of the surface from the radiation values represented by such images.

In this chapter, we present the formulation of the radiation/temperature relationship, then the heat-flow equation, which describes the behavior of temperature within soil or rocks, and the mathematical tools for solving the heat-flow equation in order to analyze remotely sensed data. This is followed by a formal discussion of atmospheric effects on satellite observations. These interfere with direct interpretation of satellite thermal-infrared data, and techniques are available to compensate for them. Then the state of

Figure 1   An 11 μm image from the Landsat Thematic Mapper. The area is approximately 15 km^2 centered in the southeast United States.

knowledge of emissivity is discussed, as our lack of precise experimental data causes residual uncertainty in our estimation of surface conditions from satellite/aircraft data. In the thermal infrared, multispectral data may be used for image analysis as for geological interpretation, without the need for formal analysis. The general prospect for the thermal infrared is reviewed at the end of the chapter.

Although the thermal-infrared spectral region covers the range from approximately 3 to 100 μm, only the spectral intervals around 3.7 μm (3.5–4 μm) and between 8 and 14 μm are of interest for remote sensing of the Earth's surface. From general studies of radiation, we know that the atmosphere is moderately transparent to radiation only in these spectral regions, called ''windows,'' whereas it is fairly opaque in the intervening wavelengths. The wavelength boundaries of the spectral windows may be extended somewhat by observations from a low-flying aircraft for which the opacity of the atmosphere to the surface is less than for satellite observations, but aircraft systems are not available on a regular basis and the implications of these broader spectral intervals will not be discussed in this chapter. Although the full range of the thermal-infrared spectrum may be explored by handheld radiometers or in the laboratory, these alternatives are not applicable for large-area studies, which are the principal domain of remote sensing.

The opaque spectral intervals in the range 4–8 and 14–100 μm are not without interest: satellite measurements of radiation at these wavelengths provide information on the temperature and humidity profile between the Earth's surface and the stratosphere. Such data are used for initialization of numerical weather-prediction models. The effects associated with surface absorption of the sun's visible radiation, and trapping of atmospheric radiation in the infrared, produce a considerable relative warming of the Earth's surface. This ''greenhouse'' effect is similar to the radiation effect experienced in greenhouses. These atmospheric radiation effects are significant for climatic studies because human activity is changing the atmospheric composition, thereby affecting the radiation balance. These topics lie within the discipline of meteorology.

II RELATIONSHIPS AMONG RADIATION, TEMPERATURE, AND SURFACE EMISSIVITY

In remote sensing, we are concerned with interpreting measurements of radiation from a distant source. In the thermal infrared, we relate this measurement to the kinetic temperature (e.g., from a thermometer) of the radiating body, which initially is considered to be a perfect radiator or black body. Thus, the relationship between radiance and temperature is given by Planck's law.

$$B_\lambda = \frac{C_1}{\lambda^5 \left(e^{C_2/\lambda T} - 1 \right)} \tag{1}$$

where B is the spectral radiance in units W m^{-2} μm^{-1}, at wavelength λ (expressed in micrometers), C_1 and C_2 are physical constants ($C_1 = 3.74 \times 10^8$, $C_2 = 1.439 \times 10^4$), and T, in degrees K, represents the physical temperature of the object. In practice, real objects are not ideal blackbodies, with the significance of the quality factor, emissivity, being discussed shortly. A general background of remote sensing and a discussion of the physical units of radiation measurements have been given by Slater (1980).

For interpretation of remotely sensed data, we generally require the inverse function because the sensor measures radiance from which we derive the value of temperature implied by Planck's law. This inverse relationship is obtained by the simple algebraic transformation of Eq. 1. Because real objects are not ideal blackbodies, it is customary to use subscripts to indicate apparent or radiance temperature as opposed to the actual temperature of the object.

$$T_{BB} = \frac{C_2}{\log \left(1 + \lambda^5 B / C_1 \right)} \tag{2}$$

As mentioned previously, there exist two atmospheric windows in the thermal infrared. These windows differ in the sense that by day, both radiation emitted from the Earth and reflected sunlight contribute significantly to radiation observed in the 3.7 μm window, whereas the contribution of reflected sunlight may be neglected compared to Earth radiation in the 8–14 μm window. To show this, we compute the ratio of solar radiance to that emitted by the Earth, where the ratio is reduced by the $1/R^2$ factor as the solar flux expands from the sun. We assume representative temperatures of 6000 K for the sun, 300 K for the Earth, and a reflectivity of 0.1, so that at 3.7 μm,

$$\frac{\text{reflected}}{\text{emitted}} = \frac{0.1 \times B_{3.7}(6000 \text{ K})}{B_{3.7}(300 \text{ K})} \times \left(\frac{R_{\text{sun}} = 6{,}900{,}000 \text{ km}}{D_{\text{earth-sun}} = 150{,}000{,}000 \text{ km}} \right)^2 = 1 \tag{3}$$

whereas at 11 μm,

$$\frac{\text{reflected}}{\text{emitted}} = \frac{0.1 \times B_{11}(6000 \text{ K})}{B_{11}(300 \text{ K})} \times \left(\frac{R_{\text{sun}}}{D_{\text{earth-sun}}} \right)^2 = 0.001 \tag{4}$$

During the day, the near equality of reflected sunlight and emitted thermal radiation at 3.7 μm causes uncertainty in the interpretation of the data, since the energy fluxes add to produce the measured radiance. For this reason, daylight observations have not been analyzed in this spectral interval. This ambiguity of reflected and emitted radiation could, in principle, be removed. Reflectance ρ is related to the quality factor for blackbodies, emissivity ε, by the general relation

$$\varepsilon_\lambda + \rho_\lambda = 1 \tag{5}$$

Thus, the ambiguity of reflected and emitted radiation at 3.7 μm could be resolved through determination of emissivity at 3.7 μm by nighttime radiation measurements, where physical temperature may be inferred from 11 μm data. Then the resulting value of emissivity, together with daytime temperature estimates from the 11 μm interval, would permit inference of the emitted radiation at 3.7 μm. This sequence is simple in principle, but subject to many errors in practice. Although suitable data are available from the NOAA meteorological satellites, this analysis has not been carried out at the present time. In the 10–14 μm window, the interpretation of radiance measurements is more direct, and we focus our attention on this spectral interval. Matson and Dozier (1981) have developed a method for estimating the temperature of small very hot areas within a uniform temperature background through analysis of combined 3.7 and 11 μm data from meteorological satellites. This analysis utilizes the differing power-law dependence of the two spectral regions, which is discussed shortly.

Before proceeding, we present the description of nonblackbody effects. The Planck formulation is an idealization of the true radiative properties of material, as the radiance of a body at kinetic temperature T is reduced by the emissivity factor according to

$$L_\lambda = \varepsilon_\lambda B_\lambda \qquad (6)$$

where L is the measured radiance. Emissivity depends on the substance and varies with wavelength, ranging from approximately 0.5 to nearly 1.0 over the range of common materials and thermal-infrared wavelengths. In the 10–14 μm window, emissivity values are generally in the range 0.8 to 0.99, so that the following simplification is appropriate for many calculations. This allows reasonably accurate numerical comparison of many radiation quantities. One may equate the Planck radiation to a power law with temperature

$$B(\lambda,\ T) \approx T^{n_\lambda} \qquad (7)$$

where n, which does not vary greatly over temperature and wavelength ranges of interest, is evaluated from the derivative of Eq. 7.

$$\frac{1}{B}\frac{dB}{dT} = \frac{n_\lambda}{T} \qquad (8)$$

Values of n for the two atmospheric windows are illustrated in Figures 2 and 3.

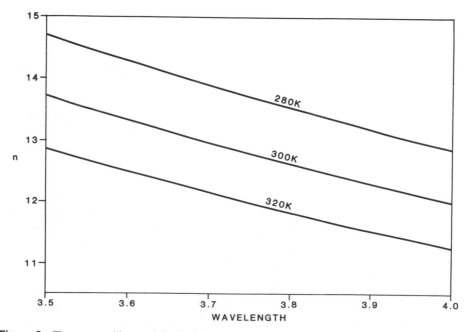

Figure 2   The power n illustrated for the 3.7 μm atmospheric window for typical values of temperature.

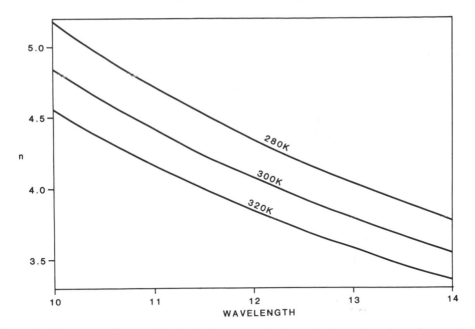

Figure 3   The power n illustrated for the 10–14 μm atmospheric window for typical values of temperature.

Equation 8 and the values of n from the figures are useful for deriving a relationship between changes of physical temperature, small variations of emissivity, and observed radiances, as given by

$$\frac{\delta L}{L} = \frac{\delta \varepsilon}{\varepsilon} + \frac{\delta B}{B} = \frac{\delta \varepsilon}{\varepsilon} + \frac{n_\lambda \delta T}{T} \tag{9}$$

For example, using this relationship, we can estimate the temperature error resulting from an error in the value assigned to emissivity. Thus, for a given radiance, i.e., with $\delta L = 0$, $\varepsilon \approx 1$, and an emissivity error of 0.01, the resulting temperature error at 300 K and wavelength 11 μm is $T\,\delta\varepsilon/n_\lambda = 0.7$ K, where $n_\lambda = 4.5$ was obtained from Figure 3.

One must distinguish between radiances measured at a particular wavelength and the total emitted radiant flux. The total radiant energy emitted by an object is found by integrating Eq. 1 over wavelength. This leads to the Stefan-Boltzmann law, where the emissivity is the wavelength averaged value,

$$R_{\text{emitted}} = \varepsilon \sigma T^4 \tag{10}$$

and σ is the Stefan-Boltzmann constant, 5.67×10^{-8} W m^{-2} K^{-4}. However, emissivity is not constant with wavelength, so that in the analysis of thermal infrared data, we must take care to distinguish between the value of emissivity appropriate for energy calculations and the value that applies to radiometric measurements in the atmospheric windows. Representative values of wavelength-averaged emissivity have been given by Sellers (1965).

III THE HEAT-FLOW EQUATION

Although measurements in the thermal infrared can provide estimates of the temperature (at the surface) of the land or water area being viewed, the resulting temperature values are seldom of interest by themselves. Instead, we seek to relate temperatures to physical properties at the Earth's surface. Is a local surface area cool because of evaporation, or because the overlying atmosphere is cold, or because of shadowing by clouds from the sun's radiation, etc? Often these questions can be answered in a qualitative sense through general knowledge of the conditions at the time and location of the measurement. However, for more precise inference of surface properties, it is necessary to obtain additional information about the surface and the local atmospheric conditions, and to process these data by means of a model, simple or complicated, according to the result desired. The starting point for a quantitative description is the heat-flow equation, which describes the behavior of temperature T in a solid such as rock or soil. Together with a surface boundary condition, the heat-flow equation relates observed surface temperatures to the energy flux exchanged with the atmosphere. The heat-flow equation is given by

$$\rho c \frac{\partial T}{\partial t} = \frac{\partial}{\partial z} \left(\kappa \frac{\partial T}{\partial z} \right) \tag{11}$$

where temperature is a function of depth z, measured downward from the surface, and time t, ρ is the density (kg m^{-3}), c is the heat capacity ($\text{J kg}^{-1} \text{K}^{-1}$), and κ is the thermal conductivity ($\text{J m}^{-1} \text{K}^{-1} \text{sec}^{-1}$). For remote-sensing applications, this equation is solved to yield T at the Earth's surface. The underlying soil/rock may be regarded as a one-dimensional half space ($z > 0$). Generally, surface-temperature variations extend only to a depth of some tens of centimeters over a day, or a few meters over the course of a year. Both distances are so small that effects of horizontal variability are negligible and the one-dimensional treatment is satisfactory. The solution of Eq. 11 requires a surface boundary condition, which may be provided either by specification of the value of temperature as a function of time at the surface $z = 0$ or else by a condition on the surface energy flux. This latter condition is appropriate for our discussion, with the condition being given by specifying the value of the ground heat flux G at the surface.

$$G(z, t) = -\kappa \frac{\partial T}{\partial z} \tag{12}$$

A thorough treatment of Eqs. 11 and 12, excluding numerical methods, has been given by Carslaw and Jaeger (1959). Representative values for density, heat capacity, and thermal conductivity are given there, and in many other sources. Most older references use the calorie as a unit of thermal energy, which is inconsistent with modern terminology. Table 1 presents values of these variables in SI units.

In order to develop remote-sensing applications, we require formal methods of solution for these equations.

III.A A Periodic Solution of the Heat-Flow Equation

Although limited in generality, the following procedure yields a solution that gives us physical insight into the behavior of temperature of the Earth's surface. We assume the form

$$T(z, t) = T_0 + T_1(k, \omega) \exp(-i\omega t + ikz) \tag{13}$$

where ω and k are as yet unspecified, and T_1, which is in general complex, represents the amplitude of the periodic variability, which we may associate with the heating and cooling of the diurnal cycle.

Substituting this function into Eq. 11 and canceling factor $T_1(k, \omega)$, we find

$$-i\omega\rho c = -k^2\kappa$$

$$k = (\omega\rho c i/\kappa)^{1/2}$$

$$k_1 = (\omega\rho c/\kappa)^{1/2}(1 + i)/2^{1/2}$$

$$k_2 = (\omega\rho c/\kappa)^{1/2}(1 - i)/2^{1/2} \tag{14}$$

where we have solved for k because ω is assumed given, e.g., a reasonable value is $\omega = \Omega = (2\pi/1 \text{ day}) = 7.27 \times 10^{-5} \text{ s}^{-1}$. If the two possible solutions for k are substituted into Eq. 13, it is apparent that one grows exponentially with increasing depth z, and thus is not physically proper, whereas the other value for k, containing the factor $1 + i$, decreases with depth below the surface, as is physically proper. The resulting solution is given by

$$T(z, t) = T_0 + T_1(k_1, \omega) \exp\left\{-i\left[\omega t - (\omega\rho c/2\kappa)^{1/2} z\right]\right\} \exp\left[-(\omega\rho c/2\kappa)^{1/2} z\right] \tag{15}$$

representing a sinusoidally varying wave that changes phase and decreases in amplitude with increasing depth below the Earth's surface. The coefficient of z may be related to an effective damping length, $L = (2\kappa/\omega\rho c)^{1/2}$, representative values of which are given in Table 1. This description by Eq. 15 is qualitatively correct, requiring specification of a value for T_1. As a crude approximation, we might consider setting the ground-heat flux variation equal to one-half the maximum solar flux. If solar heating and radiative cooling represented accurately the driving force at the surface, then this solution would be adequate for many purposes. However, this would be a great oversimplification. In order to establish the magnitude of the temperature wave, it is necessary to consider the energy balance at the Earth's surface, providing physically based values in the surface boundary condition.

TABLE 1 Representative Values for Heat Capacity, Thermal Conductivity, and Density for Common Materials

Substance	Density (Kg m^{-1})	Specific Heat $(\text{J m}^{-3} \text{ °C}^{-1})$	Thermal Conductivity $(\text{J m}^{-1} \text{ s}^{-1} \text{ °C}^{-1})$	Diurnal Heat Capacity $(\text{J m}^{-2} \text{ °C}^{-1})$	Damping Length (m)
Quartz	2650	730	8.4	34.4	0.35
Clay minerals	2650	730	2.9	20.2	0.20
Sandstone	2300	960	2.5	20.0	0.18
Wet soil	1750	1000	1.4	13.3	0.15
Water (no mixing)	1000	1000	0.60	6.6	0.13
Dry sand	1650	790	0.26	5.0	0.07

III.B   Energy Balance

For remote sensing, a realistic solution of the heat-flow equation requires a detailed description of the surface energy balance, which is a statement of the conservation of energy. The heat flux into the ground, G, equals the net flux of radiation from the atmosphere, R_{net}, minus the transfer from the ground to the atmosphere by sensible heat flow, H, and the latent heat flux of evaporation, LE. Net radiation flux equals the difference between the downward solar flux, i.e., the solar constant as adjusted for atmospheric effects, R_{sun}, the downward radiation flux from the atmosphere, R_{atm}, and the upward emitted energy from the Earth's surface, $R_{emitted}$, as described by the Stefan-Boltzmann law. Thus,

$$G = R_{net} - H - LE \tag{16}$$

$$R_{net} = R_{sun} + R_{atm} - R_{emitted} \tag{17}$$

For computation, these symbolic terms must be replaced by either explicit functions of known parameters or else by measured experimental values. Solar radiation is the dominant energy flux causing temperature changes at the Earth's surface. The solar flux, omitting for the moment the effect of the atmosphere, may be written as

$$R_{sun} = S(\sin \delta \sin \varphi + \cos \delta \cos \varphi \cos \Omega t) \qquad R_{sun} \geq 0 \tag{18}$$

where S is the solar constant, 1375 W m^{-2}, φ is the latitude of the location on the Earth, δ is the solar declination, t is the time measured from local noon, and $\Omega = (2\pi/1 \text{ day}) = 7.27 \times 10^{-5}$ s^{-1}. In order to complete the mathematical development, we defer the presentation of physically based expressions for the other energy fluxes. Instead, we assume a simple linear form with surface temperature T_s for the other terms of the surface energy flux, leading to the expression

$$G = R_{sun}V(1 - a) - A - BT \tag{19}$$

where A and B are constants that will be replaced by physical variables in a later section, V represents the transmittance of the atmosphere, typically of order 0.75 for vertical viewing, and a is the surface albedo. Then A and B, together with ρ, c, and κ, represent the quantities that one can, in principle, seek to determine from remote measurements of surface temperature.

When the actual time dependence of solar flux is inserted, the single periodic term used in Eq. 13 is generalized into a sum of terms having respectively higher and higher frequency dependence. From Fourier analysis, the amplitude of the temperature variation may be separated into components T_n by assuming

$$T = T_{mean} + \sum_{n=1}^{\infty} T_n e^{-in\Omega t} e^{ik_n z} \tag{20}$$

so that after substituting Eq. 20 into Eq. 19, multiplying both sides by cos $m\Omega t$, with m an integer, and integrating from 0 to 24 h, we find

$$-ik\kappa T_n = -BT_n + SV(1 - a) \cos \delta \cos \varphi \int_{t_1}^{t_2} \frac{dt \, e^{im\omega t}}{24 \text{ hours}} \cos n\omega t \tag{21}$$

where $t_1 = -\cos^{-1}(-\tan \delta \tan \varphi)$ and $t_2 = \cos^{-1}(\tan \delta \tan \varphi)$. Thus, the solution for temperature at the surface is given by evaluating the time integral and then separating real and imaginary parts:

$$T_s(t) = T_{mean} + SV(1 - a) \sum_{n=1}^{\infty} \frac{C_n \cos (n\Omega t - \varphi_n)}{[nD^2 + B^2 + (2BDn)^{1/2}]^{1/2}} \qquad (22)$$

where

$$\varphi_n = \cot^{-1}\left[1 + B\left(\frac{2}{nD}\right)^{1/2}\right] \qquad (23)$$

$$C_1 = \frac{1}{\pi}\left[\sin \delta \sin \varphi \, (1 - \tan^2 \delta \tan^2 \varphi)^{1/2}\right. \qquad (24)$$

$$\left. + \cos \delta \cos \varphi \cos^{-1}(-\tan \delta \tan \varphi)\right]$$

$$C_n = \frac{2}{\pi(n^2 - 1)}(n \sin nx \cos x - \cos nx \sin x) \qquad n > 1 \qquad (25)$$

and

$$x = \cos^{-1}(-\tan \delta \tan \varphi) \qquad (26)$$

Here D, an effective thermal resistance to temperature change, or diurnal heat capacity, given by $(\Omega \rho c \kappa)^{1/2}$ (W m^{-2} K^{-1}), is the product of the commonly used parameter thermal inertia $(\rho c \kappa)^{1/2}$ with peculiar units (J m^{-2} K^{-1} s$^{-1/2}$), and the square root of the Earth's rotation frequency Ω. We note that the temperature value far below the surface is T_{mean}. We also note that the solution of Eq. 22 depends only on the product $\rho c \kappa$, rather than on the individual factors (cf. Table 1). Thus, sets of values for ρ, c, and κ with a given product may not be distinguished using remote observations. This basic formulation has been used by Watson (1975) and Price (1977). The solution is capable of reproducing the major aspects of the surface-temperature cycle, although the assumption that A and B are time independent is unrealistic so that quantitative values are not very accurate. However, the solution is well suited for sensitivity studies, and with some additional effort, a very reasonable approximate form can be obtained (Hechinger et al., 1982). From this point, improving the utility of the solutions of the heat-flow equation for remote-sensing applications is a matter of specifying more precisely the various factors corresponding to $A + BT$, and improving the solution method, until a result of the desired accuracy is obtained.

IV   THE INVERSE PROBLEM

To this point, we have considered only the prediction of surface temperature, given a statement of the driving forces of the energy balance, such as solar input, and as yet unspecified formulas for the other surface fluxes. However, the remote-sensing problem is an inverse problem: one measures the radiance from a distance, derives the implied temperature from Planck's law, and then seeks to derive some physical property of the surface that is consistent with the temperature value observed. We shall not describe in detail procedures for solving this inverse problem, as the techniques must be selected according to the type and number of remote-sensing observations available, and also according to the numerical accuracy required for the application considered. The approach is, in principle, straightfoward. One measures the temperature at a location two or more times, t_1, t_2, etc., and then relates these values to the temperature values as predicted in Eq. 22. At this point, the solution is still presented symbolically in terms of A, B, T_{mean}, and D. Thus,

$$T_s(t_1) = f(A, B, D, T_{mean}, t_1) \qquad T_s(t_2) = f(A, B, D, T_{mean}, t_2) \qquad (27)$$

where more than two equations may be appropriate (two equations specify the pair of variables T_{mean} and D uniquely: we assume A and B are known parameters). However, B is related to not one, but a number of physical variables, as discussed later. The equations represented by Eqs. 27 are solved for the physical variables of interest.

A single example will be given because of its simplicity. If two thermal-infrared observations are acquired at approximately the peak of the diurnal cycle, which generally

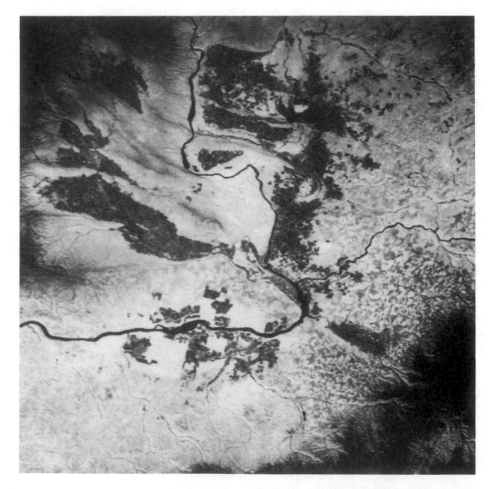

Figure 4   A midday (13:30 local time) thermal-infrared image of an area in southern Washington state. Darker regions are cool and brighter areas are hot. The area is approximately 250 km^2.

occurs around 1:30 p.m. local time, and at 2:30 a.m., or nearly the time of minimum temperature, then we can compute the temperature change T_s (13:30) $-$ T_s (2:30) using Eq. 22. In the temperature-difference expression, all the higher-order terms above $n = 1$ approximately cancel out, and T_{mean} is eliminated, leaving the simple formula

$$T_s(13:30) - T_s(2:30) \simeq \frac{2SV(1 - a)c_1}{[D + B + BD(2^{1/2})]^{1/2}} \tag{28}$$

The result may be solved for D, the diurnal heat capacity, as a function of the remote measurements (Price, 1977). Representative values of D are given in Table 1. Equation 28 is of specific interest because of a satellite program, the Heat Capacity Mapping Mission (Price, 1980), which acquired experimental data at these times for research purposes. Figures 4 and 5 illustrate data from that mission, providing evidence that thermal infrared data can discriminate among different surface conditions (Price, 1982).

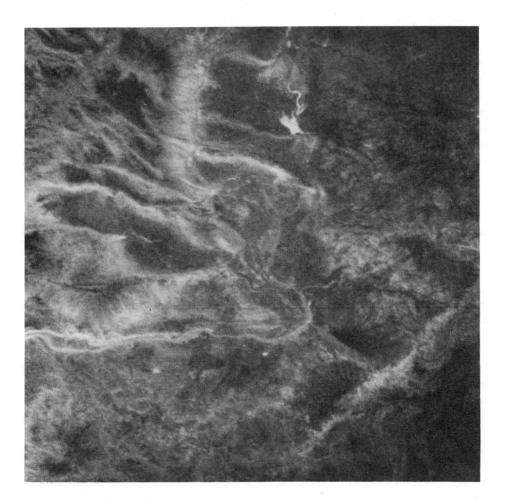

Figure 5 An early morning image taken at 2:30 local time, a few hours before that of Figure 4. Note the reversal of contrast in some areas.

In recent years, the NOAA meteorological satellites are approaching these observation times, implying that this approach may have some general utility.

The previous example illustrates a simple methodology for completing the chain of calculations from remote thermal-infrared measurements to values of a surface-related parameter. Evidently, one class of improvements is to be found through a more rigorous solution of the heat flow-equation and boundary conditions. This is discussed later. The second generalization is required when an analytic solution, such as found in Eqs. 22–26, is not available, so that temperature is given as a set of numerical values. In this case, the problem of interpreting observed temperature values is solved by first computing a multidimensional table of values of temperature for sets of values of the surface variables. Then the inversion problem becomes a task of searching the table of computed temperatures for values matching those observed, then reading off the parameter values that produced the temperature entry (Kahle et al., 1976). The dimensionality of the table is not limited in principle, i.e., temperature values may be predicted as functions of surface albedo, emissivity, moistness, heat capacity, thermal conductivity, etc. The only restriction is that from a given number of temperature measurements, one can solve for only (at most) the equivalent number of causitive surface variables.

From analysis of the general behavior of temperature as a function of surface variables, we know that a small change in a given parameter yields a small resultant change in the temperature value. For this reason, it is often useful to replace a set of look-up tables by an equivalent polynomial that expresses surface temperature as a linear or quadratic expression of the surface variables (Carlson et al., 1981). This expression is obtained by a least-squares fit of the assumed expression to the look-up table values. As another alternative Wetzel et al. (1984) have derived a formula that represents several surface variables as functions of sets of temperature measurements in the early hours after dawn. All these methods must be tailored to the number and timing of the remote-sensing observations.

V THE INITIAL-VALUE PROBLEM AND THE LAPLACE TRANSFORM

In many respects, the description of the Earth's temperature as a periodic function, with periodicity of 24 hours, is an adequate first approximation, especially in light of the many factors that must be included for significantly improved accuracy. However, the mathematical formulation to this point is not complete, as may be shown by consideration of a related class of solutions, corresponding to an initial-value problem in which the initial-temperature distribution is specified as an arbitrary function of depth.

The basis for seeking a more general solution derives because nature is not periodic, contrary to the previous assumption. One day is not exactly like the next, due to variations in radiative, latent, and sensible heat fluxes associated with changing winds, cloud cover, etc. In other words, the atmosphere changes with varying weather patterns over periods of minutes, hours, and days. The mathematical implications of this fact imply a weakness in the previous mathematical formalism: the description must contain a variable for specifying the temperature far below the Earth's surface. Thus, in principle, one may select an arbitrarily high value for the temperature under ground, leading to a constant flux of energy toward the surface, as though geothermal heating was providing a heat source, or, conversely, one might select a low value, in which case the Earth's surface would, for some period of time, be very cold, with a net flow of energy toward the Earth's center.

In order to investigate this problem, we use the Laplace-transform method. The theoretical aspects of this method need not be discussed for application to the heat-flow equation, as the formalism is very straightforward in this case. Procedurally, we multiply both sides of Eq. 11 by the factor $e^{i\omega t}e^{-ikz}$, then integrate over time t from zero to infinity, and over depth z from zero to infinity, assuming both integrals vanish at the extreme limit. Integration by parts yields the initial-value terms, which are of interest here, as well as the periodic terms discussed previously.

$$\rho c\left[-i\omega T^* + T(k, t = 0)\right] = \kappa\left[-k^2 T^* - \frac{\partial T}{\partial z}(\omega)\Big|_{z=0} + ikT(z = 0, \omega)\right] \quad (29)$$

where, by definition, $T^*(k, \omega)$ is the transform of the desired solution, and $T(k, t = 0)$ is the Laplace transform of the depth dependence of the initial temperature profile. One of the two surface terms on the right side must be selected to provide the upper (energy-balance) condition at the surface. In the remote-sensing case, the boundary condition on the flux, $G = -\kappa\, \partial T/\partial z$ provides the appropriate solution, and the other term, $T(z = 0, \omega)$, is discarded. If Eq. 19 is used for the surface flux, then this component provides the periodic solution obtained in the previous section. However, the $T(k, t = 0)$ expression is of interest here, because it describes the evolution with time of the initial temperature distribution. Keeping only this term, and calling this part of the total solution the transient component with subscript t, we find

$$T_t^*(k, \omega) = T(k, t = 0)/(i\omega\rho c - k^2\kappa) \quad (30)$$

The details of solving for $T_t(z, t)$ from $T_t(k, \omega)$ need not be given, except to say that the time aspect may be inverted first, yielding the contribution form a pole at $i\omega = k^2\kappa/\rho c$. Then the inverse of the space transform produces the final (formal) solution:

$$T_t(z, t) = \frac{1}{2\pi}\int dk\; e^{ikz}\, e^{-k^2\kappa t/\rho c}\, T(k, t = 0) \quad (31)$$

This equation represents an integral that decreases rapidly with time, leaving the periodic solution that was found earlier. Generally, if an improper temperature distribution or value $T(k, t = 0)$ is selected beneath the surface, this produces a rapid change of surface temperature during a brief initial period. The duration and magnitude of this transient depend on the details of the initial temperature distribution. There is no strict quantitative rule for distinguishing such cases, but, generally, if the initialization does not represent a realistic situation, one finds a discrepancy between the assumed initial surface temperature and the value 24 hours later, (Wetzel et al., 1984). Although it is possible to guess a reasonable value for the temperature well below the surface, a formal procedure is preferable: by solving the energy-balance equation with the mean ground heat flow (over a day) set to zero, i.e., $\langle G\rangle = 0$, which implies

$$\langle H\rangle + \langle LE\rangle + \langle R_{net}\rangle = 0 \quad (32)$$

and a value of $\langle T_s \rangle$. Then we set the temperature far below the surface, T_{mean}, equal to $\langle T_s \rangle$, so that the value of G will vanish when averaged over a day. Thus, the solution is assured to be nearly periodic. Note that it is not correct to set a deep subsurface temperature equal to the mean air temperature, as a freely evaporating surface may be cooler than the air throughout day and night.

VI MICROMETEOROLOGY AND THE SURFACE-ENERGY BALANCE

The description of the energy exchanges at the Earth's surface represents a major subdiscipline of meterology, while the single term LE, representing evapotranspiration, is of great interest to agriculture and water-use specialists. Studies of these topics began decades before the initiation of remote sensing, and available literature would fill a moderate size library. Unfortunately, there are no magic formulas that summarize the current state of knowledge: the world is too complex for universal approximations. Excellent reviews have been provided in the text by Sellers (1965), which contains useful tables of values and constants as well as basic formulation, and Carlson and Boland (1978), and Carlson (1986), who select particular expressions for the terms in the energy-balance equation, then substitute them into a complete mathematical formulation and derive estimates of surface parameters from satellite data. The possibility of obtaining some of this information using remote sensing is discussed in Chapter 16. In this selection we select commonly used expressions for R_{atm}, H, and LE.

$$R_{\text{atm}} = \varepsilon \sigma T_{\text{air}}^4 \left(a_B + b_B \, q_{\text{air}}^{1/2} \right) \tag{33}$$

$$H = \rho_{\text{air}} c_p \frac{T_s - T_{\text{air}}}{r} \tag{34}$$

$$LE = \rho_{\text{air}} L \frac{M q_s - q_{\text{air}}}{r} \tag{35}$$

$$r = \frac{\left[\ln \left(z_1/z_0 \right) \right]^2}{K^2 u} \tag{36}$$

where a_B and b_B are radiation constants (Jensen, 1973); and q_{air}, T_{air}, and u are the humidity, air temperature, and wind speed, respectively, at standard height z_1 (2 m); and ρ_{air} is the density of air. Additional variables are c_p, the heat capacity of air; L, the specific heat of evaporation; and factors pertaining to energy exchange in r, i.e., z_0 is a roughness length that is characteristic of the surface type, and K is the von Karman constant, 0.4. The important factor moisture availability, M, represents the percent saturation q_s of the surface at temperature T_s. When $M = 0$, the surface is perfectly dry, whereas $M = 1$ corresponds to a freely evaporating saturated surface. Factor r, representing the atmospheric resistance at neutral atmospheric stability, is in general a much more complicated function of atmospheric variables. The previous formulas were chosen for their simplicity, since there is no agreed formulation that is suitable for general conditions, as when a satellite instrument views mixed surface cover over an area of thousands of square kilometers. For vegetation, resistance r is modified by the addition of a stomatal resistance term, which is also not well understood. In fact, the radiation measurements of remote sensing provide a powerful new tool for studying the surface energy balance.

By using these expressions and meteorological data, it is possible to simulate the cycle of surface temperature with considerable realism and accuracy. This statement applies to rock and soil surfaces, as the description of temperature behavior for a surface covered by vegetation is still being developed. For example, the complexity of energy exchanges *within* a vegetation canopy is extreme, with the radiation processes discussed earlier in this book as a simple example (Taconet et al., 1986). However, it is useful to describe the basic properties of the solutions, which apply across all situations. Figures 6 and 7 illustrate computed temperature curves based on different values of diurnal heat capacity *D* and surface-moisture availability *M*. These two parameters have the greatest effect in determining surface-temperature variations, and thus are the variables that one may reasonably infer using remote observations. It is evident that all curves have the same general shape, with *D*, the resistance to temperature change, tending to influence mainly the amplitude of the day-to-night temperature variation, whereas *M*, which controls the surface-evaporation tendency, affects the mean surface temperature as well as the amplitude of the diurnal cycle. Similar figures by Carlson and Boland (1978) and Watson (1975) illustrate that the parameters of the surface-energy budget all tend to produce variations similar to Figures 6 and 7. This encourages those who carry out sensitivity studies, i.e., comparison of data from ground locations differing in only one or two parameters yields understandable results. However, it makes the inverse problem more difficult, as an observed variation of surface temperature from one site to another may be due to a difference of any of several factors, alone or in combination. Thus, one can

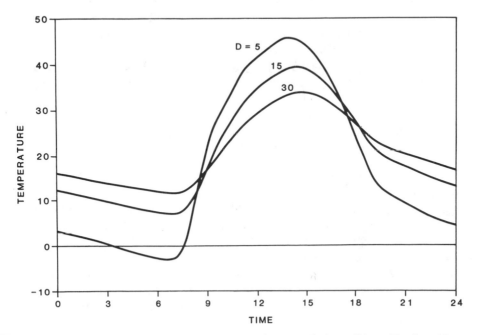

Figure 6 Computed surface temperatures for representative atmospheric conditions. The diurnal heat capacity takes on three values, while all other parameters are fixed.

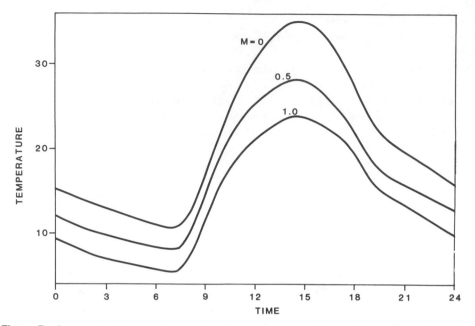

Figure 7 Computed surface temperatures for representative atmospheric conditions. The surface-moisture availability takes on three values, while all other parameters are fixed.

specify sets of values for groups of parameters that produce the same observed temperature values.

The mathematical difference in complexity resulting from replacing $A + BT$ in Eq. 19 by the formulas of Eqs. 33–36 is very great. Not only does the equation contain more parameters, but the atmospheric variables T_{air}, u, and q_{air} depend on time due to the variation of meteorological conditions, which vary both from place to place and from moment to moment in an irregular fashion. This implies at once that an analytic solution, as Eq. 22, is not possible, and numerical procedures must be used, or else formal simplification is required. For example, by combining Eqs. 16, 17, and 19, one can derive

$$A + BT = H + LE + R_{\text{emitted}} - R_{\text{atm}} \qquad (37)$$

where the expressions on the right are given by Eqs. 33–35 and 10. Then physically based formulas for A and B can be derived by averaging Eq. 37 over a day, and by equating B, the derivative of the left side with respect to temperature, with the average over a day of the temperature derivative of the right-hand side. This formalism has been explored by Price (1977) and Raffy and Becker (1985a, 1985b). The solution by numerical procedures is not difficult, as finite-difference methods for solving the heat-flow equation are readily available (Gerald, 1973). The other essential for computer solution is the surface boundary condition, which must be satisfied at each time step. This requires only the use of a root finder, which determines the value of $T(z = 0) = T_s$ obtained as the solution of

$$G(T_s) - R_{\text{net}}(T_s) - H(T_s) - LE(T_s) = 0$$

However, from the point of view of data required, one must also have available explicit values of the meteorological variables, which are not readily found over areas of thousands of square kilometers, as is appropriate for remote sensing. In principle, such values may be found as an output product of a weather-prediction program, but this implies a sophisticated data-handling capability that is not available for most potential users of remotely sensed data. Another middle ground is found by taking as given the large-scale meteorological variability, such as from a weather-forecast model, then blending this with the heat-flow equation through a formulation that includes effects in the lowest 1–2 km of the atmosphere, the planetary boundary layer. This relatively optimum procedure has been utilized in only a few studies, such as by Carlson et al. (1981).

VII THE FORCE-RESTORE APPROXIMATION

While the solution of the heat-flow equation is not difficult using either analytic or numerical methods, another simpler equation has been developed that appears to represent adequately the behavior of surface temperature. This equation has been obtained by comparing solutions given in the previous sections with those from an ordinary differential equation:

$$\frac{dT_s}{dt} = \left(\frac{2\rho\omega}{\kappa c}\right)^{1/2} G(z = 0, t) - \Omega(T_s - T_{\text{mean}}) \tag{38}$$

Equation 38 replaces the rather complicated behavior of temperature beneath the surface by a layer with temperature T_{mean}, which exchanges energy with the Earth's surface. We note that Eq. 38 includes both the properties of the periodic solutions through the term ΩT_s and the initial-value problem with the dependence on initial temperature distribution simulated by the term ΩT_{mean}. In fact, this approximation appears to provide useful accuracy, reducing the complexity of the numerical solution (Deardorff, 1978; Lin, 1980).

VIII CORRECTING THERMAL-INFRARED OBSERVATIONS FOR ATMOSPHERIC EFFECTS

Although the atmosphere is reasonably transparent at thermal-infrared wavelengths, the absorption and reemission of radiation produces appreciable effects when converted to surface parameters through the procedures developed in this chapter. Furthermore, the influence of the atmosphere is not a constant term (see Chapter 9) that may be added or subtracted from remotely sensed temperatures, since it also acts to reduce the apparent contrast of remotely sensed surface temperatures. These effects vary with atmospheric conditions. As a result, the atmospheric corrections must be derived on a case-by-case basis for the conditions prevailing at the time of the remote measurements. Data for the computation may be obtained in several ways, but first an understanding of the basic processes of radiative transfer is required (Liou, 1980). In the infrared region of the spectrum, the effect of absorbing materials is twofold: first, the absorber reduces the radiation field passing through the atmosphere, and, second, this absorber reradiates some of this radiation according to its own blackbody temperature. Thus, with $L_\lambda(z)$, the intensity of radiation at wavelength λ and height z above the surface (note the change of sign of direction as compared to the discussion of the heat-flow equation; the two problems may be addressed independently)

$$\frac{dL_\lambda}{ds} = -k_\lambda \rho [L_\lambda + B_\lambda(T)] \tag{39}$$

where k_λ is the absorption coefficient of the material, ρ is its concentration in the atmosphere, T is its temperature, and s is the path length. For observations at nadir angle θ, $s = z/\cos\theta$. This equation expresses the decrease of transmitted radiation due to absorption and scattering, and the reemission by the absorber according to its local temperature. A formal solution can be obtained by integrating Eq. 39 with respect to path length, where the concentration and temperature of the absorber must be given for numerical computation. In the thermal infrared (10–14 μm), the principle atmospheric absorber is water vapor, for which the concentration and temperature profile may be obtained from standard meteorological soundings. At numerous weather stations around the world, balloons are released twice per day with instruments that transmit the atmospheric temperature, pressure, and moisture content to a ground receiver. These values can be used in the derivation of the temperature correction of satellite data for atmospheric absorption. Thus,

$$L_\lambda = L_\lambda(s = 0)e^{-\tau_\lambda(s_T, 0)} + \int_0^{s_T} ds' \, e^{-\tau_\lambda(s_T, s')} B[T(s')] k_\lambda \rho(s') \tag{40}$$

where τ is the optical depth,

$$\tau_\lambda(s_T, s') = \int_{s'}^{s_T} ds'' \, k_\lambda \rho(s'') \tag{41}$$

and s_T is the height of the sensor, or else the effective top of the atmosphere for a satellite instrument. This solution relates the satellite-measured radiance to surface radiance temperature and the atmospheric profile of moisture and temperature. The height integral is often changed to a pressure integral using the change of variables implied by the hydrostatic equation $dp/dz = -\rho g$, where g is the acceleration of gravity. In practice, the integration is readily carried out, given atmospheric data, by general-purpose codes, such as Lowtran (Kneizys et al., 1980), or by specialized codes that have been written for this application (e.g. Price, 1983). Temperature and moisture profiles can also be estimated by specialized satellite instruments, "sounders," which obtain measurements of radiation in the opaque as well as "window" regions of the thermal-infrared spectrum. The resulting multispectral thermal-infrared data can be analyzed or inverted to yield the required atmospheric temperature and moisture profiles, and, if desired, the atmospheric correction to window channel measurements of surface temperature (Davis and Tarpley, 1983). For the meteorological satellites, this presents the advantage that the sounder data are acquired simultaneously with the window channel thermal data. The difficulty is associated with coordinating the processed results, which are oriented toward use in weather-forecast models, with the higher spatial resolution data, and the need to geographically colocate the data of different types and spatial scales.

Realizing this problem, NOAA has placed on recent satellites the capability for simultaneous 1 km resolution measurements at both 11 and 12 μm. At 12 μm, the atmospheric influence is slightly greater than at 11 μm. Thus, two measurements allow solution of a pair of equations to produce point-by-point estimates of surface temperature without the need for meteorological sounding data. The following equations, developed originally for evaluation of sea surface temperatures, illustrate the procedure. Given Eq. 40, stated at two different wavelengths λ_1 and λ_2, one can expand L and B about the surface radiance temperature T_s and treat the absorption constant k as a small quantity, so that $e^{-\tau} \approx 1 - \tau$, to produce

$$T_{BB1} = T_s(1 - \tau_1) + k_1 \int_0^{ST} dz\, \rho T(z_T - z) \tag{42}$$

$$T_{BB2} = T_s(1 - \tau_2) + k_2 \int_0^{ST} dz\, \rho T(z_T - z) \tag{43}$$

Since $\tau_\lambda = k_\lambda \int_0^{ST} ds\, \rho(s)$, we can solve to eliminate the integrals in each equation:

$$T_s = T_{BB1} + \frac{T_{BB1} - T_{BB2}}{k_2/k_1 - 1} \tag{44}$$

and the problem appears to be solved, assuming the ratio k_2/k_1 is known accurately. However, this ratio is not well established, being of order 1.3 for $\lambda_1 = 10.8$ μm, $\lambda_2 = 11.9$ μm, and the influence of surface emissivity must be considered. Using the formalism of Eqs. 8 and 9,

$$T_s = T_{BB\lambda} \left(\frac{n_\lambda - \varepsilon_\lambda + 1}{n_\lambda} \right) \tag{45}$$

so that

$$T_s = T_{BB1} + 3.3(T_{BB1} - T_{BB2})\frac{5.5 - \varepsilon_1}{4.5} + 0.75T_{BB1}(\varepsilon_1 - \varepsilon_2) \tag{46}$$

In fact, when ε is less than 1, downward radiation is reflected upward, where the reflection coefficient is $1 - \varepsilon$ (Eq. 5). Although this reflected component alters the coefficient of $\varepsilon_1 - \varepsilon_2$ in Eq. 46, this effect may be neglected here. As discussed later, the wavelength dependence of emissivity of common materials is not well known, so that any results are subject to considerable uncertainty: at 300 K, an error of 0.01 in the difference between ε_1 and ε_2 implies an error of 2 K in the estimate of surface temperature (Price, 1984, which contains a sign error).

IX   SEA SURFACE TEMPERATURE

The estimation of sea surface temperature from thermal-infrared data has received attention since the early 1970s. Sea surface temperature is important for meteorological and climatic studies, because the oceans, which cover 70% of the Earth's surface, have an enormous heat-storing capacity and thus can exchange large amounts of energy with the atmosphere. It is instructive to relate the difficulties of land-temperature measurements to the physical differences that make the ocean-temperature problem so much simpler. Questions of data compositing, cloud filtering to eliminate contaminated data, and empirical adjustment of radiative-transfer formulations have been addressed for sea surface temperature estimation. These methods may well be applied to land-temperature data, once the more important factors discussed in this chapter are under control.

As a primary simplification, the ocean temperature is essentially constant over a day, so that considerations of time dependence, as in Eq. 11, may be neglected. The large heat-storing capacity results both from the high intrinsic heat capacity of water, and because wind-induced vertical mixing carries heat down to considerable depths, often 100 m. The end result is an equivalent heat capacity orders of magnitude greater than for land surfaces. This limits the diurnal temperature cycle to barely measurable values in most locations, and restricts the typical annual temperature cycle to a few degrees Celsius. However, even the very limited seasonal and interannual variability of sea surface temperature is sufficient to cause significant meteorological effects, and an accuracy better than 1°C is sought.

For a second difference, as compared to land, sea surface temperature is relatively constant over scales of tens of kilometers, so that averaging and compositing techniques are satisfactory for obtaining highly accurate values, even if scattered clouds and noise in the data tend to reduce the accuracy of individual temperature estimates. As a third simplification, the ocean is virtually black in the visible and near-infrared spectral regions, so that identification of clouds is relatively easy over oceans as compared to land. Evidently, the most difficult situations involve very thin uniform cloud cover, e.g., cirrus, and very small (subresolution element) cumulus cloud, both of which may alter temperature estimates from thermal-infrared data. Finally, the absolute value and spectral dependence of emissivity of sea water are known, as opposed to land. The net result is that sophisticated data-processing techniques have been developed for the derivation of sea surface temperatures from satellite thermal-infrared observations (Walton, 1987). In the future, these techniques may become applicable to the land-temperature and analysis problem.

X EMISSIVITY IN THE THERMAL INFRARED

Historically, little attention has been given to spectral values of emissivity in the thermal infrared. Geologists have measured spectral behavior for some rocks and minerals (Lyon, 1972; 1975), but until recently, there has been little use for such data. Only with the acquisition of high-quality data beginning in 1987 has it been feasible to obtain high-quality emissivity data over extended areas.

Becker et al. (1981, 1985, 1986) have carried out laboratory and field measurements, finding that geometric factors, such as shadowing, influence observed radiances in the field as well as the properties of the material itself. Such data are needed for high accuracy in the determination of thermal properties and evaportranspiration over large areas. Fortunately, it appears that emissivity of about 0.96–0.97 may be assumed in agricultural areas, so that even given some uncertainty regarding absolute values, the spatial variation of the effect is small. The significance of emissivity variations in dry rocky areas is still not known (Goward, 1986). Recently, the National Aeronautics and Space

Administration has constructed a multiband radiometer that acquires data from an aircraft in a number of spectral intervals in the 8–13 μm region of the spectrum (see also Chapter 12). Since relatively little is known about the spectral behavior of materials in the 8–13 μm wavelength band, no satisfactory procedure has been developed for analyzing such data, e.g., to estimate mineral composition. As a result, much of the effort with the data has been devoted to image enhancement followed by photointerpretation. This is consistent with the general use of imagery by geologists, with the digital nature of radiometric data providing the opportunity for sophisticated image-enhancement techniques (Gillespie et al., 1986; Niblack, 1986). Theoretical studies of multispectral properties of materials in this spectral regime are just beginning (Takashima and Masuda, 1987). It should be clear that the uncertainty due to lack of knowledge of emissivity propagates through any of the techniques described in this chapter, so that high accuracy may not be expected for any temperature-related quantities, except, of course, sea surface temperature.

XI SUMMARY

Quantitative interpretation of remotely sensed thermal-infrared data is complicated by the many physical factors that influence observed temperatures. We list in approximate order of importance the various factors that affect the derivation of useful information about surface conditions:

1. The ability of the surface to support evaporation, or in the case of vegetation, evapotranspiration, influences greatly both the daily average surface temperature and day-to-night temperature range. This effect generally may be ascribed to a surface-moistness factor, or to a resistance to evapotranspiration, e.g., a stomatal resistance in the case of vegetation.

2. Moisture in the atmosphere affects the radiation balance at the surface by modifying downward radiation, while also modifying the value of satellite-observed radiances. Without a correction for atmospheric effects, any satellite-derived results will be subject to considerable uncertainty. This correction must be obtained for quantitative use of satellite data.

3. The temperature of air near the ground affects the transfer of sensible heat from ground to atmosphere. While very important, this variable can generally be estimated reasonably well from conventional meteorological data. For highly accurate interpretation of thermal measurements, local air-temperature measurements are required, or else values must be obtained from an atmospheric prediction model.

4. Surface wind speed and surface roughness combine to produce a major effect on the exchange of sensible and latent heat with the atmosphere. This difficult problem becomes even more complex when vegetative surfaces are considered. The general description of heat and moisture transfer in a plant canopy still lies in the distant future. In addition, atmospheric humidity influences the magnitude of the moisture flux exchange between the Earth's surface and the atmosphere.

5. The heat-storing capacity of soil or rocks affects the day-to-night temperature changes observed by remote-sensing instruments. This factor is of some importance in dry areas, but of rather less significance in moist areas, where evapo-

transpiration decreases the energy available to drive a large day-to-night temperature cycle.

6. Surface albedo determines the fraction of the sun's energy that is available to drive evaporation and near-surface heat storage. Albedo is readily estimated from satellite measurements, such as from the NOAA Advanced Very High Resolution Radiometers or the LANDSAT Thematic Mappers.

7. Surface emissivity, integrated over wavelength, affects radiative heat loss to the atmosphere, whereas emissivity in the atmospheric window influences the interpretation of measured radiances in terms of physical temperatures. These factors are poorly known on a regional basis.

8. Topography, i.e., attitude with respect to the sun, and openness or sheltering from the wind, modify the nominal energy balance of a flat surface.

It is not clear that the increased accuracy of the physical formulation, of the input data, or of the numerical-solution methods will make a significant contribution to the

Figure 8 A midafternoon thermal image from the NOAA-7 meteorological satellite. This image is more representative than the preceding, illustrating scattered cumulus clouds (small black spots), atmospheric moisture (dark blotches running generally top left to bottom right), and ground features.

large-area problem, i.e., the interpretation of thermal-infrared measurements at a spatial resolution of 100 m to 10 km over areas of order 10^4 to 10^6 km. Specifying micrometeorological data over such scales is difficult to imagine, unless data from smaller scales are aggregated through a geobased information-type system. Construction of such regional-scale data bases of land type, topography, radiative properties, etc. is perhaps more important than the refinement of well-developed solution methods such as those of Carlson et al. (1981); Taconet et al. (1986), and the development of better understanding of micrometeorological fluxes.

From this list of factors, it is not surprising that the use of remote sensing in the thermal infrared has not achieved operational status over the Earth's land surfaces. The problems are apparent in a typical thermal infrared image as from the NOAA AVHRR (Figure 8). However, the topic is of increasing importance, as mankind's alteration of substantial areas of the globe produces effects on the local and global exchange of heat and moisture with the atmosphere. These factors ultimately control our physical environment and our ability to produce food and fiber to sustain ourselves. The thermal-infrared spectral region offers the best hope of monitoring global surface conditions on a regular cost-effective basis. It is clear that research in this technical discipline will improve our understanding of climatic processes near the Earth's surface.

LIST OF SYMBOLS

STANDARD ALPHABETIC SYMBOLS

Symbol	Description
a	Surface albedo [unitless]
B_λ	Planck's radiance [W m^{-2} μm^{-1}]
c	Heat capacity [J kg^{-1} K^{-1}]
D	Diurnal heat capacity $= (\Omega \rho c \lambda)^{1/2}$ [W m^{-2} K^{-1}]
G	Ground heat flux [J m^{-2} s^{-1}]
H	Sensible heat flux [J m^{-2} s^{-1}]
J	Normal component of solar flux atop the atmosphere [W m^{-2}]
k	Fourier variable [m^{-1}]
k_λ	Spectral absorption coefficient [m^2 kg^{-1}]
L	Radiance [W m^{-2} μm^{-1}]
LE	Latent heat flux in atmosphere [J m^{-2} s^{-1}]
M	Moisture availability [unitless]
s	Atmospheric path length [m]
T	Temperature [degrees K]
T_{BB}	Blackbody or radiance temperature [degrees K]
T_{mean}	Temperature far below surface that results in no net heat flux [degrees K]
V	Vertical atmospheric transmittance in the visible [unitless]
z	Distance (vertical) [m]

STANDARD GREEK SYMBOLS

Symbol	Description
ε_λ	Emissivity [unitless]
κ	Thermal conductivity [J m^{-1} K^{-1} s^{-1}]

Symbol	Description
λ	Wavelength [m]
ρ	Density [kg m^{-3}]
σ	Stefan-Boltzmann constant [W m^{-2} K^{-4}]
τ	Optical depth [unitless]
ω	Fourier/Laplace variable [s^{-1}]
Ω	Earth rotation rate [rad sec^{-1}]

REFERENCES

Becker, R., W. Ngai, and M. P. Stoll (1981). An active method for measuring thermal infrared effective emissivities: Implications and perspectives for remote sensing. *Adv. Space Res.* **1**:193–210.

Becker, F., P. Ramanantsizehena, and M. P. Stoll (1985). Angular variation of the bidirectional reflectance of bare soils in the thermal infrared band. *Appl. Opt.* **24**:365–375.

Becker, F., F. Nerry, P. Ramanantsizehena, and M. P. Stoll (1986). Mesures d'emissivite angulaire par reflexion dans l'infrarough thermique—implications pour la télédection. *Int. J. Remote Sens.* **7**:1715–1762.

Carlson, T. N. (1986). Regional-scale estimates of surface moisture availability and thermal inertia using remote thermal measurements. *Remote Sens. Rev.* **1**:197–247.

Carlson, T. N. and F. E. Boland (1978). Analysis of urban-rural canopy using a surface heat flux/temperature model. *J. Appl. Meteorol.* **17**:998–1013.

Carlson, T. N., J. K. Dodd, S. G. Benjamin, and J. N. Cooper (1981). Remote estimation of surface energy balance, moisture availability and thermal inertia. *J. Appl. Meteorol.* **20**:67–87.

Carslaw, H. S., and J. C. Jaeger (1959). *Conduction of Heat in Solids*, 2nd ed. Oxford University Press, London.

Davis, P. A., and J. D. Tarpley (1983). Estimation of shelter temperatures from operational satellite sounder data. *J. Clim. Appl. Meteorol.* **22**:369.

Deardorff, J. W. (1978). Efficient prediction of ground surface temperature and moisture with inclusion of a layer of vegetation. *J. Geophys. Res.* **83**:1889–1903.

Gerald, C. F. (1973). *Applied Numerical Analysis*. Addison-Wesley, Reading, Massachusetts, pp. 218–235.

Gillespie, A. R., A. B. Kahle, and R. E. Walker (1986). Color enhancement of highly correlated images. I. Decorrelation and HSI contrast stretches. *Remote Sens. Environ.* **20**:209–235.

Goward, S. (Ed.) (1986). *Commercial Applications and Scientific Research Requirements for Thermal-Infrared Observations of Terrestrial Surfaces*. NASA/Goddard Space Flight Center, Greenbelt, Maryland, pp. 62–102.

Hechinger, E., M. Raffy, and F. Becker (1982). Comparison between accuracies of a new discretization method and an improved Fourier method to evaluate heat transfers between soil and atmosphere. *J. Geophys. Res.* **87**:7325–7339.

Jensen, M. E. (1973). *Consumptive Use of Water and Irrigation Water Requirements*. American Society of Civil Engineers, New York, pp. 26, 27.

Kahle, A. B., A. R. Gillespie, and A. F. H. Goetz (1976). Thermal inertia mapping: A new geologic mapping tool. *Geophys. Res. Lett.* **3**:26–28.

Kneizys, F. X., E. P. Shettle, W. O. Gallery, J. H. Chetwynd, L. W. Abreu, J. E. A. Selby, R. W. Fenn, and R. A. McClatchey (1980). *Atmospheric Transmittance: Computer Code Lowtran 5*, AFGL-TR-80-0067. Air Force Geophysics Laboratory, Hanscom Air Force Base, Massachusetts.

Lin, J. D. (1980). On the force-restore method for prediction of ground surface temperatures. *J. Geophys. Res.* **85:**3251–3254.

Liou, K.-N. (1980). *An Introduction to Atmospheric Radiation.* Academic Press, New York.

Lyon, R. J. P. (1972). Infrared spectral emittance in geologic mapping: Airborne spectrometer data from Pisgah Crater, CA. *Science* **175:**983–985.

Lyon, R. J. P. (1975). Reflectance and emittance of terrain in the mid-infrared (6 to 25 μm) region. In *Infrared and Raman Spectroscopy of Lunar and Terrestrail Minerals* (C. Karr, Jr., Ed.), pp. 165–196. Academic Press, New York.

Matson, M., and J. Dozier (1981). Identification of subresolution high temperature sources using a thermal IR sensor. *Photogramm. Eng. Remote Sens.* **47:**1311–1318.

Niblack, W. (1986). *An Introduction to Digital Image Processing.* Prentice-Hall, Englewood Cliffs, New Jersey.

Price, J. C. (1977). Thermal inertia mapping: A new view of the earth. *J. Geophys. Res.* **82:**2582–2590.

Price, J. C. (Ed.) (1980). *Heat Capacity Mapping Mission User's Guide.* NASA/Goddard Space Flight Center, Greenbelt, Maryland.

Price, J. C. (1982). Estimation of regional scale evapotranspiration through analysis of satellite thermal-infrared data. *IEEE Trans. Geosci. Remote Sens.* **GE-20:**286–292.

Price, J. C. (1983). Estimating surface temperatures from satellite thermal infrared data—A simple formulation for the atmospheric effect. *Remote Sens. Environ.* **13:**353–361.

Price, J. C. (1984). Land surface temperature measurements from the split window channels of the NOAA 7 advanced very high resolution radiometer. *J. Geophys. Res.* **89:**7231–7237.

Raffy, M., and F. Becker (1985a). A stable iterative procedure to obtain soil surface parameters and fluxes from satellite data. *IEEE Trans. Geosci. Remote Sens.* **GE-24:**327–333.

Raffy, M., and F. Becker (1985b). An inverse problem occuring in remote sensing in the thermal infrared bands and its solutions. *J. Geophys. Res.* **90:**5809–5818.

Sellers, W. D. (1965). *Physical Climatology.* University of Chicago Press, Chicago, Illinois.

Slater, P. N. (1980). *Remote Sensing, Optics and Optical Systems.* Addison-Wesley, Reading, Massachusetts.

Taconet, O., R. Bernard, and D. Vidal-Madjar (1986). Evapotranspiration over an agricultural region using a surface flux/temperature model based on NOAA-AVHRR data. *J. Appl. Clim. Meteorol.* **25:**284–307.

Takashima, T., and K. Masuda (1987). Emissivities of quartz and Sahara dust powders in the infrared region (7–17 μm). *Remote Sens. Environ.* **23:**51–63.

Walton, C. (1987). *The AVHRR/HIRS Operational Method for Satellite Based Seas Surface Temperature Determination*, National Environmental Satellite and Data Information Service, U.S. Department of Commerce, Washington, D.C.

Watson, K. (1975). Geologic applications of thermal infrared images. *Proc. IEEE* **63:**128–137.

Wetzel, P. J., D. Atlas, and R. Woodward (1984). Determining soil moisture from geosynchronous satellite infrared data: A feasibility study. *J. Clim. Appl. Meteorol.* **23:**375–391.

16

ESTIMATING SURFACE ENERGY-BALANCE COMPONENTS FROM REMOTELY SENSED DATA

WILLIAM P. KUSTAS

Hydrology Laboratory
Beltsville Agricultural Research Center
Agricultural Research Service
United States Department of Agriculture
Beltsville, Maryland

RAY D. JACKSON

United States Water Conservation Laboratory
Agricultural Research Service
United States Department of Agriculture
Phoenix, Arizona

and

GHASSEM ASRAR

Headquarters
National Aeronautics and Space Administration
Washington, D.C.

I INTRODUCTION

The boundary between the Earth's gaseous atmosphere and its solid and liquid phases is a complex surface at which impinging radiation from the sun and the atmosphere is either reflected or absorbed. The absorbed portion, called net radiation (R_n), is the difference between incoming radiation at all wavelengths and reflected short-wavelength (≈ 0.15 to 4 μm) and both reflected and emitted long-wavelength (> 4 μm) radiation. The amount of radiation reflected and emitted is dependent on the composition of the surface, whereas the incoming radiation is independent of surface composition. Remote-sensing techniques are ideal for measuring the reflected and emitted components of net radiation, especially since these techniques can produce images delineating the spatial distribution of the components.

The net radiant energy absorbed by the surface is dissipated by conduction into the surface, convection to the atmosphere, by utilization as latent heat of evaporation, and, in the case of vegetated surfaces, by photosynthesis. The magnitude of the energy conducted into the surface, generally called the soil heat flux (G), ranges from about 10 to about 50% of net radiation, depending on the amount of vegetative cover. Convection to the atmosphere, called the sensible heat flux (H), may either warm or cool the surface, depending on whether the surface is cooler or warmer than the surrounding air. The magnitude of H depends on many factors (as will be discussed later) and is perhaps the most difficult to estimate by remote means. The energy used to evaporate water from the surface (LE) is usually not directly measured, but is obtained as the residual when the net radiation is balanced with the dissipation terms according to the law of conservation of energy. For hydrologic purposes, it is this term that is of most interest. The amount of energy used in photosynthesis is usually small in comparison to the other terms, and is frequently neglected.

A limitation of conventional techniques for measuring the surface energy balance is that they are essentially point measurements that are representative only of the surrounding area within which the magnitude of each component of the energy balance is nearly the same. An advantage of these techniques is that the time course of changes in the energy balance can be readily documented. Remote-sensing techniques have the advantage of evaluating components of the energy balance over reasonably large areas, at a spatial scale dependent upon the resolution of the sensing devices. However, they suffer from the disadvantage of being an instantaneous measurement at a point in time. Under relatively stable atmospheric conditions, the instantaneous measurements can be extrapolated to daily values, with a reasonable degree of confidence. The instantaneity remains a serious limitation under transient atmospheric conditions.

In this chapter, we present an overview of the principles involved in using remotely sensed radiometric measurements to evaluate components of the surface energy balance. In the following sections, the energy-balance equation is briefly discussed, followed by a detailed description of methods for evaluating its components and some experimental results.

II SURFACE ENERGY-BALANCE

The energy-balance equation applicable to most land surfaces can be written as

$$R_n = G + H + LE \qquad (1)$$

Historically, Eq. 1 has been used to evaluate the latent heat flux (LE), commonly called evapotranspiration. In the literature, evapotranspiration is frequently referred to as ET. Although LE and ET are used interchangeably, LE is in units of energy, whereas ET is usually used with units of volume per unit area.

As stated in the introduction, LE is generally evaluated as a residual in Eq. 1, with R_n, G, and H measured. The evaluation of LE over large areas can best be made by using remote-sensing methods in conjunction with ground-based measurements. In the following sections, the evaluation of R_n, G, and H, using a combination of remote and ground-based measurements is reviewed. Possible sources and magnitudes of error are discussed.

III NET RADIATION

Net radiation can be expressed as the sum of four major components, i.e.,

$$R_n = R_{sd} - R_{su} + R_{\ell d} - R_{\ell u} \qquad (2)$$

where R_{sd} is the downward shortwave radiation (0.15–4 μm) from the sun and atmosphere, R_{su} is the reflected shortwave radiation by the surface, $R_{\ell d}$ is the longwave radiation (>4 μm) emitted from the atmosphere toward the surface, and $R_{\ell u}$ is the longwave radiation emitted from the surface into the atmosphere. As shortwave radiation from the sun passes through the atmosphere, a portion is absorbed or scattered by aerosols, gases, and particulates; therefore, at the Earth's surface, R_{sd} is composed of both direct solar and diffuse-sky radiation. Data to estimate R_{sd} is available from meteorological and agricultural organizations throughout the world. It can be measured fairly accurately with pyranometers. With a measure of air temperature at the standard meteorological height (called the screen height $\simeq$ 2 m above ground surface), $R_{\ell d}$ can be estimated with sufficient accuracy for many practical applications. But even under stable synoptic weather conditions, accurate estimates of R_{su} and $R_{\ell u}$ are difficult to obtain because they are highly dependent on surface conditions. Fortunately, remote-sensing techniques provide a means to integrate the variability in surface conditions and to obtain reasonable values of the upward radiative components.

III.A Incoming Components

III.A.1 Direct and Diffuse

A measurement of R_{sd} is easily obtained with a wide-band pyranometer. Under clear skies and stable weather conditions, this measure of R_{sd} can be extended over a relatively large area, that is, on the order of 100 km^2.

III.A.2 Atmospheric Long-Wave

The determination of $R_{\ell d}$ is fairly straightforward; however, it is a function of several variables, i.e.,

$$R_{\ell d} = \varepsilon_a \sigma T_a^4 \qquad (3)$$

where ε_a is the atmospheric emissivity, T_a the air temperature at screen height, and σ is the Stefan-Boltzmann constant (5.7×10^{-8} W m^{-2} K^{-4}). For cloud-free conditions, both theoretical (e.g., Brutsaert, 1975) and empirical (e.g., Idso, 1981) expressions exist for the determination of ε_a. Brutsaert's expression shows that both humidity and temperature influence the value of ε_a. Satterlund's (1979) comparison of empirical and theoretical expressions revealed that errors from either approach may be on the order of 50 W m^{-2} on the basis of using a daily mean temperature. Hatfield et al. (1983b) used hourly data from 15 locations at various latitudes and heights above mean sea level to compare expressions from Brunt (1932), Swinbank (1963), Brutsaert (1975), Idso and Jackson (1969), and Idso (1981). They found that models that included vapor pressure performed best. These models estimated $R_{\ell d}$ with less than a 5% error or < 15 W m^{-2}. Figure 1 shows values of $R_{\ell d}$ predicted by Brutsaert's (1975) equation, that is,

$$R_{\ell d} = 1.24 \left(e_a / T_a \right)^{1/7} \sigma T_a^4 \qquad (4)$$

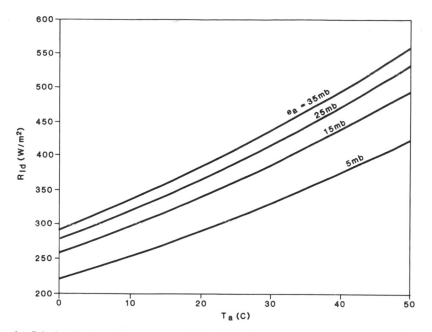

Figure 1   Calculated values of long-wave radiation emitted toward the surface (R_{ld}) as a function of air temperature (T_a) with Eq. 4 and four values of vapor pressure (e_a).

where e_a is the vapor pressure in millibars. Figure 1 illustrates that both water vapor and temperature can significantly affect the value of R_{ld}. Thus, extrapolating point estimates of R_{ld} to areas on the order of 10 km^2 may cause unacceptable errors in regions where large differences exist in atmospheric conditions. Fritschen and Nixon (1967) studied this problem by measuring T_a and e_a along a transect through desert and irrigated areas of the lower San Joaquin Valley in California. The difference in R_{ld} calculated over the two surfaces was about 15 W m^{-2}, a fairly small error. This result suggests that, for stable weather conditions, relatively small errors are incurred in extrapolating point estimates of R_{ld}.

III.B   Outgoing Components

III.B.1   Reflected Solar Radiation

The reflected solar radiation is usually expressed as $\alpha_s R_{su}$, where α_s is defined as the surface albedo. It is equal to the ratio R_{su}/R_{sd}. The albedo is an integral value over all wavelengths. For most natural surfaces, the fractional amount of R_{sd} reflected is a function of solar altitude. A table of mean albedo values given in Brutsaert (1982) for a variety of surface types (i.e., from deep water to dry snow) reveals that the α_s value can vary by ± 0.10 (on average) for a particular surface type. As a result, errors in R_{su} can be around 50 W m^{-2} for daily R_{sd} values and up to 100 W m^{-2} for instantaneous midday values.

Use of remotely sensed data in the visible and infrared regions may improve this estimate considerably since the reflected radiation is measured. However, multispectral radiometers measure only part of the total energy reflected from a surface. In fact, a radiometer with seven spectral bands in which six emulate the reflectance bands of the thematic mapper (TM) radiometer on Landsat-5 will include less than 50% of the total energy spectrum. Consequently, to determine the total reflected radiation, the fractional contribution that each sensor measures must be summed and multiplied by an appropriate conversion factor.

A number of studies concerning the estimation of surface albedo from spectral reflectance measurements has been reported. The objective of these reports was to relate the radiance measured by remote-sensing radiometers with solid angle field of view for discrete wavelength bands to all-wavelength hemispherical albedo recorded by a conventional radiometer (i.e., pyranometer). Three approaches have been reported in the literature. In the first, the surface reflectance is used directly as a substitute (Weisnet and Matson, 1983), or related empirically to surface albedo (Pinker et al., 1985). In the second, radiative-transfer models were used to compute surface albedo by integrating the radiant exitant energy from the surface in all possible directions over the 2π surface hemisphere (Kimes et al., 1987). In the third approach, a combination of measured surface reflectance and radiative-transfer models were used to compute the surface albedo (Irons et al., 1988). Jackson (1984) employed a radiative-transfer model to determine the irradiance at the Earth's surface under several atmospheric scattering and absorption conditions. A variety of surface conditions from bare soil to full canopy were used to obtain the relationship between the spectral reflectance distributions (14 in total) and the calculated irradiance data. A ratio of the radiation values from the multispectral radiometer to the total reflected solar radiation was calculated. This ratio, called the partial/total (P/T) ratio, is dependent on surface conditions when the response of all channels are summed. However, Jackson developed a procedure that allowed the values of P/T to be relatively unaffected by surface conditions. The method is simple and can be applied to any radiometer as long as its response functions and calibration factors are known. For example, a four-band instrument that emulates the Landsat Multispectral Scanner System (MSS) will produce P/T values essentially independent of surface type when the first three channels are summed and added to 0.6 times the output of the fourth channel. Each radiometer will have a different P/T ratio. The ratio for the MSS-band instrument was about 0.5.

Once the P/T value is known for a given radiometer, the total R_{su} can be calculated in several steps. First, the voltage response of the radiometer is multiplied by the calibration factor, which yields the reflected energy for each channel. Second, the proper channels are summed and divided by the P/T ratio. This yields an estimate of the total reflected solar radiation. Figure 2 is a comparison between R_{su} estimated with multispectral data obtained at screen height to measured values using an inverted pyranometer. This result at ground level is encouraging, but comparisons using airborne radiometers must also be performed to ensure that large errors are not incurred due to atmospheric effects.

III.B.2 Emitted Long-wave Radiation

The value of $R_{\ell u}$ is usually obtained by assuming that the surface under consideration can be treated as an infinitely deep grey body having a uniform emissivity ε_s and temperature T_s. Therefore, $R_{\ell u}$ can be estimated by an expression similar to Eq. 3; i.e.,

$$R_{\ell u} = \varepsilon_s \sigma T_s^4 \tag{5}$$

If in Eq. (5), one uses $T_a \simeq T_s$ and $\varepsilon_s \simeq 1$, significant errors may result. For example, a dry bare soil or a partially vegetated surface may be 20–30 °C warmer than T_a, whereas a well-irrigated full-canopy covered field in an arid region may be 10 °C cooler. Figure 3 illustrates the magnitude of such errors on $R_{\ell u}$.

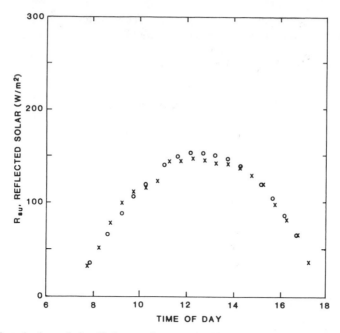

Figure 2 Reflected solar radiation (R_{su}) over wheat measured by pyranometers (x) and an eight-band radiometer (○). (Adapted from Jackson, 1984.)

Thermal-infrared radiometers are normally calibrated with reference to a blackbody assumed to have $\varepsilon_s = 1$. With the known temperature of the blackbody and the voltage response of the instrument, a calibration factor is determined. The calibration factor is employed in one of two ways. Either it is used to convert the radiometer voltage response directly into energy values, avoiding the need for ε_s, or it is used in the calculation of

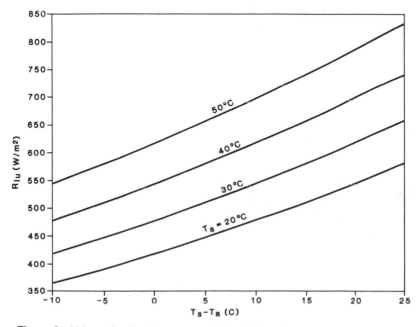

Figure 3 Values of emitted long-wave radiation R_{lu} from Eq. 5 for four values of T_s.

the apparent temperature; this would require ε_s to convert it to the actual T_s. In the latter case, the apparent surface temperature is converted back to energy units and requires the identical value of ε_s to be used in the calibration calculations. Consequently, knowledge of the emissivity is not required to estimate $R_{\ell u}$.

III.C   Estimation of Net Radiation with Remotely Sensed Data

With the two incoming radiation values measured from the ground and with the upward components determined remotely, R_n can be estimated at watershed and regional scales. Comparison with ground-based radiometers has shown good agreement with measured values (Jackson et al., 1985). Figure 4 illustrates such a comparison. More recently, Jackson et al. (1987) determined the upward components with a four-band multispectral radiometer and a single-band thermal-infrared thermometer mounted on a small aircraft flying at an altitude of about 150 m. Figure 5 illustrates the accuracy in instantaneous values of R_n determined with remotely sensed and meteorological data relative to ground measurements using net radiometers over cotton, wheat, and alfalfa fields. Figures 4 and 5 show similar scatter between multispectral estimates of R_n using ground- and airborne-based radiometers and values measured by hemispherical net radiometers. Thus, a relatively accurate measure of R_n, at least under clear-sky and stable atmospheric conditions, is feasible not only at field scales, but also at regional scales with ground-based measurements of R_{sd} and $R_{\ell d}$ and remotely sensed estimates of R_{su} and $R_{\ell u}$.

III.D   Errors in the Estimate of Net Radiation

Errors caused by atmospheric absorption and scattering of radiation can be appreciable in the measured values of reflected solar radiation and surface temperature given by a radiometer at satellite altitudes. For reflected radiation, the errors found by Jackson (1984) appear to be no greater than 10%. A detailed discussion of atmospheric effects on remotely sensed data is discussed in Chapter 9. Errors in surface temperatures appear

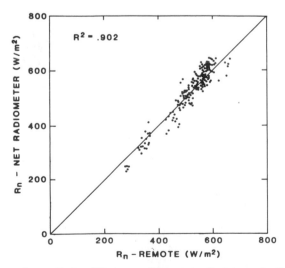

Figure 4   Comparison of net radiation (R_n) measured by a net radiometer and estimated by ground-based remote measurements over six cultivars of well-watered wheat. (Jackson et al., 1985.)

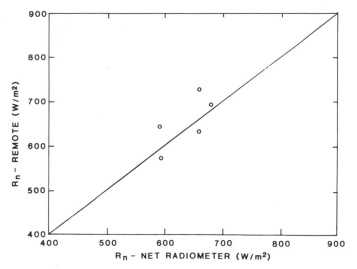

Figure 5 Instantaneous values of net radiation determined by the remote method versus values measured by net radiometers over three crops. [Reprinted from Jackson et al. (1987) by permission of Springer-Verlag, New York.]

more serious. Kiang (1982) showed errors of 30% or more can be expected with T_s obtained from radiometers at satellite altitudes.

The errors in measuring reflected solar radiation are associated with differences in the calculated P/T ratio for clear, turbid, wet, and dry atmospheres. Jackson (1984) found a difference of about 2% in the P/T ratio for clear and turbid atmospheres. The difference between wet and dry atmospheres was larger, reaching about 7%. From these results, it appears that the error could reach around 10% for ground- and airborne-based radiometers at low altitudes (i.e., 100 m). The magnitude of such an error on the overall radiation balance is relatively small since midday values for most vegetated surfaces during the growing season are on the order of 100 to 200 W m^{-2}. More work is needed to determine what the effects are for satellite-based estimates.

The magnitude of the error in surface temperature was estimated by Kiang (1982) using LOW-TRAN3, a sophisticated atmospheric correction model (Kneizys et al., 1980), for water-vapor absorption in the thermal channel of the Thematic Mapper (TM6). He showed that the radiance received by TM6 can vary by more than 30%, depending upon surface temperature and the amount of precipitable water in the atmosphere. For normal ranges in surface temperature, an error of 15–20% is more likely. This percent error will produce an uncertainty in the magnitude of $R_{\theta u}$ between 50 and 100 W m^{-2}, which is quite significant. Even more disconcerting is the fact that as precipitable water increases, the differences between temperature radiances decreases. This is illustrated in Figure 6.

Computationally simpler atmospheric correction models compared to the LOW-TRAN type have been formulated (Price, 1984). Price used radiation-transfer theory to show that regional (i.e., areas of 100–300 km^2) estimates of T_s may be in error by 2–3 °C. This would not be critical in the estimate of R_n, but could be serious for estimating H. For a detailed discussion, see Chapter 15.

Although Figures 4 and 5 give some confidence in the remote-sensing approach for evaluating R_n relatively close to the Earth's surface, it is of interest to consider the

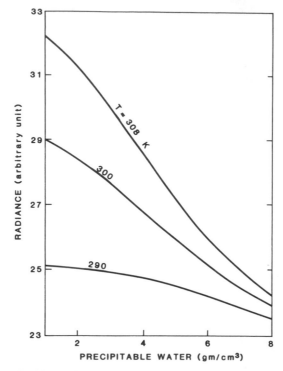

Figure 6   Radiance received by TM band 6 compared to amount of precipitable water in the atmosphere for a range of surface temperatures. [Adapted from Kiang (1982). © 1982 IEEE.]

TABLE 1   Errors in the Estimate of Net Radiation with the Remote Method

Wheat	Bare Soil
$R_{sd} = 975$ W m^{-2}	$R_{sd} = 975$ W m^{-2}
$T_a = 30$ °C	$T_a = 30$ °C
$T_s = 25$ °C	$T_s = 50$ °C
$e_a = 15$ mbars	$e_a = 10$ mbars
$R_{su} = 150$ W m^{-2}	$R_{su} = 300$ W m^{-2}
$R_{ld} = 386$ W m^{-2}	$R_{ld} = 373$ W m^{-2}
$R_{lu} = 448$ W m^{-2}	$R_{lu} = 618$ W m^{-2}
$R_n = 763$ W m^{-2}	$R_n = 430$ W m^{-2}
Errors	*Errors*
$0.20(R_{lu}) \sim 100$ W m^{-2}	$0.20(R_{lu}) \sim 130$ W m^{-2}
$0.10(R_{su}) \sim 20$ W m^{-2}	$0.10(R_{su}) \sim 30$ W m^{-2}
$0.05(R_{ld}) \sim 20$ W m^{-2}	$0.05(R_{ld}) \sim 20$ W m^{-2}
Total ~ 140 W m^{-2}	Total ~ 180 W m^{-2}
Percent error: $\dfrac{140}{760} \simeq 16\%$	$\dfrac{180}{430} \simeq 40\%$

overall error that may occur using satellite data without atmospheric corrections. Table 1 summarizes the magnitudes of various errors in the components used to calculate R_n (cf. Eq. 2). Generally, about a 20% error in R_n may occur over vegetative surfaces. Over bare soil, the error may be 40%. In both cases, the major component of the error comes from $R_{\theta\mu}$, which can be reduced substantially by methods already discussed. Thus, by accounting for atmospheric effects, differences of around 10% from ground measurements of R_n can be expected.

IV   SOIL HEAT FLUX

Over land surfaces covered by vegetation, the daily value of G, the soil heat flux, is often an order of magnitude smaller than the other terms in Eq. 1. However, neglecting G for hourly values, for example, may lead to errors of the order of 100 W m^{-2}.

The soil heat flux G has commonly been measured by a combination of soil calorimetry and the measurement of heat flux at some depth in the soil, i.e.,

$$G_{z_1} = \int_{z_1}^{z_2} C_s(z) \frac{\partial T}{\partial t} \, dz + G_{z_2} \tag{6}$$

where z_1 is zero depth (the soil surface), and G_{z_2} is the heat-flux density measured by a soil heat-flow transducer at depth z_2 (usually, 5–10 cm below the surface), $C_s(z)$ is the volumetric heat capacity of the soil, and $\partial T / \partial t$ is the time rate of change of soil temperature between z_1 and z_2. The integral is evaluated at a minimum of one level above z_2 (see Clothier et al., 1986). Although this approach works quite well when heat-flux plates are properly installed (Fuchs and Tanner, 1968; Idso, 1972), formidable problems arise when these measurements are extrapolated to larger areas.

This is exactly the area where remote sensing may prove useful. Aerial average values of G may be practical considering the work of Fuchs and Hadas (1972), who found that G may be taken as a constant of R_n for bare soil, that is,

$$G = C_G R_n \tag{7}$$

For both moist and dry soils, they found $C_G \sim 0.3$, but the dry soil (much more than moist) displayed some hysteresis in the diurnal relationship of Eq. 7. Idso et al. (1975) found C_G varied with moisture content, ranging from 0.5 to 0.3 for dry and wet soils, respectively. Again, hysteresis was observed. Unpublished data for a wheat crop analyzed by Reginato et al. (1985) showed that instantaneous values of G, measured near midday, yielded $C_G \sim 0.1$ for both bare soils and sparse canopies under moist conditions. From this result, they assumed for full-canopy cover, $C_G < 0.1$ and made C_G a linear function of crop height; this gave $C_G \simeq 0.05$ at maturity.

These few studies suggest that C_G may also have to be estimated by a combination of ground-based and remotely sensed information, as was R_n. Such an approach was recently explored by Clothier et al. (1986), who investigated C_G over two alfalfa plots of differing irrigation treatments. They found that the water content of the soil did not have a significant affect on the value of C_G. Yet the midday values (1100–1400 h local time) varied from 0.3 for alfalfa stubble to 0.1 for dense green alfalfa cover. The vari-

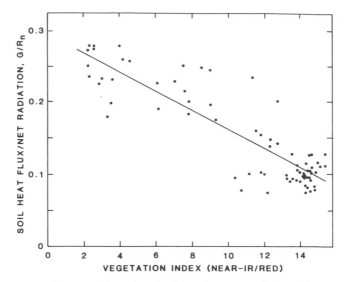

Figure 7   The average of seven midday ratios of soil heat flux to net radiation (G/R_n) versus the vegetative index (near-IR/Red). (Adapted from Clothier et al., 1986.)

ability of C_G was related to a remotely sensed vegetation index (i.e., near-IR/Red), which is shown in Figure 7. The correlation coefficient $R \sim 0.9$ was significant. This problem warrants further study over different surfaces to evaluate how general the relationship such as the one illustrated in Figure 7 may be.

V   SENSIBLE HEAT FLUX

With remotely sensed data used to determine the surface temperature, only two other meteorological variables are required to estimate the surface sensible heat flux; they are windspeed u and T_a at screen height. The equation commonly employed is a bulk transfer equation written as an analogue to Ohm's law:

$$H = (T_s - T_a)\rho C_p / r_{ah} \tag{8}$$

where ρC_p is the volumetric heat capacity (J m^{-3} K^{-1}), and r_{ah} is the resistance to heat transfer. The resistance term can be expressed as

$$r_{ah} = \frac{\left\{\ln\left[(z - d_o)/z_{oh}\right] - \psi_s\right\}\left\{\ln\left[(z - d_o)/z_{om}\right] - \psi_m\right\}}{k^2 u} \tag{9}$$

where d_o, z_{om}, and z_{oh} are the displacement height, roughness length for momentum, and scalar roughness for heat (all in meters), respectively. The symbol k is von Karman's constant, u is the wind speed (m/s) at height z, and ψ_s and ψ_m are stability corrections for heat and momentum, respectively.

V.A Estimation of Parameters in the Resistance to Heat Transfer

Historically, the stability functions have been determined by the Monin–Obukhov similarity theory (e.g., Paulson, 1970). Lately, these original expressions have been simplified and written as a function of a Richardson number, e.g.,

$$R_i = \frac{g(z - d_o)(T_s - T_a)}{T_a u^2} \tag{10}$$

This allows the use of available meteorological and remotely sensed data to calculate ψ_s and ψ_m without having to perform iterations (see Kanemasu et al., 1979; Mahrt and Ek, 1984; Choudhury et al., 1986a).

Reasonable estimates of z_{om} and d_o for vegetation have been obtained with several empirical relationships as long as the surface is uniformly covered and fairly flat. Data have shown that for a wide variety of vegetation, these terms can be estimated from plant height (h) data. Monteith (1973) reported that

$$d_o = \tfrac{2}{3}h \tag{11a}$$

and

$$z_{om} = \tfrac{1}{8}h \tag{11b}$$

yielded satisfactory estimates. Difficulties arise in the estimation of these terms when the surface is not completely covered, as is the case of row crops early in the growing season. Studies have shown that z_{om} and d_o are a function of canopy density (Verma and Barfield, 1979; Hatfield et al., 1985; Hatfield, 1987), but that the relationships are plant-specific.

Even for homogeneous surfaces, there is a distinction made between roughness lengths for scalars (i.e., heat, water vapor, CO_2, etc.) and momentum. The main reason for different roughness lengths is that the transfer processes existing in close proximity to the vegetation are not the same. In this region, heat and water-vapor transfer occur by diffusion, whereas momentum transfer occurs not only by viscous forces, but also by pressure forces (Thom, 1972). This results in an added resistance, also called a bluff body correction, to the transfer of scalars.

Experimental results suggest that the roughness length for heat (z_{oh}) can be taken as a fraction of z_{om} for most vegetated surfaces (Garratt and Hicks, 1973; Garratt and Francey, 1978). For most practical purposes, $z_{oh} = z_{om}/7$. However, over a heterogeneous surface, Garratt (1978) found $z_{oh} \simeq z_{om}/12$. Brutsaert and Kustas (1985) obtained a similar relationship for water-vapor roughness (z_{ov}) over hilly terrain. Thus, over non-homogeneous surfaces, a constant fraction of z_{om} may not be appropriate.

Under many circumstances, the error incurred by not using the appropriate fraction of z_{om} for the scalar roughness may not be significant (Thom and Oliver, 1977). They pointed out that wake diffusion effects described by Thom et al. (1975) will frequently exist because measurement heights of u and T_a will be relatively close to the vegetation. This will probably offset the difference between z_{om} and z_{oh}, resulting in $z_{om} \simeq z_{oh}$ being a reasonable approximation.

However, the overwhelming experimental evidence (Brutsaert, 1982) of $z_{oh} < z_{om}$ suggests a slight correction will generally improve the results. In fact, Heilman and Kanemasu (1976) noted that the inclusion of a bluff body correction improved their estimates of ET. A similar conclusion was drawn recently by Choudhury et al. (1986b). As a compromise between the position of taking $z_{oh} = z_{om}$ and $z_{oh} = z_{om}/10$ for non-homogeneous surfaces, the average ($z_{oh} = z_{om}/5$) is proposed.

V.B   Limitations to the One-Dimensional Resistance Equation

Raupach (1979) and Hicks et al. (1979) presented evidence that displacement heights for heat and water vapor are different than for momentum; thus, Eq. 9 should have at least two displacement heights, d_{oh} and d_{om}. Denmead (1984) was critical of the one-dimensional approach as expressed by Eq. 9. He gave experimental evidence that showed that the source/sink distributions of heat and water vapor vary seasonally and diurnally; hence, r_{ah} is not uniquely defined. A further complication was revealed by Paw U and Meyers (1987). Using a second-order closure model, they showed that the magnitude of d_{oh} was a function of stability. These findings briefly presented here are not intended to discourage the use of Eq. 9. They suggest that the limitations to the approach of using Eqs. 8 and 9 to calculate H may be important under circumstances where a one-dimensional approach is a gross simplification (Denmead and Bradley, 1987). In Chapter 17, a two-dimensional model is presented along with some of its applications.

VI   LATENT HEAT FLUX

With R_n, G, and H estimated with remotely sensed data combined with a few meteorological variables, the latent heat flux may be solved with Eq. 1. This approach has produced reliable estimates of LE, but in most cases, R_n and LE were the major components in Eq. 1 and the surface was generally homogeneous with essentially 100% canopy cover. Examples of the correlation with measured values of LE are given in Figure 8 using ground-based remote-sensing information and in Figure 9 using remote measurements taken 150 m above the surface (Jackson et al., 1987). Even under homogeneous full-canopy cover conditions, a difference between the estimated and measured values of LE with this approach can be significant. This error is commonly the result of an inaccurate determination of H, which becomes pronounced when H and LE are of the same order of magnitude. In other words, when $H \approx LE$ and there are unreliable estimates of the variables in Eqs. 8 and 9, Eq. 1 will yield relatively large deviations from measured values of LE. Consequently, an investigation into the possible errors in H and the resulting effects on the prediction of LE with Eq. 1 must be performed.

VI.A   Errors in the Estimate of Latent Heat Flux

Generally, errors in LE caused by measurement errors of the variables used in Eq. 8 increase with increasing wind speed and increasing z_{om} (Reginato et al., 1985). Moreover, errors in the stability correction to r_{ah} may also be appreciable under low wind-speed conditions (Hatfield et al., 1983a). The errors discussed in what follows pertain to surface temperature and roughness parameters.

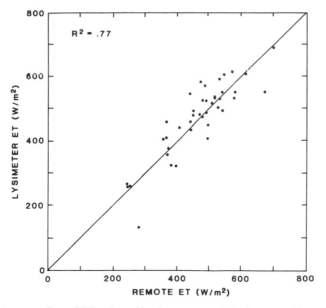

Figure 8 Instantaneous values of ET estimated by the remote method using ground-based instruments compared to lysimeter estimates (20 min averages). [Reprinted by permission of the publisher from Reginato et al. (1985). Copyright 1985 by Elsevier Science Publishing Co., Inc.]

The magnitude of the error in H and inevitably in LE is made evident by taking the derivative of Eq. 8 with respect to $\Delta T(\equiv T_s - T_a)$ and r_{ah}. This yields two equations:

$$\frac{\partial H}{\partial \Delta T} = \frac{\rho C_p}{r_{ah}} \tag{12a}$$

and

$$\frac{\partial H}{\partial r_{ah}} = \frac{-\Delta T \rho C_p}{r_{ah}^2} \tag{12b}$$

For the sensitivity analysis, two types of vegetative surfaces are used. One has $h \sim 1$ m (e.g., wheat crop) and the other has $h \sim 10$ m (e.g., forest plantation). The wind speed at $z - d_o \sim 2$ m is 2 m s^{-1}, $\rho C_p \sim 10^3$ J m^{-3} K^{-1} and near neutral conditions are assumed; i.e., $\psi_s \simeq \psi_m \simeq 0$. For the smaller vegetation, $r_{ah} \simeq 40$ s m^{-1} and Eq. 12a yields 25 W m^{-2} K^{-1}. For the larger vegetation, $r_{ah} \simeq 5$ s m^{-1}, which gives, with Eq. 12a, 200 W m^{-2} K^{-1}.

As mentioned earlier, accounting for atmospheric effects may still lead to errors in T_s of 2–3 °C from satellite altitudes. This error may be compounded as a result of using an incorrect emissivity in Eq. 5 (Huband and Monteith, 1986). Typical green-plant emissivities range from 0.97 to 0.99 (Oke, 1980); for soils, it may be less than 0.97. The temperature uncertainty associated with ±0.01 unit of green-vegetation emissivity is ±0.07 °C (Phinney and Arp, 1979). However, some of the uncertainty associated with the change in surface emissivity may be offset by the uncertainties associated with

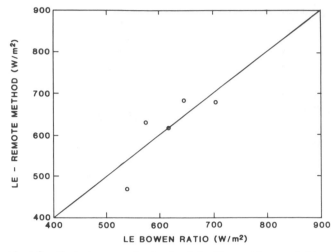

Figure 9 Latent heat flux (instantaneous values) estimated with remotely sensed data and meteorological observations for three crops versus ground-based measurements by the Bowen ratio method. [Adapted from Jackson et al. (1987) by permission of Springer-Verlag, New York.]

atmospheric attenuation of long-wavelength radiation. This is due to the longer pathlength between the surface and the airborne or spaceborne sensor systems, which contains larger water-vapor content that further attenuates the upwelling thermal energy. This reduces the effective emissivity of the surface (Price, 1984; see also Chapter 15).

Asrar et al. (1988) presented a detailed uncertainty analysis of the factors that affect in situ and airborne surface temperature measurements in a tallgrass prairie. They found the aggregate uncertainty for the two cases to be comparable in magnitude (i.e., ± 1.5 °C). Thus, an inaccuracy of 4 °C in the estimate of T_s is probably an upper limit under most conditions.

A ± 4 °C error in T_s results in H deviating by ± 100 W m^{-2} for the shorter (i.e., $h \sim 1$ m) vegetation. For the tall (i.e., $h \sim 10$ m) vegetation, $\Delta T = \pm 4$ °C produces a ± 800 W m^{-2} disagreement. The importance of surface roughness on the resulting errors in H and LE is clearly revealed. In addition, the result over tall vegetation suggests that Eqs. 8 and 9 may not be practical to estimate H.

The error in H due to incorrectly estimating r_{ah} is determined with Eq. 12b and $-\Delta T \rho C_p \simeq 3000$ J m^{-3}. For the vegetation with $h \sim 1$ m Eq. 12b produces 2 W m^{-1} s^{-1}, a fairly small error. Indeed, a factor of 2 difference in the estimate of z_{om} will yield an error in H of less than 50 W m^{-2}. Again, it is the larger roughness that is much more sensitive to errors in r_{ah}. Even if $r_{ah} = 10$ s m^{-1}, an upper value for forests given by McNaughton and Jarvis (1983), Eq. 12b gives 120 W m^{-1} s^{-1}. Therefore, the use of Eqs. 8 and 9 over surfaces with large roughness obstacles is prone to large errors in H and, as a result, will yield significant errors in the estimate of LE.

VI.B Regional Estimates of Latent Heat Flux

Throughout this discussion, the use of radiometers at satellite altitudes to obtain necessary information has been the major focal point. Jackson (1985) discussed this approach for estimating regional ET. But larger areas encompassed by satellite images require extrapolation of point measurements of meteorological data that may lead to significant errors in the estimate of the energy balance. Large errors will likely occur in r_{ah} and ΔT when extrapolating measurements of u and T_a over areas on the order of 100 km² because of changes in surface conditions (i.e., moisture, surface roughness, etc.).

Nevertheless, Klaassen and van den Berg (1985) have shown that elevating the reference height of meteorological measurements of u and T_a from standard screen height to 50 m produced reliable *ET* estimates when used with data from a NOAA-7 satellite overpass. They made use of the fact that the coupling between air temperature and local surface fluxes decreases with height and, therefore, values of T_a and u at 50 m were related to H and *LE* at a larger scale. The height of 50 m was assumed to be the upper limit in the application of surface-layer similarity theory (Wyngaard, 1983). This is also called the constant flux layer, where the mean profiles of wind speed and temperature are logarithmically related; hence, the measured values of u and T_a at 2 m were extrapolated to their value at 50 m. They discovered that calculating the variations in T_a at 50 m with a one-dimensional Lagrangian model described by Tennekes (1973) produced the estimates of *LE* close to measured values (i.e., ±10%).

Regional estimates of *ET* over complex surfaces may be obtained from mean profiles of wind and temperature over the first several kilometers above the ground. These data, obtained from radiosondes, will reflect surface conditions over tens of kilometers upwind. Following the mean profile derivations of Zilitinkevich and Deardorff (1974), Arya and Wyngaard (1975), Yamada (1976) and others, the resistance equation for bulk heat transfer over the boundary layer (cf. Eq. 8) can be written

$$H = \rho C_p (\theta_s - \hat{\theta}) / R_{bh} \qquad (13)$$

where θ is the potential temperature, R_{bh} is the atmospheric boundary-layer (ABL) resistance, θ_s is the potential temperature at the surface using T_s determined radiometrically, and $\hat{\theta}$ is the value of potential temperature for the ABL below the inversion. The boundary-layer resistance is written in analogous form to Eq. 9:

$$R_{bh} = \frac{\left\{ \ln\left[(Z - d_o)/z_{oh}\right] - \psi_{sb} \right\} \left\{ \ln\left[(Z - d_o)/z_{om}\right] - \psi_{sm} \right\}}{k^2 U} \qquad (14)$$

where Z is the ABL height scale, ψ_{mb} and ψ_{sb} are the ABL correction functions, and U is the velocity scale for the ABL. Estimates of $\hat{\theta}$, Z, and U using radiosonde data over the ocean and level land surfaces and the functional form of the correction functions for Eq. 14 have been published (e.g., Melgarejo and Deardorff, 1974, 1975; Yamada, 1976; Arya, 1977; Garratt and Francey, 1978). Under unstable conditions, the data suggest ψ_{sb} and ψ_{mb} can be represented by the Monin–Obhukhov functions for the surface layer (Mawdsley and Brutsaert, 1977; Brutsaert and Chan, 1978). This suggestion was supported by Garratt et al. (1982) for ψ_{mb}. Therefore, Eqs. 9 and 14 may differ only in the magnitude of the height and velocity values used, or, in other words, the reference level. The parameterization of the turbulent transport in the ABL is implicit in Eqs. 13 and 14, which are obviously an oversimplification of the actual exchange processes. Other ABL formulations that better model the actual turbulent transport of constituents should also be examined and tested (Brutsaert, 1986). An example of the limitation in the bulk transfer or resistance approach in modeling the turbulent boundary layer is atmospheric flow over large roughnesses like hilly terrain or forests (e.g., McNaughton and Jarvis, 1983; Brutsaert and Kustas, 1987).

Further work is needed to determine the applicability of resistance-type formulations for the ABL and to develop other models that can be used with existing meteorological data. Progress in the application of ABL formulations like the kind represented by Eqs. 13 and 14 can only occur with well-designed field experiments. The interdisciplinary field experiment, FIFE, (Schmugge and Sellers, 1985) at the Konza Prairie in Kansas is of the type needed to understand how the ABL behaves over different ecosystems relative to the energy balance at the Earth's surface.

VI.C   Estimation of Daily Evapotranspiration from Instantaneous Values

Remotely sensed data are essentially instantaneous; thus, LE estimated by Eq. 1 will also be an instantaneous value. One method to infer daily ET from one time-of-day value requires a remote measurement near midday and the assumption that LE generally follows solar radiation during daylight hours (Jackson et al., 1983). Jackson et al. (1983) using a midday value of solar radiation, S_i, under clear-sky conditions and the daily total, S_d, showed that the ratio S_d/S_i could be calculated from

$$S_d/S_i = 2N/[\pi \sin(\pi t/N)] \tag{15}$$

where variable t is the time starting at sunrise, and N is the day length in hours. Using the premise that $LE_d/LE_i = S_d/S_i$, they measured LE_i (a 20 min average) at five locations and for four crops with weighing lysimeters and solved for LE_d with Eq. 15. This value was compared to the total value measured by the lysimeter. Figure 10 shows the results for cloud-free days, which supports the expression given in Eq. 15 and the premise of a strong correlation between LE and S. Even for cloudy days, the approach does not yield appreciable scatter between measured and estimated ET, as illustrated in Figure

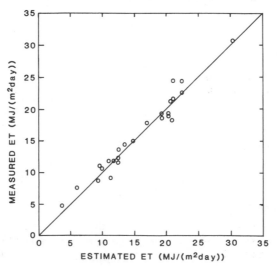

Figure 10   Daily ET estimated from latent heat flux measured by lysimeters around midday and Eq. 15 in comparison to daily totals for four crops. (Jackson et al., 1983.)

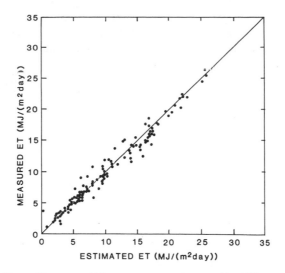

Figure 11 Same as Figure 10 except midday measurements increased by 10% to compensate for nighttime evaporation; only wheat data from Arizona are plotted and about 65% of the data contain various amounts of cloud cover. (Jackson et al., 1983.)

11. Figure 12 shows that the use of an airborne radiometer (Jackson et al., 1987) to obtain LE_i over three crops can produce acceptable estimates of LE_d with the previous approach (i.e., Eq. 15). Nevertheless, these results were obtained under stable weather conditions devoid of unsteady synoptic weather situations (i.e., frontal activity). Other reports have shown that instantaneous measurements made around midday yielded acceptable values for daily ET (Seguin and Itier, 1983; Nieuwenhuis et al., 1985; Seguin et al., 1986).

Clearly, for irrigation scheduling and other practical applications, daily estimates of ET are preferred. With the present technology, one or two remote measurements in 24

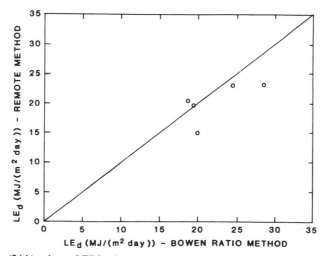

Figure 12 Daily (24 h) values of ET for three crops estimated from the instantaneous values determined by the remote method and Eq. 15 with the assumption $LE_d/LE_i = S_d/S_i$ compared to daily totals from the Bowen ratio method. [Adapted from Jackson et al. (1987) by permission of Springer-Verlag, New York.]

hours over a region at satellite altitudes is feasible. An approach that in some way simulates the diurnal cycle of the surface energy balance seems essential for inferring daily values from instantaneous measurements. Values around midday, and if possible around midnight, would seem to provide the most capability for successfully modeling the diurnal cycle of the surface-energy balance and estimating *ET* on a regular basis.

VII FUTURE DIRECTIONS

Interest in evaluating the surface-energy balance by use of spaceborne data began prior to the launch of the first Earth observation satellite. This interest has continued and, in fact, increased as more sophisticated sensors have been placed in orbit. The problems are formidable, but when one considers the importance of solving the practical problems involved in evaluating evapotranspiration on a regional scale, the necessity of the effort becomes apparent.

Some progress has been made in evaluating components of the surface-energy balance using remotely sensed data. On a local scale, with full-cover crops, surface fluxes determined using remotely sensed data compared favorably with values measured with standard ground-based techniques. At present, the remote method does not adequately account for fluxes over areas with partial canopy cover. This problem is serious and must be addressed on a regional as well as local scale.

Solving the many problems involved with evaluating the surface-energy balance from spaceborne sensors is the focus of a multinational multidisciplinary program called the International Satellite Land Surface Climatology Project (ISLSCP) (Schmugge and Sellers, 1985). The ISLSCP program includes a series of intensive field measurements, one being the First ISLSCP Field Experiment (FIFE), which was conducted on the Konza Prairie in Kansas during 1987. Results from intensive multidisciplinary experiments such as FIFE will hopefully produce the necessary knowledge from which operational systems could be devised that would provide routine information concerning the Earth's surface-energy balance.

List of Symbols

STANDARD ALPHABETICAL SYMBOLS

Symbol	Definition
C_s	Volumetric soil heat capacity [J m^{-3} K^{-1}]
d_o	Displacement height [m]
ET	Evapotranspiration
e_a	Air-vapor pressure [mbars]
g	Acceleration due to gravity [m s^{-2}]
G	Soil heat-flux density [W m^{-2}]
h	Canopy height [m]
H	Sensible heat-flux density [W m^{-2}]
IR	Infrared reflectance
k	Von Karman's constant, ~ 0.4
LE	Latent heat-flux density [W m^{-2}]
N	Daylength [h]

Symbol	Definition
P/T	Ratio of multiwavelength bands to total reflected solar radiation
r_{ah}	Resistance to heat transfer [s m^{-1}]
R	Correlation coefficient
R_{bh}	Atmospheric boundary layer (ABL) resistance [s m^{-1}]
R_i	Richardson number
$R_{\ell d}$	Downward long-wavelength radiation flux density [W m^{-2}]
$R_{\ell u}$	Upward long-wavelength radiation flux density [W m^{-2}]
R_n	Net radiation flux density [W m^{-2}]
R_{sd}	Downward short-wavelength radiation flux density [W m^{-2}]
R_{su}	Upward short-wavelength radiation flux density [W m^{-2}]
S_d	Daily solar radiation [J]
S_i	Midday solar radiation [J]
t	Time [h]
T_a	Air temperature [K]
T_s	Surface temperature [K]
u	Wind speed [m s^{-1}]
U	Scaled velocity for atmospheric boundary layer [m s^{-1}]
z	Reference height [m]
z_{oh}	Roughness length for heat [m]
z_{om}	Roughness length for momentum [m]
z_{ov}	Roughness length for water vapor [m]
Z	Boundary-layer height scale [m]

STANDARD GREEK SYMBOLS

Symbol	Definition
α_s	Surface albedo
ΔT	$T_s - T_a$ [K or °C]
ε_a	Emissivity of atmosphere
ε_s	Emissivity of surface
θ	Potential temperature [K]
π	Constant, 3.14 . . .
ρ	Air density [Kg m^{-3}]
σ	Stefan-Boltzmann constant [5.7×10^{-8} W m^{-2} K^{-4}]
ψ_m	Stability correction parameter for momentum
ψ_s	Stability correction parameter for heat
ψ_{mb}	Boundary-layer stability correction parameter for momentum
ψ_{sb}	Boundary-layer stability correction parameter for heat

REFERENCES

Arya, S. P. S. (1977). Suggested revisions to certain boundary layer parameterization schemes used in atmospheric models. *Mon. Weather Rev.* **105**:215–227.

Arya, S. P. S., and J. C. Wyngaard (1975). Effect of baroclinicity on wind profiles and the geostrophic drag law for the convective planetary boundary layer. *J. Atmos. Sci.* **32**:767–778.

Asrar, G., R. R. Harris, R. L. Lapitan, and D. I. Copper (1988). Surface radiative temperatures of the burned and unburned areas of a tallgrass prairie. *Remote Sens. Environ.* **24:** 447–457.

Brunt, D. (1932). Notes on radiation in the atmosphere. *Q. J. R. Meteorol. Soc.* **58:**389–418.

Brutsaert, W. (1975). On a derivable formula for long-wave radiation from clear skies. *Water Resour. Res.* **11:**742–744.

Brutsaert, W. H. (1982). *Evaporation into the Atmosphere.* Reidel, Dordrecht, The Netherlands.

Brutsaert, W., (1986). Catchment-scale evaporation and the atmospheric boundary layer. *Water Resour. Res.* **22:**395–495.

Brutsaert, W., and F. K.-F. Chan (1978). Similarity functions D for water vapor in the unstable atmospheric boundary layer. *Boundary-Layer Meteorol.* **14:**441–456.

Brutsaert, W., and W. P. Kustas (1985). Evaporation and humidity profiles for neutral conditions over rugged hilly terrain. *J. Clim. Appl. Meteorol.* **24:**915–923.

Brutsaert, W., and W. P. Kustas (1987). Surface water vapor and momentum fluxes under unstable conditions from a rugged-complex area *J. Atmos. Sci.* **44:**421–431.

Choudhury, B. J., S. B. Idso, and R. J. Reginato (1986a). Analysis of a resistance-energy balance method for estimating daily evaporation from wheat plots using one-time-of-day infrared temperature observations. *Remote Sens. Environ.* **19:**253–268.

Choudhury, B. J., R. J. Reginato, and S. B. Idso (1986b). An analysis of infrared temperature observations over wheat and calculation of latent heat flux. *Agric. For. Meteorol.* **37:**75–88.

Clothier, B. E., K. L. Clawson, P. J. Pinter, Jr., M. S. Moran, R. J. Reginato, and R. D. Jackson (1986). Estimation of soil heat flux from net radiation during the growth of alfalfa. *Agric. For. Meteorol.* **37:**319–329.

Denmead, O. T. (1984). Plant physiological methods for studying evapotranspiration: Problems of telling the forest from the trees. *Agric. Water Manage.* **8:**167–189.

Denmead, O. T., and E. F. Bradley (1987). On scalar transport in plant canopies. *Irrig. Sci.* **8:**131–149.

Fritschen, L. J., and P. R. Nixon (1967). Microclimate before and after irrigation. In *Ground Level Climatology* (R. H. Shaw, Ed.), Publ. No. 86. American Association for the Advancement of Science, Washington, D.C.

Fuchs, M., and A. Hadas (1972). The heat flux density in a nonhomogeneous bare loessial soil. *Boundary-Layer Meteorol.* **3:**191–200.

Fuchs, M., and C. B. Tanner (1968). Calibration and field test of soil heat flux plates. *Soil Sci. Soc. Am. Proc.* **32:**326–328.

Garratt, J. R. (1978). Transfer characteristics for a heterogeneous surface of large aerodynamic roughness. *Q. J. R. Meteorol. Soc.* **104:**491–502.

Garratt, J. R., and R. J. Francey (1978). Bulk characteristics of heat transfer in the unstable, baroclinic atmospheric boundary layer. *Boundary-Layer Meteorol.* **15:**399–421.

Garratt, J. R., and B. B. Hicks (1973). Momentum, heat and water vapour transfer to and from natural and artificial surfaces. *Q. J. R. Meteorol. Soc.* **99:**680–687.

Garratt, J. R., J. C. Wyngaard, and R. J. Francey (1982). Winds in the atmospheric boundary layer—prediction and observation. *J. Atmos. Sci.* **39:**1307–1316.

Hatfield, J. L. (1987). Evaluation of aerodynamic resistance for partial canopies. *18th Conf. Agric. For. Meteorolol., 8th Conf. Biometeorol. Aerobiol.*, West Lafayette, Indiana, American Meteorological Society, pp. 6–7.

Hatfield, J. L., A. Perrier, and R. D. Jackson (1983a). Estimation of evapotranspiration at one time-of-day using remotely sensed surface temperatures. *Agric. Water Manage.* **7:**341–350.

Hatfield, J. L., R. J. Reginato, and S. B. Idso (1983b). Comparison of long-wave radiation calculation methods over the United States. *Water Resour. Res.* **19:**285–288.

Hatfield, J. L., D. F. Wanjura, and G. L. Barker (1985). Canopy temperature response to water stress under partial canopy. *Trans. ASAE* **28**:1607–1611.

Heilman, J. L., and E. T. Kanemasu (1976). An evaluation of a resistance form of the energy balance to estimate evapotranspiration. *Agron. J.* **68**:607–611.

Hicks, B. B., G. D. Hess, and M. L. Wesely (1979). Analysis of flux-profile relationships above tall vegetation . . . an alternative view. *Q. J. R. Meteorol. Soc.* **105**:1074–1077.

Huband, N. D. S., and J. L. Monteith (1986). Radiative surface temperature and energy balance of a wheat canopy. I. Comparison of radiative and aerodynamic canopy temperature. *Boundary-Layer Meteorol.* **36**:1–17.

Idso, S. B. (1972). Calibration of soil heat flux plates by a radiation technique. *Agric. Meteorol.* **10**:467–471.

Idso, S. B. (1981). A set of equations for full spectrum and 8–14 μm and 10.5–12.5 μm thermal radiation from cloudless skies. *Water Resour. Res.* **17**:295–304.

Idso, S. B., and Jackson, R. D. (1969). Thermal radiation from the atmosphere. *J. Geophys. Res.* **74**:3397–3403.

Idso, S. B., J. K. Aase, and R. D. Jackson (1975). Net radiation—soil heat flux relations as influenced by soil water content variations. *Boundary-Layer Meteorol.* **9**:113–122.

Irons, J. R., K. J. Ranson, and C. S. T. Daughtry (1988). Estimating Big-Bluestem albedo from directional reflectance measurements. *Remote Sens. Environ.* **25**:185–199.

Jackson, R. D. (1984). Total reflected solar radiation calculated from multi-band sensor data. *Agric. For. Meteorol.* **33**:164–175.

Jackson, R. D. (1985). Evaluating evapotranspiration at local and regional scales. *Proc. IEEE* **73**:1086–1095.

Jackson, R. D., J. L. Hatfield, R. J. Reginato, S. B. Idso, and P. J. Pinter, Jr. (1983). Estimates of daily evapotranspiration from one-time-of-day measurements. *Agric. Water Manage.* **7**:351–362.

Jackson, R. D., P. J. Pinter, Jr., and R. J. Reginato (1985). Net radiation calculated from remote multispectral and ground station meteorological data. *Agric. For. Meteorol.* **35**:153–164.

Jackson, R. D., M. S. Moran, L. W. Gay, and L. H. Raymond (1987). Evaluating evaporation from field crops using airborne radiometry and ground-based meteorological data. *Irrig. Sci.* **8**:81–90.

Kanemasu, E. T., M. L. Wesely, B. B. Hicks, and J. L. Heilman (1979). Techniques for calculating energy and mass fluxes. In *Modification of the Aerial Environment of Crops.* (B. J. Barfield and J. F. Gerber, Eds.), American Society of Agricultural Engineers, St. Joseph, Michigan, pp. 156–182.

Kiang, R. K. (1982). Atmospheric effects on TM measurements: Characterization and comparison with effects on MSS. *IEEE Trans. Geosci. Remote Sens.* **GE-20**:365–370.

Kimes, D. S., P. J. Sellers, and D. J. Diner (1987). Extraction of spectral hemispherical reflectance (albedo) of surface from nadir directional reflectance data. *Int. J. Remote Sens.* **8**:1727–1746.

Klaassen, W., and W. van den Berg (1985). Evapotranspiration derived from satellite observed surface temperatures. *J. Clim. Appl. Meteorol.* **24**:412–424.

Kneizys, F. X., E. P. Shettle, W. O. Gallery, L. H. Chetwynd, L. W. Abreu, J. E. A. Selby, R. W. Fenn, and R. A. McClatchey (1980). *Atmospheric Transmittance/Radiance: Computer Code Lowtran 5*, Environ. Res. Pap. No. 697. Air Force Geophysics Laboratory, Hanscom Air Force Base, Massachusetts, pp. 57–60.

Mahrt, J., and M. Ek (1984). The influence of atmospheric stability on potential evaporation. *J. Clim. Appl. Meteorol.* **23**:222–228.

Mawdsley, J. A., and W. Brutsaert (1977). Determination of regional evapotranspiration from upper air meteorological data. *Water Resour. Res.* **13:**539–548.

McNaughton, K. G., and P. G. Jarvis (1983). Predicting effects of vegetation changes on transpiration and evaporation. In *Water Deficits and Plant Growth* (T. Kozlowski, Ed.). Academic Press, New York, pp. 1–46.

Melgarejo, J. W., and J. W. Deardorff (1974). Stability functions for the boundary-layer resistance laws based upon observed boundary-layer heights. *J. Atmos. Sci.* **31:**1324–1333.

Melgarejo, J. W., and J. W. Deardorff (1975). Revision to stability functions for the boundary-layer resistance laws, based upon observed boundary-layer heights. *J. Atmos. Sci.* **32:**837–839.

Monteith, J. L. (1973). *Principles of Environmental Physics.* Arnold, London.

Nieuwenhuis, G. J. A., E. H. Smidt, and H. A. M. Thunnissen (1985). Estimation of regional evapotranspiration of arable crops from thermal infrared images. *Int. J. Remote Sens.* **6:**1319–1334.

Oke, T. R. (1980). *Boundary Layer Climates.* Methune, New York.

Paulson, C. A. (1970). The mathematical representation of wind speed and temperature profiles in the unstable atmospheric surface layer. *J. Appl. Meteorol.* **9:**857–861.

Paw U, K. T., and T. P. Meyers (1987). Zero plane displacement and roughness length variation with stability as predicted by a turbulence closure model. *18th Conf. Agric. For. Meteorol., 8th Conf. Biometeorol. Aerobiol.*, West Lafayette, Indiana, American Meteorological Society, pp. 4–5.

Phinney, D. E., and G. K. Arp (1979). *Emissivity Corrections for Satellite Derived Radiometric Data*, Tech. Memo. LEC-12394. Lockheed, Houston, Texas.

Pinker, R. T., J. A. Ewing, and J. D. Tarpley (1985). The relationship between the planetary and surface net radiation. *J. Clim. Appl. Meteorol.* **24:**1262–1268.

Price, J. C. (1984). Land surface temperature measurements from the split window channels of NOAA 7 Advanced Very High Resolution Radiometer. *J. Geophys. Res.* **89:**7231–7237.

Raupach, M. R. (1979). Anomalies in flux-gradient relationships over forest. *Boundary-Layer Meteorol.* **16:**467–486.

Reginato, R. J., R. D. Jackson, and P. J. Pinter, Jr. (1985). Evapotranspiration calculated from remote multispectral and ground station meteorological data. *Remote Sens. Environ.* **18:**75–89.

Satterlund, D. R. (1979). An improved equation for estimating long-wave radiation from the atmosphere. *Water Resour. Res.* **15:**1649–1650.

Schmugge, T. J., and P. J. Sellers (1985). The first international satellite land-surface climatology project (ISLSCP) field experiment (FIFE). *Proc. 3rd Int. Colloq. Spectral Signatures Objects Remote Sens.*, Les Arcs, France, ESA SP-247, pp. 321–325.

Seguin, B., and B. Itier (1983). Using midday surface temperature to estimate daily evaporation from satellite thermal IR data. *Int. J. Remote Sens.* **4:**371–383.

Seguin, B., J. P. Lagouarde, and Y. Kerr (1986). Estimation of regional evaporation using midday surface temperature from satellite thermal IR data. *Proc. ISLSCP Conf.*, Rome, Italy, European Space Agency Spec. Pub. 248, pp. 339–344.

Swinbank, W. C. (1963). Long-wave radiation from clear skies. *Q.J.R. Meteorol. Soc.* **89:**339–348.

Tennekes, H. (1973). A model for the dynamics of the inversion above a convective boundary layer. *J. Atmos. Sci.* **30:**558–567.

Thom, A. S. (1972). Momentum, mass and heat exchange of vegetation. *Q. J. R. Meteorol. Soc.* **98:**124–134.

Thom, A. S., and H. R. Oliver (1977). On Penman's equation for estimating regional evaporation. *Q. J. R. Meteorol. Soc.* **103**:345–357.

Thom, A. S., J. B. Stewart, H. R. Oliver, and J. H. C. Gash (1975). Comparison of aerodynamic and energy budget estimates of fluxes over a pine forest. *Q. J. R. Meteorol. Soc.* **101**:93–105.

Verma, S. B., and B. J. Barfield (1979). Aerial and crop resistances affecting energy transport. In *Modification of the Aerial Environment of Crops* (B. J. Barfield and J. F. Gerber, Eds.) American Society of Agricultural Engineers, St. Joseph, Michigan, pp. 230–248.

Weisnet, D. R., and M. Matson (1983). Remote sensing of weather and climate. In *Manual of Remote Sensing* (R. N. Colwell, J. E. Estes, and G. A. Thorley, Eds.), 2nd ed., Vol. II. American Society of Photogrammetry, Falls Church, Virginia.

Wyngaard, J. C. (1983). Lectures on the planatary boundary layer. In *Mesoscale Meteorology: Theories, Observations and Models* (D. K. Lilly and T. Gal-Chen, Eds.), D. Reidel Publ., pp. 603–650.

Yamada, T. (1976). On the similarity functions A, B and C of the planetary boundary layer. *J. Atmos. Sci.* **33**:781–793.

Zilitinkevich, S. S., and J. W. Deardorff (1974). Similarity theory for the planetary boundary layer of time-dependent depth. *J. Atmos. Sci.* **31**:1449–1452.

17

ESTIMATING EVAPORATION AND CARBON ASSIMILATION USING INFRARED TEMPERATURE DATA: VISTAS IN MODELING

BHASKAR J. CHOUDHURY

Hydrological Sciences Branch
Goddard Space Flight Center
National Aeronautics and Space Administration
Greenbelt, Maryland

I INTRODUCTION

Radiative temperature at varied spatial and temporal resolutions could be measured rather efficiently and accurately by infrared radiometers. These data can be used in estimating evaporation and carbon assimilation (instantaneous values and/or relative temporal variation), in irrigation management and assessing dry-matter yield of agricultural fields, and regional and global studies of water balance and net primary productivity. Remote sensing is the only suitable approach for large-area estimates of evaporation at a high temporal resolution. Mathematical models are imperative for interpreting infrared temperatures or for estimating fluxes (evaporation and carbon assimilation) from infrared temperatures. This chapter discusses some simple models relating surface temperature to the fluxes and, to play with the models, representative parameter values are provided that go with the model equations. For a better appreciation of this subject, the reader should consult studies by Monteith and Szeicz (1962), Tanner (1963), Monteith (1981), and Jackson (1982).

One- and two-layer steady-state models are discussed in fair detail, because they are more easily adapted to remote-sensing applications than multilayer canopy-soil models. The canopy is visualized as a ''big leaf'' in describing heat-exchange and physiological characteristics (Sinclair et al., 1976), that is, for example, leaf stomatal resistances of all leaves within a canopy will be aggregately described by the canopy resistance. The dynamic aspects of the models are also discussed. Surface-temperature observations might be used directly as a model input or as model calibration data for estimating evaporation (Jackson, 1985; Camillo and Gurney, 1986). In either case, models require additional data such as atmospheric conditions at a reference height above the soil surface and parameters characterizing the coupling between the virtual surfaces of sensible and latent heat and the air at the reference height. Due attention is given to aerodynamic and boundary-layer resistances and the resistance for vapor diffusion out of leaf stomata (the canopy resistance). Since the boundary-layer air is coupled to the surface, it follows that the reference-height atmospheric variables may not be specified independently of the surface-characteristics and infrared-temperature observations. Thus, point observations of atmospheric variables need to be spatially extrapolated to be consistent with the spatial resolution of infrared-temperature observations. There might be some uncertainties in prescribing the atmospheric variables or the coupling parameters. Hence, some results of a model-sensitivity analysis are discussed. Empirical relations, known as the "nonwater-stressed baselines" for infrared foliage temperature of well-watered crops and the "crop-water-stress index" when a crop is running short of water, are then analyzed.

Dry-matter accumulation by a crop is a direct consequence of carbon assimilation (photosynthesis). The pathway of carbon flow into the leaves (and other green parts?) from the atmosphere is virtually identical to that of water-vapor outflow (an exception would be peristomatal transpiration). Hence, there are rather conservative relationships between crop evaporation and carbon assimilation. These relationships are evaluated in the third section in connection with infrared foliage temperature.

In all case studies presented in this chapter, the soil-hydraulic and thermal-conductivity parameters of agricultural fields have been found to have a high spatial variability. Under progressive drying, this variability alone will introduce a spatial variability in crop and soil evaporations, and hence in surface temperatures. Nonuniform irrigation or rainfall are likely to add to the spatial variability of evaporation and surface temperature. Thus, visual appearance of uniformity of an agricultural field during certain periods might give a false perception of uniformity of evaporation and other fluxes. Although the variance and coefficient of variation of foliage temperature might be useful in irrigation management, the spatial variability of soil temperature has to be recognized in interpreting infrared surface-temperature observations for irrigation purposes. In the fourth section, some results of spatial variability of infrared temperature over agricultural fields are given. For coarse-resolution satellite data, we shall illustrate, via empirical and semiempirical relations, the major effect of the spatial variability of vegetation density in determining the spatial variability of surface temperature. The dispersion of surface temperature remained after accounting for the spatial variability of vegetation density might be interpreted in terms of the spatial variabilities of surface wetness and local energy balance. In the last section, we highlight our findings and reflect on the needs for future studies in this field of remote sensing.

II EVAPORATION AND SURFACE TEMPERATURE

II.A One-Layer Heat-Balance Models

One-layer heat-balance models are by far the simplest models of the soil-plant-atmosphere continuum. In these models, no *a priori* distinction is made between soil and foliage temperatures. The basic equation (Monteith, 1981) states the balance of net supply of heat by radiation (R_n; W m^{-2}) to losses by latent heat (λE; W m^{-2}), sensible heat (C; W m^{-2}) and the flux of heat into the soil (G; W m^{-2}):

$$R_n = \lambda E + C + G \tag{1}$$

where λ is the latent heat of vaporization (2.47×10^9 J m^{-3}). Throughout this chapter, the heat losses due to photosynthesis and storage within the canopy will be ignored. Roughly 2 to 3% of R_n is used for photosynthesis during the day, and the storage flux is generally less than 50 W m^{-2} (see Monteith, 1976). It should be noted that while the heat loss for photosynthesis is largely in phase with R_n, the storage flux is generally not in phase with R_n.

In terms of potential difference and resistance, the equations for λE and C are

$$\lambda E = \frac{\rho c_p}{\gamma} \frac{e_o^* - e_a}{r_E + r_o} \tag{2}$$

$$C = \rho c_p \frac{T_o - T_a}{r_H} \tag{3}$$

where ρ and c_p are, respectively, density and specific heat of air at constant pressure ($\rho c_p \approx 1200$ J m^{-3} K^{-1}); γ is the psychrometric constant (≈ 0.066 kPa K^{-1}); T_a and e_a are, respectively, air temperature (K) and vapor pressure (kPa) at the reference height; T_o and e_o^* are, respectively, surface aerodynamic temperature (K) and saturated vapor pressure (kPa) at T_o; r_E and r_H are, respectively, the aerodynamic resistances (s m^{-1}) for exchanges of latent and sensible heat between the surface and the air at reference height; and r_o is the surface resistance (s m^{-1}) for diffusive transfer of water vapor from the location of vapor source (for example, leaf stomatal cavities) to the location of sensible heat exchange (for example, foliage surface). We shall assume that r_E and r_H are equal. In Eqs. 2 and 3, it is assumed that the locations of sensible and latent heat are at identical temperature T_o and that the vapor source is at saturation, that is, the surfaces for sensible and latent heat exchanges are located at the height of $d + Z_o'$ (symbols defined later, see Eqs. 7 and 8) from the soil surface. These crucial assumptions are justified for crop evaporation, where foliage surface and stomatal cavities are almost at identical temperatures, and the stomatal cavities are saturated; however, they are less readily justified for a drying soil, because of temperature and moisture gradients within the soil (Fuchs and Tanner, 1967; Choudhury and Monteith, 1988). Camillo and Gurney (1986) found that the introduction of a surface resistance in a detailed soil-physics model improved agreement with the observed soil evaporation and surface temperature. It is pertinent to recognize that the definition of a surface might not be precise for a soil-vegetation system. For partially vegetated soils, the surfaces for sensible and latent heat exchanges might be at different heights (i.e., r_E and r_H might differ significantly) and at different temperatures (Kustas et al., 1989). For example, the source of sensible heat might be located close to the soil surface when it is hot and dry, whereas the source of latent heat might be located close to the top of vegetation (the total evaporation is essentially equal to crop evaporation). More research is needed to understand heat exchange by partially vegetated soils.

Rearranging Eq. 1 and using Eq. 3, we arrive at a model for estimating λE as (Soer, 1980; Heilman et al., 1981; Hatfield, 1983; Hatfield et al., 1984a, among others):

$$\lambda E = R_n - G - \rho c_p (T_o - T_a)/r_H \tag{4}$$

where T_o is assumed to be the temperature measured by infrared radiometers. Differences among one-layer models (e.g., Soer, 1980; Hatfield et al., 1984a) are essentially in methods used to compute different terms in Eq. 4. Choudhury et al. (1986a) provide stability-corrected equations for r_H. Under stable conditions ($T_o < T_a$), an exact equation for r_H was obtained as

$$r_H = \left[\ln \left(\frac{Z - d}{Z_o} \right) - \psi_* \right] \left[\ln \left(\frac{Z - d}{Z_o'} \right) - \psi_* \right] \Big/ (k^2 u) \tag{5a}$$

where

$$\psi_* = \left[b - (b^2 - 4ac)^{1/2} \right] / 2a \tag{5b}$$

$$a = 1 + \eta \tag{5c}$$

$$b = \ln \left(\frac{Z - d}{Z_o'} \right) + 2\eta \ln \left(\frac{Z - d}{Z_o} \right) \tag{5d}$$

$$c = \eta \left[\ln \left(\frac{Z - d}{Z_o} \right) \right]^2 \tag{5e}$$

$$\eta = 5(Z - d)g(T_o - T_a)/T_a u^2 \tag{5f}$$

When ψ_* is less than -5, ψ_* should be set to -5. For unstable and neutral conditions ($T_o \geqslant T_a$), the equation is

$$r_H = \ln \left(\frac{Z - d}{Z_o} \right) \ln \left(\frac{Z - d}{Z_o'} \right) \Big/ k^2 u (1 + \eta)^{3/4} \tag{6}$$

where k is the von Karman constant ($= 0.4$), u is the wind speed (m s^{-1}) at the reference height Z, g is the acceleration due to gravity (9.8 m s^{-2}), d is the zero-plane displacement (m), and Z_o' is the roughness height (m) for sensible heat exchange. The height of sensible heat exchange has been related to the corresponding height for momentum exchange (Z_o) by Kustas et al. (1989) as

$$\ln (Z_o / Z_o') = \frac{\rho c_p u (T_o - T_a)}{3C} - \frac{3}{2} \tag{7}$$

But, Garratt (1978) obtained the relation $Z_o' = Z_o/7$. In Eq. 7, T_o is the infrared-temperature observation from nadir, and C has to be calculated iteratively. A summary of observations provided by Kondo and Kawanaka (1986) suggests that Z_o/Z_o' might vary between 10 and 10^5. The aerodynamic resistance for heat exchange could be a factor of 2 or more higher than that for momentum exchange when $Z_o/Z_o' \geqslant 10$. (The aerodynamic resistance for momentum might be obtained from Eqs. 5 and 6 by substituting Z_o for Z_o'.) As a first approximation, one may assume $Z_o = 10Z_o'$. Note that T_o in Eq. 3 is defined to be the temperature at the height of $d + Z_o'$.

Choudhury and Monteith (1988) used the simulation results of Shaw and Pereira (1982) to parameterize the zero-plane displacement d and the momentum roughness height Z_o. The simulation results might be expressed by the equations:

$$d = H \left[\ln (1 + X^{1/6}) + 0.03 \ln (1 + X^6) \right] \tag{8a}$$

and

$$Z_o = \begin{cases} Z_{os} + 0.28 \, HX^{1/2} & \text{for } 0 \leqslant X \leqslant 0.2 \\ 0.3H \left(1 - \dfrac{d}{H}\right) & \text{for } 0.2 \leqslant X \leqslant 2 \end{cases} \qquad (8b)$$

where

$$X = 0.2L \qquad (8c)$$

the crop height (m) is H, and the soil surface roughness (m) is Z_{os}. The crop height provides an appropriate dimension for specifying d and Z_o, and this height is considered to be an independent parameter. While no relationship is prescribed between H and L, it is physically realistic to assume a limit of $H = 0$ when $L = 0$. Thus, Eqs. 8 predict for bare soils ($L = 0$), $Z_o = Z_{os}$ and $d = 0$; and for a mature crop (say, $L = 4$), $Z_o/H = 0.10$ and $d/H = 0.67$. Typical values for soil roughness (Z_{os}) would be in the range of 0.005 to 0.02 m. Observed Z_o and d values summarized by Monteith (1973) for a moderately dense uniform stand of agricultural crops and by Jarvis et al. (1976) for coniferous forests are substantially in agreement with Eqs. 8. However, considerably less is known about Z_o and d for sparse canopies (Hatfield et al., 1985; Kondo and Yamazawa, 1986; Kustas et al., 1989). Approximate roughness heights for momentum, based on terrain description, for large-scale studies are given in Table 1. Recent studies on turbulence close to the surface suggest that the reference height Z should be chosen to be about $d + 20Z_o$ (Tennekes, 1973; Wieringa, 1986).

A wide range of importance has been assigned to G in estimating λE. For example, Gurney and Hall (1983) used a detailed soil-physics model for G in computing λE using satellite-observed infrared temperatures, and Hatfield et al. (1984a) assumed $G = 0$ in computing λE using ground-based infrared-temperature observations over partially and

TABLE 1 Generalized Roughness Heights for Momentum Exchange $(Z_o)^a$

Terrain	Z_o (m)	r_H (s m^{-1})
Mud flats	0.006	45
Marshland	0.03	30
Pasture	0.07	25
Heath	0.10	22
Agriculture	0.17	20
Orchards, bushlands	0.35	15
Forest	0.75	10
Open flat terrain; grasses, few obstacles	0.03	30
Low crops; occasional large obstacles	0.10	22
High crops; scattered obstacles	0.25	17
Suburb	1.0	10

[a]The calculated resistances for heat transfer are for neutral conditions and assuming a wind speed of 15 m s^{-1} at the reference height $Z = 50$ m (assumed to be the top of the constant flux layer) and $Z_o = 10Z'_o$ from Eq. 6.

Source: Wieringa (1986).

almost totally covered soil for various crops (alfalfa, cotton, soybean, sorghum, and tomato). Hatfield et al. (1984a) found good agreement between computed and observed λE. Observations of G reviewed by Choudhury et al. (1987) suggest that G might be 25 to 50% of R_n for bare soils, whereas under a crop cover, G might be 5 to 20% of R_n. By calculating G as the residual component of Eq. 1 from observations over one variety of wheat crop, Choudhury et al. (1987) established the following empirical relationship for G during the daytime hours (which ignores all hysteresis effects) (Figure 1):

$$G = 0.4R_s \tag{9a}$$

where R_s is the net radiation at the soil surface, given by

$$R_s = R_n \exp(-0.5L) \tag{9b}$$

and L is the canopy leaf-area index.

Clothier et al. (1986) used spectral reflectances in the visible and near-infrared to estimate the soil heat flux (Figure 2). One might reconcile the results in Figures 1 and 2 by noting that the spectral reflectances have been correlated with the leaf-area index (e.g., Tucker, 1979). Thus, measurements of L or spectral reflectances might be used to estimate G.

The equations for r_H and G given before could be used to estimate λE from Eq. 4 if surface temperature, crop height, leaf-area index, net radiation, air temperature, and wind speed are known. Note that the change in λE per unit change in $T_o - T_a$ is more pronounced under unstable as compared to stable conditions, due to nonlinear dependence of r_H on $T_o - T_a$. Thus, errors in T_o are more detrimental in estimating λE under unstable conditions as compared to stable conditions (Table 2). The surface temperature is not calculated, but the observed infrared temperatures are used for T_o. Equating T_o to the infrared temperature (either directly or after correction for emissivity) without alter-

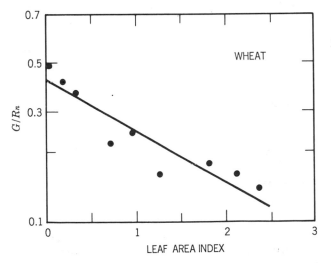

Figure 1 The ratio of soil heat flux to net radiation above the canopy of a Pavon wheat around midday hours is shown as a function of green leaf area index. The fitted equation (Eq. 9) is also shown. The soil heat was calculated as the residual component of a surface energy balance equation. (After Choudhury et al., 1987.)

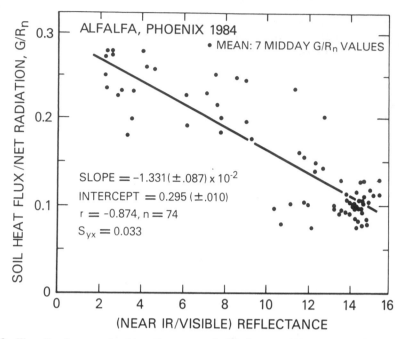

Figure 2   The ratio of measured soil heat flux to net radiation above an alfalfa canopy is shown as a function of simple ratio of near-infrared to visible reflectances. The fitted equation is also shown. This figure illustrates a remote-sensing approach to estimating the soil heat flux. (After Clothier et al., 1986.)

ing the definition of aerodynamic resistance (Eq. 5) must be recognized as a major assumption in estimated λE. As will be discussed later, infrared thermometers might not provide a measurement of T_o.

Note that the model for estimating λE, as described by Eq. 4, does not require any *a priori* knowledge about surface resistance r_o. However, if T_o is to be calculated as a solution of Eq. 1, then one must specify r_o. Alternately, the resistances might be used

TABLE 2   The Aerodynamic Resistance for Heat Exchange (r_H) and the Sensible Heat Flux (C) for Several Values of Surface-Air-Temperature Difference ($T_o - T_a$)

	$U = 1.5$ (m s^{-1})		$U = 3$ (m s^{-1})	
$T_o - T_a$ (K)	r_H (s m^{-1})	C (W m^{-2})	r_H (s m^{-1})	C (W m^{-2})
3	68	53	40	90
2	73	33	41	59
1	79	15	42	29
0	86	0	43	0
−1	97	−12	44	−27
−2	114	−21	46	−53
−3	137	−26	47	−77

[a]The reference height for atmospheric variables is $Z = 2$ m, the surface roughnesses are $Z_o = 0.05$ m and $Z_o' = 0.005$ m, the zero-plane displacement is $d = 0.3$ m, and air temperature is $T_a = 300$ K. Note that the C values are generally not symmetric with respect to $T_o - T_a$; for identical magnitudes of $T_o - T_a$, the magnitudes of C are larger under unstable conditions. Thus, errors in $T_o - T_a$ under unstable conditions would introduce larger errors in C.

as adjustable parameters for matching the calculated temperatures with the observations. Once the matching is satisfactory, the latent heat flux follows from Eq. 2 (Elkington and Hogg, 1981; Price, 1982; Carlson et al., 1984; Trent-Serafini, 1985; Camillo and Gurney, 1986, among others).

For partially vegetated soils, Jordan and Ritchie (1971) have proposed a simple model for surface resistance r_o as

$$\frac{1}{r_o} = \frac{f_v}{r_v} + \frac{1 - f_v}{r_{\text{soil}}} \tag{10}$$

where f_v is the fractional vegetated area observed from nadir, and r_v and r_{soil} are, respectively, the diffusive resistances for water vapor transport out of stomata (vegetation or canopy resistance) and through the dry surface-soil layer. The soil resistance (s m^{-1}) might be calculated according to Camillo and Gurney (1986) as

$$r_{\text{soil}} = -805 + 4140 \, (\theta_{\text{sat}} - \theta) \tag{11}$$

where θ_{sat} is the volumetric wetness at saturation (m^3/m^3), and θ is the volumetric wetness in the surface 5 mm layer; any negative value of r_{soil} must be assumed to be zero. A more physically based model for soil resistance has been discussed by Choudhury and Monteith (1988). The general validity of Eq. 10 is uncertain at present (see discussions of two-layer models). In the next section, we shall discuss in some detail the calculation of canopy resistance and more physically based models for evaluating evaporation from partially vegetated soils. For now, we note that the canopy resistance would be in the range of 50–200 s m^{-1} for grasses and crops, and 75–250 s m^{-1} for forests during the midday hours.

Direct comparisons of λE calculated from Eq. 4 with the observed λE (by lysimeters) have been done by several investigators. Figure 3 illustrates the results of Hatfield et al. (1984a) for several crops under 80% or more ground cover, Figure 3(a), and for only 15% ground cover, Figure 3(b). The data points are fairly well distributed for both ground-cover cases, and the correlation coefficients between calculated and observed λE are, respectively, 0.96 for 80% or more ground cover and 0.74 for 15% ground cover. Figures 4(a) to (d) from Choudhury et al. (1987) show comparisons for wheat (cv. Ciano) from bare soil to $L = 4.7$, and Table 3 summarizes the regression results. The correlation coefficients range from 0.91 to 0.98. Both Figures 3 and 4 demonstrate the viability of using Eq. 4 to estimate λE. However, it is pertinent to note that Hatfield et al. (1984a) had assumed $G = 0$, whereas Choudhury et al. (1987) used Eq. 9 for G, which was partly calibrated using the observed λE. Thus, neither of the previous comparisons provide an *ab initio* validation for λE derived from Eq. 4 using infrared-temperature observations.

The aerodynamic resistance decreases by a factor of 2 or more in going from a uniform stand of short vegetation to forests (Table 1). It is instructive to assess the effect of these differing resistances on the surface-air temperature difference. Clearly, if these surfaces have identical sensible heat fluxes, then the surface-air temperature difference would vary in exact proportion to their respective resistances (Eq. 3). Concurrent observations of the heat-balance fluxes by McCaughey (1985) show that during middays in summer, $R_n = 500$ W m^{-2}, $\lambda E = 280$ W m^{-2}, $C = 180$ W m^{-2}, and $G = 40$ W m^{-2} for a short (approximately 1 m high) vegetation, whereas $R_n = 600$ W m^{-2}, $\lambda E =$

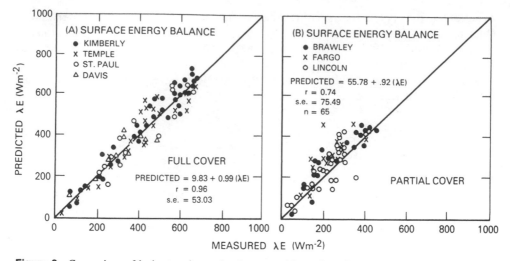

Figure 3 Comparison of lysimeter observed and computed latent heat fluxes from different locations and for different crops. For both cases of (a) almost complete ground cover and (b) very little ground cover, the estimates from an energy balance equation using infrared temperature observations are shown to agree quite well with the observed fluxes. (After Hatfield et al., 1984a.)

335 W m^{-2}, $C = 245$ W m^{-2}, and $G = 20$ W m^{-2} for a forest (approximately 20 m higher, mostly conifers). The radiative temperatures, calculated from long-wave radiation, were about 302 K for the forest and 312 K for the short vegetation. Note that the short vegetation had a higher surface temperature but a lower sensible heat flux when compared to the forest, which might be understood in terms of a higher aerodynamic resistance for a short vegetation as compared to that for a forest. The sensible heat flux for forests during summer midday hours under clear skies might be in the range 150 to 300 W m^{-2}, which, for an aerodynamic resistance of 10 s m^{-1}, corresponds to surface-air temperature differences of 1.3 to 2.5 K (Denmead, 1969; Jarvis et al., 1976; Murphy et al., 1981; Verma et al., 1986). The sensible heat flux around midday hours during summer over a uniform stand of short agricultural crop might be in the range of -100 to 250 W m^{-2}, which for an aerodynamic resistance of 25 s m^{-1}, corresponds to surface-air temperature differences between -2.1 and 5.2 K. Thus, the range of surface-air temperature difference is smaller over a forest as compared to that over a short vegetation. However, one must appreciate that for equivalent errors in estimating T_o, the resulting errors in estimating C would be larger over forests as compared to that over a short vegetation, due primarily to smaller values of the aerodynamic resistance for forests.

Equation 1 has been used quite extensively as the land surface boundary condition in the atmospheric general circulation models, GCMs (see a recent review by Carson, 1981). The surface temperature is calculated as the solution of Eq. 1 with varied approximations to each term in this equation. The soil heat flux has been approximated, for example, as (a) $G = 0$, (b) $G = C/3$, and (c) $G = \delta R_n$, where δ is a proportionality constant

Figure 4 Comparison of lysimeter observed and computed latent heat fluxes for Ciano wheat. The observed reference-height atmospheric variables and infrared surface temperature (from nadir) are also shown. The sequence (a) to (d) is for different days of the year (DOY), varying in the green-leaf-area index and crop height. A numerical summary of these results is given in Table 1. (After Choudhury et al., 1987.)

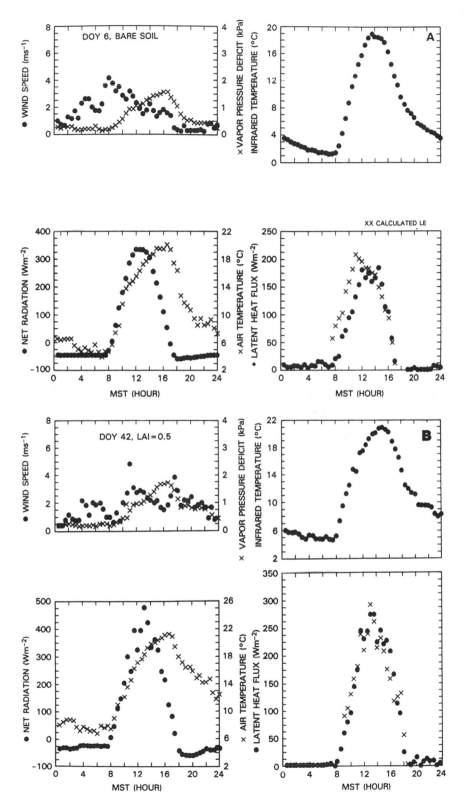

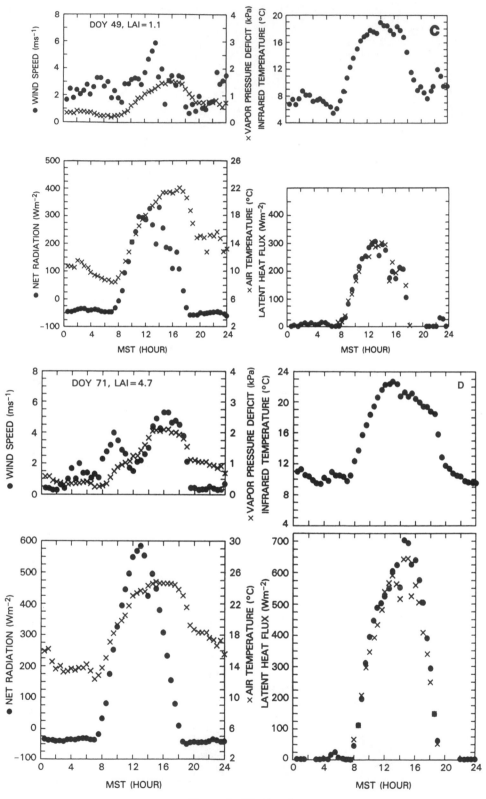

Figure 4 (*Continued*)

TABLE 3 Results of Linear Regression between the Computed (Y) and Lysimeter-Observed (X) λE: $Y = a + bX$

Day of the Year	N^a	a (W m^{-2})	b	σ^b (W m^{-2})	r^c	Daytime Evaporation (mm)		
						Observed	Computed	% Error
6	20	12	1.06	29	0.93	1.6	2.0	25
7	20	−1	1.11	24	0.95	1.6	1.7	6
42	22	−8	0.99	29	0.96	2.5	2.5	0
43	22	−4	0.92	37	0.94	2.7	2.5	−7
49	21	−3	1.05	22	0.98	2.9	2.9	0
50	21	−5	0.85	30	0.98	4.2	3.6	−14
56	23	−13	1.03	68	0.91	3.8	3.8	0
70	24	−25	0.98	83	0.93	6.0	6.0	0
71	24	−28	0.88	101	0.91	7.3	6.7	−8

[a]The number of observations.
[b]The standard deviation.
[c]The correlation coefficient.

Source: Choudhury et al. (1987).

(Bhumralkar, 1975). It is pertinent to note that once T_o is found as the solution to Eq. 1, λE and other terms are evaluated easily. The surface resistance has been generally ignored in GCMs. While some of these approximations might appear crude, one must appreciate the difficulties in modeling the land surface boundary condition in a GCM by realizing that each "point" in a GCM represents a 10^5 km^2 area or larger, and the values of all model parameters must be quantified with a fair degree of accuracy for the entire Earth's surface and for all time during a year. Clearly, such a task is far from being trivial.

Over the last several years, however, highly sophisticated parameterizations of land surface fluxes have been designed for inclusion in the GCMs (Deardorff, 1978; Dickinson, 1984; Mahrt et al., 1984; Sellers et al., 1986). A key factor in this sophistication has been the distinction of fluxes and temperatures for soil and vegetation, which is not done in Eq. 1. Vegetation morphology and its physiological characteristics are such that under both water-unlimited and water-limited conditions, the fluxes of latent and sensible heat for vegetation differ, sometimes rather drastically, from that for soil. For prescribed atmospheric conditions and an initial soil wetness, the transition time from water-unlimited to water-limited soil evaporation is determined to a large extent by the soil-hydraulic parameters, and, in contrast, this time for crop evaporation is determined most strongly by the plant rooting and stomatal characteristics. In general, this transition time is shorter for an exposed soil as compared to that for vegetation due to higher moisture availability to the roots, and consequently soil temperature begins to increase sooner than that for a crop. Because of the marked contrast that generally exists between soil and foliage temperatures, it was desirable to distinguish these two temperatures in a GCM.

For interpreting infrared-temperature observations for latent heat flux, it is also desirable to distinguish soil and foliage temperatures. Thus, the need to develop and study two-layer (soil/vegetation) models for remote-sensing applications. Additional advantage of separating soil and foliage temperatures is that crop evaporation and foliage temperature can be related directly to carbon assimilation, and consequently dry-matter accumulation might be inferred from remotely sensed temperature.

II.B Two-Layer Heat-Balance Models

A potential resistance network illustrating a two-layer model (soil and vegetation) under steady-state conditions is shown in Figure 5. Detailed analysis of analogous resistance networks has been done by Deardorff (1978), Dickinson (1984), and Shuttleworth and Wallace (1985), and an extension has been studied by Choudhury and Monteith (1988). Taconet et al. (1986) present a sensitivity analysis and application of a Deardorff (1978) type model to estimate evaporation using infrared-temperature data. These models are more suitable than one-layer models for studying surface temperature and evaporation from sparse vegetation canopies (as is generally the case with semiarid natural vegetation and dryland agriculture). Mathematical analysis of these networks is, however, considerably more complex than that for one-layer models. The basic equations for analyzing the network in Figure 5 follow. It is important to note that these networks explicitly consider the canopy microclimate, as it is created via complex interactions between soil, vegetation, and atmospheric conditions at the reference height. Choudhury and Monteith (1988) demonstrated the profound impact of this interaction in sparse canopies by calculating 50% or more *crop* evaporation when the soil surface is dry, as compared to that when the soil surface is wet. One should appreciate that a hot and dry soil surface creates a canopy microclimate that is conducive to a higher crop evaporation. Also, a hot-dry soil increases the foliage temperature compared to the case of a wet-cool soil, as will be illustrated.

The network shown in Figure 5 might be considered to have six unknown potentials: temperature and vapor pressure at the soil surface (T_s and e_s, respectively), for vegetation (T_v and e_v^*), and for air within the canopy at the effective height of heat exchange (T_b and e_b). One might identify T_b with T_o in Eq. 3. Then the calculation of these potentials requires six equations. Two of these equations are energy balance at the soil surface and for vegetation. The net radiation for soil and vegetation are R_s and R_v. Expressing the latent and sensible heat fluxes by equations analogous to Eqs. 2 and 3, one can write these energy-balance equations as follows.

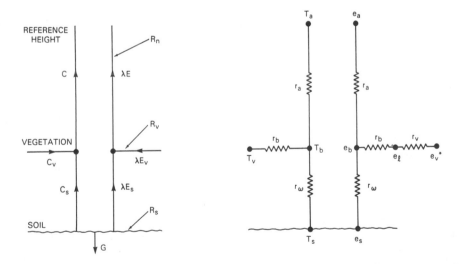

Figure 5 Fluxes and potential-resistance network for a two-layer heat-balance model. The conservation equations for the fluxes and calculation of the unknown potentials are discussed in the text (see Eq. 12). (Adaptation from Choudhury and Monteith, 1988.)

For soil,

$$R_s = \frac{\rho c_p}{\gamma r_\omega} (e_s - e_b) + \frac{\rho c_p}{r_\omega} (T_s - T_b) + G \tag{12a}$$

For vegetation,

$$R_v = \frac{\rho c_p}{\gamma} \frac{e_v^* - e_b}{r_v + r_b} + \frac{\rho c_p}{r_b} (T_v - T_b) \tag{12b}$$

where r_ω, r_b, and r_a are, respectively, the resistances ($s\ m^{-1}$) for heat exchange between the soil surface and the air within the canopy, between the foliage surface and the air within the canopy, and between the air within the canopy and the air at the reference height; and r_v is the canopy resistance ($s\ m^{-1}$) for diffusion of water vapor from stomatal cavities to the foliage surface. (Note that $R_s + R_v = R_n$, and one can use Eq. 9b to estimate R_s.)

Since the stomatal cavities are assumed to be saturated at temperature T_v, we obtain the third equation relating e_v^* to T_v (Teten's equation; see Murray, 1967)

$$e_v^* = 0.611 \exp \frac{17.27(T_v - 273.2)}{T_v - 35.86} \tag{12c}$$

The vapor pressure at the soil surface can be related to the soil surface temperature by the following thermodynamic equation (see Camillo and Gurney, 1986):

$$e_s = e_s^* \exp \frac{g\psi_s}{R'T_s} \tag{12d}$$

where e_s^* is defined by Eq. 12c, with T_s replacing T_v; R' is the gas constant for water vapor ($= 461\ m^2\ s^{-2}\ K^{-1}$); g is the acceleration due to gravity ($= 9.8\ m\ s^{-2}$); and ψ_s is the soil-water potential (in units of metric head, m). If ψ_s is assumed to be a prescribed variable, then Eq. 12d is the fourth equation.

The final two equations are conservation of latent and sensible heat fluxes for the soil-plant-atmosphere continuum:

$$\frac{\rho c_p}{\gamma r_a} (e_b - e_a) = \frac{\rho c_p}{\gamma r_\omega} (e_s - e_b) + \frac{\rho c_p}{\gamma} \frac{e_v^* - e_b}{r_v + r_b} \tag{12e}$$

$$\frac{\rho c_p}{r_a} (T_b - T_a) = \frac{\rho c_p}{r_\omega} (T_s - T_b) + \frac{\rho c_p}{r_b} (T_v - T_b) \tag{12f}$$

The aerodynamic resistances, adapted from Choudhury and Monteith (1988), based on an exponential wind profile within the canopy, are

$$r_b = \frac{50\alpha}{L[1 - \exp(-\alpha/2)]} \left[\frac{\omega}{u_H}\right]^{1/2} \tag{13}$$

$$r_\omega = \frac{H \exp(\alpha)}{\alpha K_H} \left\{ \exp(-\alpha Z_{os}/H) - \exp[-\alpha(d + Z_o)/H] \right\} \tag{14}$$

and

$$r_a = r_{ao}\phi \tag{15}$$

where ω is the leaf width (m), and u_H is the wind speed (m s^{-1}) at crop height, given by (extrapolation of a log profile with an approximate adjustment for roughness sublayer)

$$u_H = 1.5u \ln \left(\frac{H - d}{Z_o}\right) \Big/ \ln \left(\frac{Z - d}{Z_o}\right) \tag{16}$$

The eddy diffusivity K_H (m^2 s^{-1}) at the crop height is given by

$$K_H = \frac{1.5k^2(H - d)u}{\ln \left[(Z - d)/Z_o\right]} \tag{17}$$

The aerodynamic resistance r_{ao} (s m^{-1}) for momentum exchange under neutral conditions is given by (Thom and Oliver, 1977)

$$r_{ao} = \frac{\left\{\ln \left[(Z - d)/Z_o\right]\right\}^2}{k^2 u} \tag{18}$$

The stability correction term ϕ for the aerodynamic resistance is given by

$$\phi = \quad \begin{array}{ll} 1/(1 + \eta)^2 & \text{for stable atmosphere} \\ 1/(1 + \eta)^{3/4} & \text{for unstable atmosphere} \end{array} \tag{19}$$

The damping coefficient for eddy diffusivity and wind speed within the canopy (an exponential damping was assumed in deriving Eqs. 13 and 14) is α, and η is defined by Eq. 5f. Cionco (1972) found α to be proportional to [(flexibility) (leaf area) (density)]$^{1/3}$, and its value was in a rather limited range of 0.3–3 (roughly 0.5 for a citrus orchard, 1.0–2.0 for semirigid crops like mature corn and sunflower, whereas for flexible crops like immature corn, wheat, and oats, α depended upon the wind speed at crop height u_H (m s^{-1}) as $\alpha = 1.5 + 0.6u_H$). For moderately dense canopies, Cowan's (1968) analysis suggests the following equation:

$$\alpha = \frac{1}{\left(\dfrac{d}{H}\right) \ln \left(\dfrac{H - d}{Z_o}\right)} \tag{20}$$

If $d = 0.65H$ and $Z_o = 0.13H$, then $\alpha = 2.2$. The stomatal diffusive resistance r_v will be discussed later. Knowing the resistances, soil-water potential, R_s (or R_v), and the weather variables at the reference height, one can solve the six simultaneous equations (12a to 12f) for the six unknowns (temperatures and vapor pressures in Figure 5). Analytic solutions are possible under simplifying assumptions (see Dickinson, 1984; Shuttleworth and Wallace, 1985; Choudhury and Monteith, 1988). By matching the calculated aerodynamic temperature T_o with the infrared-temperature observations, one may infer the latent heat flux. Again, it is a major assumption to interpret infrared temperature as being the aerodynamic temperature without altering the defining equation for aerodynamic resistance.

Two limiting cases of Eqs. 12a–12f are worth noting. These cases are for very little vegetation ($L \to 0$) and for fairly dense vegetation ($L \gg 1$). In the limit $L \to 0$, we find $d \to 0$ and $Z_o \to Z_{os}$. Thus, in this limit,

$$r_{ao} = \frac{[\ln (Z/Z_{os})]^2}{k^2 u} \tag{21a}$$

$$r_\omega \to 0 \tag{21b}$$

and

$$r_b = \frac{100}{L}\left(\frac{\omega}{u_H}\right) \to \infty \tag{21c}$$

From Eqs. 21b and 21c, it is clear that temperature T_b and vapor pressure e_b of air within the canopy are more strongly coupled to soil surface condition than to the foliage. Therefore, the canopy would experience a microclimate determined quite strongly by the soil. When the soil surface is dry and hot, the canopy ambient air would have a higher temperature and a higher saturation deficit, as compared to the case of cool and wet soil. A hot and dry crop microclimate is clearly to enhance crop evaporation, if the crop is not running short of water and the stomata are not sensitive to the saturation deficit (see Choudhury and Monteith, 1988, and a further discussion to follow).

For dense canopies ($L \gg 1$), we find

$$r_\omega \approx \frac{90 \ln \left(\dfrac{Z - d}{Z_o}\right)}{u} \tag{22a}$$

$$r_b \approx \frac{50\alpha}{L}\left(\frac{\omega}{u_H}\right)^{1/2} \tag{22b}$$

Since r_ω is a factor of 5 or more larger than r_{ao} and r_b progressively decreases as L increases, the canopy air temperature and vapor pressure become progressively more coupled to the foliage and decoupled from the soil. As a consequence, the vegetation determines its own microclimate for evaporation. Additionally, soil evaporation is reduced considerably, even when the surface is wet, because of reduced net radiation and a higher resistance for heat exchange, r_ω. The total evaporation is essentially equal to crop evaporation for dense canopies.

The difference between foliage temperature and air temperature at the reference height for wet and dry soils is illustrated in Figure 6. These results are based upon the simulation model of Choudhury and Monteith (1988), and the crop is not running short of water. The reference-height (1.5 m) meteorologic variables are identical on each day [at 1300 h, the values are $R_n = 785$ W m^{-2}, $T_a = 24.8$ °C, $u = 3$ m s^{-1}, and D (the air vapor-pressure deficit) $= 1.6$ kPa] and the wheat crop height and leaf-area index are increasing (Choudhury and Monteith, 1988). The crop height increased from 0.11 to 0.29 m, while the leaf-area index increased from 0.3 to 2.8. The canopy ambient air is

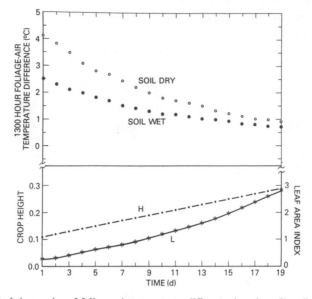

Figure 6 Simulated time series of foliage–air temperature difference based on Choudhury and Monteith's (1988) model (upper portion) and the time series of crop height and leaf-area index used in the simulation (lower portion). The vegetation (wheat) is assumed to be transpiring at the potential rate (i.e., unstressed). When the soil is dry, the foliage temperature is higher compared to that when the soil is wet, primarily because the canopy ambient air is warmer under drying soil conditions.

warmer when the soil is dry as compared to when the soil is wet. As a consequence, the foliage temperature is warmer when the soil is dry as compared to when the soil is wet. This difference in foliage temperatures is more than 1 °C when L is less than 0.5. The difference in foliage temperatures for wet and dry soils decreases as L increases, because, as discussed previously, the canopy ambient-air temperature progressively is decoupled from the soil. In dry-land agriculture, where L is generally less than 3, the foliage temperature would be higher when the soil surface is dry, even though the crop is not running short of water. This indeed was observed by Hatfield et al. (1985). The foliage temperature would decrease soon after any irrigation or rainfall due purely to the soil surface being wet and thus modifying the crop microclimate. It is therefore clear that for partially vegetated soils, the crop microclimate should be considered in interpreting any remotely sensed temperature.

A major effect of canopy microclimate in determining evaporation is illustrated in Figure 7, which is based upon the simulation model of Choudhury and Monteith (1988). When the roots permeate deep soil layers so that the crop is not running short of water, crop evaporation is significantly enhanced when the soil surface is dry and hot, again as a result of modification of the canopy microclimate by the drying soil.

The resistance network in Figure 5 does not specifically address water extraction by plant roots and the effect of soil-water tension (within the root zone) on the leaf stomatal behavior. Thus, changes in foliage temperature while a crop is running short of water were not specifically considered by Choudhury and Monteith (1988). These considerations are addressed in some detail in models developed by Soer (1980), Nieuwenhuis (1981), Choudhury (1983), Choudhury and Federer (1984), Choudhury and Idso (1984, 1985), and Choudhury et al. (1986b). Note, however, that none of these TERGRA-type

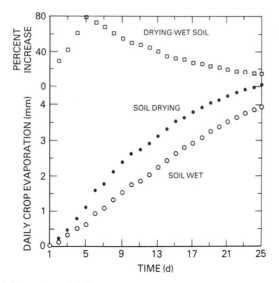

Figure 7 Simulated time series of daily total crop (wheat) evaporation when the soil surface is wet and when it is drying. The upper portion shows the percent increase in crop evaporation under drying soil conditions as a result of the canopy microclimate being hot and dry. Day 1 in Figure 6 corresponds to day 5 in this figure. (Choudhury and Monteith, 1988.)

models (after Soer, 1980) considered the details of canopy microclimate and, therefore, the results are likely to be applicable to moderately dense vegetation (L greater than about 2). A general outline of the approach considered by TERGRA-type models follows, along with a discussion of the effect of canopy microclimate. Since soil energy balance is not separated from the crop energy balance, one might consider these models to be an effective one-layer model.

Figure 8 shows a resistance network illustrating water extraction by plant roots and the coupling of stomatal response to the root-zone soil water tension $\overline{\psi}_s$ under steady-state conditions. The model recognizes the location of water vapor during evaporation to be the stomatal cavities, but the source of water to be the soil root zone. The plant cells conducting water from the root zone to the stomatal cavities are assumed to be rigid. (The dynamical aspects of plant water balance are discussed in Choudhury and Federer, 1984.) Thus, if there is no *net* gain or loss of water for the vegetation during the evaporation process, then the water balance of vegetation demands that the rate of water extracted by plant roots (E_e) must be equal to the rate of crop evaporation (E_v). From Eq. 2, the rate of crop evaporation (m s^{-1}) is given by

$$E_v = \frac{\rho c_p}{\lambda \gamma} \frac{e_v^* - e_a}{r_v + r_H} \tag{23a}$$

According to van den Honert (1948), the rate of water extraction (m s^{-1}) by plant roots is given by

$$E_e = \frac{\overline{\psi}_s - \psi_l}{R_{\text{soil}} + R_{\text{plant}}} \tag{23b}$$

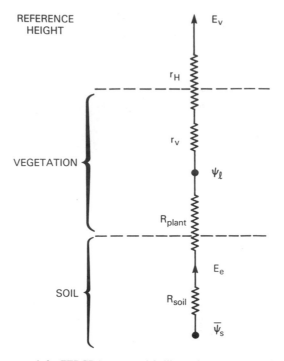

Figure 8 A resistance network for TERGRA-type models illustrating water extraction by plant roots, transport through xylem, diffusion through stomata, and then turbulent transport to the atmospheric reference height. The rate of water extraction by the plant roots (E_e) is equated to the rate of crop evaporation (E_v). The net water content of the vegetation is assumed to remain unchanged during the transpiration process. The symbols are defined in the text (Eqs. 23). The canopy is visualized as a big leaf. (Sinclair et al., 1976.)

where ψ_l is the leaf-water potential (m), and R_{soil} and R_{plant} are, respectively, the resistances (s) for water flow from the bulk soil root zone to the root surface and from the root surface to leaf stomata. One must appreciate that while water extraction by roots and its transport through xylem are mandatory for plant-nutrient requirements, the daily transpiration is greatly in excess of nutrient needs. Transpiration appears to be more closely linked with the carbon gain via photosynthesis.

By noting that $\overline{\psi}_s$, ψ_l, R_{soil}, and R_{plant} are "bulk" soil-plant parameters and recognizing that ψ_l is variable within a canopy, the roots permeate a soil volume and the soil-water tension is variable within this volume. The root-zone soil water tension $\overline{\psi}_s$ is a weighted mean tension within the root zone, and the weighting function has been assessed to be the rooting density (Federer, 1979). Some caution might be exercised in interpreting tension-probe or neutron-probe data for soil water in terms of the mean root-zone water potential that is effective in transpiration when root distribution is not considered. A semiempirical equation for calculating the resistance for water flow within the soil (in units of seconds) is given by Feddes and Rijtema (1972) as

$$R_{soil} = \frac{0.0013}{Z_r K(\overline{\psi}_s)} \tag{24a}$$

where Z_r is the rooting depth (m), and $K(\overline{\psi}_s)$ is the soil hydraulic conductivity (m s^{-1}). The rooting depth systematically increases with crop growth (Taylor and Klepper, 1978;

Borg and Grimes, 1986). For a mature crop in a fine soil, Z_r might be 1 to 1.5 m (Mason et al., 1983; Hamblin and Hamblin, 1985; Borg and Grimes, 1986). The hydraulic conductivity varies with texture (Clapp and Hornberger, 1978). The proposed empirical equation for hydraulic conductivity is (Clapp and Hornberger, 1978)

$$K(\overline{\psi}_s) = K_s(\psi_{\text{sat}}/\overline{\psi}_s)^{2 + 3/b} \tag{24b}$$

where K_s, ψ_{sat}, and b are empirical soil-texture-dependent constants (Table 4).

The plant resistance (R_{plant}) might be separated into two components; the resistance for water entry into the roots and the resistance for water flow through the roots and stem into the leaves. For most crops and grasses, the resistance for water flow through the roots and stem is about an order of magnitude less than that for water entry into the roots, but for trees, the stem resistance might not be negligible (Hellkvist et al., 1974). The resistance for water entry into the roots is inversely proportional to the effective rooting density (effective, because all roots and different portions along the length of a root generally do not participate equally in the water extraction; see Taylor and Klepper, 1978), and thus this resistance would be expected to decrease with crop growth (i.e., increasing rooting density) if other physiological changes do not occur for the roots. Observed R_{plant} values range from 10^8 to 10^9 s, and sometimes decrease during a day with increasing evaporation (Hellkvist et al., 1974; Taylor and Klepper, 1978; Choudhury et al., 1986b). For a mature crop, R_{plant} might be about 5 ($\pm$2) $\times$ 10^8 s. Some results of sensitivity analysis for R_{plant} are discussed later.

The water balance of a crop is ensured by equating Eqs. 23a and 23b as

$$\frac{\rho c_p}{\lambda \gamma} \frac{e_v^* - e_a}{r_v + r_H} = \frac{\overline{\psi}_s - \psi_l}{R_{\text{soil}} + R_{\text{plant}}} \tag{25}$$

To solve Eq. 25, two additional equations are defined: (1) a canopy energy-balance equation, which would provide the foliage temperature T_v at which the saturation vapor pressure e_v^* should be calculated; and (2) an empirical equation prescribing the dependence of canopy (stomatal) resistance r_v on the leaf-water potential ψ_l to couple r_v with the soil-water tension.

The canopy energy-balance equation is

$$R_v = \frac{\rho c_p}{\gamma} \frac{e_v^* - e_a}{r_v + r_H} + \frac{\rho c_p(T_v - T_a)}{r_H} \tag{26}$$

TABLE 4 Representative Parameter Values for Soil Hydraulic Conductivity

Texture	K_s (m s^{-1})	ψ_{sat} (m)	b
Loamy sand	1.8×10^{-4}	0.03	4.4
Sandy loam	3.5×10^{-5}	0.07	4.9
Loam	7.0×10^{-6}	0.15	5.4
Clay loam	2.5×10^{-6}	0.36	8.5
Clay	1.3×10^{-6}	0.20	11.4

Source: After Clapp and Hornberger (1978).

where the symbols have been defined previously. (In comparing Eq. 26 with Eq. 12b, one notes that the canopy microclimate variables e_b and T_b have been replaced by the reference-height air variables e_a and T_a, and r_b has been replaced by r_H.)

The canopy stomatal resistance depends upon many factors, other than the leaf-water potential. These factors include irradiance, saturation deficit, leaf-area index, foliage temperature, leaf age, and leaf mineral nutrients (for example, nitrogen, phosphorus, potassium). The irradiance within the photosynthetically active region (PAR, 0.4 to 0.7 μm) would perhaps be the most appropriate variable for describing the stomatal response, although total solar irradiance and net radiation have also been used. Considering net radiation and leaf-water potential, Choudhury and Idso (1985) obtained the canopy resistance (s m^{-1}) for wheat (cv. Anza) as

$$r_v = \frac{1000\left[1 + (-\psi_l/231)^{5.5}\right]}{L + 0.025R_v} \tag{27}$$

Since canopy resistance plays a central role in crop evaporation and hence its temperature, a short review of this resistance follows. We shall discuss some results for natural grass and wheat canopies based on the solutions of Eqs. 25–27. It is important to note that while empirical relationships are used to describe stomatal resistances, significant advances have been made in understanding why stomata functions the way they do (Cowan and Farquhar, 1977; Cowan, 1982).

The canopy resistance of a crop might be viewed as leaf stomatal resistances of all leaves within a canopy joined in parallel. If the canopy is divided into n strata of leaves, where, for a strata $j (j = 1, 2, \ldots n)$, the leaf-area index is L_j and leaf stomatal resistance is $r_{l,j}$, then the canopy resistance might be expressed as

$$\frac{1}{r_v} = \sum_{j=1}^{n} \frac{L_j}{r_{l,j}} \tag{28a}$$

One may visualize a strata to be of roughly constant irradiance in applying Eq. 28a. It is convenient to express Eq. 28a in terms of canopy conductance c_v and leaf stomatal conductance $c_{l,j}$, which are simply the inverse of the corresponding resistances, as

$$c_v = \sum_{j=1}^{n} L_j c_{l,j} \tag{28b}$$

The dependence of leaf stomatal conductance on various factors, say, photosynthetically active (PAR) irradiance S, stomata-to-leaf ambient-air saturation deficit D' ($\cong e_v' - e_b$), temperature T_v, and leaf-water potential ψ_l could be expressed as

$$c_l = f_1(S)f_2(D')f_3(T_v)f_4(\psi_l) \tag{29}$$

where $f_1, f_2, f_3,$ and f_4 are generic empirical functions. Although Eq. 29 does not show any dependence on nutrients, the major role of nitrogen in determining the conductance has been documented in Schulze and Hall (1982) and Chapin et al. (1987). The conductance generally increases with increasing nitrogen. A convenient function for PAR dependence is

$$f_1(S) = c_{max} + (c_o - c_{max}) \exp(-k_1 S) \tag{30a}$$

where c_o and c_{max} are, respectively, the minimum and maximum conductances (m s^{-1}), k_1 is the damping exponent (W^{-1} m^2), and S is PAR irradiance on the leaf (W m^{-2}). Representative values of c_o, c_{max}, and k_1 are given in Table 5 for grasses and deciduous trees under unstressed conditions. The rate of light saturation is lowest for C_4 grasses (maize and sorghum), followed by C_3 grasses (wheat and barley), and then deciduous trees (oak and maple). There are very few observations for the stomatal conductance in the dark (c_o), and the c_o values in Table 4 are somewhat uncertain. However, c_o is generally a factor of 10 to 20 less than c_{max}.

The stomatal response to saturation deficit could be approximated by a linear function (Choudhury and Monteith, 1986) as

$$f_2(D') = 1 - D'/D'_m \tag{30b}$$

where D'_m represents the saturation deficit at which the conductance would go to zero, which is roughly 8 kPa for the grasses and 7 kPa for several deciduous trees (Hinckley et al., 1978). For D' less than 2 or 3 kPa, however, the stomatal conductance is relatively insensitive to D', and, therefore, some caution should be exercised in applications of Eq. 30b. This stomatal response might be of significance for short crops under partially vegetated soils.

The dependence of conductance on the mean crop temperature T_v might be expressed as

$$f_2(T_v) = \frac{1}{1 + \left(\dfrac{T_v - T_{opt}}{15}\right)^4} \tag{30c}$$

TABLE 5 Representative Light-Response Parameters for Leaf Stomatal Conductance (mm s^{-1}) and Net Assimilation (mg CO$_2$ m^{-2} s^{-1}) for Unstressed Plants[a]

Crop	Conductance			Carbon Assimilation			Source
	c_{max}	c_o	k_1	A_{max}	A_o	k_1	
C_3 grasses	10	0.6	0.012	1.0	−0.05	0.012	Nobel (1983, p. 454); Jones (1983, p. 159); Turner et al. (1978)
C_4 grasses	8	0.3	0.006	2.0	−0.15	0.006	Nobel (1983, p. 454); Coyne and Bradford (1985); Turner et al. (1978)
Deciduous trees	6	0.8	0.03	0.45	−0.06	0.03	Hinckley et al. (1978); Hicks and Chabot (1985)

[a]See text for the meaning of the symbols and the fitted equation (Eq. 30a for conductance, and an analogous equation for net assimilation).

where T_{opt} is the optimum temperature at which the conductance is maximum, which for c_3 grasses and deciduous trees is about 296 K, and for c_4 grasses, about 308 K (see Jones, 1983; Nobel, 1983). The conductance is assumed to be half that at the optimum temperature, when T_v departs from the optimum value by 15 K. The optimum temperature and the functional dependence of conductance on temperature are quite variable. Coyne and Bradford (1985) found for C_4 grasses that the conductance increases linearly with temperature up to T_{opt} and decreases beyond T_{opt}. Same species grown in different environments have different T_{opt} and temperature dependence, clearly, a response to adaptation. While Eq. 30c catches the essence of temperature dependence of stomatal conductance, its applicability to a particular crop at a particular time of a day (transient temperature conditions) is very uncertain. Under most operating conditions, the crop temperature may be assumed to remain fairly close to the optimum temperature, and $f_3(T_v)$ may be assumed to be 1.0.

The dependence of stomatal conductance on leaf-water potential might be described as (Fisher et al., 1981)

$$f_4(\psi_l) = \frac{1}{1 + (\psi_l/\psi_{1/2})^q} \tag{30d}$$

where q is a parameter, and $\psi_{1/2}$ is the value of leaf-water potential at which the conductance decreases by half of its maximum value. Both q and $\psi_{1/2}$ are quite variable, both among crops and for a particular crop, depending upon previous rainfall or irrigation history. The range of ψ_l over which the conductance is most sensitive to the water status is determined by the value of q; for $q = 1$, a linear decrease, whereas for q values of 10 or higher, a threshold-type response is obtained. For C_3 grasses, q is roughly 5 and $\psi_{1/2}$ roughly -150 m (Hansen, 1974; Ripley and Saugier, 1978; Jones et al., 1980; Kemp and Williams, 1980; Choudhury and Idso, 1985), whereas for C_4 grasses, $q = 1$ and $\psi_{1/2} = -70$ m might be used (Ackerson and Krieg, 1977; Henson et al., 1982), although highly threshold-type response has also been observed (Turner, 1974; Sala et al., 1981). For deciduous trees, $\psi_{1/2}$ might be -150 and $q = 15$ (a highly threshold type) or $\psi_{1/2} = -190$ and $q = 8$ (a rather steep variation) (Federer, 1980; Hinckley et al., 1975; Ginter-Whitehouse et al., 1983).

With leaf stomatal conductance defined by Eqs. 29 and 30, we have to sum (integrate) these conductances for all leaves within a canopy to obtain the canopy stomatal conductance (Eq. 28b). The stomatal conductances of all leaves within a canopy are far from being identical; differences in leaf age and leaf microclimate are factors contributing to the differences in the leaf conductance. As a first approximation, one may assume that the irradiance on the leaf will be the major factor contributing to the differences in leaf conductance within a canopy; the sunlit leaves will have higher conductance than the shaded leaves (Turner and Begg, 1973; Aston, 1984). Then the canopy stomatal conductance can be written as the sum of conductances of all sunlit and shaded leaves— a premise of the "big-leaf" model (Sinclair et al., 1976). Eq. 28b can be written as

$$c_v = L_{sun}c_l^{sun} + L_{shade}c_l^{shade} \tag{31}$$

where L_{sun} and L_{shade} are, respectively, sunlit and shaded leaf-area indices, and c_l^{sun} and c_l^{shade} are the corresponding leaf conductances. The sunlit and shaded leaf-area indices of a canopy, whose leaf angles are uniformly distributed (a good approximation for grasses and deciduous trees), are

$$L_{\text{sun}} = 2\mu_o[1 - \exp(-L/2\mu_o)] \tag{32a}$$
$$L_{\text{shade}} = L - L_{\text{sun}} \tag{32b}$$

where L is the canopy leaf-area index, and μ_o is the cosine of solar zenith angle ξ ($\mu_o = \cos \xi$).

The sunlit and shaded leaf conductances can be expressed, using Eq. 29, as

$$c_l^{\text{sun}} = f_1(S_{\text{sun}})f_2(D')f_3(T_v)f_4(\psi_l) \tag{33a}$$
$$c_l^{\text{shade}} = f_1(S_{\text{shade}})f_2(D')f_3(T_v)f_4(\psi_l) \tag{33b}$$

where $f_1(S_{\text{sun}})$ and $f_1(S_{\text{shade}})$ are the light-response function (Eq. 30a) to be evaluated at irradiances, respectively, on the sunlit (S_{sun}) and the shaded (S_{shade}) leaves (i.e., substitute S_{sun} and S_{shade} in place of S in Eq. 30a).

Fairly detailed models for calculating the irradiances within a canopy (and hence on a leaf) may be found in Ross (1981), Choudhury (1987), and elsewhere in this book. Because scattering of PAR by leaves is rather negligible, one may use the following approximate equations for estimating S_{sun} and S_{shade} under clear skies based upon a single-scattering approximation to the radiative-transfer equation:

$$S_{\text{sun}} = 0.5S_0 + S_{\text{shade}} \tag{34a}$$

$$S_{\text{shade}} = \frac{1}{L}\left\{ S_d(1 - e^{-L}) + \left(\frac{\omega_o S_o \mu_o}{2}\right)\left[1 - \frac{e^{-L} - 2\mu_o \exp(-L/2\mu_o)}{1 - 2\mu_o}\right]\right\} \tag{34b}$$

where S_o is the PAR corresponding to the direct solar irradiance on a surface perpendicular to the beam (roughly 420 W m^{-2}, varying with the atmospheric transmittance), S_d is the diffuse PAR irradiance (roughly 80 W m^{-2} under clear skies, varying with μ_o and atmospheric transmittance), and ω_o is the single-scattering albedo (roughly, 0.2, but varying somewhat with crop species). Note that $S_o\mu_o$ is the direct solar irradiance on a horizontal surface at the top of a canopy. For a canopy with $L = 3$ and a solar zenith angle $\xi = 30°$ ($\mu_o = 0.87$), the mean irradiance on the shaded leaves is about 35 W m^{-2}), whereas the irradiance on the sunlit leaves is about 245 W m^{-2}. Thus, the dependence of stomatal conductance on irradiances greater than about 250 W m^{-2} would be largely inconsequential in determining canopy conductance. To complete the example, we note that for $L = 3$ and $\mu_o = 0.87$, the sunlit and shaded leaf-area indices are (from Eq. 32), respectively, 1.4 and 1.6. Again, by considering these leaf-area indices, it is clear that the dependence of leaf stomatal conductance on irradiance less than 250 W m^{-2} would largely determine the canopy stomatal conductance. Additionally, the magnitude of leaf stomatal conductance in the dark (c_o) would play a progressively important role in determining the canopy conductance when the irradiance is low or the leaf-area index L is high. The maximum canopy conductance (high L and high irradiance) would roughly be equal to $[2\mu_o f_1(250) + (L - 2\mu_o)c_o]$, which for C_3 grasses and deciduous trees is roughly $[Lc_o + 2\mu_o(c_{\max} - c_o)]$. For low values of L, one might obtain $L_{\text{sun}} \simeq L$ and $L_{\text{shade}} \simeq 0$, and thus the maximum canopy conductance would be $Lf_1(290)$ or $LC_{\max}$ for most C_3 plants. While the canopy conductance initially increases with L, this linear dependence is not maintained for high L values due to shading.

To summarize this discussion, we note that Eqs. 25, 26, and 31 provide the basic equations for TERGRA-type models. The atmospheric conditions at the reference height and leaf stomatal, rooting, soil-texture, and water-potential parameters have to be prescribed for the solution of this type of model.

The results of TERGRA-model simulation by Soer (1980) for a grassland in The Netherlands are shown in Figure 9. The diurnal trends of the simulated and observed temperatures are in remarkable agreement, although the measured temperatures are systematically higher than the simulated values. The uncertainty associated with the surface long-wave emissivity was suggested to be responsible for this systematic difference.

Comparisons of another TERGRA-type model with the observations over wheat at Phoenix, Arizona, are shown in Figures 10(*a*) and (*b*), and the pertinent weather data in Fig. 10(*c*) (Choudhury and Idso, 1985). These figures show concurrent observations of evaporation measured by lysimeters and canopy temperature measured by a nadir viewing infrared radiometer over well-watered and water-stressed plots. The leaf-area indices for well-watered and water-stressed plots were, respectively, 6.2 and 6.0, and the crop height was about 0.9 m. The calculated evaporations are seen to agree well with the observations, but the calculated canopy temperatures are generally higher than the observations. The discrepancy with respect to canopy temperatures is particularly noticeable for the well-watered crop, and this discrepancy is not easily understood from energy-balance considerations when the infrared temperatures are assumed to be identical to the temperatures in the energy-balance equation. The simulated higher canopy temperatures should have resulted in lower computed evaporation as compared to the

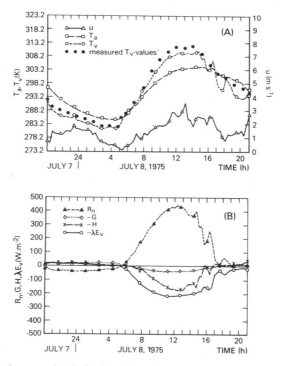

Figure 9 Comparison of measured and simulated foliage temperature for a grassland using a TERGRA model. (*a*) The reference-height atmospheric variables are given and (*b*) the computed latent, sensible and soil heat fluxes are also shown. (After Soer, 1980.)

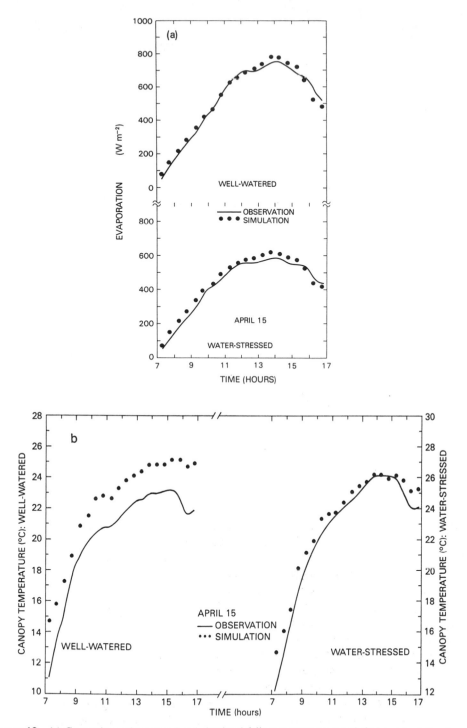

Figure 10 (*a*) Comparison of measured and simulated foliage temperatures and (*b*) latent heat flux for a wheat crop in Phoenix. (After Choudhury and Idso, 1985.) (*c*) The reference-height weather conditions are shown. For the well-watered crop, the simulated latent heat fluxes agreed better than the foliage temperatures with the corresponding observations. There are uncertainties in interpreting infrared-temperature observations as being the temperatures that satisfy an energy-balance equation (see Figure 11).

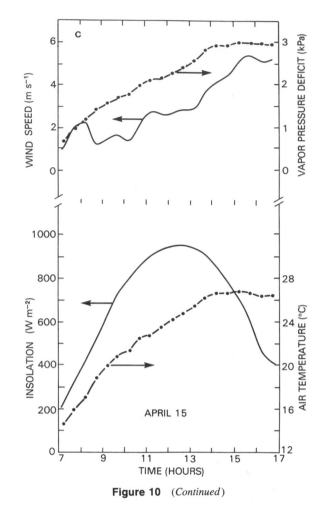

Figure 10   (*Continued*)

observations; this, however, was not the case. One may conclude that the infrared temperatures may not always provide the temperature that satisfies the energy-balance equation (Figure 11 from Choudhury et al., 1986a; Huband and Monteith, 1986a,b).

The previous comparisons show the potential utility of TERGRA-type models for estimating evaporation. It is pertinent to recognize that under well-watered conditions, some crop-specific parameters, namely, the plant and canopy resistances, play important roles in determining evaporation, and uncertainties in these parameters could introduce errors in the computed evaporation (Table 6, Choudhury et al., 1986b). In general, when the canopy resistance or the plant resistance is erroneously estimated to be higher than its actual value, then evaporation would be computed to be lower than its actual value. In remote-sensing applications of these models, one may not have *a priori* knowledge of these parameters, and thus one has to be cautious in interpreting the results obtained.

A basic premise in incorporating the plant water-balance equation (eq. 25) within the TERGRA-type models is that the effect of soil-water potential on the stomatal resistance is mediated via the leaf-water potential (see Eqs. 27 and 29). If osmotic adjustment is an important consideration, then we might use turgor pressure in lieu of the leaf-water potential for expressing stomatal resistance, and define a supplementary equation, relating the turgor pressure and the leaf-water potential, for the water-balance equation.

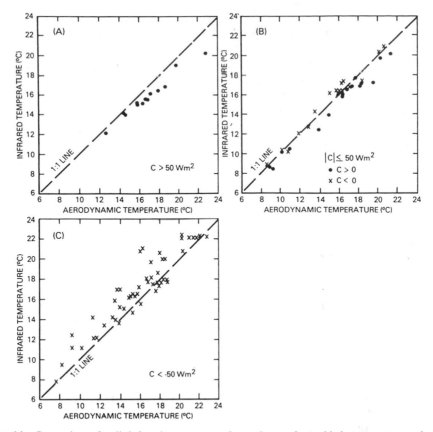

Figure 11 Comparison of nadir infrared temperatures observed over wheat with the temperatures calculated as the solution of a one-layer energy-balance equation; the aerodynamic temperature is defined by Eq. 3. Only for near-neutral conditions the calculated temperatures agreed with the infrared temperature data, as shown in (*b*). (After Choudhury et al., 1986a.) A more detailed analysis of this problem has been done by Huband and Monteith (1986a, b).

TABLE 6   Results of Sensitivity Analysis (Percent Error) in Computing the Daily Total Evaporation from a Stressed Wheat Plot Using One-Time-of-Day (1300 h) Infrared-Temperature (T_{eff}) Observations in a TERGRA-Type Model

Day of the Year	Observed L	% Error Resulting from					
		$L + 0.5$	$L - 0.5$	$T_{eff} + 0.5$	$T_{eff} - 0.5$	$r_v{}^a$	$R_{plant}{}^a$
104	6.3	−2.2	3.3	2.2	−4.4	15.0	1.3
105	6.0	−2.3	2.3	5.8	−5.8	17.9	1.9
106	5.6	−5.5	1.4	2.7	−4.1	13.0	0.5
108	4.7	−2.5	3.2	7.1	−2.5	11.0	2.2
109	4.3	−3.6	4.8	3.6	−6.0	10.6	2.3
112	3.3	−5.2	10.3	12.1	−9.5	12.6	1.9
113	3.0	−9.1	9.6	7.3	−5.5	8.7	2.0
116	2.1	−12.2	17.4	5.7	−3.6	5.7	6.5
118	1.6	−22.2	27.8	4.4	−11.1	4.7	9.2

[a]The canopy and plant resistances were altered as per differing observations for wheat.

Source: Adopted from Choudhury et al. (1986b).

However, several experiments and recent thoughts are that the soil-water potential directly affects the stomatal resistance (Gollan et al., 1985; Turner, 1986). That is, the function $f_4(\psi_l)$ in Eq. 29 should instead be $f_4(\overline{\psi}_s)$, and then one has to prescribe $\overline{\psi}_s$ only to solve for the canopy temperature from the canopy energy-balance equation (Eq. 26). The consequent redundancy of the plant water-balance equation (Eq. 25) would considerably simplify the TERGRA-type models and also facilitate the solution of the system of Eqs. 12. The evaporation and surface temperature could be better estimated by considering the canopy microclimate, particularly in sparse canopies.

In the following, we shall discuss how soil and foliage temperatures could be combined to estimate an effective surface temperature, which might be used to understand infrared-temperature data.

II.C Effective Surface Temperature

Two-layer models recognize the difference in soil and foliage temperatures, which has important implications for interpreting infrared-radiometer observations. As a first approximation, the effective temperature measured by a radiometer might be considered as a weighted sum of soil and foliage temperatures, the weighting factors being the fractional areas of soil and vegetation. Thus, the effective surface temperature, assuming identical emissivities for soil and vegetation, may be written as

$$T_{\text{eff}} = f_s T_s + (1 - f_s) T_v \tag{35}$$

where f_s is the fractional soil area viewed by the radiometer. (For a more accurate analysis of effective temperature, see Becker, 1981.) If the foliage elements are assumed to be randomly distributed above the soil surface (note that this is not a good description for row crops or clump vegetation), then the fractional soil area that will be viewed by a radiometer from a zenith angle θ_r ($\mu_r = \cos \theta_r$) is

$$f_s = \exp \left(-g'L/\mu_r \right) \tag{36}$$

where g' is determined by the angular distribution of foliage elements, and for uniform leaf-angle distribution, $g' = 0.5$.

Soil and foliage temperatures are roughly equal soon after an irrigation or rainfall. In general, however, the soil is considerably warmer than the foliage; the difference in temperatures could be more than 25 K. For a given g' and L, we see from Eq. 36 that f_s has the highest value for $\theta_r = 0$ (i.e., nadir observations) and it decreases as θ_r increases. As a consequence, T_{eff} will, in general, decrease as θ_r increases because the radiometer would "see" a progressively lower fraction of the warmer soil surface. (We shall assume that the angular variation of T_{eff} is due solely to Eq. 36, but T_s and T_v, when observed by an infrared radiometer, might also vary with the angle of observation.) The angle of the radiometer plays an important role in the observed infrared temperature, and hence in its interpretation. If T_{eff}, as determined by an infrared radiometer, is used in Eq. 4 as a substitute for T_o, then the calculated λE would have a pronounced *angular* dependence, and, in general, the calculated λE will increase as the radiometer view angle increases. It is quite possible that T_{eff} measured by a nadir viewing infrared radiometer might result in *negative* values of λE during the *daytime* hours, when T_{eff} is substituted for T_o in Eq. 4, when L is low and the soil surface is hot (Kustas et al., 1989). That λE depends upon the radiometer view angle when the observed T_{eff} is substituted for T_o in Eq. 4 is the central dichotomy of the aerodynamic (T_o) and the radiative temperature measured by a radiometer (T_{eff}). In one sense, Eq. 4 would generally provide the minimum value of λE when nadir viewing radiometer observations (T_{eff}) are substituted for T_o. For realistic estimates of λE using T_{eff} in Eq. 4 one has to redefine the aerodynamic resistance (Kustas et al., 1989).

Consider concurrent radiometer observations at two angles, for example, at nadir (T_{eff}^o) and at 45° (T_{eff}^1). From Eqs. 35 and 36, the difference of the two temperatures ($\Delta T = T_{\mathrm{eff}}^1 - T_{\mathrm{eff}}^o$) for a uniform leaf-angle distribution ($g' = 0.5$) is given by

$$\Delta T = (T_v - T_s) \left[\exp\left(-0.5L\right) - \exp\left(-0.71L\right) \right] \tag{37}$$

Figure 12 illustrates ΔT as a function of fractional ground cover, calculated as $1 - \exp(-0.5L)$, together with the observations of Hatfield (1979) for wheat. In the observations and in the calculations, L is the implicit variable, and we assumed $T_s - T_v = 15$ K, which probably varied throughout the observations. That T_{eff} varies with θ_r and that T_v is smaller than T_s are largely the factors determining the shape of ΔT. Clearly, λE calculated from Eq. 4 using T_{eff}^o and T_{eff}^1 might have differed by 80 W m^{-2} ($\Delta T = 2$ K, $r_H = 30$ s m^{-1}) when the ground cover was 0.5. Again, the nadir observations of surface temperature are likely to give the minimum value of λE from Eq. 4 for partially vegetated soils, because, in this case, the radiometer "sees" the highest fraction of warmer soil area for a given L (see Eq. 36). The leaf-angle distribution, which determines g' in Eq. 36, also has an important effect on the angular variation of infrared temperatures; a nadir viewing radiometer would see a higher fraction of soil area for an erectophile canopy as compared to that for a planophile canopy.

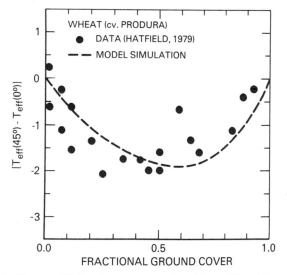

Figure 12 Observed difference of infrared temperatures viewed from 45° and from nadir as a function of ground cover (after Hatfield, 1979) and the simulated relation by Eq. 37. This figure clearly demonstrates that the infrared temperatures vary strongly with the angle of observation under partial ground cover conditions.

II.D Dynamical Aspects of Surface Temperature

The steady-state equations for surface heat balance might be extended to include an explicit temporal variation due to finite heat capacities of soil and vegetation. The heat capacity of short vegetation is very low, and, therefore, the steady-state equations are good approximations. The heat capacity of soil is not negligible, which has an important bearing for soil surface temperature (for example, hysteresis) and hence soil evaporation (Deardorff, 1978; Lin, 1980; Camillo and Gurney, 1986; Taconet et al., 1986). A rather simple but fairly accurate set of equations describing the temporal variation of (soil) surface (T_s) and deep-soil ($\overline{T}_s$) temperatures are (Lin, 1980)

$$\alpha \frac{\partial T_s}{\partial t} = \frac{2G}{c_s d} - \omega_\phi (T_s - \overline{T}_s) \tag{38a}$$

$$\frac{\partial \overline{T}_s}{\partial t} = \frac{G}{\left[c_s d (365\pi)^{1/2} \right]} \tag{38b}$$

where

$$\alpha = 1 + \delta/d \tag{38c}$$

$$d = (2\kappa/\omega_\phi)^{1/2} \tag{38d}$$

δ is the thickness of soil surface layer (m), κ is the diffusivity (m^2 s^{-1}), c_s is the heat capacity (W m^{-3} s K^{-1}), ω_ϕ is the diurnal frequency (s), and G is the soil heat flux (W m^{-2}), as given by Eq. 12a. The solution of Eq. 38a provides the temporal variation of T_s during a day, and the solution of Eq. 38b provides the day-to-day variation of the deep-soil temperature. One can combine the set of Eqs. 12 with Eqs. 38 to form a coupled soil-plant-atmosphere model for studying the heat balance. For a comprehensive model, one should also add the dynamic equations for soil-water potential (see Camillo and Gurney, 1986).

Application of infrared temperatures to estimate soil evaporation using coupled dynamic equations for temperature and moisture has been described by Camillo et al. (1983) and Camillo and Gurney (1986). In the latter study, they adjust soil-surface roughness, soil resistance, and saturation hydraulic conductivity to fit surface temperature, wetness, and soil evaporation. After calibration, this procedure can be used to estimate the daily total evaporation by fitting the simulation model only to the surface-temperature observations, as has been done by Camillo et al. (1983). Figure 13, from Camillo and Gurney (1986), shows observed (at Phoenix) and fitted surface temperatures and the predicted rate of soil evaporation. The agreement between surface temperatures is quite well, except for a systematic difference during the afternoon. The estimated rate of evaporation is not as good, but the agreement for daily total evaporation might be satisfactory. For a soil-vegetation system, one might use a steady-state energy-balance equation (for example, Eq. 26) to compute the foliage temperature and a dynamic equation of the form of Eq. 38 for soil temperature. A numerical scheme to simulate the thermal regimes of soil together with the thermal and moisture regimes of the atmospheric boundary layer has been discussed by Novak and Black (1985). Such coupled schemes would be useful in computing the daily total latent heat flux from one-time-of-day observations of surface temperature and the reference-height atmospheric variables.

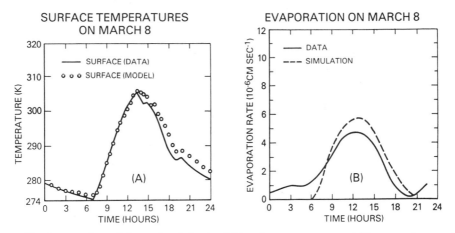

Figure 13 A comparison of observed and simulated (*a*) surface temperatures and (*b*) evaporation rates for bare soils using coupled dynamic equations for temperature and moisture. The model parameters were adjusted to fit the observed surface temperatures, and the evaporation rates are as predicted by the model. The soil surface resistance (s cm^{-1}) used in the model is given by Eq. 11. The approach used by Camillo and Gurney (1986) might be useful in estimating the daily total evaporation from one- or two-time-of-day observations of surface temperature. (After Camillo and Gurney, 1986.)

II.E Semiempirical Relations

In 1982, Idso published results of an experimental program in which he showed that for 26 well-watered plant canopies, the difference of foliage temperature as measured by an infrared radiometer and air temperature at the height of roughly 1.5 m was linearly related to the air vapor-pressure deficit at the height of roughly 1.5 m. These linear relationships, differing only in slope and intercept for different species, were called the "nonwater-stressed baselines." Figure 14, extracted from Idso (1982), illustrates these observed baselines. Then, in 1983, Idso analyzed the implication of these baselines for stomatal behavior and crop evaporation. These studies are summarized in this section, together with a rather simple technique to quantify water stress and evaporation from water-stressed canopies developed by Idso et al. (1981a) and Jackson et al. (1981), the so-called crop-water-stress-index. This technique for quantifying water stress received very high praise from O'Toole et al. (1984) when they compared several methods for quantifying water stress for rice. The crop-water-stress index might be a more valuable index for scheduling irrigation than the stress-day index (Hiler et al., 1974) and the stress-degree-day index (Jackson et al., 1977).

The aerodynamic equation for latent heat flux from an well-watered canopy (Eq. 23a) is

$$\lambda E_v^o = \frac{\rho c_p}{\gamma} \frac{e_{v,o}^* - e_{a,o}}{r_v^o + r_H^o} \tag{39a}$$

and the energy-balance equation (Eq. 26) is

$$R_v^o = \lambda E_v^o + \rho c_p (T_v^o - T_a^o)/r_H^o \tag{39b}$$

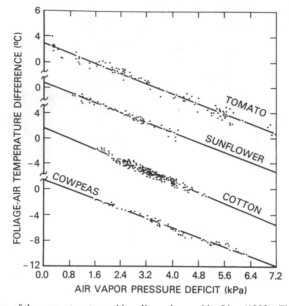

Figure 14 Examples of the nonwater-stressed baselines observed by Idso (1982). The foliage temperatures were measured by a handheld infrared thermometer at about a 45° angle for both sunlit and shaded sides of a canopy, and the time series of all observations during a day were smoothed by a three-point slide-rule method. Whereas net radiation and wind speed were quite variable, the data consistently showed that 80% or more of the variability of foliage–air temperature difference (at the reference height) could be explained in terms of air vapor-pressure deficit.

where superscript/subscript o denotes measurements under well-watered conditions. (It is recognized that the aerodynamic resistance and the reference-height atmospheric variables over a well-watered canopy might differ from those over a stressed canopy.) If the saturated vapor pressure at the foliage temperature ($e_{v,o}^*$) is linearized as

$$e_{v,o}^* = e_{a,o}^* + \Delta(T_v^o - T_a^o) \tag{39c}$$

then one can verify that the foliage–air temperature difference that would satisfy Eqs. 39a–39c (assuming r_v^o and r_H^o do not depend upon T_v^o) is given by

$$T_v^o - T_a^o = \frac{(r_v^o + r_H^o)R_v^o}{\rho c_p(1 + \Delta/\gamma + r_v^o/r_H^o)} - \frac{D^o}{\gamma(1 + \Delta/\gamma + r_v^o/r_H^o)} \tag{40}$$

where D^o is the air vapor-pressure deficit ($= e_{a,o}^* - e_{a,o}$).

Since the empirical equation for nonwater-stressed baselines is given by (assuming the infrared foliage temperature and T_v^o in Eq. 39 are identical)

$$T_v^o - T_a^o = \alpha - \beta D^o \tag{41}$$

where α and β are crop-specific empirical parameters, Idso (1983) stipulates that for Eq. 41 to be consistent with Eq. 40, parameters α and β should be given as follows:

$$\alpha = \frac{R_v^o(r_v^o + r_H^o)}{\rho c_p(1 + \Delta/\gamma + r_v^o/r_H^o)} \tag{42a}$$

$$\beta = \frac{1}{\gamma(1 + \Delta/\gamma + r_v^o/r_H^o}$$ (42b)

From Eqs. 42a and 42b, we obtain

$$\frac{1}{r_v^o + r_H^o} = \frac{\gamma\beta R_v^o}{\rho c_p \alpha}$$ (43)

Then, by using Eqs. 39c, 41, and 43 in Eq. 39a, the latent heat flux from a well-watered canopy is obtained as

$$\lambda E_v^o = \frac{\lambda R_v^o \beta}{\rho c_p \alpha} (D^o + \alpha\Delta - \beta\Delta D^o)$$ (44)

Thus, the evaluation of λE_v^o requires R_v^o, D^o, a rough knowledge of air temperature (to calculate Δ), and the empirical constants α and β. From Idso's (1982) observations for 26 sunlit crops, we calculate $\alpha = 2.8 \pm 1.8$ and $\beta = 2.1 \pm 0.45$. Whereas the general validity and accuracy of Eq. 44 are somewhat uncertain at present, the equation certainly contains the important environmental parameters, R_v^o and D^o, which are expected to determine evaporation from well-watered crops. The values of α and β given before show that the intercepts (α) of the nonwater-stressed baseline are more variable among crops than the slopes (β) of the baseline.

Note that if instead of constraining α and β by Eqs. 42a and 42b, one simply equates Eqs. 40 and 41, i.e.,

$$\alpha - \beta D^o = \frac{(r_v^o + r_H^o)R_v^o}{\rho c_p(1 + \Delta/\gamma + r_v^o/r_H^o)} - \frac{D^o}{\gamma(1 + \Delta/\gamma + r_v^o/r_H^o)}$$ (45)

then the canopy resistance is obtained as

$$r_v^o = \frac{\rho c_p[(1 + \Delta/\gamma)\alpha - (D^o/\gamma)(1 + \beta\gamma - \beta\Delta)]}{R_v^o - (\rho c_p/r_H^o)(\alpha - \beta D^o)}$$ (46)

That r_v^o depends upon r_H^o, D^o, and air temperature (through Δ) is a rather puzzling result, unless there exists a strong correlation between r_H^o, R_v^o, air temperature, and D^o. This correlation is also suggested by stipulating that the right-hand sides of Eqs. 42a and 42b are constants.

A further point worth noting is that the derivation of Eq. 40 contains assumptions that r_H^o and r_v^o are independent of T_v^o. Only under near-neutral or high wind-speed conditions can one assume r_H^o to be independent of T_v^o (see Eqs. 5 and 6), and r_v^o would be independent of T_v^o if leaf stomata are not sensitive to the air humidity (see Eq. 30b and Korner, 1985). The interpretation of constants in Eq. 41 in terms of Eq. 40 is not, therefore, rigorous and might not be generally valid. Although the physical basis of Eq. 41 is not fully understood at present, the practical utility of the equation should not be underestimated.

To quantify water stress, Jackson et al. (1981) defined a crop-water-stress index (CWSI) as

$$CWSI = 1 - \frac{\lambda E_v}{\lambda E_v^o}$$ (47a)

where λE_v and λE_v^o are, respectively, the latent heat fluxes from water-stressed and from well-watered canopies under identical atmospheric conditions at the reference height. Idso et al. (1981a) then defined CWSI in terms of the infrared foliage temperature as (Figure 15):

$$\text{CWSI} = \frac{T_v - T_v^o}{T_v^{\text{max}} - T_v^o} \tag{47b}$$

where T_v^o and T_v are, respectively, the infrared foliage temperatures of well-watered and water-stressed canopies, and T_v^{max} is the maximum attainable foliage temperature under the prevailing atmospheric conditions (nonevaporating crop). One can use Eq. 41 for T_v^o (knowing D^o) and calculate T_v^{max} as

$$T_v^{\text{max}} = T_a^o + \alpha + \beta D'' \tag{47c}$$

where D'' is the difference of saturated vapor pressures (Eq. 12c) calculated at temperatures $(T_a^o + \alpha)$ and T_a^o. [Note that T_v^{max} might also be obtained as the solution of the energy-balance equation, $R_v = \rho c_p (T_v^{\text{max}} - T_a)/r_H$.]

Combining Eqs. 47a and 47b, one obtains the equation for calculating the latent heat flux from water-stressed canopies based on infrared foliage temperatures as

$$\lambda E_v = \lambda E_v^o \left(\frac{T_v^{\text{max}} - T_v}{T_v^{\text{max}} - T_v^o} \right) \tag{47d}$$

with λE_v^o given by Eq. 44. Apart from crop-specific constants α and β, one needs air temperature, vapor-pressure deficit, net radiation, and infrared foliage temperature to calculate the latent heat flux. The major savings in the data requirements is that the

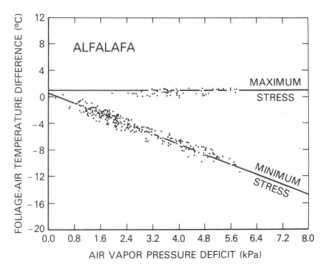

Figure 15 Observed minimum and maximum foliage–air temperature difference for an alfalfa canopy. The minimum foliage–air temperature difference curve is essentially the nonwater-stressed baseline. Identification of these minimum and maximum differences led to the definition of the crop-water-stress index; see Eq. 47b. (After Idso et al., 1981a.)

canopy resistance (r_v), crop height, and wind speed (needed for r_H) are not needed to calculate λE_v. One should not undermine this savings in any practical application.

II.F Some Field Data

One-layer heat-balance models, as summarized by Eq. 4, provide the most direct approach to estimating the latent heat flux using infrared-temperature observations. We have quoted several studies discussing such applications. For the ardent and the curious, we have tabulated some observations to play with varied energy-balance equations, including Eq. (4). (These observations were provided by Dr. R. J. Reginato of the U.S. Water Conservation Laboratory in Phoenix and his colleagues.) Two sets of observations are over well-watered wheat, and the third set is over semiarid natural vegetation. In all cases, the infrared-temperature data are from nadir viewing infrared radiometers.

The observations over Pavon wheat (Tables 7 and 8) were collected in Phoenix on March 30 and March 31, 1983. The leaf-area index and canopy height were, respectively, 6.0 and 0.75 m. The meteorologic data (R_n, T_a, u, and D) were collected at 1.5 m above the soil surface, the soil heat flux (G) was measured at a depth of 0.05 m, and the latent heat flux (λE) was measured by a lysimeter. The leaf stomatal conductance c (mm s^{-1}) increased almost linearly with the photosynthetically active irradiance S (W

TABLE 7 Observations over Pavon Wheat at Phoenix, Arizona, on March 30, 1983 from 0100 to 2400 h

R_n (W m^{-2})	λE (W m^{-2})	G (W m^{-2})	T_{eff} (°C)	T_a (°C)	u (m s^{-1})	D (kPa)
−45	0	−26	6.7	13.5	0.9	0.6
−48	0	−22	7.0	12.5	0.9	0.5
−55	9	−14	7.7	11.6	1.6	0.3
−50	4	−17	7.2	11.1	1.4	0.3
−50	0	−21	6.9	11.3	1.2	0.3
−51	0	−23	6.2	9.9	1.2	0.3
−25	2	−13	7.5	10.4	1.7	0.3
116	60	3	11.7	13.3	2.5	0.4
286	215	18	15.7	16.0	2.2	0.6
462	383	29	17.6	18.5	2.6	1.0
589	460	40	19.9	20.8	2.1	1.3
668	553	50	20.7	22.3	2.0	1.5
656	540	60	21.8	23.8	1.9	1.7
601	617	60	21.6	24.7	2.2	2.0
503	695	29	21.5	25.7	2.9	2.2
343	592	18	21.3	26.0	2.7	2.3
160	465	9	20.7	26.2	3.3	2.4
1	286	1	19.8	25.8	3.0	2.4
−55	58	−12	14.7	23.6	1.0	1.9
−48	9	−23	10.9	21.3	0.5	1.5
−53	2	−25	9.8	20.0	0.5	1.4
−48	2	−25	9.2	18.5	1.0	1.1
−49	13	−21	8.9	16.3	1.3	0.9
−52	0	−18	9.2	15.4	1.4	0.7

TABLE 8 Observations over Pavon Wheat at Phoenix on March 31, 1983 from 0100 to 2400 h

R_n (W m^{-2})	λE (W m^{-2})	G (W m^{-2})	T_{eff} (°C)	T_a (°C)	u (m s^{-1})	D (kPa)
-55	0	-13	9.8	14.7	1.8	0.5
-58	0	-11	10.0	13.9	2.2	0.4
-56	11	-13	9.4	13.0	2.4	0.4
-57	9	-12	9.0	12.6	1.9	0.3
-58	22	-14	8.7	12.0	2.3	0.3
-57	13	-16	8.0	11.1	2.3	0.2
-29	0	-13	9.0	11.6	2.8	0.3
118	93	3	12.7	14.3	3.1	0.4
276	228	17	16.7	16.9	1.6	0.6
466	376	28	18.4	19.9	2.8	1.1
593	518	37	20.4	22.7	2.0	1.4
644	497	50	21.8	25.0	2.2	1.7
723	723	57	21.4	25.8	3.9	2.2
615	906	41	23.3	27.0	6.8	2.7
503	873	20	23.8	27.3	7.4	3.0
322	770	6	24.7	28.3	9.9	3.6
142	981	5	24.1	27.7	9.2	3.5
-18	561	3	22.8	26.6	8.1	3.2
-82	174	0	20.8	24.3	4.8	2.7
-78	125	-7	17.5	21.5	4.4	2.1
-73	116	-6	17.2	20.7	4.2	2.0
-77	101	-6	17.3	20.4	4.7	2.0
-73	112	-7	15.2	19.4	2.4	1.7
-63	58	-25	9.6	16.8	1.3	1.3

TABLE 9 Observations at Owens Valley, California, over Semiarid Natural Vegetation on June 2, 1986 between 1000 to 1530 h[a]

R_n (W m^{-2})	λE (W m^{-2})	G (W m^{-2})	T_{eff} (°C)	T_a (°C)	u (m s^{-1})	D (kPa)
540	70	165	35.1	29.4	2.5	2.9
540	70	170	35.7	29.1	2.4	2.8
640	80	160	42.9	31.8	1.7	3.5
650	80	150	42.3	32.0	1.6	3.5
650	80	150	49.7	35.8	1.4	4.7
680	85	160	49.8	35.4	1.5	4.6
680	85	170	51.8	36.8	1.5	5.2
680	85	170	50.6	36.7	2.4	5.1
610	80	160	51.3	34.4	1.0	4.6
610	80	160	51.1	34.4	1.0	4.6
520	70	110	45.8	35.7	1.5	5.0
520	70	110	45.9	35.7	1.5	5.0

[a]The infrared temperatures are corrected for surface emissivity.

m^{-4}) measured on top of the canopy: $c_l = 0.6 + 0.045S$. Thus, the midday values of c_l were in the range 14 to 18 mm s^{-1}.

The observations over natural vegetation (Table 9) were collected at Owens Valley, California, on June 2, 1986. The major species include saltbush, rabbitbush, greasewood, alkali sacaton, and Russian thistle. The vegetation appeared vigorous at the time of the observations. The vegetation height was about 1 m, the leaf-area index was not measured, and the fractional vegetation cover was about 0.3. The meteorologic data were collected at the height of 2 m above the soil surface, the soil heat flux was measured at a depth of 0.01 m, and the latent heat fluxes were determined by a Bowen ratio instrument. The midday leaf stomatal conductances ranged between 4 and 6 mm s^{-1}.

One should note the highly stable conditions developing during the afternoons over the wheat plot, whereas highly unstable conditions prevailed over the natural vegetation.

III CARBON ASSIMILATION

While estimates of evaporation are of direct interest in irrigation scheduling, estimates of carbon assimilation (net photosynthesis) are of direct relevance in assessing the crop yield and net productivity. The existence of an intimate relationship between crop evaporation and dry-matter accumulation was established more than 70 years ago through pioneering experiments by Briggs and Shantz (see Tanner and Sinclair, 1983), and a recent review by Zelitch (1982) shows the existence of a close relationship between the rates of dry-matter accumulation and carbon assimilation. It is clear from the discussion in the previous section that if infrared temperature can be used to estimate crop evaporation, it should also be possible to estimate carbon assimilation and dry-matter accumulation.

In this section, we shall discuss the carbon-flow pathway as a resistance network, analogous to the resistance network for vapor-flow pathway (Figure 5), establish a relationship between crop evaporation and carbon assimilation, and then establish the relationships between infrared foliage temperature, carbon assimilation, and dry-matter accumulation.

III.A A Resistance Network

A potential/flux resistance network for carbon flow is shown in Figure 16. The fluxes (mg CO$_2$ m^{-2} s^{-1}) of carbon dioxide (CO$_2$) from the air at reference height and from soil to the canopy ambient air are, respectively, F_a and F_s, and A_v is the rate of carbon assimilation by the canopy. The canopy is visualized as a big leaf (Sinclair et al., 1976). Analogous to the vapor pressures in Figure 5, the CO$_2$ mass densities (mg CO$_2$ m^{-3}) at the reference height, in the canopy ambient air, and at the soil surface are, respectively, χ_a, χ_b, and χ_s, and these densities at the "leaf" (the big-leaf) surface and within the stomatal cavities are, respectively, χ_1 and χ_i (no distinction should be made between the subscripts i and v used in Figures 16 and 5 for vapor pressure and CO$_2$ concentration since they refer to the same location; i happens to be the conventional subscript for CO$_2$ density within stomata as for intercellular). The transfer resistances from the reference height and from the soil surface to the canopy ambient air, r_a and r_ω, respectively, are identical to those for water vapor (because the mass transfer is via turbulence), and the canopy boundary layer and canopy stomatal resistances, r_b' and r_v', respectively, are

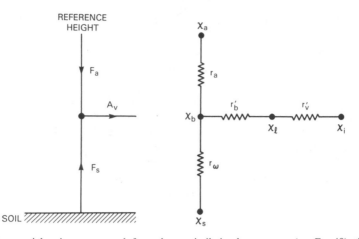

Figure 16 A potential-resistance network for carbon assimilation by a canopy (see Eq. 49); the fluxes are shown on the left. A prime is used to differentiate the diffusive resistances for CO_2 exchange from the corresponding resistances for H_2O exchange (Figure 5). The turbulent transfer resistances are identical for CO_2 and H_2O. The CO_2 mass densities in assimilation play the role of water-vapor concentration (the air vapor pressure) in transpiration. The canopy is visualized as a big leaf. (Sinclair et al., 1976.)

different from those for water vapor due primarily to the differences in molecular weights of CO_2 and H_2O (the mass transfer is by molecular diffusion). There is, however, a constant proportional relationship between r'_b and r_b, and r'_v and r_v, namely (Jones, 1983)

$$r'_b = 1.39r_b \tag{48a}$$

$$r'_v = 1.64r_v \tag{48b}$$

and the total diffusive resistance for CO_2 $(r'_b + r'_v)$ is approximately related to the corresponding resistance for H_2O as (Eqs. 13 and 28)

$$(r'_b + r'_b) = 1.6(r_b + r_v) \tag{48c}$$

Now, analogous to the rate of evaporation, we define the rate of net carbon assimilation by a canopy, A_v, ignoring the hindrance of carbon flow into stomata caused by the water-vapor outflow through the stomatal pores (Jarman, 1974) as

$$A_v = \frac{\chi_b - \chi_i}{1.6(r_b + r_v)} \tag{49a}$$

Considering the evidence for leaves (Ramos and Hall, 1982, among others) that χ_i has a constant proportional relationship to χ_1, it is perhaps more convenient to express A_v as

$$A_v = \frac{\chi_1 - \chi_i}{1.64r_v} \tag{49b}$$

Since mass conservation demands

$$\frac{\chi_b - \chi_1}{1.39r_b} = \frac{\chi_1 - \chi_i}{1.64r_v} \tag{49c}$$

we can write Eq. 49b as

$$A_v = \frac{\chi_b(1 - f_i)}{1.64r_v + 1.39r_b(1 - f_i)} \tag{49d}$$

where the "constant" f_i is given by

$$f_i = \chi_i/\chi_1 \tag{49e}$$

Using mass conservation, one can relate χ_b to the concentrations at the reference height (χ_a) and at the soil surface (χ_s) to obtain the rate of net carbon assimilation as

$$A_v = \frac{(1/r_a + 1/r_\omega)^{-1}(1 - f_i)(\chi_a/r_a + \chi_s/r_\omega)}{1.64r_v + [1.39r_b + (1/r_a + 1/r_\omega)^{-1}](1 - f_i)} \tag{49f}$$

where r_a and r_ω are defined by Eqs. 15 and 14, respectively.

Evaluation of A_v requires a knowledge of the resistances (r_v, r_a, and r_ω), the concentrations (χ_a and χ_s), and the ratio of intercellular to ambient-air CO_2 concentrations, f_i. At moderate or high irradiance, f_i for a single leaf has been observed to be rather constant, roughly, 0.7 for unstressed C_3 plants and 0.3 for unstressed C_4 plants (Ramos and Hall, 1982) although varying somewhat with the saturation deficit (Wong and Dunin, 1987). At low irradiances, f_i increases, reaching a value of 1.0 at the light compensation point (2 to 10 W m^{-2} PAR), whereas in the dark, f_i is about 1.1 (Ball and Berry, 1982). For canopies visualized as a big leaf, the dependence of f_i on irradiance is quite a bit stronger than that for a single leaf, because of the decreasing effective irradiance on a "big leaf" with increasing leaf-area index. Thus, the light compensation point for a canopy shifts toward higher irradiances compared to that for a single leaf. One can calculate the irradiance dependence of f_i for a canopy by integrating the light-response function for assimilation by a leaf (see Table 5) and the canopy stomatal conductance, as illustrated previously, and evaluating the ratio of canopy assimilation and conductance, e.g., Eq. 49b. Nevertheless, at fairly high irradiances, f_i for a well-watered C_3 crop canopy would be about 0.7, and for a well-watered C_4 canopy, about 0.3. One can use Eq. 49f to simulate canopy assimilation, and study its relationship to crop evaporation and temperature by solving Eq. 12. Note that the close correspondence of transfer resistances for CO_2 and H_2O (e.g., Figures 5 and 16, and Eq. 48) is essentially responsible for close relationships between evaporation and carbon assimilation. After discussing conversion of CO_2 assimilation to dry matter, we shall present a relationship between evaporation, assimilation, and crop temperature considering only the diffusion across stomatal cavities.

The key aspects of conversion from the rate of CO_2 assimilation (A_v) to the rate of dry-matter accumulation (Y; mg dry matter m^{-2} s^{-1}) have been discussed by Tanner and

Sinclair (1983). The equation is

$$Y = 0.5\{A_v - R\} \tag{50}$$

where R is the rate of respiration (mg CO_2 m^{-2} s^{-1}) during the daytime by the plant parts that do not participate in CO_2 assimilation (for example, roots and perhaps all of the stems). The coefficient 0.5 in Eq. (50) results from the conversion of CO_2 to hexose (0.68) and from hexose to carbohydrates (0.75), resulting the overall conversion of (0.68 $\times$ 0.75 =) 0.5. Note that Y includes both above- and below-ground plant parts. If only the above-ground biomass is measured, then one should multiply this biomass by 1.2 to 1.5 to account for root biomass.

When integrated over a whole day (24 h), the accumulated dry matter $\overline{Y}$ (mg dry matter m^{-2} d^{-1}) is related to the integrated (daytime) assimilation $\overline{A}_v$ (mg CO_2 m^{-2} d^{-1}) as

$$\overline{Y} = 0.5\{\overline{A}_v - \overline{R}\} \tag{51}$$

where $\overline{R}$ includes the respiration by the nonassimilating plant parts during the daytime and by all plant parts, including leaves, during the night since A_v is assumed to be zero during the night.

Some observations (e.g., King and Evans, 1967; Biscoe et al., 1975; Denmead, 1976; Farquhar and Richards, 1984; Gent and Kiyomoto, 1985) suggest that for agricultural crops, a rough proportionality exists between assimilation and respiration, e.g., $\overline{R} = 0.30(\pm 0.1) A_v$. Thus, for a rough estimate,

$$\overline{Y} = 0.35(\pm 0.05)\overline{A}_v \tag{52}$$

Experimental evidence justifying a close relationship between the rates of dry-matter accumulation and carbon assimilation (as illustrated by Eq. 52) has been reviewed recently by Zelitch (1982).

III.B   Intrinsic Transpiration Efficiency

If we consider only the diffusion across stomata of CO_2 and H_2O, then the rate of assimilation (mg CO_2 m^{-2} s^{-1}) is given by

$$A_v = \frac{\chi_1(1 - f_i)}{1.64 r_v} \tag{53}$$

and the rate of crop evaporation (m s^{-1}) would be given by

$$E_v = \frac{\rho c_p}{\lambda \gamma} \frac{e_v^* - e_1}{r_v} \tag{54a}$$

To compare evaporation and assimilation, it is convenient to have them in identical dimension. Thus, we write E_v in units of mg H_2O m^{-2} s^{-1} as

$$E_v = \frac{\rho M_v}{P M_a} \frac{e_v^* - e_1}{r_v} \tag{54b}$$

where ρ is the density of air (1.2×10^6 mg m^{-3}), M_a is the molecular weight of air (28.96×10^3 mg mol^{-1}), M_v is the molecular weight of water vapor (18×10^3 mg mol^{-1}), and P is the barometric air pressure (kPa).

Eliminating r_v from Eqs. 53 and 54b, we get the dimensionless ratio (A_v/E_v), called the transpiration (or water-use) efficiency (mg CO_2/mg H_2O) as

$$\frac{A_v}{E_v} = \frac{\chi_1(1 - f_i)P}{(1.2 \times 10^6)(e_v^* - e_1)} \tag{55}$$

The numerator on the right-hand side of Eq. 55 is a rather conservative quantity, and the major determinant of transpiration efficiency is the saturation deficit in the numerator (f_i) and in the denominator of the right-hand side of Eq. (55) (see Tanner and Sinclair, 1983). Thus, under well-watered conditions, and also under moderate stress, one finds that the foliage temperature and canopy microclimate (which sets the value of $e_v^* - e_1$) would determine, to a large extent, the fractional carbon gain and dry matter produced per unit of water transpired. The conservative nature of transpiration efficiency has been demonstrated for agricultural crops as well as for natural vegetation (cf. Hellmuth, 1971; Caldwell et al., 1977; Verma et al., 1986). Considering the uncertainties in prescribing χ_1, f_i, and e_1, and measurement errors in A_v and E_v, one may use the following equations in practical applications (assume $\chi_1 = 650$ mg m^{-3}, $P = 100$ kPa, and, under high irradiances, $f_i = 0.3$ for C_4 and $f_i = 0.7$ for C_3 plants).

For C_3 plants,

$$\frac{A_v}{E_v} = \frac{0.016}{D} \tag{56a}$$

For C_4 plants,

$$\frac{A_v}{E_v} = \frac{0.038}{D} \tag{56b}$$

where D is the daytime air vapor-pressure deficit (kPa) at the reference height.

In terms of daily accumulation of dry matter ($\overline{Y}$) and daily crop evaporation ($\overline{E}_v$), the ratio (called the dry-matter water-use efficiency) is obtained (using Eq. 52) as follows.

For C_3 plants,

$$\frac{\overline{Y}}{\overline{E}_v} = \frac{0.006}{\overline{D}} \tag{57a}$$

For C_4 plants,

$$\frac{\overline{Y}}{\overline{E}_v} = \frac{0.013}{\overline{D}} \tag{57b}$$

where $\overline{D}$ is the daytime mean saturation deficit (kPa). Although we considered only the diffusion of CO_2 and H_2O across the stomatal pores in deriving the previous equations, the final results (Eqs. 56 and 57) describe observations reasonably well. For example, Figure 17 from Day et al. (1978) shows that evaporation from barley, when normalized

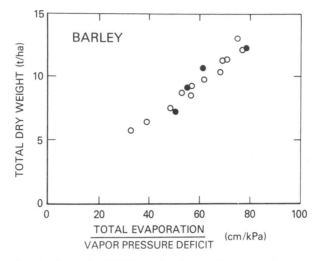

Figure 17   Observations by Day et al. (1978) for barley showing that evaporation, when normalized by the air vapor-pressure deficit, could be uniquely related to the dry-matter accumulation. A theoretical basis of this relationship is discussed in the text (Eq. 57). Remote sensing of evaporation, together with the air vapor-pressure deficit, might provide an estimate of dry-matter production. The open and solid circles represent the data for two separate trials.

by the air vapor-pressure deficit, is linearly related to dry-matter accumulation, as would be predicted by Eq. 57. A more detailed analysis of Eq. 57 has been given by Tanner and Sinclair (1983). This agreement is partly because of the measurement uncertainties in the transpiration efficiency (20 to 30%) and also because the canopy resistance is generally a factor of 2 or 3 larger than the other transfer resistances for well-watered agricultural crops (the factor is larger than 3 for moderately stressed crops and forests). Thus, stomata exert the major control in the diffusion of gases at least for well-watered plants (Cowan and Farquhar, 1977). Assimilation under high water-stress conditions is determined more by the biochemical kinetics than by carbon diffusion through stomata.

If infrared-temperature observations can be used to estimate crop evaporation, then carbon assimilation and dry-matter accumulation could be obtained from Eqs. 56 and 57. In the next section, we show how infrared-temperature observations might be used to estimate carbon assimilation.

III.C   Semiempirical Relations with Infrared Temperatures

If we combine Eqs. 47a and 56, then the effect of water stress on carbon assimilation by a canopy would be described by the equation:

$$A_v = \xi E_v^o (1 - \text{CWSI})/D \tag{58}$$

where ξ is 0.016 for C_3 and is 0.038 for C_4 crops.

Although Eq. 58 is approximate, Figure 18, from Idso et al. (1982), does show that carbon assimilation by *leaves* is a highly linear function of CWSI calculated using infrared temperature (Eq. 47b). We note in Figure 18 that $A_v = 0$ at CWSI of 0.9 instead of being at CWSI = 1, as predicted by Eq. 58. At high water stress, the biochemical kinetics, which determine the intercellular CO_2 concentration, is substantially modified;

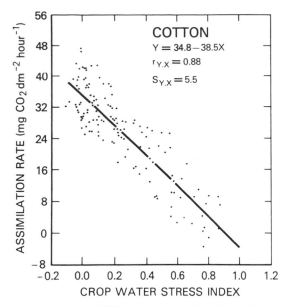

Figure 18   Observations of Idso et al. (1982) for cotton showing the relation between the rate of carbon assimilation and crop-water-stress index calculated using infrared-temperature data. The assimilation rate decreases with increasing water stress, and the crop-water-stress index could explain roughly 75% of the variability of carbon assimilation through a linear regression.

the intercellular CO_2 concentration increases and, hence, ξ in Eq. 58 decreases, which shifts the intercept from CWSI = 1 to a lower value (Berry and Bjorkman, 1980). A more realistic evaluation of A_v was done by Choudhury (1986). Figure 19 compares the computed A_v for a cotton *canopy* with the observed assimilation by cotton *leaves*. Since carbon assimilation by a full canopy is larger than that by individual leaves, we see the departure from the 1 : 1 line in Figure 19.

One can compute E_v^o and CWSI using Eqs. 44 and 47b, and thus quantify carbon assimilation by water-stressed crops using infrared-temperature observations. The practical value of quantifying assimilation by infrared radiometers is quite immense, because large areas could be sampled rapidly by an easy-to-use instrument. Estimation of dry-matter accumulation or economic yield is often more desirable than assimilation. But assimilation and dry-matter accumulation are highly correlated, as has been demonstrated previously (cf. Eq. 50). Indeed, several observations have demonstrated the potential of quantifying yield using infrared-temperature data. In the following three separate experiments, these are briefly described.

Figure 20, from Idso et al. (1981b), shows grain yield of wheat as a function of CWSI. These observations are from six differentially irrigated plots of Produra wheat in Phoenix. The CWSI was calculated daily using the infrared-temperature observations about 1.5 h past solar noon. These daily CWSI values were averaged over the reproductive growth period of wheat, from heading to senescence. The relationship between the grain yield and CWSI is quite strong, but the authors cautioned that the constants of this empirical relationship may be unique to the particular crop and the prevailing climate in Phoenix. This caution is quite appropriate because the theoretical results presented illustrated the important roles of stomatal resistance and atmospheric conditions in determining assimilation.

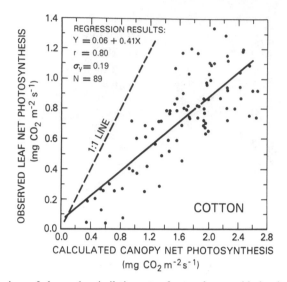

Figure 19 A comparison of observed assimilation rate of cotton leaves with the simulation using semiempirical relations developed from nonwater-stressed baselines of infrared foliage temperature (Eqs. 43, 44, and 58). For details, see Choudhury (1986). The assimilation rate for a canopy (per-unit-ground-area basis) should be higher than that for a leaf (per-unit-leaf-area basis) when the leaf-area index (which is the ratio of total one-side leaf area per unit ground area) is greater than 1. The 1:1 line is shown in the figure to demonstrate that the calculated assimilation rates for the cotton canopy are higher than those observed for a leaf.

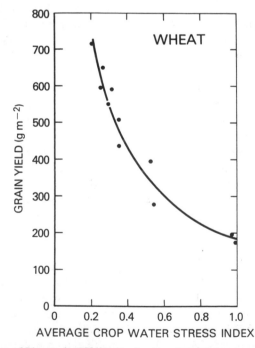

Figure 20 Observations of Idso et al. (1981b) showing the relation between grain yield of Produra wheat from six plots and the average crop-water-stress index (from the infrared temperature). The decrease in grain yield with increasing water stress might be understood in terms of a decrease in the assimilation rate (cf. Figure 18).

Pinter et al. (1983) studied seed and lint yields of cotton (cv. Deltapine 70) as a function of CWSI. These observations, Figure 21(a), are from eight differentially irrigated plots (four irrigation regimes were replicated) in Phoenix. The CWSI was calculated daily from the infrared-temperature measurements about 1.5 h past noon, Figure 21(b). These daily values of CWSI were averaged over an 88-day period, from the appearance of first square (flower bud) until two weeks past the final irrigation. Over the observed range of 0.12 to 0.34 for the CWSI, the relationship between the yield and the CWSI was found to be highly linear.

Smith et al. (1985) complemented the observations of Idso et al. (1981b) by showing a highly linear relationship between grain yield of wheat (cv. Avocet) and the mean foliage–air temperature difference (Figure 22). These observations of Smith et al. (1985)

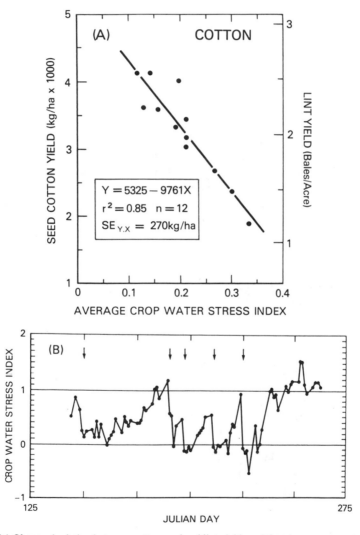

Figure 21 (a) Observed relation between cotton seed and lint yields and the average crop-water-stress index under different irrigation schedulings, and (b) the time series of the crop-water-stress index of a cotton plant under one irrigation scheduling used to compute an average value. Arrows indicate the days when the plot was irrigated. The results in (a) are from eight differentially irrigated plots. (After Pinter et al., 1983.)

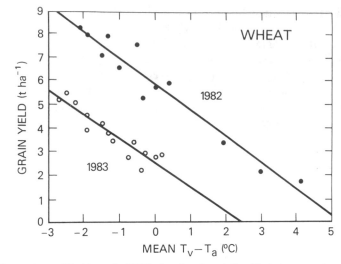

Figure 22 Observations of Smith et al. (1985) showing the relationship between grain yield of wheat for two years and the mean foliage–air temperature difference. Differing intercepts of the linear relations were explained to be due to the differing lengths of the growing season for the two years.

are from Griffith, Australia, for two consecutive years, with different frequency of irrigation and time of sowing. Infrared foliage temperature measurements were taken daily around solar noon, and the daily foliage–air temperature difference was averaged from jointing to maturity. For both years, the grain yield is seen to be linearly related to the mean foliage–air temperature difference; the major difference between the years is in the intercept of the linear relationship. The authors explained much of this difference as being due to the length of the growing season (i.e., the duration of assimilation period). A longer growing season resulted in higher yield at identical mean foliage–air temperature difference.

The previous theoretical analysis and observations demonstrate the feasibility of estimating carbon assimilation and dry-matter production (or yield) using infrared-temperature data. The close correspondence of resistances for carbon and water-vapor exchange, which leads to conservative relationships between evaporation and carbon assimilation, has been exploited to estimate carbon assimilation using the infrared-temperature data. It is important to know the foliage temperature, rather than the composite soil–foliage temperature (Eq. 35), to apply the analysis given. Hot soil temperature, when included in the infrared-temperature observations, might give higher values of CWSI and hence a lower rate of assimilation.

Another important aspect of the observed infrared temperatures is its spatial variability. The foliage temperature could vary spatially due to the differential soil-water stress. However, the spatial variability of infrared temperatures could also be due to a different fractional soil area viewed by the radiometer (Eq. 35). In the next section, we shall discuss the spatial variability of infrared temperatures.

IV SPATIAL VARIABILITY

Surface temperatures measured by infrared radiometers at different locations within a field are quite likely to show variability, even when there is no instrument noise and the measurements are concurrent. The major factors responsible for this variability would be (1) atmospheric conditions at the reference height, particularly, wind speed, (2) fractional soil and foliage areas viewed by the radiometer, and (3) differential soil drying due to the spatial variability of soil hydraulic parameters and roughness and resulting crop stress. The separation of observed surface-temperature variability into components caused by these factors may not always be straightforward. Except for factor (2), temperature variability is due to the variability of surface energy balance and, hence, would be an indicator of the spatial variability of evaporation. In the following, we shall discuss the observed spatial variabilities by ground-based and satellite-borne infrared radiometers.

IV.A Ground Data

Vauclin et al. (1982) present a detailed analysis of the spatial structure of infrared temperature over a bare field for three consecutive days after irrigation. All temperatures were normally distributed along the transects of observation. The spatial structure could be described by a first-order autoregressive process (Figure 23). An understanding of the spatial structure of infrared temperatures is needed to design the sampling strategy to compute the mean surface temperature over a field from point observations. The mean temperature can then be used to assess the average rate of evaporation and, hence, any need for irrigation.

Aston and van Bavel (1972) proposed that the spatial variability of foliage temperature might be used to do irrigation scheduling. Clearly, irrigation scheduling based on some measurements of plant parameters should be superior to those based only upon meteorologic or soil parameters. Clawson and Blad (1982) and Hatfield et al. (1984b) tested this theory for irrigation scheduling for corn and sorghum, respectively. A theoretical analysis of the spatial variability of temperature has been done by Choudhury and Federer (1984).

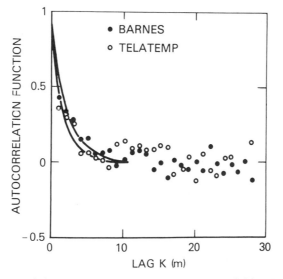

Figure 23 Spatial autocorrelation function of infrared temperatures at varied lag distances along a transect of bare field in Davis, California. The observations were by two radiometers (Barnes and Teletemp) differing in the viewing spatial resolutions. Exponential functions, which are characteristic of a first-order autogressive process, are shown. The spatial variability of infrared temperatures was found to be a first-order autogressive process. (Vauclin et al., 1982.)

Observations of Clawson and Blad (1982) were done in Nebraska using five plots of sprinker irrigated corn. An infrared thermometer was used to measure canopy (foliage) temperatures daily at 1400 h solar time. The thermometer was held atop a 4 m portable ladder, and aimed at different plots to "view" only the foliage. The view angle for different plots was variable, but they did not observe any view-angle dependence for the foliage temperature. From neutron-probe observations of soil moisture in the top 1.7 m soil layer, they designated one plot as the control "well-watered" plot, and this plot was irrigated weekly. In other plots, the irrigation scheduling was done using the *average* foliage temperature over a plot and also by the canopy-temperature variability (CTV), which was defined as the difference of maximum and minimum infrared temperatures observed over a plot. Irrigation was done when the average temperature was either 1 or 3 °C higher than the temperature for the control plot. In one plot, the irrigation was done when CTV = 0.8 °C.

The effectiveness of the irrigation scheduling might be judged by comparing the grain yields, total applied water (rain plus irrigation), and by the grain yield water-use efficiency (which is the ratio of grain yield to applied water). These values are given in Table 10. The data for the 3 °C stressed plot are not included because this plot was accidentally irrigated twice during the study period. Both canopy-temperature-based irrigation schedulings (CTV and 1 °C stressed) have markedly higher water-use efficiencies as compared to that for the well-watered control plot, albeit at the expense of a reduced yield. For the CTV trial, the grain yield is reduced by roughly 5%, but the water-use efficiency is increased by roughly 32%. For the 1 °C stressed trial, the grain yield is reduced by 18%, but the water-use efficiency is increased by 38%. Clawson and Blad (1982) concluded that "using canopy temperature variability to initiate irrigation has the potential for significant water savings. . . . "

Observations of Hatfield et al. (1984b) were done at Davis using differentially irrigated fields of grain sorghum. They measured canopy temperature near solar noon at 1 m intervals along 85 m long transects using a handheld infrared thermometer at a 30° angle from horizontal. They found no correlation of temperatures at 1 m interval, and the distribution was normal with respect to the mean temperature along the transect. During the drying cycle from one irrigation to the next, the variance of temperature along a transect increased when 60% or more of the available water within the top 1.5 m soil layer was depleted (Figure 24). They concluded that the temperature-variability method may be used to evaluate the distribution efficiency of irrigation water.

Further observations of temperature variability are needed to understand its usefulness in quantifying crop stress and irrigation scheduling. The possibility of quantifying the spatial variability of evaporation using the infrared-temperature data also has to be investigated.

TABLE 10   The Grain Yield, Applied Water, and Water-Use Efficiency in the Irrigation Trials over Corn

Trial	Grain Yield (kg ha^{-1})	Applied Water (mm)	Water-Use Efficiency (kg ha^{-1} mm^{-1})
Well watered	7575	558	13.6
Canopy-temperature variability	7202	402	17.9
1 °C	6197	329	18.8

Source: Clawson and Blad (1982).

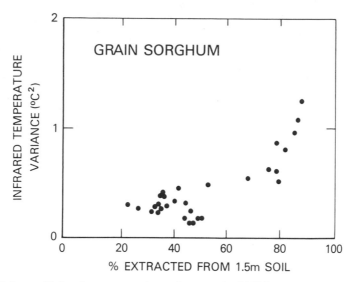

Figure 24 Variance of infrared temperature observed at an angle of 30° from the horizontal over a sorghum field as a function of water extracted from the 1.5 m surface layer. The variance was found to increase sharply when 60% of the available water was depleted from the 1.5 m surface layer. Such an analysis of variance might be useful in irrigation scheduling. (Hatfield et al., 1984b.)

IV.B Satellite Data

During the past several years, highly sophisticated parameterizations of land surface temperature and latent and sensible heat fluxes have been developed for inclusion in the atmospheric general circulation models (cf. Sellers et al., 1986). A key factor in this sophistication has been the distinction of fluxes and temperatures for soil and vegetation. The area of the smallest land surface unit, called a grid box, varies between 10^4 to 10^5 km². The existence of spatial heterogeneity within any grid box must be considered a norm rather than the exception. Considering the large number of parameters (for example, stomatal resistance, rooting characteristics, plant height, soil wetness, and fractional ground cover) that could determine the spatial variability of temperature and fluxes within a grid box, it is important to establish the relative importance of these parameters in determining the effective temperature and fluxes. Analysis of satellite data might provide an understanding of the parameters determining the surface temperature at a coarse spatial resolution.

In this section, we shall present some infrared-temperature observations by a space-borne sensor, namely, the Advanced Very High Resolution Radiometer (AVHRR) on board the NOAA-7 satellite. Observations by this radiometer are available at five spectral bands: two in the visible to near-infrared region (0.58–0.68 μm and 0.73–1.1 μm) and three in the infrared region (3.55–3.93 μm, 10.3–11.3 μm, and 11.5–12.5 μm). The radiometer scans at $\pm 56°$ about the nadir, with daily global coverage at 0230 and 1430 h local solar time. Although the nadir resolution is 1.1 km, the readily accessible global data are considered to have a nadir resolution of 4 km because they are produced by averaging four pixels of every third scan line (Kidwell, 1985). This coarse resolution data is quite suitable for multitemporal global studies. The visible and near-infrared data have been used successfully to quantify vegetation density and productivity (e.g., Prince and Tucker, 1986). The data from infrared bands can be used to study the relationship between surface temperature and vegetation density. In the following, we shall discuss this relationship for a $4° \times 2°$ (latitude $\times$ longitude) area over the U.S. southern Great Plains ($31.5°$ to $35.5°$N, $98°$ to $100°$W) for two consecutive mostly clear-sky days (September 4 and 5, 1982), following rainfalls on the previous two days.

The radiative temperatures T_1 and T_2, respectively, at $10.3–11.3$ μm and $11.5–12.5$ μm bands were computed from the raw data counts using the calibration coefficients as discussed by Kidwell (1985). From these radiative temperatures, the surface temperature T_{eff} was calculated following Price (1984) as (see also Becker, 1987)

$$T_{eff} = T_1 + 3.33(T_1 - T_2) \tag{59}$$

From the raw data counts in the visible ($0.58–0.68$ μm) and near-infrared ($0.73–1.1$ μm) bands together with the calibration coefficients (embedded in the data stream within the magnetic tapes), we calculated the reflectances for these bands, A_1 and A_2, respectively, following Kidwell (1985). Then, to describe the vegetation density, we calculated the simple ratio (SR) and normalized difference (ND), as discussed by Tucker (1979):

$$SR = A_2/A_1 \tag{60a}$$

$$ND = (A_2 - A_1)/(A_2 + A_1) \tag{60b}$$

The results of daytime (1430 h local solar time) surface temperatures (Eq. 59) and vegetation indices (Eq. 60), aggregated over $0.5° \times 0.5°$ areas, are shown in Figure 25 for September 4 and Figure 26 for September 5, 1982. The results of linear regression analysis, given in these figures, show that either of the two vegetation indices could explain roughly 80% of the spatial variability of surface temperature for both days. The surface temperature is lower when the vegetation indices and, hence, vegetation density are higher (cf. Goward et al., 1985). The intercept and slope of the linear regressions are higher on September 5 as compared to those on September 4. Meteorologic data showed that the maximum air temperature over the study area was identical on both days ($34 \pm 1°$ C). However, one would expect the soil surface to be drier and warmer on September 5 as compared to September 4. Field studies, satellite data, and simulations (e.g., Huete et al., 1985; Choudhury, 1987; Ormsby et al., 1987) show a high linear correlation between the vegetation indices and the fractional ground cover (Figure 27). Considering this correlation and Eq. 35, it appears that the fractional ground cover could be the major determinant of the effective surface temperature at coarse spatial resolutions. One might also infer this correlation between the fractional ground cover and the effective surface temperature from the aircraft observations of Hatfield et al. (1982). Since on a per-unit-ground-area basis, the ratio of crop and soil evaporations is generally greater than 1, it follows that a lower surface temperature would be an indicator of higher evaporation. One should contrast the linear correlation shown in Figures 14 and 25 (or 26) by noting that they correspond, respectively, to foliage temperature and to a composite soil-foliage temperature.

We discussed before the spatial variabilities of infrared temperatures observed by ground-based and satellite-borne radiometers. The spatial variability of foliage temperature over agricultural fields has been studied with the aim of efficient use of irrigation water. This variability could arise from variable root-zone soil water either due to differential soil drying as a result of the spatial variability of soil hydraulic parameters or

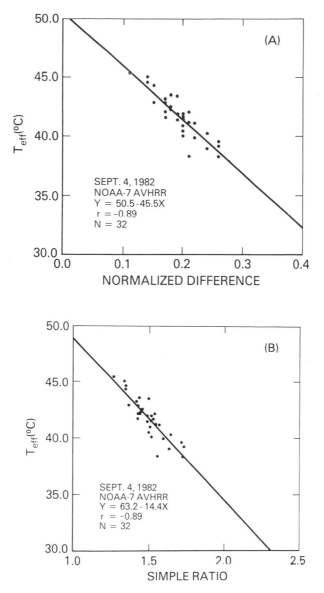

Figure 25 Observed relation between the surface temperature and (*a*) the normalized difference, and (*b*) the simple ratio for September 4, 1982 based upon NOAA-7 Advanced Very High Resolution Radiometer data. Results of linear regression are also shown.

due to nonuniform application of irrigation water and the resulting effect on the crop energy balance, as shown by Choudhury and Federer (1984) using a TERGRA-type model. Field data also show a rather pronounced effect of radiometer look angle on the measured temperature (cf. Figure 12), but to what extent the statistical attributes of spatial variability (for example, the variance) are dependent upon the look angle needs further research. Nevertheless, irrigation trials based on indices of spatial variability of foliage temperature have given promising results. For coarse-resolution satellite data,

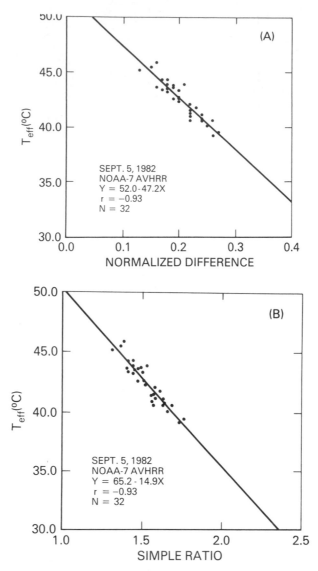

Figure 26   Observed relation between the surface temperature and (*a*) the normalized difference, and (*b*) the simple ratio for September 5, 1982 based upon NOAA-7 Advanced Very High Resolution Radiometer data. Results of linear regression are also shown.

the spatial variability of surface temperature could be primarily due to the spatial variability of fractional vegetation cover (Figures 25–27). The surface temperature increases as the fractional vegetation cover decreases, because the soil surface during summer is generally much warmer than vegetation. Certainly, the foliage and soil temperatures had spatial variabilities within themselves, but these variabilities appear to be of second-order importance in understanding the spatial variabilities of coarse-resolution infrared-temperature data from satellites. Also, the radiometer look angle, spatial heterogeneity of vegetation type (grasses, crops, and forests), and local energy balance in going from one pixel to the next would give somewhat different temperatures, even though the frac-

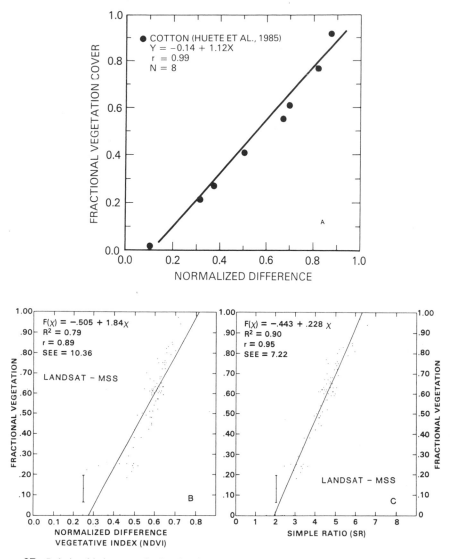

Figure 27 Relationship between the fractional ground cover and the normalized difference vegetation index: (*a*) for a cotton crop observed by Huete et al. (1985), and (*b*) based on LANDSAT MSS data by Ormsby et al. (1987). (*c*) The relation between the fractional vegetation cover and the simple ratio for the LANDSAT data is shown. The results of linear regression are also shown. The results from these figures together with Eq. 35 provide an understanding of the satellite data given in Figures 25 and 26. The surface temperature would be lower under higher fractional vegetation cover because vegetation is cooler than soil.

tional ground cover might be identical. But these variabilities also appear to be of second-order importance, as compared to the variability of surface temperature due to variable fractional ground cover. Thus, in going from fine to coarse spatial resolutions, we see a changing hierarchy of parameters determining the spatial variability of surface temperature. Further research is needed for a quantitative assessment of this hierarchy for infrared temperatures and its implication for estimating latent heat flux using infrared-temperature data. For further discussions of satellite data, see Chapter 15 of this book.

V CONCLUSION

This chapter covered discussion of some energy-balance models for interpreting infrared-temperature observations and for estimating evaporation and carbon assimilation from the infrared-temperature data. Past applications of the infrared-temperature observations have been almost exclusively for agricultural crops. One-layer energy-balance models (as summarized by Eq. 4) provide the most direct and perhaps most promising approach for estimating the latent heat flux using the infrared-temperature data, and the rate of carbon assimilation might be estimated from Eq. 58 if the observed temperature could be partitioned into soil and foliage components. Analysis of two-layer models provides an understanding of the canopy microclimate and its effect on crop and soil evaporations. These models could be used to calculate soil and foliage temperatures and to provide an understanding of the infrared-temperature data in terms of soil and foliage temperatures.

Very little has been done to evaluate large-scale (regional, continental, and global) fluxes using multitemporal global infrared-temperature data available from spaceborne sensors. Infrared-temperature observations have been stated to provide a remote-sensing approach to estimating fluxes. Although it is important to test the methodology in the field, the full potential of this approach is yet to be realized from the analysis of satellite data. The frontier of large-scale studies using infrared-temperature data remains wide open.

LIST OF IMPORTANT SYMBOLS

STANDARD ALPHABETICAL SYMBOLS

Symbol	Definition
$A_v, \overline{A}_v$	Rate of carbon assimilation per unit ground area [mg CO_2 m^{-2} s^{-1}]; for the daily total assimilation [mg CO_2 m^{-2} d^{-1}]
c_l	Leaf stomatal conductance [m s^{-1}]
c_p	Specific heat of air at constant pressure [J kg^{-1} K^{-1}]
c_v	Canopy stomatal conductance [m s^{-1}]
C	Sensible heat flux [W m^{-2}]
$D, D', D°, \overline{D}$	Vapor-pressure deficit; for canopy to air; for that at the reference height over well-watered crop; for the daytime mean at the reference height [kPa]
E, E_v, E_s	Rate of evaporation per unit ground area; for crop; and for soil [m s^{-1}]
d	Zero-plane displacement [m]
e_a	Air vapor pressure at the reference height [kPa]
e_o^*	Saturated vapor pressure at temperature T_o [kPa]
e_s^*	Saturated vapor pressure at temperature T_s [kPa]
f_i	Ratio of CO_2 concentrations within the stomatal cavities and at the leaf surface
f_s	Fractional ground cover viewed by a radiometer
g	Acceleration due to gravity [m s^{-2}]
g'	Leaf-angle distribution parameter
G	Soil heat flux [W m^{-2}]

Symbol	Definition
H	Crop height [m]
k	Von Karman constant
L	Canopy leaf-area index
P	Barometric air pressure at the reference height [kPa]
r_E	Aerodynamic resistance for vapor exchange [s m^{-1}]
r_H	Aerodynamic resistance for sensible heat exchange [s m^{-1}]
r_o	Surface resistance for diffusion of water vapor [s m^{-1}]
R	Rate of respiration per unit ground area [mg CO$_2$ m^{-2} s^{-1}]
R_s, R_v, R_n	Net radiation for soil (s), foliage (v), and above the canopy (n) [W m^{-2}]
R'	Gas constant [m^2 s^{-2} K^{-1}]
$R_{\text{soil}}, R_{\text{plant}}$	Resistances for water flow within the soil (s) and through the plant, respectively
$S_{\text{sun}}, S_{\text{shade}}$	Photosynthetically active irradiance at the leaf surface for sunlit and shaded leaves, respectively [W m^{-2}]
T_a, T_b	Air temperature at the reference height and within the canopy at the height of heat exchange, respectively [K]
T_o	Aerodynamic temperature for the surface [K]
T_s, T_v	Aerodynamic temperatures for soil and foliage, respectively [K]
T_{eff}	Infrared surface temperature (K)
u	Wind speed at the reference height [m s^{-1}]
u_H	Wind speed at the canopy height [m s^{-1}]
Z	Reference height [m]
Z_o, Z_o'	Surface roughness heights for momentum and heat exchange, respectively [m]
Z_{os}	Aerodynamic roughness height for soil surface [m]

STANDARD GREEK SYMBOLS

Symbol	Definition
γ	Phychrometer constant [kPa K^{-1}]
Δ	Gradient of saturated vapor-pressure curve with respect to temperature [kPa K^{-1}]
θ	Volumetric soil wetness [m^3/m^3]
λ	Latent heat of evaporation [J kg^{-1}]
μ_o	Cosine of solar zenith angle
μ_r	Cosine of the zenith angle of an infrared radiometer
ρ	Density of air [kg m^{-3}]
χ_a	Density of CO$_2$ at the reference height [mg m^{-3}]
χ_b	Density of CO$_2$ in air within a canopy [mg m^{-3}]
χ_s	Density of CO$_2$ at the soil surface [mg m^{-3}]
χ_1	Density of CO$_2$ at the leaf surface [mg m^{-3}]
χ_i	Density of CO$_2$ within the stomatal cavities [mg m^{-3}]
ψ_l	Leaf-water potential [m]
ψ_s	Matric potential of the surface soil water [m]
ψ_s	Root-zone soil-water potential [m]
ω	Leaf width [m]
ω_o	Single-scattering albedo of a leaf for photosynthetically active irradiance

REFERENCES

Ackerson, R. C., and D. R. Krieg. Stomatal and non-stomatal regulation of water use in cotton, corn, and sorghum. *Plant Physiol* **60:**850–853.

Aston, A. R. (1984). Evaporation from eucalyptus growing in a weighing lysimeter: Test of the combination equations. *Agric. For. Meteorol.* **31:**241–249.

Aston, A. R., and C. H. M. van Bavel. Soil surface water depletion and leaf temperatures. *Agron. J.* **64:**368–373.

Ball, J. T., and J. A. Berry (1982). The C_i/C_s ratio: A basis for predicting stomatal control of photosynthesis. *Year Book—Carnegie Inst. Washington* **81:**88–92.

Becker, F. (1981) *Angular Reflectivity and Emissivity of Natural Media in the Thermal Infrared Bands.* Signatures Spectrales d'Objects en Télédetection, INRA. Toulouse, France.

Becker, F. (1987). The impact of spectral emissivity on the measurement of land surface temperature from satellite. *Int. J. Remote Sens.* **8:**1509–1522.

Berry, J. A., and O. Bjorkman (1980). Photosynthetic response and adaptation to temperature in higher plants. *Annu. Rev. Plant Physiol* **31:**491–543.

Bhumralkar, C. M. (1975). Numerical experiments on the computation of ground surface temperature in an atmospheric general circulation model. *J. Appl. Meteorol* **14:**1246–1258.

Biscoe, P. V., R. K. Scott, and J. L. Monteith (1975). Barley and its environment. III. Carbon budget of the stand. *J. Appl. Ecol.* **12:**269–293.

Borg, H., and D. W. Grimes (1986). Depth development of roots with time: An empirical description. *Tran. ASAE* **29:**194–197.

Caldwell, M. M., R. W. White, R. T. Moore, and L. B. Camp (1977). Carbon balance, productivity, and water use of cold-winter desert shrub communities dominated by C_3 and C_4 species. *Oecologia* **29:**275–300.

Camillo, P. J., and R. J. Gurney (1986). A resistance parameter for bare-soil evaporation models. *Soil Sci.* **141:**95–105.

Camillo, P. J., R. J. Gurney, and T. J. Schmugge (1983). A soil and atmosphere boundary layer model for evapotranspiration and soil moisture studies. *Water Resour. Res.* **19:**371–380.

Carlson, T. N., F. G. Rose, and E. M. Perry (1984). Regional-scale estimates of surface moisture availability from GOES infrared satellite measurements. *Agron. J.* **76:**972–979.

Carson, D. J. (1981). Current parameterization of land surface processes in atmospheric General Circulation Models. In *Land-Surface Processes in Atmospheric General Circulation Models* (P. S. Eagleson, Ed.), Cambridge University Press, London and New York, pp. 67–108.

Chapin, F. S., III, A. J. Bloom, C. B. Field, and R. H. Waring (1987). Plant responses to multiple environmental stress. *BioScience* **37:**49–57.

Choudhury, B. (1983). Simulating the effects of weather variables and soil water potential on a corn canopy temperature. *Agric. Meteorol.* **29:**169–182.

Choudhury, B. J. (1986). An analysis of observed linear correlations between net photosynthesis and a canopy-temperature-based plant water stress index. *Agric. For. Meteorol.* **36:**323–333.

Choudhury, B. J. (1987). Relationships between vegetation indices, radiation absorption and net photosynthesis evaluated by a sensitivity analysis. *Remote Sens. Environ.* **22:**209–233.

Choudhury, B. J., and C. A. Federer (1984). Some sensitivity results for corn canopy temperature and its spatial variation induced by soil hydraulic heterogeneity. *Agric. For. Meteorol.* **31:**297–317.

Choudhury, B. J., and S. B. Idso (1984). Simulating sunflower canopy temperatures to infer root-zone soil water potential. *Agric. For. Meteorol.* **31:**69–78.

Choudhury, B. J., and S. B. Idso (1985). An empirical model for stomatal resistance of wheat. *Agric. For. Meteorol.* **36:**65–82.

Choudhury, B. J., and J. L. Monteith (1986). Implications of stomatal response to saturation deficit for the heat balance of vegetation. *Agric. For. Meteorol.* **36**:215–225.

Choudhury, B. J., and J. L. Monteith (1988). A four-layer model for the heat budget of homogeneous land surfaces. *Q. J. R. Meteorol. Soc.* **114**:373–398.

Choudhury, B. J., R. J. Reginato and S. B. Idso (1986a). An analysis of infrared temperature observations over wheat and calculation of latent heat flux. *Agric. For. Meteorol.* **37**:75–88.

Choudhury, B. J., S. B. Idso, and R. J. Reginato (1986b). Analysis of a resistance-energy balance method for estimating daily evaporation from wheat plots using one-time-of-day infrared temperature observations. *Remote Sens. Environ.* **19**:253–268.

Choudhury, B. J., S. B. Idso, and R. J. Reginato (1987). Analysis of an empirical model for soil heat flux under a growing wheat crop for estimating evaporation by an infrared temperature based energy balance equation. *Agric. For. Meteorol.* **39**:283–297.

Cionco, R. M. (1972). A wind-profile index for canopy flow. *Boundary-Layer Meteorol.* **3**:255–263.

Clapp, R. B., and G. M. Hornberger (1978). Empirical equations for some soil hydraulic properties. *Water Resour. Res.* **14**:601–604.

Clawson, K. L., and B. L. Blad (1982). Infrared thermometry for scheduling irrigation of corn. *Agron. J.* **74**:311–316.

Clothier, B. E., K. L., Clawson, P. J. Pinter, Jr., M. S. Moran, R. J. Reginato, and R. D. Jackson (1986). Estimation of soil heat flux from net radiation during growth of alfalfa. *Agric. For. Meteorol* **37**:319–329.

Cowan, I. R. (1968). Mass, heat and momentum exchange between stands of plants and their atmospheric environment. *J. R. Meteorol. Soc.* **94**:523–544.

Cowan, I. R. (1982). Regulation of water use in relation to carbon gain in higher plants. In *Physiological Plant Ecology II.* (O. L. Lange, P. S. Nobel, C. B. Osmond, and H. Ziegler, Eds.). Springer-Verlag, New York, pp. 589–613.

Cowan, I. R., and G. D. Farquhar (1977). Stomatal function in relation to leaf metabolism and environment. *Symp. Soc. Exp. Bot.* **31**:471–505.

Coyne, P. I., and J. A. Bradford (1985). Comparison of leaf gas exchange and water use efficiency in two eastern gamagrass accessions. *Crop Sci.* **25**:65–75.

Day, W., B. J. Legg, B. K., French, A. E. Johnston, D. W. Lawlor, and W. de C. Jeffers (1978). A drought experiment using mobile shelters. *J. Agric. Sci.* **91**:599–623.

Deardorff, J. W. (1978). Efficient prediction of ground surface temperature and moisture, with inclusion of a layer of vegetation. *J. Geophys. Res.* **83**:1889–1903.

Denmead, O. T. (1969). Comparative micrometeorology of a wheat field and a forest of *Pinus Radiata. Agric. Meteorol* **6**:357–371.

Denmead, O. T. (1976). Temperate cereals. In *Vegetation and the Atmosphere.* (J. L. Monteith, Ed.). Academic Press, New York, pp. 1–31.

Dickinson, R. E. (1984). Modeling evapotranspiration for three-dimensional global climate model. *Geophys. Monogr.* **29**:58–72.

Elkington, M. D., and J. Hogg (1981). The characterization of soil moisture content and actual evapotranspiration from crop canopies using thermal infrared remote sensing. In *Remote Sensing in Geological and Terrain Studies* (J. A. Allan and M. Bradshaw, Ed.). Remote Sensing Society, London, pp. 69–90.

Farquhar, G. D., and R. A. Richards (1984). Isotopic composition of plant carbon correlates with water-use efficiency of wheat genotypes. *Aust. J. Plant Physiol.* **11**:539–552.

Feddes, R. A., and P. E. Rijtema (1972). Water withdrawal by plant roots. *J. Hydrol.* **17**:33–59.

Federer, C. A. (1979). A soil-plant-atmosphere model for transpiration and availability of soil water. *Water Resour. Res.* **15**:555–562.

Federer, C. A. (1980). Paper birch and white oak saplings differ in response to drought. *For. Sci.* **26**:313–324.

Fisher, M. J., D. A. Charles-Edwards, and M. M. Ludlow (1981). An analysis of the effects of repeated short-term water deficits on stomatal conductance to carbon dioxide and leaf photosynthesis by the legume. *Macroptilium atropurpureum* cv. Sirato. *Aust. J. Plant Physiol* **8**:347–357.

Fuchs, M., and C. B. Tanner (1967). Evaporation from a drying soil. *J. Appl. Meteorol.* **6**:852–857.

Garratt, J. R. (1978). Transfer characteristics for a heterogeneous surface of large aerodynamic roughness. *Q. J. R. Meteorol. Soc.* **104**:491–502.

Gent, M. P. N., and R. K. Kiyomoto (1985). Comparison of canopy and flag leaf net carbon dioxide exchange of 1920 and 1977 New York winter wheats. *Crop Sci.* **25**:81–86.

Ginter-Whitehouse, D. K., T. M. Hinckley, and S. G. Pallardy (1983). Spatial and temporal aspects of water relations of three tree species with different vascular anatomy. *For. Sci.* **29**:317–329.

Gollan, T., N. C. Turner, and E. -D. Schulze (1985). The responses of stomata and leaf gas exchange to vapor pressure deficits and soil water content. III. In the sclerophyllous woody species *Nerium oceander. Oecologia* **65**:356–362.

Goward, S. N., G. D. Cruickshanks, and A. S. Hope (1985). Observed relation between thermal emission and reflected spectral radiance of a complex vegetated landscape. *Remote Sens. Environ.* **18**:137–146.

Gurney, R. J., and D. K. Hall (1983). Satellite derived surface energy balance estimates in the Alaskan sub-arctic. *J. Clim. Appl. Meteorol.* **22**:115–125.

Hamblin, A. P., and J. Hamblin (1985). Root characteristics of some temperate legume species and varieties on deep, free-draining entisols. *Aust. J. Agric. Res.* **36**:63–72.

Hansen, G. K. (1974). Resistance to water flow in soil and plants, plant water status, stomatal resistance and transpiration of Italian Ryegrass, as influenced by transpiration demand and soil water depletion. *Acta Agric. Scand.* **24**:83–92.

Hatfield, J. L. (1979). Canopy temperatures: The usefulness and reliability of remote measurements. *Agron. J.* **71**:889–892.

Hatfield, J. L. (1983). Evapotranspiration obtained from remote sensing method. *Adv. Irrig.* **2**:395–416.

Hatfield, J. L., J. P. Millard, and R. C. Goettleman (1982). Variability of surface temperature in agricultural fields of central California. *Photogramm. Eng. Remote Sens.* **48**:1319–1325.

Hatfield, J. L., R. J. Reginato, and S. B. Idso (1984a). Evaluation of canopy temperature-evapotranspiration models over various crops. *Agric. For Meteorol* **32**:41–53.

Hatfield, J. L., M. Vauclin, S. R. Vieira, and R. Bernard (1984b). Surface temperature variability patterns within irrigated field. *Agric. Water Manage.* **8**:429–437.

Hatfield, J. L., D. F. Wanjura, and G. L. Becker (1985). Canopy temperature response to water stress under partial canopy. *Trans. ASAE* **28**:1607–1611.

Heilman, J. L., W. E. Heilman, and D. G. Moore (1981). Remote sensing of canopy temperature and incomplete cover. *Agron. J.* **73**:403–406.

Hellkvist, J., G. P. Richards, and P. G. Jarvis (1974). Vertical gradients of water potential and tissue water relations in Sitka spruce trees measured with the pressure chamber. *J. Appl. Ecol.* **11**:637–667.

Hellmuth, O. (1971). Eco-physiological studies on plants in arid and semi-arid regions in western Australia. III. Comparative studies on photosynthesis, respiration and water relations of ten arid zone and two semi-arid zone plants under winter and later summer climatic conditions. *J. Ecol.* **59**:225–260.

Henson, I. E., Alagarswamy, F. R. Bidinger, and V. Mahalakshmi (1982). Stomatal responses of pearl millet (*Pennisetum americanum* (L.) Leeke) to leaf water status and environmental factors in the field. *Plant Cell Environ.* **5**:65–74.

Hicks, D. J., and Chabot, B. F. (1985). Deciduous forest. In *Physiological Ecology of North American Plant Communities* (B. F. Chabot and H. A. Mooney, Eds.). Chapman & Hall, New York, pp. 257–277.

Hiler, E. A., T. A. Howell, R. B. Lewis, and R. P. Boos (1974). Irrigation timing by the stress day index method. *Trans. ASAE* **17**:393–398.

Hinckley, T. M., M. O. Schroeder, J. E. Roberts, and D. N. Bruckerhoff (1975). Effect of several environmental variables and xylem pressure potential on leaf surface resistance in white oak. *For. Sci.* **21**:201–211.

Hinckley, T. M., R. G. Aslin, R. R. Aubuchon, C. L. Metcalf, and J. E. Roberts (1978). Leaf conductance and photosynthesis in four species of the oak-hickory forest type. *For. Sci.* **24**:73–84.

Huband, N. D. S. and J. L . Monteith (1986a). Radiative surface temperature and energy balance of a wheat canopy. I. Comparison of radiative and aerodynamic temperatures. *Boundary-Layer Meteorol* **36**:1–17.

Huband, N. D. S., and J. L. Monteith (1986b). II. Estimating fluxes of sensible and latent heat. *Boundary-Layer Meteorol.* **36**:107–116.

Huete, A. R., R. D. Jackson, and D. F. Post (1985). Spectral response of a plant canopy with different soil backgrounds. *Remote Sens. Environ.* **17**:37–53.

Idso, S. B. (1982). Non-water-stressed baselines: A key to measuring and interpreting plant water stress. *Agric. Meteorol.* **27**:59–70.

Idso, S. B. (1983). Stomatal regulation of evaporation from well-watered plant canopies: A new synthesis. *Agric. Meteorol.* **29**:213–217.

Idso, S. B., R. D. Jackson, P. J. Pinter, Jr., R. J. Reginato, and J. L. Hatfield (1981a). Normalizing the stress-degree-day parameter for environmental variability. *Agric. Meteorol.* **24**:45–55.

Idso, S. B., R. J. Reginato, and J. W. Radin (1982). Leaf diffusion resistance and photosynthesis in cotton as related a foliage temperature based plant water stress index. *Agric. Meteorol.* **27**:27–34.

Idso, S. B., R. J. Reginato, R. D. Jackson, and P. J. Pinter, Jr. (1981b). Measuring yield-reducing plant water potential depressions in wheat by infrared thermometry. *Irrig. Sci.* **2**:205–212.

Jackson, R. D. (1982). Canopy temperature and crop water stress. *Adv. Irrig.* **1**:43–85.

Jackson, R. D. (1985). Evaluating evapotranspiration at local and regional scales. *Proc. IEEE* **73**:1086–1097.

Jackson, R. D., R. J. Reginato, and S. B. Idso (1977). Wheat canopy temperature: A practical tool for evaluating water requirements. *Water Resour. Res.* **13**:651–656.

Jackson, R. D., S. B. Idso, R. J. Reginato, and P. J. Pinter, Jr. (1981). Canopy temperature as a crop water stress indicator. *Water Resour. Res.* **17**:1133–1138.

Jarman, P. D. (1974). The diffusion of carbon dioxide and water vapor through stomata. *J. Exp. Bot.* **25**:927–936.

Jarvis, P. G., G. B. James, and J. J. Landsberg (1976). Coniferous forests. In *Vegetation and the Atmosphere* (J. L. Monteith. Ed.). Academic Press, New York, pp. 171–240.

Jones, H. G. (1983). *Plant and Microclimate.* Cambridge University Press, New York.

Jones, M. B., E. L. Leafe, and W. Stiles (1980). Water stress in field-grown perennial ryegrass. II. Its effect on leaf water status, stomatal resistance and leaf morphology. *Ann. Appl. Biol.* **96**:103–110.

Jordan, W. R., and J. T. Ritchie (1971). Influence of soil water stress on evaporation, root absorption, and internal water status of cotton. *Plant Physiol.* **48**:783–788.

Kemp, P. R., and G. J. Williams, III (1980). A physiological basis for niche separation between *Agropyron Smithii* (C_3) and *Bouteloua gracilis* (C_4). *Ecology* **61**:846–858.

Kidwell, K. B. (1985). *NOAA Polar Orbiter Data Users Guide.* U.S. Department of Commerce, Satellite Data Service Division, Washington, D.C.

King, R. W., and L. T. Evans (1967). Photosynthesis in artificial communities of wheat, lucerne and subterranean clover plants. *Aust. J. Biol. Sci.* **20**:623–635.

Kondo, J., and A. Kawanaka (1986). Numerical study on the bulk heat transfer coefficient for a variety of vegetation types and densities. *Boundary-Layer Meteorol* **37**:285–296.

Kondo, J., and H. Yamazawa (1986). Aerodynamic roughness over an inhomogeneous ground surface. *Boundary-Layer Meteorol.* **35**:331–348.

Korner, C. (1985). Humidity responses of forest trees: Precautions in thermal scanning surveys. *Arch. Meteorol. Geophys. Bioklimatol., Ser. B* **36**:83–98.

Kustas, W. P., B. J. Choudhury, M. S. Moran, L. W. Gay, R. J. Reginato, and R. D. Jackson (1989). Estimating sensible heat flux using infrared temperatures over partially vegetated soils. *Agric. For. Meteorol.*

Lin, J. D. (1980). On the force-restore method for prediction of ground surface temperature. *JGR, J. Geophys. Res.* **85**:3251–3254.

Mahrt, L., H. Pan, J. Paumier, and I. Troen (1984). *A Boundary Layer Parameterization for a General Circulation Model*, AFGL-TR-84-0063. Air Force Geophysics Laboratory, Hanscom. Air Force Base, Massachusetts.

Mason, W. K., W. S. Meyer, R. C. G. Smith, and H. D. Barrs (1983). Water balance of three irrigated crops on fine-textured soils of the Riverine Plain. *Aust. J. Agric. Res.* **34**:183–191.

McCaughey, J. H. (1985). A radiation and energy balance study of mature forest and clear-cut sites. *Boundary-Layer Meteorol.* **32**:1–24.

Monteith, J. L. (1973). *Principles of Environmental Physics.* Arnold, London.

Monteith, J. L. (1976). *Vegetation and the Atmosphere*, Vol. 2. Academic Press, New York.

Monteith, J. L. (1981). Evaporation and surface temperature. *Q. J. R. Meteorol. Soc.* **107**:1–27.

Monteith, J. L., and G. Szeicz (1962). Radiative temperature in the heat balance of natural surfaces. *Q. J. R. Meteorol. Soc.* **88**:496–507.

Murphy, C. E., Jr., J. F. Shubert, and A. H. Dexter (1981). The energy and mass exchange characteristics of a Loblolly pine plantation. *J. Appl. Ecol.* **18**:271–281.

Murray, F. W. (1967). On the computation of saturation vapor pressure. *J. Appl. Meteorol.* **6**:203–204.

Nieuwenhuis, G. J. A. (1981). Applications of HCMM satellite and airplane reflection and heat maps in agro-hydrology. *Adv. Space Res.* **1**:71–86.

Nobel, P. S. (1983). *Biophysical Plant Physiology and Ecology.* Freeman, San Francisco, California.

Novak, M. D., and Black, T. A. (1985). Theoretical determination of the surface energy balance and thermal regimes of bare soils. *Boundary-Layer Meteorol.* **33**:313–333.

Ormsby, J. P., B. J. Choudhury, and M. Owe (1987). Vegetation spatial variability and its effect on vegetation indices. *Int. J. Remote Sens.* **8**:1301–1306.

O'Toole, J. C., N. C. Turner, O. P. Namuco, M. Dingkuhn, and K. A. Gomez (1984). Comparison of some crop water stress measurement methods. *Crop Sci.* **24**:1121–1128.

Pinter, P. J., Jr., K. E. Fry, G. Guinn, and J. R. Mauney (1983). Infrared thermometry: A remote sensing technique for predicting yield in water-stressed cotton. *Agric. Water Manage.* **6**:385–395.

Price, J. C. (1982). On the use of satellite data to infer surface fluxes at meteorological scales. *J. Appl. Meteorol.* **21**:1111–1122.

Price, J. C. (1984). Land surface temperature measurements from the split window channels of the NOAA-7 advanced very high resolution radiometer. *J. Geophys. Res.* **89**:7231–7237.

Prince, S. D., and Tucker, C. J. (1986). Satellite remote sensing of rangelands in Botswana. II. NOAA AVHRR and herbaceous vegetation. *Int. J. Remote Sens.* **7**:1555–1570.

Ramos, C., and A. E. Hall (1982). Relationship between leaf conductance, intercellular CO_2 partial pressure and CO_2 uptake rate in two C_3 and C_4 plant species. *Photosynthetica* **16**:343–355.

Ripley, E. A., and B. Saugier (1978). Biophysics of a natural grassland evaporation. *J. Appl. Ecol.* **15**:459–479.

Ross, J. (1981). *The Radiation Regime and Architecture of Plant Stands.* Junk Publ., Boston, Massachusetts.

Sala, O. E., W. K. Lauenroth, W. J. Parton, and M. J. Trlica (1981). Water status of soil and vegetation in a shortgrass steppe. *Oecologia* **48**:327–331.

Schulze, E.-D., and A. E. Hall (1982). Stomatal response, water loss and CO_2 assimilation rates of plants in contrasting environments. In *Physiological Plant Ecology II* (O. L. Lange, P. S. Nobel, C. B. Osmond, and H. Ziegler, Eds.). Springer-Verlag, New York, pp. 181–230.

Sellers, P. J., Y. Mintz, Y. C. Sud, and A. Dalcher (1986). A simple biosphere model (SiB) for use within general circulation models. *J. Atmos. Sci.* **43**:505–531.

Shaw, R. H., and A. R. Pereira (1982). Aerodynamic roughness of a plant canopy: A numerical experiment. *Agric. Meteorol.* **26**:51–65.

Shuttleworth, W. J., and J. S. Wallace (1985). Evaporation from sparse crop–an energy combination theory. *Q. J. R. Meteorol. Soc.* **111**:839–855.

Sinclair, T. R., C. E. Murphy, and K. R. Knoerr (1976). Development and evaluation of simplified models for simulating canopy photosynthesis and transpiration. *J. Appl. Ecol.* **13**:813–829.

Smith, R. C. G., H. D. Barrs, J. L. Steiner, and M. Stapper (1985). Relationship between wheat yield and foliage temperature: Theory and its application to infrared measurements. *Agric. For. Meteorol.* **36**:129–143.

Soer, G. J. R. (1980). Estimating regional evapotranspiration and soil moisture conditions using remotely sensed crop surface temperatures. *Remote Sens. Environ.* **9**:27–45.

Taconet, O., R. Bernard, and D. Vidal-Madjar (1986). Evapotranspiration over an agricultural region using a surface flux/temperature based on NOAA-AVHRR data. *J. Clim. Appl. Meteorol.* **25**:284–307.

Tanner, C. B. (1963). Plant temperatures. *Agron. J.* **55**:210 211.

Tanner, C. B., and T. R. Sinclair (1983). Efficient water use in crop production: Research or research? *Limitations to Efficient Water Use in Crop Production* (H. M. Taylor, W. R. Jordan, and T. R. Sinclair, Eds.). American Society of Agronomy, Madison, Wisconsin, pp. 1–28.

Taylor, H. M., and Klepper, B. (1978). The role of rooting characteristics in the supply of water to plants. *Adv. Agron.* **30**:99–128.

Tennekes, H. (1973). The logarithmic wind profile. *J. Atmos. Sci.* **30**:234–238.

Thom, A. S., and H. R. Oliver (1977). On Penman's equation for estimating regional evaporation. *Q. J. R. Meteorol. Soc.* **103**:345–357.

Trent-Serafini, Y. V. Le. (1985). Estimation of regional scale evapotranspiration using satellite data. In *Advanced Technology for Monitoring and Processing Global Environmental Data.* Remote Sensing Society, London, pp. 99–107.

Tucker, C. J. (1979). Red and photographic infrared linear combinations for monitoring vegetation. *Remote Sens. Environ.* **8**:127–150.

Turner, N. C. (1974). Stomatal behavior and water status of maize, sorghum, and tobacco under field conditions. II. At low soil water potential. *Plant Physiol.* **53**:360–365.

Turner, N. C. (1986). Adaptation to water deficits: A changing perspective. *Aust. J. Plant Physiol.* **13**:175–190.

Turner, N. C., and J. E. Begg (1973). Stomatal behavior and water status of maize, sorghum, and tobacco under field conditions. I. At high soil water potential. *Plant Physiol.* **51**:31–36.

Turner, N. C., J. E. Begg, H. M. Rawson, S. D. English, and A. B. Hearn (1978). Agronomic and physiological responses of soybean and sorghum crops to water deficit. III. Components of leaf water potential, leaf conductance, $^{14}CO_2$ photosynthesis, and adaptation to water deficit. *Aust. J. Plant Physiol.* **5**:179–194.

van den Honert, T. H. (1948). Water transport as a catenary process. *Discuss. Faraday Soc.* **3**:146–153.

Vauclin, M., S. R. Vieira, R. Bernard, and J. L. Hatfield (1982). Spatial variability of surface temperature along two transects of a bare soil. *Water Resour. Res.* **18**:1677–1686.

Verma, S. B., D. D. Baldocchi, D. E. Anderson, D. R. Matt, and R. J. Clement (1986). Eddy fluxes of CO_2, water vapor, and sensible heat over a deciduous forest. *Boundary-Layer Meteorol.* **36**:71–91.

Wieringa, J. (1986). Roughness-dependent geographical interpolation of surface wind averages. *Q. J. R. Meteorol. Soc.* **112**:867–889.

Wong, S. C. and F. X. Dunin (1987). Photosynthesis and transpiration of trees in a Eucalypt forest stand: CO_2, light and humidity responses. *Aust. J. Plant. Physiol.* **14**:619–632.

Zelitch, I. (1982). The close relationship between net photosynthesis and crop yield. *BioScience* **32**:796–802.

18

FUTURE DIRECTIONS FOR REMOTE SENSING IN TERRESTRIAL ECOLOGICAL RESEARCH

DIANE E. WICKLAND

Terrestrial Ecosystems Program
National Aeronautics and Space Administration
Washington, D.C.

I INTRODUCTION

Future research in Earth remote sensing will be directed toward understanding how the planet functions as an integrated system. In the past few decades, we have learned that there are many complex interactions among the various components of the Earth: the geosphere, hydrosphere, cryosphere, atmosphere, and biosphere. We have become aware that humans are effecting changes in certain of these components that are of truly global significance. These discoveries have lead to an increased recognition that we must develop a unified multidisciplinary approach to the study of how the Earth functions as a whole and of how it might be expected to change on time scales of relevance to humans.

The Earth is the only planet known to support life. This life is both dependent on the unique characteristics of the planet and capable of influencing those characteristics. The planet and its biota have evolved together throughout their mutual histories and will continue to do so. The abundance, distribution, and function of terrestrial biota affect the Earth's energy balance, climate, biochemistry, and cycling of chemical elements and water [National Research Council (NRC), 1986b]. Thus, knowledge of the role of terrestrial biota in global processes and of the effects of global processes on terrestrial biota is critical to understanding how the Earth functions as a system—in fact, as an ecosystem.

Assessing the role of terrestrial biota in processes of global significance will require interdisciplinary cooperation, a global perspective, and new research strategies. Remote sensing provides the global perspective and is destined to be a key tool in developing new research strategies and methodologies. In fact, the maturation of remote sensing as a research tool has made possible the comprehensive pursuit of certain global questions for the first time.

It is important to remember that passive optical remote-sensing instruments only detect surface states. In heavily vegetated terrain, only the vegetation canopy is observed, whereas in sparsely vegetated terrain, various combinations of the soil or rock substrate, the litter layer, the vegetation understory, and the vegetation canopy are observed. Thus, optical remote sensing of terrestrial ecosystems is limited to evaluation of those ecosystem properties that can be measured or inferred through knowledge of vegetation canopy characteristics. Fortunately, the vegetation canopy is a sensitive indicator of a number of important ecosystem characteristics (Waring et al., 1986). Remote-sensing observations of surface states can only be used to infer processes, or rates, through comparison of multiple observations over time or through relating the measured biophysical state to a known dynamical process.

Before considering the future of remote sensing and its associated technology, it is worth assessing the scientific context for the application of remote sensing to global ecological problems. The previous chapters in this book have presented in great detail the state-of-the-art in instrumentation and research for optical remote sensing. In the discussion that follows, the research issues for the future will be described. The well-understood applications of remote sensing for ecosystems studies will be summarized. Satellite and airborne sensors presently in use will be described; airborne sensors are particularly relevant, because today's airborne research sensors are often precursors of tomorrow's satellite sensors. Finally, plans for future satellite sensors will be reviewed.

II   PLANS FOR FUTURE RESEARCH

II.A   A Global Emphasis

II.A.1   *International Geosphere-Biosphere Program*

Planning for future research to study the Earth as a system has resulted in a number of reports outlining priority issues and detailing the necessary remote-sensing observations [National Research Council (NRC), 1982, 1985, 1986a, b; National Aeronautics and Space Administration (NASA), 1984, 1986c, 1987a, 1988; Rasool, 1987]. Foremost among these national and international plans is the program approved in 1986 by the International Council of Scientific Unions (ICSU), called the International Geosphere-Biosphere Program (IGBP): A Study of Global Change (Rasool, 1987; Malone, 1986). The goal for IGBP is

> To describe and understand the interactive physical, chemical, and biological processes that regulate the total Earth system, the unique environment for life, the changes that are occurring in this system, and the manner in which these changes are influenced by human actions.

The essential feature of the IGBP, or Global Change Program, will be its focus on interdisciplinary studies of the interactive processes that affect changes in the global environment; these studies will involve relevant elements of the earth and biological sciences, but will transcend the bounds of any one discipline. Priority will fall on those areas in each of the fields involved that show the greatest promise of elucidating interactions that might lead to significant change in the next 100 years, that most affect the biosphere, and that are most susceptible to human perturbation (NRC, 1986a).

The potential role of remote sensing in the Global Change Program has been detailed in a special report to ICSU prepared by the Committee on Space Research (COSPAR) (Rasool, 1987). Near- and longer-term actions necessary for the study of global change were recommended in this report. The near-term actions include the validation of current remotely sensed data sets, the transfer of demonstrated new technology to operations, and the identification and filling of gaps in the current observation system. The requisite longer-term actions include development of new platforms and instruments for polar orbit, deployment of a second generation of geostationary satellites, and development of new instruments and techniques to be used in ground-based validation networks and data systems. The remote observation system for the initial phase of the Global Change Program, presumed to start in the early 1990s, will consist of continued operational space measurements (from geostationary and polar orbits), special research and new operational missions (already well underway in their planning), *in-situ* validation and calibration studies, and improved data systems.

II.A.2 Earth System Science

In the United States, advance planning by several different research groups and government agencies has led to the definition of research activities with similar goals and objectives to those of the IGBP. The idea of studying the Earth as a system and attempting to understand how humans are changing its state and function has captured worldwide attention. It is clear, given the potentially disastrous, but not well understood, consequences of such environmental changes as increases in global atmospheric carbon dioxide or decreases in stratospheric ozone, that the time to begin such work is now.

A program called Earth System Science: A Program for Global Change has been recommended by a NASA advisory committee to focus future Earth science research for NASA, the National Oceanic and Atmospheric Administration (NOAA), and the National Science Foundation (NSF) (NASA, 1986c, 1988). The goal of Earth System Science is

> To obtain a scientific understanding of the entire Earth System on a global scale by describing how its component parts and their interactions have evolved, how they function, and how they may be expected to continue to evolve on all time scales.

The challenge to Earth System Science is

> To develop the capability to predict those changes that will occur in the next decade to century, both naturally and in response to human activity.

The central approach of Earth System Science is to divide the study of Earth processes by time scale, rather than by discipline. The Earth must be viewed as a dynamical system, described by a collection of variables that specify its state and the associated rules for inferring how a given state will evolve. Through this approach, Earth System Science will concentrate on describing change through global observations, understanding change through transforming observed patterns into understanding of processes, simulating change through conceptual and qualitative models of the Earth system, and predicting change on the time scale of decades to centuries through the development and verification of Earth system models (NASA, 1986c). The remote observation program for Earth System Science is essentially the same as the one defined for the IGBP and is described in more detail in the next section.

II.A.3 The Earth Observing System

The Earth Observing System (Eos) is a science mission, the stated goal of which is

> To advance the understanding of the entire Earth system on the global scale through de-

veloping a deeper understanding of the components of that system, the interactions among them, and how the Earth system is changing.

Eos is planned to be an evolutionary program with an observing capability that is built up on orbit over time and is maintained for up to 15 years to provide a long-term data base. Current plans call for the launch of two instrumented space station-derived polar-orbiting platforms; one will be launched in 1995 by NASA and the other will be launched in 1997 by the European Space Agency (ESA). A third platform is being planned for an earliest possible launch by NASA in 1997, and a fourth platform is being planned for launch in 1998 by Japan. Plans are for the NOAA operational sensors of the 1990s to be part of Eos. The platforms will be shared by research and operational facility instruments as well as by additional scientific instruments selected through NASA, ESA, and Japanese Announcements of Opportunity. In addition, certain Eos instruments may be deployed on the 28.5-degree-inclination low-altitude-orbit manned space station. The platforms and instruments will be designed to accommodate servicing in orbit. The Earth surface observing optical sensors planned for Eos are described in Table 1 (NASA, 1987a).

TABLE 1   Future Optical Satellite Sensors for Earth Surface Observations

Sensor	Number of Channels	Spectral Range	Spatial Resolution (IFOV)[o]	Swath Width
JERS-1 VNIR[a]	TBD[b]	TBD	25 m	150 km
IRS LISS-I[c]	4	0.45–0.86 μm	73 m	148 km
IRS LISS-II	4	0.45–0.86 μm	36.5 m	148 km
Eos AMRIR[d]	11	0.65–12.0 μm	0.50 $\pm$ 0.25 km	3100 km
Eos AVNIR[e]	TBD	TBD	TBD	TBD
Eos GLRS[f]	3	1.064, 0.532, and 0.355 μm	80–320 m	Profiler
Eos HIRIS[g]	192	0.4–2.5 μm	30 m	30 km
Eos HRIS[h]	$\geq 10^i$	0.45–2.70 μm	10–40 m	20–60 km
Eos ITIR[j]	11	0.85–11.70 μm	15 and 60 m[k]	30 km
Eos MERIS[l]	$\geq 10^i$	0.39–1.05 μm	$\leq$ 0.5 km	$\geq$ 1000 km
Eos MODIS-N[m]	35+	0.4–12.5 μm	0.5–1.0 km	1500 km
Eos MODIS-T[m]	64	0.4–1.1 μm	1 km	1500 km
Eos OCTS[n]	TBD	TBD	TBD	TBD

[a]Japanese Earth Remote Sensing Satellite, Visible and Near Infrared Radiometer.
[b]To be decided.
[c]Indian Remote Sensing Satellite, Linear Imaging Self-Scanner Sensor.
[d]Advanced Medium Resolution Imaging Radiometer (NOAA Operational Payload).
[e]Advanced Visible and Near Infrared Radiometer (Japan).
[f]Geodynamics Laser Ranging System (ranging and altimetric) (NASA).
[g]High-Resolution Imaging Spectrometer (NASA).
[h]High-Resolution Imaging Spectrometer (ESA).
[i]Minimum of ten selectable channels throughout the spectral range; spectral bandwidths 5–20 nm in visible and near-infrared and 10–40 nm in the shortwave-infrared region (HRIS only).
[j]Intermediate Thermal Infrared Radiometer (Japan).
[k]Spatial resolution is 15 m for the near- and shortwave-infrared and 60 m for the thermal-infrared regions.
[l]Medium-Resolution Imaging Spectrometer (ESA).
[m]Moderate-Resolution Imaging Spectrometer; N = nadir and T = tilt (NASA).
[n]Ocean Color and Temperature Sensor (Japan).
[o]Instantaneous Field of View.

II.A.4   *Mission to Planet Earth*

In order to satisfy fully the observational requirements of the Earth System Science and Global Change programs, a more comprehensive observational program than just Eos must be put in place. It is believed that a comprehensive approach to observing the full spectrum of processes that comprise the Earth system will require enlarged concentrations of orbiting remote-sensing instruments with mission lifetimes of a decade or more. Advance planning coordinated by NASA has led to the concept of a "Mission to Planet Earth" (Malone, 1986). Mission to Planet Earth requires that we build upon the understanding developed through the current operational weather and commercial land remote-sensing satellites. Approved and planned near-term Earth observing missions must be conducted as planned. Certain midterm missions that require unique orbits, such as a tropical rainfall measurement mission (TRMM), also are called for in Mission to Planet Earth. At least four platforms in sun-synchronous polar orbits are required to cover both morning and afternoon equator crossing-time requirements and to accommodate the full suite of required remote-sensing instruments. Eos will provide this key component. Mission to Planet Earth also requires at least five geostationary platforms to provide full coverage of the Earth up to 45-degree latitudes. Presumably, NASA, ESA/Eumetsat, and Japan will be interested in providing these geostationary platforms, perhaps as soon as the late 1990s. Thus, the Mission to Planet Earth concept encompasses the full breadth of Earth observations required for studies of global change.

II.B   Ecological Research Issues: Detection of State and Dynamics

Remote-sensing research conducted since the advent of satellite sensors has concentrated on the use of broad-band radiance measurements to characterize land cover classes and to determine the state and, to a lesser extent, the dynamics of above-ground vegetation. A major emphasis has been on the use of vegetation indices (various ratios of red and infrared radiance values) for estimating leaf area index (LAI) and, through their relationship to LAI, above-ground biomass and productivity. The satellite sensors used in these studies have been the Landsat Multispectral Scanner System (MSS), the Landsat Thematic Mapper (TM), the NOAA polar orbiters' Advanced Very High Resolution Radiometer (AVHRR), and, more recently, the High Resolution Visible (HRV) sensor on the French Système Probatoire d'Observacion de la Terre (SPOT) satellite. A summary of the characteristics of these instruments is provided in Table 2.

Ecologists are about to become significant users of remotely sensed data for the first time. Until recently, they have focused primarily on developing quantitative understanding of natural physiological, population, community, and ecosystem processes at relatively fine spatial scales. Remote sensing of terrestrial ecosystems, on the other hand, has been largely descriptive and qualitative; and when quantitative, it has focused on relatively coarse spatial scales. Further, satellite remote-sensing research programs in the 1960s and 1970s were devoted to understanding a restricted set of patterns and processes in highly managed and controlled (i.e., agricultural) ecosystems, rather than natural representative global ecosystems.

We are now at a point where ecological and remote-sensing priorities are converging. Remote-sensing techniques have matured and are becoming more quantitative. Interests in remote-sensing applications have shifted to the examination of important natural ecosystems. Ecologists, concurrently, have become more interested in studying ecosystem pattern and process at larger spatial scales. It seems clear that both research communities have been responding to the challenge to understand global change.

TABLE 2   Present Satellite Optical Sensors

Satellite/ Sensor	Number of Channels	Spectral Range	Spatial Resolution (IFOV)[i]	Swath Width
Landsat MSS[a]	4	0.5–1.1 μm	80 m	185 km
Landsat TM[b]	7	0.45–12.5 μm	30 m (VIS-IR)[c] 120 m (thermal IR)[d]	185 km
NOAA AVHRR[e]	5	0.58–12.5 μm	1.1 km (LAC)[f] 4 km (GAC)[g]	2700 km
SPOT HRV[h]	4	0.5–0.9 μm (multispectral)	20 m	60 km
		0.5–0.73 μm (panchromatic)	10 m	60 km

[a]Multispectral Scanner System.
[b]Thematic Mapper.
[c]Visible and Infrared.
[d]Infrared.
[e]Advanced Very High Resolution Radiometer.
[f]Local Area Coverage.
[g]Global Area Coverage.
[h]High Resolution Visible.
[i]Instantaneous Field of View.

In order to document and understand global changes and their effects on ecosystems, we must first understand the state and dynamics of relatively unperturbed terrestrial ecosystems and their natural variations over year to century time scales. It is only after natural patterns and variations can be accounted for that we will be able to detect deviations that may be indicative of directional change. There are many aspects of natural ecosystem function that are not yet understood (NRC, 1986a). Thus, a vigorous program of basic ecosystems research must be conducted in parallel with studies of global change in order to acquire the requisite baseline data set.

The subset of IGBP and Earth System Science research goals and objectives that are of particular relevance for terrestrial ecology, summarized in Table 3, can be categorized as falling into four major topic areas: landscape patterns and processes, vegetation change, biogeochemical cycles, and global energy balance (Arvidson et al., 1985). These categories are somewhat artificial, but do serve to focus our attention on the classes of research activity required. Global change provides the overriding rationale for the focus on these topic areas and is inextricably entwined in the discussion of each.

II.B.1 Landscape Pattern and Process

Landscapes have been defined as heterogeneous land areas composed of clusters of interacting ecosystems (Forman and Godron, 1986). Landscapes may be as small as a few kilometers or as large as many thousands of kilometers in area. Landscapes are composed of a mosaic of land cover units (e.g., habitats, vegetation types, water bodies, cultural features), shaped by differences in topography, microclimate, substrate characteristics, and disturbance or management history. The natural patterns of these regional mosaics and the processes that maintain them have not been investigated in any detail until recently. Landscape ecology is a relatively new subdiscipline devoted to the study of structure, function, and change in landscapes (Forman and Godron, 1986). Remote

TABLE 3   Global Change Goals and Objectives for Terrestrial Ecology

Landscape Pattern and Process

To measure the distribution and areal extent of terrestrial ecosystems.

To measure the biomass densities in the various terrestrial ecosystems.

Vegetation Change

To measure change in the distribution and abundance of terrestrial vegetation. This includes change in areal extent; change in biomass density; changes in biological diversity; changes in production, respiration, and decomposition; and changes in the biogeochemical cycles for all terrestrial biomes.

Biogeochemical Cycling Processes

To understand the biogeochemical cycling of carbon, nitrogen, phosphorus, sulfur, and trace metals.

To quantify our knowledge of production and decomposition processes both regionally and globally, to determine the factors that control them, and to understand their annual cycle and year-to-year variation.

To measure the magnitudes of the terrestrial sources and sinks for the key biogeochemicals: carbon, nitrogen, sulfur, and phosphorus. Particular attention should be given to radiatively and chemically important tropospheric trace gases [CO_2, CO, CH_4, and other hydrocarbons, N_2O, NH_3, $(CH_3)_2S$, H_2S, OCS, and SO_2] and aerosols.

To measure the transport of sediments and nutrients from the land to inland waters and the oceans.

To predict on time scales of decades to centuries changes in the Earth's biogeochemistry that can affect climate, biological productivity, and human health.

Global Energy Balance

To measure the spatial distribution and amounts of fresh-water runoff, soil moisture, precipitation, snow and ice, and evapotranspiration over the Earth.

To quantify the influences of changes in land surface evaporation, albedo, and roughness on local and regional climate.

To obtain data on the details of the biological inputs into micro–mesoscale climate processes.

To quantify the interactions between the vegetation, soil, and topographic characteristics of the land surface and the components of the hydrologic cycle.

---

sensing will prove to be a fundamental research tool for this new subdiscipline. In addition, studies in landscape ecology are likely to contribute significantly to our understanding of regional and global ecosystem dynamics (Swanson et al., 1988).

One of the basic requirements for most landscape, regional, or global ecosystem studies will be to have accurate measurements, or estimates, of the amount of land surface covered by each land cover type. The synoptic coverage provided by satellite sensors has already proven to be ideal for determination of the areal extent, distribution, and abundance of land cover types, but improvements in the extent of available coverage and its level of detail will be required, and are expected, in the future. It also will be important to develop a better understanding of the inherent limitations for classification of remotely sensed imagery. Only after areal extent and pattern are documented and properly understood can the study of process within a regional context be attempted. Routine monitoring and mapping of land cover classes and vegetation activity also can be expected to provide a key means of integrating ecosystem information across spatial scales.

Land Cover and Vegetation Type Classification. Much progress has been made in analyzing land use and in classifying land cover and vegetation types using Landsat, SPOT, and AVHRR data. The level of detail that can be identified and the accuracy of such analyses depend on the nature of the region to be analyzed as well as the spectral, spatial, temporal, and radiometric resolution of the sensor. Single and multidate Landsat data have been used routinely to classify vegetation community types with approximately 70–90% accuracies (Botkin et al., 1984; Ustin et al., 1986; Franklin et al., 1986; Butera, 1983; Morrissey et al., 1986). Single and multidate AVHRR data have been used to produce continental-scale classifications of major land cover types (Tucker et al., 1985a; Clark et al., 1986; Malingreau and Tucker, 1987). Thus, satellite data at a variety of scales of resolution can provide consistent and timely information on global vegetation patterns and processes.

There is considerable evidence that canopy architecture determines, to a large extent, the angular distribution of radiances above the canopy (Simmer and Gerstl, 1985; Goel and Deering, 1985; Kimes, 1983; Kirchner et al., 1982). Thus, angular radiance data should contain information about canopy structure. Such information would, at the very least, improve vegetation classification accuracies. The "hot spot", defined as the narrow intensity peak of reflected light observed in the reverse solar direction, has been shown to carry information about plant stand architecture that can be used for crop identification (Gerstl, 1986). Structural information is likely to prove useful in the estimation of biomass, canopy surface roughness, and canopy condition, as well as for vegetation identification (Gerstl, 1986; Kirchner et al., 1982).

Landscape Pattern to Infer Process. In general, vegetation patterns are not uniquely related to particular ecosystem processes; in fact, different processes may yield similar vegetation patterns. However, in certain situations, pattern and process are sufficiently linked that the former can be used as an indicator of the latter. Once detailed ground and remote-sensing studies have clearly established that such a relationship exists, it should be possible to use changes in pattern to infer changes in processes, perhaps even quantitatively.

Nested multistage sampling of landscape patterns using multiple optical sensors, which acquire data at differing spatial scales, will provide a powerful means for developing a global understanding of landscape patterns and processes. The spatial pattern of vegetation units will be used to integrate information among differing spatial scales. For example, in order to transfer detailed site-specific information to a regional and global context, and vice versa, it will be necessary to develop a hierarchy of nested models addressing relevant processes at all spatial scales. Spatial pattern then may provide the key to transferring the results obtained from a model at one scale to the parameter set of a model operating at the next scale up or down. In the simplest case, spatial pattern can be used to aggregate or sum the data for one specific area over all occurrences of that type of area (Bartlett et al., 1986; Woodwell et al., 1986).

II.B.2 *Vegetation Change*

Understanding global vegetation change will require both an understanding of the effects of changes in external environmental conditions on vegetation and an understanding of the effects of changes in vegetation patterns and processes on the environment, including the oceans and atmosphere. The effects on vegetation of such global perturbations as climatic warming associated with increased concentrations of greenhouse gases in the atmosphere must be understood. The effects on vegetation of such regional phenomena as air pollution, acid precipitation, deforestation, and large-scale changes in land use have potentially global significance as well, and must be understood.

Research to understand vegetation change seems to consist of two components. The first is related to assessing the patterns of change over time and space and the second is related to developing a detailed understanding of the symptomology and physiology of change in vegetation.

Change Detection. Retrospective analysis of the current record of Landsat MSS data since 1972, Landsat TM data since 1982, and AVHRR data since 1981 can provide a synoptic assessment of recent change in vegetation patterns for selected regions of the Earth's surface. In the future, analysis of calibrated multidate remotely sensed data sets will be a powerful technique for the study of vegetation change. Change-detection analyses will identify large-scale patterns of change, and it will be possible to estimate the rates and magnitude of changes.

Large-scale studies of change in landscapes are beginning to be conducted using the historical remote-sensing record (Hall et al., 1987; Tucker et al, 1985b, 1986b; Malingreau and Tucker, 1987; Sader and Joyce, 1988). Change-detection studies are documenting natural and anthropogenic patterns of change throughout the world. Because this type of investigation will be so important for global change research, a few examples are cited in the next sections. Continued progress in analyzing multidate satellite imagery will depend upon the existence of calibrated data and readily available easy-to-use image-processing techniques for image coregistration, change detection, and pattern analysis.

LANDSCAPE DYNAMICS AND SECONDARY SUCCESSION. Secondary succession patterns and processes in the boreal forest of northern Minnesota have been studied through analysis of two Landsat MSS scenes collected in 1973 and 1983 (Hall et al., 1987). A stochastic description of the key life-cycle states of community landscape components was generated. A transition matrix describing the probability of change from one successional state to another was calculated based on classifications of the two Landsat scenes. The ten-year observations of this boreal forest region indicated that there were considerable successional changes at the landscape component level in an ecosystem that has been relatively stable over several centuries. Managed areas within the region were much more dynamic and heterogeneous than the wilderness areas. Stochastic descriptions such as this can provide input and verification for models of community development, landscape dynamics, and ecosystem stability (Hall et al., 1987).

PRIMARY PRODUCTION AND DROUGHT. Changes in primary production related to regional drought have been monitored using the AVHRR for the Sahelian region of Africa (Tucker et al., 1985b, 1986b). Tucker et al. (1985b) reported strong correlations between the integrated AVHRR normalized difference vegetation index (NDVI) data and end-of-season above-ground dry biomass as sampled on the ground in the Senegalese Sahel. An enormous decline in production, ranging from 1093 kg/ha in 1980 to 55 kg/ha in 1985, was observed. Vegetation production, estimated by the integrated NDVI, across the entire Sahel was higher in 1985 (Tucker et al., 1986b). These studies demonstrate the utility of the coarse-resolution AVHRR for monitoring rangeland productivity patterns and providing synoptic information on the progression of regional drought. Interannual comparisons of productivity for the region have been used to identify areas of greatest change and where more detailed satellite or ground observations must be made.

AMAZONIAN DEFORESTATION. NDVI and thermal data of 1.1 km resolution from the AVHRR have been used to document large-scale deforestation in the southern part of the Amazon Basin. Data from 1982, 1984, and 1985 were used to generate estimates of fire occurrence and frequency, which then were used to identify general disturbance areas. NDVI data also were used to identify areas more likely to be subject to human disturbances. Brightness temperatures in the 3.55–3.93 μm region were found to be significantly higher for cleared areas than for forested areas, and supervised classifications based on these data were used to estimate the total cleared area in each general disturbance area. Total disturbance area and total deforested area for the combined states of Acre, Rondonia, and Mato Grosso were estimated at 165,742 km^2 and 89,573 km^2, respectively. These data confirm predictions of near-exponential increases in the rate of deforestation for these areas (Malingreau and Tucker, 1987).

Stress and Condition Assessment. It has long been recognized that remote-sensing observations can provide information for assessing vegetation growth and condition, or relative vigor (Bauer, 1985). Vegetation indices (e.g., the NDVI) are indicators of relative growth and vigor of green vegetation. Various stress agents cause visible and shortwave-infrared reflectances to increase and near-infrared reflectance to decrease. Spectral indices or variables estimated from spectral data (e.g., LAI), in combination with meteorological and soils data, have been used to assess crop growth and yield (Bauer, 1985). Such spectral indices also have been developed for natural vegetation and, therefore, can be used in the study of vegetation change. For example, recent research into the effects of acid deposition has produced regional estimates of the extent and severity of damage in spruce-fir forests of the northeastern United States using an index derived from shortwave- and near-infrared radiances (Vogelmann and Rock, 1986).

Reflectance of vegetation in the shortwave-infrared region of the spectrum (1.3–2.6 μm) is dominated by strong water absorptions. Tucker (1980) demonstrated that for this spectral region, the absorption characteristics of a leaf can be simulated by the absorption of an equivalent water thickness. Leaf reflectance is inversely related to the total leaf water content in the shortwave infrared (Gates et al., 1965; Bauer, 1985). This relationship has been exploited in some comparative studies using broad-band radiance values in a qualitative fashion; but, because comparisons of absolute reflectance over time must be made, quantification of leaf or canopy water content has not been possible. Better calibration of sensors could improve this situation in the future.

It should be noted that a number of field studies with crop species have failed to find a useful predictive relationship between leaf shortwave-infrared reflectance and leaf water status (Thomas et al., 1971; Kimes et al., 1981; Goward, 1985). It has been suggested that the relationship between shortwave-infrared reflectance and canopy water content may saturate at relatively low canopy water content levels and that internal leaf structure also plays a strong role in determining shortwave-infrared reflectance (Thomas et al., 1971; Hoffer and Johannsen, 1969; Goward, 1985). Thus, it seems that our understanding of the water content information contained in shortwave-infrared reflectance data is not yet complete.

Ratios of near-infrared and shortwave-infrared reflectance have been used to generate moisture-stress indices, which provide an indication of leaf and canopy moisture status when compared with leaf water potential measurements (Rock et al., 1985). A leaf water content index using Landsat TM bands 4 (0.76–0.90 μm) and 5 (1.55–1.75 μm) has been proposed for estimating the average relative water content of plant canopies under certain known conditions (Hunt et al., 1987). Just how useful these indices will be when applied to a variety of vegetation types or ecological conditions is not yet known.

Remotely sensed thermal-infrared data have been used to identify stressed vegetation; stress frequently results in closure of the stomata and, hence, higher leaf and canopy temperatures (Sader, 1986). It has been noted that radiant temperatures indicate the degree of stress at a particular time, whereas reflectance measurements indicate the effects of stress over time (Bauer, 1985).

The exact location of the sharp rise in vegetation reflectance between 0.68 and 0.72 μm, the red edge, changes with vegetation condition. Its location shifts about 7–10 nm to shorter wavelengths, in the so-called blue shift of the red edge, in heavy metal stressed vegetation (Horler et al., 1980; Collins et al., 1983; Goetz et al., 1983). More recent research suggests that a blue shift of the red edge results from a wide variety of stress factors (Rock et al., 1986; Westman and Price, 1987). The issue of the specificity of the blue shift is very much in question. Nonetheless, the blue shift of the red edge seems likely to remain a useful generic indicator of stress in vegetation. Collins (1978) also reported a shift to longer wavelengths that correlated with the time of fruiting in wheat.

II.B.3 *Biogeochemical Cycles*

In addition to water and oxygen, four elements—carbon, nitrogen, phosphorus, and sulfur—are of particular importance in the study of the Earth (NRC, 1985). The biogeochemical cycles represent the fluxes of material and energy among the various components of the Earth system. The reservoirs for these elements and the rates of transfer among them are determined to a large extent by the nature and abundance of the Earth's biota. Thus, the biogeochemical cycles of carbon, nitrogen, phosphorus, and sulfur are inextricably linked; a perturbation in the cycle of one element leads to changes in the function of the other cycles (NRC, 1986a; Moore and Bolin, 1986). Important goals for IGBP and Earth System Science are to identify and quantify the biotic reservoirs of carbon, nitrogen, phosphorus, and sulfur, to quantify the fluxes among them, and to quantify their exchanges with the oceanic and atmospheric reservoirs.

Human activities have begun to have major effects on the biogeochemical reservoirs and cycles. The well-documented, steady increase in atmospheric carbon dioxide concentration attributable to fossil-fuel combustion is one of the best known examples. Already, present atmospheric carbon dioxide concentrations are well outside their recent historic distribution (NRC, 1985). Tropospheric methane concentrations are increasing annually by 1% (NRC, 1985; Mooney et al., 1987). The rate of nitrogen fixation by humans now approaches that of algae and bacteria. Human production of sulfur gases is comparable to that of volcanoes, and humans are the primary eroders of phosphate rock (Slobodkin, 1984). We have no evidence as to how natural or managed ecosystems will respond to these simultaneous changes. We do know that the effects can range from enhanced productivity or competitive ability to substantial mortality (Mooney et al.,

1987). Future studies must focus on developing an understanding of ecosystem response to change.

Understanding the sources, sinks, and dynamics of biotic trace gases is of particular importance because of the effects they can have on chemical and physical processes in the atmosphere. Biogenic trace gases are capable of altering the Earth's energy budget, the concentrations of oxidants in the troposphere, and the absorption of ultraviolet radiation in the stratosphere (Mooney et al., 1987). They are equally important for what they can indicate concerning ecosystem function. The production and consumption of trace gases in terrestrial ecosystems is indicative of the occurrence and magnitude of certain physiological processes and ecosystem fluxes. Thus, future studies to identify the biotic sources and sinks of radiatively and chemically important trace gases also should reveal new understanding of fundamental ecological processes.

Remotely sensed data on land cover pattern can be used to identify and estimate the areal extent of the sources and sinks for various biogeochemical compounds. Remote sensing can also provide data on the existence and duration of environmental conditions required for the occurrence of certain biogeochemical processes. There also is evidence that high spectral resolution reflectance data can be used to estimate plant canopy biochemical concentrations (Peterson et al., 1985, 1987).

Many surface materials, particularly rock-forming minerals, have characteristic absorption features that are 20–40 nm or more wide. These absorption features are related to electronic transitions and vibrational processes (Goetz et al., 1985b). So far, the only absorption features of terrestrial vegetation that have been consistently noted and understood are relatively broad features for pigments in the blue and red regions and for water in the shortwave infrared. Algae, due to their varied pigments, have a number of distinctive absorption features in the visible wavelength region (Perry, 1986). Lichens also have unique reflectance and absorption features in the visible wavelength region; absorptions in the blue region appear to be due to the presence of unique pigments in the algal component of the lichen (Petzold and Rencz, 1975; Petzold and Goward, 1988). For higher plants, small absorption features observed in the near-infrared region may be related to differences in leaf structure or hydration state, but no consistent understandable relationship has yet been described (Goetz et al., 1985a, b).

The lack of unique narrow absorption features in vegetation spectra is very likely because plants are highly structured mixtures of many organic compounds. Various of the constituent organic compounds do have characteristic absorption maxima corresponding to electronic transitions or fundamental stretching of organic bonds. Harmonics and overtones of these fundamental stretching frequencies are found throughout the infrared region of the spectrum. In a whole plant, these subtle absorption features are superimposed on each other and on the strong first-order absorption effects due to water and pigments (Peterson et al., 1985; Marten et al., 1985). Statistical analysis methods can be used to develop relationships between high spectral resolution data and canopy biochemical concentrations (Marten et al., 1985; Norris et al., 1976). Laboratory studies and preliminary aircraft-based studies have demonstrated strong correlations between canopy lignin and nitrogen concentrations and high spectral resolution leaf or canopy reflectance data (Norris et al., 1976; Peterson et al., 1985; Wessman et al., 1987; also see the discussion of AVIRIS and Chapter 10.)

Carbon Cycle. Photosynthetic fixation of carbon is the fundamental process that captures energy to support life on Earth. The processes of photosynthesis, respiration, and decomposition are key to understanding the fluxes of elements and energy through the

Earth system. Yet, the carbon cycle is not well understood; we have not been able even to balance the annual carbon budget (NRC, 1985; Detwiler and Hall, 1988). The role of terrestrial ecosystems is uncertain in at least two areas related to carbon storage (NRC, 1985). First, the net effect of changes in land use on atmospheric carbon dioxide concentrations is not known. Second, the response of ecosystems, in terms of productivity, decomposition, and carbon storage, to changes in external factors is not understood. Understanding carbon storage is important because changes in ecosystem net carbon storage patterns control the feedback from carbon dioxide absorption by plants to the atmosphere (Mooney et al., 1987). In addition, the sources, rates of flux, and controls for emissions into the atmosphere of carbon monoxide, methane, higher hydrocarbons, isoprene, and terpenes must be better understood (NRC, 1985).

PRODUCTION. Applied remote-sensing research in the 1960s and 1970s accumulated a vast body of information on the utility of broad-band optical data for the assessment of vegetation condition and the prediction of above-ground biomass production and agricultural yields (Tucker et al., 1981; Curran, 1983; Asrar et al., 1984; Botkin et al., 1984; Hogg, 1986). A strong positive nonlinear relation exists between LAI, green biomass, and several vegetation indices. For most vegetation types, this relationship is nearly linear for single-sided LAI values of up to 5 or 7 (Bauer, 1985). However, results of studies in western coniferous forests suggest that the relationship may remain linear for LAI values of up to 10 or 12 for some ecosystems (Running et al., 1986).

Vegetation indices have been widely used in ecological and agronomic studies as surrogates for LAI despite the fact that it has been well established that other leaf and surface properties influence the vegetation index. Leaf-angle distribution, leaf optical properties, canopy structure, and, frequently, background reflectances also contribute to the overall radiances that are used to calculate a vegetation index (Curran, 1983; Richardson and Wiegand, 1977). Thus, a certain amount of judgment must be exercised when using a vegetation index to estimate vegetation properties in the fashion previously described. Vegetation indices will continue to be used to estimate local and global LAI and productivity patterns; however, in the future, it would be extremely desirable to have estimates of accuracy associated with them.

Recent research has shown that the integrated vegetation index can be related directly to primary productivity (Tucker et al., 1985b, 1986a; Sellers, 1985; Goward et al., 1985b). The vegetation index is near-linearly related to the percentage of photosynthetically active radiation that a canopy may absorb and, thus, is an index of canopy photosynthetic capacity (Sellers, 1985). The time integral over a growing season of the vegetation index is an estimate of net primary productivity (Goward et al., 1985b).

It will soon be possible to use the historical AVHRR record to make comparisons of annual global and regional productivity patterns from year to year. This will enable us to quantify natural interannual variability. Directional trends of change in total production or the rate of production should be good, sensitive indicators of environmental variations. Such studies may provide the first insights into how regional and global vegetation productivity patterns are changing with changes in the global climate.

Research relating the AVHRR NDVI to absorbed photosynthetically active radiation and canopy resistance has laid the foundatation for the use of the NDVI as input to mechanistic models of ecosystem productivity (Sellers, 1985; Asrar et al., 1984). Satellite-derived estimates of biomass and LAI have already been successfully used as inputs to such models (Running et al., 1986; Peterson et al., 1985).

Measurement of biomass per unit area has been set as one of the goals for an IGBP

(NRC, 1986a). Biomass density can be used in the estimation of the size of the reservoirs of biogeochemical compounds in organisms. Vegetation indices will probably continue to be used to estimate above-ground biomass, but it is hoped that greater effort will be taken to understand and quantify the errors associated with the approach. In the future, estimation of biomass density should be greatly improved through the combination of remotely sensed canopy chemical and structural information. For example, vegetation index data combined with estimates of canopy height from lidar, radar, or bidirectional measures of canopy structure may yield improved estimates of biomass density.

Lignin concentration is an indicator of the carbon quality of leaf tissues. Tissues with higher lignin concentrations are usually less palatable to predators and are more difficult for microorganisms to decompose. The higher the lignin concentration in leaf litter, the more nitrogen is immobilized per unit carbon respired. Thus, lignin concentrations can be used to infer rates of nitrogen turnover in litter and soils (Melillo et al., 1982). Relative lignin concentration in the canopy also can be used as an indicator of site fertility (Melillo et al., 1982; Waring et al., 1986). Therefore, the preliminary research indicating that canopy lignin concentrations can be estimated using remotely sensed high spectral resolution reflectance data has important implications for understanding canopy carbon quality and site fertility (Wessman et al., 1987).

High spectral resolution data have been successfully used to estimate ocean chlorophyll concentrations and productivity. This has been accomplished by measurement of absorption in narrow-wavelength regions corresponding to algal pigment absorption wavelengths (Perry, 1986). This technique should be explored for estimating chlorphyll concentrations in fresh-water aquatic ecosystems as well as in terrestrial vegetation now that appropriate narrow-wavelength data can be acquired over land (see discussion of AVIRIS). Chlorophyll concentrations are correlated with photosynthetic rate and gross productivity. Estimates of chlorophyll concentrations in terrestrial vegetation may prove to be useful for improving our interpretation of vegetation index data, and should yield additional insight into vegetation condition.

RESPIRATION AND DECOMPOSITION. It seems likely that decomposition, which occurs primarily in the soil, and respiration will never be measured directly by means of remote sensing. However, we may be able to estimate some of the controls on these processes through remote sensing. Respiration and decomposition are both controlled by temperature. Knowledge of surface and soil temperatures may prove useful for making general estimates of the rates of these processes in certain climatic regions.

It might be possible to infer indirectly something about respiration through remotely sensed estimates of biomass, measurements of surface temperature over the seasons, estimates of soil moisture over the seasons, and knowledge of the ecosystem functional type. Such inferences never have been attempted. Similarly, decomposition rates also might be indirectly inferred using such information along with information on litter and soil biochemistry.

However, if canopy lignin and nitrogen concentrations can be estimated using high spectral resolution reflectance data acquired at approximately the time of leaf fall, then decomposition rates could be fairly directly estimated. The ratio of lignin to nitrogen in leaf litter determines the rate at which it decomposes and the rate at which nitrogen is released from organic compounds in a given climatic region (Melillo et al., 1982; Waring et al., 1986). In particular, nitrogen mineralization, a key parameter controlling ecosystem properties in temperate regions, could be estimated (Wessman et al., 1987).

OTHER EXCHANGES WITH THE ATMOSPHERE AND OCEANS. Remote sensing can be used to estimate the size of various biotic reservoirs of key biogeochemicals. Knowledge of ecosystem areal extent and biomass content can be combined with information about average nutrient concentrations to derive the size of the reservoir.

Methane is produced in the anaerobic soils of wetlands. Remote estimates of wetlands extent and the duration of flooding in them have been quite useful for extrapolating point measurements of methane emissions for a given climatic regime to regional estimates of methane production (Bartlett et al., 1986).

Monitoring of coastal wetlands and estuaries may provide important information on the fluxes of carbon and nutrients to the oceans. It may be possible to estimate carbon export from wetlands to the ocean through comparison of multitemporal estimates of biomass density for the wetlands or through monitoring of water turbidity and color over time (Butera et al., 1984). It also may be possible to use ocean color measurements to determine whether the nutrients associated with these fluxes contribute to marine primary productivity (see Chapter 11).

Fire is a means by which biochemical compounds are released from ecosystems to the atmosphere. Substantial amounts of carbon dioxide and other carbon compounds (e.g., methane and carbon monoxide) are released by fires annually (Mooney et al., 1987). Remotely sensed thermal-infrared data have been used for detecting fires. The AVHRR, due to its twice-a-day repeat coverage, can provide thermal data useful for estimating the extent of large-scale biomass combustion. AVHRR thermal data have been used to estimate fire frequency and to identify areas of clearing in the tropics (Malingreau and Tucker, 1987). The high temperature contrast enables easy detection, even of subpixel-size fires, during day and night, and through dense smoke (Sader, 1986). It may become possible to use remotely sensed information about the extent and quality of fires to develop regional estimates of trace gas inputs to the atmosphere attributable to fires.

Nitrogen, Phosphorus, and Sulfur Cycles. The role of remote sensing in understanding the nitrogen, phosphorus, and sulfur cycles could be limited to identifying ecosystems where certain processes or rates of processes may be occurring. Evolution of sulfur gases, as for methane, occurs to a large extent in wetlands, where reducing conditions prevail in the soils. Remote estimates of wetlands extent and moisture regime should be very helpful for identifying potential source areas, for determining their areal extent, and for estimating the duration of suitable conditions for trace-gas production.

NITROGEN CYCLE. The biological processes of nitrogen fixation and denitrification are what close the atmospheric nitrogen cycle. We do not understand what controls the balance between biological and lightning fixation and denitrification (NRC, 1985, 1986a). Mineral nitrogen (NH_4^+, NO_2^-, and NO_3^-) is a major plant nutrient, and its availability is known to limit productivity in many global ecosystems, particularly in temperate regions (Vitousek, 1982). Decomposition processes and fires release a number of volatile nitrogen-containing compounds to the atmosphere (e.g., NH_3, NO_2, NO, N_2O, and N_2). Of particular importance, is N_2O, which is a greenhouse gas and comparatively inert. It ultimately reaches the stratosphere, where it decomposes into N_2 and NO. NO, of course, is known to play a role in the photochemical reactions associated with the destruction of stratospheric ozone (NRC, 1985).

Humans are presently modifying the nitrogen cycle in a major way. Cultivation of legumes and the production of artificial fertilizers are now responsible for over half of the annual global nitrogen fixation. Further, high-temperature combustion is responsible for 10–20% of the total flux of fixed nitrogen to the atmosphere. These nitrogen oxides are responsible for regional changes in atmospheric chemistry, biological productivity, and the acidity of precipitation (NRC, 1985).

As described previously, high spectral resolution data may provide information on canopy chlorophyll, lignin, and nitrogen concentrations. Classification of ecosystems according to their relative canopy chlorophyll or nitrogen content would be extremely useful. Chlorophyll concentrations are related to gross productivity and vegetation vigor. Nitrogen is frequently the limiting nutrient in terrestrial ecosystems, and, therefore, knowledge of canopy nitrogen concentration would permit assessment of ecosystem productivity and other processes that are closely related to nitrogen concentration. The ratio of lignin to nitrogen could also be used to infer nitrogen mineralization rates in certain ecosystems, and, perhaps, even potential rates of nitrogen-bearing trace gas emissions.

PHOSPHORUS CYCLE. Phosphorus is another major plant nutrient. It is relatively abundant, but exists principally as insoluble minerals, unavailable for uptake by plants. Phosphorus is often the limiting nutrient in lake and tropical ecosystems. The major exchange processes for phosphorus are associated with dissolved and particulate transport in rivers, weathering processes, and diagenesis in soils and sediments. Thus, there are important linkages between the hydrologic cycle and the phosphorus cycle (NRC, 1985). Human mining of phosphate for fertilizer has resulted in a major increase in the mobility of phosphorus. Much of this phosphorus is transferred from the fertilized land to the rivers and oceans, where it may be lost to marine sediments. Again, it has been suggested that a wide variety of organic compounds, including those containing phosphorus, may be identified using high spectral resolution reflectance data (Peterson et al., 1985). Very little research has been conducted on this topic with respect to phosphorus.

Phosphorus is often the limiting nutrient in fresh-water ecosystems. Eutrophication of fresh-water lakes associated with increased inputs of phosphorus has been well documented (Odum, 1971). Remote sensing of algal productivity and phenological patterns in fresh-water lakes could be used as a means of monitoring for changes in phosphorus status.

Monitoring of erosion, riverine transport of materials, and coastal wetlands and estuaries could provide some insight into rates of transfer of phosphorus-bearing materials to the ocean, but the problems to solve would be many. Detailed ground measurements are likely to remain the best source of information on phosphorus cycling. Remotely sensed pattern may be useful for extrapolating point measurements to larger scales.

SULFUR CYCLE. Sulfur is a major plant nutrient, but rarely limits productivity in natural ecosystems. Sulfur exists in a variety of oxidation states and is cycled among them by the biota, volcanoes, combustion, and atmospheric reactions. Sulfur enters the atmosphere as a result of biological processes in wetlands and the ocean surface waters, through combustion of fossil fuels, and through volcanic eruptions. The contributions from the natural versus the anthropogenic sources are about equal. The anthropogenically induced increases in atmospheric concentrations of sulfur have been manifest in large-scale increases in acid deposition in many regions. It will be important to learn more about how the biology and geochemistry of terrestrial ecosystems are affected by sulfuric acid deposition (NRC, 1985). Remote sensing of acid-deposition-damaged ecosystems can play a role here. Since biological production of gaseous sulfides (i.e., H_2S, OCS, $(CH_3)_2S$, and CS_2) occurs in wetlands, remote sensing of the areal extent and duration of flooding for wetlands should be useful in estimating the amounts and rates of production.

II.B.4 Global Energy Balance

The energy balance of the Earth is dependent on a number of surface properties and processes. The physical and chemical properties of terrestrial vegetation determine the albedo of large portions of the Earth's surface. As discussed in the previous section, indirect, but substantial, effects on the energy budget result from biotic inputs of trace gases into the atmosphere. The role of terrestrial ecosystems is important in the energy-exchange processes of the global hydrologic cycle, primarily through evapotranspiration and other controls on water routing.

Land Surface Climatology. The physical interactions of vegetation with climate are of importance in the global energy balance. Vegetation alters the amount of intercepted radiation and alters boundary-layer characteristics (Mooney et al., 1987). Changes in vegetation type or extent in a region can alter the albedo of the region. Such changes have the potential to lead to alterations in regional weather patterns and long-term climate (NRC, 1986b). Changes in the relative proportions of land cover types in a region, or changes in the pattern of differing land cover types in a region may also affect weather and climate patterns (Forman and Godron, 1986). Thus, remotely sensed information on the pattern of landscapes may be necessary to infer likely changes in regional climate-related processes. In addition, if climate changes, its effects on vegetation patterns are likely to cause changes that feed back into regional climate patterns. Initial responses to change may not be indicative of the long-term ultimate response; several cycles of change and feedback may be necessary before any sort of equilibrium is reached.

In order to estimate properly the Earth's radiation budget, the total hemispherical albedo of all land cover components and vegetation types must be known. Most surface materials are anisotropic reflectors (Gerstl and Simmer, 1986). Therefore, nadir-viewing sensors cannot measure the complete bidirectional reflectance distribution function (BRDF). The relationship between the directional reflectance they do measure and the total hemispherical reflectance is only beginning to be addressed (Kimes and Sellers, 1985; Brest and Goward, 1987). Satellite-borne multidirectional sensors will be necessary to obtain improved estimates of global albedos.

The International Satellite Land Surface Climatology Project (ISLSCP) is an internationally sponsored research program for study of the complex and closely coupled land surface–atmosphere interactions. Its major objective is to develop methodologies for deriving quantitative information concerning land surface climatological variables from satellite observations of radiation reflected and emitted by the Earth. Such quantitative information is required to monitor global-scale change of the land surface caused by variations in climate or by human activity, to develop further mathematical models designed to predict or simulate climate at various time scales, and to permit inclusion of land surface climatological variables in diagnostic and empirical studies of climate variations. The land surface variables of primary interest are vegetation cover, albedo, surface temperature, and radiation budget (Rasool, 1987).

Present ISLSCP research is divided into two major activities. The first, the ISLSCP Retrospective Analyses Project (IRAP), is focused on the evaluation of satellite data collected since 1972 to determine their usefulness for detecting climate-related fluctuations or human-induced changes in the surface of the Earth. The second activity is devoted to the development and validation of methods to convert satellite-observed radiances to climatological variables through major field programs. The First ISLSCP Field Experiment (FIFE), begun in 1987, has collected an extensive set of ground, aircraft, and satellite measurements of land surface properties over a 15 km^2 grassland site in Kansas. FIFE was designed to obtain the relevant satellite- and ground-based data for the validation of algorithms relating land surface parameters to satellite radiances (Rasool, 1987; Sellers et al., 1988). FIFE should significantly advance our knowledge of the role of remote sensing in estimating land surface and terrestrial ecosystems properties. Additional large-scale experiments will be planned in the future in order to extend the results of FIFE to larger scales and different types of ecosystems.

Hydrologic Cycle. The global water cycle is, perhaps, the most fundamental of the biogeochemical cycles. It influences all other biogeochemical cycles. The hydrologic cycle shapes weather and climate through its direct effects on atmospheric chemistry and global circulation. Much of the heat that drives atmospheric circulation is derived from phase changes of water (NRC, 1985). One of these important phase changes is accomplished through evapotranspiration. Evapotranspiration is a process that is as important to understanding ecosystem function as it is to understanding regional energy balance and water cycling. Of course, it is intimately linked to photosynthesis because both processes are mediated by the leaf stomata.

EVAPOTRANSPIRATION. The latent heat flux associated with evaporation is the dominant energy-exchange process in vegetated landscapes. Therefore, large-scale changes in evapotranspiration could have major climatic consequences. For example, there is concern that widespread clearing in the Amazon will change the region's water and energy balance. A significant fraction, perhaps as much as 50% of the rainfall in the Amazon is derived from evapotranspiration from the forest itself (Salati, 1987). The effects of Amazonian deforestation on regional weather patterns and atmospheric circulation should be monitored closely, as should evapotranspiration of all vegetation, including primary and secondary forests, in the region. This large-scale human-induced manipulation of a natural ecosystem will provide considerable insight into the feedbacks of vegetation and hydrological processes to climate.

A limited number of research studies have attempted to use remotely sensed thermal-infrared data to estimate regional evapotranspiration. Presently, it is not possible to estimate all of the parameters necessary to calculate evapotranspiration using current remote-sensing techniques. Modeling strategies, combined with remotely sensed data and ground measurements, represent the best approaches to date (Jackson, 1985; Hope et al., 1986). In modeling studies of regional evapotranspiration, thermal-infrared observations of surface temperature are used to calculate the sensible heat flux and then to determine the latent heat flux from the energy-balance equation (Jackson et al., 1983; Jackson, 1985; Gurney and Camillo, 1984; Goward et al., 1985a; Hope et al., 1986). At least two measurements of surface temperature under clear-sky conditions are required during a diurnal cycle for most of these approaches (Jackson, 1985).

The 1978 Heat Capacity Mapping Mission (HCMM) provided the first opportunity to measure surface temperatures from space twice a day (Price, 1980). Observations from the HCMM thermal scanner have been examined in conjunction with a Landsat-derived vegetation index for their theoretical suitability in parameterizing surface moisture availability for evapotranspiration. Results indicate the approach may be useful, but additional research and verification is needed (Hope et al., 1986).

Data from the TIROS Operational Vertical Sounder (TOVS) system on the NOAA polar orbiting satellites have been used to obtain surface air temperature estimates (Davis and Tarpley, 1983). Other aspects of surface climate necessary for the estimation of evapotranspiration can be estimated using meteorological satellite data (Yates et al., 1983; Running et al., 1987). Thermal data from airborne sensors also are being applied to studies of evapotranspiration (Sader, 1986; Luvall and Holbo, 1987; also see discussion of TIMS).

It seems that remotely sensed measurements of surface temperature, moisture conditions, and vegetation cover could be used to estimate evapotranspiration. The vegetation index can be used as an indicator of the potential for evapotranspiration as well as for carbon fixation (Sellers, 1985; Hope et al., 1986). Models that incorporate measurements of surface temperature and the vegetation index to estimate evapotranspiration should be developed. A great deal of research employing present capabilities can be conducted to improve our abilities to estimate evapotranspiration.

WATER ROUTING AND WATERSHED DYNAMICS. Vegetation plays a major role in controlling the routing of water at the surface. Vegetation influences the amount of precipitation that reaches the soil surface, it influences the amount of water that enters the soil profile, and it influences the amount of water available for evaporation (Mooney et al., 1987). Remotely sensed estimates of vegetation density, height, and moisture status may provide useful information for predicting certain components of water routing in terrestrial ecosystems.

Remotely sensed information on surface temperatures, meteorological conditions, topography, and land cover type could be used to model patterns and processes within drainage basins and to estimate the hydrologic balance for large watersheds. Such studies would require considerable ground information to complement the remotely sensed data.

III PLANS FOR FUTURE TECHNOLOGY: EXPECTED ADVANCES FOR ECOLOGICAL RESEARCH

III.A Present Airborne Sensors: Testbeds of the Future

Airborne sensors will continue to be important in remote-sensing research. Aircraft offer flexibility and timeliness in data acquisition and permit development and testing of instruments, algorithms, and scientific understanding in advance of the launch of a new satellite instrument. Aircraft are used to conduct unique remote-sensing experiments, to underfly satellite sensors in nested sampling strategies, and to evaluate experimental prototype satellite instruments. Current aircraft capabilities are often precursors to future satellite capabilities. Research conducted using new aircraft sensors frequently represents the state-of-the-art in remote sensing. Data from a number of operational and experimental airborne sensors are currently being acquired and analyzed. Many of these sensors are operated by NASA; they are listed in Table 4 and described briefly in the following sections.

Airborne multispectral scanners and thematic mapper simulators (TMS) have proven extremely useful for a wide variety of research applications. While originally developed

TABLE 4   NASA Operational and Experimental Airborne Imaging Optical Sensors

Sensor	Aircraft Platform	Number of Channels	Spectral Range	Spectral Sampling Interval	Spatial Resolution (IFOV)[h]	Swath Width
AOCI[a]	U-2/ER-2	10	0.44–12.28 μm	20 and 60 nm; and 3.86 μm[b]	50 m	33 km
ASAS[c]	Skyvan/ C-130	30	0.45–0.88 μm	14 nm	0.85 mrad[d]	Varies[d]
AVIRIS[e]	U-2/ER-2	220	0.41–2.45 μm	9.4–9.7 nm	20 m	11 km
TIMS[f]	Learjet/ C-130	6	8.2–12.2 μm	0.4–1.0 μm	2.5 mrad[d]	Varies[d]
TMS[g] (NS-001)	C-130	8	0.45–12.3 μm	0.06–1.4 μm	2.5 mrad[d]	Varies[d]
TMS[g] (Daedalus)	U-2/ER-2	11	0.42–14.0 μm	0.02–5.5 μm	30 m	17 km

[a]Airborne Ocean Color Imager.
[b]Visible and near-infrared = 20 nm; near- and shortwave-infrared = 60 nm; thermal infrared = 3.86 μm.
[c]Advanced Solid-State Array Spectrometer.
[d]Swath width and spatial resolution can be varied widely by flying the sensor at different altitudes.
[e]Airborne Visible-Infrared Imaging Spectrometer.
[f]Thermal Infrared Multispectral Scanner.
[g]Thematic Mapper Simulator.
[h]Instantaneous Field of View.

to learn about the utility of broad-band radiance measurements in advance of the Landsat satellites, they remain useful for basic research in a variety of fields and as a component in nested sampling strategies involving the Landsat sensors and the AVHRR. Airborne thematic mapper simulators can provide finer spatial resolution and greater spectral coverage than the satellite sensors, and they can be deployed quickly and at times when satellite coverage is not possible. A number of commercial and research multispectral scanners and thematic mapper simulators are in current use.

The Airborne Visible-Infrared Imaging Spectrometer (AVIRIS), flown on a NASA U-2 or ER-2 aircraft, acquires a complete reflectance spectrum within the wavelength region of 0.41–2.45 μm with a 10 nm sampling interval, yielding 220 spectral bands. AVIRIS was flown to collect research data for the first time in 1987. Its precursor, the Airborne Imaging Spectrometer (AIS), gathered similar data in the 0.9–2.4 μm region during 1982–1986. Much of our current understanding of AVIRIS's capabilities comes from studies of AIS data.

High spectral resolution data will be necessary to resolve the blue shift of the red edge, which can be used for assessing vegetation stress. Early research with field instruments and experience with the AIS suggested that the visible channels of AVIRIS might be able to provide such information about vegetation stress (Goetz et al., 1985a). The spectral resolution of AVIRIS (about 20 nm) is such that it may not be able to detect any but the most pronounced of shifts in the location of the red edge. However, AVIRIS should be able to provide information on vegetation vigor through measurements of reflectance in pigment absorption regions.

It is anticipated that the additional information in AVIRIS's 220 spectral bands, relative to the 7 bands of TM, will enhance our ability to identify vegetation types. The additional spectral channels may provide sufficient information to add a level of detail to vegetation classifications derived from AVIRIS data that was heretofore not possible with TM data. Early studies with the AIS have already demonstrated that very detailed classifications of vegetation community types can be achieved using high spectral resolution data (Goetz et al., 1985a).

AVIRIS data currently are being used to evaluate the utility of high spectral resolution data for a variety of ecological applications, including identification of vegetation communities, assessment of canopy condition or vigor, assessment of canopy water status, and estimation of concentrations of canopy biochemical constituents.

The Thermal Infrared Multispectral Scanner (TIMS) acquires data in six broad channels spanning the region from 8.2–12.2 μm with a thermal sensitivity of approximately 0.1–0.2 K. The TIMS can be flown on the NASA Learjet and the NASA C-130. Although it was designed as a multispectral instrument for geological studies, to detect emissivity minima related to the silicate composition of rocks, it also can be used as an extremely sensitive and accurate radiometer for the measurement of surface temperatures.

Very few studies of ecosystem properties have been conducted using TIMS. The studies that have been conducted indicate a need for a great deal more research on the thermal properties of vegetation. Surface temperatures derived from TIMS data have been used to assess various components of the surface energy balance (Sader, 1986; Luvall et al., 1986). Canopy temperatures, as estimated by TIMS, currently are being used to calculate evapotranspiration in instrumented watersheds (Luvall and Holbo, 1987). Soil surface temperatures have been related to soil moisture content, and TIMS has been used to qualitatively assess variations in soil moisture content (Pelletier et al., 1985).

NASA has been developing an experimental instrument, called the Advanced Solid-State Array Spectrometer (ASAS), to measure multidirectional reflectance properties of the surface. ASAS has 30 channels in the wavelength region 0.45–0.88 μm, each approximately 14 nm wide. It can point $\pm 45°$ fore and aft of nadir. ASAS will permit measurement of radiance at a sufficient variety of angles to estimate the BRDF. ASAS also is being used to develop understanding of fine scale differences between directional reflectance and hemispherical reflectance.

The Airborne Ocean Color Imager (AOCI) is actually an optional reconfiguration of one of NASA's airborne multispectral scanners. It is flown on the NASA U-2 or ER-2 aircraft. The AOCI was designed for oceanographic remote sensing to simulate the spectral characteristics of a proposed second-generation instrument to follow the Coastal Zone Color Scanner on the Nimbus-7 satellite. It has eight relatively narrow bands in the visible part of the spectrum and one band each in the near- and thermal-infrared portions of the spectrum. Its ground resolution is about 50 m. The AOCI is of interest for terrestrial ecological studies because it can be flown over fresh-water lakes to study phytoplankton dynamics and other surface-water characteristics.

Airborne light detection and ranging (LIDAR) systems have been used to measure forest canopy profiles and tree height (Nelson et al., 1984). In one study, a pulsed laser unit was flown at low altitude in a profiling mode to assess the stand structure of an eastern deciduous forest canopy. Laser estimates of canopy height differed from photogrammetrically measured tree heights by less than 1 m (Nelson et al., 1984). This laser system is limited by its power to flight altitudes of less than 305 m. In the future, satellite laser systems may be developed to measure canopy height from space. Canopy height would be an extremely useful parameter to factor into the estimation of vegetation biomass and surface roughness.

Laser profiling can also be used to assess terrain elevation (Krabill et al., 1980). Future satellite-borne lasers for measuring global topography with appropriate horizontal and vertical resolutions would provide an extremely important data base for many terrestrial ecological and hydrological studies. Although there are no definite plans at present to collect such a data set, various options for compiling global digital topographic data sets, including radar as well as laser options, are being considered both nationally and internationally.

III.B   Plans for Future Satellite Sensors

National, international, and commercial entities are planning satellite optical sensors to be launched in the late 1980s and early 1990s. Many of these will offer broad-band data similar to those of the Landsat sensors, and thus should provide additional global coverage and continuity of the data type. They include enhanced commercial versions of the Landsat satellites, additional SPOT satellites, an Indian satellite, and a Japanese satellite. Eos, planned for the mid- and late 1990s, will carry several optical sensors for land surface observations. Proposed attributes for these new sensors are summarized in Table 1. Details concerning these sensors and the wide variety of potential scientific applications of data from them can be found in several publications (NASA, 1984, 1986b, 1987b; Rasool, 1987). Those optical sensors of particular importance to terrestrial ecological research will be described in the following sections.

III.B.1   Near-Term Instruments

One of the major requirements for future global change research is that there be long-term continuous global observations of a baseline set of Earth properties. Thus, continuity of current satellite data sets is of primary importance. The AVHRR-derived vegetation index record started in 1979 is one of these key data sets. AVHRR or improved AVHRR-like sensors will continue to be flown on the operational polar-orbiting meteorological satellites for the foreseeable future. However, more attention must be given to their pre- and post-launch calibration and to their intercalibration. Care also must be taken to ensure that these data are properly preserved and archived, as described in previous chapters.

The Landsat MSS record began in 1972 with the launch of ERTS-1 (Landsat-1). The Landsat TM record began in 1982 with the launch of Landsat-4. As of January, 1988, Landsats-4 and -5 were still functional and MSS and TM data could be collected from at least one of them during any given period of time. Planning for Landsats-6 and -7 has suffered a number of delays; the launch of Landsat-6 is currently planned for 1991. Thus, there is potential for a gap in Landsat coverage prior to the launch of Landsat-6. With its launch, the continuity of TM data will very likely be assured into the mid-1990s.

Landsat-6 will not carry an MSS. However, data similar to that of Landsat's MSS will be available into the 1990s. The French SPOT satellite series will provide such continued coverage, as will the future Japanese and Indian optical sensors listed in Table 1. However, to use these data sets together most effectively, and to be able to use them with Landsat data for retrospective studies back to 1972, we require detailed information on the unique attributes of the sensors, the nature of their data, and their calibration. In addition, we must be able to relate their calibration to that of the historical sensor of interest.

III.B.2 *Eos Instruments*

The next generation of research optical sensors will be deployed on Eos. Two proposed facility instruments, the Moderate-Resolution Imaging Spectrometer (MODIS) and the High-Resolution Imaging Spectrometer (HIRIS), will provide data of great relevance for terrestrial ecological research.

MODIS will be composed of two instrument packages, MODIS-T (tilt) for fore and aft off nadir viewing and MODIS-N (nadir), which will not view the Earth's surface off nadir. Both will have 500–1000 m spatial resolution and two-day global repeat coverage. MODIS-N will have at least 35 spectral bands of varying widths within the range of 0.4–12.0 μm to measure properties of the land, oceans, and atmosphere. MODIS-N will be used to obtain a wide variety of terrestrial information, and it will very likely replace the AVHRR as the research instrument for global ecological observations. MODIS-T will have 64 spectral bands, each 10 nm wide, covering the range 0.4–1.0 μm. It will be capable of pointing $\pm 60°$ fore and aft. MODIS-T will be used primarily for ocean research. In terrestrial studies, MODIS-T will be used to examine the BRDF for large targets and for determining atmospheric characteristics necessary for atmospheric corrections (NASA, 1986b).

HIRIS will have a 30 m spatial resolution and acquire 192 continuous 10 nm wide spectral bands in the wavelength region of 0.4–2.5 μm. It is the logical successor to AVIRIS. HIRIS will be able to point and acquire images up to $+60°/-30°$ downtrack and $\pm 24°$ crosstrack. Its repeat cycle is 16 days, but the pointing capability will permit more frequent coverage of particular sites. HIRIS data will be used to identify surface materials, measure the BRDF of small targets, and perform detailed studies of fine-scale ecological processes (NASA, 1987b).

Although this discussion has been directed toward a consideration of optical sensors, it would be extremely remiss to fail to mention the multisensor approach to Earth system research that is likely to dominate the future. Information about the Earth's surface obtained from all wavelength regions capable of being sensed remotely will be necessaary to develop understanding of global processes. Microwave sensors, both active and passive, will provide critical information about surface structure, texture, moisture content, and dielectric properties (NASA, 1987c, d). Because of its insensitivity to cloud cover and solar illumination, radar will extend our capability for observing surface patterns and processes. Future remote-sensing research must emphasize the development of methods for integrating information in data sets from all wavelength regions. The combination of optical and microwave data seems particularly challenging. The fundamentally different information to be obtained from each new wavelength region promises to enhance substantially our ability to identify surface phenomena and to monitor and predict Earth system processes.

III.C Plans for Future Data Systems

Future data-processing systems will have to deal with increased data rates, increased spectral and spatial resolutions, a wider variety of user products, more complex operational applications, and the problem of broader product dissemination (MacDonald, 1987). This is an inevitable consequence of advances both in remote-sensing technology and the scientific understanding of the Earth system. As research progresses from discovery and exploration to interpretation and consolidation, it becomes increasingly team oriented and multidisciplinary in character (Ludwig, 1987). This progression is accompanied by an increased need for tightly integrated, but distributed, information systems.

Planned data volumes for any of the proposed Eos sensors are far beyond any we have previously handled. The variety, complexity, and volume of data to be produced by the Eos instruments will present major challenges in mission operations, data transmission, data processing, and long-term data management and maintenance. Therefore, a data and information system has been included as a fundamental part of the total plan for Eos. The Eos data and information system will be a complete research information system that will transcend the traditional mission system and include additional capabilities such as maintaining long-term time-series data bases and providing access by Eos researchers to non-Eos data (NASA, 1986a). Its core will be an electronic information network that will allow access to the full suite of Eos system capabilities. The multidisciplinary integrative Earth system science research of the 1990s will depend heavily on the success of the Eos data and information system.

As more complex multitemporal multisource data sets become available, more sophisticated analysis techniques will be required to produce synthetic, integrative results and interpretations (Swain, 1985). In addition, a wide variety of currently disparate sources of information must be gathered together and the technologies by which they can be integrated must be developed and rendered easy to use (Goward, 1989). Key techniques include geographic information systems (GIS), automated image registration, change detection, pattern analysis, and elements derived from the field of artificial intelligence (Swain, 1985; Estes, 1985). There are a number of problems related to the use of GIS that must be solved. These include resolving the raster-vector conflict that so limits the integration of remotely sensed digital data in GIS; developing means of relating information at various spatial scales and realistically interpolating between point data values; establishing standards for data products, data formats, documentation of data sets, and calibration and validatation of data sets; and resolving the cartographic complexities of preserving location information when incorporating data layers mapped to differing projections (Goward, 1989).

The technology for accomplishing many of these tasks is available now, and the current rate of progress in this field promises greatly enhanced technological capabilities for the future. However, it seems quite possible that the technology may far outstrip the scientists' ability to utilize it effectively. Terrestrial ecologists, and in fact, land scientists in general, must become more sophisticated in their use of advanced computational facilities and information systems. The recent spread of GIS capabilities in the community is one favorable development in this regard. NASA's Pilot Land Data System (PLDS) should provide further opportunities for the land research community to learn how to conduct research in an advanced information systems environment (Estes et al., 1985).

III.D Expected Advances in Capabilities

Advances in remote sensing and associated capabilities are expected in a number of areas. Most importantly, globally consistent data sets from which spatial, temporal, and spectral ecological information can be derived will be made available for the first time (NRC, 1986b). In addition, fundamentally new types of information may become available from sensors which measure such phenomena as off-nadir reflectance, polarization, or fluorescence. Advances in ecological modeling capabilities and in the handling, processing, and integration of data are also expected to contribute significantly to progress in terrestrial ecological research. The expected advances in each of these areas are summarized in the following sections.

III.D.1 Temporal Coverage

Ecosystems are inherently temporally dynamic, and any effort to study them must attempt to capture these dynamics. One great advance in remote sensing that the future will bring is increased temporal coverage of the Earth's land surface. Until recently, we have not been able to obtain data for a given site more frequently than every 16 days at finer than 1 km resolution. This, combined with problems of cloud cover, has limited our ability to acquire appropriate temporal and phenological information for land cover classification, and has made studies of most local-scale ecosystems processes nearly impossible. The successful application of the relatively coarse spatial scale daily AVHRR data to studies of global and regional vegetation dynamics is an excellent example of what can be done with frequent coverage (Tucker et al., 1985b, 1986a). Integration over the growing season of monthly composites of daily measurements of the NDVI has been necessary to produce one estimate of annual productivity for global ecosystems (Tucker et al., 1985b; Goward et al., 1985b). SPOT now allows more frequent repeat coverage, but with variations in viewing angles. In the Eos era, MODIS will provide two-day repeat coverage with moderate spatial and spectral resolution and HIRIS's pointing capabilities will permit relatively frequent high-resolution coverage of selected sites.

III.D.2 Spectral Resolution

Advances in high spectral resolution remote sensing can be expected to lead to improvements in our ability to identify vegetation, to assess its biochemical composition, and to evaluate its condition. It may be possible to identify uniquely, not just classify, vegetation types and to diagnose, not just detect, vegetation condition (Goetz et al., 1983, 1985a). Many distinct vegetation types are composed of mixtures of species; thus, unique identification by means of remote sensing of types may prove to be quite challenging. Nonetheless, we may be able to improve considerably our abilities to classify vegetation using only remotely sensed data. Extremely detailed and accurate classifications of forest communities have been achieved using data from low spectral resolution sensors collected under optimal viewing conditions (Rock, 1982; Lang et al., 1984).

If reports of unique spectral characteristics for the effects of various stress agents can be verified, then we may be able to diagnose, or at least narrow the range of possibilities for, the stress causal agent. A great deal of fundamental research on the biochemistry of stress and the biophysics of canopy–light interactions will be required before these phenomena are fully understood.

Continued research to understand the use of high spectral resolution data for the prediction of canopy biochemical concentrations, particularly canopy lignin and nitrogen concentrations, is expected to lead to extremely significant ecological applications. If this research is successful, we will be able to assess several fundamentally important ecosystem characteristics and processes (i.e., litter carbon quality, decomposition rates, nitrogen mineralization, relative site fertility) using remote-sensing techniques. This will remove a major barrier to efforts to scale up from site-specific understanding to regional- and global-scale understanding of biogeochemical cycling.

III.D.3   Off-Nadir Viewing

Accurate land surface albedos will be measured once instruments that can point off nadir in all directions have been placed in orbit (see discussions of ASAS, MODIS, and HIRIS and Chapter 2). Improved understanding of land surface heterogeneity and its effect on climate should follow.

Bidirectional reflectance data from multidirectional sensors also may provide information about canopy and landscape structural properties as well as information that can improve our ability to identify vegetation types.

III.D.4   Polarization

A significant percentage of the radiation incident on vegetation canopies never enters the leaves, but is reflected specularly. This specularly reflected light is polarized, and the magnitude of the polarized reflection depends on the angle of the incident light on the leaf, on the optical index of refraction of the cuticle wax, and on the surface roughness characteristics of the cuticle wax (Vanderbilt et al., 1985).

These findings suggest that if the polarization of reflected light could be measured, it would be possible to calculate the specular component and distinguish it from the diffuse component (Bauer, 1985). We might then be able to isolate the diffuse component, which contains information on leaf internal structure and chemistry. Further, polarization measurements should contain information about the leaf surfaces and their geometric arrangement in the canopy. It is well known that leaf cuticle properties vary with species, developmental state, and environmental conditions. Thus, fundamentally new information about vegetation canopies that may be of use for identification, classification, or condition assessment could be obtained by measuring polarization (Vanderbilt et al., 1985).

To date, only laboratory and field measurements of polarization have been made. No satellite sensors to measure polarization are being planned, but the possibility of a future polarization sensor should not be ruled out.

III.D.5   Laser-Induced Fluorescence

Laboratory studies of laser-induced fluorescence in terrestrial vegetation have demonstrated a strong relationship between photosynthetic efficiency and amount of fluorescence (Chappelle et al., 1984). Further, species groups with differing types of photosynthetic mechanisms have characteristic fluorescence patterns (Chappelle et al., 1985; Chappelle and Williams, 1986). Changes in the degree of fluorescence may prove to be sensitive early indicators of vegetation stress. To date, measurements of laser-induced fluorescence in terrestrial vegetation have not been made successfully from aircraft. Future airborne or spaceborne systems seem more than just a few years away, but should not be discounted given the current rapid pace of advance in laser technology.

III.D.6   Vegetation Classifications and Maps

Coarse-resolution remotely sensed data, most likely averaged or integrated over several years, will be used to generate global and regional maps of land cover types and vegetation units. Tucker et al.'s (1985a) classification of African vegetation is an excellent example of a first attempt along these lines. Finer-scale remotely sensed data will be used to generate highly accurate maps of local and regional areas. In many cases, these maps should greatly surpass in accuracy more conventionally prepared maps. In other cases, they will be used to modify or refine conventional maps. These maps will serve as vital baseline data sets for studies of global change. Any studies requiring extrapolation of local understanding to regional or global scales will require such maps to derive areal extent of the relevant land cover types.

Future analyses of landscape patterns using advanced sensors or more sophisticated algorithms are likely to identify new classes of vegetation functional units. These classes would be based on criteria other than those traditionally used to classify vegetation. Such criteria could be canopy chemistry type, canopy structural configuration, or canopy process rates. Such new classes would be extremely useful for extrapolating information over large regions and estimating global impacts.

III.D.7 Multisensor Data Sets

It is very likely that in the near future, the current suite of experimental airborne sensors will be flown concurrently to acquire simultaneous or near-simultaneous data sets that can be used to develop multisensor information extraction techniques. This simulation of Eos-like data also will provide an opportunity for scientists to conduct the first coordinated multisensor analyses of ecosystems pattern and process. Such studies should be conducted over very carefully selected study sites, or regions, where important ecological questions can be addressed at the same time. A future FIFE-like experiment would provide an ideal opportunity to bring together all appropriate airborne sensors.

III.D.8 Calibration

Progress in analyzing multidate satellite imagery will require that the satellite data be well-calibrated and intercalibrated. Data from one date to the next from a given satellite sensor must be comparable, data from different sensors in the same instrument series must be comparable, and data from different types of optical sensors must be comparable. Researchers must insist that sensor calibration be performed routinely and that calibration data and information about calibration precision and accuracy be readily obtainable. Current satellite sensors can be calibrated and monitored for change over time by periodic measurement of a ground target of known spectral radiance and by periodic underflights with calibrated aircraft sensors using known targets.

Future satellite sensors should have redundant in-flight means of calibration and also should periodically measure known targets and be regularly underflown by calibrated airborne sensors (Slater, 1985). All calibrations should be traceable to National Bureau of Standard (NBS) standard sources. Without such care in calibration and intercalibration of optical sensors, it will be impossible to determine whether a perceived change from one satellite image to another is due to actual changes on the ground or to changes in the satellite sensor. Separate from the sensor calibration, but equally important to the multidate comparisons, is the development and implementation of good atmospheric corrections for satellite data sets (see Chapter 9). Scientists must be able to compare surface reflectance values and not just satellite-observed radiances.

III.D.9 Modeling

The development of mathematical models to simulate and predict ecosystem function has long been an important component of terrestrial ecological research. In recent years, considerable effort has been expended in the development of models that require as drivers, either exclusively or predominantly, remotely sensed inputs. Once developed, such models for large vegetation stands or regions can be initiated and run without requiring extensive field measurement programs for their parameterization. In addition, field data collection can then be reserved as a means of model verification or validation. One such model, described by Peterson et al. (1985), requires as input LAI, microclimatic conditions, canopy nitrogen content, canopy phosphorus content, and canopy lignin concentration. Model outputs are net primary production, decomposition, gaseous water exchange, and nitrogen and phosphorus cycling. As remote-sensing capabilities improve, and as our biophysical understanding of remotely sensed data increases, mechanistic remote-sensing driven models are going to be in great demand.

As ecologists approach the study of ecosystem processes at larger scales, they are beginning to develop models that operate at various scales and the means to link them. Detailed information about site-specific processes must be synthesized and integrated into mesoscale and global models. Similarly, outputs from global climate process models must be input into detailed ecosystem models. Models that operate at various scales will be necessary to integrate information across scales; and models that operate at intermediate scales may serve as important means of working between local and global scales. Nested hierarchical mechanistic models, as biologically and physically realistic as possible, will be required to achieve these goals. Remotely sensed spatial pattern may provide the key for linking models of differing scales.

III.D.10   Data Handling and Processing

Sophisticated data and information systems and advanced computer processing capabilities are expected to be available for future terrestrial ecological research. New methods for gathering and integrating data from a wide variety of sources, in a wide variety of formats, are expected. Current activities in the development of computer workstations and information systems should lead to increasingly easy-to-use interfaces between the scientist and the computer. If scientists are to make progress in understanding integrated Earth system processes, this will be essential. Scientists must be freed from having to invest inordinate amounts of time in learning how to use specialized computer systems. Advances in computerized image processing and analysis techniques for handling and analyzing large regional data sets are expected to facilitate future landscape studies. Active research in the develoment of improved GIS is expected to yield more powerful capabilities. The use of GIS to analyze many layers of digital imagery and supporting data sets is expected to be widespread in the future.

IV   THE EOS ERA AND BEYOND

In the 1990s and beyond, we should have a much better understanding of the capabilities and limitations of remote-sensing technology for identifying, monitoring, and assessing terrestrial ecosystem structure and function. We should be in the midst of learning how to perform multidisciplinary studies of important global processes using data from a variety of remote-sensing instruments. Techniques for merging optical and microwave data should be available. Models that can accept remotely sensed inputs to predict global ecosystem function should be in routine use.

By the mid-1990s, we should have acquired some information about the ability of terrestrial ecosystems to adapt to the sorts of anthropogenic changes that have already been made to the Earth system. A number of human-induced ecosystem ''experiments'' should be available for detailed observation and assessment. We should be gaining a better impression of the resilience or susceptibility to change of certain ecosystems and of the processes most likely to be affected by certain types of perturbations. Clues may even be available as to the kinds of feedbacks ecosystems are likely to give to climate and biogeochemical cycles.

One challenge that is likely to remain is that of learning how to manage the planetary ecosystem such that its habitability for humans and the biota essential to their survival is not jeopardized.

ACKNOWLEDGMENTS

Few of the words and ideas presented in this chapter are the author's own. Many have come from national and international planning activities and reports. The National Research Council reports on the IGBP (1986a) and a strategy for earth science from space in the 1980s and 1990s (NRC, 1982, 1985) were drawn upon extensively, as were NASA's Earth System Science reports. The author's perspectives on the use of remote sensing have been developed through close interaction with NASA's Terrestrial Ecosystems Program Advance Planning Group and represent a synthesis, or, perhaps, an amalgamation, of the ideas raised in that group's meetings. Many of the ideas and words that appear in this chapter can be traced to Sam Goward, Richard Waring, Berrien Moore III, Piers Sellers, Dave Simonett, Pamela Matson, Steve Running, Forrest Hall, Jim Smith, Jim Tucker, Dave Peterson, Barry Rock, Jim Lawless, Bob Murphy, Jerry Melillo, or Francis Bretherton. Their intellectual leadership is most gratefully acknowledged. Special thanks are due to Sam Goward for his many and varied inputs.

REFERENCES

Arvidson, R. E., D. M. Butler, and R. E. Hartle (1985). Eos: The Earth observing system of the 1990s. *Proc. IEEE* **73**(6):1025–1030.

Asrar, G., M. Fuchs, E. T. Kanemasu, and J. L. Hatfield (1984). Estimating absorbed photosynthetic radiation and leaf area index from spectral reflectance in wheat. *Agron. J.* **76**:300–306.

Bartlett, K. B., D. S. Bartlett, D. I. Sebacher, R. C. Harriss, D. P. Brannon, C. A. Clark, and J. M. Hartman (1986). Remote sensing and the assessment of methane emissions from the Florida Everglades. Program 4th Int. Cong. Ecol., Syracuse, New York, p. 85.

Bauer, M. E. (1985). Spectral inputs to crop identification and condition assessment. *Proc. IEEE* **73**(6):1071–1085.

Botkin, D. B., J. E. Estes, R. M. MacDonald, and M. V. Wilson (1984). Studying the Earth's vegetation from space. *BioScience* **34**(8):508–514.

Brest, C. L., and S. N. Goward (1987). Deriving surface albedo measurements from narrow band satellite data. *Int. J. Remote Sens.* **8**(3):351–367.

Butera, M. K. (1983). Remote sensing of wetlands. *IEEE Trans. Geosci. Remote Sens.* **GE-21**(3):383–392.

Butera, M. K., J. A. Browder, and A. L. Frick (1984). A preliminary report on the assessment of wetland productive capacity from a remote-sensing-based model—a NASA/NMFS joint research project. *IEEE Trans. Geosci. Remote Sens.* **GE-22**(6):502–511.

Chappelle, E. W., and D. L. Williams (1986). Laser-induced fluorescence (LIF) from plant foliage. *Proc. Int. Geosci. Remote Sens. Symp. (IGARSS '86)*, Zurich, Switzerland, pp. 1591–1598.

Chappelle, E. W., F. M. Wood, J. E. McMurtrey, III, and W. W. Newcomb (1984). Laser-induced fluorescence of green plants. 1. A technique for the remote detection of plant stress and species differentiation. *Appl. Opt.* **23**(1):134–138.

Chappelle, E. W., F. M. Wood, W. W. Newcomb, and J. E. McMurtrey, III (1985). Laser-induced fluorescence of green plants. 3. LIF spectral signatures of five major plant types. *Appl. Opt.* **24**(1):74–80.

Clark, C. A., R. B. Cate, M. H. Trenchard, J. A. Boatright, and R. M. Bizzell (1986). Mapping and classifying large ecological units. *BioScience* **36**(7):476–478.

Collins, W. (1978). Remote sensing of crop type and maturity. *Photogramm. Eng. Remote Sens.* **44**:43–55.

Collins, W., S. H. Chang, G. Raines, F. Canney, and R. Ashley (1983). Airborne biogeochemical mapping of hidden mineral deposits. *Econ. Geol.* **78**:737–749.

Curran, P. J. (1983). Multispectral remote sensing for the estimation of green leaf area index. *Philos. Trans. R. Soc. London, Ser. A* **309**:257–270.

Davis, P. A., and J. D. Tarpley (1983). Estimation of shelter temperatures from operational satellite sounder data. *J. Appl. Meteorol.* **22**:369–376.

Detwiler, R. P., and C. A. S. Hall (1988). Tropical forests and the global carbon cycle. *Science* **239**:42–47.

Estes, J. E. (1985). Geographic applications of remotely sensed data. *Proc. IEEE* **73**(6):1097–1107.

Estes, J. E., J. L. Star, P. J. Cressy, and M. Devirian (1985). Pilot Land Data System. *Photogramm. Eng. Remote Sens.* **51**(6):703–709.

Forman, R. T. T., and M. Godron (1986). *Landscape Ecology*. Wiley, New York.

Franklin, J., T. L. Logan, C. E. Woodcock, and A. H. Strahler (1986). Coniferous forest classification and inventory using Landsat and digital terrain data. *IEEE Trans. Geosci. Remote Sens.* **GE-24**(1):139–149.

Gates, D. M., H. J. Keegan, J. C. Schleter, and V. P. Weidner (1965). Spectral properties of plants. *Appl. Opt.* **4**:11–20.

Gerstl, S. A. W. (1986). Off-nadir optical remote sensing from satellites for vegetation identification. *Proc. Int. Geosci. Remote Sens. Symp. (IGARSS '86)*, Zurich, Switzerland, pp. 1457–1460.

Gerstl, S. A. W., and C. Simmer (1986). Radiation physics and modeling for off-nadir satellite-sensing of non-Lambertian surfaces. *Remote Sens. Environ.* **20**:1–29.

Goel, N. S., and D. W. Deering (1985). Evaluation of a canopy reflectance model for LAI estimation through its inversion. *IEEE Trans. Geosci. Remote Sens.* **GE-23**(5):674–684.

Goetz, A. F. H., B. N. Rock, and L. C. Rowan (1983). Remote sensing for exploration: An overview. *Econ. Geol.* **78**(4):573–590.

Goetz, A. F. H., G. Vane, J. E. Solomon, and B. N. Rock (1985a). Imaging spectrometry for Earth remote sensing. *Science* **228**:1147–1153.

Goetz, A. F. H., J. B. Wellman, and W. L. Barnes (1985b). Optical remote sensing of the Earth. *Proc. IEEE* **73**(6):950–969.

Goward, S. N. (1985). Shortwave infrared detection of vegetation. *Adv. Space Res.* **5**(5):41–50.

Goward, S. N. (1989). Geographic information systems and data integration. In *Global Change: A Geographical Approach* (V. M. Kotlyakov and G. White, Eds.). International Geographical Union (in press).

Goward, S. N., G. D. Cruickshanks, and A. S. Hope (1985a). Observed relation between thermal emission and reflected spectral radiance of a complex vegetated landscape. *Remote Sens. Environ.* **18**:137–146.

Goward, S. N., C. J. Tucker, and D. G. Dye (1985b). North American vegetation patterns observed with the NOAA-7 Advanced Very High Resolution Radiometer. *Vegetatio* **64**:3–14.

Gurney, R. J., and P. J. Camillo (1984). Modelling daily evapotranspiration using remotely sensed data. *J. Hydrol.* **69**:305–324.

Hall, F. G., D. E. Strebel, S. J. Goetz, K. D. Woods, and D. B. Botkin (1987). Landscape pattern and successional dynamics in the boreal forest. *Proc. Int. Geosci. Remote Sens. Symp. (IGARSS '87)*, Ann Arbor, Michigan, pp. 473–482.

Hoffer, R. M., and D. J. Johannsen (1969). Ecological potentials in spectral signature analysis. In *Remote Sensing in Ecology* (P. L. Johnson, Ed.). University of Georgia Press, Athens, pp. 1–16.

Hogg, H. C. (Ed.) (1986). Special issue on Agriculture and Resources Inventory Surveys Through Aerospace Remote Sensing (AgRISTARS). *IEEE Trans. Geosci. Remote Sens.* **GE-24**(1):1–184.

Hope, A. S., D. E. Petzold, S. N. Goward, and R. M. Ragan (1986). Simulated relationships between spectral reflectance, thermal emissions, and evapotranspiration of a soybean canopy. *Water Resour. Bull.* **22**(6):1011–1019.

Horler, D. N. H., J. Barber, and A. R. Barringer (1980). Effects of heavy metals on the absorbance and reflectance spectra of plants. *Int. J. Remote Sens.* **1**:121–136.

Hunt, E. R., Jr., B. N. Rock, and P. S. Nobel (1987). Measurement of leaf relative water content by infrared reflectance. *Remote Sens. Environ.* **22**:429–435.

Jackson, R. D. (1985). Evaluating evapotranspiration at local and regional scales. *Proc. IEEE* **73**(6):1086–1096.

Jackson, R. D., J. L. Hatfield, R. J. Reginato, S. B. Idso, and P. J. Pinter (1983). Estimation of daily evapotranspiration from one-time-of-day measurements. *Agric. Water Manage.* **7**:351–362.

Kimes, D. S. (1983). Dynamics of directional reflectance factor distributions for vegetation canopies. *Appl. Opt.* **22**(9):1364–1372.

Kimes, D. S., B. L. Markham, C. J. Tucker, and J. E. McMurtrey, III (1981). Temporal relations between spectral response and agronomic variables of a corn canopy. *Remote Sens. Environ.* **11**:401–411.

Kimes, D. S., and P. J. Sellers (1985). Inferring hemispherical reflectance of the Earth's surface for global energy budgets from remotely sensed nadir or directional radiance values. *Remote Sens. Environ.* **18**:205–223.

Kirchner, J. A., D. S. Kimes, and J. E. McMurtrey, III (1982). Variation of directional reflectance factors with structural changes of a developing alfalfa canopy. *Appl. Opt.* **21**(20):3766–3774.

Krabill, W. B., J. G. Collins, R. N. Swift, and M. L. Butler (1980). Airborne laser topographic mapping results from initial joint NASA/U.S. Army Corps of Engineers experiment. *NASA Tech. Memo.* **NASA TM-73287**:1–33.

Lang, H. R., J. B. Curtis, and J. S. Kovacs (1984). Lost River, West Virginia, petroleum test site report. In *The Joint NASA/Geosat Test Case Project: Final Report* (M. J. Abrams, J. E. Conel, and H. R. Lang, Eds.). AAPG Bookstore, Tulsa, Oklahoma, pp. 12-i–12-96.

Ludwig, G. H. (1987). Distributed information systems for space remote sensing. *Proc. Int. Geosci. Remote Sens. Symp. (IGARSS '87)*, Ann Arbor, Michigan, pp. 885–886.

Luvall, J. C., J. E. Anderson, and H. R. Holbo (1986). Thermal infrared energy processes in western coniferous forest canopies. *Program 4th Int. Congr. Ecol.*, Syracuse, New York, p. 222.

Luvall, J. C., and R. Holbo (1987). Using the TIMS to estimate evapotranspiration from a white pine forest. *Assoc. Southeast. Biol. Bull.* **34**(2):69.

MacDonald, J. (1987). Concepts for Eos data processing. *Proc. Int. Geosci. Remote Sens. Symp. (IGARSS '87)*, Ann Arbor, Michigan, p. 895.

Malingreau, J. P., and C. J. Tucker (1987). The contribution of AVHRR data for measuring and understanding global processes: Large scale deforestation in the Amazon basin. *Proc. Int. Geosci. Remote Sens. Symp. (IGARSS '87)*, Ann Arbor, Michigan, pp. 443–448.

Malone, T. F. (1986). Mission to Planet Earth: Integrating studies of global change. *Environment* **28**(8):6–11.

Marten, G. C., J. S. Shenk, and F. E. Barton, II (Eds.) (1985). *Near Infrared Reflectance Spectroscopy (NIRS): Analysis of Forage Quality*, U.S. Dep. Agric. Handb. No. 643. USDA, Washington, D.C.

Melillo, J. M., J. D. Aber, and J. F. Muratore (1982). Nitrogen and lignin control of hardwood leaf litter decomposition dynamics. *Ecology* **63**(3):621–626.

Mooney, H. A., P. M. Vitousek, and P. A. Matson (1987). Exchange of materials between terrestrial ecosystems and the atmosphere. *Science* **238**:926–932.

Moore, B., III and B. Bolin (1986). The oceans, carbon dioxide, and global climate change. *Oceanus* **29**(4):9–15.

Morrissey, L. A., L. L. Strong, and D. H. Card (1986). Mapping permafrost in the boreal forest with Thematic Mapper satellite data. *Photogramm. Eng. Remote Sens.* **52**(9):1513–1520.

National Aeronautics and Space Administration (NASA) (1984). Earth observing system: Science mission requirements working group report. *NASA Tech. Memo.* **NASA TM-86129**:1–51.

National Aeronautics and Space Administration (NASA) (1986a). *Earth Observing System*, Vol. IIa, Data Inf. Syst. Panel Rep. NASA, Washington, D.C.

National Aeronautics and Space Administration (NASA) (1986b). *Earth Observing System*, Vol. IIb, MODIS Instrum. Panel Rep. NASA, Washington, D.C.

National Aeronautics and Space Administration (NASA) (1986c). *Earth System Science: Overview*. NASA, Washington, D. C.

National Aeronautics and Space Administration (NASA) (1987a). *Earth Observing System*, Vol. II, From pattern to process: The strategy of the Earth observing system. NASA, Washington, D.C.

National Aeronautics and Space Administration (NASA) (1987b). *Earth Observing System*, Vol. IIc, HIRIS Instrum. Panel Rep. NASA, Washington, D.C.

National Aeronautics and Space Administration (NASA) (1987c). *Earth Observing System*, Vol. IIe, HMMR Instrum. Panel Rep. NASA, Washington, D.C.

National Aeronautics and Space Administration (NASA) (1987d). *Earth Observing System*, Vol. IIf, SAR Instrum. Panel Rep. NASA, Washington, D.C.

National Aeronautics and Space Administration (NASA) (1988). *Earth System Science: A Closer View*. NASA, Washington, D. C.

National Research Council (NRC) (1982). *A Strategy for Earth Science from Space in the 1980's*, Part I. National Academy Press, Washington, D. C.

National Research Council (NRC) (1985). *A Strategy for Earth Science from Space in the 1980's and 1990's*, Part II. National Academy Press, Washington, D. C.

National Research Council (NRC) (1986a). *Global Change in the Geosphere-Biosphere: Initial Priorities for an IGBP*. National Academy Press, Washington, D. C.

National Research Council (NRC) (1986b). *Remote Sensing of the Biosphere*. National Academy Press, Washington, D. C.

Nelson, R., W. Krabill, and G. MacLean (1984). Determining forest canopy characteristics using airborne laser data. *Remote Sens. Environ.* **15**:201–212.

Norris, K. H., R. F. Barnes, J. E. Moore, and J. S. Shenk (1976). Predicting forage quality by infrared reflectance spectroscopy. *J. Anim. Sci.* **43**(4):889–897.

Odum, E. P. (1971). *Fundamentals of Ecology*. Saunders, Philadelphia, Pennsylvania.

Pelletier, R. E., M. C. Ochoa, and B. F. Hajek (1985). Agricultural application for Thermal Infrared Multispectral Scanner data. In *Machine Processing of Remotely Sensed Data*. Purdue University, West Lafayette, Indiana, pp. 321–328.

Perry, M. J. (1986). Assessing marine primary production from space. *BioScience* **36**(7):461–467.

Peterson, D.L., P. A. Matson, J. G. Lawless, J. D. Aber, P. M. Vitousek, and S. W. Running (1985). Biogeochemical cycling in terrestrial ecosystems: Modeling, measurement, and remote sensing. *Proc. 36th Int. Astronaut. Fed. Congr.*, Stockholm, Sweden.

Peterson, D. L., J. D. Aber, P. A. Matson, D. H. Card, N. Swanberg, C. Wessman, and M. Spanner (1987). Remote sensing of forest canopy and leaf biochemical contents. *Remote Sens. Environ.* **23**:85–108.

Petzold, D. E., and S. N. Goward (1988). Reflectance spectra of subarctic lichens. *Remote Sens. Environ.* **24**:481–492.

Petzold, D. E., and A. N. Rencz (1975). The albedo of selected subarctic surfaces. *Arct. Alp. Res.* **7**:393–398.

Price, J. C. (1980). The potential of remotely sensed infrared data to infer surface soil moisture and evaporation. *Water Resour. Res.* **16**:787–795.

Rasool, S. I. (Ed.) (1987). Potential of remote sensing for the study of global change. *Adv. Space Res.* **7**(1):1–97.

Richardson, A. J., and C. L. Wiegand (1977). Distinguishing vegetation from soil background information. *Photogramm. Eng. Remote Sens.* **43**:1541–1552.

Rock, B. N. (1982). Mapping of deciduous forest cover using simulated Landsat-D TM data. *Proc. Int. Geosci. Remote Sens. Symp. (IGARSS '82)*, Munich, Germany, pp. 3.1–3.5.

Rock, B. N., J. E. Vogelmann, D.L. Williams, A. F. Vogelmann, and T. Hoshizaki (1986). Remote detection of forest damage. *BioScience* **36**(7):439–445.

Rock, B. N., D. L. Williams, and J. E. Vogelmann (1985). Field and airborne spectral characterization of suspected acid deposition damage in red spruce (*Picea rubens*) from Vermont. In *Machine Processing of Remotely Sensed Data*. Purdue University, West Lafayette, Indiana, pp. 71–81.

Running, S. W., D. L. Peterson, M. A. Spanner, and K. B. Teuber (1986). Remote sensing of coniferous forest leaf area. *Ecology* **67**(1):273–276.

Running, S. W., R. R. Nemani, and R. D. Hungerford (1987). Extrapolation of synoptic meteorological data in mountainous terrain and its use for simulating forest evapotranspiration and photosynthesis. *Can. J. For. Res.* **17**(6):472–483.

Sader, S. A. (1986). Analysis of effective radiant temperatures in a Pacific Northwest forest using Thermal Infrared Multispectral Scanner data. *Remote Sens. Environ.* **19**:105–115.

Sader, S. A., and A. T. Joyce (1988). Deforestation rates and trends in Costa Rica—1940 to 1983. *Biotropica* **20**(1):11–19.

Salati, E. (1987). The forest and the hydrological cycle. In *The Geophysiology of Amazonia* (R. E. Dickinson, Ed.). Wiley, New York, pp. 273–296.

Sellers, P. J. (1985). Canopy reflectance, photosynthesis and transpiration. *Int. J. Remote Sens.* **6**(8):1335–1372.

Sellers, P. J., F. G. Hall, G. Asrar, D. E. Strebel, and R. E. Murphy (1988). The First ISLSCP Field Experiment (FIFE). *Bull. Am. Meteorol. Soc.* **69**(1):22–27.

Simmer, C., and S. A. W. Gerstl (1985). Remote sensing of angular characteristics of canopy reflectances. *IEEE Trans. Geosci. Remote Sens.* **GE-23**(5):648–658.

Slater, P. N. (1985). Radiometric considerations in remote sensing. *Proc. IEEE* **73**(6):997–1011.

Slobodkin, L. B. (1984). Toward a global perspective. *BioScience* **34**(8):484–485.

Swain, P. H. (1985). Advanced interpretation techniques for Earth data information systems. *Proc. IEEE* **73**(6):1031–1039.

Swanson, F. J., T. K. Kratz, N. Caine, and R. G. Woodmansee (1988). Landform effects on ecosystem patterns and processes. *BioScience* **38**(2):92–98.

Thomas, J. R., L. N. Namken, G. F. Oerther, and R. G. Brown (1971). Estimating leaf water content by reflectance measurements. *Agron. J.* **63**:845–847.

Tucker, C. J. (1980). Remote sensing of leaf water content in the near infrared. *Remote Sens. Environ.* **10**:23–32.

Tucker, C. J., B. N. Holben, J. H. Elgin, and J. E. McMurtrey, III (1981). Remote sensing of total dry-matter accumulation in winter wheat. *Remote Sens. Environ.* **11**:171–189.

Tucker, C. J., J. R. G. Townshend, and T. E. Goff (1985a). African land-cover classification using satellite data. *Science* **227**:369–375.

Tucker, C. J., C. L. Vanpraet, M. J. Sharman, and G. van Ittersum (1985b). Satellite remote sensing of total herbaceous biomass production in the Senegalese Sahel: 1980–1984. *Remote Sens. Environ.* **17**:233–249.

Tucker, C. J., I. Y. Fung, C. D. Keeling, and R. H. Gammon (1986a). Relationship between atmospheric CO_2 variations and a satellite-derived vegetation index. *Nature (London)* **319**:195–199.

Tucker, C. J., C. O. Justice, and S. D. Prince (1986b). Monitoring the grasslands of the Sahel 1984–1985. *Int. J. Remote Sens.* **7**(11):1571–1581.

Ustin, S. L., J. B. Adams, C. D. Elvidge, M. Rejmanek, B. N. Rock, M. O. Smith, R. W. Thomas, and R. A. Woodward (1986). Thematic Mapper studies of semiarid shrub communities. *BioScience* **36**(7):446–452.

Vanderbilt, V. C., L. Grant, and C. S. T. Daughtry (1985). Polarization of light scattered by vegetation. *Proc. IEEE* **73**(6):1012–1024.

Vitousek, P. M. (1982). Nutrient cycling and nutrient use efficiency. *Am. Nat.* **119**:553–572.

Vogelmann, J. E., and B. N. Rock (1986). Assessing forest decline in coniferous forests of Vermont using NS-001 Thematic Mapper simulator data. *Int. J. Remote Sens.* **7**(10):1303–1321.

Waring, R. H., J. D. Aber, J. M. Melillo, and B. Moore, III (1986). Precursors of change in terrestrial ecosystems. *BioScience* **36**(7):433–438.

Wessman, C. A., J. D. Aber, and D. L. Peterson (1987). Estimating key forest ecosystem parameters through remote sensing. *Proc. Int. Geosci. Remote Sens. Symp. (IGARSS '87)*, Ann Arbor, Michigan, pp. 1189–1193.

Westman, W. E., and C. V. Price (1987). Remote detection of air pollution stress to vegetation: Laboratory-level studies. *Proc. Int. Geosci. Remote Sens. Symp. (IGARSS '87)*, Ann Arbor, Michigan, pp. 451–456.

Woodwell, G. M., R. A. Houghton, T. A. Stone, and A. B. Park (1986). Changes in the area of forests in Rondonia, Amazon basin, measured by satellite imagery. In *The Changing Carbon Cycle: A Global Analysis* (J. R. Trabalka and D. Reichle, Eds.). Springer-Verlag, New York, pp. 242–257.

Yates, H. W., J. D. Tarpley, S. R. Schneider, D. F. McGinnis, and R. A. Scofield (1983). The role of meteorological satellites in agricultural remote sensing. *Remote Sens. Environ.* **14**:219–234.

INDEX